The Greek Alphabet

Alpha	A	α		Iota	I	ι		Rho	P	ρ
Beta	B	β		Kappa	K	κ		Sigma	Σ	σ
Gamma	Γ	γ		Lambda	Λ	λ		Tau	T	τ
Delta	Δ	δ		Mu	M	μ		Upsilon	Υ	υ
Epsilon	E	ϵ		Nu	N	ν		Phi	Φ	ϕ
Zeta	Z	ζ		Xi	Ξ	ξ		Chi	X	χ
Eta	H	η		Omicron	O	o		Psi	Ψ	ψ
Theta	Θ	θ		Pi	Π	π		Omega	Ω	ω

Abbreviations for Units

A	ampere		lb	pound
Å	angstrom (10^{-10} m)		L	liter
atm	atmosphere		m	meter
Btu	British thermal unit		MeV	mega-electron volts
Bq	becquerel		Mm	megameter (10^6 m)
C	coulomb		mi	mile
°C	degree Celsius		min	minute
cal	calorie		mm	millimeter
Ci	curie		ms	millisecond
cm	centimeter		N	newton
dyn	dyne		nm	nanometer (10^{-9} m)
eV	electron volt		pt	pint
°F	degree Fahrenheit		qt	quart
fm	femtometer, fermi (10^{-15} m)		rev	revolution
ft	foot		R	roentgen
Gm	gigameter (10^9 m)		Sv	seivert
G	gauss		s	second
Gy	gray		T	tesla
g	gram		u	unified mass unit
H	henry		V	volt
h	hour		W	watt
Hz	hertz		Wb	weber
in	inch		y	year
J	joule		yd	yard
K	kelvin		μm	micrometer (10^{-6} m)
kg	kilogram		μs	microsecond
km	kilometer		μC	microcoulomb
keV	kilo-electron volts		Ω	ohm

Physics For Scientists and Engineers

Third Edition

Volume 1

Physics

For Scientists and Engineers

Paul A. Tipler

Worth Publishers

Physics for Scientists and Engineers, Third Edition, Volume 1

Paul A. Tipler

Copyright © 1991, 1982, 1976 by Worth Publishers, Inc.

Printed in the United States of America

Library of Congress Catalog Card Number: 89-52166

Extended Version (Chapters 1–42) ISBN: 0-87901-432-6

Standard Version (Chapters 1–35) ISBN: 0-87901-430-X

Volume 1 (Chapters 1–17) ISBN: 0-87901-433-4

Volume 2 (Chapters 18–42) ISBN: 0-87901-434-2

Printing: 5   6   7   Year: 99 98 97

Development Editors: Valerie Neal and Steven Tenney

Design: Malcolm Grear Designers

Art Director: George Touloumes

Production Editor: Elizabeth Mastalski

Production Supervisor: Sarah Segal

Layout: Patricia Lawson

Photographs: Steven Tenney, John Schultz of PAR/NYC, and Lana Berkovich

Line Art: York Graphic Services and Demetrios Zangos

Composition: York Graphic Services

Printing and binding: R. R. Donnelley and Sons

Cover: Supersonic Candlelight. A stroboscopic color schlieren or shadow picture taken at one-third microsecond exposure shows a supersonic bullet passing through the hot air rising above a candle. Schlieren pictures make visible the regions of nonuniform density in air. Estate of Harold E. Edgerton/ Courtesy of Palm Press.

Worth Publishers

33 Irving Place

New York, NY 10003

For Claudia

Preface

Preface

This third edition of *Physics*, now titled *Physics for Scientists and Engineers*, is a textbook for the standard two- or three-semester introductory physics course for engineering and science majors. It is assumed that the student has taken, or is concurrently taking, calculus. The book is divided into six parts: Mechanics, Oscillations and Waves, Thermodynamics, Electricity and Magnetism, Optics, and Modern Physics. It is available both in the standard version of 35 chapters, including two on modern physics (Relativity and The Origins of Quantum Theory), and an extended version of 42 chapters (available in one volume or two), which has seven additional chapters on modern physics (Quantum Mechanics, Atoms, Molecules, Solids, Nuclei, Elementary Particles, and Astrophysics and Cosmology), of which six are new.

SI units are used throughout the text. Except for a few problems on the conversion of force units (for example, pounds to newtons) in Chapter 4 and in the discussion of R factors for insulation materials in Chapter 16, all the worked examples, exercises, and problems are given in SI units.

My primary goals in writing this text have been

1. To provide a balanced introduction to the most important concepts and phenomena in classical and modern physics in a way that reflects the beauty and excitement of physics and also provides a solid foundation for further study.

2. To present physics in a logical and coherent manner that is interesting and accessible to all students.

3. To help students build self-confidence in their understanding of physics and in their problem-solving skills.

4. To stimulate students by exposing them to some of the many contemporary applications of and developments in physics in everyday life, in today's technology, and in the study of the cosmos.

Some of the features of this new edition are described below.

Streamlining and Consolidation

Every chapter has been extensively revised to make the presentation more concrete and to provide clear, logical, and succinct introductions to the central ideas of physics. For example,

> Work and energy are treated in a single chapter (Chapter 6) that discusses both the application of the work–energy theorem and conservation of mechanical energy to the solving of mechanics problems, and the generalized law of the conservation of energy. The difficulties that arise in the application of the work–energy theorem to work done on extended objects is discussed fully, as is the conversion of internal energy to mechanical energy. The concept of pseudowork is presented in Chapter 7 after the center of mass has been introduced. The treatment of escape speed has been moved to Chapter 10 (Gravity).

The discussion of rotational motion is now covered in a single chapter (Chapter 8) and has been reorganized so that angular momentum and rolling bodies can be discussed before the full vector treatment of rotation.

The chapter on oscillations (Chapter 12) has been moved so that it immediately precedes the chapters on mechanical waves. Care has been taken in the discussion of resonance to match the equations in it to those in Chapter 28 (Alternating-Current Circuits) describing resonance in *LRC* circuits. Many of the concepts of waves are introduced in Chapter 13 (Waves on Strings), with Chapter 14 being devoted entirely to sound waves. Reflection, refraction, interference, and diffraction are discussed qualitatively as applying to all types of waves, but all detailed calculations are deferred to the chapters on optics (Chapters 30 through 33).

The thermodynamics unit has been rewritten and is now organized into three chapters (Chapter 15 through 17). The introduction to temperature has been made more physical, and the section on the kinetic-theory interpretation of temperature has been expanded by the inclusion of the Maxwell–Boltzmann distribution.

The unit on electricity and magnetism (Chapters 18 through 29) has been extensively revised with particular attention to enhancing its accessibility. Electric charge, electric fields due to discrete charges, and the behavior of point charges and dipoles in an electric field are covered in Chapter 18 (Electric Fields I). In Chapter 19 (Electric Fields II), the fields due to continuous charge distributions are calculated using Coulomb's law, Gauss's law, or both. Expanded coverage of the classical model of conduction, together with a qualitative discussion of contemporary theory in Chapter 22 (Electric Current), complements the more detailed introduction to the band theory of solids in Chapter 39 of the extended edition.

In Chapter 25 (Sources of the Magnetic Field), the Biot–Savart law is given in terms of the field of a moving charge as well as that of a current element. The field is then calculated for a current loop, a solenoid, and a straight wire before Ampère's law is discussed. The material on magnetic flux has been moved to Chapter 26 (Magnetic Induction), the coverage of the magnetization of a bar magnet has been moved to Chapter 27 (Magnetism in Matter), and the discussion of Maxwell's displacement current has been moved to Chapter 29 (Maxwell's Equations and Electromagnetic Waves).

In the optics unit (Chapters 30 through 33), the material on the eye, microscopes, and telescopes is now in a short, new chapter on optical instruments (Chapter 32), which also includes a new section on the camera. A discussion of rainbows has been added to Chapter 30 (Light). The diffraction of light is first covered qualitatively and is then given a full mathematical treatment using phasors in Chapter 33 (Interference and Diffraction).

Because of this streamlining and consolidation, the length of the book has not increased significantly despite the substantial number of new worked examples, problems, photographs, and figures.

Modern Physics and Contemporary Applications

In addition to the separate chapters on modern physics (two in the standard version and nine in the extended version), the coverage of modern and applied physics has been greatly expanded throughout this edition. A section on numerical methods to support computer-based problem solving has been written for Chapter 5 (Newton's Laws II). A set of computer problems that supplement the end-of-chapter problems is available.

A solid foundation for the study of modern physics has been provided in the presentation of classical physics. For example, wave packets and group velocity are introduced in Chapter 14 (Sound) so that these concepts will be available for the study of electron waves in Chapter 36 (Quantum Mechanics) in the extended edition. Similarly, the Maxwell–Boltzmann energy distribution is given in Chapter 15 (Temperature) so that it can be used in the discussion of the electrical properties of solids in Chapter 39, and vacuum-tube diodes and triodes are discussed in Chapter 28 (Alternating-Current Circuits) to provide a basis for the treatment of semiconductor devices, also in Chapter 39.

The nine chapters on modern physics in the extended version of this edition provide an introduction to quantum physics and to the physics of atoms, molecules, solids, nuclei, and elementary particles, plus a concluding chapter on astrophysics and cosmology. Although the material in these chapters is necessarily more descriptive than that in the previous chapters, the important ideas of modern physics are often introduced by the study of specific, simple problems. For example, the concepts of quantum numbers and degeneracy arise naturally in the study of the three-dimensional infinite square well in Chapter 36. Similarly, the Pauli exclusion principle arises from symmetry considerations in the study of two identical particles in a one-dimensional infinite square well in the same chapter. Later in Chapter 39, the infinite square well is again used to introduce the concept of Fermi energy needed to understand the electrical and thermal properties of solids.

New to this edition are hundreds of photographs, most of them in color, that complement and amplify the text and provide effective examples of contemporary applications of physics.

Seventeen guest essays, of which fourteen are new, have been written for the enjoyment and enlightenment of students and instructors. These include:

Examples, Exercises, and Problems

The understanding of physics and the development of problem-solving skills are enhanced by the extensive and integrated use of examples, in-text exercises, and graded sets of problems. Of the 434 worked examples in the 42-chapter extended version, 290 are new. Nearly all of the examples are numerical and have been written to ensure correspondence between the examples and the end-of-chapter problems (especially those at the intermediate level). Worked examples are often paired with a numerical in-text exercise (with its answer given immediately) that asks the student to perform a simple calculation to extend the example and reinforce understanding. Many of the simple, single-step examples in the second edition have been converted into in-text exercises, and many of the nonnumerical examples have been converted into text material under an appropriate subhead. Despite these conversions, the total number of worked examples has been increased by about 50 percent.

Problems at the end of each chapter are grouped into three levels of difficulty. Level I problems are relatively easy, single-step problems. They are keyed to the appropriate sections of the chapter so that the student can quickly find help if needed. Level II problems require a more sophisticated understanding and are not divided by section. Great effort has been invested in providing an extensive selection of these intermediate level problems. Level III problems are the most challenging and will be of value for more advanced students. A typical chapter has about 35 problems at Level I, 20 at Level II, and 12 at Level III. The number of intermediate level problems has been increased by more than 40 percent.

Mathematics

The teaching of introductory physics is complicated by the wide range of student backgrounds and abilities. Some have had both physics and calculus courses in high school; others have had no high school physics and are just beginning their first calculus course. To accommodate this diversity of backgrounds, a general review of mathematics (including algebra, geometry, trigonometry, complex numbers, and calculus) is provided in Appendix A.

In the text itself, the mathematical sophistication required increases gradually so that those students who are studying calculus concurrently will not be at a disadvantage. New mathematical methods and concepts are introduced as dictated by the physics, with the introduction of a new method or concept immediately followed by an example. The derivative and integral are presented in Chapter 2 (Motion in One Dimension) as an intrinsic component of the discussion of instantaneous velocity and displacement. Vector dot products are introduced in Chapter 6, where they are needed for the discussion of work and energy, and cross products are introduced in Chapter 8 in conjunction with the discussion of torque and rotational motion. A few sections that are more mathematically sophisticated are optional and are located at the ends of chapters, where they do not interrupt the flow of ideas and can easily be omitted.

The gradual increase in mathematical level through the text, combined with the relatively easy in-text exercises and numerous Level I problems, should help the least prepared students to build confidence. At the same time, the broad spectrum of complexity in the problem sets, culminating in the Level III problems, ensures that all students can be challenged and stimulated at a level appropriate to their abilities.

Ease of Review

Several pedagogical features will aid students as they review the material just covered. Important equations, laws, and tables are highlighted by a color screen. Margin heads are provided for quick reference. Key terms are introduced in boldface type, defined in the text, and listed in the review section of each chapter. Thought questions immediately follow some sections within each chapter. These include routine questions that can be easily answered from the preceding text as well as open-ended questions that can serve as a basis for classroom discussion.

Each chapter concludes with a summary, which lists the important laws and results that were discussed in the chapter, along with the equations that will be most useful in solving problems. Following each chapter is a list of suggested further readings, a review section, and the graded sets of problems. The review section contains a list of learning objectives, a list of key terms that the student should be able to identify and define, and a set of true–false questions.

Use of Color

Color has been used to improve the pedagogical effectiveness of the illustrations, most of which have been redrawn to enhance their clarity and dimensionality. Students will appreciate, for example, the way color is used to distinguish between vectors and their components, between force and velocity vectors in the same diagram, or between conductor surfaces and gaussian surfaces. The conventions used in the illustrations, such as the placement of force vectors in a diagram, have been chosen with pedagogical effectiveness as the primary goal.

Comparison with the Second Edition

Those instructors familiar with my previous edition should be reassured that the features that contributed to its usefulness have not been diminished in this extensive revision. All of the many changes have been made to enhance and complement what was already sound. A few of the changes are given here to expedite familiarization with this edition for those who have taught from the second edition:

The list of learning objectives, formerly at the beginning of each chapter, now appears in the review section following the summary.

The reviews of algebra and trigonometry have been moved from Chapter 1 to Appendix A.

Relative velocity is presented in Chapter 3 (Motion in Two and Three Dimensions):

Each of Newton's three laws of motion is discussed in a separate section in Chapter 4 (Newton's Laws I). The basic interactions are then listed with a brief qualitative discussion. In Chapter 4, applications are limited to problems involving a single particle. A brief discussion of rolling friction has been included in Chapter 5 (Newton's Laws II).

Static equilibrium is now treated separately in a new chapter (Chapter 9).

Discussions of the Maxwell–Boltzmann distribution and the heat pump have been added to the thermodynamics unit, while the Clausius inequality and the proof of the existence of entropy have been deleted.

Chapter 29 (Maxwell's Equations and Electromagnetic Waves) now includes material on Maxwell's displacement current, energy and momentum in electromagnetic waves, and the electromagnetic spectrum.

The Lorentz transformation is treated early in Chapter 34 and is used to simplify the discussion of clock synchronization and simultaneity.

Seven new chapters on modern physics have been added to the two in the standard version to complete the modern physics unit in the extended version. This material can easily serve as the basis of a one-quarter or one-semester course in modern physics.

A number of deletions have been made to streamline the text or to make space for the coverage of contemporary topics. These include reaction thresholds, the vector nature of angular displacement, gravitational field lines, the buoyant force in accelerated reference frames, the capacitance of an isolated sphere, the betatron, the Wheatstone bridge, and the magnetic vector **H**.

Acknowledgements

Many people have contributed to this edition. I would like to thank everyone who used the first or second edition and offered comments and suggestions. Your letters have been valuable and welcome.

Ralph Llewellyn (University of Central Florida) wrote the exciting and informative chapter on astrophysics and cosmology (Chapter 42) in the extended edition.

Many new and interesting end-of-chapter problems were provided by Howard Miles (Washington State University), Lawrence B. Golden and James Klein (Pennsylvania State University), Robert Rundel (Mississippi State University), and William E. Fasnacht (United States Naval Academy). Among them, they have also provided one of the independently worked sets of solutions for all of the problems in the text.

Lawrence Golden and James Klein also wrote the Computer Investigations supplement.

James Walker (Washington State University) prepared the answers listing at the end of the text, produced the elegant solutions that are published in the accompanying complete *Solutions Manual* (for instructors), and also offered many valuable suggestions for improving and clarifying the end-of-chapter problems.

Robin Macqueen (University of British Columbia) contributed the Suggestions for Further Reading for each chapter.

The accuracy of the numerical calculations in the examples and exercises has been expertly checked by Robert Weidman (Michigan Technological University), Chelcie Liu (City College of San Francisco), and Edward Brown (Manhattan College). Professor Brown also offered many helpful suggestions in his reviews of the end-of-chapter problems.

Gene Mosca (United States Naval Academy) has applied his profound understanding of the teaching of physics to the revision of the *Study Guide* originally written by Granvil C. Kyker.

David Mills (College of the Redwoods) has produced a test bank of about 3000 conceptual and numerical multiple-choice questions.

The *Instructors Resource Manual* has been prepared by Robert Allen (Inver Hills Community College), John Davis (University of Washington), John Risley (North Carolina State University), James Walker (Washington State University), Vicki Williams (Pennsylvania State University), and myself. It provides a comprehensive selection of demonstrations, listings of commercially available and public-domain software programs suitable for use in teaching introductory physics, a film and video guide, suggested homework assignments, critical-thinking questions, and a complete answers listing for all end-of-chapter problems.

Roger Clapp (University of South Florida), Manuel Gómez-Rodríguez (University of Puerto Rico, Río Piedras), John Russell (Southeastern Massachusetts University), and Jim Smith (University of Illinois, Champaign–

Urbana) reviewed the entire second edition; John Russell also reviewed parts of the second draft manuscript, and Jim Smith reviewed the entire first draft for the third edition. Brooke Pridmore (Clayton State College) reviewed the second draft manuscript; Grant Hart (Brigham Young University) reviewed major portions of the second edition, as well as both drafts of the manuscript; and James Brown (The Colorado School of Mines) reviewed major portions of the second edition and the first draft of the manuscript. Their substantive and constructive comments and suggestions were a continual source of inspiration over the many months of this project.

Many other instructors have provided extensive and invaluable reviews. They have all made a deeply appreciated and fundamental contribution to the quality of this revision, and I therefore wish to thank:

Edward Brown, *Manhattan College*

James Brown, *The Colorado School of Mines*

Walter Borst, *Texas Technological University*

Christopher Cameron, *University of Southern Mississippi*

Bob Coakley, *University of Southern Maine*

Andrew Coates, *University College, London, United Kingdom*

Miles Dressler, *Washington State University*

Allin Gould, *John Abbott College C.E.G.E.P. Canada*

Dennis Hall, *University of Rochester*

Grant Hart, *Brigham Young University*

Jerold Izatt, *University of Alabama*

Alvin Jenkins, *North Carolina State University*

Lorella Jones, *University of Illinois, Champaign–Urbana*

Michael Kambour, *Miami-Dade Junior College*

Patrick Kenealy, *California State University at Long Beach*

Doug Kurtze, *Clarkson University*

Lui Lam, *San Jose State University*

Chelcie Liu, *City College of San Francisco*

Robert Luke, *Boise State University*

Stefan Machlup, *Case Western Reserve University*

Eric Matthews, *Wake Forest University*

Konrad Mauersberger, *University of Minnesota, Minneapolis*

Duncan Moore, *University of Rochester*

Gene Mosca, *United States Naval Academy*

Elizabeth Nickles, *Albany College of Pharmacy*

Harry Otteson, *Utah State University*

Jack Overley, *University of Oregon*

Larry Panek, *Widener University*

Malcolm Perry, *Cambridge University, United Kingdom*

Brooke Pridmore, *Clayton State College*

Arthur Quinton, *University of Massachusetts, Amherst*

John Risley, *North Carolina State University*

Robert Rundel, *Mississippi State University*

Michael Simon, *Housatonic Community College*

Richard Smith, *Montana State University*

Larry Sorenson, *University of Washington*

Thor Stromberg, *New Mexico State University*

Edward Thomas, *Georgia Institute of Technology*

Colin Thomson, *Queens University, Canada*

Gianfranco Vidali, *Syracuse University*

Brian Watson, *St. Lawrence University*

Robert Weidman, *Michigan Technological University*

Stan Williams, *Iowa State University*

Thad Zaleskiewicz, *University of Pittsburgh, Greensburg*

George Zimmerman, *Boston University*

Finally, I would like to thank everyone at Worth Publishers for their help and encouragement, and particularly Steven Tenney, Valerie Neal, Betsy Mastalski, Wendy Schechter, Anne Vinnicombe, George Touloumes, and Sarah Segal.

Berkeley, California
February 1991

Paul Tipler

Supplements

For Students

Study Guide

Volume 1 (Chapters 1–17) and
Volume 2 (Chapters 18–42)

Each chapter contains:

Key ideas and equations

Possible pitfalls

Questions and answers

True/False questions

Problems with detailed solutions, followed by a similar problem with the answer only

Student's Solutions Manual

Contains worked solutions to selected odd-numbered problems, totaling about 25% of the text problems.

Computer Investigations

An introduction to the application of computers to physics. Spreadsheets and programs, including 25 that are supplied for the student, are used to model, visualize, and calculate. No previous computer literacy is assumed. Explanatory text describes spreadsheets, rudiments of BASIC, numerical methods, and elements of equation-solver software. The approximately 110 problems that follow are divided into three levels: Level I problems supply a spreadsheet or program and require variation of input parameters; Level II problems indicate a prototype spreadsheet or program to be modified by the student; Level III problems require the student to create the needed spreadsheet or program. An accompanying IBM or Mac disk contains spreadsheets and programs. Answers to selected problems are supplied.

For Instructors

Solutions Manual

Volume 1 (Chapters 1–17) and
Volume 2 (Chapters 18–42)

Contains complete worked-out solutions to all problems in the text.

Test Bank

Volume 1 (Chapters 1–17) and
Volume 2 (Chapters 18–42)

Approximately 3000 multiple-choice questions spanning all sections of the text. Each question is identified by topic and noted as factual, conceptual, or numerical.

Computerized Test-Generation System

(IBM and Macintosh)

A data base comprising the questions in the Test Bank.

Instructor's Resource Manual

Contains:

Demonstrations

Film and videocassette guide

Software guide

Critical thinking questions

Answers to all end-of-chapter problems

Transparencies

Volume 1 (Chapters 1–17) and
Volume 2 (Chapters 18–42)

Approximately 150 full-color acetates of figures and tables from the text, with type enlarged for projection.

About the Author

Paul Tipler was born in the small farming town of Antigo, Wisconsin, in 1933. He graduated from high school in Oshkosh, Wisconsin, where his father was Superintendent of the Public Schools. He received his BS at Purdue University in 1955 and his PhD at the University of Illinois in 1962, where he studied the structure of nuclei. He taught for one year at Wesleyan University in Connecticut while writing his thesis. He then moved to Oakland University in Michigan, where he was one of the original members of the physics department, playing a major role in developing the physics curriculum. During the next 20 years, he taught nearly all the physics courses and wrote the first and second editions of his widely used textbooks *Modern Physics* (1969, 1978) and *Physics* (1976, 1982). In 1982 he moved to Berkeley, California, where he now resides and where he wrote *College Physics* (1987). In addition to physics, his interests include music, hiking, and camping, and he is an accomplished jazz pianist and poker player.

Contents in Brief

Extended Version

Chapters 1–42:

Mechanics; Oscillations and Waves; Thermodynamics; Electricity and Magnetism; Optics; and nine chapters of Modern Physics

Standard Version

Chapters 1–35:

Mechanics; Oscillations and Waves; Thermodynamics; Electricity and Magnetism; Optics; and two chapters of Modern Physics

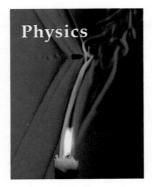

Volume 1

Chapters 1–17:

Mechanics; Oscillations and Waves; and Thermodynamics

Volume 2

Chapters 18–42:

Electricity and Magnetism; Optics; and nine chapters of Modern Physics

Contents

Part 1

Mechanics 19

Part **2**   Oscillations and Waves   367

Part 3

Thermodynamics   485

Physics For Scientists and Engineers

Chapter 1

Systems of Measurement

Tycho Brahe (1546–1601) with his large brass quadrant for measuring the position of planets and stars.

We have always been curious about the world around us. Since the beginning of recorded thought, we have sought ways to impose order on the bewildering diversity of the events we observe. This search for order has taken a variety of forms: one is religion, another is art, and a third is science. Although the term "science" has its origins in the Latin verb meaning "to know," science has come to mean not merely knowledge but specifically knowledge of the natural world. Most important, this knowledge is organized in a systematic and rational way.

We generally think of science as being divided into several separate but interrelated fields. Biology, for example, is the study of living organisms. Chemistry deals with the interaction of elements and compounds. Geology is the study of the earth. Astronomy is the study of the solar system, the stars, and galaxies, and the universe as a whole. Physics deals with matter and energy, with the principles that govern the motion of particles and waves, with the interactions of particles, and with the properties of molecules, atoms, and atomic nuclei, and of larger-scale systems such as gases, liquids, and solids. Some consider physics to be the most fundamental science because it is the basis of all the other fields of science.

1-1   Units

The laws of physics express relationships between physical quantities, such as length, time, force, energy, and temperature. Thus, the ability to define such quantities precisely and measure them accurately is a requisite of physics. The measurement of any physical quantity involves comparing it with some precisely defined unit value of the quantity. For example, to measure the distance between two points, we compare that distance with a standard unit of distance, such as the meter. The statement that a certain distance is "25 meters" means that it is 25 times the length of the unit meter. That is, a standard meterstick fits into the distance 25 times. It is important to include the unit "meter" along with the number "25" in expressing a distance because there are other units of distance, such as feet or miles, in common use. To say that a distance is "25" is meaningless. The magnitude of any physical quantity must include both a number and a unit.

All physical quantities can be expressed in terms of a small number of fundamental units. For example, speed is expressed in terms of a unit of length and a unit of time, such as meters per second or miles per hour. Many of the quantities that we will be studying, such as force, momentum, work, energy, and power, can be expressed in terms of three fundamental quantities—length, time, and mass. The choice of standard units for these fundamental quantities determines a system of units. The system used universally in the scientific community is the **International System (SI).**[*] In the SI the standard unit of length is the meter, the standard unit of time is the second, and the standard unit of mass is the kilogram.

The standard unit of length, the meter (abbreviated m), was originally indicated by two scratches on a bar made of a platinum-iridium alloy kept at the International Bureau of Weights and Measures in Sèvres, France. This length was chosen so that the distance between the equator and the North Pole along the meridian through Paris would be 10 million meters (Figure 1-1). The standard meter is now defined in terms of the speed of light as the distance traveled by light in empty space during a time of 1/299,792,458 second. (This makes the speed of light exactly 299,792,458 m/s.) The standard meter is used to construct secondary standards that are used to calibrate measuring rods throughout the world.

The standard unit of time, the second (s), was originally defined in terms of the rotation of the earth as $\frac{1}{60} \times \frac{1}{60} \times \frac{1}{24}$ of the mean solar day. The second is now defined in terms of light. All atoms, after absorbing energy, emit light with wavelengths and frequencies that are characteristic of the particular element. There is a particular frequency and wavelength associated with each energy transition within the atom. As far as we know, these frequencies remain constant. The second is defined so that the frequency of the light from a certain transition in cesium is 9,192,631,770 cycles per second. With these definitions, the fundamental units of length and time are accessible to laboratories throughout the world.

The standard unit of mass, the kilogram (kg), which equals 1000 grams (g), is defined to be the mass of a standard body, also kept at Sèvres. We discuss the concept of mass in detail in Chapter 4. We will see that the weight of an object at a given point on earth is proportional to its mass. Thus, masses of ordinary size can be compared by weighing them. A duplicate of the standard 1-kg body is kept at the National Bureau of Standards in Gaithersburg, Maryland.

In our study of thermodynamics and electricity, we will need three more fundamental physical units: the unit of temperature, the kelvin (K) (formerly the degree Kelvin); the unit for the amount of a substance, the mole (mol);

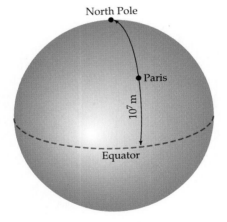

Figure 1-1 The meter was originally chosen so that the distance from the equator to the North Pole along the meridian through Paris would be 10^7 m.

[*]SI stands for Système International. Complete definitions of the SI units are given in Appendix B.

(b)

(a) Water clock used to measure time intervals in the 13th century. (b) Modern portable cesium clock.

(a)

and the unit of current, the ampere (A). There is another fundamental unit, the candela (cd), for luminous intensity, which we will have no occasion to use in this book. These seven fundamental units—the meter (m), the second (s), the kilogram (kg), the kelvin (K), the ampere (A), the mole (mol), and the candela (cd)—constitute the SI.

The units of every physical quantity can be expressed in terms of the fundamental SI units. Some frequently used combinations are given special names. For example, the SI unit of force, $kg \cdot m/s^2$, is called a newton (N). Similarly, the SI unit of power, $kg \cdot m^2/s^3 = N \cdot m/s$, is called a watt (W).

Prefixes for common multiples and submultiples of SI units are listed in Table 1-1. These multiples are all powers of 10, and so such a system is called a **decimal system.** The decimal system based on the meter is called the **metric system.** The prefixes can be applied to any SI unit; for example, 0.001 second is 1 millisecond (ms); 1,000,000 watts is 1 megawatt (MW).

Another decimal system still in use but gradually being replaced by the SI is the **cgs system,** based on the *c*entimeter, *g*ram, and *s*econd. The centimeter is defined as 0.01 m. The gram is now defined as 0.001 kg. Originally, the gram was defined as the mass of one cubic centimeter of water. (The kilogram was then the mass of 1000 cubic centimeters or one liter of water.)

In another system of units used in the United States, the **U.S. customary system,** a unit of force, the pound, is chosen to be a fundamental unit. The pound is defined in terms of the gravitational attraction of the earth at a particular place for a standard body. The unit of mass is then defined in terms of the pound. (We will see in Chapter 4 that mass is a better choice for a fundamental unit than force because mass is an intrinsic property of an object that is independent of its location relative to the earth.) The fundamental unit of length in this system is the foot (ft). The foot is defined as exactly one-third of a yard (yd), which is now defined in terms of the meter:

$$1 \text{ yd} = 0.9144 \text{ m} \qquad\qquad 1\text{-}1$$

$$1 \text{ ft} = \tfrac{1}{3} \text{ yd} = 0.3048 \text{ m} \qquad\qquad 1\text{-}2$$

This makes the inch (in) exactly 2.54 cm. The fundamental unit of time is the second, which is defined as it is in the SI. This system is not a decimal system. It is less convenient than the SI or other decimal systems because common multiples of its units are not powers of 10. For example, 1 yd = 3 ft and 1 ft = 12 in. Relationships between the units in the U.S. customary system and those in the SI are given in Appendix D.

Table 1-1	**Prefixes for Powers of 10**	
Multiple	Prefix	Abbreviation
10^{18}	exa	E
10^{15}	peta	P
10^{12}	tera	T
10^{9}	giga	G
10^{6}	mega	M
10^{3}	kilo	k
10^{2}	hecto†	h
10^{1}	deka†	da
10^{-1}	deci†	d
10^{-2}	centi†	c
10^{-3}	milli	m
10^{-6}	micro	μ
10^{-9}	nano	n
10^{-12}	pico	p
10^{-15}	femto	f
10^{-18}	atto	a

†The prefixes hecto (h), deka (da), and deci (d) are not powers of 10^3 or 10^{-3} and are rarely used. The other prefix which is not a power of 10^3 or 10^{-3} is centi (c), now used only with the meter, as in 1 cm = 10^{-2} m.

(a)

(b)

(*a*) Laser beams from the Macdonald Observatory used to measure the distance to the moon. The distance can be measured to within a few centimeters by measuring the time for the beam to go to the moon and return after reflecting off a corner mirror (*b*) placed on the moon by the Apollo 14 astronauts.

Questions

1. What are the advantages and disadvantages of using the length of your arm for a standard length?
2. A certain clock is known to be consistently 10 percent fast compared with the standard cesium clock. A second clock varies in a random way by 1 percent. Which clock would make a more useful secondary standard for a laboratory? Why?

1-2   Conversion of Units

We have said that the magnitude of a physical quantity must include both a number and a unit. When such quantities are added, subtracted, multiplied, or divided in an algebraic equation, the unit can be treated like any other algebraic quantity. For example, suppose you wish to find the distance traveled in 3 hours (h) by a car moving at a constant rate of 80 kilometers per hour (km/h). The distance is the product of the speed v and the time t:

$$x = vt = \frac{80 \text{ km}}{\text{h}} \times 3 \text{ h} = 240 \text{ km}$$

We cancel the unit of time, the hours, just as we would any algebraic quantity to obtain the distance in the proper unit of length, the kilometer. This method of treating units makes it easy to convert from one unit to another. Suppose we want to convert our answer of 240 km to miles (mi). We use the fact that

$$1 \text{ mi} = 1.61 \text{ km}$$

If we divide each side of this equation by 1.61 km, we obtain

$$\frac{1 \text{ mi}}{1.61 \text{ km}} = 1$$

Since any quantity can be multiplied by 1 without changing its value, we can now change 240 km to miles by multiplying by the factor (1 mi)/(1.61 km):

$$240 \text{ km} = 240 \text{ km} \times \frac{1 \text{ mi}}{1.61 \text{ km}} = 149 \text{ mi}$$

The factor (1 mi)/(1.61 km) is called a **conversion factor.** All conversion factors have a value of 1 and are used to convert a quantity expressed in one unit of measure into its equivalent in another unit of measure. By writing out the units explicitly and canceling them, we need not think about whether we multiply by 1.61 or divide by 1.61 to change kilometers to miles because the units that remain tell us whether we have chosen the correct or incorrect factor.

Example 1-1

What is the equivalent of 90 km/h in meters per second and in miles per hour?

We use the facts that 1000 m = 1 km, 60 s = 1 min, and 60 min = 1 h to convert kilometers per hour to meters per second. We multiply the quantity 90 km/h by a set of conversion factors each having the value 1 so that the value of the speed is not changed:

$$\frac{90 \text{ km}}{\text{h}} \times \frac{1000 \text{ m}}{1 \text{ km}} \times \frac{1 \text{ h}}{60 \text{ min}} \times \frac{1 \text{ min}}{60 \text{ s}} = 25 \text{ m/s}$$

To convert this speed into miles per hour, we use the conversion factor (1 mi)/(1.61 km) = 1.

$$\frac{90 \text{ km}}{\text{h}} \times \frac{1 \text{ mi}}{1.61 \text{ km}} = 55.9 \text{ mi/h}$$

Exercise

What is the equivalent of 65 mi/h in meters per second?
(Answer: 29.1 m/s)

1-3   Dimensions of Physical Quantities

The area of a surface is found by multiplying one length by another. For example, the area of a rectangle of sides 2 m and 3 m is $A = (2 \text{ m})(3 \text{ m}) = 6 \text{ m}^2$. The unit of area is square meters. Because area is the product of two lengths, it is said to have the dimensions of length times length or length squared, often written L^2. The idea of dimensions is easily extended to non-geometric quantities. For example, speed is said to have the dimensions of length divided by time, or L/T. The dimensions of other quantities, such as force or energy, are written in terms of the fundamental quantities of length, time, and mass.

Adding two physical quantities makes sense only if the quantities have the same dimensions. For example, we cannot add an area to a speed and obtain a meaningful sum. If we have an equation like

$$A = B + C$$

the quantities A, B, and C must all have the same dimensions. The addition of B and C also requires that the dimensions of these quantities be expressed in the same units. For example, if B is an area of 500 in^2 and C is an area of 4 ft^2, we must either convert B into square feet or C into square inches in order to find the sum of the two areas.

We can often find mistakes in a calculation by checking the dimensions or units of the quantities in the calculation. Suppose, for example, we mistakenly use the formula $A = 2\pi r$ for the area of a circle. We can see immediately that this cannot be correct because the right side of the equation, $2\pi r$, has dimensions of length whereas area should have dimensions of length squared. For another example, consider this formula for distance x:

$$x = vt + \tfrac{1}{2}at$$

where t is the time, v is the speed, and a is the acceleration, which (as we will see) has dimensions of L/T^2. We can see by looking at the dimensions of each quantity that this formula cannot be correct. Since x has dimensions of length, each term on the right side of the equation must also have dimensions of length. The term vt has dimensions of length, but the dimensions of $\tfrac{1}{2}at$ are $(L/T^2)T = L/T$. Since this term does not have the correct dimensions, the formula cannot be correct. Dimensional consistency is a necessary but not a sufficient condition for an equation to be correct. An equation can have the correct dimensions in each term and still not describe any physical situation.

1-4   Scientific Notation

Handling very large or very small numbers is simplified by using scientific notation. In this notation, a number is written as a product of a number between 1 and 10 and a power of 10, such as 10^2 ($= 100$) or 10^3 ($= 1000$). For example, the number 12,000,000 is written 1.2×10^7; the distance from the earth to the sun, about 150,000,000,000 m, is written 1.5×10^{11} m. The number 11 in 10^{11} is called the **exponent.** For numbers smaller than 1, the exponent is negative. For example, $0.1 = 10^{-1}$ and $0.0001 = 10^{-4}$. The diameter of a virus, which is about 0.00000001 m, is written 1×10^{-8} m.

In multiplication, exponents are added; in division, they are subtracted. Some simple examples follow that show the use of these rules.

$$10^2 \times 10^3 = 100 \times 1000 = 100,000 = 10^{2+3} = 10^5$$

Similarly,

$$\frac{10^2}{10^3} = \frac{100}{1000} = \frac{1}{10} = 10^{2-3} = 10^{-1}$$

In scientific notation, 10^0 is defined to be 1. We can see this as follows. Suppose we divide 1000 by itself. We have

$$\frac{1000}{1000} = \frac{10^3}{10^3} = 10^{3-3} = 10^0 = 1$$

Example 1-2

Using scientific notation, compute (a) 120×6000 and (b) 3,000,000/0.00015.

(a) $(1.20 \times 10^2)(6.00 \times 10^3) = (1.20)(6.00) \times 10^{2+3} = 7.20 \times 10^5$

(b) $\dfrac{3.00 \times 10^6}{1.50 \times 10^{-4}} = \dfrac{3.00}{1.50} \times 10^{6-(-4)} = 2.00 \times 10^{10}$

The two minus signs in (b) arise because -4 is subtracted from 6.

Exercise

Using scientific notation, calculate $(2.50 \times 10^7)(1.90 \times 10^{-3})$.
(Answer: 4.75×10^4)

In raising a power to another power, the exponents are multiplied. For example,

$$(10^2)^4 = 10^2 \times 10^2 \times 10^2 \times 10^2 = 10^8$$

Example 1-3

A liter is the volume of a cube that is 10 cm by 10 cm by 10 cm. Find the volume of a liter in cubic centimeters and in cubic meters.

The volume V of a cube of side L is L^3.

$$V = L^3 = (10 \text{ cm})(10 \text{ cm})(10 \text{ cm}) = 10^3 \text{ cm}^3$$

To convert to cubic meters (m^3), we use $1 \text{ cm} = 10^{-2} \text{ m}$,

$$10^3 \text{ cm}^3 = 10^3 \text{ cm}^3 \times \left(\frac{10^{-2} \text{ m}}{1 \text{ cm}}\right)^3 = 10^3 \text{ cm}^3 \times \frac{10^{-6} \text{ m}^3}{1 \text{ cm}^3} = 10^{-3} \text{ m}^3$$

Note that raising the conversion factor (which equals 1) to the third power does not change its value and enables us to cancel the units.

Care must be taken when adding or subtracting numbers that are written in scientific notation when the exponents don't match. Consider, for example,

$$(1.200 \times 10^2) + (8 \times 10^{-1}) = 120.0 + 0.8 = 120.8$$

To find the sum without first converting both numbers into ordinary decimal form, it is sufficient to rewrite either of the numbers so that its power of 10 is the same as that of the other. For example, we can find the sum by writing $1.200 \times 10^2 = 1200 \times 10^{-1}$ and then adding:

$$(1200 \times 10^{-1}) + (8 \times 10^{-1}) = 1208 \times 10^{-1} = 120.8$$

If the exponents are very different, then one of the numbers is much smaller than the other and can often be neglected in addition or subtraction. For example,

$$(2 \times 10^6) + (9 \times 10^{-3}) = 2,000,000 + 0.009$$

$$= 2,000,000.009 \approx 2 \times 10^6$$

where the symbol $\approx$ means "is approximately equal to."

1-5 Significant Figures and Orders of Magnitude

Many of the numbers in science are the result of measurement and are therefore known only within the limits of some experimental uncertainty. The magnitude of the uncertainty depends on the skill of the experimenter and on the apparatus used and often can only be estimated. A rough indication of the uncertainty in a measurement is implied by the number of digits used. For example, if we say that a table is 2.50 m long, we are implying that its length is probably between 2.495 m and 2.505 m. That is, we know the length to within about $\pm 0.005 \text{ m} = \pm 0.5 \text{ cm}$ of the stated length. If we used a meterstick with millimeter markings and measured the table length carefully, we might estimate that we could measure the length to $\pm 0.5 \text{ mm}$ rather than $\pm 0.5 \text{ cm}$. We would indicate this precision by using four digits, such as 2.503 m, to give the length. A reliably known digit (other than a zero used to locate the decimal point) is called a **significant figure.** The number 2.50 has three significant figures; 2.503 has four. The number 0.00103 has three significant figures. (The first three zeros are not significant figures but merely locate the decimal point.) In scientific notation, this number is expressed as 1.03×10^{-3}. A common student error, particularly since the advent of hand calculators, is to carry many more digits in an answer than are warranted. Suppose, for example, that you measure the area of a circular playing field by pacing off the radius and using the formula $A = \pi r^2$. If you estimate the radius to be 8 m by pacing and use a 10-digit calculator to compute the area, you obtain $\pi(8\text{ m})^2 = 201.0619298\text{ m}^2$. The digits after the decimal point are misleading about the accuracy with which you know the area. Since you found the radius by pacing, you might expect that your measurement is accurate to only about 0.5 m. That is, the radius could be as great as 8.5 m or as small as 7.5 m. If the radius is 8.5 m, the area is $\pi(8.5\text{ m})^2 = 226.9800692\text{ m}^2$, whereas if the radius is 7.5 m, the area is $\pi(7.5\text{ m})^2 = 176.714587\text{ m}^2$. A general rule to follow when multiplying or dividing several numbers is

> The number of significant figures in the result of multiplication or division is no greater than the least number of significant figures in any of the factors.

In the previous example, the radius of the playing field is known to only one significant figure, so the area is also known only to one significant figure. It should be written as $2 \times 10^2\text{ m}^2$, which implies that the area is somewhere between 150 m² and 250 m².

The precision of the sum or difference of two measurements is only as good as the precision of the least precise of the two measurements. A general rule to follow is

> The result of the addition or subtraction of two numbers has no significant figures beyond the last decimal place where both of the original numbers had significant figures.

Example 1-4

Find the sum of 1.040 and 0.2134.

The first number, 1.040, has only three significant figures beyond the decimal point, whereas the second, 0.2134, has four. According to our rule, the sum can have only three significant figures beyond the decimal point. We thus obtain

$$1.040 + 0.2134 = 1.253$$

Exercise

Apply the appropriate rule for determining the number of significant figures to calculate the following: (a) 1.58×0.03, (b) $1.4 + 2.53$, and (c) $(2.34 \times 10^2) + 4.93$.   [Answers: (a) 0.05, (b) 3.9, (c) 2.39×10^2]

In a textbook, it is cumbersome to write every number with the proper number of significant figures. Most examples and exercises will be done with data to three (or sometimes four) significant figures, but occasionally we will say, for example, that a table top is 3 ft by 8 ft rather than taking the time and space to say it is 3.00 ft by 8.00 ft. Any data you see in an example or exercise can be assumed to be known to three significant figures unless otherwise indicated.

(a) (b) (c)

(d)

In doing rough calculations or comparisons, we sometimes will round off a number to the nearest power of 10. Such a number is called an **order of magnitude.** For example, the height of a small insect, say an ant, might be 8×10^{-4} m $\approx 10^{-3}$ m. We would say that the order of magnitude of the height of an ant is 10^{-3} m. Similarly, though the height of most people is about 2 m, we might round that off and say that the order of magnitude of the height of a person is 10^0 m. By this we do not mean to imply that a typical height is really 1 m but that it is closer to 1 m than to 10 m or to $10^{-1} = 0.1$ m. We might say that a typical human being is 3 orders of magnitude taller than a typical ant, meaning that the ratio of their heights is about 1000 to 1. An order of magnitude does not provide any digits that are reliably known. It may be thought of as having no significant figures.

Tables 1-2 to 1-4 give some typical order-of-magnitude values for some lengths, masses, and time intervals encountered in physics.

(e)

Table 1-2 Order of Magnitude of Some Lengths

Length	m
Radius of proton	10^{-15}
Radius of atom	10^{-10}
Radius of virus	10^{-7}
Radius of giant amoeba	10^{-4}
Radius of walnut	10^{-2}
Height of human being	10^0
Height of highest mountains	10^4
Radius of earth	10^7
Radius of sun	10^9
Earth-sun distance	10^{11}
Radius of solar system	10^{13}
Distance to nearest star	10^{16}
Radius of Milky Way galaxy	10^{21}
Radius of visible universe	10^{26}

Table 1-3 Order of Magnitude of Some Masses

Mass	kg
Electron	10^{-30}
Proton	10^{-27}
Amino acid	10^{-25}
Hemoglobin	10^{-22}
Flu virus	10^{-19}
Giant amoeba	10^{-8}
Raindrop	10^{-6}
Ant	10^{-2}
Human being	10^2
Saturn 5 rocket	10^6
Pyramid	10^{10}
Earth	10^{24}
Sun	10^{30}
Milky Way galaxy	10^{41}
Universe	10^{52}

(a) Benzene molecules of the order of 10^{-10} m diameter, as seen in a scanning tunneling microscope. (b) Chromosomes measuring of the order of 10^{-6} m across, as seen in a scanning electron microscope. (c) Distances familiar in our everyday macroscopic world. The height of the girl is of the order of 10^0 m and that of the mountain is of the order of 10^4 m. (d) Earth with a diameter of the order of 10^7 m, as seen from space in the Apollo Mission. (e) Andromeda galaxy with diameter of the order of 10^{21} m.

Table 1-4 **Order of Magnitude of Some Time Intervals**	
Interval	s
Time for light to cross nucleus	10^{-23}
Period of visible-light radiation	10^{-15}
Period of microwaves	10^{-10}
Half-life of muon	10^{-6}
Period of highest audible sound	10^{-4}
Period of human heartbeat	10^{0}
Half-life of free neutron	10^{3}
Period of earth's rotation (day)	10^{5}
Period of revolution of earth (year)	10^{7}
Lifetime of a human	10^{9}
Half-life of plutonium 239	10^{12}
Lifetime of a mountain range	10^{15}
Age of earth	10^{17}
Age of universe	10^{18}

In many cases, the order of magnitude of a quantity can be estimated using reasonable assumptions and simple calculations. The physicist Enrico Fermi was a master at computing approximate answers to questions that seemed impossible to answer because of limited available information. Such questions are often called **Fermi questions**. The following is an example of a Fermi question.

Example 1-5

What thickness of rubber tread is worn off an automobile tire as it travels 1 km (0.6 mi)?

We assume the tread thickness of a new tire is 1 cm. This may be off by a factor of 2 or so, but 1 mm is certainly too small and 10 cm is too large. Since tires have to be replaced after about 60,000 km (about 37,000 mi), we will assume that the tread is completely worn off after 60,000 km. In other words, the rate of wear is 1 cm of tire per 60,000 km of travel. The thickness that is worn per kilometer is therefore

$$\frac{1 \text{ cm}}{60,000 \text{ km}} = 1.7 \times 10^{-5} \text{ cm/km} \approx 0.2 \ \mu\text{m/km}$$

Summary

1. The magnitudes of physical quantities (for example, length, time, force, and energy) are expressed as a number times a unit.

2. The fundamental units in the International System (SI) are the meter (m), the second (s), the kilogram (kg), the kelvin (K), the ampere (A), the mole (mol), and the candela (cd). Every physical quantity can be expressed in terms of these fundamental units.

3. Units in equations are treated just like any other algebraic quantity.

4. The two sides of an equation must have the same dimensions.

5. Conversion factors, which are always equal to 1, provide a convenient method for converting from one unit to another.

6. Very small and very large numbers are most easily handled when written as a number between 1 and 10 times a power of 10. This is called scientific notation. When multiplying two numbers, the exponents are added; when dividing, the exponents are subtracted. When a number containing an exponent is itself raised to an exponent, the exponents are multiplied.

7. The number of significant figures in the result of multiplication or division is no greater than the least number of significant figures in any of the factors. The result of the addition or subtraction of two numbers has no significant figures beyond the last decimal place where both of the original numbers had significant figures.

8. A number rounded to the nearest power of 10 is called an order of magnitude. The order of magnitude of a quantity can often be estimated using reasonable assumptions and simple calculations.

The Fermi Solution

Hans Christian von Baeyer
The College of William and Mary

At twenty-nine minutes past five, on a Monday morning in July 1945, the world's first atom bomb exploded in the desert sixty miles northwest of Alamogordo, New Mexico. Forty seconds later, the blast's shock wave reached the base camp, where scientists stood in stunned contemplation of the historic spectacle. The first person to stir was the Italian-American physicist Enrico Fermi, who was on hand to witness the culmination of a project he had helped begin.

Before the bomb detonated, Fermi tore a sheet of notebook paper into small bits. Then, as he felt the first quiver of the shock wave spreading outward through the still air, he released the shreds above his head. They fluttered down and away from the mushroom cloud growing on the horizon, landing about two and half yards behind him. After a brief mental calculation, Fermi announced that the bomb's energy had been equivalent to that produced by ten thousand tons of TNT. Sophisticated instruments also were at the site. Analyses of their

Hans Christian von Baeyer is a professor of physics at the College of William and Mary in Williamsburg, Virginia. After attending high school in Canada, he graduated from Columbia

University and earned his Ph.D. in physics at Vanderbilt University. His research specialty is elementary particle theory, with particular emphasis on classical relativistic mechanics— a problem that should have been solved 70 years ago but wasn't. He is a fellow of the American Physical Society, and a recipient of the American Institute of Physics Science Writing Award. This essay originally appeared in his bimonthly column entitled "PHYSIKA" that appears in the journal *The Sciences*. Professor von Baeyer likes to run in 10-km races and finishes in the top half of the field.

readings of the shock wave's velocity and pressure, which took several weeks to complete, confirmed Fermi's instant estimate. (It is not known for sure how Fermi did it, but what he did was probably this: by measuring the speed with which air was pushed out by the explosion, he estimated the total kinetic energy dissipated by the atmosphere, and then divided that by the amount of energy liberated by a ton of TNT.)

The bomb-test team was impressed, but not surprised, by this brilliant bit of scientific improvisation. Enrico Fermi's genius was known throughout the world of physics. In 1938, he won a Nobel Prize for his work in elementary particle physics, and, four years later, in Chicago, produced the first sustained nuclear chain reaction, thereby ushering in the age of atomic weapons and commercial nuclear power. No other physicist of his generation, and no one since, had been at once a masterful experimentalist and a leading theoretician. In miniature, the bits of paper and the analysis of their motion exemplified this unique combination of gifts.

Like all virtuosos, Fermi had a distinctive style. His approach to physics brooked no opposition; it simply never occurred to him that he might fail to find the solution to a problem. His scientific papers and books reveal a disdain for embellishment—a preference for the most direct, rather than the most intellectually elegant, route to an answer. When he reached the limits of his cleverness, Fermi completed a task by brute force.

To illustrate this approach, imagine that a physicist must determine the volume of an irregular object—say, the earth, which is slightly pear-shaped. Without some kind of formula, one might feel stymied; there are several ways of getting one. Consulting a mathematician is one way, but finding one with enough knowledge and interest to be of help is usually difficult. Searching through the mathematical literature is another, though time-consuming and probably fruitless, because the ideal shapes that interest mathematicians often do not match those of the irregular objects found in nature. Finally the physicist could consider setting aside all other research in order to derive the formula from basic mathematical principles, but, of course, anyone who wanted to devote vast amounts of time to theoretical geometry wouldn't have become a physicist.

Alternatively, the physicist could do what Fermi would have done—compute the volume *numerically*. Instead of relying on a formula, this could per-

Continued

Enrico Fermi (1901–1954).

remotest notion what the answer might be. And one feels certain that too little information exists to find a solution. Yet, when the problem is broken down into subproblems, each one answerable without the help of experts or reference books, an estimate can be made, either mentally or on the back of an envelope, that comes remarkably close to the exact solution.

Suppose, for example, that one wants to determine the earth's circumference without looking it up. If one knows that New York and Los Angeles are separated by about three thousand miles and that the time difference between the two coasts is three hours. Three hours corresponds to one-eighth of a day, and a day is the time it takes the planet to complete one revolution. Therefore, the circumference of the earth can be estimated to be eight times three thousand, or twenty-four thousand miles. At the equator, the earth's circumference is in fact 24,902 miles. In John Milton's words:

> so easy it seemed
> Once found, which yet unfound most
>   would have thought
> Impossible.

Fermi problems might seem to resemble the brainteasers that appear among the back pages of airline magazines and other popular publications (given three containers that hold eight, five, and three liters, respectively, how do you measure out a single liter?), but the two genres differ significantly. The answer to a Fermi problem, in contrast to that of a brainteaser, cannot be verified by logical deduction alone and is always approximate. (To determine precisely the circumference of the earth, it is necessary that the planet actually be measured.) Then, too, solving a Fermi problem requires a knowledge of facts not mentioned in the statement of the problem. (In contrast, the decanting puzzle above contains all the information necessary for its solution.)

These differences mean that Fermi problems are more closely tied to the physical world than are mathematical puzzles, which rarely have anything practical to offer physicists. By the same token, Fermi problems are reminiscent of the ordinary dilemmas that nonphysicists encounter every day of their lives. Indeed, Fermi problems, and the way they are solved, are not only essential to the practice of physics but also valuable for the lesson they give in the art of living.

How many piano tuners are there in Chicago? The whimsical nature of this question, the improb-

haps involve mentally dividing the planet into a number of small cubes, each with a volume easily determined by multiplying the length times the width times the height, and then adding together the answers to these more tractable problems (increasing the number of these cubes will increase the accuracy of the calculation). This method yields only an approximate solution, but because it does not depend on any unknown formulas, it is sure to produce the desired result; this is what mattered to Fermi. With the introduction, after the Second World War, of computers and, later, pocket calculators, numerical computation has become standard procedure in physics.

The technique of dividing difficult problems into small, manageable ones applies to many problems besides those amenable to numerical computation. Fermi excelled at this rough-and-ready method. To pass it on to his students, he developed a type of question that has become associated with his name. A Fermi problem has a characteristic profile. Upon first hearing it, one doesn't have even the ability that anyone knows the answer, and the fact that Fermi posed it to his classes at the University of Chicago have elevated it to the status of legend. There is no standard solution (that's exactly the point), but anyone can make assumptions that quickly lead to an approximate answer. Here is one way: If the population of metropolitan Chicago is three million, an average family consists of four people, and one-third of all families own pianos, there are 250,000 pianos in the city. If each piano is tuned every ten years, there are 25,000 tunings a year. If each tuner can service 4 pianos a day, 250 days a year, for a total of 1000 tunings a year, there must be about 25 piano tuners in the city. The answer is not exact; it could be as low as 10 or as high as 50. But, as the yellow pages of the telephone directory attest, it is definitely in the ballpark.

Fermi's intent was to show that although at the outset even the answer's order of magnitude is unknown, one can proceed on the basis of different assumptions and still arrive at estimates that fall within the range of the answer. The reason is that, in any string of calculations, errors tend to cancel out one another. Every sixth, rather than third, family might be assumed to own a piano, but then pianos might just as well be tuned every five, rather than ten, years. It is just as improbable that all of one's errors will be underestimates (or overestimates) as it is that all the throws in a series of coin tosses will be heads (or tails). The law of probabilities dictates that deviations from the correct assumptions will tend to compensate for one another, so the final results will converge toward the right number. An essential caveat is that one must be cautious that no possible source of bias favors deviations in one direction over another.

Of course, the Fermi problems that physicists face deal with atoms and molecules more often than pianos. To answer them, one needs to commit to memory a few basic magnitudes, such as the approximate diameter of a typical atom or the number of molecules in a thimbleful of water. Equipped with such facts, one can estimate, in a fashion similar to that given in Example 1-5, the amount by which a car tire is worn down in one revolution during normal driving conditions. Assume that the tread is about 1 cm thick and that it wears off in 60,000 km of driving. If 1 cm is divided by the number of revolutions made by a typical tire, with its typical circumference, in 60,000 km, the answer is roughly one molecular diameter.

Another Fermi problem illustrates the immensity of the number of atoms and molecules that surround us. The task is to prove a theorem called "Caesar's Last Breath," which states that with every breath you take, you inhale a single molecule of the air that Julius Caesar exhaled as he died. Behind the statement lurks a qualification and several assumptions. The qualification is that, since the theorem is based on averages, it can only be approximately true. If you take three or four breaths without encountering one of Caesar's air molecules, don't be disappointed; later you may inhale several at once. One of the assumptions is that over the last two millennia no air molecules have been added to the atmosphere or been taken out of circulation by becoming attached to other molecules, which isn't true. Also taken for granted is that Caesar's last breath has had a chance to mingle evenly throughout the entire atmosphere, which isn't quite realistic either. But if both assumptions *were* tenable, the theorem would be true and could also be applied to the dying gasp of Attila, Socrates, or Jesus.

The point of the parable is that, although the earth's atmosphere is vast, the number of molecules exhaled in a single human breath is also vast. In order to prove the theorem, you estimate the total volume of the atmosphere, and then divide it by the volume of air in your lungs. The answer is the number of lungsful of air in the atmosphere, and this turns out to be approximately equal to the estimated number of molecules in Caesar's last breath. Both are roughly equal to one-tenth of Avogadro's number, 6×10^{23}, a number that physicists carry around in their heads.

More momentous Fermi problems might concern energy policy (the number of solar cells required to produce a certain amount of electricity), environmental quality (the amount of acid rain caused annually by coal consumption in the United States), or the arms race. Prudent physicists—those who want to avoid false leads and dead ends—operate according to a long-standing principle: Never start a lengthy calculation until you know the range of values within which the answer is likely to fall (and, equally important, the range within which the answer is *un*likely to fall). These physicists attack every problem as if it were a Fermi problem, estimating the order of magnitude of the result before engaging in an investigation.

Physicists also use Fermi problems to communicate with one another. Whether gathering in university hallways, convention-center lobbies, or French restaurants, as they describe, say, new experiments

Continued

or discuss unfamiliar subjects, they often first survey the lay of the land, staking out, in a numerical way, the perimeter of the problem at hand. Those accustomed to tackling Fermi problems approach the experiment or subject as if it were their own, demonstrating their understanding of the underlying physical principles by performing rough calculations. If the conversation turns to a new particle accelerator, for example, physicists might estimate the strength of the magnets it requires; if the subject is the structure of a novel crystal, they could calculate the spacing between its atoms. The goal is to arrive at a reasonable answer using the least effort. It is this spirit of independence, which he himself possessed in ample measure, that Fermi sought to instill by posing his unconventional problems.

Questions about atom bombs, piano tuners, automobile tires, particle accelerators, and crystal structure have little in common. But the manner in which they are answered is the same in every case and can be applied to questions outside the realm of physics. Whether the problem concerns cooking, automobile repair, or personal relationships, there are two basic types of responses. The fainthearted turn to authority—reference books, bosses, expert consultants, physicians, or ministers—whereas the independent of mind delve into that private store of common sense and limited factual knowledge that everyone possesses, make reasonable assumptions, and derive their own, admittedly approximate, solutions. To be sure, it would be foolish to practice neurosurgery at home, but mundane challenges, such as chili from scratch, replacing a water pump, or resolving a family quarrel, can often be sorted out with nothing more than logic, common sense, and patience.

Not everyone trusts these informal approaches. For example, some people were probably skeptical of Fermi's analysis of a two-billion-dollar bomb test that he carried out with the aid of a handful of confetti. Such an attitude demonstrates less, perhaps, about their knowledge of the problem than about their attitude toward life.

Enrico Fermi and Niels Bohr on a stroll.

Ultimately, the value of dealing with the problems of science, or everyday life, the way Fermi did lies in the rewards one gains for making independent discoveries and inventions. It doesn't matter whether the discovery is as momentous as the determination of the yield of an atom bomb or as insignificant as an estimate of the number of piano tuners in a midwestern city. Looking up the answer, or letting someone else find it, actually impoverishes one; it robs one of the pleasure and pride that accompany creativity and deprives one of an experience that, more than anything else in life, bolsters self-confidence. Self-confidence, in turn, is the essential prerequisite for solving Fermi problems. Thus, approaching personal dilemmas as Fermi problems can become, by a kind of chain reaction, a habit that enriches life.

Suggestions for Further Reading

McMahon, Thomas A., and John Tyler Bonner: "The Physics of Dimensions," Chapter 3 of *On Size and Life*, Scientific American Books, New York, 1983.

Submarines, airplanes, and stringed instruments are used as examples to illustrate the principles of dimensional analysis. This prepares the reader for later chapters in which the importance of scale to living things is considered.

Root-Bernstein, Robert S.: "Setting the Stage for Discovery: Breakthroughs Depend on More Than Luck," *The Sciences*, May/June 1988, p. 26.

The author's view of the nature of scientific discovery, illustrated with examples from the life sciences and astronomy.

von Baeyer, Hans Christian: "New Measures of Matter," *The Sciences*, May/June 1988, p. 2.

How the atom is replacing man as the measure of all things: a personal view of the significance of the redefinition of the second and meter.

von Baeyer, Hans Christian: "How Fermi Would Have Fixed It," *The Sciences*, September/October 1988, p. 2.

Enrico Fermi's approach to obtaining approximate solutions, illustrated with entertaining examples.

von Baeyer, Hans Christian: "The Inverse Problem," *The Sciences*, November/December 1985, p. 13.

Two approaches to deducing general principles from observations in physics: working backwards versus guessing and working forwards.

von Baeyer, Hans Christian: "Rainbows, Whirlpools, and Clouds: A Field Guide to the Physics of a Summer's Day," *The Sciences*, July/August 1984, p. 24.

An introduction to physics as a way of thinking about and investigating the world around you.

Review

A. Objectives: After studying this chapter you should:

1. Be able to define the units of length, time, and mass.

2. Know what is meant by SI units, U.S. customary units, and cgs units.

3. Know what conversion factors are, and be able to use them to convert from one kind of unit to another.

4. Know what is meant by the dimensions of a quantity.

B. Define, explain, or otherwise identify the following:

Units
SI units
Metric system
U.S. customary system
Conversion factor
Dimensions
Significant figure

C. True or false: If the statement is true, explain why it is true. If it is false, give a counterexample, that is, an example that contradicts the statement.

1. Two quantities to be added must have the same dimensions.

2. Two quantities to be multiplied must have the same dimensions.

3. All conversion factors have the value 1.

Problems

Level I

1-1 Units

1. Write the following using the prefixes listed in Table 1-1 and the abbreviations listed on the inside cover. For example, 10,000 meters = 10 km. (*a*) 1,000,000 watts; (*b*) 0.002 gram; (*c*) 3×10^{-6} meter; (*d*) 30,000 seconds.

2. Write the following without using prefixes: (*a*) 40 μW; (*b*) 4 ns; (*c*) 3 MW; (*d*) 25 km.

3. Write out the following (which are not SI units) without using any abbreviations. For example, 10^3 meters = 1 kilometer. (*a*) 10^{-12} boo; (*b*) 10^9 low; (*c*) 10^{-6} scope; (*d*) 10^{-18} boy; (*e*) 10^6 phone; (*f*) 10^{-9} goat; (*g*) 10^{12} bull.

4. In the following equations, the distance x is in meters, the time t is in seconds, and the velocity v is in meters per second. What are the SI units of the constants C_1 and C_2? (*a*) $x = C_1 + C_2 t$; (*b*) $x = \frac{1}{2} C_1 t^2$; (*c*) $v^2 = 2 C_1 x$; (*d*) $x = C_1 \cos C_2 t$; (*e*) $v = C_1 e^{-C_2 t}$. *Hint:* The arguments of trigonometric functions and exponentials must be dimensionless. The "argument" of $\cos \theta$ is θ and that of e^x is x.

5. If x is in feet, t is in seconds, and v is in feet per second, what are the units of the constants C_1 and C_2 in each part of Problem 4?

1-2 Conversion of Units

6. From the original definition of the meter in terms of the distance from the equator to the North Pole, find in meters (*a*) the circumference and (*b*) the radius of the earth. (*c*) Convert your answers to (*a*) and (*b*) from meters into miles.

7. Complete the following: (a) 100 km/h = _____ mi/h; (b) 60 cm = _____ in; (c) 100 yd = _____ m.

8. In the following, x is in meters, t is in seconds, v is in meters per second, and the acceleration a is in meters per second squared. Find the SI units of each combination: (a) v^2/x; (b) $\sqrt{x/a}$; (c) $\frac{1}{2}at^2$.

9. Find the conversion factor to convert from miles per hour into kilometers per hour.

10. (a) Find the number of seconds in a year. (b) If one could count $1 per second, how many years would it take to count 1 billion dollars (1 billion = 10^9)? (c) If one could count one molecule per second, how many years would it take to count the molecules in a mole? (The number of molecules in a mole is Avogadro's number, $N_A = 6.02 \times 10^{23}$.)

11. Complete the following: (a) 1.296×10^5 km/h^2 = _____ km/h · s; (b) 1.296×10^5 km/h^2 = _____ m/s^2; (c) 60 mi/h = _____ ft/s; (d) 60 mi/h = _____ m/s.

12. There are 1.057 quarts in a liter and 4 quarts in a gallon. (a) How many liters are there in a gallon? (b) A barrel equals 42 gallons. How many cubic meters are there in a barrel?

13. There are 640 acres in a square mile. How many square meters are there in one acre?

14. Sometimes a conversion factor can be derived from the knowledge of a constant in two different systems. (a) The speed of light in a vacuum is 186,000 mi/s = 3×10^8 m/s. Use this fact to find the number of kilometers in a mile. (b) The weight of 1 ft^3 of water is 62.4 lb. Use this and the fact that 1 cm^3 of water has a mass of 1 g to find the weight in pounds of a 1-kg mass.

1-3 Dimensions of Physical Qualities

15. What are the dimensions of the constants in each part of Problem 4?

16. The SI unit of force, the kilogram-meter per second squared (kg·m/s^2), is called the newton (N). Find the dimensions and the SI units of the constant G in Newton's law of gravitation $F = Gm_1m_2/r^2$.

17. An object on the end of a string moves in a circle. The force exerted by the string depends on the mass of the object, its speed, and the radius of the circle. What combination of these variables gives the correct dimensions (ML/T^2) for the force?

18. Kepler's third law relates the period of a planet to its radius r, the constant G in Newton's law of gravitation ($F = Gm_1m_2/r^2$), and the mass of the sun M_s. What combination of these factors gives the correct dimensions for the period of a planet?

1-4 Scientific Notation

19. Express each of the following as a decimal number without using power-of-10 notation: (a) 3×10^4; (b) 6.2×10^{-3}; (c) 4×10^{-6}; (d) 2.17×10^5.

20. Write the following in scientific notation in watts, seconds, and meters: (a) 3.1 GW = _____ W; (b) 10 pm = _____ m; (c) 2.3 ks = _____ s; (d) 4 μs = _____ s.

21. Calculate the following, round off to the correct number of significant figures, and express your result in scientific notation: (a) $(1.14)(9.99 \times 10^4)$; (b) $(2.78 \times 10^{-8}) - (5.31 \times 10^{-9})$; (c) $12\pi/(4.56 \times 10^{-3})$; (d) $27.6 + (5.99 \times 10^2)$.

22. Calculate the following, round off to the correct number of significant figures, and express your result in scientific notation: (a) $(200.9)(569.3)$; (b) $(0.000000513)(62.3 \times 10^7)$; (c) $28{,}401 + (5.78 \times 10^4)$; (d) $63.25/(4.17 \times 10^{-3})$.

23. A cell membrane has a thickness of about 7 nm. How many cell membranes would it take to make a stack 1 in high?

24. Calculate the following, round off to the correct number of significant figures, and express your result in scientific notation: (a) $(2.00 \times 10^4)(6.10 \times 10^{-2})$; (b) $(3.141592)(4.00 \times 10^5)$; (c) $(2.32 \times 10^3)(1.16 \times 10^8)$; (d) $(5.14 \times 10^3) + (2.78 \times 10^2)$; (e) $(1.99 \times 10^2) + (9.99 \times 10^{-5})$.

25. Perform the following calculations and round off the answers to the correct number of significant figures: (a) $3.141592654 \times (23.2)^2$; (b) $2 \times 3.141592654 \times 0.76$; (c) $\frac{4}{3}\pi \times (1.1)^3$; (d) $(2.0)^5/(3.141592654)$.

26. The sun has a mass of 1.99×10^{30} kg. The sun is mostly hydrogen, with only a small amount of heavier elements. The hydrogen atom has a mass of 1.67×10^{-27} kg. Estimate the number of hydrogen atoms in the sun.

Level II

27. The angle subtended by the moon's diameter at a point on the earth is about 0.524° (Figure 1-2). Use this and the fact that the moon is about 384 Mm away to find the diameter of the moon. (The angle θ subtended by the moon is approximately D/r_m, where D is the diameter of the moon and r_m is the distance to the moon.)

Figure 1-2 Problem 27.

28. The astronomical unit is defined as the mean distance from the earth to the sun, namely, 1.496×10^{11} m. The parsec is the radial length at which one astronomical unit of arc length subtends an angle of 1 second (Figure 1-3). The lightyear is the distance that light travels in one year. (a) How many parsecs are there in one astronomical unit? (b) How many meters are there in a parsec? (c) How many meters are there in a lightyear? (d) How many astronomical units are there in a lightyear? (e) How many lightyears are there in a parsec?

Figure 1-3 Problem 28.

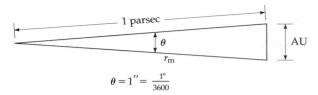

$$\theta = 1'' = \frac{1°}{3600}$$

29. The United States imports 6 million barrels of oil per day. This imported oil provides about one-fourth of our total energy needs. A barrel fills a drum that stands about 1 m high. (*a*) If the drums are laid end to end, what is the length in kilometers of the oil imported each day? (*b*) The largest tankers hold about a quarter million barrels. How many tanker loads per year would supply our imported oil? (*c*) If oil costs $20 a barrel, how much do we spend for imported oil each year?

30. Every year the United States generates 160 million tons of municipal solid waste and a grand total of 10 billion tons of solid waste of all kinds. If one allows one cubic meter of volume per ton of waste, how many square miles of area at an average height of 10 m is needed for landfills each year?

31. An iron nucleus has a radius of 5.4×10^{-15} m and a mass of 9.3×10^{-26} kg. (*a*) What is its mass per unit volume in kilograms per cubic meter? (*b*) If the earth has the same mass per unit volume, what would its radius be? (The mass of the earth is 5.98×10^{24} kg.)

32. If the universe is to stop expanding and begin contracting, its average density must be at least 6×10^{-27} kg/m^3. (*a*) How many electrons are needed in a cubic meter to produce the critical density? (*b*) How many protons per cubic meter would produce the critical density? (The mass of an electron and of a proton are listed on the inside cover.)

33. Observational estimates of the density of the universe yield an average of about 2×10^{-28} kg/m^3. (*a*) If a 100-kg football player had his mass uniformly spread out in a sphere to match the estimate for the average density of the universe, what would be the radius of the sphere? (*b*) Compare this radius with the earth–moon distance of 3.84×10^8 m.

34. Beer and soft drinks are sold in aluminum cans. The mass of a typical can is about 0.018 kg. (*a*) Estimate the number of aluminum cans used in the United States in one year. (*b*) Estimate the total mass of aluminum in a year's consumption from these cans. (*c*) Recycled aluminum sells for about $1 per kilogram. How much is this year's accumulation of aluminum cans worth?

Level III

35. The following table gives experimental results for the measurement of the period of motion T of an object of mass m suspended on a spring versus the mass of the object. These data are consistent with a simple equation expressing T as a function of m of the form $T = Cm^n$,

where C and n are constants and n is not necessarily an integer. (*a*) Find n and C. (There are several ways to do this. One way is to guess the value of n and check by plotting T versus m^n on graph paper. If your guess is right, the plot will be a straight line. Another way is to plot log T versus log m. The slope of the straight line on this plot is n.) (*b*) Which data points deviate the most from a straight-line plot of T versus m^n?

Mass m, kg	0.10	0.20	0.40	0.50	0.75	1.00	1.50
Period T, s	0.56	0.83	1.05	1.28	1.55	1.75	2.22

36. The table that follows gives the period T and orbit radius r for the motions of four satellites orbiting a dense, heavy asteroid.

Period T, y	0.44	1.61	3.88	7.89
Radius r, Gm	0.088	0.208	0.374	0.600

(*a*) These data can be fitted by the formula $T = Cr^n$. Find C and n. (*b*) A fifth satellite is discovered to have a period of 6.20 y. Find the radius for the orbit of this satellite, which fits the same formula.

37. The period T of a simple pendulum depends on the length L of the pendulum and the acceleration of gravity g (dimensions L/T^2). (*a*) Find a simple combination of L and g that has the dimensions of time. (*b*) Check the dependence of the period T on the length L by measuring the period (the time for a complete to-and-fro swing) of a pendulum for two different values of L. (*c*) The correct formula relating T to L and g involves a constant that is a multiple of π and cannot be obtained by the dimensional analysis of part (*a*). It can be found by experiment as in part (*b*) if g is known. Using the value $g = 9.81$ m/s^2 and your experimental results from part (*b*), find the formula relating T to L and g.

38. A projectile fired at an angle of 45° travels a total distance R, called the range, which depends only on the initial speed v and the acceleration of gravity g (dimensions L/T^2). Using dimensional analysis, find how R depends on the speed and on g.

39. A ball thrown horizontally from a height H with speed v travels a total horizontal distance R. (*a*) Do you expect R to increase or decrease with increasing H? With increasing v? (*b*) From dimensional analysis, find a possible dependence of R on H, v, and g.

40. The earth's atmosphere pushes down on the surface of the earth with a force of 14.7 lb for each square inch of the earth's surface. What is the weight in pounds of the earth's atmosphere? (The radius of the earth is about 6370 km.)

41. Each binary digit is termed a bit. A series of bits grouped together is called a word. An eight-bit word is called a byte. Suppose a computer disk has a capacity of 100 megabytes. (*a*) How many bits can be stored on the disk? (*b*) Estimate the number of typical books that can be stored on the disk.

Part 1 Mechanics

Modern version of Galileo's legendary experiment in which he dropped a heavy ball and a light ball from the Leaning Tower of Pisa. In this demonstration, a feather and an apple fall together in a large vacuum chamber with the same acceleration due to gravity. The light bar at the top of the photograph is a portion of a trap door, which initially held the apple and feather at rest, and then dropped them simultaneously.

Chapter 2

Motion in One Dimension

This jet-fuel powered car on the Bonneville Salt Flats deploys its chute to slow it down after a qualifying run of 243.3 km/h.

The description of the motion of objects is an important part of the description of the physical universe. Indeed, it was central to the development of science from Aristotle to Galileo. The laws of how things fall were developed long before Newton described why things fall. One of the earliest scientific puzzles concerned the apparent motion of the sun across the sky and the seasonal motion of the planets and stars. A great triumph of newtonian mechanics was the discovery that the motion of the earth and the other planets around the sun could be explained in terms of a force of attraction between the sun and the planets.

In this chapter and the next, we will be concerned with the description of motion (kinematics) without worrying about its causes. (We will consider the causes of motion in Chapter 4 when we study Newton's laws.) We will confine our discussion in this chapter to motion in one dimension, that is, motion along a straight line. A simple example of one-dimensional motion is a car moving along a flat, straight, narrow road. For such restricted motion, there are only two possible directions, which we distinguish by designating one positive and the other negative.

To simplify our discussion of motion, we will start with objects whose position can be described by locating one point. Such an object is called a

particle. One tends to think of a particle as a very small object, but actually no size limit is implied by the word "particle." For example, it is sometimes convenient to consider the earth to be a particle moving around the sun in a nearly circular path. (Certainly, when viewed from a distant planet or a distant galaxy, the earth looks like a point.) In such cases, we are interested only in the motion of the center of the earth, so we ignore the size of the earth and its rotation. In some astronomical problems, the entire solar system or even a whole galaxy is sometimes treated as a particle. When we are analyzing the rotation or the internal structure of an object, we can no longer treat it as a single particle. But our study of particle motion is useful even in these cases because any object, no matter how complex, can be treated as a collection or "system" of particles.

2-1 Speed, Displacement, and Velocity

We are all familiar with the concept of speed. We define the **average speed** of a particle as the ratio of the total distance traveled to the total time taken:

$$\text{Average speed} = \frac{\text{total distance}}{\text{total time}}$$

The SI unit of average speed is meters per second (m/s), and the U.S. customary unit is feet per second (ft/s). The familiar, everyday unit of speed in the United States is miles per hour (mi/h). Internationally, the more common unit is kilometers per hour (km/h). If you drive 200 km in 5 h, your average speed is (200 km)/(5 h) = 40 km/h. The average speed tells you nothing about the details of the trip. You may have driven at a steady rate of 40 km/h for the whole 5 h, or you may have driven faster part of the time and slower the rest of the time, or you may have stopped for an hour and then driven at varying rates during the other 4 h.

The concept of velocity is similar to that of speed but differs because it includes the *direction* of motion. To understand this concept, we first introduce the idea of displacement. Let us set up a coordinate system by choosing some reference point on a line for the origin O. To every other point on the line we assign a number x that indicates how far the point is from the origin. The value of x depends on the unit (feet, meters, or whatever) chosen to measure the distance. The sign of x depends on its position relative to the origin O. The usual convention we choose is that points to the right of the origin are given positive values and those to the left, negative values.

Figure 2-1 shows a car (which we can treat as a particle) that is at position x_1 at some time t_1 and at position x_2 at time t_2. The change in the position of the particle, $x_2 - x_1$, is called the **displacement** of the particle. It is customary to use the Greek letter Δ (capital delta) to indicate the change in quantity. Thus, the change in x is written Δx:

$$\Delta x = x_2 - x_1 \qquad\qquad \text{2-1} \qquad\qquad \textit{Displacement defined}$$

The notation Δx (read "delta x") stands for a single quantity, the change in x. It is not a product of Δ and x any more than $\cos\theta$ is a product of cos and θ.

$$\Delta x = x_2 - x_1$$

Figure 2-1 When the car travels from point x_1 to point x_2, its displacement is $\Delta x = x_2 - x_1$.

Velocity is the rate at which the position changes. The **average velocity** of the particle is defined as the ratio of the displacement Δx to the time interval $\Delta t = t_2 - t_1$:

Average velocity defined

$$v_{av} = \frac{\Delta x}{\Delta t} = \frac{x_2 - x_1}{t_2 - t_1} \qquad 2\text{-}2$$

Note that the displacement and the average velocity may be either positive or negative, depending on whether x_2 is greater or less than x_1. A positive value indicates motion to the right and a negative value indicates motion to the left.

Example 2-1

A snail is at $x_1 = 18$ mm at $t_1 = 2$ s and is later found at $x_2 = 14$ mm at $t_2 = 7$ s. Find the displacement and the average velocity of the snail for this time interval.

By the definition, the displacement of the snail is

$$\Delta x = x_2 - x_1 = 14 \text{ mm} - 18 \text{ mm} = -4 \text{ mm}$$

and the average velocity is

$$v_{av} = \frac{\Delta x}{\Delta t} = \frac{x_2 - x_1}{t_2 - t_1} = \frac{14 \text{ mm} - 18 \text{ mm}}{7 \text{ s} - 2 \text{ s}} = \frac{-4 \text{ mm}}{5 \text{ s}} = -0.8 \text{ mm/s}$$

The displacement and average velocity are negative, indicating that the snail moved to the left.

Note that the unit millimeters per second is included as part of the answer for the average velocity found in Example 2-1. Since there are many other possible choices for units of length (for example, feet, inches, miles, lightyears) and time (for example, hours, days, years), it is essential to include the unit with a numerical answer. The statement "the average velocity of a particle is -3" is meaningless.

Example 2-2

How far does a car go in 5 min if its average velocity is 80 km/h during this time interval?

We are interested in the displacement during a time interval of 5 min. From Equation 2-2, the displacement Δx is given by

$$\Delta x = v_{av} \, \Delta t$$

Since the time is given in minutes and the average velocity in kilometers per hour, we must convert either the time to hours or the average velocity to kilometers per minute. Doing the latter, we obtain

$$v_{av} = \frac{80 \text{ km}}{1 \text{ h}} \times \frac{1 \text{ h}}{60 \text{ min}} = \frac{4 \text{ km}}{3 \text{ min}}$$

Then

$$\Delta x = \frac{4 \text{ km}}{3 \text{ min}} \times 5 \text{ min} = 6.67 \text{ km}$$

Exercise

A runner runs with an average velocity of 0.25 km/min along a straight line. How long does it take to run a distance of 10 km?   (Answer: 40 min)

Example 2-3

A runner runs 100 m in 12 s, then turns around and jogs 50 m back toward the starting point in 30 s. What are her average *speed* and average *velocity* for the total trip?

The total distance traveled is 100 m + 50 m = 150 m, and the total time taken is 42 s. The average speed is therefore $(150 \text{ m})/(42 \text{ s})$ = 3.57 m/s. Note that this is not the average of her running and jogging speeds because she ran for 12 s but jogged for 30 s. To find the average velocity, we first find the total displacement. If her starting point x_1 is taken to be 0, then her final point x_2 is 50 m, so her total displacement is $x_2 - x_1 = 50$ m. The average velocity is then

$$v_{av} = \frac{\Delta x}{\Delta t} = \frac{50 \text{ m}}{42 \text{ s}} = +1.19 \text{ m/s}$$

Again, this is not the average of her running velocity (+8.33 m/s) and her jogging velocity (−1.67 m/s) because the times are different.

Figure 2-2 shows a graph of x versus t for some arbitrary motion along the x axis. Each point on the curve has an x value, which is the location of the particle at a particular time, and a t value, which is the time the particle was at that location. On the graph, we have drawn a straight line between the position labeled P_1 and the position labeled P_2. The displacement $\Delta x = x_2 - x_1$ and the time interval $\Delta t = t_2 - t_1$ between these points are indicated. The line between P_1 and P_2 is the hypotenuse of the triangle having sides Δx and Δt. The ratio $\Delta x / \Delta t$ is called the **slope** of this straight line. In geometric terms, it is a measure of the steepness of the straight line in the graph. For a given interval Δt, the steeper the line, the greater the value of $\Delta x / \Delta t$. Since the slope of this line is just the average velocity for the interval Δt, we have a geometric representation of average velocity.

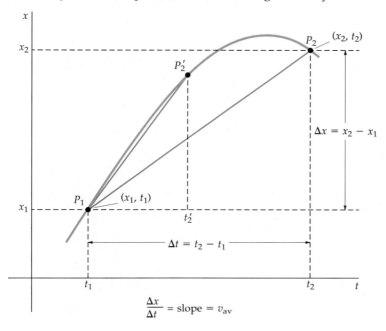

Figure 2-2 Graph of x versus t for a particle moving in one dimension. The points P_1 and P_2 are connected by a straight line. The average velocity is the slope of this line, $\Delta x / \Delta t$, which depends on the time interval, as indicated by the fact that the line from P_1 to P_2' has a greater slope than the line from P_1 to P_2.

> The average velocity is the slope of the straight line connecting the points (x_1, t_1) and (x_2, t_2).

Unless the velocity is constant, the average velocity will depend on the time interval on which it is based. For example, in Figure 2-2, if we chose a smaller time interval by choosing a time t_2' closer to t_1, the average velocity would be greater, as indicated by the greater steepness of the line connecting points P_1 and P_2'.

Questions

1. What sense, if any, does the following statement make? "The average velocity of the car at 9 A.M. was 60 km/h."

2. Is it possible for the average velocity for some interval to be zero even though the average velocity for a shorter interval included in the first interval is not zero? Explain.

3. What is the approximate average velocity of the race cars during the Indianapolis 500?

2-2   Instantaneous Velocity

At first glance, to define the velocity of a particle at a single instant, that is, at a specific time, might seem impossible. At a time t_1, the particle is at a single point x_1. If it is at a single point, how can it be moving? On the other hand, if it is not moving, shouldn't it stay at the same point? This is an age-old paradox, which can be resolved when we realize that, to observe motion and thus define it, we must look at the position of the object at more than one time. It is then possible to define the velocity at an instant by using a limiting process.

The fastest running animal, the cheetah, can reach speeds of greater than 100 km/h.

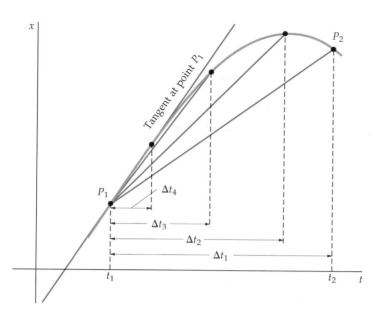

Figure 2-3 Graph of x versus t from Figure 2-2. As the time interval beginning at t_1 is decreased, the average velocity for that interval approaches the slope of the line tangent to the curve at time t_1. The instantaneous velocity is defined as the slope of this line.

Figure 2-3 is the same x-versus-t curve as Figure 2-2, showing a sequence of time intervals, $\Delta t_1, \Delta t_2, \Delta t_3, \ldots$, each smaller than the previous one. For each time interval Δt, the average velocity is the slope of the straight line appropriate for that interval. The figure shows that, as the time interval becomes smaller, the straight lines get steeper, but they never incline more than the tangent to the curve at point t_1. We define the slope of this tangent as the **instantaneous velocity** at the time t_1.

> The instantaneous velocity at some time is the slope of the line tangent to the x-versus-t curve at that time.

It is important to realize that the displacement Δx depends on the time interval Δt. As Δt approaches zero, Δx does too (as can be seen from Figure 2-3), and the ratio $\Delta x/\Delta t$ approaches the slope of the line tangent to the curve. Since the slope of the tangent is the limit of the ratio $\Delta x/\Delta t$ as Δt approaches zero, we can restate our definition as follows:

> The instantaneous velocity is the limit of the ratio $\Delta x/\Delta t$ as Δt approaches zero.

$$v = \lim_{\Delta t \to 0} \frac{\Delta x}{\Delta t} = \frac{\text{slope of the line tangent}}{\text{to the } x\text{-versus-}t \text{ curve}}$$ 2-3   *Instantaneous velocity defined*

This limit is called the **derivative** of x with respect to t. In usual calculus notation, the derivative is written dx/dt:

$$v = \lim_{\Delta t \to 0} \frac{\Delta x}{\Delta t} = \frac{dx}{dt}$$ 2-4

This slope may be positive (x is increasing) or negative (x is decreasing); correspondingly, the instantaneous velocity may be positive or negative in one-dimensional motion. The magnitude of the instantaneous velocity is called the **instantaneous speed.**

Example 2-4

The position of a particle is given by the curve shown in Figure 2-4. Find the instantaneous velocity at the time $t = 2$ s. When is the velocity greatest? When is it zero? Is it ever negative?

In the figure, we have sketched the line tangent to the curve at time $t = 2$ s. The slope of this line is measured from the figure to be (4.5 m)/(3 s) = 1.5 m/s. Thus $v = 1.5$ m/s at time $t = 2$ s. According to the figure, the slope (and therefore the velocity) is greatest at about $t = 4$ s. The velocity is zero at times $t = 0$ and $t = 6$ s, as indicated by the fact that the tangent lines at these times are horizontal and so have zero slopes. After $t = 6$ s, the curve has a negative slope, indicating that the velocity is negative. (The slope of the line tangent to a curve at a given point is often referred to merely as the "slope of the curve" at that point.)

Figure 2-4   Graph of x versus t for Example 2-4. The instantaneous velocity at time $t = 2$ s can be found by measuring the slope of the line tangent to the curve at that time.

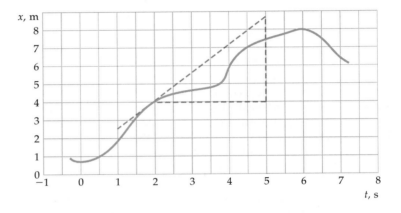

Example 2-5

The position of a stone dropped from rest from a cliff is given by $x = 5t^2$, where x is in meters measured downward from the original position at $t = 0$, and t is in seconds. Find the velocity at any time. (We omit explicit indication of the units to simplify the notation.)

The curve corresponding to $x = 5t^2$ is shown in Figure 2-5. Tangent lines have been drawn for three different times, t_1, t_2, and t_3. The slopes of these lines differ. As time passes, the slope of the curve increases, indicating that the instantaneous velocity is increasing with time. We can compute the velocity at some time t by computing the derivative dx/dt directly from the definition in Equation 2-3. At time t, the position is

$$x(t) = 5t^2$$

At a later time, $t + \Delta t$, the position is $x(t + \Delta t)$, which is given by

$$x(t + \Delta t) = 5(t + \Delta t)^2 = 5[t^2 + 2t\, \Delta t + (\Delta t)^2]$$
$$= 5t^2 + 10t\, \Delta t + 5(\Delta t)^2$$

The displacement for this time interval is thus

$$\Delta x = x(t + \Delta t) - x(t) = [5t^2 + 10t\, \Delta t + 5(\Delta t)^2] - 5t^2$$
$$= 10t\, \Delta t + 5(\Delta t)^2$$

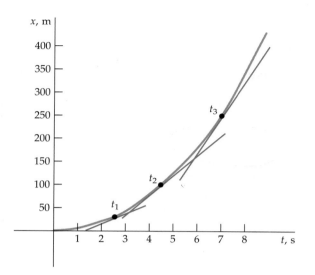

Figure 2-5   Graph of the function $x(t) = 5t^2$ for Example 2-5. Tangent lines are drawn at times t_1, t_2, and t_3. The slopes of these tangent lines increase steadily from t_1 to t_2 to t_3, indicating that the instantaneous velocity increases steadily with time.

The average velocity for this time interval is

$$v_{av} = \frac{\Delta x}{\Delta t} = \frac{10t\,\Delta t + 5(\Delta t)^2}{\Delta t} = 10t + 5\,\Delta t$$

As we consider shorter and shorter time intervals, Δt approaches zero, so the second term, $5\,\Delta t$, approaches zero while the first term, $10t$, remains unaltered. The instantaneous velocity at time t is thus

$$v = \lim_{\Delta t \to 0} \frac{\Delta x}{\Delta t} = 10t$$

For this example, the instantaneous velocity is proportional to the time.

Notice that if we set Δt in Equation 2-3 directly equal to 0, then Δx would also be 0 and the ratio $\Delta x / \Delta t$ would not be defined. However, by starting with an equation that represents how x varies with t, we have calculated exactly the limit of $\Delta x / \Delta t$ as Δt approaches zero.

It is instructive to examine the limiting process numerically by computing the average velocity for smaller and smaller time intervals. Table 2-1 gives the average velocity for Example 2-5 for $t = 2$ s and various time intervals Δt, each smaller than the previous one. The table shows that for very small time intervals, the average velocity is very nearly equal to the instantaneous velocity, which is 20 m/s. The difference between $\Delta x / \Delta t$ and $\lim_{\Delta t \to 0}(\Delta x / \Delta t)$ can be made arbitrarily small by choosing a sufficiently small Δt.

It is important to distinguish carefully between average velocity and instantaneous velocity. By custom, the word "velocity" alone is assumed to mean instantaneous velocity.

Table 2-1   Displacement and Average Velocity for Various Time Intervals Δt Beginning at $t = 2$ s for the Function $x = 5t^2$

Δt, s	Δx, m	$\Delta x / \Delta t$, m/s
1.00	25	25
0.50	11.25	22.5
0.20	4.20	21.0
0.10	2.05	20.5
0.05	1.0125	20.25
0.01	0.2005	20.05
0.005	0.100125	20.025
0.001	0.020005	20.005
0.0001	0.00200005	20.0005

Questions

4.   If the instantaneous velocity does not change, will the average velocities for different intervals differ?

5.   If $v_{av} = 0$ for some time interval Δt, must the instantaneous velocity v be zero at some point in the interval? Support your answer by sketching a possible x-versus-t curve that has $\Delta x = 0$ for some interval Δt.

2-3   Acceleration

When the instantaneous velocity of a particle is changing with time, as in Example 2-5, the particle is said to be *accelerating*. The **average acceleration** for a particular time interval $\Delta t = t_2 - t_1$ is defined as the ratio $\Delta v/\Delta t$, where $\Delta v = v_2 - v_1$ is the change in the instantaneous velocity for that time interval:

Average acceleration defined

$$a_{\text{av}} = \frac{\Delta v}{\Delta t} \qquad\qquad 2\text{-}5$$

The dimensions of acceleration are length divided by $(\text{time})^2$. Convenient units are meters per second per second, more compactly written meters per second squared (m/s^2), or feet per second squared (ft/s^2). For example, if we say that a particle is accelerating at $5 \ \text{m/s}^2$, we mean that, if it begins at rest, after 1 s it is traveling with a velocity of 5 m/s, after 2 s it is traveling with a velocity of 10 m/s, after 3 s it is traveling with a velocity of 15 m/s, and so on.

The **instantaneous acceleration** is the limit of the ratio $\Delta v/\Delta t$ as Δt approaches zero. If we plot the velocity versus time, the instantaneous acceleration at time t is defined as the slope of the line tangent to the curve at that time:

Instantaneous acceleration defined

$$a = \lim_{\Delta t \to 0} \frac{\Delta v}{\Delta t} = \begin{array}{l}\text{slope of the line tangent} \\ \text{to the } v\text{-versus-}t \text{ curve}\end{array} \qquad\qquad 2\text{-}6$$

The acceleration is thus the derivative of the velocity with respect to time. The calculus notation for this derivative is dv/dt. Since velocity is the derivative of the position x with respect to t, the acceleration is the second derivative of x with respect to t, which is usually written d^2x/dt^2. We can see the reason for this notation by writing the acceleration as dv/dt and replacing v with dx/dt:

$$a = \frac{dv}{dt} = \frac{d(dx/dt)}{dt} = \frac{d^2x}{dt^2} \qquad\qquad 2\text{-}7$$

If the velocity is constant, the acceleration is zero because $\Delta v = 0$ for all time intervals. In this case, the slope of the corresponding x-versus-t curve does not change. In Example 2-5 we found that for the position function $x = (5 \ \text{m/s}^2)t^2$, the velocity increases linearly with the time according to $v = (10 \ \text{m/s}^2)t$. In this case, the acceleration is constant and equal to $10 \ \text{m/s}^2$, which is the slope of the corresponding v-versus-t curve.

We will see in Chapter 4 that the acceleration is directly proportional to the net force on the particle.

Example 2-6

A fast car can accelerate from 0 to 90 km/h in 5 s. What is the average acceleration during this period? Compare this with the free-fall acceleration due to gravity, which is $9.81 \ \text{m/s}^2$.

From Equation 2-5, we have for the average acceleration

$$a_{\text{av}} = \frac{\Delta v}{\Delta t} = \frac{90 \ \text{km/h}}{5 \ \text{s}} = 18 \ \text{km/h·s}$$

To compare this with the acceleration due to gravity, we convert to meters per second squared using 1 h = 3600 s = 3.6 ks. Then

$$\frac{18 \text{ km}}{\text{h·s}} \times \frac{1 \text{ h}}{3.6 \text{ ks}} = 5 \text{ m/s}^2$$

This is about half the free-fall acceleration due to gravity.

Exercise

A car is traveling at 45 km/h at time $t = 0$. It accelerates at a constant rate of 10 km/h·s. How fast is it traveling at $t = 2$ s? (Answer: 65 km/h)

Example 2-7

The position of a particle is given by $x = Ct^3$, where C is a constant with the unit m/s^3. Find the velocity and acceleration as functions of time.

As in Example 2-5, we can compute the velocity by computing the derivative dx/dt directly from its definition as $\lim\limits_{\Delta t \to 0} (\Delta x / \Delta t)$ (Equation 2-3). At some time t the position is $x(t) = Ct^3$. At a later time $t + \Delta t$, the position is

$$x(t + \Delta t) = C(t + \Delta t)^3 = Ct^3 + 3Ct^2\,\Delta t + 3Ct(\Delta t)^2 + C(\Delta t)^3$$

The displacement is thus

$$\Delta x = x(t + \Delta t) - x(t) = C(t + \Delta t)^3 - Ct^3$$

$$= Ct^3 + 3Ct^2\,\Delta t + 3Ct(\Delta t)^2 + C(\Delta t)^3 - Ct^3$$

$$= 3Ct^2\,\Delta t + 3Ct(\Delta t)^2 + C(\Delta t)^3$$

The average velocity for this time interval is

$$v_{\text{av}} = \frac{\Delta x}{\Delta t} = 3Ct^2 + 3Ct(\Delta t) + C(\Delta t)^2$$

As $\Delta t \to 0$, the two terms containing Δt approach zero and the term $3Ct^2$ remains unchanged. The instantaneous velocity at time t is thus

$$v = \lim_{\Delta t \to 0} \frac{\Delta x}{\Delta t} = 3Ct^2$$

We find the acceleration by repeating the process and finding the derivative of v with respect to t, which is the second derivative of x with respect to t. We omit some of the algebra in this derivation because it is similar to that just shown. The change in velocity for the time interval from t to $t + \Delta t$ is

$$\Delta v = 3C(t + \Delta t)^2 - 3Ct^2 = 6Ct\,\Delta t + 3C(\Delta t)^2$$

We divide this expression by Δt to obtain the average acceleration for this interval:

$$a_{\text{av}} = \frac{\Delta v}{\Delta t} = 6Ct + 3C\,\Delta t$$

The instantaneous acceleration is thus

$$a = \lim_{\Delta t \to 0} \frac{\Delta v}{\Delta t} = 6Ct$$

In this example the acceleration is not constant but increases with time.

Strobe photograph of a falling apple at 60 flashes per second. The acceleration of the apple is indicated by the greater spacing of the images at the bottom.

In the examples so far, we have computed derivatives directly from the definition by taking the appropriate limit explicitly. It is useful to examine various properties of the derivative and to develop rules that allow us to calculate the derivatives of many functions quickly without applying the definition each time. Appendix Table A-4 contains a list of such rules, followed by brief discussions of their origins. We will use these rules without comment and leave detailed study of them to your calculus course.

Rule 7 in Appendix Table A-4 is used so often that we will repeat it here. If x is a simple power function of t, such as

$$x = Ct^n$$

where C and n are any constants, the derivative of x with respect to t is given by

$$\frac{dx}{dt} = \frac{d}{dt}(Ct^n) = Cnt^{n-1} \qquad 2\text{-}8$$

We have already seen applications of this rule. For example, for the position function of Example 2-7, $x = Ct^3$, we found $v = dx/dt = 3Ct^2$ and $a = dv/dt = 6Ct$.

Questions

6. Give an example of a motion for which the velocity is negative but the acceleration is positive; for example, sketch a graph of v versus t.

7. Give an example of a motion for which both the acceleration and the velocity are negative.

8. Is it possible for a body to have zero velocity and nonzero acceleration?

2-4   Motion with Constant Acceleration

The motion of a particle that has constant acceleration is common in nature. For example, near the surface of the earth all unsupported objects fall vertically with constant acceleration due to gravity provided that air resistance can be neglected. The acceleration due to gravity is designated by g and has the approximate value

$$g = 9.81 \text{ m/s}^2 = 32.2 \text{ ft/s}^2$$

A constant acceleration means that the slope of the v-versus-t curve is constant; that is, the velocity varies linearly with time. If the velocity is v_0 at time $t = 0$, its value v at a later time t is given by

$$v = v_0 + at \qquad 2\text{-}9$$

If the particle starts at x_0 at time $t = 0$ and its position is x at time t, the displacement $\Delta x = x - x_0$ is given by

$$\Delta x = v_{av}t$$

(This is the same as Equation 2-2, with t replacing Δt because we have chosen the initial value of t to be zero.) For constant acceleration, the velocity

"It goes from zero to 60 in about 3 seconds."

© 1987 Sidney Harris

Constant acceleration, v versus t

varies linearly with time and the average velocity is the mean value of the initial and final velocities, as shown in Figure 2-6.* If v_0 is the initial velocity and v the final velocity, the average velocity is $\frac{1}{2}(v_0 + v)$.

$$v_{av} = \tfrac{1}{2}(v_0 + v)$$

2-10   *Constant acceleration, v_{av}*

The displacement is then

$$\Delta x = v_{av}t = \tfrac{1}{2}(v_0 + v)t$$

2-11

Figure 2-6 Average velocity for constant acceleration.

We can eliminate v by substituting $v = v_0 + at$ from Equation 2-9:

$$\Delta x = \tfrac{1}{2}(v_0 + v)t = \tfrac{1}{2}(v_0 + v_0 + at)t = v_0t + \tfrac{1}{2}at^2$$

2-12a

The position function is thus

$$x = x_0 + v_0t + \tfrac{1}{2}at^2$$

2-12b   *Constant acceleration, x versus t*

We can eliminate the time t from Equations 2-9 and 2-12a and obtain a relation among the displacement Δx, the acceleration a, and the initial and final velocities. Solving Equation 2-9 for $t = (v - v_0)/a$ and substituting this result into Equation 2-12a, we obtain

$$\Delta x = v_0t + \tfrac{1}{2}at^2 = v_0\left(\frac{v - v_0}{a}\right) + \tfrac{1}{2}a\left(\frac{v - v_0}{a}\right)^2$$

Multiplying each side by a and writing out the terms, we obtain

$$a\,\Delta x = v_0v - v_0^2 + \tfrac{1}{2}v^2 - v_0v + \tfrac{1}{2}v_0^2 = \tfrac{1}{2}v^2 - \tfrac{1}{2}v_0^2$$

or

$$v^2 = v_0^2 + 2a\,\Delta x$$

2-13   *Constant acceleration, v versus x*

Equation 2-13 is useful, for example, if we want to find the final velocity of a ball dropped from rest from some height x and we are not interested in the time the fall takes.

We will now look at some examples of constant-acceleration problems. The first step in solving such a problem is to choose a convenient coordinate system. When possible, we choose the origin to be the location of the particle at $t = 0$ so that $x_0 = 0$. The choice of the positive direction for the x axis determines the positive direction for the velocity and acceleration. Although this choice is arbitrary, a judicious choice for the positive direction can make the solution of a problem easier. For example, if we have a problem in which we drop a ball from some height, it is easiest to choose the downward direction as positive. Then the acceleration is positive (since the acceleration due

*This is not true if the acceleration is not constant.

to gravity is always downward), and the velocity is also positive since the ball moves only downward. On the other hand, if we throw a ball upwards, it is usually more convenient to choose the upward direction as positive. The acceleration is then negative, and the velocity is positive when the ball is moving upward and negative when it is moving downward.

The next step in solving a constant-acceleration problem is to list the information given in terms of an equation. For example, we would write $v_0 = 0$ if the initial velocity is given as zero. Next, list the quantities to be found; for example, $v = ?$ if the final velocity is asked for. Then choose the equation (from Equations 2-9 to 2-13) that contains the given and unknown quantities, and solve for the unknown quantity. When possible, it is always a good idea to solve a problem in two different ways to check your solution. Finally, you should check your answer to see if it is reasonable. For example, if a ball is thrown upward and you find that it travels 4500 m before reaching its highest point, you have probably made a numerical error.

Example 2-8

A ball is thrown upward with an initial velocity of 30 m/s. If its acceleration is 10 m/s^2 downward, how long does it take to reach its highest point, and what is the distance to the highest point? (In this example, we approximate the acceleration of gravity to be 10 m/s^2 to simplify the calculations.)

We choose the origin at the initial position of the ball and take the upward direction to be positive. The given quantities are $v_0 = 30$ m/s and $a = -10$ m/s^2. The acceleration is negative because it is in the downward direction. As the ball moves upward (v is positive), the velocity decreases from its initial value until it is zero. When the velocity is zero, the ball is at its highest point. It then begins to fall, and the velocity becomes negative, indicating that the ball is moving downward. We wish to find the time $t = ?$ when the velocity v is zero. Equation 2-9 contains the unknown time t in terms of the given quantities v_0, a, and v:

$$v = v_0 + at$$

$$0 = 30 \text{ m/s} + (-10 \text{ m/s}^2)t$$

$$t = \frac{30 \text{ m/s}}{10 \text{ m/s}^2} = 3.0 \text{ s}$$

Note that the units work out correctly.

We can find the distance traveled from Equation 2-11. Since the initial velocity is +30 m/s and the final velocity is 0, the average velocity for the upward motion is 15 m/s. The distance traveled is then

$$\Delta x = v_{av}t = (15 \text{ m/s})(3.0 \text{ s}) = 45 \text{ m}$$

We can also find Δx from Equation 2-12a, but the calculation is somewhat more difficult and there is more of a chance of making a mistake. We will use this method to check our result, however.

$$\Delta x = v_0 t + \tfrac{1}{2}at^2$$

$$= (30 \text{ m/s})(3.0 \text{ s}) + \tfrac{1}{2}(-10 \text{ m/s}^2)(3.0 \text{ s})^2$$

$$= +90 \text{ m} - 45 \text{ m}$$

$$= 45 \text{ m}$$

Example 2-9

What is the total time the ball in Example 2-8 is in the air?

We could guess the answer to be 6 s since, by symmetry, if it takes 3.0 s to rise 45 m, it should take the same time to fall 45 m. This is correct. We can also find the time from Equation 2-12a by setting $\Delta x = 0$.

$$\Delta x = v_0 t + \tfrac{1}{2}at^2 = 0$$

DISPLACEMENT   *FROM START TO END*

Factoring, we obtain

$$t(v_0 + \tfrac{1}{2}at) = 0$$

The two solutions are $t = 0$ and

$$t = -\frac{2v_0}{a} = -\frac{2(30 \text{ m/s})}{-10 \text{ m/s}^2} = 6 \text{ s}$$

The solution $t = 0$ corresponds to the initial condition that the ball was at x_0 at $t = 0$.

Figures 2-7a and b show x-versus-t and v-versus-t curves for the ball in Examples 2-8 and 2-9. Note that at the time 3.0 s, the velocity of the ball is zero but the slope of the v-versus-t curve is not. The slope of the v-versus-t curve has the value -10 m/s^2 at this time and at all other times because the acceleration is constant. Note that when the ball is moving upward, its velocity is positive and decreasing so its speed is decreasing. When it is moving downward, its velocity is negative and decreasing, so its speed is increasing.

Figure 2-7 (*a*) Graph of *x* versus *t* for the ball thrown into the air in Examples 2-8 and 2-9. The curve is a parabola, $x = (30 \text{ m/s})t - \tfrac{1}{2}(10 \text{ m/s}^2)t^2$. (*b*) Graph of *v* versus *t* for the same ball. The velocity decreases steadily from its initial value of 30 m/s to its final value of −30 m/s just before the ball hits the ground. At time $t = 3.0$ s, when the ball is at its highest point, the velocity is zero, but its rate of change is -10 m/s^2, the same as at any other time.

(a)

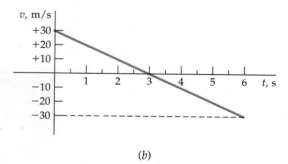

(b)

Exercise

A car accelerates from rest at a constant rate of 8 m/s². (*a*) How fast is it going after 10 s? (*b*) How far has it gone after 10 s? (*c*) What is its average velocity for the interval $t = 0$ to $t = 10$ s? [Answers: (*a*) 80 m/s, (*b*) 400 m, (*c*) 40 m/s]

Example 2-10

A car traveling at 30 m/s (about 67 mi/h) brakes to a stop. If the acceleration is −5 m/s², how far does the car travel before stopping? This distance is called the **stopping distance.**

In this example, we choose the original direction of motion to be positive. The stopping distance will then also be positive, but the acceleration will be negative. (An acceleration that causes the speed to decrease is sometimes called *deceleration*.) In this problem we are given the initial velocity $v_0 = 30$ m/s, the final velocity $v = 0$, the acceleration $a = -5$ m/s^2, and we wish to find the distance traveled $\Delta x = ?$. We do not need to know the time it takes for the car to stop, so Equation 2-13 is the most convenient. Setting $v = 0$ in Equation 2-13, we have

$$v^2 = v_0^2 + 2a\,\Delta x = 0$$

$$0 = (30 \text{ m/s})^2 + 2(-5 \text{ m/s}^2)\,\Delta x$$

$$\Delta x = 90 \text{ m}$$

Note that this is a considerable distance. The force that produces this deceleration is that of friction between the car tires and the road. On wet pavement or gravel, the frictional force is smaller and the magnitude of the acceleration is less than 5 m/s^2, giving an even greater stopping distance.

Example 2-11

What is the stopping distance under the same conditions as Example 2-10 if the car is initially traveling at 15 m/s?

From Equation 2-13 with $v = 0$, we see that the stopping distance is proportional to the square of the initial speed. If we double the speed, the stopping distance is increased by a factor of 4. Similarly, if we halve the initial speed, the stopping distance is reduced by a factor of 4. The stopping distance at 15 m/s is therefore one-fourth that at 30 m/s, or $\frac{1}{4}(90 \text{ m}) = 22.5$ m.

Sometimes, even when the acceleration is not constant, valuable insight can be gained about the motion of an object by assuming that the constant-acceleration formulas still apply to it and analyzing what happens in such an ideal case.

(a) The two-mile long linear accelerator at Stanford University is used to accelerate electrons and positrons in a straight line to nearly the speed of light.
(b) Cross section of the electron beam as shown on a television screen monitor.

(a)   (b)

Example 2-12

A car traveling 100 km/h (about 62 mi/h) crashes into a concrete wall, which does not move. How long does it take the car to come to rest, and what is its acceleration?

In this example, it is not accurate to treat the car as a particle because different parts of it have different accelerations. Moreover, the accelerations are not constant. Nevertheless, let us assume that the crashing of the car can be described by the constant acceleration of a point particle. We need more information in order to find either the stopping time or the acceleration. The information missing is the stopping distance. We can estimate this from our practical knowledge. The center of the car certainly moves less than half the length of the car. (Otherwise the car would be flattened.) A reasonable estimate for the stopping distance is probably between 0.5 and 1.0 m. Let us use 0.75 m as our estimate. We can then find the time it takes the car to stop from $\Delta x = v_{av} \Delta t$, with $v_{av} = \frac{1}{2}v_0 = 50$ km/h $= 14$ m/s. (Since we are making estimates, two significant figures are sufficient.) Then

$$\Delta t = \frac{\Delta x}{v_{av}} = \frac{0.75 \text{ m}}{14 \text{ m/s}} = 0.054 \text{ s}$$

Since the car is brought from $v_0 = 100$ km/h $= 28$ m/s to rest in this time, the acceleration is

$$a = \frac{\Delta v}{\Delta t} = \frac{0 - 28 \text{ m/s}}{0.054 \text{ s}} = -520 \text{ m/s}^2$$

To get a feeling for the magnitude of this acceleration, we note that it is more than 50 times the acceleration due to gravity.

Example 2-13

A car is traveling 80 km/h in a school zone. A police car starts from rest just as the speeder passes it and accelerates at a constant rate of 8 km/h·s. (a) When does the police car catch the speeding car? (b) How fast is the police car traveling when it catches the speeder?

This problem is more difficult because there are two moving objects. We take the origin to be at the original position of both cars, with the positive direction in the direction of motion, and define $t = 0$ to be when the speeder passes the police car.

(*a*) Since the car is traveling with constant velocity, its position x_c is given by Equation 2-12*b*, with $x_0 = 0$ and $a = 0$:

$$x_c = v_0 t = (80 \text{ km/h})t$$

The position of x_p of the police car is given by

$$x_p = \tfrac{1}{2}at^2 = \tfrac{1}{2}(8 \text{ km/h·s})t^2$$

We find the time when the two cars are at the same position by setting $x_c = x_p$ and solving for *t*:

$$(80 \text{ km/h})t = \tfrac{1}{2}(8 \text{ km/h·s})t^2$$

or

$$t(4t \text{ km/h·s} - 80 \text{ km/h}) = 0$$

The two solutions are

$$t = 0$$

corresponding to the initial conditions and

$$t = \frac{80 \text{ km/h}}{4 \text{ km/h·s}} = 20 \text{ s}$$

The police car thus catches the speeder at time $t = 20$ s.

(*b*) The velocity of the police car is given by Equation 2-9, with $v_0 = 0$:

$$v_p = at = (8 \text{ km/h·s})t$$

At $t = 20$ s, the velocity of the police car is

$$v_p = (8 \text{ km/h·s})(20 \text{ s}) = 160 \text{ km/h}$$

At this time, the speed of the police car is twice that of the speeder. This must be true because the average velocity of the police car is half its final velocity, and since both cars cover the same distance in the same time, they must have equal average velocities. Figure 2-8 shows *x*-versus-*t* curves for the two cars.

Questions

9. Two boys standing on a bridge throw two rocks straight down into the water below. They throw the rocks at the same time, but one hits the water before the other. How can this be if the rocks have the same acceleration?

10. A ball is thrown straight up. What is its velocity at the top of its flight? What is its acceleration at that point?

Figure 2-8 Graphs of *x* versus *t* for the speeding car (x_c) and the police car (x_p) for Example 2-13. The curve for the police car has a zero slope at $t = 0$ because the police car starts from rest. It catches the speeding car at $t = 20$ s. Note that at that time the slope of $x_p(t)$ is greater than that of $x_c(t)$. At $t = 20$ s, the police car is traveling twice as fast as the speeding car.

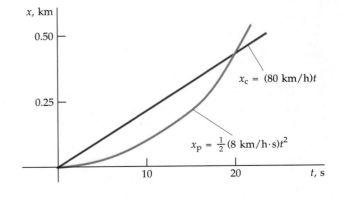

2-5   Integration

We have learned how to obtain the velocity and acceleration functions from a given position function by differentiation. The inverse problem is to find the position function x given the velocity v or the acceleration a. To do this we use a procedure called **integration.** Consider, for example, the problem of finding the velocity and position from a given acceleration. If we know the acceleration as a function of time, we can find the velocity by finding the function $v(t)$ whose derivative is the acceleration. For example, if the acceleration is constant,

$$\frac{dv}{dt} = a$$

the velocity is the function of time that when differentiated equals this constant. One such function is

$$v = at$$

This is not the most general expression for v that will satisfy the relation $dv/dt = a$, however. In particular, we can add any constant to at without changing the value of its derivative with respect to time. Calling this constant v_0, we have

$$v = at + v_0$$

The constant v_0 is the initial velocity. The position function x is that function whose derivative is the velocity:

$$\frac{dx}{dt} = v = v_0 + at$$

We can treat each term separately. The function whose derivative is a constant v_0 is $v_0 t$ plus any constant. The function whose derivative is at is $\frac{1}{2}at^2$ plus any constant. (This result is easily verified by differentiation.) Combining these results and writing x_0 for the combined arbitrary constants, we have for the position function

$$x = x_0 + v_0 t + \tfrac{1}{2}at^2$$

Whenever we find a function from its derivative, we must include an arbitrary constant in the general function. Since we go through the integration process twice to find $x(t)$ from the acceleration, two constants arise. These constants are usually determined from the velocity and position at some given time, which is usually chosen to be $t = 0$. They are therefore called the **initial conditions.** The problem "given $a(t)$, find $x(t)$" is called the **initial-value problem.** The solution depends on the form of the function $a(t)$ and on the values of v and x at some particular time. This problem is particularly important in physics because the acceleration of a particle is determined by the forces acting on it. Thus, if we know the forces acting on a particle and the position and velocity of the particle at some particular time, we can find its position uniquely at all other times.

The problem of integration is related to the problem of finding the area under a curve. Consider the case of a constant velocity v_0. The change in position Δx during some time interval Δt is just the velocity times the time interval:

$$\Delta x = v_0 \, \Delta t$$

This is the area under the v-versus-t curve, as can be seen in Figure 2-9. This

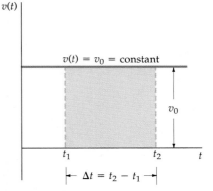

Shaded area $= v_0 \, \Delta t = v_0(t_2 - t_1)$
$\qquad\qquad\quad = v_0 t_2 - v_0 t_1$
$\qquad\qquad\quad = x_2 - x_1 = \Delta x$

Figure 2-9 The displacement for the interval Δt equals the area under the velocity-versus-time curve for that interval. For $v(t) = v_0 = $ constant, the displacement equals the area of the rectangle shown.

graphical interpretation of the displacement as the area under the v-versus-t curve is true both for cases in which the velocity is constant and for cases in which it varies, as in Figure 2-10. In this case, the area under the curve can be approximated by first dividing the time interval into a number of small intervals, Δt_1, Δt_2, and so on, and drawing a set of rectangular

Figure 2-10   Graph of a general $v(t)$-versus-t curve. The displacement for the interval Δt_i is approximately $v_i \, \Delta t_i$, which is indicated by the shaded rectangular area. The total displacement from t_1 to t_2 is the area under the curve for this interval, which can be approximated by summing the areas of the rectangles.

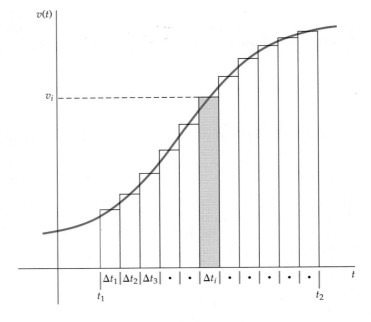

areas. The area of the shaded rectangle is $v_i \, \Delta t_i$, which is approximately equal to the displacement Δx_i of the particle during the interval Δt_i. We can make the approximation as good as we wish by taking enough rectangles and making each Δt very small. The sum of the rectangular areas is therefore approximately the sum of the displacements during the time intervals and is approximately equal to the total displacement from time t_1 to t_2.

Mathematically, we write this as

$$\Delta x \approx \sum_i v_i \, \Delta t_i$$

where the Greek letter Σ (capital sigma) stands for "sum." For the limit of smaller and smaller time intervals, this sum equals the area under the curve, which equals the displacement. This limit is called an **integral** and is written

$$\Delta x = \lim_{\Delta t_i \to 0} \sum_i v_i \, \Delta t_i = \int_{t_1}^{t_2} v \, dt \qquad \text{2-14}$$

It is helpful to think of the integral sign $\int$ as an elongated S indicating a sum. The limits t_1 and t_2 indicate the initial and final values of the variable t.

The change in velocity for some time interval can similarly be interpreted as the area under the a-versus-t curve for that interval. This is written

$$\Delta v = \lim_{\Delta t_i \to 0} \sum_i a_i \, \Delta t_i = \int_{t_1}^{t_2} a \, dt \qquad \text{2-15}$$

The average velocity has a simple geometric interpretation in terms of the area under a curve. Consider the v-versus-t curve in Figure 2-11. The displacement Δx during the time interval $\Delta t = t_2 - t_1$ is indicated by the shaded area. By the definition of the average velocity for this interval (Equation 2-2), the displacement is the product of v_{av} and Δt:

$$\Delta x = v_{av}\,\Delta t$$

The average velocity is indicated in Figure 2-11 by the horizontal line, which is drawn such that the area under it from t_1 to t_2 equals the area under the actual v-versus-t curve.

Example 2-14

A particle moves from rest with constant acceleration a. Show by calculating the area under the v-versus-t curve that the average velocity for some time interval beginning at $t = 0$ equals half the final velocity.

Figure 2-12 shows the v-versus-t curve for this problem. The displacement from $t = 0$ to some final time t is indicated by the shaded area. The area of this triangle is $\frac{1}{2}v_f\,\Delta t$, where v_f is the final velocity. The displacement is

$$\Delta x = \tfrac{1}{2}v_f\,\Delta t = v_{av}\,\Delta t$$

The average velocity is thus half the final velocity for this time interval.

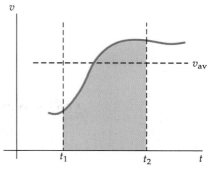

Figure 2-11 Geometric interpretation of average velocity. By definition, $v_{av} = \Delta x/\Delta t$. Thus, the rectangular area $v_{av}(t_2 - t_1)$ must equal the displacement in the time interval $t_2 - t_1$. It follows that the rectangular area $v_{av}(t_2 - t_1)$ and the shaded area under the curve must be equal.

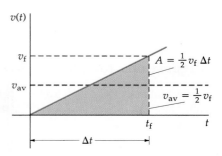

Figure 2-12 Proof that $v_{av} = \tfrac{1}{2}v_f$ for a particle starting from rest and moving with constant acceleration. The displacement equals the area under the curve, which equals $\tfrac{1}{2}v_f\,\Delta t$. It also equals $v_{av}\,\Delta t$, so $v_{av} = \tfrac{1}{2}v_f$.

Summary

1. Average velocity is the ratio of the displacement Δx to the time interval Δt:

$$v_{av} = \frac{\Delta x}{\Delta t}$$

2. The instantaneous velocity v is the limit of this ratio as the time interval approaches zero. This is the derivative of x with respect to t:

$$v = \lim_{\Delta t \to 0} \frac{\Delta x}{\Delta t} = \frac{dx}{dt}$$

The instantaneous velocity is represented graphically as the slope of the x-versus-t curve. In one dimension, both average and instantaneous velocity can be either positive or negative. The magnitude of the instantaneous velocity is called the speed.

3. The average acceleration is the ratio of the change in velocity Δv to the time interval Δt:

$$a_{av} = \frac{\Delta v}{\Delta t}$$

The instantaneous acceleration is the limit of this ratio as the time interval approaches zero. Instantaneous acceleration is the derivative of v with respect to t, which is the second derivative of x with respect to t:

$$a = \frac{dv}{dt} = \frac{d^2x}{dt^2}$$

The instantaneous acceleration is represented graphically as the slope of the v-versus-t curve.

4. In the special case of constant acceleration, the following formulas hold:

$$v = v_0 + at$$

$$\Delta x = v_{av}t = \tfrac{1}{2}(v_0 + v)t$$

$$\Delta x = x - x_0 = v_0 t + \tfrac{1}{2}at^2$$

$$v^2 = v_0^2 + 2a\,\Delta x$$

A common example of motion with constant acceleration is the motion of an object near the surface of the earth in free fall under the influence of gravity. In this case, the acceleration of the body is directed downward and has the magnitude $g = 9.81$ m/s^2 = 32.2 ft/s^2.

5. The displacement is represented graphically as the area under the v-versus-t curve. This area is the integral of v over time from some initial time t_1 to some final time t_2 and is written

$$\Delta x = \lim_{\Delta x_i \to 0} \sum_i v_i\,\Delta t_i = \int_{t_1}^{t_2} v\,dt$$

Similarly, the change in velocity for some time is represented graphically as the area under the a-versus-t curve.

Suggestions for Further Reading

Drake, Stillman: "Galileo's Discovery of the Law of Free Fall," *Scientific American*, May 1973, p. 84.

The author's reexamination of unpublished manuscripts clarifies the development of Galileo's ideas about motion.

Magie, W. F.: *A Source Book in Physics*, McGraw-Hill, New York and London, 1935.

This useful book contains short life histories and extracts from the important works of the great physicists from the time of Galileo to the year 1900.

U.S. Department of Transportation, National Highway Traffic Safety Administration: "Acceleration and Passing Ability: A Comparison of Acceleration and Passing Ability for 1974 Passenger Cars and Motorcycles," *Consumer Aid Series*, vol. 1, part 3, 1974.

A tabulation of the time and distance required for vehicles sold in the United States to pass a 55-foot-long truck.

U.S. Department of Transportation, National Highway Traffic Safety Administration: "Brakes: A Comparison of Braking Performance for 1974 Passenger Cars and Motorcycles," *Consumer Aid Series*, vol. 4, part 1, 1974.

A tabulation of stopping distance from 60 mi/h, under specified conditions, for vehicles sold in the United States.

Review

A. Objectives: After studying this chapter you should:

1. Be able to define displacement, velocity, and acceleration.

2. Be able to distinguish between velocity and speed.

3. Be able to calculate the instantaneous velocity from a graph of position versus time.

4. Be able to state the important equations relating displacement, velocity, acceleration, and time that apply when the acceleration is constant, and use them to work problems.

5. Be able to calculate the displacement of a particle from the v-versus-t curve and the change in the velocity of a particle from the a-versus-t curve by finding the areas under the appropriate curves.

B. Define, explain, or otherwise identify:

Particle
Speed
Displacement
Average velocity
Instantaneous velocity
Slope
Derivative
Average acceleration
Instantaneous acceleration
Initial-value problem
Integral

C. True or false: If the statement is true, explain why it is true. If it is false, give a counterexample.

1. The equation $\Delta x = v_0 t + \frac{1}{2}at^2$ is true for all motion in one dimension.

2. If the velocity is zero at an instant, the acceleration must also be zero at that instant.

3. If the acceleration is zero, the body cannot be moving.

4. If the acceleration is zero, the x-versus-t curve is a straight line.

5. The equation $\Delta x = v_{av} \, \Delta t$ holds for all motion in one dimension.

Problems

Unless instructed otherwise, use $g = 9.81 \ m/s^2 = 32.2 \ ft/s^2$ for the acceleration due to gravity.

Level I

2-1 Speed, Displacement, and Velocity

1. A runner runs 2 km in 5 min and then takes 10 min to walk back to the starting point. (*a*) What is the average velocity for the first 5 min? (*b*) What is the average velocity for the time spent walking? (*c*) What is the average velocity for the total trip? (*d*) What is the average speed for the total trip?

2. Work Problem 1 if the runner walked only halfway back in 10 min and then stopped.

3. A particle is at $x = +5$ m at $t = 0$, $x = -7$ m at $t = 6$ s, and $x = +2$ m at $t = 10$ s. Find the average velocity of the particle during the intervals (*a*) $t = 0$ to $t = 6$ s, (*b*) $t = 6$ s to $t = 10$ s, and (*c*) $t = 0$ to $t = 10$ s.

4. A driver begins a 200-km trip at noon. (*a*) She drives nonstop and arrives at her destination at 5:30 P.M. Calculate her average velocity for the trip. (*b*) She drives for 3 h, rests for $\frac{1}{2}$ h, and continues driving, arriving at 5:30 P.M. Calculate her average velocity. (*c*) After resting 2 h, she drives back home, taking 6 h for the return trip. What is her average velocity for the total round trip? (*d*) What is her displacement?

5. A car travels in a straight line with an average velocity of 80 km/h for 2.5 h and then with an average velocity of 40 km/h for 1.5 h. (*a*) What is the total displacement for the 4-h trip? (*b*) What is the average velocity for the total trip?

6. As you drive into Kentucky, you notice that the mile marker on the interstate is 325. You drive straight through to mile marker 0 on the other side of the state in 6 h. You are then struck by a desire for fried chicken, so you turn around and drive back 25 mi for a snack. This return takes 30 min. (*a*) What was your average speed in miles per hour for the 350-mi trip? (*b*) What was your average velocity for the trip? (*c*) What was your average velocity for the last 25 mi?

7. (*a*) How long does it take a supersonic jet flying at 2.4 times the speed of sound to fly across the Atlantic, which is about 5500 km wide? Take the speed of sound to be 350 m/s. (*b*) How long does it take a subsonic jet flying at 0.9 times the speed of sound to make the same trip? Assuming it takes 2 h at each end of the trip to get to or from the airport and to check or pick up your baggage, what is your average speed from your home to your final destination for (*c*) the supersonic jet and (*d*) the subsonic jet?

8. As you drive along the freeway, your speedometer quits working. You measure your speed by measuring the time it takes to travel between mile markers. (*a*) How many seconds should elapse between mile markers if your average speed is 55 mi/h? (*b*) What is your average speed in miles per hour if the time between mile markers is 45 s?

9. Light travels at a speed of $c = 3 \times 10^8$ m/s. (*a*) How long does it take light to travel from the sun to the earth, a distance of 1.5×10^{11} m? (*b*) How long does it take light to travel from the moon to the earth, a distance of 3.84×10^8 m? (*c*) A lightyear is a unit of distance equal to that traveled by light in 1 year. Find the equivalent distance of 1 lightyear in kilometers and in miles.

10. The nearest star, Proxima Centauri, is 4.1×10^{13} km away. (*a*) How much time is required for a light signal sent from the earth to reach Proxima Centauri? (*b*) How many years would it take a spaceship traveling with a speed of $10^{-4} c$ to reach the nearest star? (See Problem 9.)

11. A car making a 100-km journey covers the first 50 km at 40 km/h. How fast must it cover the second 50 km to average 50 km/h?

2-2 Instantaneous Velocity

12. For each of the four graphs of x versus t in Figure 2-13, indicate whether (*a*) the velocity at time t_2 is greater than, less than, or equal to the velocity at time t_1 and (*b*) the speed at time t_2 is greater than, less than, or equal to the speed at time t_1.

Figure 2-13   Graphs of x versus t for Problem 12.

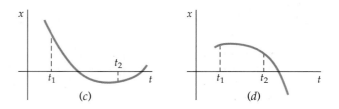

13. From the graph of x versus t in Figure 2-14, (a) find the average velocity between the times $t = 0$ and $t = 2$ s. (b) Find the instantaneous velocity at $t = 2$ s by measuring the slope of the tangent line indicated.

Figure 2-14   Graph of x versus t with a tangent line drawn at $t = 2$ s for Problem 13.

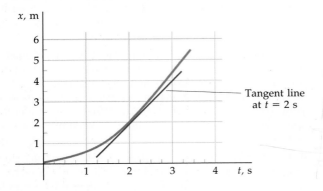

14. For the graph of x versus t in Figure 2-15, find the average velocity for the time intervals $\Delta t = t_2 - 0.75$ s when t_2 is 1.75, 1.5, 1.25, and 1.0 s. What is the instantaneous velocity at $t = 0.75$ s?

Figure 2-15   Graph of x versus t for Problem 14.

15. For the graph of x versus t shown in Figure 2-16, (a) find the average velocity for the interval $t = 1$ s to $t = 5$ s. (b) Find the instantaneous velocity at $t = 4$ s. (c) At what time is the velocity of the particle zero?

Figure 2-16   Graph of x versus t with a tangent line drawn at $t = 4$ s for Problem 15.

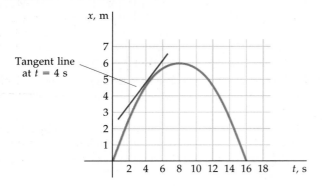

2-3 Acceleration

16. A car is traveling at 45 km/h at time $t = 0$. It accelerates at a constant rate of 10 km/h·s. (a) How fast is the car going at $t = 1$ s? At $t = 2$ s? (b) What is its speed at a general time t?

17. At $t = 5$ s an object is traveling at 5 m/s. At $t = 8$ s its velocity is -1 m/s. Find the average acceleration for this interval.

18. State whether the acceleration is positive, negative, or zero for each of the position functions $x(t)$ in Figure 2-17.

Figure 2-17   Graphs of x versus t for Problem 18.

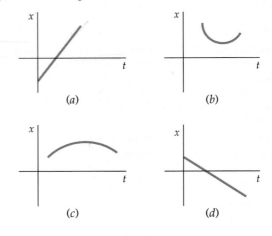

2-4 Motion with Constant Acceleration

19. A car accelerates from rest at a constant rate of 8 m/s^2. (a) How fast is it going after 10 s? (b) How far has it gone after 10 s? (c) What is its average velocity for the interval $t = 0$ to $t = 10$ s?

20. An object with initial velocity of 5 m/s has a constant acceleration of 2 m/s^2. When its speed is 15 m/s, how far has it traveled?

21. An object with constant acceleration has velocity $v = 10$ m/s when it is at $x = 6$ m and $v = 15$ m/s when it is at $x = 10$ m. What is its acceleration?

22. An object has constant acceleration $a = 4$ m/s^2. Its velocity is 1 m/s at $t = 0$, when it is at $x = 7$ m. How fast is it moving when it is at $x = 8$ m? At what time is this?

23. How long does it take for a particle to travel 100 m if it begins from rest and accelerates at 10 m/s^2? What is the velocity when it has traveled 100 m? What is the average velocity for this time?

24. A ball is thrown upward with initial velocity of 20 m/s. (a) How long is the ball in the air? (b) What is the greatest height reached by the ball? (c) When is the ball 15 m above the ground?

25. Suppose a rifle fires a bullet straight up with a muzzle speed of 300 m/s. Neglecting air friction, how high will the bullet rise?

26. The minimum distance for a controlled stop, starting from 98 km/h, for a certain car is 50 m for level braking.

Find the acceleration, assuming it to be constant, and express your answer as a fraction of the free-fall acceleration due to gravity. How long does it take to stop?

27. A world-land-speed car can decelerate at about $1g$. (That is, the magnitude of a is g.) (*a*) How long does it take the car to come to a stop from a record-setting run of 885 km/h? (*b*) How far does it travel while stopping?

2-5 Integration

28. The velocity of a particle is given by $v = 6t + 3$, where t is in seconds and v is in meters per second. (*a*) Sketch $v(t)$ versus t, and find the area under the curve for the interval $t = 0$ to $t = 5$ s. (*b*) Find the general position function $x(t)$. Use it to calculate the displacement during the interval $t = 0$ to $t = 5$ s.

29. The velocity of a particle in meters per second is given by $v = 7 - 4t$, where t is in seconds. (*a*) Sketch $v(t)$ versus t, and find the area between the curve and the t axis from $t = 2$ s to $t = 6$ s. (*b*) Find the position function $x(t)$ by integration, and use it to find the displacement during the interval $t = 2$ s to $t = 6$ s. (*c*) What is the average velocity for this interval?

30. Figure 2-18 shows the velocity of a particle versus time. (*a*) What is the magnitude in meters of the area of the rectangle indicated? (*b*) Find the approximate displacement of the particle for the 1-s intervals beginning at $t = 1$ s and $t = 2$ s. (*c*) What is the approximate average velocity for the interval from $t = 1$ s to $t = 3$ s?

Figure 2-18 Graph of v versus t for Problems 30 and 41.

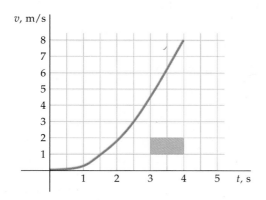

Level II

31. Figure 2-19 shows the position of a particle versus time. Find the average velocity for the time intervals a, b, c, and d indicated in the figure.

32. The position of a particle depends on the time according to $x = (1 \text{ m/s}^2)t^2 - (5 \text{ m/s})t + 1$ m. (*a*) Find the displacement and average velocity for the interval $t = 3$ s to $t = 4$ s. (*b*) Find a general formula for the displacement for the time interval from t to $t + \Delta t$. (*c*) Find the instantaneous velocity for any time t.

33. The height of a certain projectile is related to time by $y = -5(t - 5)^2 + 125$, where y is in meters and t is in sec-

onds. (*a*) Sketch y versus t for $t = 0$ to $t = 10$ s. (*b*) Find the average velocity for each of the 1-s time intervals between integral time values from $t = 0$ to $t = 10$ s. Sketch v_{av} versus t. (*c*) Find the instantaneous velocity as a function of time.

34. A particle moves with velocity given by $v = 8t - 7$, where v is in meters per second and t is in seconds. (*a*) Find the average acceleration for the 1-s intervals beginning at $t = 3$ s and $t = 4$ s. (*b*) Sketch v versus t. What is the instantaneous acceleration at any time?

35. The position of a particle versus time is given by

t, s	0	1	2	3	4	5	6	7	8	9	10	11
x, m	0	5	15	45	65	70	60	−30	−50	−50	−55	−55

Plot x versus t, and draw a smooth curve $x(t)$. Indicate the times or time intervals for which (*a*) the speed is greatest, (*b*) the speed is least, (*c*) the speed is zero, (*d*) the speed is constant, (*e*) the acceleration is positive, and (*f*) the acceleration is negative.

36. The position of an object is related to time by $x = At^2 - Bt + C$, where $A = 8$ m/s^2, $B = 6$ m/s, and $C = 4$ m. Find the instantaneous velocity and acceleration as functions of time.

Figure 2-19 Graph of x versus t for Problem 31.

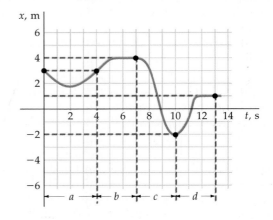

37. A ball is dropped from a height of 3 m and rebounds from the floor to a height of 2 m. (*a*) What is the velocity of the ball just as it reaches the floor? (*b*) What is the velocity just as it leaves the floor? (*c*) If it is in contact with the floor for 0.02 s, what are the magnitude and direction of its average acceleration during this interval?

38. The velocity of a particle in meters per second is given by $v = 7t + 5$, where t is in seconds. Find the general position function $x(t)$.

39. The acceleration of a certain rocket is given by $a = Ct$, where C is a constant. (*a*) Find the general position function $x(t)$. (*b*) Find the position and velocity at $t = 5$ s if $x = 0$ and $v = 0$ at $t = 0$ and $C = 3$ m/s^3.

40. Figure 2-20 shows the acceleration of a particle versus time. (*a*) What is the magnitude of the area of the rectangle indicated? (*b*) The particle starts from rest at $t = 0$. Find the velocity at $t = 1$, 2, and 3 s by counting the rectangles under the curve. (*c*) Sketch the curve $v(t)$ versus t from your results for part (*b*), and estimate how far the particle travels in the interval $t = 0$ to $t = 3$ s.

41. The equation of the curve shown in Figure 2-18 is $v = 0.5t^2$ m/s. Find the displacement of the particle for the interval $t = 1$ s to $t = 3$ s by integration, and compare this answer with your answer for Problem 30. Is the average velocity equal to the mean of the initial and final velocities for this case?

42. A car traveling at a constant speed of 20 m/s passes an intersection at time $t = 0$, and 5 s later another car passes the same intersection traveling 30 m/s in the same direction. (*a*) Sketch the position functions $x_1(t)$ and $x_2(t)$ for the two cars. (*b*) Find when the second car overtakes the first. (*c*) How far from the intersection have the cars traveled when this happens?

43. Hare and Tortoise begin a 10-km race at time $t = 0$. Hare runs at 4 m/s and quickly outdistances Tortoise, who runs at 1 m/s (about 10 times faster than a tortoise can actually run but convenient for this problem). After running 5 min, Hare stops and falls asleep. His nap lasts 135 min. He awakes and again runs at 4 m/s but loses the race. Plot *x*-versus-*t* curves for Hare and Tortoise on the same axes. At what time does Tortoise pass Hare? How far behind is Hare when Tortoise crosses the finish line? How long can Hare nap and still win the race?

44. A particle moves with a constant acceleration of 3 m/s². At $t = 4$ s, it is at $x = 100$ m. At $t = 6$ s, it has a velocity $v = 15$ m/s. Find its position at $t = 6$ s.

45. Figure 2-21 shows the position of a car plotted as function of time. At which of the times t_0 to t_7 is (*a*) the velocity negative, (*b*) the velocity positive, (*c*) the velocity zero, (*d*) the acceleration negative, (*e*) the acceleration positive, and (*f*) the acceleration zero?

46. It has been found that galaxies are moving away from the earth at a speed that is proportional to their distance from the earth. This discovery is known as Hubble's law. The speed of a galaxy a distance r away is given by $v = Hr$, where H is Hubble's constant, which equals 1.58×10^{-18} s^{-1}. What is the speed of a galaxy (*a*) 5×10^{22} m from earth and (*b*) 2×10^{25} m from earth? (*c*) If each of these galaxies has traveled with constant speed, how long ago were they both located at the same place as the earth?

47. For each of the graphs in Figure 2-22, indicate (*a*) at what times the acceleration of the object is positive, negative, and zero; (*b*) at what times the acceleration is constant; and (*c*) at what times the instantaneous velocity is zero.

48. In the Blackhawk landslide in California, a mass of rock and mud fell 460 m down a mountain and then traveled 8 km across a level plain on a cushion of compressed air. Assume that the mud dropped with the free-fall acceleration due to gravity and then slid horizontally with constant deceleration. (*a*) How long did the mud take to drop the 460 m? (*b*) How fast was it traveling when it reached

Figure 2-20   Graph of *a* versus *t* for Problem 40.

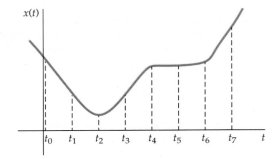

Figure 2-21   Graph of the position of a car versus time for Problem 45.

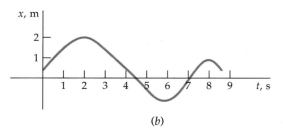

Figure 2-22   Graphs of (*a*) *v* versus *t* and (*b*) *x* versus *t* for Problem 47.

the bottom? (*c*) How long did the mud take to slide the 8 km horizontally?

49. A police car chases a speeder who is traveling at 125 km/h. The maximum speed of the police car is 190 km/h. The police car travels from rest with a constant acceleration of 8 km/h·s until its speed reaches 190 km/h. It then moves at constant speed. (*a*) When does the police car catch the speeder if it starts just as the speeder passes? (*b*) How far has each car traveled? (*c*) Sketch $x(t)$ for each car.

50. When the police car of Problem 49 is traveling at 190 km/h and is 100 m behind the speeder (traveling at 125 km/h), the speeder sees the police car and slams on his brakes, locking the wheels. (*a*) Assuming that each car can brake at 6 m/s^2 and that the driver of the police car brakes as soon as she sees the brake lights of the speeder, that is, with no reaction time, show that the cars collide. (b) At what time after the speeder applies his brakes do they collide? (c) The interval between the time the police officer sees the speeder's brake lights and the time she puts on her own brakes is called her *reaction time T*. Discuss how reaction time affects this problem.

51. A glider moves along an air track with constant acceleration a. It is projected from the start of the track ($x = 0$) with an initial velocity v_0. At time $t = 8$ s, it is at $x = 100$ cm and is moving along the track at velocity $v = -15$ cm/s. Find the initial speed v_0 and the acceleration a.

52. A load of bricks is being lifted by a crane at the steady velocity of 5 m/s when one brick falls off 6 m above the ground. Describe the motion of the free brick by sketching $x(t)$. (a) What is the greatest height the brick reaches above the ground? (b) How long does it take to reach the ground? (c) What is its speed just before it hits the ground?

53. A student running at 9 m/s is 40 m behind Eddy when Eddy starts from rest on his motor bike with an acceleration of 0.9 m/s^2. (a) How long does it take for the student to catch Eddy? (b) What is the time interval during which the student remains ahead of Eddy?

54. An automobile accelerates from rest at 2 m/s^2 for 20 s. The speed is then held constant for 20 s, after which there is an acceleration of -3 m/s^2 until the automobile stops. What is the total distance traveled?

55. A flower pot falls from a ledge of an apartment building. A person in an apartment below, who just happens to have a stopwatch handy, notices that it takes 0.2 s for the pot to fall past his window, which is 4 m high. How far above the top of the window is the ledge from which the pot fell?

56. A pilot jumps out of a burning plane without a parachute. She reaches a speed of 120 km/h prior to impact. The pilot can survive a 35g deceleration. She falls onto a stack of loose hay. Assuming uniform deceleration to the bottom of the haystack, how high a stack of hay is needed for the pilot to barely survive?

57. A bolt comes loose from the bottom of an elevator that is moving upward at a speed of 6 m/s. The bolt reaches the bottom of the elevator shaft in 3 s. (a) How high up was the elevator when the bolt came loose? (b) What is the speed of the bolt when it hits the bottom of the shaft?

58. An archer fires an arrow that produces a loud noise when it strikes its target. The average velocity of the arrow is 150 m/s. The archer hears the impact exactly 1 s after firing the arrow. If the velocity of sound is 340 m/s, how far away is the target?

Level III

59. Sketch a single v-versus-t curve in which there are points or segments for which (a) the acceleration is zero and constant while the velocity is not zero; (b) the acceleration is zero but not constant; (c) the velocity and acceleration are both positive; (d) the velocity and acceleration are both negative; (e) the velocity is positive and the acceleration is negative; (f) the velocity is negative and the acceleration is positive; and (g) the velocity is zero, but the acceleration is not.

60. When a car traveling at speed v_1 rounds a corner, the driver sees another car traveling at a slower speed v_2 a distance d ahead. (a) If the maximum acceleration the driver's brakes can provide is a, show that the distance d must be greater than $(v_1 - v_2)^2/2a$ if a collision is to be avoided. (b) Evaluate this distance for $v_1 = 90$ km/h, $v_2 = 45$ km/h, and $a = 6$ m/s^2. (c) Estimate or measure your reaction time, and calculate the effect it would have on the distance found in part (b).

61. A passenger is running at his maximum velocity of 8 m/s to catch a train. When he is a distance d from the nearest entry to the train, the train starts from rest with a constant acceleration $a = 1.0$ m/s^2 away from him. (a) If $d = 30$ m and the passenger keeps running, will he be able to jump onto the train? (b) Sketch the position function $x(t)$ for the train, choosing $x = 0$ at $t = 0$. On the same graph sketch $x(t)$ for the passenger for various initial separation distances d, including $d = 30$ m and the critical separation distance d_c such that he just catches the train. (c) For the critical separation distance d_c, what is the speed of the train when the passenger catches it? What is the train's average speed for the time interval from $t = 0$ until he catches it? What is the value of d_c?

62. A train starts from a station with a constant acceleration of 0.40 m/s^2. A passenger arrives at the track 6.0 s after the end of the train left the very same point. What is the slowest constant speed at which she can run and catch the train? Sketch curves for the motion of the passenger and the train as functions of time.

63. Ball A is dropped from the top of a building at the same instant that ball B is thrown vertically upward from the ground. When the balls collide, they are moving in opposite directions, and the speed of A is twice the speed of B. At what fraction of the height of the building does the collision occur?

64. Solve Problem 63 if the collision occurs when the balls are moving in the same direction and the speed of A is 4 times that of B.

65. Figure 2-23 is a graph of v versus t for a particle moving along a straight line. The position of the particle at time $t = 0$ is $x_0 = 5$ m. (a) Find x for various times t by counting squares, and sketch x versus t. (b) Sketch the acceleration a versus t.

Figure 2-23 Graph of v versus t for Problem 65.

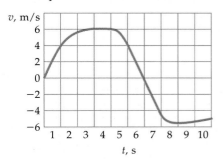

66. Figure 2-24 shows a plot of x versus t for a body moving along a straight line. Sketch rough graphs of v versus t and a versus t for this motion.

Figure 2-24   Graph of x versus t for Problem 66.

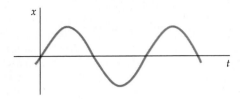

67. The position of a body oscillating on a spring is given by $x = A \sin \omega t$, where A and ω are constants and have the values $A = 5$ cm and $\omega = 0.175$ s^{-1}. Sketch x versus t for times from $t = 0$ to $t = 36$ s. (*a*) Measure the slope of your graph at $t = 0$ to find the velocity at this time. (*b*) Calculate the average velocity for a series of intervals beginning at $t = 0$ and ending at $t = 6, 3, 2, 1, 0.5$, and 0.25 s. (*c*) Compute dx/dt and find the velocity at time $t = 0$.

68. A car has a maximum acceleration a, which remains constant to high speeds, and it has a maximum deceleration $2a$. The car has to travel a short distance L, starting and ending at rest, in the minimum time T. (The distance is so short that the car can never reach top speed.) After what fraction of L should the driver move her foot from the gas pedal to the brake, and what fraction of the time for the trip has elapsed at this point?

69. Table 2-2 lists some world track records for short sprints. A simple model of sprinting assumes that a sprinter starts from rest, accelerates with constant acceleration a for a short time T, and then runs with constant speed $v_0 = aT$. According to this model, for times t greater than T, the distance x varies linearly with time. (*a*) Make a graph of distance x versus time t from the data in the table. (*b*) Set up an equation for x versus T according to the simple model described, and show that for $t > T$, x can be written $x = v_0(t - \frac{1}{2}T)$. (*c*) Connect the points on your graph with a straight line, and determine the slope and the intercept of the line with the time axis. From the fact that the slope is v_0 and the intercept is $\frac{1}{2}v_0 T$, compute the acceleration a. (*d*) The record for $x = 200$ m is 19.5 s. Discuss the applicability of this simple model to races of 200 m or more.

	x	
yd	m	t, s
50		5.1
	50	5.5
60		5.9
	60	6.5
100		9.1
	100	9.9

70. A Sprint ABM missile can accelerate at $100g$. If an ICBM is detected at an altitude of 100 km moving straight down at a constant speed of 3×10^4 km/h and the ABM missile is launched to intercept it, at what time and altitude will the interception take place? *Note:* You can neglect the acceleration of gravity in this problem. Why?

71. The acceleration of a particle falling under the influence of gravity and a resistive force, such as air resistance, is given by

$$a = \frac{dv}{dt} = g - bv$$

where g is the free-fall acceleration due to gravity and b is a constant that depends on the mass and shape of the particle and on the medium. Suppose the particle begins with zero velocity at time $t = 0$. (*a*) Discuss qualitatively how the speed v varies with time from your knowledge of the rate of change dv/dt given by this equation. What is the value of the velocity when the acceleration is zero? This is called the *terminal speed*. (*b*) Sketch the solution $v(t)$ versus t without solving the equation. This can be done as follows: At $t = 0$, v is zero and the slope is g. Sketch a straight-line segment, neglecting any change in slope for a short time interval. At the end of the interval the velocity is not zero, so the slope is less than g. Sketch another straight-line segment with a smaller slope. Continue until the slope is zero and the velocity equals the terminal velocity.

72. Suppose acceleration is a function of x, where $a(x) = 2x$ m/s^2. (*a*) If the velocity at $x = 1$ m is zero, what is the speed at $x = 3$ m? (*b*) How long does it take to travel from $x = 1$ m to $x = 3$ m?

73. Suppose that a particle moves in a straight line such that, at any time t, its position, velocity, and acceleration all have the same numerical value. Give the position x as a function of time.

74. A typical automobile has a maximum deceleration of about 7 m/s^2; and the typical reaction time to engage the brakes is 0.50 s. A school board sets the speed limit in a school zone to meet the condition that all cars should be able to stop in a distance of 4 m. (*a*) What maximum speed should be allowed for a typical automobile? (*b*) What fraction of the 4 m is due to the reaction time?

75. An object moving in a straight line doubles its velocity each second for the first 10 s. Let the initial speed be 2 m/s. (*a*) Sketch a smooth function $v(t)$ that gives the velocity. (*b*) What is the average velocity over the first 10 s?

76. Suppose a fast-moving object encounters resistance such that its velocity is reduced by one-half for each second that elapses. Let the initial velocity be 1000 m/s. (*a*) Sketch a smooth function $v(t)$ that gives the velocity. (*b*) What is the average velocity over the first 10 s?

Chapter 3

Motion in Two and Three Dimensions

Illuminated fountains, St. Louis Missouri. The jets follow parabolic paths like those followed by projectiles.

We now extend our description of the motion of a particle to two and three dimensions. In these cases, displacement, velocity, and acceleration are quantities that have direction in space as well as magnitude. We call such quantities **vectors**. In future chapters, we will encounter many other vector quantities, such as force, momentum, and the electric field. Quantities that have magnitude only and no associated direction, such as distance, speed, mass, or temperature, are called **scalars.**

In this chapter we will investigate the properties of vectors in general and those of displacement, velocity, and acceleration in particular. Many of the interesting features of motion in three dimensions are also present in motion in two dimensions. Since two-dimensional motion is more easily illustrated on paper or a blackboard, most of our examples will be limited to it. Two important special cases, projectile motion and circular motion, will be discussed in detail.

3-1   The Displacement Vector and Vector Addition

If you ask someone where the post office is and she says it is 10 blocks away, you would likely ask in what direction before you set off to find it. It makes quite a difference whether the post office is 10 blocks east, or 10 blocks north, or 6 blocks west and 8 blocks south (and therefore 14 blocks walking but 10 blocks "as the crow flies"). The quantity that gives the straight-line distance and the direction from one point in space to another is a directed line segment called the **displacement vector.** We use displacement vectors to illustrate results for vectors in general because, by definition, **vectors** *are quantities with magnitude and direction that add and subtract like displacements.* A vector is represented graphically by an arrow whose direction is the same as the direction of the vector and whose length is proportional to the magnitude of the vector.

Figure 3-1 shows the path of a particle that moves from point P_1 to a second point P_2 and then to a third point P_3. The displacement from point P_1 to point P_2 is represented by the arrow **A**. Note that the displacement **A** does not depend on the path taken by the particle as it moves from P_1 to P_2 but only on the endpoints P_1 and P_2. A second displacement from P_2 to P_3 is indicated by the arrow **B**. The resultant displacement from P_1 to P_3 is represented by the arrow **C**. The resultant displacement vector **C** is the sum of the two successive displacements **A** and **B**:

$$C = A + B \qquad\qquad 3\text{-}1$$

Any two vectors (whose units are the same) may be added graphically in this way, by placing the tail of one arrow at the head of the other.

Figure 3-1   The addition of vectors. The displacement **C** is equivalent to the successive displacements **A** and **B**; that is, **C** = **A** + **B**.

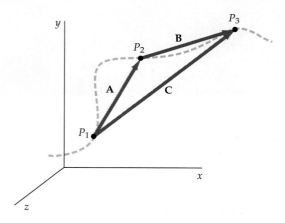

The resultant vector is then from the tail of the first vector to the head of the second.

We denote vector quantities with boldface type, as in **A**. We will use this convention throughout the text to distinguish vectors from scalar quantities, which are denoted with italic type. (In handwriting, we indicate a vector by drawing an arrow over the symbol as in $\vec{A}$.) The magnitude of a vector **A** is written $|\mathbf{A}|$ or simply A. Ordinarily, the magnitude of a vector has physical units. For example, a displacement vector has a magnitude that can be expressed in feet, meters, or any other unit of distance.

Note that the sum of the magnitudes of **A** and **B** does not equal the magnitude of **C** unless **A** and **B** are in the same direction. That is, **C** = **A** + **B** does not imply that $C = A + B$.

Example 3-1

A man walks 3 km east and then 4 km north. What is his resultant displacement?

The two displacements and the resultant displacement are shown in Figure 3-2. Since these three vectors form a right triangle, we can find the magnitude of the resultant displacement using the pythagorean theorem, which states

$$C^2 = A^2 + B^2 = (3 \text{ km})^2 + (4 \text{ km})^2 = 25 \text{ km}^2$$
$$C = \sqrt{25 \text{ km}^2} = 5 \text{ km}$$

To describe the resultant displacement, we need to give its direction as well as its magnitude. If θ is the angle between the east axis and the resultant displacement, we have from Figure 3-2

$$\tan \theta = \frac{4 \text{ km}}{3 \text{ km}} = 1.33$$

We find θ from trigonometric tables or from a calculator that has trigonometric functions:

$$\theta = \tan^{-1} 1.33 = 53.1°$$

The resultant displacement is therefore 5 km at 53.1° north of east.

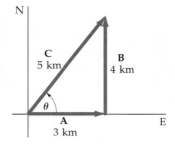

Figure 3-2 Displacement vectors for Example 3-1. The magnitude of the resultant displacement vector **C** can be found from the pythagorean theorem.

Example 3-2

A man walks 3 km east and then 4 km at 60° north of east. What is his resultant displacement?

The displacement vectors for this example are shown in Figure 3-3. In this case, the triangle formed by the three vectors is not a right triangle, so we cannot use the pythagorean theorem to find the resultant displacement. In the next section, we will learn how to find the resultant vector for a case like this using vector components. For now, we find the resultant graphically by drawing each of the displacements to scale and measuring the resultant displacement with a ruler. For example, if we draw the first displacement vector 3 cm long and the second 4 cm long, we find the resultant vector to be about 6 cm long. Thus, the magnitude of the resultant displacement is 6 km. The angle between the resultant displacement and the east direction can be measured with a protractor. It is about 35°.

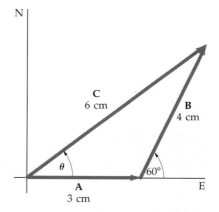

Figure 3-3 Displacement vectors for Example 3-2. Since **A** and **B** are not perpendicular, the pythagorean theorem cannot be used to find the magnitude of **C**. Instead, C can be found graphically.

Since vectors are defined only by their magnitude and direction, two vectors are equal if they have the same magnitude and the same direction, independent of their points of origin. Graphically, two vectors are equal if they have the same length and are parallel to one another. Thus, all the vectors in Figure 3-4 are equal. Figure 3-5*a* shows two vectors **A** and **B**

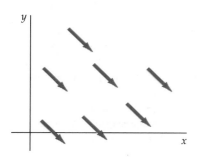

Figure 3-4 Vectors are equal if their magnitudes and directions are the same. All the vectors in this figure are equal.

Figure 3-5 (*a*) Vectors **A** + **B** = **C**. In (*b*) the vector **B** has been moved parallel to itself so that **A** and **B** are tail to tail. The resultant **C** = **A** + **B** is along the diagonal of the parallelogram formed when **A** and **B** are placed tail to tail. We see from this figure that the order of addition makes no difference; that is, **A** + **B** is the same as **B** + **A**.

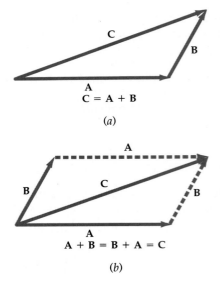

(*a*)

(*b*)

whose sum is **C**. In Figure 3-5*b*, we have moved vector **B** parallel to itself so that it originates at the same point as **A**. The resultant vector **C** lies along the diagonal of the parallelogram formed by **A** and **B**. The graphical addition of two vectors by placing them tail to tail and finding the diagonal of the parallelogram formed is known as the **parallelogram method of vector addition**. From Figure 3-5*b*, we can see that it makes no difference in which order we add two vectors, that is, **A** + **B** = **B** + **A**.

We can subtract vector **B** from vector **A**, as shown in Figure 3-6*a*, by simply adding −**B**, which has the same magnitude as **B** but points in the opposite direction. The result is **C** = **A** + (−**B**) = **A** − **B**. Another method of subtraction, illustrated in Figure 3-6*b*, is to draw the two vectors **A** and **B** tail to tail and then note that the vector **C** = **A** − **B** is that vector that must be added to **B** to obtain the resultant vector **A**.

Figure 3-6 Subtraction of vectors. (*a*) Here **C** = **A** − **B** is found by adding −**B** to **A**. (*b*) An alternative method of finding **A** − **B** is to find the vector **C** that when added to **B** gives the vector **A**.

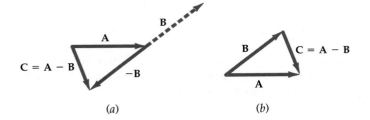

(*a*)   (*b*)

Questions

1. Can the displacement of a particle have a magnitude that is less than the distance traveled by the particle along its path? Can its magnitude be more than the distance traveled? Explain.

2. Give an example in which the distance traveled is a significant amount yet the corresponding displacement is zero.

3-2   Addition of Vectors by Components

We add or subtract vectors analytically by first breaking down the vectors into their components. The component of a vector is the projection of that vector on a line in space that is found by dropping a perpendicular from the head of the vector to the line, as shown in Figure 3-7. An important example is the projection of a vector on an axis of a rectangular coordinate system.

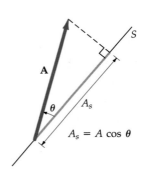

Figure 3-7 The component A_s of a vector **A** along a line in space is found by dropping the perpendicular from the head of the vector to the line.

Such a projection is called a **rectangular component** of the vector. Figure 3-8 shows a vector **A** lying in the xy plane. It has the rectangular components A_x and A_y. In general, the components can be either positive or negative. For example, if **A** points in the negative x direction, A_x is negative. If θ is the angle between vector **A** and the x axis, we see from the figure that

$$\tan \theta = \frac{A_y}{A_x} \qquad\qquad 3\text{-}2a$$

$$\sin \theta = \frac{A_y}{A} \qquad\qquad 3\text{-}2b$$

$$\cos \theta = \frac{A_x}{A} \qquad\qquad 3\text{-}2c$$

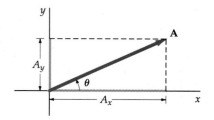

Figure 3-8   The rectangular components of the vector **A** are related to the magnitude A and the angle θ by $A_x = A \cos \theta$ and $A_y = A \sin \theta$.

where A is the magnitude of **A**. We can therefore determine the components of **A** analytically from the magnitude A and the angle θ by

$$A_x = A \cos \theta \qquad\qquad 3\text{-}3$$

and

$$A_y = A \sin \theta \qquad\qquad 3\text{-}4$$

Conversely, if we know the components A_x and A_y, we can find the angle θ from Equation 3-2a and the magnitude A from the pythagorean theorem:

$$A = \sqrt{A_x^2 + A_y^2} \qquad\qquad 3\text{-}5$$

Figure 3-9 illustrates the use of components in the addition of two vectors **A** and **B** that lie in the xy plane. The rectangular components of each vector and those of the sum **C** = **A** + **B** are shown in the figure. We can see from the figure that **C** = **A** + **B** implies both

$$C_x = A_x + B_x \qquad\qquad 3\text{-}6a$$

and

$$C_y = A_y + B_y \qquad\qquad 3\text{-}6b$$

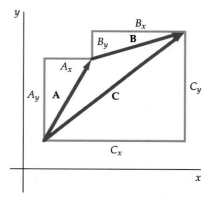

Figure 3-9   The x and y components of the vectors **A**, **B**, and **C** = **A** + **B**. We see from this figure that $C_x = A_x + B_x$ and $C_y = A_y + B_y$.

Exercise

A car travels 20 km at an angle of 30° north of west. Assuming the x axis points east and the y axis points north as in Figure 3-10, find the x and y components of the displacement vector of the car.
(Answers: $A_x = -17.3$ km, $A_y = +10$ km)

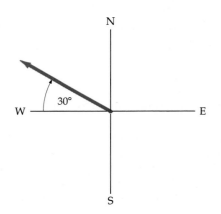

Figure 3-10   Geometry for the exercise.

Example 3-3

Work Example 3-2 using vector components.

In this example, a man first walked east 3 km. If **A** represents his displacement and we take the x axis to be in the easterly direction, the components of **A** are

$$A_x = 3 \text{ km}$$

and

$$A_y = 0$$

He next walked 4 km at 60° north of east. The vector **B** representing this displacement has the components

$$B_x = (4 \text{ km}) \cos 60° = (4 \text{ km})(0.5) = 2 \text{ km}$$

and

$$B_y = (4 \text{ km}) \sin 60° = (4 \text{ km})(0.866) = 3.46 \text{ km}$$

The components of the resultant displacement are thus

$$C_x = A_x + B_x = 3 \text{ km} + 2 \text{ km} = 5 \text{ km}$$

and

$$C_y = A_y + B_y = 0 + 3.46 \text{ km} = 3.46 \text{ km}$$

We obtain the magnitude of the resultant displacement **C** from the pythagorean theorem:

$$C^2 = C_x^2 + C_y^2 = (5 \text{ km})^2 + (3.46 \text{ km})^2 = 37.0 \text{ km}^2$$

So

$$C = \sqrt{37.0 \text{ km}^2} = 6.1 \text{ km}$$

The angle between **C** and the x axis is found from

$$\tan \theta = \frac{C_y}{C_x} = \frac{3.46 \text{ km}}{5 \text{ km}} = 0.692$$

The angle θ is

$$\theta = \tan^{-1} 0.692 = 34.7°$$

This agrees with our results for Example 3-2 within the accuracy of our measurement in that example.

3-3 Unit Vectors and Multiplication of Vectors by Scalars

A vector **A** may be multiplied by a scalar s. The result is the vector **B** = s**A**, which points in the direction of **A** and has a magnitude of sA. The dimensions of **B** are the dimensions of s multiplied by the dimensions of **A**.

We can express a vector conveniently in terms of its components by making use of unit vectors. A **unit vector** is a dimensionless vector that is defined to have the magnitude of 1 and points in some specified direction. For example, let **i**, **j**, and **k** be unit vectors that point in the x, y, and z directions, respectively. The vector A_x**i** is the product of the component A_x and the unit vector **i**. It is a vector that is parallel to the x axis (or antiparallel if A_x is negative) and has a magnitude $|A_x|$. A general vector **A** can thus be written as the sum of three vectors, each of which is parallel to a coordinate:

$$\mathbf{A} = A_x\mathbf{i} + A_y\mathbf{j} + A_z\mathbf{k} \qquad \text{3-7}$$

The vector sum of Equation 3-7 is illustrated in Figure 3-11. The addition of two vectors **A** and **B** can be written in terms of unit vectors as

$$\mathbf{A} + \mathbf{B} = (A_x\mathbf{i} + A_y\mathbf{j} + A_z\mathbf{k}) + (B_x\mathbf{i} + B_y\mathbf{j} + B_z\mathbf{k})$$

$$= (A_x + B_x)\mathbf{i} + (A_y + B_y)\mathbf{j} + (A_z + B_z)\mathbf{k} \qquad \text{3-8}$$

Exercise

Given the two vectors $\mathbf{A} = (4 \text{ m})\mathbf{i} + (3 \text{ m})\mathbf{j}$ and $\mathbf{B} = (2 \text{ m})\mathbf{i} - (3 \text{ m})\mathbf{j}$, find (a) A, (b) B, (c) $\mathbf{A} + \mathbf{B}$, and (d) $\mathbf{A} - \mathbf{B}$. [Answers: (a) $A = 5$ m, (b) $B = 3.61$ m, (c) $\mathbf{A} + \mathbf{B} = (6 \text{ m})\mathbf{i}$, (d) $\mathbf{A} - \mathbf{B} = (2 \text{ m})\mathbf{i} + (6 \text{ m})\mathbf{j}$]

The general properties of vectors that we have discussed are summarized in Table 3-1.

Questions

3. How would you subtract two vectors using the component method?

4. Can a component of a vector have a magnitude greater than the magnitude of the vector? Under what circumstances can a component of a vector have a magnitude equal to the magnitude of the vector?

5. Can a vector be equal to zero and still have one or more components not equal to zero?

6. Are the components of $\mathbf{C} = \mathbf{A} + \mathbf{B}$ necessarily larger than the corresponding components of either **A** or **B**?

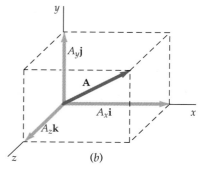

Figure 3-11   (a) The unit vectors **i**, **j**, and **k** in a rectangular coordinate system. (b) The vector **A** can be written in terms of the unit vectors as $\mathbf{A} = A_x\mathbf{i} + A_y\mathbf{j} + A_z\mathbf{k}$.

Table 3-1   Properties of Vectors

Property	Explanation	Figure	Component representation
Equality	**A** = **B** if $|\mathbf{A}| = |\mathbf{B}|$ and their directions are the same		$A_x = B_x$ $A_y = B_y$ $A_z = B_z$
Addition	**C** = **A** + **B**		$C_x = A_x + B_x$ $C_y = A_y + B_y$ $C_z = A_z + B_z$
Negative of a vector	$\mathbf{A} = -\mathbf{B}$ if $|\mathbf{B}| = |\mathbf{A}|$ and their directions are opposite		$A_x = -B_x$ $A_y = -B_y$ $A_z = -B_z$
Subtraction	**C** = **A** − **B**		$C_x = A_x - B_x$ $C_y = A_y - B_y$ $C_z = A_z - B_z$
Multiplication by a scalar	$\mathbf{B} = s\mathbf{A}$ if $|\mathbf{B}| = s|\mathbf{A}|$ and the direction of **B** is the same as the direction of **A**		$B_x = sA_x$ $B_y = sA_y$ $B_z = sA_z$

3-4   The Velocity Vector

Suppose you are driving a car at 50 km/h as indicated by the speedometer reading and are headed south as indicated by a compass. The speedometer gives the speed, magnitude of the velocity, and the compass gives the direction of the velocity. The instantaneous-velocity vector is a vector that points in the direction of motion and has a magnitude equal to the speed of the car. It equals the rate of change of the displacement vector.

Figure 3-12 shows a particle moving along some curve in space. (The curve represents the physical path or trajectory actually followed by the particle. It should not be confused with the x-versus-t plots in the previous chapter.) We define the position of the particle by means of its vector displacement from the origin O. This displacement is called the **position vector r**. If the particle is at point (x, y), the position vector is

$$\mathbf{r} = x\mathbf{i} + y\mathbf{j} \qquad\qquad 3\text{-}9$$

Figure 3-12 A particle moving along some arbitrary curve in space, with position vectors $\mathbf{r}_1$ and $\mathbf{r}_2$ at two different times t_1 and t_2. The displacement vector $\Delta\mathbf{r}$ is the difference in the position vectors, $\Delta\mathbf{r} = \mathbf{r}_2 - \mathbf{r}_1$.

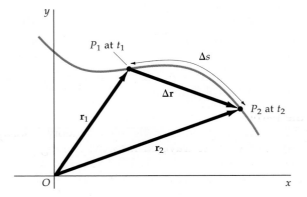

At some time t_1, the particle is at point P_1. This is characterized by the position vector $\mathbf{r}_1$ drawn from the origin to P_1. At some later time t_2, it is at point P_2 and has the position vector $\mathbf{r}_2$. The displacement vector is the change in the position vector:

$$\Delta\mathbf{r} = \mathbf{r}_2 - \mathbf{r}_1 \qquad\qquad 3\text{-}10$$

(This is analogous to our one-dimensional definition in Chapter 2, where the displacement is the change in the position coordinate x.) The new position vector $\mathbf{r}_2$ is the sum of the original position vector $\mathbf{r}_1$ and the displacement vector $\Delta\mathbf{r}$, as shown in the figure. The ratio of the displacement vector to the time interval $\Delta t = t_2 - t_1$ is the **average-velocity vector:**

Average-velocity vector defined

$$\mathbf{v}_{av} = \frac{\Delta\mathbf{r}}{\Delta t} \qquad\qquad 3\text{-}11$$

We note from Figure 3-12 that the magnitude of the displacement vector is not equal to the actual distance traveled Δs as measured along the curve. It is, in fact, less than this distance (unless the particle travels in a straight line between points P_1 and P_2). However, if we consider smaller and smaller time intervals, as indicated in Figure 3-13, the magnitude of the displacement approaches the actual distance traveled by the particle along the curve, and the direction of $\Delta\mathbf{r}$ approaches the direction of the line tangent to the

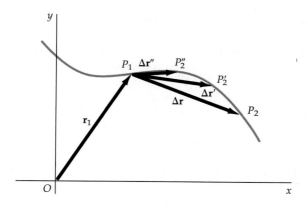

Figure 3-13 As smaller and
smaller time intervals are considered, the magnitude of the
displacement vector approaches
the actual distance traveled along
the curve, and the direction of the
displacement vector approaches
the direction of the line that is
tangent to the curve at point P_1.

curve at point P_1. We define the **instantaneous-velocity vector** as the limit of
the average-velocity vector as the time interval Δt approaches zero:

$$\mathbf{v} = \lim_{\Delta t \to 0} \frac{\Delta \mathbf{r}}{\Delta t} = \frac{d\mathbf{r}}{dt}$$

3-12   *Instantaneous-velocity vector
defined*

The instantaneous-velocity vector is thus the derivative of the position vector with respect to time. Its direction is along the line tangent to the curve
traveled by the particle in space. It therefore points in the direction of motion of the particle. The magnitude of the instantaneous velocity is the speed
ds/dt, where s is the distance traveled measured along the curve.

   To calculate the derivative in Equation 3-12, we must express the position vector in terms of its components as in Equation 3-9:

$$\Delta \mathbf{r} = \mathbf{r}_2 - \mathbf{r}_1 = (x_2 - x_1)\mathbf{i} + (y_2 - y_1)\mathbf{j}$$

$$\mathbf{v} = \lim_{\Delta t \to 0} \frac{\Delta \mathbf{r}}{\Delta t} = \lim_{\Delta t \to 0} \frac{\Delta x\, \mathbf{i} + \Delta y\, \mathbf{j}}{\Delta t} = \lim_{\Delta t \to 0} \left(\frac{\Delta x\, \mathbf{i}}{\Delta t} \right) + \left(\frac{\Delta y\, \mathbf{j}}{\Delta t} \right)$$

or

$$\mathbf{v} = \frac{dx}{dt}\, \mathbf{i} + \frac{dy}{dt}\, \mathbf{j}$$

Example 3-4

A sailboat has initial coordinates $(x_1, y_1) = (100 \text{ m}, 200 \text{ m})$. Two minutes
later, it has the coordinates $(x_2, y_2) = (120 \text{ m}, 210 \text{ m})$. What are the components, magnitude, and direction of its average velocity for this
2.00-min interval?

$$v_{x,\text{av}} = \frac{x_2 - x_1}{\Delta t} = \frac{120 - 100 \text{ m}}{2.00 \text{ min}} = 10.0 \text{ m/min}$$

$$v_{y,\text{av}} = \frac{y_2 - y_1}{\Delta t} = \frac{210 - 200 \text{ m}}{2.00 \text{ min}} = 5.0 \text{ m/min}$$

$$v_{\text{av}} = \sqrt{(v_{x,\text{av}})^2 + (v_{y,\text{av}})^2} = \sqrt{10.0^2 + 5.0^2} = \sqrt{125} = 11.2 \text{ m/min}$$

$$\tan \theta = \frac{v_{y,\text{av}}}{v_{x,\text{av}}} = \frac{5.0 \text{ m/min}}{10.0 \text{ m/min}} = 0.500$$

$$\theta = \tan^{-1} 0.500 = 26.6°$$

3-5   The Acceleration Vector

The **average-acceleration vector** is defined as the ratio of the change in the instantaneous-velocity vector $\Delta\mathbf{v}$ to the time interval Δt:

Average-acceleration vector defined

$$\mathbf{a}_{av} = \frac{\Delta\mathbf{v}}{\Delta t}$$

3-13

The **instantaneous-acceleration vector** is the limit of this ratio as the time interval approaches zero. That is, the instantaneous-acceleration vector is the derivative of the velocity vector with respect to time:

Instantaneous-acceleration vector defined

$$\mathbf{a} = \lim_{\Delta t \to 0} \frac{\Delta\mathbf{v}}{\Delta t} = \frac{d\mathbf{v}}{dt}$$

3-14

To calculate the instantaneous acceleration, we express $\mathbf{v}$ in terms of its rectangular coordinates:

$$\mathbf{v} = v_x\mathbf{i} + v_y\mathbf{j} = \frac{dx}{dt}\,\mathbf{i} + \frac{dy}{dt}\,\mathbf{j}$$

Then

$$\mathbf{a} = \frac{dv_x}{dt}\,\mathbf{i} + \frac{dv_y}{dt}\,\mathbf{j} = \frac{d^2x}{dt^2}\,\mathbf{i} + \frac{d^2y}{dt^2}\,\mathbf{j}$$

It is particularly important to note that the velocity vector may be changing in magnitude, direction, or both. If the velocity vector is changing in any way, the particle is accelerating. We are perhaps most familiar with acceleration in which the velocity changes in magnitude, that is, in which the speed changes. However, a particle can be traveling with constant speed and still be accelerating if the direction of the velocity vector is changing. A particularly important example of this is circular motion, which will be discussed in Section 3-8. This acceleration is just as real as when the speed is changing.

Example 3-5

A car is traveling east at 60 km/h. It rounds a curve, and 5 s later it is traveling north at 60 km/h. Find the average acceleration of the car.

Figure 3-14 shows the initial and final velocity vectors $\mathbf{v}_1 = 60$ km/h $\mathbf{i}$ and $\mathbf{v}_2 = 60$ km/h $\mathbf{j}$. The change in velocity

$$\Delta\mathbf{v} = \mathbf{v}_2 - \mathbf{v}_1 = 60 \text{ km/h } \mathbf{j} - 60 \text{ km/h } \mathbf{i}$$

is also shown in the figure. (Note that we have drawn the figure such that $\mathbf{v}_1 + \Delta\mathbf{v} = \mathbf{v}_2$.) The average acceleration is given by

$$\mathbf{a}_{av} = \frac{\Delta\mathbf{v}}{\Delta t} = \frac{60 \text{ km/h } \mathbf{j} - 60 \text{ km/h } \mathbf{i}}{5 \text{ s}}$$

$$= 12 \text{ km/h·s } \mathbf{j} - 12 \text{ km/h·s } \mathbf{i}$$

The magnitude of the average acceleration is

$$a = \sqrt{a_x^2 + a_y^2} = \sqrt{(-12 \text{ km/h·s})^2 + (12 \text{ km/h·s})^2} = 17.0 \text{ km/h·s}$$

Note that in this example, the car accelerates even though the speed does not change.

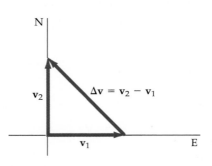

Figure 3-14   Velocity vectors for Example 3-5.

Questions

7. For an arbitrary motion of a given particle, does the direction of the velocity vector have any particular relation to the direction of the position vector?

8. Give examples in which the directions of the velocity and position vectors are (a) opposite, (b) the same, and (c) mutually perpendicular.

9. How is it possible for a particle moving at constant speed to be accelerating? Can a particle with constant velocity be accelerating at the same time?

10. Is it possible for a particle to round a curve without accelerating?

11. Can the velocity vector change direction without changing magnitude? If so, give an example.

3-6   Relative Velocity

The velocity of an object is sometimes measured relative to a coordinate system that is itself moving relative to another coordinate system. For example, suppose that a person is walking on a railroad car with velocity $\mathbf{v}_{pc}$ relative to the car while the car is moving with velocity $\mathbf{v}_{cg}$ relative to the ground, as shown in Figure 3-15. The velocity of the person relative to the ground $\mathbf{v}_{pg}$ is the sum of these two velocities:

$$\mathbf{v}_{pg} = \mathbf{v}_{pc} + \mathbf{v}_{cg}$$

3-15   *Relative velocity*

Figure 3-15   (a) Relative to the railroad car, the person has velocity $\mathbf{v}_{pc}$. (b) The velocity of the car relative to the ground is $\mathbf{v}_{cg}$. (c) Relative to the ground, the person's velocity is $\mathbf{v}_{pg} = \mathbf{v}_{pc} + \mathbf{v}_{cg}$, where $\mathbf{v}_{cg}$ is the velocity of the car relative to the ground.

(a)

(b)

(c)

The addition of relative velocities is done in the same way as the addition of displacements, either graphically by placing the tail of one velocity vector at the head of the other or analytically using vector components.

Figure 3-16 Velocity vectors for Example 3-6.

Example 3-6

A river flows from west to east with a speed of 3 m/s. A boy swims north across the river with a speed of 2 m/s relative to the water. What is the velocity of the boy relative to the shore?

Figure 3-16 shows the velocity vectors for this problem. The velocity of the boy relative to the shore is the vector sum of the velocity of the boy relative to the water $\mathbf{v}_{bw}$ and the velocity of the water relative to the shore $\mathbf{v}_{ws}$, as shown in the figure. The magnitude of this velocity is

$$v = \sqrt{v_{bw}^2 + v_{ws}^2} = \sqrt{(2 \text{ m/s})^2 + (3 \text{ m/s})^2}$$

$$= \sqrt{13 \text{ m}^2/\text{s}^2} = 3.61 \text{ m/s}$$

The direction is at an angle θ to the shore:

$$\tan \theta = \frac{v_{bw}}{v_{ws}} = \frac{2 \text{ m/s}}{3 \text{ m/s}} = 0.667$$

$$\theta = \tan^{-1} 0.667 = 33.7°$$

3-7   Projectile Motion

An interesting application of motion in two dimensions is that of a projectile, which is a body that is launched into the air and is then allowed to move freely. The motion of a projectile is complicated by air resistance, the rotation of the earth, and variations in the acceleration due to gravity. For simplicity, we will neglect these complications. The projectile then has a constant acceleration directed vertically downward with a magnitude $g = 9.81 \text{ m/s}^2 = 32.2 \text{ ft/s}^2$. In projectile motion, the horizontal and vertical components of the motion are independent. For example, consider a ball that is thrown from a platform car that is moving horizontally with constant velocity, as illustrated in Figure 3-17. If the ball is thrown straight upwards relative to the car, it moves to its highest point, which depends on its initial vertical velocity, and then returns (Figure 3-17a). This type of vertical motion with constant downward acceleration was discussed in Chapter 2. It has nothing to do with the horizontal motion of the car relative to the ground. The horizontal motion of the ball relative to the ground is motion with constant velocity, the velocity of the car. It has nothing to do with the vertical motion of the ball. Relative to the ground, the ball follows a parabolic path (Figure 3-17b), which is characteristic of projectile motion.

Figure 3-17 (a) A ball is thrown vertically into the air by a person riding on a cart. Relative to the cart, the ball moves vertically upwards and then falls back to its original position. (b) Relative to the ground, the ball has an initial horizontal velocity equal to the velocity of the cart and therefore moves along a parabolic path.

Path seen by observer on cart

(a)

Path seen by observer on ground

(b)

Consider a particle that is launched with some initial velocity that has both vertical and horizontal components relative to a fixed origin. If we take the y axis to be vertical with the positive direction upward and the x axis to be horizontal with the positive direction in the direction of the original horizontal component of the projectile's velocity, we have for the acceleration

$$a_y = -g \qquad \text{3-16a}$$

and

$$a_x = 0 \qquad \text{3-16b}$$

Suppose we launch a projectile from the origin with an initial speed v_0 at an angle of θ with the horizontal axis (Figure 3-18). The initial velocity then has the components

$$v_{0x} = v_0 \cos \theta \qquad \text{3-17a}$$

$$v_{0y} = v_0 \sin \theta \qquad \text{3-17b}$$

Since there is no horizontal acceleration, the x component of the velocity is constant:

$$v_x = v_{0x} \qquad \text{3-18a}$$

The y component varies with time according to

$$v_y = v_{0y} - gt \qquad \text{3-18b}$$

(This is the same as Equation 2-9 with $a = -g$.) The components of the displacement of the projectile are

$$\Delta x = v_{0x} t \qquad \text{3-19a}$$

$$\Delta y = v_{0y} t - \tfrac{1}{2} g t^2 \qquad \text{3-19b}$$

(Compare this equation for y with Equation 2-12a.)

Figure 3-18 The components of the initial velocity of a projectile are $v_{0x} = v_0 \cos \theta$ and $v_{0y} = v_0 \sin \theta$, where θ is the angle between $\mathbf{v}_0$ and the horizontal x axis.

Projectile motion

Projectile motion

Projectile motion

Example 3-7

A ball is thrown into the air with an initial velocity of 50 m/s at 37° to the horizontal. Find the total time the ball is in the air and the total horizontal distance it travels using the approximation $g = 10$ m/s².

The components of the initial velocity vector are

$$v_{0x} = (50 \text{ m/s}) \cos 37° = 40 \text{ m/s}$$

$$v_{0y} = (50 \text{ m/s}) \sin 37° = 30 \text{ m/s}$$

The total time the ball is in the air can be found from Equation 3-19b by setting $\Delta y = 0$ and solving for t:

$$\Delta y = v_{0y} t - \tfrac{1}{2} g t^2 = t(v_{0y} - \tfrac{1}{2} g t) = 0$$

This has the solutions $t = 0$, corresponding to the initial conditions, and

$$t = \frac{2 v_{0y}}{g} = \frac{2 \, (30 \text{ m/s})}{10 \text{ m/s}^2} = 6 \text{ s}$$

Multiflash photograph of a ball thrown into the air. The position of the ball is recorded at approximately 0.43-s intervals.

This is the same result we found in Example 2-9, in which the vertical acceleration and initial vertical velocity of the ball were the same as in this example. As in that example, the total time the ball is in the air is twice the time t_1 it takes the ball to reach its highest point, which can be found by setting $v_y = 0$ in Equation 3-18b:

$$v_y = v_{0y} - gt_1 = 0$$

$$t_1 = \frac{v_{0y}}{g} = \frac{30 \text{ m/s}}{10 \text{ m/s}^2} = 3 \text{ s}$$

Since the ball moves horizontally with a constant velocity of 40 m/s, the total horizontal distance traveled is

$$\Delta x = v_{0x}t = 40 \text{ m/s} \times 6 \text{ s} = 240 \text{ m}$$

This distance is called the **range** of the projectile.

Figure 3-19 shows the height y versus the time t for this example. This curve is identical to that in Figure 2-7a for Examples 2-8 and 2-9 because the vertical accelerations and vertical velocities are the same for all three examples Since the projectile moves 40 m horizontally during each second, we can interpret this curve as a graph of y versus x if we change the horizontal axis from a time scale to a distance scale by multiplying the time values by 40 m/s. The curve of y versus x is a parabola.

The general equation for the path $y(x)$ can be obtained from Equations 3-19a and b by eliminating the variable t between these equations. Choosing $x_0 = y_0 = 0$ and using $t = x/v_{0x}$ in Equation 3-19b, we obtain

$$y = v_{0y}\left(\frac{x}{v_{0x}}\right) - \frac{1}{2}g\left(\frac{x}{v_{0x}}\right)^2$$

or

$$y = \left(\frac{v_{0y}}{v_{0x}}\right)x - \frac{1}{2}\left(\frac{g}{v_{0x}^2}\right)x^2 \qquad 3\text{-}20$$

This equation is of the form $y = ax + bx^2$, which is the equation for a parabola that passes through the origin. Figure 3-20 shows the path of a projectile with the velocity vector and its components indicated at several points.

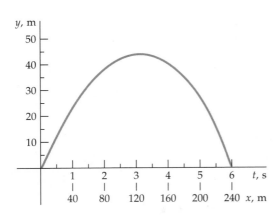

Figure 3-19   Graph of y versus t and y versus x for the ball in Example 3-7. The time scale can be converted into a horizontal-distance scale by multiplying each time by 40 m/s because x is related to t by $x = (40 \text{ m/s})t$.

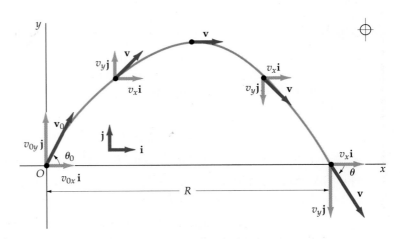

Figure 3-20   Path of a projectile with the velocity vector and its rectangular components shown at several points. The total horizontal distance traveled is the range R.

Exercise

A bullet is fired horizontally from a height of 2 m with an initial speed of 200 m/s. How far does it travel before striking the ground? Use $g = 9.81$ m/s^2. (Answer: 129 m)

In Example 3-7, we found the range of the projectile by finding the total time the projectile was in the air from the vertical component of the velocity and then using this time and the horizontal component of the velocity to find the horizontal distance traveled. This method works whether or not the projectile lands at the same elevation as that from which it is projected.

Example 3-8

The ball in Example 3-7 is projected with the same initial velocity but from a cliff that is 55 m above a plane below (Figure 3-21). Where does the ball land?

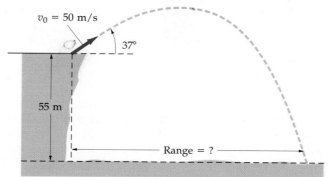

$v_0 = 50$ m/s

37°

55 m

Range = ?

Figure 3-21   A ball thrown from a cliff in Example 3-8. The range is found by first finding the total time the ball is in the air and then multiplying it by the x component of the ball's velocity.

In this case, the time it takes for the ball to reach its maximum height is again 3 s, but the time of fall is longer because the ball falls a greater distance. We first find the maximum height the ball reaches, as we did in Example 2-8. The vertical component of velocity is initially 30 m/s and at the ball's greatest height it is zero. The average upward velocity for the 3 s it takes the ball to reach its maximum height is therefore 15 m/s, and the maximum height is $y = (15$ m/s$)(3$ s$) = 45$ m. After reaching this maximum height, the ball falls 45 m back to the level of its projection plus 55 m to the plane below, so the total distance of fall is 45 m + 55 m = 100 m. The time for the ball to fall from "rest" (the vertical component of velocity is zero at the top of the path) a distance of 100 m is found from the constant-acceleration formula:

$$\Delta y = v_{0y}t - \tfrac{1}{2}gt^2$$
$$-100 \text{ m} = 0 - \tfrac{1}{2}(10 \text{ m/s}^2)t^2$$

or

$$t = \sqrt{20 \text{ s}^2} = 4.5 \text{ s}$$

The total time the ball is in the air is thus 3 s + 4.5 s = 7.5 s. The horizontal distance covered in this time is

$$x = v_{0x}t = (40 \text{ m/s})(7.5 \text{ s}) = 300 \text{ m}$$

We could have used Equation 3-19b to find the time directly without considering the problem in two parts. Since the ball lands 55 m below its starting point, we set $\Delta y = -55$ m. The time is then found from

$$\Delta y = v_{0y}t - \tfrac{1}{2}gt^2$$
$$-55 \text{ m} = (30 \text{ m/s})t - \tfrac{1}{2}(10 \text{ m/s}^2)t^2$$

There are two solutions for this quadratic equation, $t = -1.5$ s and $t = +7.5$ s. The negative time corresponds to the time of 1.5 s it would take for the ball to reach its starting point at $y = 0$ if it were projected from $y = -55$ m with an initial speed such that the component of its vertical velocity component would be 30 m/s when it reached $y = 0$, as shown in Figure 3-22.

Figure 3-22 The solution $t = -1.5$ s, found by setting $\Delta y = -55$ m in Example 3-8, describes the situation that would occur if the ball were projected from $y = -55$ m such that it reaches $y = 0$ at time $t = 0$. This mathematical solution is discarded because it does not describe the given problem.

For the special case in which the initial and final elevations are equal, we can derive a general formula for the range of a projectile in terms of its initial speed and the angle of projection. The time it takes the projectile to reach its maximum height is found by setting the vertical component of its velocity equal to zero:

$$v_y = v_{0y} - gt = 0$$

or

$$t = \frac{v_{0y}}{g}$$

The range R is then the distance traveled in twice this time:

$$R = 2v_{0x}\left(\frac{v_{0y}}{g}\right) = \frac{2v_{0x}v_{0y}}{g}$$

In terms of the initial speed v_0 and the angle of projection θ, the range is

$$R = \frac{2(v_0 \cos \theta)(v_0 \sin \theta)}{g} = \frac{2v_0^2 \sin \theta \cos \theta}{g}$$

This can be further simplified by using the trigonometric identity for the sine of twice an angle (Appendix A):

$$\sin 2\theta = 2 \sin \theta \cos \theta$$

We then have

Range of a projectile

$$R = \frac{v_0^2}{g} \sin 2\theta \qquad\qquad 3\text{-}21$$

Since the maximum value of $\sin 2\theta$ is 1 when $2\theta = 90°$ or $\theta = 45°$, the range is a maximum and equal to v_0^2/g when $\theta = 45°$.

Equation 3-21 is useful for finding the range for projectile problems when the initial and final elevations are the same. Note, however, that the formula could not be used in Example 3-8 because the elevations were not the same. More important, Equation 3-21 is useful because we can learn something from it about the dependence of the range in the initial angle of projection. For example, we have just seen that the range is maximum when the angle of projection is 45°.

We note that the horizontal distance traveled is the product of the initial horizontal component of the velocity v_{0x} and the time the projectile is in the air, which is in turn proportional to v_{0y}. The maximum range occurs when the horizontal and vertical components are equal, which means that the angle of projection is 45°. In some practical applications, other considerations are important. For example, in the shot put, the initial and final elevations are not equal because the ball is projected from an initial height of about 2 m rather than from ground level, which is where the ball lands. The extra initial height increases the time the shot is in the air. In this case, the range is maximum when v_{0x} is somewhat greater than v_{0y}, that is, when the angle of projection is somewhat smaller than 45° (Figure 3-23). Studies of the form of the best shot putters show that the maximum range occurs with an angle of projection of about 42°. With artillery shells, air resistance must be taken into account to predict the range accurately. Air resistance reduces the range for a given angle of projection. It also decreases the optimum angle of projection slightly.

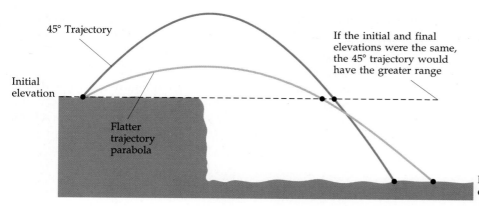

45° Trajectory

If the initial and final elevations were the same, the 45° trajectory would have the greater range

Initial elevation

Flatter trajectory parabola

Final elevation

Figure 3-23   If a projectile lands at an elevation lower than the elevation of projection, the range is maximum when the projection angle is slightly less than 45°.

According to our analysis of projectile motion, an object dropped from a height h above the ground will hit the ground in the same time as one projected horizontally from the same height. In each case, the distance the object falls is given by $y = \frac{1}{2}gt^2$ (measuring y downward from the original height). This remarkable fact can easily be demonstrated. It was first commented upon during the Renaissance by Galileo Galilei (1564–1642), the first person to give the modern, quantitative description of projectile motion we have discussed. In the following, Galileo illustrates the validity of treating the horizontal and vertical components of a projectile's motion as independent motions:*

Galileo Galilei (1564-1642)

Let the ship be motionless and the fall of the stone from the mast take two pulse beats. Then cause the ship to move, and drop the same stone from the same place; from what has been said, it will take two pulse beats to arrive at the deck. In these two pulse beats, the ship will have gone, say, twenty yards so that the natural motion of the stone will have been a diagonal line much longer than the

*Galileo Galilei, *Dialogue Concerning Two Chief World Systems—Ptolemaic and Copernican*, pp. 154–155, trans. Stillman Drake, University of California Press, Berkeley, 1953; reprinted by permission of The Regents of the University of California.

Comparison of a ball dropped with one projected horizontally. The vertical position is independent of the horizontal motion.

first straight and perpendicular one, which was merely the length of the mast; nevertheless, it will have traversed this distance in the same time. Now assuming the ship to be speeded up still more, so that the stone in falling must follow a diagonal line very much longer still than the other, eventually the velocity of the ship may be increased by any amount while the falling rock will describe always longer and longer diagonals, and still pass over them in the same two pulse beats. Similarly, if a perfectly level cannon on a tower were fired parallel to the horizon, it would not matter whether a small charge or a great one was put in, so that the ball would fall a thousand yards away, four thousand, or six thousand, or ten thousand or more; all these shots would require equal times and each time would be equal to that which the ball would have taken in going from the mouth of the cannon to the ground if it were allowed to fall straight down without any other impulse.

Example 3-9

A park ranger with a tranquilizer dart gun wishes to shoot a monkey hanging from a branch of a tree. The ranger aims right at the monkey, not realizing that the dart will follow a parabolic path and thus fall below the monkey. The monkey, however, seeing the dart leave the gun, lets go of the branch and drops out of the tree, expecting to avoid the dart. Show that the monkey will be hit regardless of the initial speed of the dart so long as the speed of the dart is great enough for it to travel the horizontal distance to the tree before hitting the ground and the monkey drops the instant the dart is fired.

This situation is sometimes demonstrated using a target suspended by an electromagnet. When the dart leaves the gun, the circuit is broken and the target falls. Let the horizontal distance to the tree be x and the original height of the monkey be h, as shown in Figure 3-24. The dart is projected at an angle given by $\tan \theta = h/x$. If there were no gravity, the dart would reach the height h in the time t it takes for it to travel the horizontal distance x:

$$y = v_{0y}t = h$$

$$t = \frac{x}{v_{0x}}$$

However, because of gravity, the dart has a vertical acceleration downward. In time $t = x/v_{0x}$, the dart reaches a height y given by

$$y = v_{0y}t - \tfrac{1}{2}gt^2 = h - \tfrac{1}{2}gt^2$$

Figure 3-24   The monkey and dart from Example 3-9. The height of the dart at any time is $y = v_{0y}t - \tfrac{1}{2}gt^2$. The height of the monkey at any time is $h - \tfrac{1}{2}gt^2$, which is the same as that of the dart because $h = v_{0y}t$. The dart will therefore always hit the monkey if the monkey drops the instant the dart is fired.

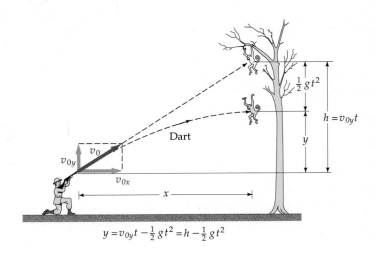

$$y = v_{0y}t - \tfrac{1}{2}gt^2 = h - \tfrac{1}{2}gt^2$$

This is lower than h by $\frac{1}{2}gt^2$, which is exactly the distance the monkey falls in the same time. In the usual lecture demonstration, the initial speed of the dart is varied so that for a large v_0 the target is hit very near its original height and for a small v_0 it is hit just before it reaches the floor.

Questions

12. What is the acceleration of a projectile at the top of its flight?

13. Can the velocity of an object change direction while its acceleration is constant in both magnitude and direction? If so, give an example.

3-8   Circular Motion

Circular motion is common in nature and in our everyday experience. The earth revolves in a nearly circular orbit around the sun; the moon, around the earth. Wheels rotate in circles, cars travel in circular arcs as they round corners, and so on. In this section, we consider a particle moving in a circle with constant speed. In everyday usage, we might say that since the speed is constant the particle is not accelerating. But we have defined acceleration as the rate of change of the velocity vector, and when a particle moves in a circle its velocity vector is continually changing direction.

Newton was one of the first to recognize the importance of circular motion. He showed that when a particle moves with constant speed v in a circle of radius r, it has an acceleration with a magnitude of v^2/r directed toward the center of the circle. This acceleration is called **centripetal acceleration.**

Consider a satellite that is moving in a circular orbit around the earth. Why doesn't the satellite fall toward the earth? The answer is not that there is no force of gravity acting on the satellite. At 200 km above the surface of the earth, the gravitational force acting on a satellite is about 94 percent of what it would be were the satellite at the earth's surface. The satellite does "fall" toward the earth, but because of its tangential velocity, it continually

The rings of Saturn as photographed by Voyager 1 in 1980. The rings consist of ice or ice-covered rock in circular orbits close to the planet.

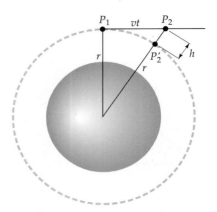

Figure 3-25 Satellite moving with speed v in a circular orbit of radius r about the earth. If the satellite did not accelerate toward the earth, it would move in a straight line from point P_1 to P_2 in time t. Because of its acceleration, it falls a distance h in this time. For small t, $h = \frac{1}{2}(v^2/r)t^2 = \frac{1}{2}at^2$.

misses the earth. To see this, consider Figure 3-25. If the satellite were not accelerating, it would move from point P_1 to P_2 in some time t. Instead, it arrives at point P_2' on its circular orbit. Thus, in a sense, the satellite "falls" the distance h shown. If we take the time t to be very small, the points P_2 and P_2' are approximately on a radial line, as shown in the figure, and we can use the approximation that h is much smaller than the orbit radius r. (From the figure, we can see that the smaller we make the time t, the shorter the distance vt and the smaller h will be for any given orbital radius r.) We can then calculate h from the right triangle with the sides vt, r, and $r + h$. Since $r + h$ is the hypotenuse of the right triangle, the pythagorean theorem gives

$$(r + h)^2 = (vt)^2 + r^2$$

$$r^2 + 2hr + h^2 = v^2t^2 + r^2$$

or

$$h(2r + h) = v^2t^2$$

For very short times t, h will be much less than r, so we can neglect h compared with $2r$ for the term in the parentheses. Thus,

$$2rh \approx v^2t^2$$

or

$$h \approx \frac{1}{2}\left(\frac{v^2}{r}\right)t^2$$

Comparing this with the constant-acceleration expression $h = \frac{1}{2}at^2$, we see that the magnitude of the acceleration of the satellite is

Centripetal acceleration

$$a = \frac{v^2}{r} \qquad\qquad 3\text{-}22$$

The direction of the acceleration is inward toward the center of the circle.

We can show that this result holds in general for circular motion with constant speed by considering the position and velocity vectors (Figure 3-26). Here the initial velocity vector $\mathbf{v}_1$ is perpendicular to the initial position vector $\mathbf{r}_1$. A short time later, the velocity is $\mathbf{v}_2$, which is perpendicular to $\mathbf{r}_2$. The angle between the velocity vectors $\Delta\theta$ is the same as that between the position vectors because both the position and velocity vectors must move through equal angles to remain mutually perpendicular. If we take the time interval to be very small, the magnitude of the displacement $|\Delta\mathbf{r}|$ is approximately equal to the distance traveled along the arc Δs. The average acceleration is the ratio of the change in velocity $\Delta\mathbf{v} = \mathbf{v}_2 - \mathbf{v}_1$ to the time interval Δt. From the figure we see that for very small Δt, the change in velocity (and therefore the average acceleration) is approximately perpendicular to the velocity vectors and is directed toward the center of the circle. We can find the magnitude of the average acceleration from the angle $\Delta\theta$ in the similar triangles in Figure 3-26. We have

$$\Delta\theta = \frac{\Delta s}{r} \approx \frac{|\Delta\mathbf{v}|}{v}$$

where r is the radius of the circle and v is the speed of the particle. If we now use the fact that the distance is given by $\Delta s = v\,\Delta t$, we have

$$\Delta\theta = \frac{v\,\Delta t}{r} \approx \frac{|\Delta\mathbf{v}|}{v}$$

or

$$\frac{|\Delta\mathbf{v}|}{\Delta t} \approx \frac{v^2}{r}$$

It is often convenient to describe the motion of a particle moving in a circle with constant speed in terms of the time required for one complete revolution, called the period T. If the radius of the circle is r, the particle travels a distance $2\pi r$ during one period, so its speed is related to the radius and the period by

$$v = \frac{2\pi r}{T} \qquad\qquad 3\text{-}23$$

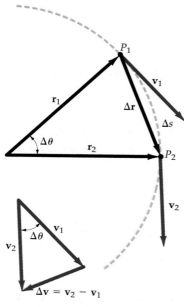

Figure 3-26 Position and velocity vectors for a particle moving in a circle. The angle $\Delta\theta$ between $\mathbf{v}_1$ and $\mathbf{v}_2$ is the same as the angle between $\mathbf{r}_1$ and $\mathbf{r}_2$. For very small time intervals, the change in velocity $\Delta\mathbf{v}$ is approximately perpendicular to $\mathbf{v}$ and points inward toward the center of the circle.

Example 3-10

A tether ball moves in a horizontal circle of radius 2 m. It makes one revolution in 3 s. Find its acceleration.

To find the acceleration we must first find the speed, which is given by Equation 3-23:

$$v = \frac{2\pi r}{T} = \frac{2\pi(2\ \text{m})}{3\ \text{s}} = 4.19\ \text{m/s}$$

The magnitude of the ball's acceleration is then

$$a = \frac{v^2}{r} = \frac{(4.19\ \text{m/s})^2}{2\ \text{m}} = 8.78\ \text{m/s}^2$$

This is about 0.9 times the acceleration due to gravity g. The direction of the acceleration is radially inward, towards the center of the circle.

Example 3-11

A car rounds a curve of radius 30 m. If the maximum centripetal acceleration that can be provided by friction is 5 m/s², what is the maximum speed of the car in kilometers per hour?

From Equation 3-22, we have

$$\frac{v^2}{r} = a$$

$$v_{\max} = \sqrt{ra_{\max}}$$

$$= \sqrt{(30\ \text{m})(5\ \text{m/s}^2)} = 12.2\ \text{m/s}$$

To convert this speed to kilometers per hour, we use 1 h = 3600 s and 1 km = 1000 m:

$$12.2\ \frac{\text{m}}{\text{s}} \times \frac{3600\ \text{s}}{1\ \text{h}} \times \frac{1\ \text{km}}{1000\ \text{m}} = 44\ \text{km/h}$$

Exercise

A car rounds a curve of radius 40 m at 48 km/h. What is its centripetal acceleration? (Answer: 4.44 m/s²)

Example 3-12

A satellite moves at constant speed in a circular orbit about the center of the earth near the surface of the earth. If its acceleration is 9.81 m/s², what is its speed, and how long does it take to make one complete revolution?

This acceleration is the same as that for any body falling freely near the surface of the earth. We take the radius of the earth, about 6370 km, to be the approximate radius of the orbit. (For actual satellites that are put into orbit a few hundred kilometers above the earth's surface, the orbital radius will be slightly greater, so the acceleration will be slightly less than 9.81 m/s² because of the decrease in the gravitational force with distance from the center of the earth.) The speed of the satellite can be found from Equation 3-22:

$$v^2 = rg = (6370 \text{ km})(9.81 \text{ m/s}^2)$$

$$v = 7.91 \text{ km/s}$$

The period is related to the speed by Equation 3-23:

$$T = \frac{2\pi r}{v} = \frac{2\pi(6370 \text{ km})}{7.91 \text{ km/s}} = 5060 \text{ s} = 84.3 \text{ min}$$

Figure 3-27 is a drawing from Newton's *System of the World* that illustrates the connection between projectile motion and satellite motion. Newton's description of it reads*

That by means of centripetal forces the planets may be retained in certain . . . orbits, we may easily understand, if we consider the motions of projectiles; for a stone that is projected is by the pressure of its own weight forced out of the rectilinear path, which by the initial projection alone it should have pursued, and made to describe a curved line in the air; and through that crooked way is at last brought down to the ground; and the greater the velocity is with which it is projected, the farther it goes before it falls to the earth. We may therefore suppose the velocity to be so increased, that it would describe an arc of 1, 2, 5, 10, 100, 1000 miles before it arrived at the earth, till at last, exceeding the limits of the earth, it should pass into space without touching it.

System of the World, trans. Andrew Motte, 1728, University of California Press, Berkeley, 1960; reprinted by permission of The Regents of the University of California.

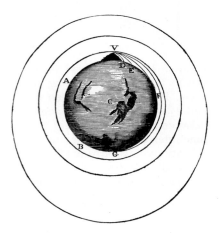

Figure 3-27 Drawing from Newton's *System of the World*, published in 1728, illustrating the connection between projectile motion and satellite motion.

The space shuttle orbits freely after disengagement from the booster rockets shown at the left.

If a particle moves in a circle with a speed that is varying, there is a component of acceleration that is tangent to the circle as well as the centripetal acceleration inward. The tangential component of the acceleration is simply the rate of change of the speed, whereas the radially inward component has the magnitude v^2/r. For any general motion along a curve, we can treat a portion of the curve as an arc of a circle (Figure 3-28). The particle then has centripetal acceleration v^2/r toward the center of curvature, and if the speed is changing, it has tangential acceleration of magnitude dv/dt.

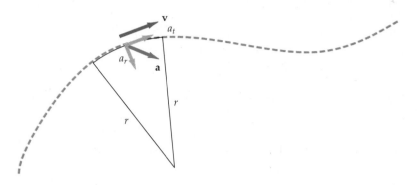

Figure 3-28   A particle moving along an arbitrary curve can be considered to be moving in a circular arc during a small time interval. The instantaneous-acceleration vector has a component a_r of magnitude v^2/r toward the center of curvature of the arc and a component a_t of magnitude dv/dt that is tangential to the curve in the direction of motion.

Summary

1. Quantities that have both magnitude and direction, such as displacement, velocity, and acceleration, are vector quantities.

2. Vectors can be added graphically by placing the tail of one vector at the head of the other and drawing the resultant vector from the tail of the first to the head of the second. Subtracting a vector **B** is the same as adding $-$**B**, where $-$**B** is a vector with a magnitude equal to B but in the opposite direction.

3. Vectors can be added analytically by first finding the components of the vectors given by

$$A_x = A \cos \theta$$

$$A_y = A \sin \theta$$

where θ is the angle between **A** and the x axis. The x component of the resultant vector is the sum of the x components of the individual vectors, and the y component is the sum of the y components of the individual vectors.

4. The position vector **r** points from an arbitrary origin to a particle's position. In the time interval Δt, **r** changes by Δ**r**. The velocity vector **v** is the rate of change of the position vector. Its magnitude is the speed, and it points in the direction of motion, tangent to the curve along which the particle is traveling. The instantaneous-velocity vector is given by

$$\mathbf{v} = \lim_{\Delta t \to 0} \frac{\Delta \mathbf{r}}{\Delta t} = \frac{d\mathbf{r}}{dt}$$

5. The acceleration vector is the rate of change of the velocity vector. The instantaneous-acceleration vector is given by

$$\mathbf{a} = \lim_{\Delta t \to 0} \frac{\Delta \mathbf{v}}{\Delta t} = \frac{d\mathbf{v}}{dt}$$

A particle is accelerating if its velocity vector is changing in magnitude or direction or both.

6. If a particle moves with velocity $\mathbf{v}_{pA}$ relative to a coordinate system A, which is in turn moving with velocity $\mathbf{v}_{AB}$ relative to another coordinate system B, the velocity of the particle relative to B is

$$\mathbf{v}_{pB} = \mathbf{v}_{pA} + \mathbf{v}_{AB}$$

7. In projectile motion, the horizontal and vertical motions are independent. The horizontal motion has constant velocity that is equal to the horizontal component of the original velocity:

$$v_x = v_{0x} = v_0 \cos \theta$$

$$\Delta x = v_{0x}t$$

The vertical motion is the same as motion in one dimension with constant acceleration due to gravity g downward:

$$v_y = v_{0y} - gt$$

$$\Delta y = v_{0y}t - \tfrac{1}{2}gt^2$$

The total distance traveled by the projectile, called the range R, is found by first finding the total time the projectile is in the air and then multiplying this time by the constant horizontal component of the velocity. For the special case in which the initial and final elevations are the same, the range is related to the projection angle θ by

$$R = \frac{v_0^2}{g} \sin 2\theta$$

and is maximum at $\theta = 45°$.

8. When an object moves in a circle with constant speed, it is accelerating because its velocity is changing in direction. This acceleration is called centripetal acceleration, and it points toward the center of the circle. The magnitude of the centripetal acceleration is

$$a = \frac{v^2}{r}$$

where v is the speed and r is the radius of the circle.

Suggestions for Further Reading

Drake, Stillman, and James MacLachan: "Galileo's Discovery of the Parabolic Trajectory," *Scientific American*, March 1975, p. 102.

Galileo knew that a falling body with a horizontal component of velocity describes a parabola 30 years before he published the fact, according to unpublished manuscripts. (See also Chapter 2 references.)

Review

A. Objectives: After studying this chapter, you should

1. Be able to add and subtract vectors graphically.

2. Be able to obtain the components of vectors and use them to add and subtract vectors.

3. Be able to express arbitrary vectors in terms of unit vectors.

4. Be able to combine relative velocities to express the velocity of a particle relative to a coordinate system that is itself moving relative to another coordinate system.

5. Know that in projectile motion the horizontal and vertical motions are independent, and be able to use this fact in working projectile problems.

6. Know that when a particle moves in a circle with constant speed it has centripetal acceleration of magnitude v^2/r directed towards the center of the circle.

B. Define, explain, or otherwise identify:

Vectors
Scalars
Displacement vector
Parallelogram method of vector addition
Rectangular component of a vector
Unit vector
Position vector
Velocity vector
Acceleration vector
Relative velocity
Range
Centripetal acceleration

C. True or false: If the statement is true, explain why it is true. If it is false, give a counterexample.

1. The magnitude of the sum of two vectors must be greater than the magnitude of either vector.

2. The instantaneous-velocity vector is always in the direction of motion.

3. The instantaneous-acceleration vector is always in the direction of motion.

4. If the speed is constant, the acceleration must be zero.

5. If the acceleration is zero, the speed must be constant.

6. It is impossible to go around a curve without acceleration.

7. The time required for a bullet fired horizontally to reach the ground is the same as if it were dropped from rest from the same height.

Problems

Level I

3-1 The Displacement Vector and Vector Addition

1. A bear walks northeast for 10 m and then east for 10 m. Show each displacement graphically, and find the resultant displacement vector.

2. (*a*) A man walks along a circular arc from the position $x = 5$ m, $y = 0$ to a final position $x = 0$, $y = 5$ m. What is his displacement? (*b*) A second man walks from the same initial position along the x axis to the origin and then along the y axis to $y = 5$ m and $x = 0$. What is his displacement?

3. A circular path has a radius of 10 m. An xy coordinate system is established such that the center of the circle is on the positive y axis and the circle passes through the origin. A woman starts at the origin and walks around the path at a steady speed, returning to the origin exactly 1 min after she started. (*a*) Find the magnitude and direction of her displacement from the origin 15, 30, 45, and 60 s after she starts. (*b*) Find the magnitude and direction of her displacement for each of the four successive 15-s intervals of her walk. (*c*) How is her displacement for the first 15 s related to that for the second 15 s? (*d*) How is her displacement for the second 15-s interval related to that for the last 15-s interval?

4. For the two vectors **A** and **B** shown in Figure 3-29, find the following graphically: (*a*) **A** + **B**, (*b*) **A** − **B**, (*c*) 2**A** + **B**, (*d*) **B** − **A**, (*e*) 2**B** − **A**.

Figure 3-29   Problems 4 and 9.

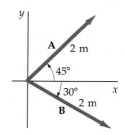

5. A scout walks 2 km due east from camp, then turns left and walks 2 km along the arc of a circle centered at the campsite, and finally walks 1 km directly toward camp. (*a*) How far is the scout from camp? (*b*) What is the direction of the scout's position as measured from the campsite? (*c*) What is the ratio of the final displacement to the total distance walked?

3-2 Addition of Vectors by Components

6. A hiker sets off at 8 A.M. on level terrain. At 9 A.M. she is 2 km due east of her starting point. At 10 A.M. she is 1 km northwest of where she was at 9 A.M. At 11 A.M. she is 3 km due north of where she was at 10 A.M. (*a*) Make a drawing showing these successive displacements as vectors, with the tail of each new vector being at the head of the previous one. What are the magnitudes and directions of these displacements? (Specify the direction of a vector by giving the angle it makes with the eastward direction.) (*b*) What are the north and east components of these displacements? (*c*) How far is the hiker from her starting point at 11 A.M.? In what direction? (*d*) Draw the three displacement vectors to scale and add them graphically. Do these successive straight lines represent the actual path the hiker followed? Is the distance she walked the sum of the lengths of the three displacement vectors?

7. Find the rectangular components of the vectors that lie in the xy plane, have a magnitude of A, and make an angle θ with the x axis, as shown in Figure 3-30, for the following values of A and θ: (*a*) $A = 10$ m, $\theta = 30°$; (*b*) $A = 5$ m, $\theta = 45°$; (*c*) $A = 7$ km, $\theta = 60°$; (*d*) $A = 5$ km, $\theta = 90°$; (*e*) $A = 15$ km/s, $\theta = 150°$; (*f*) $A = 10$ m/s, $\theta = 240°$; and (*g*) $A = 8$ m/s^2, $\theta = 270°$.

Figure 3-30   Problem 7.

8. A plane is inclined at an angle of 30° with the horizontal. Choose the x axis pointing down the slope of the plane and the y axis perpendicular to the plane. Find the x and y components of the acceleration of gravity, which has the magnitude 9.81 m/s^2 and points vertically down.

9. The displacement vectors **A** and **B** shown in Figure 3-29 both have a magnitude of 2 m. (*a*) Find their x and y components. (*b*) Find the components, magnitude, and direction of the sum **A** + **B**. (*c*) Find the components, magnitude, and direction of the difference **A** − **B**.

3-3 Unit Vectors and Multiplication of Vectors by Scalars

10. Find the magnitude and direction of the following vectors: (*a*) **A** = 5**i** + 3**j**, (*b*) **B** = 10**i** − 7**j**, and (*c*) **C** = −2**i** − 3**j** + 4**k**.

11. Find the magnitude and direction of **A**, **B**, and **A** + **B** for (*a*) **A** = −4**i** − 7**j**, **B** = 3**i** − 2**j**, and (*b*) **A** = 1**i** − 4**j**, **B** = 2**i** + 6**j**.

12. A cube of side 2 m has its faces parallel to the coordinate planes with one corner at the origin. A fly begins at the origin and walks along three edges until it is at the far corner. Write the displacement vector of the fly using the unit vectors **i**, **j**, and **k**, and find the magnitude of this displacement.

13. Describe the following vectors using the unit vectors **i** and **j**: (*a*) a velocity of 10 m/s at an angle of elevation of 60°; (*b*) a vector **A** of magnitude $A = 5$ m and $\theta = 225°$; and (*c*) a displacement from the origin to the point $x = 14$ m, $y = -6$ m.

14. For the vector **A** = 3**i** + 4**j**, find any three other vectors **B** that also lie in the xy plane and have the property that $A = B$ but **A** ≠ **B**. Write these vectors in terms of their components and show them graphically.

15. Two vectors **A** and **B** lie in the xy plane. Under what conditions does the ratio A/B equal A_x/B_x?

16. If **A** = 5**i** − 4**j** and **B** = −7.5**i** + 6**j**, write an equation relating **A** to **B**.

3-4 The Velocity Vector

17. A vector **A**(t) has a constant magnitude but is changing direction in a uniform way. Draw the vectors **A**($t + \Delta t$) and **A**(t) for a small time interval Δt, and find the difference $\Delta \mathbf{A} = \mathbf{A}(t + \Delta t) - \mathbf{A}(t)$ graphically. How is the direction of $\Delta \mathbf{A}$ related to **A** for small time intervals?

18. Consider the path of a particle as it moves in space. (*a*) How is the velocity vector related geometrically to the path of the particle? (*b*) Sketch a curved path, and draw the velocity vector for the particle for several positions along the path.

19. A stationary radar operator determines that a ship is 10 km south of him. An hour later the same ship is 20 km southeast of him. If the ship moved at constant speed and always in the same direction, what was its velocity during this time?

20. A particle's position coordinates (x, y) are (2 m, 3 m) at $t = 0$; (6 m, 7 m) at $t = 2$ s; and (13 m, 14 m) at $t = 5$ s.

(*a*) Find $\mathbf{v}_{av}$ from $t = 0$ to $t = 2$ s. (*b*) Find $\mathbf{v}_{av}$ from $t = 0$ to $t = 5$ s.

21. The position vector of a particle is given by **r** = 5t**i** + 10t**j**, where t is in seconds and **r** is in meters. (*a*) Draw the path of the particle in the xy plane. (*b*) Find **v** in component form and find its magnitude.

3-5 The Acceleration Vector

22. A ball is thrown directly upward. Consider the 2-s time interval $\Delta t = t_2 - t_1$, where t_1 is 1 s before the ball reaches its highest point and t_2 is 1 s after it reaches its highest point. Find (*a*) the change in speed, (*b*) the change in velocity, and (*c*) the average acceleration for this time interval.

23. Figure 3-31 shows the path of an automobile that is made up of segments of straight lines and arcs of circles. The automobile starts from rest at point A. After it reaches point B, it travels at constant speed until it reaches point E. It comes to rest at point F. (*a*) At the middle of each segment (AB, BC, CD, DE, and EF), what is the direction of the velocity vector? (*b*) At which of these points does the automobile have an acceleration? In those cases, what is the direction of the acceleration? (*c*) How do the magnitudes of the acceleration compare for segments BC and DE?

Figure 3-31   Problem 23.

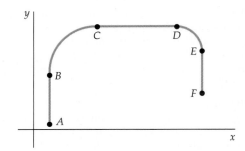

24. Initially, a particle is moving due west with a speed of 40 m/s, and 5 s later it is moving north with a speed of 30 m/s. (*a*) What was the change in the magnitude of the particle's velocity during this time? (*b*) What was the change in the direction of the velocity? (*c*) What are the magnitude and direction of $\Delta \mathbf{v}$ for this interval? (*d*) What are the magnitude and direction of $\mathbf{a}_{av}$ for this interval?

25. At $t = 0$ a particle located at the origin has a speed of 40 m/s at $\theta = 45°$. At $t = 3$ s the particle is at $x = 100$ m and $y = 80$ m with speed of 30 m/s at $\theta = 50°$. Calculate (*a*) the average velocity and (*b*) the average acceleration of the particle during this interval.

3-6 Relative Velocity

26. A swimmer heads directly across a river, swimming at 1.6 m/s relative to still water. She arrives at a point 40 m downstream from the point directly across the river, which is 80 m wide. (*a*) What is the speed of the river current? (*b*) What is the swimmer's speed relative to the shore? (*c*) In what direction should she head so as to arrive at the point directly opposite her starting point?

27. A plane flies at a speed of 250 km/h relative to still air. There is a wind blowing at 80 km/h in the northeast direction (that is, at 45° to the east of north). (*a*) In what direction should the plane head so as to fly due north? (*b*) What is the speed of the plane relative to the ground?

28. A missile fired from a jet takes off from the wing with an acceleration of 500 m/s² that lasts for 3 s. The jet's velocity is 500 m/s due east. (*a*) What are the position and velocity of the missile as observed by the jet pilot 2 s after it is fired? (*b*) What are the position and velocity of the missile as observed by a person on the ground 2 s after it is fired? (Neglect the acceleration due to gravity.)

3-7 Projectile Motion

29. A bullet is fired horizontally with an initial velocity of 245 m/s. The gun is 1.5 m above the ground. How long is the bullet in the air?

30. A supersonic transport is flying horizontally at an altitude of 20 km with a speed of 2500 km/h when an engine falls off. (*a*) How long does it take the engine to hit the ground? (*b*) How far horizontally is the engine from where it fell off when it hits the ground? (*c*) How far is the engine from the aircraft (assuming it continues to fly as if nothing had happened) when the engine hits the ground? Neglect air resistance.

31. A cannon is elevated at an angle of 45°. It fires a ball with a speed of 300 m/s. (*a*) What height does the ball reach? (*b*) How long is the ball in the air? (*c*) What is the horizontal range?

32. A projectile is launched with speed v_0 at an angle of θ_0 with the horizontal. Find an expression for the maximum height it reaches above its starting point in terms of v_0, θ_0, and g.

33. A projectile is fired with an initial velocity of 30 m/s at 60° to the horizontal. At the projectile's highest point, what is its velocity? Its acceleration?

34. A pitcher throws a fastball at 140 km/h toward home plate, which is 18.4 m away. Neglecting air resistance (not a good idea if you are the batter), find how far the ball drops because of gravity by the time it reaches home plate.

3-8 Circular Motion

35. A particle travels in a circular path of radius 5 m with a constant speed of 15 m/s. What is the magnitude of its acceleration?

36. An airplane pilot pulls out of a dive by following the arc of a circle whose radius is 300 m. At the bottom of the circle, where her speed is 180 km/h, what are the direction and magnitude of her acceleration?

37. An object travels with a constant speed v in a circular path of radius r. (*a*) If v is doubled, how is the acceleration a affected? (*b*) If r is doubled, how is a affected? (*c*) Why is it impossible for an object to travel around a perfectly sharp angular turn?

38. In Figure 3-32 the particles are traveling counterclockwise in circles of radius 5 m with speeds that may be vary-

ing. The acceleration vectors are indicated at certain times. Find the values of v and dv/dt at each of these three times.

39. A boy whirls a ball on a string in a horizontal circle of radius 1 m. How many revolutions per minute must the ball make if its acceleration toward the center of the circle is to have the same magnitude as the acceleration due to gravity?

Level II

40. A particle moves in a circle of radius 4 cm. It takes 8 s to make a complete trip. Draw the path of the particle to scale, and indicate the particles's positions at 1-s intervals. Draw the displacement vectors for these 1-s intervals. These vectors also indicate the average-velocity vectors for these intervals. Find graphically the change in the average velocity Δv for two consecutive 1-s intervals. Compare $\Delta v/\Delta t$, measured in this way, with the instantaneous acceleration computed from $a_r = v^2/r$.

41. A particle moves in the xy plane with constant acceleration. At $t = 0$ the particle is at $\mathbf{r} = 4$ m $\mathbf{i} + 3$ m $\mathbf{j}$. At $t = 2$ s the particle has moved to $\mathbf{r} = 10$ m $\mathbf{i} - 2$ m $\mathbf{j}$, and its velocity has changed to $\mathbf{v} = 5$ m/s $\mathbf{i} - 6$ m/s $\mathbf{j}$. (*a*) What is the acceleration of the particle? (*b*) What is the velocity of the particle as a function of time? (*c*) What is the position vector of the particle as a function of time?

42. A particle moves in an xy plane with constant acceleration. At time zero, the particle is at $x = 4$ m, $y = 3$ m. The acceleration is given by the vector $\mathbf{a} = 4$ m/s² $\mathbf{i} + 3$ m/s² $\mathbf{j}$. The velocity vector initially is $\mathbf{v} = 2$ m/s $\mathbf{i} - 9$ m/s $\mathbf{j}$. (*a*) Find the velocity vector at $t = 2$ s. (*b*) Find the position vector at $t = 4$ s. Give its magnitude and direction.

Figure 3-32   Problem 38.

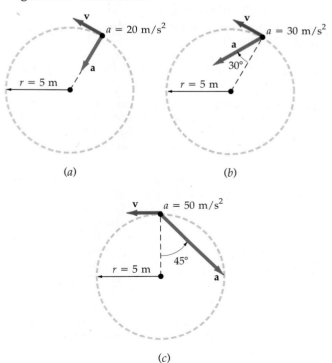

(*a*)   (*b*)

(*c*)

Figure 3-33   Problem 43.

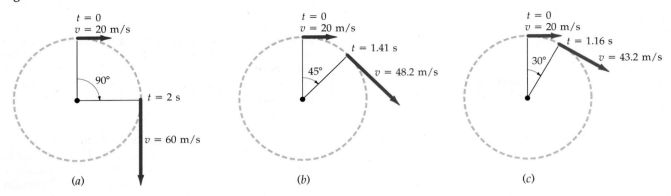

(a)   *(b)*   *(c)*

43. In parts (*a*) to (*c*) of Figure 3-33, particles are traveling in circular paths with varying speeds. The velocity vectors are shown. Find the magnitude of the average-acceleration vector between the two given positions in each case.

44. A particle travels counterclockwise with constant speed in a circular path of radius 5 m around the origin. It begins at $t = 0$ at $x = 5$ m, $y = 0$ and takes 100 s to make a complete revolution. (*a*) What is the speed of the particle? (*b*) Give the magnitude and direction of the position vector **r** at the times $t = 50$ s, $t = 25$ s, $t = 10$ s, and $t = 0$. (*c*) Find the magnitude of $\mathbf{v}_{av}$ and show its direction graphically for each of the following time intervals: $t = 0$ to $t = 50$ s, $t = 0$ to $t = 25$ s, and $t = 0$ to $t = 10$ s. (*d*) How does $\mathbf{v}_{av}$ for the interval $t = 0$ to $t = 10$ s compare with the instantaneous velocity at $t = 0$?

45. A particle has a position vector given by $\mathbf{r} = 30t\mathbf{i} + (40t - 5t^2)\mathbf{j}$, where **r** is in meters and t in seconds. Find the instantaneous-velocity and instantaneous-acceleration vectors as functions of time t.

46. A particle has a constant acceleration of $\mathbf{a} = 6\mathbf{i} + 4\mathbf{j}$ m/s². At time $t = 0$, the velocity is zero and the position vector is $\mathbf{r}_0 = (10$ m$)\mathbf{i}$. (*a*) Find the velocity and position vectors at any time t. (*b*) Find the equation of the particle's path in the xy plane, and sketch the path.

47. A stone thrown horizontally from the top of a tower hits the ground at a distance of 18 m from the base of the tower. (*a*) Find the speed at which the stone was thrown if the tower is 24 m high. (*b*) Find the speed of the stone just before it hits the ground.

48. A freight train is moving at a constant speed of 10 m/s. A man standing on a flatcar throws a ball into the air and catches it as it falls. Relative to the flatcar, the initial velocity of the ball is 15 m/s straight up. (*a*) What are the magnitude and direction of the initial velocity of the ball as seen by a second man standing next to the track? (*b*) How long is the ball in the air according to the man on the train? According to the man on the ground? (*c*) What horizontal distance has the ball traveled by the time it is caught according to the man on the train? According to the man on the ground? (*d*) What is the minimum speed of the ball during its flight according to the

man on the train? According to the man on the ground? (*e*) What is the acceleration of the ball according to the man on the train? According to the man on the ground?

49. A projectile is fired into the air from the top of a 200-m cliff above a valley (Figure 3-34). Its initial velocity is 60 m/s at 60° with the horizontal. Neglecting air resistance, where does the projectile land?

Figure 3-34   Problem 49.

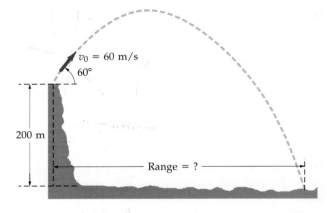

50. A car is traveling down a highway at 25 m/s. Just as the car crosses a perpendicularly intersecting crossroad, the passenger throws out a can at a 45° angle of elevation in a plane perpendicular to the motion of the car. The initial speed of the can relative to the car is 10 m/s. It is released at a height of 1.2 m above the road. (*a*) Write the initial velocity of the can (relative to the road) in terms of the unit vectors **i**, **j**, and **k**. (*b*) Where does the can land?

51. Car A is traveling east at 20 m/s. As car A crosses the intersection shown in Figure 3-35, car B starts from rest 40 m north of the intersection and moves south with a constant acceleration of 2 m/s². (*a*) What is the position of B relative to A 6 s after A crosses the intersection? (*b*) What is the velocity of B relative to A for $t = 6$ s? (*c*) What is the acceleration of B relative to A for $t = 6$ s?

Figure 3-35 Problem 51.

52. An object on the equator has an acceleration toward the center of the earth because of the earth's rotation and an acceleration toward the sun because of the earth's motion along its orbit. Calculate the magnitudes of both accelerations, and express them as fractions of the free-fall acceleration due to gravity g.

53. A baseball is thrown toward a player with an initial speed of 20 m/s at an angle of 45° with the horizontal. At the moment the ball is thrown, the player is 50 m from the thrower. At what speed and in what direction must he run to catch the ball at the same height at which it was released?

54. Find the angle of projection such that the maximum height of a projectile is equal to its horizontal range.

55. A trail bike comes to a ditch. A ramp with an incline of 10° has been built so that the bike can jump the ditch. If the bike needs to jump a horizontal distance of 7 m to clear the ditch, how fast must it be going when it leaves the ramp?

56. The position of a particle is given by the vector

$$\mathbf{r} = -10 \text{ m cos } \omega t \, \mathbf{i} + 10 \text{ m sin } \omega t \, \mathbf{j}$$

where $\omega = 2 \text{ s}^{-1}$. (a) Show that the motion is a circle. What is the radius of the circle? Does the particle move clockwise or counterclockwise around the circle? (b) What is the speed of the particle? (c) What is the time for one complete revolution?

57. A workman on the roof of a house drops his hammer, which slides down the roof at a constant speed of 4 m/s. The roof makes an angle of 30° with the horizontal, and its lowest point is 10 m from the ground. What is the horizontal distance traveled by the hammer after it leaves the roof of the house before it hits the ground?

58. A motorboat begins on a course heading toward an island that is 1 km east and 3 km north from the boat's initial position. After traveling 45 min, the driver discovers that the boat is due east of the island. He reverses his course, and 45 min later the boat is 6 km east of its starting position. (a) What is the velocity of the water current? (b) What is the velocity of the boat relative to the water for the first 45 min? (c) What is the velocity of the boat relative to the island for the first 45 min?

Level III

59. A gun shoots bullets that leave the muzzle at 250 m/s. If a bullet is to hit a target 100 m away at the level of the muzzle, the gun must be aimed at a point above the target. How far above the target is this point? (Neglect air resistance.)

60. Galileo showed that, if air resistance is neglected, the ranges for projectiles whose angles of projection exceed or fall short of 45° by the same amount are equal. Prove this.

Projectile paths for various projection angles (taken from a book on physics by Petrus van Musschenbroek in 1762).

61. A particle moves clockwise in a circle of radius 1 m with its center at $(x, y) = (1 \text{ m}, 0)$. The particle starts from rest at the origin at time $t = 0$. Its speed increases at the constant rate of $(\pi/2) \text{ m/s}^2$. (a) How long does it take the particle to travel halfway around the circle? (b) What is its speed at that time? (c) What is the direction of its velocity at that time? (d) What is its radial acceleration then? Its tangential acceleration? (e) What are the magnitude and direction of the total acceleration halfway around the circle?

62. A boy stands 4 m from a vertical wall and throws a ball (Figure 3-36). The ball leaves the boy's hand at 2 m above the ground with initial velocity $\mathbf{v} = 10 \text{ m/s } \mathbf{i} + 10 \text{ m/s } \mathbf{j}$. When the ball hits the wall, the horizontal component of its velocity is reversed and the vertical component remains unchanged. Where does the ball hit the ground?

Figure 3-36 Problem 62.

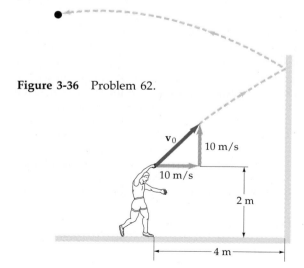

63. A baseball just clears a 3-m wall that is 120 m from home plate. If it leaves the bat at 45° and 1.2 m above the ground, what must its initial velocity be (again, making the unrealistic assumption that air resistance can be ignored)?

64. A baseball is struck by a bat, and 3 s later it is caught 30 m away. (*a*) If the baseball was 1 m above the ground when it was struck and caught, what was the greatest height it reached above the ground? (*b*) What were the horizontal and vertical components of its velocity when it was struck? (*c*) What was its speed when it was caught? (*d*) At what angle with the horizontal did it leave the bat? (Neglect air resistance.)

65. A hockey puck struck at ice level just clears the top of a glass wall that is 2.80 m high. The flight time to this point is 0.650 s, and the horizontal distance is 12.0 m. Find (*a*) the initial speed of the puck and (*b*) the maximum height it will reach.

66. For short time intervals any path can be considered to be an arc of a circle. How can the radius of curvature of a path segment be determined from the instantaneous velocity and acceleration? Consider a projectile at the top of its path. Indicate the velocity vector just before and just after this point. Is the speed changing? What is the radius of curvature of the path segment at this point?

67. The position of a particle as a function of time is

$$\mathbf{r} = 4 \text{ m} \sin 2\pi t \ \mathbf{i} + 4 \text{ m} \cos 2\pi t \ \mathbf{j}$$

where *t* is in seconds. (*a*) Show that the path of this particle is a circle of radius 4 m with its center at the origin. (*b*) Compute the velocity vector. Show that $v_x/v_y = -y/x$. (*c*) Compute the acceleration vector and show that it is in the radial direction and has the magnitude v^2/r.

68. The position of a particle is given by

$$\mathbf{r} = 3 \text{ m} \sin 2\pi t \ \mathbf{i} + 2 \text{ m} \cos 2\pi t \ \mathbf{j}$$

where *t* is in seconds. (*a*) Plot the path of the particle in the *xy* plane. (*b*) Find the velocity vector. (*c*) Find the acceleration vector and show that its direction is along **r**; that is, it is radial. (*d*) Find the times for which the speed is a maximum or minimum.

69. In Figure 3-37, a trail bike takes off from a ramp at angle θ to clear a ditch of width *x* and land on the other side, which is elevated at a height *H*. (*a*) For a given angle θ and distance *x*, what is the upper limit for *H* such that the bike has any chance of making the jump? (*b*) For *H* less than this upper limit, what is the minimum takeoff speed v_0 necessary for a successful jump? (Neglect the size of the bike, and assume that covering a horizontal distance *x* and a vertical distance *H* is sufficient to clear the ditch.)

70. Two balls are thrown with equal speeds from the top of a cliff of height *H*. One ball is thrown upward at an angle α above the horizontal. The other ball is thrown downward at an angle β below the horizontal. Show that each ball strikes the ground with the same speed, and find that speed in terms of *H* and the initial speed v_0.

71. A baseball player hits a baseball that lands in the stands 24 m above the playing field. The ball lands with a velocity of 50 m/s at an angle of 35° below the horizontal. (*a*) If the player hits the ball 1 m above the playing field, what was the velocity of the ball upon leaving the bat? (*b*) What was the horizontal distance traveled by the ball? (*c*) How long was the ball in the air? (Neglect air resistance.)

72. A steel ball is dropped from a point with (*x, y*) coordinates of (8 m, 16 m). At the same time, another ball is launched from the origin with a speed of 20 m/s at an angle of 30°. (*a*) Find the minimum distance of separation of the two balls. (*b*) At what time does this separation occur? (*c*) Give the coordinates of the two balls for the minimum separation.

73. The position vector of a particle moving in the *xy* plane is given by

$$\mathbf{r} = (10t + 5 \cos 2t) \ \mathbf{i} + (5 - 5 \sin 2t) \ \mathbf{j}$$

where **r** is in meters and *t* is in seconds. (*a*) What is the velocity of the particle? (*b*) The position vector describes a particle on the rim of a rolling wheel. Sketch the locus of points that the particle traces out as the wheel rolls along. Find the *x* values for which the particle is in contact with the surface upon which the wheel rolls. (*c*) What is the acceleration of the particle? (*d*) At what times is the particle not moving, and how is the particle on the rim related to the horizontal surface at these times?

Figure 3-37   Problem 69.

Chapter 4

Newton's Laws I

Isaac Newton (1642-1727)

Classical, or newtonian, mechanics is a theory of motion based on the ideas of mass and force and the laws connecting these physical concepts to the kinematic quantities—displacement, velocity, and acceleration—discussed in the preceding chapters. All phenomena in classical mechanics can be described using just three simple laws called Newton's laws of motion. Newton's laws relate the acceleration of an object to its mass and the forces acting on it. In this chapter, we will examine each of Newton's laws in detail, and we will carefully define the concepts of force and mass. We will see how Newton's laws can be applied to simple problems in which a single object is acted on by forces that are constant in magnitude. In Chapter 5, we will examine more general applications of Newton's laws.

A modern version of Newton's laws is

Law 1. An object continues in its initial state of rest or motion with uniform velocity unless it is acted on by an unbalanced, or net external, force. (The net force acting on an object, also called the resultant force, is the vector sum of all the forces acting on it: $\mathbf{F}_{net} = \Sigma\mathbf{F}$.)

Newton's Laws

Law 2. The acceleration of an object is inversely proportional to its mass and directly proportional to the net external force acting on it:

$$\mathbf{a} = \frac{\mathbf{F}_{net}}{m}$$

or

$$\mathbf{F}_{net} = m\mathbf{a}$$

Law 3. Forces always occur in pairs. If object *A* exerts a force on object *B*, an equal but opposite force is exerted by object *B* on object *A*.

4-1   Newton's First Law: The Law of Inertia

Newton's first law states that an object at rest or moving with constant velocity will remain at rest or will continue moving with constant velocity unless it is acted on by a net external force. This tendency is described by saying that the object has **inertia.** Accordingly, Newton's first law is often called the **law of inertia.** Prior to Galileo, it was generally thought that a force, such as a push or a pull, was necessary to keep an object moving with constant velocity. In everyday experience, if a book is pushed along a table and then let go, it slides for a while and then stops. Galileo, and later Newton, recognized that in such a situation the book is not free from external forces because friction is present. If we polish the surface of the table, the book slides farther, and the decrease in its velocity in a given time is smaller. If we support the book on a thin cushion of air (this is possible with an air table), the book will glide for a considerable time and distance with almost no perceptible change in its velocity.

Galileo studied motion by doing experiments in which he rolled balls down and up inclined planes. He found, for example, that when a ball rolled down an incline its speed increased by equal amounts in equal time intervals; that is, the acceleration of the ball was constant. Similarly, when a ball rolled up an incline, its speed decreased by equal amounts in equal time intervals. Figure 4-1 shows a ball rolling down one incline and up another.

Figure 4-1 Galileo's experiment with balls rolling down and up inclined planes. The ball rolls up the second incline to nearly its original height no matter what the angle of incline. As the angle is decreased, the ball rolls farther and farther.

The ball rolls up the second incline to nearly the same height from which it started, regardless of the relative slopes of the inclines. As the slope of the second incline is reduced, the ball rolls farther and farther. Galileo reasoned that, if he could eliminate the effects of friction, a ball rolling on a horizontal plane would roll forever with no change in speed. Newton stated these results as his first law of motion.

Note that Newton's first law makes no distinction between an object at rest and one that is moving with constant velocity. The question of whether an object is at rest or is moving with constant velocity depends on the refer-

Midair refueling. Each plane is nearly at rest relative to the other plane, although they are both moving with a very large velocity relative to the earth.

Friction is greatly reduced by a cushion of air that supports the hovercraft.

ence frame in which the object is observed. Consider a book resting on an air table in a railroad boxcar. In a coordinate system whose origin O' is attached to the boxcar (Figure 4-2a), the book is at rest. The coordinate system provides a **reference frame** for measuring the position, velocity, and acceleration of the book. We will call the reference frame attached to the boxcar S'. Now suppose the boxcar is moving along the track to the right with speed v. The book is placed on the air table so that it is again at rest relative to the boxcar. The speed v of the boxcar is measured relative to a second coordinate system whose origin O is attached to the track. In the reference frame of the track, which we will call S, the book is moving to the right with speed v (Figure 4-2b). According to Newton's first law, the book will continue to move with constant velocity in reference frame S or will remain at rest in reference frame S' unless it is acted on by a net force.

Newton's first law does not hold in all reference frames. Suppose reference frame S'' is attached to a boxcar that is accelerating along the track with an acceleration a relative to the track. Let us suppose that the boxcar starts from rest at time $t = 0$ and at that time we place a book on an air table in the

Figure 4-2 (a) A book is at rest on an air table in a boxcar in frame S'. (b) Relative to frame S, the book is moving with the speed v of the boxcar.

(a)

(b)

car. The book will remain at rest relative to the track, but it will accelerate backward with acceleration $-a$ relative to the boxcar. Thus, relative to the boxcar, the book has a horizontal acceleration with no horizontal force. If the book is to remain at rest relative to the boxcar, a horizontal force must be exerted on the book (Figure 4-3). Newton's first law does not hold in this reference frame.

Figure 4-3   A horizontal force is exerted on the book by the attached spring to give the book an acceleration a relative to frame S. Relative to frame S'' attached to the boxcar, the book is at rest even though there is a net force acting on it. Newton's first law does not apply in the reference frame S'' that is accelerating.

A reference frame in which Newton's first law holds is called an **inertial reference frame.** Any reference frame that is moving with constant velocity relative to an inertial reference frame is itself an inertial reference frame. A reference frame attached to the surface of the earth is not quite an inertial reference frame because of the small acceleration of the surface of the earth (relative to the center of the earth) due to the rotation of the earth, and because of the small centripetal acceleration of the earth itself due to its revolution around the sun. However, these accelerations are of the order of 0.01 m/s^2 or less, so to a good approximation, a reference frame attached to the surface of the earth is an inertial reference frame.

4-2   Force, Mass, and Newton's Second Law

Newton's first and second laws may be considered to be a definition of force. A **force** is an influence on an object that causes the object to change its velocity, that is, to accelerate. The direction of the force is the direction of the acceleration it causes if it is the only force acting on the object. The magnitude of the force is the product of the mass of the object and the magnitude of the acceleration produced by the force. (We will define the mass of an object shortly.) This definition of force is in accord with our intuitive idea of a force as a push or pull like that exerted by our muscles. It has been found experimentally that when two or more forces act on the same object, the acceleration of the object is the same as if it were acted on by a single force equal to the vector sum of the individual forces. That is, forces combine as vectors.

Mass is an intrinsic property of an object that measures its resistance to acceleration. The ratio of any two masses can be defined as follows. If a force F is applied to an object of mass m_1, producing acceleration a_1, then

$$F = m_1 a_1$$

If the same force is applied to a second object of mass m_2, producing acceleration a_2, then

$$F = m_2 a_2$$

Combining these equations, we have

$$F = m_1 a_1 = m_2 a_2$$

or

$$\frac{m_2}{m_1} = \frac{a_1}{a_2}$$

4-1   *Mass defined*

Thus the ratio of the masses of any two objects is defined by applying the same force to each and comparing their accelerations. This definition is in agreement with our intuitive idea of mass. For example, if one object is more "massive" than another according to our everyday use of this term, we will find that a given force produces a smaller acceleration on the more massive object. Experimentally, we find that the ratio of the accelerations a_2/a_1 produced by the same force acting on two objects is independent of the magnitude or direction of the force. It is also independent of the kind of force used; that is, it does not matter whether the force is due to springs, the pull of gravity, electric or magnetic attraction or repulsion, and so forth. We also find that if mass m_2 is found to be twice mass m_1 by direct comparison and if a third mass m_3 is found to be 4 times the mass m_1, then m_3 will be twice the mass of m_2 when the two masses are compared directly. We can therefore set up a mass scale by choosing one particular object to be the standard object and assigning it a mass of 1 unit. The international standard object is a cylinder of a platinum alloy carefully preserved at the International Bureau of Weights and Measures at Sèvres, France. The mass of the standard object is 1 **kilogram**, the SI unit of mass.* The standard object can be used to produce secondary standards by direct comparison, and the mass of any other object can then be found by comparing the acceleration produced on it by a given force with the acceleration produced on one of the secondary standards. The mass of an object is an intrinsic property of the object that does not depend on the location of the object. That is, the mass of an object remains the same whether the object is on the earth, on the moon, or in outer space.

The force required to produce an acceleration of 1 m/s² on the standard object is defined to be 1 **newton** (N). Similarly, the force that produces an acceleration of 2 m/s² on the standard object is defined to be 2 N.

Example 4-1

A given force produces an acceleration of 5 m/s² on the standard object. When the same force is applied to a second object, it produces an acceleration of 15 m/s². What is the mass of the second object, and what is the magnitude of the force?

Since the acceleration of the second object is 3 times that of the standard object under the influence of the same force, the mass of the second object is $\frac{1}{3}$ that of the standard object or 0.33 kg. The magnitude of the force is $F = (1 \text{ kg})(5 \text{ m/s}^2) = 5 \text{ N}$.

Note that we have defined the concepts of force and mass so that Newton's second law follows directly from these definitions:

$$\mathbf{F}_{net} = m\mathbf{a}$$

4-2   *Newton's second law*

---

*As mentioned in Chapter 1, the mass of the standard object, 1 kg, was originally intended to be equal to the mass of 1000 cm³ = 1 liter of water.

Newton's second law provides the link between the dynamic quantities of force and mass and the kinematic quantities of acceleration, velocity, and displacement discussed in Chapters 2 and 3. It is useful because it allows us to describe a wide variety of physical phenomena using just a few, relatively simple force laws. For example, with the addition of Newton's law of gravitational attraction between two bodies, we can describe such phenomena as the motion of the moon, the orbits of the planets around the sun, the orbits of artificial satellites, the variation in the acceleration of gravity g with altitude and with latitude, the variations in the acceleration of gravity due to the presence of mineral deposits, and the paths of ballistic missiles.

Example 4-2

A 4-kg block is at rest at $t = 0$. A single constant horizontal force F_x acts on the block. At $t = 3$ s the block has moved 2.25 m. Find the force F_x.

Since the net force acting on the block is constant, the acceleration of the block is constant and can be found from the constant acceleration formulas from Chapter 2. Using Equation 2-12 with $v_0 = 0$, we have

$$\Delta x = \tfrac{1}{2}at^2$$

$$a = \frac{2\Delta x}{t^2} = \frac{2(2.25 \text{ m})}{(3 \text{ s})^2} = 0.500 \text{ m/s}^2$$

The force is therefore

$$F_x = ma = (4 \text{ kg})(0.500 \text{ m/s}^2) = 2.00 \text{ N}$$

Example 4-3

A particle of mass 0.4 kg is subjected to two forces, $\mathbf{F}_1 = 2 \text{ N i} - 4 \text{ N j}$ and $\mathbf{F}_2 = -2.6 \text{ N i} + 5 \text{ N j}$. If the particle starts from rest at the origin at $t = 0$, find its position and velocity at $t = 1.6$ s.

The net force acting on the particle is the vector sum of the two given forces:

$$\mathbf{F}_{\text{net}} = \mathbf{F}_1 + \mathbf{F}_2 = (2 \text{ N i} - 4 \text{ N j}) + (-2.6 \text{ N i} + 5 \text{ N j})$$

$$= -0.6 \text{ N i} + 1.0 \text{ N j}$$

The acceleration of the particle is given by Newton's second law:

$$\mathbf{a} = \frac{\mathbf{F}_{\text{net}}}{m} = \frac{-0.6 \text{ N i} + 1.0 \text{ N j}}{0.4 \text{ kg}} = -1.5 \text{ m/s}^2 \text{ i} + 2.5 \text{ m/s}^2 \text{ j}$$

or

$$a_x = -1.5 \text{ m/s}^2 \quad \text{and} \quad a_y = 2.5 \text{ m/s}^2$$

Since the particle starts from rest at the origin at $t = 0$, its x and y coordinates at $t = 1.6$ s are given by

$$x = \tfrac{1}{2}a_x t^2 = \tfrac{1}{2}(-1.5 \text{ m/s}^2)(1.6 \text{ s})^2 = -1.92 \text{ m}$$

and

$$y = \tfrac{1}{2}a_y t^2 = \tfrac{1}{2}(2.5 \text{ m/s}^2)(1.6 \text{ s})^2 = 3.20 \text{ m}$$

The x and y components of the particle's velocity at $t = 1.6$ s are

$$v_x = a_x t = (-1.5 \text{ m/s}^2)(1.6 \text{ s}) = -2.40 \text{ m/s}$$

and

$$v_y = a_y t = (2.5 \text{ m/s}^2)(1.6 \text{ s}) = 4.00 \text{ m/s}$$

In vector notation, the velocity of the particle at $t = 1.6$ s is

$$\mathbf{v} = -2.40 \text{ m/s } \mathbf{i} + 4.00 \text{ m/s } \mathbf{j}$$

Exercise

A force of 3 N produces an acceleration of 2 m/s^2 on an object of unknown mass. (*a*) What is the mass of the object? (*b*) If the force is increased to 4 N, what is the acceleration? [Answers: (*a*) 1.5 kg (*b*) 2.67 m/s^2]

Questions

1. If an object has no acceleration, can you conclude that no forces are acting on it?

2. If only a single force acts on an object, must the object accelerate? Can it ever have zero velocity?

3. Is there an unbalanced force acting when (*a*) an object moves at constant speed in a circle, (*b*) an object that is moving in a straight line slows down, or (*c*) an object moves at constant speed in a straight line?

4. Is it possible for an object to round a curve without any force acting on it?

5. If a single known force acts on an object, can you tell in which direction the object will move from this information alone?

6. If several forces of different magnitudes and directions are applied to an object initially at rest, how can you predict the direction in which it will move?

7. Can you judge the mass of an object by its size? If *A* is twice as big as *B*, does that mean that $m_A = 2m_B$?

8. Can the mass of an object be negative?

4-3   The Force Due to Gravity: Weight

The force most common in our everyday experience is the force of the gravitational attraction of the earth for an object. This force is called the **weight w** of the object. If we drop an object near the surface of the earth and neglect air resistance so that the only force acting on the object is the force due to gravity (this situation is called **free fall**), the object accelerates toward the earth with acceleration of 9.81 m/s^2. At any given point in space, this acceleration is the same for all objects, independent of their masses. We call the value of this acceleration **g**. From Newton's second law, we can write the gravitational force $\mathbf{F}_g$ on an object of mass *m* as

$$\mathbf{F}_g = m\mathbf{a}$$

Using $\mathbf{a} = \mathbf{g}$ and writing **w** for the gravitational force, we have

$$\mathbf{w} = m\mathbf{g} \qquad\qquad 4\text{-}3$$

Since **g** is the same for all objects at a given point, we can conclude that the weight of an object is proportional to its mass. The vector **g** in Equation 4-3 is the force per unit mass exerted by the earth on any object and is called the

When air resistance can be neglected, balls of different mass fall with the same acceleration due to gravity.

gravitational field of the earth. (We will discuss gravitational fields in Chapter 10.) It is equal to the free-fall acceleration experienced by an object when the only force acting on it is the gravitational force of the earth. Near the surface of the earth g has the value

$$g = 9.81 \text{ N/kg} = 9.81 \text{ m/s}^2$$

Careful measurements of g at various places show that it does not have the same value everywhere. *The force of attraction of the earth for an object varies with location.* In particular, at points above the surface of the earth, the force due to gravity varies inversely with the square of the distance of the object from the center of the earth. Thus, an object weighs slightly less at very high altitudes than it does at sea level. The gravitational field also varies slightly with latitude because the earth is not exactly spherical but is slightly flattened at the poles. Thus weight, unlike mass, is not an intrinsic property of an object; that is, it is not a property of the object itself.

Although the weight of an object varies from place to place because of changes in g, this variation is too small to be noticed in most practical applications. Thus, in our everyday experience, the weight of an object *appears* to be as much a constant characteristic of the object as its mass.

Near the surface of the moon, the gravitational attraction of the earth on an object is much less than that of the moon. The force exerted on the object by the moon is usually called the weight of the object when it is near the moon. Note again that the mass of an object is the same whether it is on the earth, on the moon, or somewhere in space. Mass is a property of the object itself, whereas weight depends on the nature and distance of other objects that exert gravitational forces on the object. An example should help clarify the difference between mass and weight. Suppose you take a heavy ball such as a bowling ball to the moon. Since the ball weighs only about one-sixth as much on the moon as on earth, lifting the ball is much easier on the moon. But throwing the ball with some given horizontal velocity requires the same force on the moon as on the earth. That is, the same force is required to produce a given acceleration of the ball on the moon as on the earth because the mass of the ball is the same. The same force would also be required to produce the same acceleration of the ball in free space, far from any gravitational field.

Since, at any particular location, the weight of an object is proportional to its mass, we can conveniently compare the mass of one object with that of another by comparing their weights as long as we determine the weights at the same location.

Our sensation of our weight usually comes from other forces that balance it. For example, sitting on a chair, we feel the force exerted by the chair that balances our weight and thereby prevents us from falling to the floor. When we stand on a spring scale, our feet feel the force exerted on us by the scale. The scale is calibrated to read the force it must exert (by the compression of its springs) to balance our weight. The force that balances our weight is called our **apparent weight.** It is the apparent weight that is given by a spring scale. If there is no force to balance your weight, as in free fall, your apparent weight is zero. This condition, called **weightlessness,** is experienced by astronauts in orbiting satellites. Consider a satellite in a circular orbit near the surface of the earth with a centripetal acceleration v^2/r, where r is the orbital radius and v is the speed. The only force acting on the satellite is its weight. Thus it is in free fall with the acceleration of gravity. An astronaut in the satellite is also in free fall. The only force acting on the astronaut is his or her weight, which produces the acceleration $g = v^2/r$. Since there is no force balancing the force of gravity, the astronaut's apparent weight is zero.

Weightlessness in a space capsule. The astronauts are accelerating toward the earth in free fall with the acceleration of gravity.

Units of Force and Mass

The SI unit of mass is the kilogram. Like the second and the meter, the kilogram is a fundamental unit in SI. The unit of force, the newton, and the units for other quantities that we will study, such as momentum and energy, are derived from these three units. Since 1 N produces an acceleration of 1 m/s^2 when it acts on 1 kg, we have from $F = ma$

$$1\ N = 1\ kg \cdot m/s^2 \qquad 4\text{-}4$$

Although we will generally use SI units in this book, we need to know another system, the U.S. customary system, which is based on the foot, the second, and a force unit (the pound) and is still used in the United States. The U.S. customary system differs from SI in that a unit of force rather than a unit of mass has been chosen as a fundamental unit. The **pound** was originally defined as the weight of a particular standard object at a particular location. It is now defined as 4.448222 N. (This is the weight of an object of mass 0.45359237 kg at a point where g has the value 9.80665 m/s^2 = 32.1740 ft/s^2.) Rounding to three places, we have

$$1\ lb \approx 4.45\ N$$

Since 1 kg weighs 9.81 N, its weight in pounds is

$$9.81\ N \times \frac{1\ lb}{4.45\ N} = 2.20\ lb \qquad \qquad \textit{Weight of 1 kg}$$

The unit of mass in the U.S. customary system is that mass that will be given an acceleration of one foot per second squared when a force of one pound is applied to it. This unit, called a *slug*, is the mass of an object that weighs 32.2 pounds. We will not need to use this unit. Instead, when working problems in this system, we substitute w/g for the mass m, where w is the weight in pounds and g is the acceleration due to gravity in feet per second per second (feet per second squared).

Example 4-4

The net force acting on a 10.0-lb object is 3.00 lb. What is its acceleration?

The acceleration is the force divided by the mass:

$$a = \frac{F}{m} = \frac{F}{w/g} = \frac{3.00\ lb}{(10.0\ lb)/(32.2\ ft/s^2)} = 9.66\ ft/s^2$$

Exercise

What force is needed to give an acceleration of 3 ft/s^2 to a 5-lb block? (Answer: 0.466 lb)

Questions

9. Suppose an object was sent far out in space, away from galaxies, stars, or other bodies. How would its mass change? Its weight?

10. How would an astronaut in a condition of apparent weightlessness be aware of his or her mass?

11. Under what circumstances would your apparent weight be greater than your true weight?

12. What is your weight in newtons?

13. What is your mass in kilograms?

14. What would your weight be in pounds on the moon, where objects fall freely with an acceleration of about 5.33 ft/s^2?

4-4   Newton's Third Law

Newton's third law is sometimes called the **law of interaction.** It describes an important property of forces, that they always occur in pairs. If a force is exerted on some object A, there must be another object B exerting the force. Furthermore, if B exerts a force on A, then A must exert a force on B that is equal in magnitude and opposite in direction. For example, the earth exerts a gravitational force $\mathbf{F}_g$ on a projectile, causing it to accelerate toward the earth. According to the third law, the projectile in turn exerts a force on the earth equal in magnitude and opposite in direction. Thus, the projectile exerts a force $\mathbf{F}'_g = -\mathbf{F}_g$ on the earth toward the projectile. If this were the only force acting on the earth, the earth would accelerate toward the projectile. Since the earth has a very large mass, the acceleration it experiences due to this force is negligible and unobserved.

In discussions of Newton's third law, the words "action" and "reaction" are frequently used. If the force exerted on object A is called the action of object B upon A, then the force A exerts back on B is called the reaction of A upon B. It does not matter which force in such a pair is called the action and which the reaction. The important points are that forces always occur in action–reaction pairs, and that the reaction force is equal in magnitude and opposite in direction to the action force.

Note that the action and reaction forces can never balance each other because they act on *different objects.* This is illustrated in Figure 4-4, which shows two action–reaction pairs for a block resting on a table. The force acting downward on the block is the weight $\mathbf{w}$ due to the attraction of the earth. An equal and opposite force $\mathbf{w}' = -\mathbf{w}$ is exerted by the block on the earth. These forces are an action–reaction pair. If they were the only forces present, the block would accelerate downward because it would have only a single force acting on it. However, the table in contact with the block exerts an upward force $\mathbf{F}_n$ on it. This force balances the weight of the block. The block also exerts a force $\mathbf{F}'_n = -\mathbf{F}_n$ downward on the table. The forces $\mathbf{F}_n$ and $\mathbf{F}'_n$ are also an action–reaction pair.

Figure 4-4 Action–reaction forces. The weight $\mathbf{w}$ is the force exerted on the block by the earth. The equal and opposite reaction force exerted on the earth by the block is $\mathbf{w}' = -\mathbf{w}$. Similarly, the table exerts a force $\mathbf{F}_n$ on the block, and the block exerts an equal and opposite force $\mathbf{F}'_n$ on the table. Action–reaction forces are exerted on different objects and therefore cannot balance.

Example 4-5

A horse refuses to pull a cart. The horse reasons, "according to Newton's third law, whatever force I exert on the cart, the cart will exert an equal and opposite force on me, so the net force will be zero and I will have no chance of accelerating the cart." What is wrong with this reasoning?

Figure 4-5 is a sketch of a horse pulling the cart. Since we are interested in the motion of the cart, we have enclosed it with a dashed line and have indicated the forces acting on it. The force exerted by the horse is labeled **T**. It is exerted by the horse on the harness. (Since the harness is attached to the cart, we are considering it part of the cart.) Other forces acting on the cart are its weight **w**, the vertical support force of the ground $\mathbf{F}_n$, and the horizontal force exerted by the ground, labeled **f** (for friction). The vertical forces **w** and $\mathbf{F}_n$ balance. (We know this because we know the cart does not accelerate vertically.) The horizontal forces are **T** to the right and **f** to the left. The cart will accelerate if **T** is greater than **f**. Note that the reaction force to **T**, which we call **T'**, is exerted on the horse, not on the cart. It has no effect on the motion of the cart, but it does affect the motion of the horse. If the horse is to accelerate to the right, there must be a force **F** (to the right) exerted by the ground on the horse's feet that is greater than **T'**. This example illustrates the importance of drawing a simple diagram when solving mechanics problems. Had the horse drawn a simple diagram, he would have seen that he need only push back hard against the ground so that the ground would push him forward.

Figure 4-5 Horse pulling a cart. The cart will accelerate to the right if the force **T** exerted on it by the horse is greater than the frictional force **f** exerted on the cart by the ground. The force **T'** is equal and opposite to **T**, but because it is exerted on the *horse*, it has no effect on the motion of the cart.

(a)

(b)

(a) The tides are caused by the gravitational force exerted by the moon and the sun. Mont-Saint-Michel, France is an island when the tide is in. (b) These lightning bolts at the Kitt Peak National Observatory are the result of the electromagnetic force. (c) This hydrogen bomb explosion illustrates the strong nuclear force. (d) The weak interaction between muons and electrons is illustrated by this false-color cloud chamber photograph showing the path of a cosmic ray muon (green) and an electron (red) that is knocked out of an atom.

4-5   Forces in Nature

Fundamental Forces

All the different forces observed in nature can be explained in terms of four basic interactions that occur between elementary particles:

1. The gravitational force
2. The electromagnetic force
3. The strong nuclear force (also called the hadronic force)
4. The weak nuclear force

The gravitational force between the earth and an object near the earth's surface is the weight of the object. The gravitational force exerted by the sun on the earth and the other planets is responsible for keeping the planets in their orbits around the sun. Similarly, the gravitational force exerted by the earth on the moon keeps the moon in its nearly circular orbit around the earth. The gravitational forces exerted by the moon and the sun on the oceans of the earth are responsible for the tides. We will study the gravitational force in detail in Chapter 10.

The electromagnetic force includes both the electric and the magnetic forces. A familiar example of the electric force is the attraction between small bits of paper and a comb that has been electrified by being rubbed across hair. Although the familiar magnetic force between a magnet and iron objects seems to be very different from the electric force, the magnetic force arises when electric charges are in motion. The electromagnetic force between charged elementary particles is so much greater than the gravitational force between them that the gravitational force can nearly always be ignored. For example, the electrostatic force of repulsion between two protons is of the order of 10^{36} times the gravitational attraction between them.

The strong nuclear force occurs between elementary particles called hadrons, which include protons and neutrons, the constituents of atomic nuclei. It is responsible for holding nuclei together. For example, the two protons in the helium atom are attracted via the strong nuclear force, which more than balances the electrostatic repulsion of the protons. However, the strong nuclear force has a very short range. It decreases rapidly with separa-

(c)

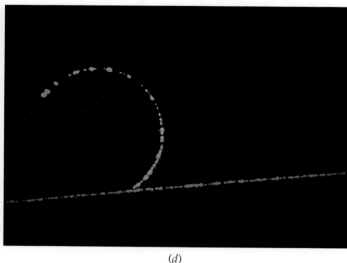

(d)

tion of the particles and is negligible when they are a few nuclear diameters apart. The weak nuclear force, which also has a short range, occurs between electrons and protons or neutrons. It is responsible for a certain kind of radioactive decay called beta decay.

The fundamental forces act between particles that are separated in space. This concept is referred to as **action at a distance.** Newton perceived action at a distance as a flaw in his theory of gravitation, but he avoided giving any other hypothesis. Indeed, in 1692 he wrote the following:*

> It is inconceivable that inanimate, brute matter should, without the mediation of something else, which is not material, operate upon, and affect other matter without mutual contact, as it must be if gravitation, in the sense of Epicurus, be essential and inherent in it. And this is one reason why I desired you would not ascribe innate gravity to me. That gravity should be innate, inherent, and essential to matter, so that one body may act upon another at a distance through a vacuum, without the mediation of anything else, by and through which their action and force may be conveyed from one to another, is to me so great an absurdity that I believe no man who has in philosophical matters a competent faculty of thinking can ever fall into it.

Today, we treat the problem of action at a distance by introducing the concept of a **field.** For example, we can consider the attraction of the earth by the sun in two steps. The sun creates a condition in space that we call the gravitational field. This field produces a force on the earth. The field is thus the intermediary agent. Similarly, the earth produces a gravitational field that exerts a force on the sun. If the earth moves to a new position, the field of the earth is changed. This change is not propagated through space instantly but rather with the speed $c = 3 \times 10^8$ m/s, which is also the speed of light. If we can neglect the time it takes for propagation of the field, we can ignore this intermediary agent and treat the gravitational forces as if they were exerted by the sun and the earth directly on each other. For example, during the 8 min it takes for propagation of the gravitational field from the earth to the sun, the earth moves through only a small fraction of its total orbit around the sun.

*Isaac Newton, Third Letter to Bentley (Feb. 25, 1692), R. and J. Dodsley, London, 1756.

Contact forces exerted by one object on another produce deformations in the objects that are often not visible. Here, the forces exerted by the C-clamp on the plastic block produce stress patterns in the block that are made visible by polarized light.

Contact Forces

Most of the everyday forces we observe on macroscopic objects are contact forces exerted by springs, strings, and surfaces in direct contact with the object. The forces are the result of molecular forces exerted by the molecules of one object on those of another. These molecular forces are themselves complicated manifestations of the basic electromagnetic force.

A spring made by winding a stiff wire into a helix is a familiar device. The force exerted by the spring when it is compressed or extended is the result of complicated intermolecular forces in the spring, but an empirical description of the macroscopic behavior of the spring is sufficient for most applications. If the spring is compressed or extended and released, it returns to its original, or natural, length, provided the displacement is not too great. There is a limit to such displacements beyond which the spring does not return to its original length but remains permanently deformed. If we allow only displacements below this limit, we can calibrate the extension or compression Δx in terms of the force needed to produce the extension or compression. It has been found experimentally that, for small Δx, the force exerted by the spring is approximately proportional to Δx and in the opposite direction. This relationship, known as **Hooke's law,** can be written

$$F_x = -k(x - x_0) = -k \, \Delta x \qquad \text{4-5}$$

where the constant k is called the **force constant** of the spring. The distance x is the coordinate of the free end of the spring or of any object attached to that end of the spring. The constant x_0 is the value of this coordinate when the spring is unstretched from its equilibrium position. There is a negative sign in Equation 4-5 because, if the spring is stretched (Δx is positive), the force F_x is negative, whereas if the spring is compressed (Δx is negative), F_x is positive (Figure 4-6). Such a force is called a **restoring force** because it tends to restore the spring to its initial configuration.

Figure 4-6 A horizontal spring attached to a block. (a) When the spring is unstretched, it exerts no force on the block. (b) When the spring is stretched such that Δx is positive, it exerts a force on the block of magnitude $k \, \Delta x$ in the negative x direction. (c) When the spring is compressed such that Δx is negative, the spring exerts a force on the block of magnitude $k \, \Delta x$ in the positive x direction.

(a)

(a)

(b)

Figure 4-7   (a) Model of a solid as consisting of atoms connected to each other by springs. (b) The elasticity of nylon arises from the shape and cross linking of its fibers shown here under polarized light.

The force exerted by a spring is similar to that exerted by one atom on another in a molecule or in a solid in the sense that, for small displacements from equilibrium, the restoring force is proportional to the displacement. It is often useful to visualize the atoms in a molecule or solid as being connected by springs (Figure 4-7). For example, if we slightly increase the separation of the atoms in a molecule and release them, we would expect the atoms to oscillate back and forth as if they were two masses connected by a spring.

If we pull on a flexible string, the string stretches slightly and pulls back with an equal but opposite force (unless the string breaks). We can think of a string as a spring with such a large force constant that the extension of the string is negligible. Because the string is flexible, however, we cannot exert a force of compression on it. When we push on a string, it merely flexes or bends.

When two bodies are in contact with each other, they exert forces on each other due to the interaction of the molecules of one object with those of the other. Consider a block resting on a horizontal table. The weight of the block pulls the block downward, pressing it against the table. Because the molecules in the table have a great resistance to compression, the table exerts a force upward on the block perpendicular, or normal, to the surface. Such a force is called a **normal force.** (The word *normal* means perpendicular.) Careful measurement would show that a supporting surface always bends slightly in response to a load, but this compression is not noticeable to the naked eye. Since the table exerts an upward force on the block, the block must exert an equal force downward on the table. Note that the normal force exerted by one surface on another can vary over a wide range of values. For example, unless the block is so heavy that the table breaks, the table will exert an upward support force on the block exactly equal to the weight of the

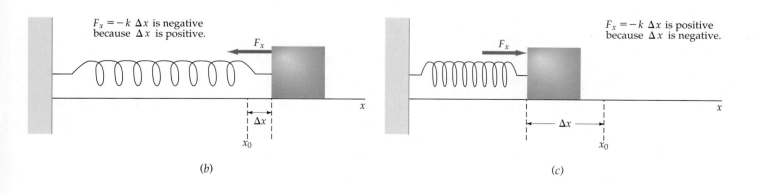

$F_x = -k \, \Delta x$ is negative because Δx is positive.

F_x

x

Δx

x_0

(b)

$F_x = -k \, \Delta x$ is positive because Δx is negative.

F_x

x

Δx

x_0

(c)

$$F_s = -k\,\Delta x = mg$$

m

mg

Figure 4-8 An object suspended from a spring. The weight of the object $\mathbf{w} = m\mathbf{g}$ is balanced by the upward force $\mathbf{F}_s$, exerted by the spring.

block no matter how large or small its weight is. Furthermore, if you press down on the block, the table will exert a support force greater than the weight of the block to prevent it from accelerating downward.

Under certain circumstances, bodies in contact will exert forces on each other that are parallel to the surfaces in contact. The parallel component of the contact force exerted by one object on another is called a **frictional force**. We will study friction in the next chapter.

Example 4-6

A spring has a force constant $k = 300$ N/m. A 4-kg object is suspended motionless from the spring, as shown in Figure 4-8. Find the extension of the spring.

This is an example of static equilibrium, which we will study in more detail in Chapter 9. Since the block is not accelerating, the net force acting on it must be zero. There are two forces acting on the block: its weight mg downward and the force of the spring F_s upward. If we take the downward direction to be positive, the weight is positive and the extension of the spring Δx is also positive. The force the spring exerts is upward and is therefore negative. Setting the sum of the forces on the block equal to zero, we obtain

$$mg + (-k\,\Delta x) = 0$$

Using $m = 4$ kg, $k = 300$ N/m, and $g = 9.81$ m/s^2 = 9.81 N/kg, we obtain

$$(4\text{ kg})(9.81\text{ N/kg}) - (300\text{ N/m})(\Delta x) = 0$$

or

$$\Delta x = \frac{(4\text{ kg})(9.81\text{ N/kg})}{300\text{ N/m}} = 0.131\text{ m}$$

Exercise

A spring of force constant 400 N/m is attached to a 3-kg block that rests on an air track so that friction is negligible. What extension of the spring is needed to give the block an acceleration of 4 m/s^2?
(Answer: 3.0 cm)

4-6   Applications to Problem Solving

Newton's laws can be used to solve a variety of mechanics problems. For instance, they allow us to determine the acceleration of a particle from a knowledge of all the forces acting on it, or to determine the forces acting on a particle given its acceleration. In this section, we will examine the application of Newton's laws to some simple problems involving motion under forces that are constant in magnitude. Careful study of these simple examples will make you aware of the content of newtonian mechanics and how it can be applied. Practical problems are generally more complex than these examples, but the methods of solving them are natural extensions of those illustrated here. Some more general examples of the applications of Newton's laws will be discussed in Chapter 5.

Consider a block of mass m resting on a frictionless horizontal table and pulled with a force $\mathbf{F}$ applied to a string, as shown in Figure 4-9. To find the motion of the block, we need to find the net force acting on it. The first step is to choose the object whose acceleration is to be determined and upon which the forces to be considered act. In the figure, a circle is drawn around

Figure 4-9 A block on a frictionless horizontal surface with a horizontal force exerted on it by a string. The first step in solving the problem is to isolate the object to be analyzed. In this case, the circle isolates the box from its surroundings.

the block to help us mentally isolate it from its surroundings. We then look for all the possible forces that act on the chosen object. Such forces may arise from the contact of the object with its surroundings, or they may be action-at-a-distance forces such as gravity.

Three significant external forces act on the block in this example. They are shown in the diagram in Figure 4-10. Such a diagram is called a **free-body diagram.** The three forces are

1. The weight of the block **w.**

2. The contact force $\mathbf{F_n}$ exerted by the table. Since we are assuming the table to be frictionless, the contact force is perpendicular to the table.

3. The contact force **T** exerted by the string.

A convenient coordinate system is also indicated in Figure 4-10. Note that the normal force $\mathbf{F_n}$ and the weight **w** are drawn with equal magnitudes. We know these forces have equal magnitudes because the block does not accelerate vertically. Since the resultant force is in the x direction and has magnitude T, Newton's second law gives

$$T = ma_x$$

The force **F** exerted by the hand on the string equals the force **T** exerted by the string on the block. We can see this by considering the free-body diagram for the string shown in Figure 4-11. The force **T'** is the force the block exerts on the string. It is equal in magnitude and opposite the direction of the force **T** exerted by the string on the block. (We have neglected the weight of the string. In actual fact, the string will sag slightly and the forces **F** and **T'** will have small vertical components, but they are so small that we can safely ignore them in this problem). Assuming the string remains taut, it must have the same acceleration as the block. If m_s is the mass of the string, applying Newton's second law to the string gives us

$$F - T' = m_s a_x$$

Figure 4-10 caption:

Figure 4-10 A free-body diagram for the block in Figure 4-9. The three significant forces acting on the box are the force exerted by the earth **w**, the normal force $\mathbf{F_n}$, and the force exerted by the string **T**.

Figure 4-11 Free-body diagram for the string in Figure 4-9. If the string is light enough that its mass can be neglected, the forces **F** and **T'** are equal in magnitude.

If the string is light enough so that we can neglect its mass, we have

$$F - T' = m_s a_x \approx 0$$

Since **T'** and **T** have the same magnitudes, **F** and **T** are equal.

Figure 4-12 shows a small segment of the string of mass Δm_s. The forces acting on this segment are $\mathbf{T_1}$ exerted by that part of the string to the right of the segment and $\mathbf{T_2}$ exerted by that part of the string to the left of the segment. Because the string has negligible mass, these two forces are equal in magnitude, $T_1 = T_2 = T$. The force T is called the **tension** in the string. Each segment of the string exerts a force T on each of its neighbors. These forces act along the string, so a light string connecting two points has a tension that has a constant magnitude throughout. This result also holds for a string that passes over a frictionless peg or a pulley of negligible mass as long as there are no tangential forces acting on the string between the two points considered.

Figure 4-12 Free-body diagram for a segment of the string of mass Δm_s. If the mass of the segment is negligible, the force $\mathbf{T_1}$ exerted by that part of the string to the right of the segment is equal in magnitude to the force $\mathbf{T_2}$ exerted by that part of the string to the left of the segment. The tension T is the magnitude of the force each segment exerts on each adjacent segment.

Even in this simple example, both kinds of applications of Newton's laws are used: The horizontal acceleration is found in terms of the given force **F**, and the vertical force $\mathbf{F}_n$ exerted by the table is found from the fact that the block remains on the table and thus $a_y = 0$. Conditions on the motion of an object, such as the requirement that the block remain on the table, are called **constraints**.

According to Newton's third law, forces always act in pairs. Figure 4-10 shows only those forces that act on the block. Figure 4-13 shows the reaction forces to those in Figure 4-10. These are the gravitational force **w'** exerted by the block on the earth, the force $\mathbf{F}_n'$ exerted by the block on the table, and the force **T'** exerted by the block on the string. Since these forces are not exerted on the block they have nothing to do with its motion. We can therefore omit them in applying Newton's second law to the motion of the block.

Figure 4-13 The reaction forces corresponding to the three forces shown in Figure 4-10. Note that these forces *do not* act on the block. **T'** acts on the string, $\mathbf{F}_n'$ acts on the table, and **w'** acts on the earth.

This simple example illustrates a general method of attack for solving problems using Newton's laws. Basically, this method consists of the following steps:

Solving problems using Newton's laws

1. Draw a neat diagram.

2. Isolate the object (particle) of interest, and draw a free-body diagram, showing each external force that acts on the object. If there is more than one object of interest in the problem, draw a separate free-body diagram for each.

3. Choose a convenient coordinate system for each object, and apply Newton's second law $\mathbf{F}_{net} = m\mathbf{a}$ in component form.

4. Solve the resulting equations for the unknowns using whatever additional information is available. The unknowns may include the masses, the components of the accelerations, or the components of some of the forces.

5. Finally, inspect your results carefully, checking to see whether they correspond to reasonable expectations. Particularly valuable is the determination of the predictions of your solution when variables are assigned extreme values. In this way you can check your work for errors.

We will now look at some examples.

Example 4-7

Find the acceleration of a block of mass m that slides down a fixed, frictionless surface inclined at an angle θ to the horizontal.

There are only two forces acting on the block, the weight **w** and the normal force $\mathbf{F_n}$ exerted by the incline (see Figure 4-14). (For real surfaces, there would be a force of friction parallel to the incline, but here we are assuming an ideal frictionless surface.) Since the two forces do not act along the same line, they cannot add to zero, so the block must accelerate. The acceleration is along the incline. This is another example of a constraint. For this problem, it is convenient to choose a coordinate frame with one axis parallel to the incline and the other perpendicular to it, as shown in the figure. Then the acceleration has only one component, a_x. For this choice, $\mathbf{F_n}$ is in the y direction, and the weight **w** has the components

$$w_x = w \sin \theta = mg \sin \theta$$
$$w_y = -w \cos \theta = -mg \cos \theta$$

4-6

where m is the mass of the block and g is the acceleration due to gravity (Figure 4-15). The net force in the y direction is $F_n - mg \cos \theta$. From Newton's second law and the fact that $a_y = 0$,

$$\sum F_y = F_n - mg \cos \theta = ma_y = 0$$

and thus

$$F_n = mg \cos \theta$$

4-7

Similarly, for the x components,

$$\sum F_x = mg \sin \theta = ma_x$$
$$a_x = g \sin \theta$$

4-8

The acceleration down the incline is constant and equal to $g \sin \theta$. It is useful to check our results at the extreme values of inclination, $\theta = 0$ and $\theta = 90°$. At $\theta = 0$, the surface is horizontal. The weight has only a y component, which is balanced by the normal force $\mathbf{F_n}$. The acceleration is zero: $a_x = g \sin 0° = 0$. At the opposite extreme, $\theta = 90°$, the incline is vertical. Then the weight has only an x component along the incline, and the normal force is zero: $F_n = mg \cos 90° = 0$. The acceleration is $a_x = g \sin 90° = g$. That is, the block is in free fall.

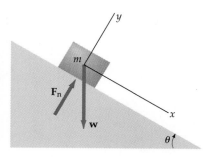

Figure 4-14   Forces acting on a block of mass m on a frictionless incline. It is convenient to choose the x axis parallel to the incline.

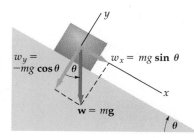

Figure 4-15   The weight of the block can be replaced by its components along the x and y axes. Since the weight vector is perpendicular to the horizontal and the negative y axis is perpendicular to the incline, the angle between these two lines is the same as the angle θ of the incline. The y component of the weight is thus $-mg \cos \theta$, and the x component is $mg \sin \theta$.

When friction is neglected, the body slides down the incline with acceleration $g \sin \theta$.

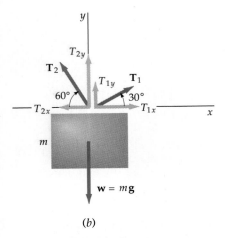

Figure 4-16   (*a*) Picture supported by two wires in Example 4-8. (*b*) Free-body diagram for the picture with the forces resolved into their *x* and *y* components.

Example 4-8

A picture weighing 8 N is supported by two wires of tension T_1 and T_2 as shown in Figure 4-16*a*. Find the tension in the wires.

Since the picture does not accelerate, the net force acting on it must be zero. The three forces acting on the picture—its weight *mg*, the tension T_1 in one wire, and the tension T_2 in the other—must therefore sum to zero. Figure 4-16*b* shows the free-body diagram for the picture with the forces resolved into their horizontal and vertical components. Since the weight has only a vertical component *mg* downward, the horizontal components of the tensions T_1 and T_2 must balance each other, and the vertical components of the tensions must balance the weight:

$$\sum F_x = T_1 \cos 30° - T_2 \cos 60° = 0$$

$$\sum F_y = T_1 \sin 30° + T_2 \sin 60° - mg = 0$$

Using $\cos 30° = \sqrt{3}/2 = \sin 60°$ and $\sin 30° = \frac{1}{2} = \cos 60°$ and solving for the tensions, we obtain

$$T_1 = \tfrac{1}{2}mg = 4 \text{ N}$$

$$T_2 = \sqrt{3}T_1 = \frac{\sqrt{3}}{2} mg = 6.93 \text{ N}$$

We will now consider two examples of the application of Newton's laws to objects moving in circles. As discussed in Chapter 3, when a particle moves with a speed *v* in a circle of radius *r*, it has an acceleration with a magnitude of v^2/r directed toward the center of the circle whether or not the speed is changing. If the speed is changing, there is also a component of acceleration tangential to the circle and equal to the rate of change of the speed.

As with any acceleration, there must be a net force in the direction of the centripetal acceleration to produce it. This net force is called the **centripetal force.** It is important to understand that the centripetal force is not a new kind of force that we have not yet studied. It is merely a name for the net inward force that must be present to provide the centripetal acceleration needed for circular motion. Whenever an object moves in a circle, there must be a net force to provide its centripetal acceleration just as there must be a net force to provide linear acceleration. The centripetal force may be due to a string, a spring, or some other type of contact force such as a normal force or friction; it may be an action-at-a-distance force such as a gravitational force; or it may be a combination of any of these familiar forces.

Example 4-9

A tetherball of mass *m* is suspended from a rope of length *L* and travels at a constant speed *v* in a horizontal circle of radius *r*. The rope makes an angle θ given by $\sin \theta = r/L$, as shown in Figure 4-17. Find the tension in the rope and the speed of the ball.

The two forces acting on the ball are the weight *mg*, which acts vertically downward, and the tension **T**, which acts along the rope. In this problem, we know that the acceleration is horizontal, directed toward the center of the circle, and of magnitude v^2/r. Thus, the vertical component of the tension must balance the weight *mg*. The horizontal component of the tension is the centripetal force. The vertical and horizontal components of $\mathbf{F}_{net} = m\mathbf{a}$ therefore give

$$T \cos \theta - mg = 0$$

or

$$T \cos \theta = mg$$

and

$$T \sin \theta = ma = \frac{mv^2}{r}$$

The tension is found directly from the first equation since θ is given. We can find the speed v in terms of the known quantities r and θ by dividing one equation by the other to eliminate T. We thereby obtain

$$\frac{T \sin \theta}{T \cos \theta} = \frac{mv^2/r}{mg}$$

$$\frac{\sin \theta}{\cos \theta} = \tan \theta = \frac{v^2}{rg}$$

so

$$v = \sqrt{rg \tan \theta}$$

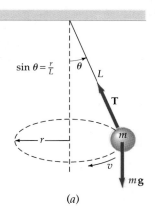

$$\sin \theta = \frac{r}{L}$$

(a)

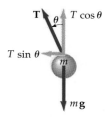

(b)

Figure 4-17   (a) Tetherball moving in a horizontal circle. (b) Free-body diagram for the ball with the force **T** resolved into its vertical and horizontal components.

Example 4-10

A pail of water is whirled in a vertical circle of radius r. If its speed is v_t at the top of the circle, find the force exerted on the water by the pail at the top of the circle. Find also the minimum value of v_t for which the water will remain in the pail.

The forces on the water at the top of the circle are shown in Figure 4-18. They are the force of gravity mg and the force $\mathbf{F}_p$ exerted by the pail. Both these forces act downward. The acceleration, which is toward the center of the circle, is also downward at this point. Newton's second law gives

$$F_p + mg = m\frac{v_t^2}{r}$$

The force exerted by the pail is therefore

$$F_p = m\frac{v_t^2}{r} - mg$$

Note that both the force of gravity and the contact normal force exerted by the pail contribute to the necessary centripetal force.

If we increase the speed of the pail, the bottom of the pail will exert a larger force on the water to keep it moving in a circle. If we decrease the speed, F_p will decrease. Since the pail cannot exert an upward force on the water, the minimum speed the water can have at the top of the circle occurs when $F_p = 0$. Then,

$$mg = m\frac{v_{t,min}^2}{r}$$

or

$$v_{t,min} = \sqrt{rg} \qquad\qquad 4\text{-}9$$

When the water is moving at this minimum speed, its acceleration at the top of the circle is $\mathbf{g}$, the free-fall acceleration of gravity, and the only force acting on the water is the gravitational attraction of the earth, the weight mg.

If you wish to try this, you might want to know how fast you need to move the pail so that you will not get wet. A reasonable value for r is

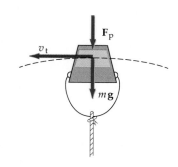

Figure 4-18   Pail of water whirled in a vertical circle. At the top of the circle the forces acting on the water are its weight mg and the force exerted by the pail $\mathbf{F}_p$. Both of these forces act downward at this point, toward the center of the circle.

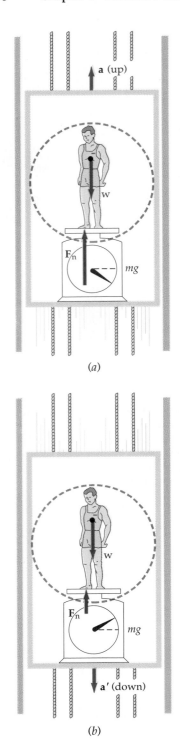

(a)

(b)

Figure 4-19 Man on a scale in an accelerating elevator. The scale indicates the man's *apparent weight*, which equals the magnitude of $\mathbf{F_n}$. (a) When the acceleration is upward, his apparent weight is greater than *mg*. (b) When the acceleration is downward, his apparent weight is less than *mg*.

1 m, assuming that your arm is 70 cm long and that there is another 30 cm from your hand to the water. The minimum speed at the top of the circle is then

$$\bar{v}_{t,min} = \sqrt{rg} = \sqrt{(1\ m)(9.81\ m/s^2)} = 3.13\ m/s$$

If you rotate the pail at constant speed, the maximum period of revolution can be found using Equation 3-23 from Chapter 3:

$$T = \frac{2\pi r}{v} = \frac{(2)(3.14)(1\ m)}{3.13\ m/s} \approx 2\ s$$

Our final example is an application of Newton's laws to an object that is at rest relative to an elevator that is itself accelerating.

Example 4-11

A man stands on a scale fastened to the floor of an elevator, as illustrated in Figure 4-19. What does the scale read when the elevator is accelerating (*a*) up and (*b*) down?

(*a*) Since the man is at rest relative to the elevator, he is also accelerating up. The forces acting on the man are $\mathbf{F_n}$ upward, exerted by the scale platform on which he stands, and **w** downward, the force of gravity. The net force is $F_n - w$ upward in the direction of the acceleration **a**. Newton's second law gives

$$F_n - w = ma$$

or

$$F_n = w + ma = mg + ma \qquad 4\text{-}10$$

The force $\mathbf{F_n'}$ exerted by the man on the scale determines the scale reading, which is his apparent weight. Since $\mathbf{F_n'}$ and $\mathbf{F_n}$ are an action–reaction pair, they are equal in magnitude. Thus, when the elevator accelerates up, the apparent weight of the man is greater than his true weight by the amount *ma*.

(*b*) For the case of the elevator accelerating downward, let us call the acceleration **a'**. In this case, the net force must be downward, implying that the weight *mg* is greater than F_n. Choosing down to be the positive direction, Newton's second law gives

$$w - F_n = ma'$$

or

$$F_n = w - ma' = mg - ma' \qquad 4\text{-}11$$

Again, the scale reading, or apparent weight, equals F_n in magnitude. In this case, the apparent weight is less than *mg*. If $a' = g$, as it would if the elevator were in free fall, the man would be apparently weightless. What if the downward acceleration of the elevator is greater than *g*? (For this to happen something in addition to gravity would have to pull down on the elevator.) Assuming that the surface of the scale is not sticky, the scale cannot exert a downward force on the man. Since the downward force on the man cannot be greater than *w*, the scale will soon leave the man. The man will have the acceleration *g*, which is less than that of the elevator, so he will eventually hit the ceiling of the elevator. Then, if the ceiling is strong enough, it can provide the force downward necessary to give him the acceleration **a'**.

Exercise

An elevator is descending to the ground floor and comes to a stop with an acceleration of magnitude 4 m/s^2. If your mass is 70 kg and you are standing on a scale in the elevator, what does the scale read as the elevator is stopping? (Answer: 967 N)

Questions

15. A picture is supported by two wires as in Example 4-8. Do you expect the tension to be greater or less in the wire that is more nearly vertical?

16. A weight is hung on a wire that was originally stretched horizontally. Can the wire remain horizontal? Explain.

17. What effect does the velocity of the elevator have on the apparent weight of the man in Example 4-11?

Summary

1. The fundamental relations of classical mechanics are contained in Newton's laws of motion:

   Law 1. An object continues in its initial state of rest or motion with uniform velocity unless it is acted on by an unbalanced, or net external, force.

   Law 2. The acceleration of an object is inversely proportional to its mass and directly proportional to the net external force acting on it:

   $$\mathbf{a} = \frac{\mathbf{F}_{net}}{m}$$

   or

   $$\mathbf{F}_{net} = m\mathbf{a}$$

   Law 3. Forces always occur in pairs. If object A exerts a force on object B, an equal but opposite force is exerted by object B on object A.

2. A reference frame in which Newton's laws hold is called an inertial reference frame. Any reference frame that is moving with constant velocity relative to an inertial reference frame is itself an inertial reference frame. A reference frame that is accelerating relative to an inertial frame is not an inertial reference frame. A reference frame attached to the earth is approximately an inertial reference frame.

3. Force is defined in terms of the acceleration it produces on a given object. A force of 1 newton (N) is that force that produces an acceleration of 1 m/s^2 on a standard object with a mass of 1 kilogram (kg).

4. Mass is an intrinsic property of an object that measures its resistance to acceleration. The mass of one object can be compared to that of another by applying an equal force to each object and measuring their accelerations. The ratio of the masses of the objects is then equal to the inverse ratio of the accelerations of the objects produced by the same force:

   $$\frac{m_1}{m_2} = \frac{a_2}{a_1}$$

   The mass of an object does not depend on the location of the object.

5. The weight **w** of an object is the force of gravitational attraction between the object and the earth. It is proportional to the mass *m* of the object and the gravitational field **g**, which also equals the free-fall acceleration of gravity:

$$\mathbf{w} = m\mathbf{g}$$

Weight is not an intrinsic property of an object. It does depend on the location of the object because **g** does.

6. All the forces observed in nature can be explained in terms of four basic interactions:

   1. The gravitational force
   2. The electromagnetic force
   3. The strong nuclear force (also called the hadronic force)
   4. The weak nuclear force

   The everyday forces we observe between macroscopic objects, such as the contact forces of support and friction and those exerted by springs and strings, are due to molecular forces that arise from the basic electromagnetic force.

7. The general method of attack for solving a problem using Newton's laws includes the following steps:

   1. Draw a neat diagram.
   2. Isolate the object (particle) of interest and draw a free-body diagram, showing each external force that acts on the object. Draw a separate free-body diagram for each object of interest.
   3. Choose a convenient coordinate system for each object, and apply Newton's second law in component form.
   4. Solve the resulting equations for the unknowns using whatever additional information is available.
   5. Check your results to see if they are reasonable. Examine your solutions when the variables are assigned extreme values.

Suggestions for Further Reading

Cohen, I. Bernard: "Isaac Newton," *Scientific American*, December 1955, p. 73.

This is a short biographical sketch of Newton.

Diggins, John Patrick: "Science and the American Experiment: How Newton's Laws Shaped the Constitution," *The Sciences*, November/December 1987, p. 28.

Newton's laws had social consequences!

Feld, Michael S., Ronald E. McNair, and Stephen R. Wilk: "The Physics of Karate," *Scientific American*, April 1979, p. 150.

An analysis of the forces that a karate expert can apply with bare hands to a block of wood or concrete.

Shamos, Morris H.: "The Laws of Motion—Isaac Newton," in *Great Experiments in Physics*, Henry Holt and Company, New York, 1959. Reprinted by Dover, 1987.

Newton's own account of the laws of motion, with editorial annotations for clarity and a biographical sketch.

Walker, Jearl: "The Amateur Scientist: In Judo and Aikido Application of the Physics of Forces Makes the Weak Equal to the Strong," *Scientific American*, July 1980, p. 150.

Many diagrams help to make this an entertaining and instructive exposition of physical principles employed in two martial arts.

Weaver, Jefferson Hane: "Newton's World," in *The World of Physics: A Small Library of the Literature of Physics from Antiquity to the Present*, Simon and Schuster, New York, 1987.

Contains selections from Newton's writings, including "Mathematical Principles of Natural Philosophy" and "Rules of Reasoning in Philosophy."

Review

A. Objectives: After studying this chapter, you should

1. Be able to discuss the definitions of force and mass and state Newton's laws of motion.

2. Be able to define an inertial reference frame.

3. Be able to distinguish between mass and weight.

4. Be familiar with the following units and know how they are defined: kilogram, newton, pound.

5. Be able to distinguish between action–reaction pairs of forces that act on different bodies and balancing forces that act on the same body.

6. Be able to apply Newton's laws in a systematic way to solve mechanics problems.

B. Define, explain, or otherwise identify:

Newton's laws	Weight
Law of inertia	Apparent weight
Inertia	Free fall
Reference frame	Weightlessness
Inertial reference frame	Newton
Force	Pound
Mass	Law of interaction
Kilogram	Action and reaction forces

Fundamental forces	Force constant
Action at a distance	Normal force
Field	Free-body diagram
Contact forces	Constraints
Hooke's law	Centripetal force

C. Write out the steps in the general method of attack for solving problems using Newton's laws of motion.

D. True or false: If the statement is true, explain why it is true. If it is false give a counterexample.

1. If there are no forces acting on an object, the body will not accelerate.

2. If an object is not accelerating, there must be no forces acting on it.

3. The motion of an object is always in the direction of the resultant force.

4. Action–reaction forces never act on the same object.

5. The mass of an object depends on its location.

6. Action equals reaction only if the objects are not accelerating.

Problems

Level I

4-1 Newton's First Law: The Law of Inertia

There are no problems for this section.

4-2 Force, Mass, and Newton's Second Law

1. An object experiences an acceleration of 4 m/s² when a certain force F_0 acts on it. (*a*) What is its acceleration when the force is doubled? (*b*) A second object experiences an acceleration of 8 m/s² under the influence of the force F_0. What is the ratio of the masses of the two objects? (*c*) If the two objects are tied together, what acceleration will the force F_0 produce?

2. An object is pulled in a straight line along a level, frictionless surface with a constant force. The increase in its speed in a 10-s interval is 5 km/h. When a second constant force is applied in the same direction in addition to the first force, the speed increases by 15 km/h in a 10-s interval. How do the magnitudes of the two forces compare?

3. A force F_0 causes an acceleration of 5 m/s² when it acts on an object of mass m. Find the acceleration of the same object when it is acted on by the forces shown in Figure 4-20*a* and *b*.

4. A certain force F_0 gives an object an acceleration of 6×10^6 m/s². Another force gives the same object an acceleration of 9×10^6 m/s². (*a*) What is the magnitude of the sec-

Figure 4-20 Forces acting on an object for Problem 3.

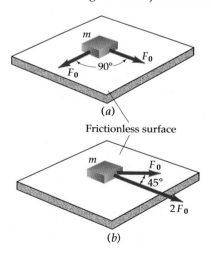

ond force? What is the acceleration of the object (*b*) if the two forces act together on the object in the same direction, (*c*) if the two forces act in opposite directions on the object, and (*d*) if the two forces are perpendicular to each other?

5. A certain force applied to a particle of mass m_1 gives it an acceleration of 20 m/s². The same force applied to a particle of mass m_2 gives it an acceleration of 30 m/s². If the two particles are tied together and the same force is applied to the combination, find the acceleration.

6. A force of 15 N is applied to an object of mass m. The object moves in a straight line, with its speed increasing by 10 m/s every 2 s. Find the mass m.

7. A 5-kg object is pulled along a frictionless horizontal surface by a horizontal force of 10 N. (*a*) If the object is at rest at $t = 0$, how fast is it moving after 3 s? (*b*) How far does it travel from $t = 0$ to $t = 3$ s?

8. A force $\mathbf{F} = 6\,\text{N}\,\mathbf{i} - 3\,\text{N}\,\mathbf{j}$ acts on an object of mass of 2 kg. Find the acceleration **a**. What is the magnitude a?

9. A single force of 10 N acts on a particle of mass m. The particle starts from rest and travels in a straight line a distance of 18 m in 6 s. Find its mass m.

10. In order to drag a 100-kg log along the ground at constant velocity, you have to pull on it with a horizontal force of 300 N. (*a*) What is the resistive force exerted by the ground? (*b*) What force must you exert if you want to give the log an acceleration of 2 m/s²?

11. The graph in Figure 4-21 shows a plot of v_x versus t for an object of mass 10 kg moving in a straight line. Make a plot of the net force acting on the object as a function of time.

Figure 4-21 Graph of velocity v_x versus time t for Problem 11.

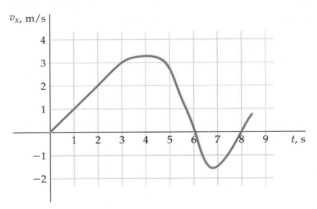

12. Figure 4-22 shows the position x versus time t of a particle moving in one dimension. During what time intervals is there a net force acting on the particle? Give the direction (+ or −) of the net force during these time intervals.

13. Figure 4-23 shows the path taken by an automobile. It consists of straight lines and arcs of circles. The automobile starts from rest at point A and accelerates until it reaches point B. It then proceeds at constant speed until it reaches point E. From point E on it slows down, coming to rest at point F. What is the direction of the net force, if there is any, on the automobile at the midpoint of each section of the path?

4-3 The Force Due to Gravity: Weight

14. Find the weight of a 50-kg girl in (*a*) newtons and (*b*) pounds.

15. Find the mass of a 175-lb man in (*a*) kilograms and (*b*) grams.

16. Find the weight of a 50-g object in newtons.

17. The gravitational force exerted by the earth on an object at height h above the earth's surface can be written

$$F = mg\,\frac{R_E^2}{(R_E + h)^2}$$

where R_E is the radius of the earth (about 6370 km) and g is the acceleration due to gravity at the earth's surface. (*a*) Find the weight of an 80-kg woman in newtons and in pounds at the earth's surface. (*b*) Find the weight of the woman (in newtons and in pounds) at a height of 300 km above the earth's surface. (*c*) What is the mass of this woman at this altitude?

4-4 Newton's Third Law

18. A 2-kg object hangs at rest from a string attached to the ceiling. (*a*) Draw a diagram showing the forces acting on the object and indicate each reaction force. (*b*) Do the same for the forces acting on the string.

19. A box slides down a frictionless inclined plane. Draw a diagram showing the forces acting on the box. For each force in your diagram, indicate the reaction force.

Figure 4-22 Graph of position x versus time t for Problem 12.

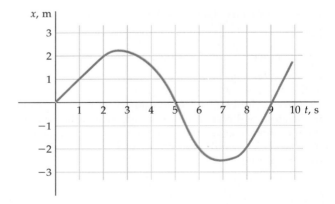

Figure 4-23 Path taken by automobile in Problem 13.

20. A 3-kg box rests on a 7-kg box that rests on a horizontal table. (*a*) Draw a diagram showing the forces acting on the 3-kg box and indicate each reaction force. (*b*) Do the same for the forces acting on the 7-kg box.

4-5 Forces in Nature

21. The moon is 3.84×10^8 m from the earth and the star Sirius is about 8.25×10^{16} m from the earth. (*a*) How long does it take for the gravitational field of the earth to propagate to the moon? (*b*) How long does it take for the gravitational field of the earth to propagate to the star Sirius?

22. A vertical spring of force constant 400 N/m is attached to a 10-kg block resting on a horizontal surface so that the spring exerts an upward force on the block. The spring is stretched by 10 cm. (*a*) What force does the spring exert on the block? (*b*) What is the force that the surface exerts on the block?

23. A 6-kg box slides on a frictionless horizontal surface. It is pulled by a horizontal spring with force constant of 800 N/m. If the spring is stretched 4 cm from its equilibrium position, what is the acceleration of the box?

4-6 Applications to Problem Solving

24. A 10-kg object is subjected to the two forces, $\mathbf{F}_1$ and $\mathbf{F}_2$, as shown in Figure 4-24. (*a*) Find the acceleration **a** of the object. (*b*) A third force $\mathbf{F}_3$ is applied so that the object is in static equilibrium. Find $\mathbf{F}_3$.

Figure 4-24   Problem 24.

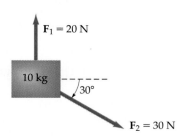

25. In Figure 4-25, the objects are attached to spring balances calibrated in newtons. Give the readings of the balances in each case, assuming that the strings are massless and the incline is frictionless.

Figure 4-25   Problem 25.

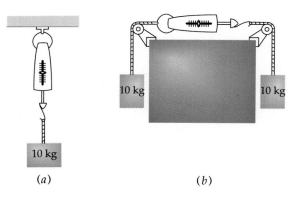

(*a*)   (*b*)

26. A vertical force **T** is exerted on a 5-kg body near the surface of the earth, as shown in Figure 4-26. Find the acceleration of the body if (*a*) $T = 5$ N, (*b*) $T = 10$ N, and (*c*) $T = 100$ N.

Figure 4-26   Problem 26.

27. A 3-kg body is acted on by a single force **F** perpendicular to the velocity of the body. The body travels in a circle of radius 2 m. It makes one complete revolution every 3 s. (*a*) What is the magnitude of the acceleration? (*b*) What is the magnitude of **F**?

28. A woman holding a 10-kg block by a cord designed to withstand a tension of 150 N steps into an elevator. When the elevator starts up, the cord breaks. What was the minimum acceleration of the elevator?

29. A 60-kg girl weighs herself by standing on a scale in an elevator. What does the scale read when (*a*) the elevator is descending at a constant rate of 10 m/s, (*b*) the elevator is accelerating downward at 2 m/s², and (*c*) the elevator is ascending at 10 m/s but its speed is decreasing by 2 m/s in each second?

Level II

30. A 4-kg object is subjected to two forces, $\mathbf{F}_1 = 2$ N **i** $-$ 3 N **j** and $\mathbf{F}_2 = 4$ N **i** $+$ 11 N **j**. The object is at rest at the origin at time $t = 0$. (*a*) What is the object's acceleration? (*b*) What is its velocity at time $t = 3$ s? (*c*) Where is the object at time $t = 3$ s?

(*c*)

(*d*)

31. A 2-kg picture is hung by two wires of equal length. Each makes an angle of θ with the horizontal, as shown in Figure 4-27. (*a*) Find the tension T for general values of θ and weight w of the picture. For what angle θ is T the least? The greatest? (*b*) If $\theta = 30°$, find the tension in the wires.

32. A box is held in position by a cable along a smooth incline (Figure 4-28). (*a*) If $\theta = 60°$ and $m = 50$ kg, find the tension in the cable and the normal force exerted by the incline. (*b*) Find the tension as a function of θ and m, and check your result for $\theta = 0°$ and $\theta = 90°$.

33. A 2-kg block hangs from a spring balance (calibrated in newtons) attached to the ceiling of an elevator (Figure 4-29). What does the balance read (*a*) when the elevator is moving up with a constant velocity of 30 m/s, (*b*) when the elevator is moving down with a constant velocity of 30 m/s, and (*c*) when the elevator is accelerating upward at 10 m/s^2? (*d*) From $t = 0$ to $t = 2$ s, the elevator moves upward at 10 m/s. Its velocity is then reduced uniformly to zero in the next 2 s, so that it is at rest at $t = 4$ s. Describe the reading on the balance during the interval $t = 0$ to $t = 4$ s.

Figure 4-29   Problem 33.

Elevator

34. A bullet of mass 1.8×10^{-3} kg moving at 500 m/s embeds itself in a large fixed piece of wood and travels 6 cm before coming to rest. Assuming that the deceleration of the bullet is constant, find the force exerted by the wood on the bullet.

35. A 1000-kg load is being moved by a crane. Find the tension in the cable that supports the load as (*a*) it is accelerated upward at 2 m/s^2, (*b*) it is lifted at constant speed, and (*c*) it moves upward but its speed decreases by 2 m/s in each second.

Figure 4-27   Problem 31.

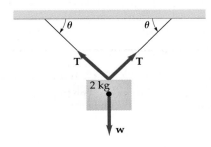

Figure 4-28   Problem 32.

36. The acceleration a versus spring length L observed when a 0.5-kg mass is pulled along a frictionless table by a single spring is shown in the following table:

L, cm	4	5	6	7	8	9	10	11	12	13	14
a, m/s^2	0	2.0	3.8	5.6	7.4	9.2	11.2	12.8	14.0	14.6	14.6

(*a*) Make a plot of the force exerted by the spring versus length L. (*b*) If the spring is extended to 12.5 cm, what force does it exert? (*c*) By how much is the spring extended when the mass suspended from it is at rest near sea level, where $g = 9.81$ N/kg?

37. Your car is stuck in a mudhole. You are alone, but you have a long, strong rope. Having studied physics, you tie the rope tautly to a tree and pull on it sideways, as shown in Figure 4-30. (*a*) Find the force exerted by the rope on the car when the angle θ is 3° and you are pulling with a force of 400 N but the car does not move. (*b*) How strong must the rope be if it takes a force of 600 N to move the car when $\theta = 3°$?

Figure 4-30   Problem 37.

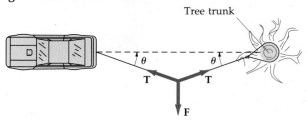

38. A rifle bullet of mass 9 g starts from rest and moves 0.6 m in the rifle barrel. The speed of the bullet as it leaves the barrel is 1200 m/s. Find the force exerted on the bullet assuming it to be constant while the bullet is in the barrel.

39. A car traveling 90 km/h comes to a sudden stop to avoid an accident. Fortunately the driver is wearing a seat belt. Using reasonable values for the mass of the driver and the time it takes the car to come to a stop, estimate the force (assuming it to be constant) exerted on the driver by the seat belt.

40. A man stands on a scale in an elevator that has an upward acceleration a. The scale reads 960 N. When he picks up a 20-kg box, the scale reads 1200 N. Find the mass of the man, his weight, and the acceleration a.

41. A horizontal force of 100 N pushes a 12-kg block up a frictionless incline that makes an angle of 25° with the horizontal. (*a*) What is the normal force that the incline exerts on the block? (*b*) What is the acceleration of the block?

42. In a tug-of-war, two boys pull on a rope, each trying to pull the other over a line midway between their original positions. According to Newton's third law, the forces exerted by each boy on the other are equal and opposite. Show with a force diagram how one boy can win.

43. A man swings his child in a circle of radius 0.75 m as shown in Figure 4-31. If the mass of the child is 25 kg and the child makes one revolution in 1.5 s, what is the magnitude and direction of the force that must be exerted by the man on the child?

Figure 4-31 Problem 43.

Level III

44. A box of mass m_1 is pulled by a force **F** exerted at the end of a rope that has a much smaller mass m_2, as shown in Figure 4-32. The box slides along a smooth horizontal surface. (*a*) Find the acceleration of the rope and box assuming them to be one object. (*b*) What is the net force acting on the rope? (*c*) Find the tension in the rope at the point where it is attached to the box. (*d*) The diagram in Figure 4-32, with the rope horizontal, is not quite correct for this situation. Correct the diagram, and state how this correction affects your solution.

Figure 4-32 Problem 44.

45. A 2-kg body rests on a smooth wedge that has an inclination of 60° (Figure 4-33) and an acceleration a to the right such that the mass remains stationary relative to the wedge. (*a*) Find a. (*b*) What would happen if the wedge were given a greater acceleration?

Figure 4-33 Problem 45.

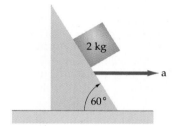

46. A student has to escape from his girlfriend's dormitory through a window that is 15.0 m above the ground. He has a heavy 20-m-long rope, but it will break when the tension exceeds 360 N and he weighs 600 N. The student will be injured if he hits the ground with a speed greater than 10 m/s. (*a*) Show that he cannot safely slide down the rope. (*b*) Find a strategy using the rope that will permit the student to reach the ground safely.

47. A 60-kg house painter stands on a 15-kg aluminum platform. A rope attached to the platform and passing over an overhead pulley allows the painter to raise herself and the platform (Figure 4-34). (*a*) To get started, she accelerates herself and the platform at a rate of 0.8 m/s². With what force must she pull on the rope? (*b*) When her speed reaches 1 m/s, she pulls such that she and the platform go up at a constant speed. What force must she exert on the rope? (Ignore the mass of the rope.)

Figure 4-34 Problem 47.

48. (*a*) Show that a point on the surface of the earth at latitude θ has an acceleration relative to the center of the earth with a magnitude of 3.37 cos θ cm/s^2. What is the direction of this acceleration? (*b*) Discuss the effect of this acceleration on the apparent weight of an object near the surface of the earth. (*c*) The free-fall acceleration of an object at sea level measured *relative to the earth's surface* is 9.78 m/s^2 at the equator and 9.81 m/s^2 at latitude $\theta = 45°$. What are the values of the gravitational field g at these points?

49. A simple accelerometer can be made by suspending a small object from a string attached to a fixed point in an accelerating object—to the ceiling of a passenger car, for example. When there is an acceleration, the object will deflect and the string will make some angle with the vertical. (*a*) How is the direction in which the suspended object deflects related to the direction of the acceleration? (*b*) Show that the acceleration a is related to the angle θ the string makes by $a = g \tan \theta$. (*c*) Suppose the accelerometer is attached to the ceiling of an automobile that brakes to rest from 50 km/h in a distance of 60 m. What angle will the accelerometer make? Will the object swing forward or backward?

50. A 65-kg girl weighs herself by standing on a scale mounted on a skateboard that is rolling down an incline as shown in Figure 4-35. Assume there is no friction so that the force exerted by the incline on the skateboard is perpendicular to the incline. What is the reading on the scale if $\theta = 30°$?

51. A small bead with a mass of 100 g slides along a semicircular wire with a radius of 10 cm that rotates about a vertical axis at a rate of 2 revolutions per second as shown in Figure 4-36. Find the values of θ for which the bead will remain stationary with respect to the rotating wire.

Figure 4-35   Problem 50.

Figure 4-36   Problem 51.

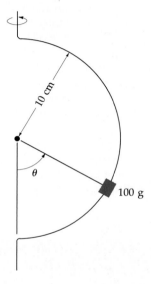

Chapter 5

Newton's Laws II

The Utah/MIT Dextrous Hand is a tendon-operated, multidegree-of-freedom dextrous hand that has multichannel touch sensing capability.

In this chapter, we give examples of the application of Newton's laws to problems that involve frictional forces and problems that involve two or more objects. We will also briefly discuss the motion of an object under the influence of drag forces, which are not constant but depend on the velocity of the object.

The method of problem solving discussed in Section 4-6 applies to many of the problems in this chapter, too. The steps involved should be reread and learned well. In particular, you should develop the habit of approaching a problem by first drawing a picture and then indicating the important forces that act on each object in a free-body diagram.

5-1   Friction

If you push on a large box resting on the floor with a small horizontal force, it is possible that the box won't move at all. The reason is that the floor is exerting a horizontal force called the force of **static friction f$_s$**, which balances the force you are exerting (Figure 5-1). This frictional force is due to the bonding of the molecules of the box and the floor at those places where the

Figure 5-1 When you try to push a large box across the floor, friction opposes the motion. The floor exerts a force of static friction that balances the applied force unless the applied force is greater than the maximum possible force of static friction.

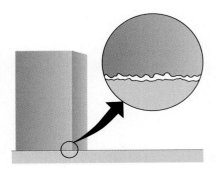

Figure 5-2 The microscopic area of contact between a box and a table is only a small fraction of the macroscopic area of contact. This fraction is proportional to the normal force exerted between the surfaces.

surfaces are in very close contact. It is in the opposite direction of the applied force. The force of static friction is somewhat like a support force in that it can adjust from zero to some maximum force $f_{s,max}$, depending on how hard you push. If you push hard enough, the box will slide across the floor. When the box is sliding, molecular bonds are continually being formed and ruptured and small pieces of the surfaces are being broken off. The result is a force of **kinetic friction f_k** (also called sliding friction) that opposes the motion. To keep the box sliding with constant velocity, you must exert a force equal in magnitude and opposite in direction to this force of kinetic friction.

We will consider static friction first. We might expect the maximum force of static friction to be proportional to the area of contact between the two surfaces. However, it has been shown experimentally that, to a good approximation, this force is independent of the area of contact and is simply proportional to the normal force exerted by one of the surfaces on the other. Figure 5-2, which includes an enlarged view of the contact between a box and a table, shows that the actual microscopic area of contact where the molecules can bond together is just a small fraction of the total macroscopic area of contact. Static friction is a complicated phenomenon that is not completely understood, even today. The following, however, is a possible model that is consistent both with our intuition and with empirical results. The maximum force of static friction is proportional to the microscopic area of contact, but this microscopic area is proportional to the total macroscopic area A and to the normal force per unit area F_n/A exerted between the surfaces. The product of A and F_n/A is thus independent of the total macroscopic area A.

Consider, for example, a 1-kg box with a side area of 60 cm² and an end area of 20 cm². When the box is on its side on a table, a small fraction of the total 60 cm² is actually in microscopic contact with the table. When the box is placed on end, the fraction of the total area actually in microscopic contact is increased by a factor of 3 because the normal force per unit area is 3 times as great. However, since the area of the end is one-third that of the side, the actual microscopic area of contact is unchanged. Thus, the maximum force of static friction $f_{s,max}$ is proportional to the normal force between the surfaces:

$$f_{s,max} = \mu_s F_n \qquad 5\text{-}1$$

where μ_s is called the **coefficient of static friction.** It depends on the nature of the surfaces of the box and the table. If we exert a horizontal force smaller than $f_{s,max}$ on the box, the frictional force will just balance this horizontal force. In general, we can write

$$f_s \leq \mu_s F_n \qquad 5\text{-}2$$

(a) Magnified section of a polished steel surface showing surface irregularities. The irregularities are about 5×10^{-5} cm high, corresponding to about several thousand atomic diameters. (b) Computer depiction derived from microscope data. It shows gold atoms (bottom) adhering to a nickel tip (top), as the tip is lifted after being in contact with the gold surface.

1 μ

10 μ

(a)

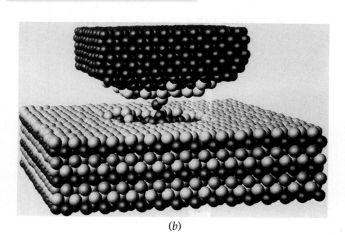

(b)

The force of kinetic friction is opposite the direction of motion. Like static friction, kinetic friction is a complicated and incompletely understood phenomenon. The **coefficient of kinetic friction** μ_k is defined as the ratio of the magnitudes of the kinetic frictional force f_k and the normal force F_n. Then

$$f_k = \mu_k F_n \qquad\qquad 5\text{-}3$$

Experimentally, it has been found that:

1. μ_k is less than μ_s.

2. μ_k depends on the relative speed of the surfaces; but for speeds in the range from about 1 cm/s to several meters per second, μ_k is approximately constant.

3. μ_k (like μ_s) depends on the nature of the surfaces in contact, but it is independent of the (macroscopic) area of contact.

We will neglect any variation in μ_k with speed and assume that it is a constant that depends only on the nature of the surfaces in contact.

Figure 5-3 shows the frictional force exerted on the box by the table as a function of the applied force. The force of friction balances the applied force until the applied force equals $\mu_s F_n$, at which point the box begins to slide. Then the frictional force is constant and equal to $\mu_k F_n$.

We can measure μ_s and μ_k for two surfaces simply by placing a block on a plane surface and inclining the plane until the block begins to slide. Let θ_c be the critical angle at which sliding starts. For angles of inclination less than this, the block is in static equilibrium under the influence of its weight $m\mathbf{g}$, the normal force $\mathbf{F}_n$, and the force of static friction $\mathbf{f}_s$ (see Figure 5-4). Choosing the x axis parallel to the plane and the y axis perpendicular to the plane, we have

$$\sum F_y = F_n - mg \cos \theta = 0$$

and

$$\sum F_x = mg \sin \theta - f_s = 0$$

We can eliminate the weight mg from these two equations by solving the first equation for $mg = F_n/\cos \theta$ and substituting this into the second equation:

$$f_s = mg \sin \theta = \frac{F_n}{\cos \theta} \sin \theta = F_n \tan \theta$$

At the critical angle θ_c, the force of static friction equals its maximum value, and we can replace f_s by $\mu_s F_n$. Then

$$\mu_s = \tan \theta_c \qquad\qquad 5\text{-}4$$

Thus, the coefficient of static friction equals the tangent of the angle of inclination at which the block just begins to slide.

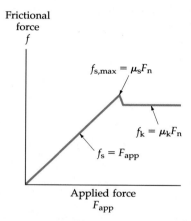

Figure 5-3   Graph of the force of friction acting on a box versus the force applied to the box. When the applied force exceeds the maximum force of static friction $\mu_s F_n$, the box slides and the friction is kinetic friction.

$\mu F_n = F_n \tan \theta$

Figure 5-4   Forces acting on a block on a rough inclined plane. At angles less than the critical angle θ_c, the frictional force balances the component $mg \sin \theta$ directed down the incline. At angles greater than θ_c the block slides down the incline. The critical angle is related to the coefficient of static friction by $\tan \theta_c = \mu_s$.

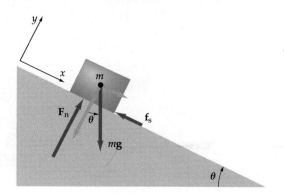

At angles greater than θ_c, the block slides down the incline with an acceleration a_x. In this case, the frictional force is $\mu_k F_n$, and we have

$$F_x = mg \sin \theta - \mu_k F_n = ma_x$$

Substituting $mg \cos \theta$ for F_n, we obtain for the acceleration

$$a_x = g(\sin \theta - \mu_k \cos \theta)$$

By measuring the acceleration a_x, we can determine μ_k for the two surfaces. Table 5-1 lists some approximate values of μ_s and μ_k for various surfaces.

Table 5-1 Approximate Values of Frictional Coefficients		
Materials	μ_s	μ_k
Steel on steel	0.7	0.6
Brass on steel	0.5	0.4
Copper on cast iron	1.1	0.3
Glass on glass	0.9	0.4
Teflon on Teflon	0.04	0.04
Teflon on steel	0.04	0.04
Rubber on concrete (dry)	1.0	0.80
Rubber on concrete (wet)	0.30	0.25
Waxed ski on snow (0°C)	0.1	0.05

Exercise

The coefficient of static friction between a car's tires and the road on a particular day is 0.7. What is the steepest angle of inclination of the road for which the car can be parked with its wheels locked and not slide down the hill?   (Answer: 35°)

Example 5-1

A box slides along a horizontal floor with an initial speed of 2.5 m/s. It comes to rest after sliding 1.4 m. Find the coefficient of kinetic friction.

The weight of the box mg is balanced by the normal force of the floor $F_n = mg$. The only horizontal force acting on the box is the force of kinetic friction, which acts opposite to the direction of motion. If we assume the velocity to be in the positive direction, the force of friction is

$$f = -\mu_k F_n = -\mu_k mg$$

The acceleration is then

$$a = \frac{f}{m} = -\mu_k g$$

Since the acceleration is constant, we may use Equation 2-13 to relate the acceleration to the total distance traveled and the initial velocity. We have

$$v^2 = v_0^2 + 2a\,\Delta x = 0$$

$$a = -\frac{v_0^2}{2\,\Delta x} = -\frac{(2.5\text{ m/s})^2}{2(1.4\text{ m})} = -2.23\text{ m/s}^2$$

The coefficient of kinetic friction is then found from

$$\mu_k = -\frac{a}{g} = -\frac{-2.23\text{ m/s}^2}{9.81\text{ m/s}^2} = 0.228$$

Example 5-2

Two children are pulled on a sled over snow-covered ground. The sled is pulled by a rope that makes an angle of 40° with the horizontal as shown in Figure 5-5. The children have a combined mass of 45 kg and the sled has a mass of 5 kg. The coefficients of static and kinetic friction are $\mu_s = 0.2$ and $\mu_k = 0.15$. Find the frictional force exerted by the ground on the sled and the acceleration of the children and sled if the tension in the rope is (*a*) 100 N and (*b*) 140 N.

Figure 5-5 Man pulling children on a sled for Example 5-2.

(*a*) The vertical and horizontal components of the tension in the rope are

$$T_y = T \sin 40° = (100 \text{ N})(0.643) = 64.3 \text{ N}$$

and

$$T_x = T \cos 40° = (100 \text{ N})(0.766) = 76.6 \text{ N}$$

The vertical forces acting on the sled are its weight acting downward, the normal force exerted by the ground acting upward, and the upward component of the tension in the rope. Since there is no vertical acceleration, the net vertical force must be zero:

$$\sum F_y = F_n + T_y - mg = 0$$

The normal force is therefore

$$F_n = mg - T_y = (50 \text{ kg})(9.81 \text{ m/s}^2) - 64.3 \text{ N}$$
$$= 490 \text{ N} - 64.3 \text{ N} = 426 \text{ N}$$

The maximum possible force of static friction is

$$f_{s,\text{max}} = \mu_s F_n = 0.2(426 \text{ N}) = 85.2 \text{ N}$$

Since the applied horizontal force $T_x = 76.6$ N does not exceed the maximum possible force of static friction, the sled remains at rest. The frictional force is therefore 76.6 N to the left to balance the horizontal force applied to the right.

There are two important points to note about this example: (1) The normal force is not equal to the weight of the children and the sled because the vertical component of the tension helps lift the sled off the ground. (2) The force of static friction is not equal to $\mu_s F_n$; it is less than this maximum possible limiting value.

(b) When the tension is increased to 140 N, its vertical and horizontal components are

$$T_y = (140 \text{ N})(\sin 40°) = 90.0 \text{ N}$$

and

$$T_x = (140 \text{ N})(\cos 40°) = 107 \text{ N}$$

The normal force is then

$$F_n = mg - T_y = 490 \text{ N} - 90.0 \text{ N} = 400 \text{ N}$$

so the maximum possible force of static friction is

$$f_{s,\text{max}} = \mu_s F_n = 0.2(400 \text{ N}) = 80.0 \text{ N}$$

Since this maximum force of static friction is less than the applied horizontal force, the sled will slide. The frictional force on the sled will thus be due to kinetic friction and will have the value

$$f_k = \mu_k F_n = 0.15(400 \text{ N}) = 60.0 \text{ N}$$

The sum of the horizontal forces (taking the rightward direction in the figure to be positive) is

$$\sum F_x = T_x - f_k = 107 \text{ N} - 60.0 \text{ N} = 47.0 \text{ N}$$

and the acceleration of the sled is

$$a_x = \frac{\sum F_x}{m} = \frac{47.0 \text{ N}}{50 \text{ kg}} = 0.940 \text{ m/s}^2$$

When a car is accelerating along a straight road, the unbalanced force that provides the acceleration is due to friction between the tires and the road. Figure 5-6 shows the forces acting on a car that is just starting to move from rest. The weight of the car is balanced by the normal force exerted on the tires. To start the car moving, the engine delivers a torque to the drive shaft that makes the wheels rotate. (We will discuss torque in Chapter 8.) If the road were frictionless, the wheels would merely spin, with the surface of the tire in contact with the road moving backward. If there is friction, and if the torque delivered by the engine is not too great, the tire will not slip on the road because of static friction. This frictional force exerted by the road on the car is in the forward direction, and it provides the acceleration needed for the car to start moving forward. If each wheel rolls without slipping, the tire surface in contact with the road is at rest relative to the road. Although the tire surface in contact with the road is moving backward with speed v relative to the axle, the axle is moving forward with speed v relative to the road as the car moves forward (Figure 5-7). The friction between the road and the tire is therefore static friction. If the engine torque is too great, the wheels will spin, in which case the surface of each tire in contact with the road will slip backward relative to the road. Then the force that accelerates the car is the force of kinetic friction, which is less than the force of static friction. We can see from this example that if we are stuck on

Figure 5-6   Forces acting on a car. As the engine turns the front wheels (for front-wheel drive), the force of static friction $\mathbf{f}_s$ prevents the tires from slipping on the road, causing the car to roll forward. If the wheels spin, the smaller force of kinetic friction pushes the car forward. (The normal forces $\mathbf{F}_n$ are not generally equal on the front and rear tires.)

ice or snow, our chances of getting free are better if we use a light touch on the accelerator pedal of the car so that the wheels do not slip.

When braking a car to a stop, the force exerted by the road on the tires may be either static friction or kinetic friction, depending on how the brakes are applied. If the brakes are applied so hard that the wheels lock, the tires will slide along the road and the stopping force will be that of kinetic friction. However, if the brakes are applied gently so that no slipping occurs between the tires and the road, the stopping force will be that of static friction.

When a wheel rolls *at constant speed* along a horizontal road without slipping, there is neither static nor kinetic friction. However, it is observed that a small force is required to keep a wheel rolling with constant velocity. As the wheel rolls, the surfaces of the wheel and road in contact must be continually peeled apart. In addition, each surface is deformed slightly so that in effect, the wheel is rolling uphill (Figure 5-8). We can define a **coefficient of rolling friction** μ_r as the ratio of the force needed to keep a wheel rolling at constant velocity on a level surface to the normal force exerted by the surface on the wheel. Typical values of μ_r are 0.01 to 0.02 for rubber tires on concrete, and 0.001 to 0.002 for steel wheels on steel rails.

Figure 5-7   When a wheel rolls without slipping, each point on the rim has a velocity of magnitude v relative to the center of the wheel, where v is the speed of the center of the wheel relative to the ground. The velocity of the point on the tire in contact with the ground is zero relative to the ground. In this figure, the dashed lines represent velocities relative to the center of the wheel, and the solid lines represent velocities relative to the ground.

Example 5-3

A car is traveling at 30 m/s along a horizontal road. The coefficients of friction between the road and the tires are $\mu_s = 0.5$ and $\mu_k = 0.3$. How far does the car travel before stopping if (*a*) the car is braked firmly so that the wheels almost slip and (*b*) the car is braked hard so that the wheels lock?

(*a*) Figure 5-9 shows the forces on the car as it is stopping. Since there is no vertical acceleration, the weight of the car mg is balanced by the normal force F_n exerted by the road:

$$\sum F_y = F_n - mg = 0$$

so

$$F_n = mg$$

Since the wheels do not slip, the horizontal force exerted by the road is that of static friction.* Taking the direction of motion as the positive direction (to the right in the figure), Newton's second law gives

$$\sum F_x = -\mu_s F_n = ma_x$$

$$a_x = -\frac{\mu_s F_n}{m} = -\frac{\mu_s mg}{m} = -\mu_s g$$

$$= -(0.5)(9.81 \text{ m/s}^2) = -4.90 \text{ m/s}^2$$

Figure 5-8   Rolling friction. When a wheel rests on a horizontal surface, both surfaces are deformed. As the wheel rolls, its center is raised slightly, just as if the wheel were rolling up a small hill.

---

*For simplicity, we assume that the weight of the car is distributed equally over all four tires and that the brakes are applied to each wheel so that the total frictional force is μ_s times the total normal force.

Figure 5-9   The force that stops a car when it brakes is the force of friction exerted by the road on the tires. If the brakes are applied gently so that the tires do not slip, the force is that of static friction. If the tires slip, the force is that of kinetic friction.

Since the acceleration is constant, we can use Equation 2-13 to find the stopping distance:

$$v^2 = v_0^2 + 2a\,\Delta x$$

$$0 = (30\text{ m/s})^2 + 2(-4.90\text{ m/s}^2)\,\Delta x$$

$$\Delta x = 91.8\text{ m}$$

(b) When the wheels lock, the force exerted by the road on the car is that of kinetic friction. Using reasoning similar to that in part (a), we obtain for the acceleration

$$a_x = -\mu_k g = -(0.3)(9.81\text{ m/s}^2) = -2.94\text{ m/s}^2$$

The stopping distance is then

$$\Delta x = \frac{-v_0^2}{2a} = \frac{-(30\text{ m/s})^2}{2(-2.94\text{ m/s}^2)} = 153\text{ m}$$

Exercise

What must the coefficient of static friction be between the road and the tires of a four-wheel-drive car if the car is to accelerate from rest to 25 m/s in 8 s?   (Answer: 0.319)

When a car rounds a curve on a horizontal road, the centripetal force is provided by the force of friction exerted by the road on the tires of the car. If the car does not slide radially, the friction is static friction.

Example 5-4

A car travels on a horizontal road in a circle of radius 30 m. If the coefficient of static friction is $\mu_s = 0.6$, how fast can the car travel without slipping?

Figure 5-10 shows the forces acting on the car. The normal force $\mathbf{F_n}$ balances the downward force due to gravity $m\mathbf{g}$. The only horizontal force is due to friction. Its maximum value is $f_{s,max} = \mu_s F_n = \mu_s mg$. In this case, the frictional force is the centripetal force. The maximum speed of the car v_{max} occurs when the frictional force equals its maximum value. From Newton's second law

$$f_{s,max} = \mu_s mg = m\,\frac{v_{max}^2}{r}$$

or

$$v_{max} = \sqrt{\mu_s gr} = \sqrt{(0.6)(9.81\text{ m/s}^2)(30\text{ m})} = 13.3\text{ m/s}$$

This speed is approximately equal to 47.8 km/h = 29.7 mi/h. If the car travels at a speed greater than 13.3 m/s, the force of static friction will not be great enough to provide the acceleration needed for the car to travel in a circle. The car will slide out away from the center of the circle; that is, it will tend to travel in a straight line.

Figure 5-10 (a) Car traveling in a horizontal circle for Example 5-4. The normal force balances the weight of the car, and the frictional force provides the centripetal acceleration. (b) As the motorcycle rounds the curve, it is tilted so that the resultant of the normal and frictional forces exerted by the road acts along the plane of the cycle.

(a)

(b)

Exercise

If a car is to travel at 60 km/h around a curve whose radius is 40 m, what is the least value for the coefficient of static friction such that the car will not slide?   (Answer: 0.71)

If a curved road is not horizontal but banked, the normal force of the road will have a component directed inward toward the center of the circle that will contribute to the centripetal force. The banking angle can be such that, for a given speed, no friction is needed for a car to make the curve.

Example 5-5

A curve of radius 30 m is banked at an angle θ as shown in Figure 5-11. Find θ for which a car can round the curve at 40 km/h even if the road is frictionless.

In this example, the normal force of the road acting on the car has a component directed inward toward the center of the circle that provides the centripetal force. From the figure, we can see that the angle between the normal force F_n and the vertical is θ, the same as the banking angle. The vertical component of the normal force $F_y = F_n \cos \theta$ must balance the weight of the car:

$$F_y = F_n \cos \theta = mg$$

The horizontal component of the normal force $F_x = F_n \sin \theta$ provides the centripetal force:

$$F_x = F_n \sin \theta = \frac{mv^2}{r}$$

If we divide this equation by the first, we can eliminate m and F_n and obtain an equation relating θ to the speed v and the radius r:

$$\frac{F_n \sin \theta}{F_n \cos \theta} = \tan \theta = \frac{v^2}{rg}$$

Substituting in $v = 40$ km/h $= 11.1$ m/s, $r = 30$ m, and $g = 9.81$ m/s^2, we obtain

$$\tan \theta = \frac{(11.1 \text{ m/s})^2}{(30 \text{ m})(9.81 \text{ m/s}^2)} = 0.419$$

$$\theta = 22.7°$$

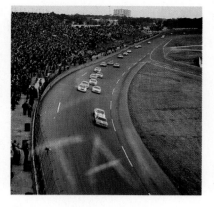

Cars on a race track that is banked so that the normal force exerted by the track on a car has a horizontal component toward the center of the circle to provide the centripetal force.

Questions

1. Various objects lie on the floor of a truck. If the truck accelerates, what force acts on the objects to cause them to accelerate?

2. Any object resting on the floor of a truck will slide if the acceleration of the truck is too great. How does the critical acceleration at which a light object slips compare with that at which a much heavier object slips?

Figure 5-11 (a) Car on banked curve for Example 5-5. The forces exerted by the road on the car are represented by F_{n1} and F_{n2}. (b) The force diagram for the car. The total normal force has the component $F_n \sin \theta$ directed towards the center of the curve, which contributes to the centripetal acceleration of the car.

A golf ball and styrofoam ball falling in air. The air resistance is negligible for the heavier golf ball, which falls with essentially constant acceleration. The styrofoam ball reaches terminal velocity quickly, as indicated by the nearly equal spacing at the bottom.

5-2   Drag Forces

When an object moves through a fluid such as air or water, the fluid exerts a retarding or **drag force** that tends to reduce the speed of the object. This drag force depends on the shape of the object, on the properties of the fluid, and on the speed of the object relative to the fluid. Like the force of friction, the drag force is very complicated. Unlike ordinary friction, the drag force increases as the speed of the object increases. For small speeds, the drag force is approximately proportional to the speed of the object; for higher speeds, it is more nearly proportional to the square of the speed.

Consider an object dropped from rest and falling under the influences of the force of gravity, which we assume to be constant, and a drag force of magnitude bv^n, where b and n are constants. We then have a constant downward force mg and an upward force bv^n (Figure 5-12). If we take the downward direction to be positive, we obtain from Newton's second law

$$F_{\text{net}} = mg - bv^n = ma \qquad\qquad 5\text{-}5$$

At $t = 0$, when the object is dropped, the speed is zero, so the retarding force is zero and the acceleration is g downward. As the speed of the object increases, the drag force increases and the acceleration is less than g. Eventually, the speed is great enough for the drag force bv^n to equal the force of gravity mg, so the acceleration is zero. The object then continues moving at a constant speed v_t, called its **terminal speed.** Setting the acceleration a in Equation 5-5 equal to zero, we obtain

$$bv_t^n = mg$$

Solving for the terminal speed, we get

$$v_t = \left(\frac{mg}{b}\right)^{1/n} \qquad\qquad 5\text{-}6$$

The larger the constant b, the smaller the terminal speed. The constant b depends on the shape of the object. A parachute is designed to make b large so that the terminal speed will be small. On the other hand, cars are designed to make b small to reduce the effect of wind resistance.

For a sky diver (with parachute closed), the terminal speed is about 60 m/s = 216 km/h. When the parachute is opened, the drag force is greater than the force of gravity and the sky diver has an upward acceleration while falling; that is, the downward speed of the sky diver decreases. The drag force then decreases until a new terminal speed, about 20 km/h, is reached.

(a)

(b)

Figure 5-12 (a) Free-body diagram showing forces on an object falling with air resistance. The upward force is the drag force due to air resistance bv^n, which depends on the speed v of the body. As the speed increases, the drag force increases until its magnitude equals the weight, after which the body falls with constant speed called the terminal speed. (b) Use of a parachute greatly reduces the terminal speed.

(*a*) The space shuttle Columbia just after landing at Dryden Flight Research Center, Edwards, California on Nov. 14, 1981. The sides of the fuselage show scorching marks from the tremendous heat generated by air resistance as the shuttle fell through the atmosphere.

(*b*) As it falls through the atmosphere, a meteor burns making it visible in the night sky.

Example 5-6

A sky diver of mass 64 kg reaches a terminal speed of 180 km/h with her arms and legs outspread. (*a*) What is the magnitude of the upward drag force on the sky diver? (*b*) If the drag force is equal to bv^2, what is the value of b?

(*a*) Since the sky diver is moving with constant velocity, the downward force on her, her weight mg, must be balanced by the upward drag force F_d:

$$F_d = mg = (64 \text{ kg})(9.81 \text{ N/kg}) = 628 \text{ N}$$

(*b*) Setting bv^2 equal to 628 N and using $v = 180$ km/h $= 50$ m/s, we obtain

$$bv^2 = 628 \text{ N}$$

$$b = \frac{628 \text{ N}}{(50 \text{ m/s})^2} = 0.251 \text{ N·s}^2/\text{m}^2 = 0.251 \text{ kg/m}$$

where we have used 1 N = 1 kg·m/s^2.

Question

3. How would you expect the value of b for air resistance to depend on the density of air?

5-3   Problems with Two or More Objects

In many mechanics problems, two or more objects are in contact or are connected by a string or spring. Such problems can be solved by treating each object separately. For each object, a free-body diagram is drawn and Newton's second law is applied. The resulting equations are then solved simultaneously for the unknown forces or accelerations. For a system of two objects, the force exerted by the first object on the second must be equal and opposite to that exerted by the second object on the first, as stated in Newton's third law.

Example 5-7

A block hangs by a massless string that passes over a frictionless peg* and is connected to another block on a frictionless table. Find the acceleration of each block and the tension in the string.

Figure 5-13 shows the important elements of this problem. The string tensions T_1 and T_2 have equal magnitudes because the string is assumed to be massless, and since the peg is assumed to be frictionless, there are no tangential forces acting on the string. Because the block on the table has no vertical acceleration, the vertical forces $\mathbf{F}_n$ and $m_1\mathbf{g}$ must balance. If a_1 is the horizontal acceleration of m_1, Newton's second law gives

$$T = m_1 a_1 \qquad\qquad 5\text{-}7$$

where $T = T_1 = T_2$ is the tension in the string. The acceleration of the hanging block is vertically downward. The forces acting on it are its weight $m_2\mathbf{g}$ downward and the tension $\mathbf{T}_2$ upward. If we take the downward direction to be positive for the acceleration a_2 of this block, Newton's second law gives

$$m_2 g - T = m_2 a_2 \qquad\qquad 5\text{-}8$$

If the connecting string does not stretch or become slack, each block must move at the same speed. This constraint allows us to simplify Equations 5-7 and 5-8. Since the speeds of the blocks are the same, the accelerations a_1 and a_2 must be equal in magnitude (but not in direction). Let us call this magnitude a. We then have

$$T = m_1 a \qquad\qquad 5\text{-}9$$

$$m_2 g - T = m_2 a \qquad\qquad 5\text{-}10$$

We solve these two equations for the two unknowns T and a by first eliminating one of the unknowns. For example, we can eliminate T by substituting $m_1 a$ from Equation 5-9 for T in Equation 5-10 (or by simply adding the two equations). We then obtain

$$m_2 g - m_1 a = m_2 a$$

or

$$a = \frac{m_2}{m_1 + m_2}\, g \qquad\qquad 5\text{-}11$$

Note that the result for the magnitude of a is the same as that for a mass $m = m_1 + m_2$ acted on by a force $m_2 g$. The result for a can be substituted

*In some laboratory experiments, the string is replaced by a tape that passes over a curved channel that supports the tape on a cushion of air blown through tiny holes in the channel. In others, the "frictionless" peg is a pulley. If the pulley has negligible mass, it acts the same as a frictionless peg. We will learn how to handle cases involving a pulley with mass in Chapter 8.

Figure 5-13 (a) The two blocks of Example 5-7. (b) Free-body diagram for the block on the table. (c) Free-body diagram for the hanging block.

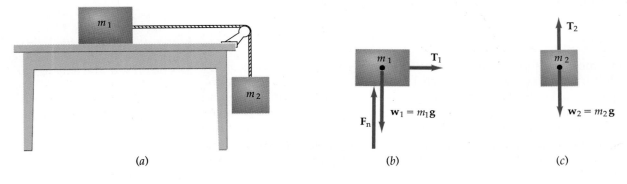

into Equation 5-9 to find T:

$$T = \frac{m_1 m_2}{m_1 + m_2} g \qquad\qquad 5\text{-}12$$

If the block on the table is very much lighter than the one hanging, we would expect the hanging block to fall with the acceleration of gravity g and the tension T to be zero. Setting $m_1 = 0$ in Equations 5-11 and 5-12, we obtain $a = g$ and $T = 0$. At the other extreme, if m_1 is very much heavier than m_2, we would expect essentially no acceleration. If $m_1 \gg m_2$, the denominators in Equations 5-11 and 5-12 are approximately equal to m_1, and we obtain

$$a \approx \frac{m_2}{m_1} g \approx 0 \qquad \text{for } m_1 \gg m_2$$

and

$$T \approx \frac{m_1 m_2}{m_1} g = m_2 g \qquad \text{for } m_1 \gg m_2$$

Exercise

(a) Find the acceleration of the blocks in Example 5-6 if the masses are $m_1 = 2$ kg and $m_2 = 8$ kg. (b) Find the acceleration if these two masses are interchanged.   [Answer: (a) $a = 0.8g = 7.85$ m/s^2, (b) $a = 0.2g = 1.96$ m/s^2]

Example 5-8

A block of mass m_1 sits on a second block of mass m_2, which in turn sits on a frictionless horizontal table as shown in Figure 5-14a. A force **F** is applied to the bottom block as shown in the figure. The coefficients of static and sliding friction between the blocks are μ_s and μ_k. (a) Find the maximum value of F for which the blocks will not slide relative to each other. (b) Find the acceleration of each block when F is greater than this value.

(a)   (b)

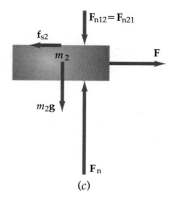

(c)

Figure 5-14 (a) A small block of mass m_1 on top of a large block of mass m_2 for Example 5-8. (b) Free-body diagram for the small block. The only horizontal force is that of friction exerted by the larger block. (c) Free-body diagram for the large block.

(a) The free-body diagrams for each block are shown in Figure 5-14b and c. The forces acting on the top block are its weight $m_1\mathbf{g}$ downward, the normal contact force $\mathbf{F}_{n21}$ upward exerted by the lower block, and the force of static friction $\mathbf{f}_{s1}$ exerted to the right by the lower block. Since there is no vertical acceleration, the vertical forces must balance. For the top block, Newton's second law gives

$$\sum F_y = F_{n21} - m_1 g = 0 \qquad\qquad 5\text{-}13a$$

and

$$\sum F_x = f_{s1} = m_1 a_1 \qquad\qquad 5\text{-}13b$$

The forces acting on the bottom block are its weight $m_2\mathbf{g}$ downward, the downward normal contact force $\mathbf{F}_{n12} = -\mathbf{F}_{n21}$ exerted by the upper block, the upward normal contact force $\mathbf{F}_n$ exerted by the table, the force of static friction $\mathbf{f}_{s2}$ exerted to the left by the upper block, and the applied force $\mathbf{F}$ to the right. Again, the vertical forces must balance. For the lower block, Newton's second law gives

$$\sum F_y = F_n - F_{n12} - m_2 g = 0 \qquad \text{5-14}a$$

and

$$\sum F_x = F - f_{s2} = m_2 a_2 \qquad \text{5-14}b$$

Since f_{s1} and f_{s2} are action–reaction forces, they are equal in magnitude, $f_{s1} = f_{s2} = f_s$. The blocks are moving together so they have the same acceleration. Thus, $a_1 = a_2 = a$. We can then eliminate f_s by adding Equations 5-13b and 5-14b and obtaining

$$F = (m_1 + m_2)a \qquad \text{5-15}$$

The greater the applied force F, the greater the acceleration a. But the acceleration of the upper block is limited because the force of static friction f_{s1} is limited. Solving Equation 5-13b for a_1, we obtain

$$a_1 = \frac{f_{s1}}{m_1} \leq \frac{\mu_s F_{n21}}{m_1}$$

But from Equation 5-13a, $F_{n21} = m_1 g$. Then

$$a \leq \frac{\mu_s m_1 g}{m_1}$$

or

$$a \leq \mu_s g$$

When a_1 has its maximum value $\mu_s g$, the force F has the maximum value for which the blocks will not slide relative to each other. Substituting $\mu_s g$ for a in Equation 5-15, we obtain

$$F_{max} = (m_1 + m_2)\mu_s g \qquad \text{5-16}$$

(b) When F is greater than this value, the blocks slide relative to each other, and the accelerations of the blocks are not equal. From Newton's second law, the horizontal component of the net force acting on the upper block is

$$f_k = m_1 a_1$$

Using $f_k = \mu_k F_{n21} = \mu_k m_1 g$, we obtain

$$\mu_k m_1 g = m_1 a_1$$

or

$$a_1 = \mu_k g \qquad \text{5-17}$$

Similarly, the horizontal component of the net force acting on the lower block is

$$F - f_k = m_2 a_2$$

$$F - \mu_k m_1 g = m_2 a_2$$

$$a_2 = \frac{F - \mu_k m_1 g}{m_2} \qquad \text{5-18}$$

Note that in part (a) we could find the acceleration of the two objects moving together by treating them as a single object of mass $m_1 + m_2$ moving under the net force $\mathbf{F}$. The vertical forces balance, and the horizontal frictional forces are equal and opposite forces that are internal to the system. However, when the upper block slides on the lower block, the accelerations are not the same, and each block must be treated separately.

5-4   Pseudoforces

Newton's laws hold only in inertial reference frames. When the acceleration of any object is measured relative to a reference frame that is itself accelerating relative to an inertial frame, the net force on the object does not equal the mass of the object times its acceleration as measured in the accelerating frame. In some cases, the object will be at rest relative to the noninertial frame even though there is obviously a net force acting on it. In other cases, the object has no forces acting on it even though it is accelerating relative to the noninertial frame. If we wish to use Newton's second law, $\mathbf{F}_{net} = m\mathbf{a}$, in such an accelerating reference frame, we must introduce fictitious forces, called **pseudoforces,** that depend on the acceleration of the reference frame. These fictitious forces are not exerted by any agent. They are introduced merely to make $\mathbf{F}_{net} = m\mathbf{a}$ work when the acceleration $\mathbf{a}$ is measured relative to the noninertial frame. However, to observers in the noninertial frame, the pseudoforces appear to be as real as other forces.

Let us first consider a railroad boxcar that is moving in a straight line along a horizontal track with a constant acceleration $\mathbf{a}_c$ relative to the track, which we assume to be in an inertial reference frame. If we drop an object in the car, it does not fall straight down but rather toward the back of the car. Relative to the car, the object has a vertical acceleration $\mathbf{g}$ and a horizontal acceleration $-\mathbf{a}_c$ (Figure 5-15). Similarly, if an object is placed on a smooth table in the car such that the net force acting on it is zero, it accelerates toward the back of the car relative to an observer in the car. Of course, from the point of view of an observer on the tracks in an inertial frame, the object does not accelerate. Instead, the car and table accelerate out from under the object. We can use Newton's second law in the reference frame of the car if we introduce a pseudoforce $\mathbf{F}_p = -m\mathbf{a}_c$ that acts on any object of mass m.

Consider, for example, a lamp hanging by a cord from the ceiling of the car. The acceleration of the lamp and the forces acting on it relative to both the inertial and the noninertial reference frames are shown in Figure 5-16.

Figure 5-15   An object is dropped inside a railroad boxcar that has an initial velocity of zero and a constant acceleration $\mathbf{a}_c$ to the right. (*a*) An observer on the ground in an inertial reference frame sees the ball fall straight downward. (*b*) An observer in the accelerating car sees the ball fall downward and toward the back of the car. He attributes the backward acceleration of the ball to a pseudoforce $-m\mathbf{a}_c$.

(*a*)   (*b*)

Observer

Figure 5-16 A lamp is hung by a cord from the roof of an accelerating railroad boxcar. (*a*) According to an observer in an inertial reference frame, the lamp accelerates to the right because there is a net force acting on it to the right, the horizontal component of the tension **T**. (*b*) In the accelerating frame, the lamp appears to be at rest and not accelerating. The force acting on it to the right is balanced by the pseudoforce $-m\mathbf{a}_c$ that must be introduced in this frame so that Newton's second law can be used.

The vertical component of the tension in the cord equals the weight of the lamp according to each observer. In the inertial frame of the track, the lamp is accelerating. This acceleration is provided by the net force due to the horizontal component of the tension in the cord. In the frame of the car, the lamp is at rest and therefore has no acceleration. This is explained by the fact that the horizontal component of the tension balances the pseudoforce $\mathbf{F}_P = -m\mathbf{a}_c$ seen to act on all objects in the car by an observer in the car.

Figure 5-17 shows another noninertial frame, a rotating platform. Each point on the platform is moving in a circle and therefore has centripetal acceleration. Thus, a frame attached to the platform is a noninertial frame. In the figure, a block at rest relative to the platform is attached to the center post by a string. According to an observer in an inertial frame, the block is moving in a circle with a speed v, so it is accelerating toward the center of the circle. This centripetal acceleration v^2/r is provided by the unbalanced force of the tension T in the string. However, according to an observer on the platform, the block is at rest and so is not accelerating. In order to use $\mathbf{F}_{net} = m\mathbf{a}$, this observer must introduce a pseudoforce with a magnitude of mv^2/r that is acting on the object radially outward to balance the string tension. This fictitious outward force, called the **centrifugal force,** appears quite real to the observer on the platform. If he wants to stand "at rest" on the platform, an inward force of this magnitude must be exerted on him (by the floor) to "balance" the outward centrifugal force. It is important to understand that the centrifugal force is a pseudoforce that exists *only in a rotating reference frame.*

Consider a satellite near the surface of the earth that is observed in an inertial frame attached to the earth. (We are neglecting the earth's rotation here.) It is incorrect to say that the satellite does not fall because the gravitational attraction of the earth "is balanced by the centrifugal force." Pseudoforces such as the centrifugal force appear only in accelerating reference frames. In the earth's frame, the satellite does "fall" toward the earth with an acceleration v^2/r produced by the single, unbalanced force of gravity acting on it. However, an observer in the satellite who considers the satellite to be at rest can use $\mathbf{F}_{net} = m\mathbf{a}$ only by introducing an outward centrifugal force to balance the force of gravity.

Figure 5-17 A block is tied to the center post of a rotating platform by a string. (*a*) An observer in an inertial frame sees the block moving in a circle with a centripetal acceleration provided by the net force **T**. (*b*) According to an observer on the platform, the block is not accelerating. Therefore, Newton's second law can be used only if a pseudoforce mv^2/r acting outward on the block is introduced to balance the tension.

Large centrifuge used for research at Sandia National Laboratories.

A second pseudoforce, which depends on the velocity of a particle, must be introduced in a rotating frame in order to use $\mathbf{F}_{net} = m\mathbf{a}$ in that frame. Called the **Coriolis force,** this force is perpendicular to the velocity of the particle (relative to the rotating frame) and causes a sideways deflection. Consider two boys standing along a radial line on a rotating platform and playing catch (Figure 5-18). If the ball is thrown outward along the radial line, an observer in an inertial frame will see the ball travel in a straight line and miss the receiver because the receiver is moving to the left (Figure 5-18a). The path of the ball relative to the rotating platform is shown in Figure 5-18b. To reach the receiver, the ball must be thrown to the left of the receiver to take into account this sideways deflection.

Figure 5-18   A boy at the center of a rotating platform throws a ball toward a friend on the edge of the platform. (a) In an inertial frame, the ball travels in a straight line and misses the receiver because the receiver moves with the platform. (b) In the rotating frame of the platform, the receiver is at rest, and the ball deflects to the right. The pseudoforce that deflects the ball from a straight line in this frame is called the Coriolis force.

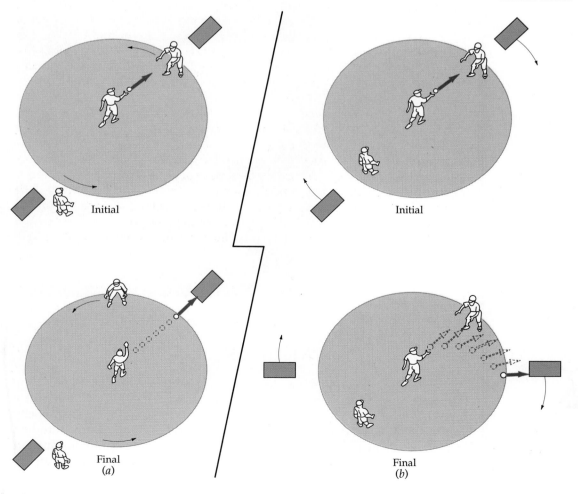

Initial

Initial

Final
(a)

Final
(b)

Figure 5-19   (*a*) If the earth were not rotating, the winds would blow inward toward a low-pressure center. (*b*) Because the earth does rotate, the Coriolis force deflects the winds to the right in the northern hemisphere, setting up a counterclockwise pattern. (In the southern hemisphere, the pattern is clockwise.)

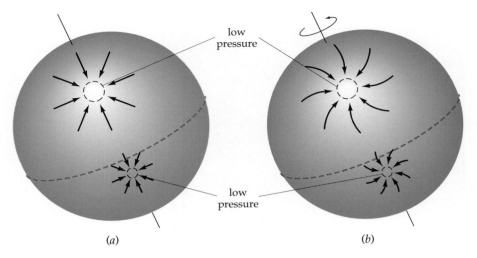

low pressure

low pressure

(*a*)

(*b*)

These two types of pseudoforce for a rotating reference frame—the centrifugal force and Coriolis force—have direct application to reference frames that are attached to the earth because of the earth's rotation. In particular, Coriolis forces are important for understanding the weather. For example, these forces are responsible for cyclones being counterclockwise in the northern hemisphere and clockwise in the southern hemisphere, when viewed from above (see Figure 5-19).

Typhoon Pat photographed over the Western Pacific by Shuttle Discovery astronauts. Shown in the photo is the counterclockwise rotation of the typhoon due to the Coriolis effect.

5-5   Numerical Methods

When we apply Newton's laws to a problem in which a particle is under the influence of various forces, we can obtain the acceleration of the particle from $\mathbf{a} = \mathbf{F}_{\text{net}}/m$. If the acceleration is constant, we can obtain the velocity and position of the particle from the constant-acceleration formulas studied in Chapter 2. In many cases, though, the acceleration is not constant but depends on the velocity or the position of the particle. For example, in the motion of a particle under the influence of a drag force, the acceleration depends on the velocity of the particle. In such cases, an analytical determination of the velocity and position of the particle may be very difficult or even impossible. However, it is always possible to find the velocity and position approximately by **numerical integration.** For problems that can be solved analytically, the analytical solution has an advantage over numerical integration in that it gives the general features of the solution, whereas numerical integration applies only to a specific case. However, numerical integration has the advantage that it can be applied to any problem no matter how complex.

By definition, the average acceleration of a particle in one dimension is

$$a_{av} = \frac{\Delta v}{\Delta t}$$

and the average velocity is

$$v_{av} = \frac{\Delta x}{\Delta t}$$

The change in the velocity for a time interval Δt is then

$$\Delta v = a_{av} \, \Delta t$$

and the displacement during this interval is

$$\Delta x = v_{av} \, \Delta t$$

Numerical integration is based on the assumption that if the time interval Δt is very small, the average values of the acceleration and velocity can be replaced by particular values chosen at some time during the interval. In **Euler's method,** a_{av} and v_{av} are replaced by the values of the acceleration and velocity at the beginning of the time interval.* The acceleration and velocity are then assumed to be constant during that time interval. Let x_0, v_0, and a_0 be the initial values of the position, velocity, and acceleration at time $t = t_0$. At some later time $t_1 = t_0 + \Delta t$, the velocity is approximately

$$v_1 = v_0 + a_0 \, \Delta t$$

and the position is approximately

$$x_1 = x_0 + v_0 \, \Delta t$$

We now use the new values v_1 and x_1 to compute the acceleration a_1 for the next interval, and we then use this acceleration to compute the new values of v and x:

$$v_2 = v_1 + a_1 \, \Delta t$$

and

$$x_2 = x_1 + v_1 \, \Delta t$$

In general, if x_i, v_i, and a_i are the values of the position, velocity, and acceleration at some time t_i, the velocity and position at time $t_{i+1} = t_i + \Delta t$ are approximately

$$v_{i+1} = v_i + a_i \, \Delta t \qquad \text{5-19}a$$

Euler's Method

$$x_{i+1} = x_i + v_i \, \Delta t \qquad \text{5-19}b$$

*Other methods of numerical integration are more accurate but less simple to use. For example, the accuracy is improved if a_{av} and v_{av} are replaced by the values that occur at the midpoint of the interval rather than those at the beginning of the interval.

To find the velocity and position at some time t, we therefore divide the time interval $t - t_0$ into a large number of smaller intervals Δt and apply Equations 5-19a and b beginning at the initial time t_0. This involves a large number of simple, repetitive calculations that are easily done on a computer.

To illustrate the use of numerical methods, let us consider a problem in which a sky diver is dropped from rest at some height under the influences of gravity and a drag force that is proportional to the square of the speed. We will find the velocity v and the distance x traveled as functions of time. The equation describing the motion of an object of mass m dropped from rest is Equation 5-5 with $n = 2$:

$$F_{net} = mg - bv^2 = ma$$

The acceleration is thus

$$a = g - \left(\frac{b}{m}\right)v^2 \qquad \qquad 5\text{-}20$$

It is convenient to write the constant b/m in terms of the terminal speed v_t. Setting $a = 0$ in Equation 5-20, we obtain

$$0 = g - \left(\frac{b}{m}\right)v_t^2$$

$$\frac{b}{m} = \frac{g}{v_t^2}$$

Substituting g/v_t^2 for b/m in Equation 5-20 gives

$$a = g\left(1 - \frac{v^2}{v_t^2}\right) \qquad \qquad 5\text{-}21$$

To solve Equation 5-21 numerically, we need numerical values for g and v_t. A reasonable terminal speed for a sky diver is 60 m/s. Using this and $g = 9.81$ m/s^2, we obtain

$$a = 9.81\left(1 - \frac{v^2}{3600}\right) \qquad \qquad 5\text{-}22$$

We have omitted the units in this equation. Since we are using SI, the unit for v is meters per second, and the unit for x is meters. If we choose $x_0 = 0$ for the initial position, the initial values are $x_0 = 0$, $v_0 = 0$, and $a_0 = g = 9.81$. To find the velocity v and position x after some time, say $t = 20$ s, we divide the time interval $0 < t < 20$ s into many small intervals Δt and apply Equations 5-19a and b. We do this by writing a computer program or by using a computer spreadsheet. Figure 5-20 shows graphs of v versus t and x versus t based on data found using a spreadsheet with $\Delta t = 0.5$ s. At $t = 20$ s, the computed values are $v = 59.97$ m/s and $x = 957.5$ m.

But how accurate are our computations? We can estimate the accuracy by running the program again using a smaller time interval. If we use $\Delta t = 0.25$ s, one-half of the value we originally used, we obtain $v = 59.92$ m/s and $x = 952.0$ m at $t = 20$ s. The difference in v is about 0.1 percent and that in x is about 0.5 percent. These are our estimates of the accuracy of the original computations.

Since the difference between the value of a_{av} for some time interval Δt and the value of a_i at the beginning of the interval becomes smaller as the time interval becomes smaller, we might expect that it would be better to use very small time intervals, say $\Delta t = 0.000000001$ s. But there are two reasons for not using very small time intervals. First, the smaller the time interval, the larger the number of calculations that are required and the longer the program takes to run. Second, the computer keeps only a fixed number of digits at each step of the calculation, so that at each step there is a round-off

(a)

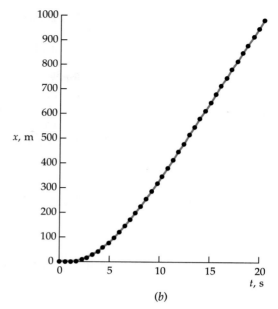

(b)

Figure 5-20 (a) Graph of v versus t for the sky diver found by numerical integration using $\Delta t = 0.5$ s. The horizontal dashed line is the terminal speed $v_t = 60$ m/s. (b) Graph of x versus t using $\Delta t = 0.5$ s.

error. These round-off errors add up. The larger the number of calculations, the more significant the total round-off errors become. When we first decrease the time interval, the accuracy improves because a_i more nearly approximates a_{av} for the interval. However, as the time interval is decreased further, the round-off errors build up and the accuracy of the computation decreases. A good rule of thumb to follow is to use no more than about 10^4 or 10^5 time intervals for the typical numerical integration.

Summary

1. When two bodies are in contact, they can exert frictional forces on each other. These frictional forces are parallel to the surfaces of the bodies at the points of contact. When the surfaces are at rest relative to each other, the frictional force is static friction, which can vary anywhere from 0 to its maximum value $\mu_s F_n$, where F_n is the normal force of contact and μ_s is the coefficient of static friction. When the surfaces are moving relative to each other, the frictional force is kinetic or sliding friction and has the value $\mu_k F_n$, where μ_k is the coefficient of kinetic friction. The coefficient of kinetic friction is slightly less than the coefficient of static friction.

2. When a body moves through a fluid such as air or water, the body experiences a drag force that opposes its motion. The drag force increases with increasing speed. If the body is dropped from rest, its speed increases until the drag force equals the force of gravity, after which it moves with a constant speed called its terminal speed. The terminal speed depends on the shape of the body and on the medium through which it falls.

3. In applying Newton's laws to problems with two or more objects, a free-body diagram should be drawn for each object. $\mathbf{F}_{net} = m\mathbf{a}$ should then be applied to each object separately.

4. Newton's laws do not hold in accelerating reference frames. However, they can be applied if certain fictitious forces or pseudoforces that depend on the acceleration of the frame are introduced.

5. If the acceleration of a particle as a function of its position or velocity is known and if the initial position and velocity are known, the position and velocity of the particle at a later time can be found by numerical integration. In this procedure, the total time interval is divided into a large number of small time intervals Δt, and the average acceleration a_{av} and average velocity v_{av} are replaced by particular values that occur at some time in the interval. In Euler's method, the acceleration and velocity are found at the beginning of a time interval and are assumed to be constant over that interval. The velocity and position at the end of an interval are found from

$$v_{i+1} = v_i + a_i \, \Delta t$$

and

$$x_{i+1} = x_i + v_i \, \Delta t$$

where x_i, v_i, and a_i are the values of the position, velocity, and acceleration at the beginning of the interval. The accuracy of the results obtained by numerical integration can be estimated by repeating the computation using a smaller time interval.

Suggestions for Further Reading

Armstrong, H. L.: "How Dry Friction Really Behaves," *American Journal of Physics*, vol. 53, no. 9, 1958, p. 910.

This one-page article discusses the dependence of friction on velocity and points out that the change from static to kinetic friction is not instantaneous but takes place gradually without discontinuity.

Brancazio, Peter J.: "Trajectory of a Fly Ball," *The Physics Teacher*, January 1985, p. 20.

This article shows why drag forces cannot be neglected in any realistic analysis of the motion of a baseball.

Gross, A. C., C. R. Kyle, and D. J. Malewicki: "The Aerodynamics of Human-Powered Land Vehicles," *Scientific American*, December 1983, p. 142.

Drag forces play an important role in determining the maximum velocity a bicycle can attain. By applying the principles of aerodynamics, a bicycle that can travel 60 miles per hour on level ground can be constructed.

von Baeyer, Hans Christian: "Physika: Einstein at the Ex," *The Sciences*, May/June 1985, p. 10.

An account of the author's "first conscious encounter with a fictitious force" (or pseudoforce) at the carnival. Complements the article "Thinking About Physics While Scared to Death (on a Falling Roller Coaster)" by Jearl Walker (see Suggestions for Further Reading, Chapter 8).

Review

A. Objectives: After studying this chapter you should:

1. Know that the maximum force of static friction and the force of kinetic friction are proportional to the normal force between the surfaces involved.

2. Be able to apply Newton's laws to problems involving frictional forces.

3. Be able to apply Newton's laws to problems involving two or more objects.

4. Be able to discuss qualitatively motion with a velocity-dependent drag force.

5. Be able to discuss qualitatively the use of Newton's laws in accelerating reference frames using pseudoforces.

6. Be able to apply Euler's method of numerical inte-

gration to solve problems in which the acceleration of a particle depends on its velocity or position.

B. Define, explain, or otherwise identify:

Static friction
Kinetic friction
Coefficient of static friction
Coefficient of kinetic friction
Coefficient of rolling friction
Drag force
Terminal speed
Pseudoforces
Centrifugal force
Coriolis force
Numerical integration
Euler's method

C. True or false: If the statement is true, explain why it is true. If it is false, give a counterexample.

1. The force of static friction always equals $\mu_s F_n$.

2. Friction always opposes the motion of an object.

3. The terminal speed of an object depends on its shape.

4. Pseudoforces exist only in noninertial reference frames.

5. The accuracy of numerical integration always increases when smaller time intervals are used.

Problems

Level I

5-1 Friction

1. A box weighing 600 N is pushed along a horizontal floor at constant velocity with a force of 250 N parallel to the floor. What is the coefficient of kinetic friction between the box and the floor?

2. The coefficient of static friction between the tires of a car and a horizontal road is $\mu_s = 0.6$. If the net force on the car is the force of static friction exerted by the road, (a) what is the maximum acceleration of the car? (b) What is the least distance in which the car can stop if it is initially traveling at 30 m/s?

3. The force that accelerates a car along a flat road is the frictional force exerted by the road on the car's tires. (a) Explain why the acceleration can be greater when the wheels do not spin. (b) If a car is to accelerate from 0 to 90 km/h in 12 s at constant acceleration, what is the minimum coefficient of friction needed between the road and tires?

4. A chair is sliding across a polished floor with an initial speed of 3 m/s. It comes to rest after sliding 2 m. What is the coefficient of kinetic friction between the floor and chair?

5. An 800-N box rests on a plane surface inclined at 30° to the horizontal. A physics student finds that she can prevent the box from sliding if she pushes on it with a force of at least 200 N parallel to the surface. (a) What is the coefficient of static friction between the box and the surface? (b) What is the greatest force that can be applied to the box parallel to the incline before the box slides up the incline?

6. A 5-kg block is held at rest against a vertical wall by a horizontal force of 100 N. (a) What is the frictional force exerted by the wall on the block? (b) What is the minimum horizontal force needed to prevent the block from falling if the coefficient of friction between the wall and the block is $\mu_s = 0.40$?

7. On a snowy day with the temperature near the freezing point, the coefficient of static friction between the tires and an icy road is 0.08. What is the maximum incline that a vehicle with four-wheel drive can climb with zero acceleration?

8. A 50-kg box that is resting on a level floor must be moved. The coefficient of static friction between the box and the floor is 0.6. One method to move the box is to push down on the box at an angle θ with the horizontal.

Another method is to pull up on the box at an angle θ with the horizontal. (a) Explain why one method is better than the other. (b) Calculate the force necessary to move the box by each method if $\theta = 30°$ and compare these results with those for $\theta = 0°$.

5-2 Drag Forces

9. What are the dimensions and SI units of the constant b in the retarding force bv^n (a) if $n = 1$, and (b) if $n = 2$?

10. A small pollution particle drops to earth in still air with a terminal speed of 0.3 mm/s. The particle has a mass of 10^{-10} g and a retarding force of the form bv. What is the value of b?

11. A Ping-Pong ball has a mass of 2.3 g and a terminal speed of 9 m/s. The retarding force is of the form bv^2. What is the value of b?

12. A sky diver of mass 60 kg can slow herself to a constant speed of 90 km/h by adjusting her form. (a) What is the magnitude of the upward drag force on the sky diver? (b) If the drag force is equal to bv^2 what is the value of b?

13. Newton showed that the air resistance on a falling object with a circular cross section should be approximately $\frac{1}{2}\rho\pi r^2 v^2$, where $\rho = 1.2$ kg/m^3 is the density of air. Find the terminal speed for a 56-kg sky diver, assuming that his cross-sectional area is equivalent to that of a disk of radius 0.30 m.

5-3 Problems with Two or More Objects

14. A hand pushes two bodies on a frictionless horizontal surface, as shown in Figure 5-21. The masses of the bodies are 2 and 1 kg. The hand exerts a force of 5 N on the 2-kg body. (a) What is the acceleration of the system? (b) What is the acceleration of the 1-kg body? What force is exerted on it? What is the origin of this force? (c) Show all the forces acting on the 2-kg body. What is the net force acting on this body?

Figure 5-21   Problem 14.

15. Two objects are connected by a massless string as shown in Figure 5-22. The incline and peg are frictionless. Find the acceleration of the objects and the tension in the string for (a) $\theta = 30°$ and $m_1 = m_2 = 5$ kg, and (b) for general values of θ, m_1, and m_2.

Figure 5-22 Problem 15.

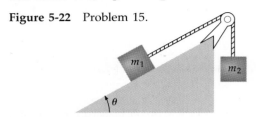

16. A 3-kg box rests on a horizontal table. It is attached to a 2-kg box by a light string, as shown in Figure 5-23. (a) What is the minimum coefficient of static friction such that the objects remain at rest? (b) If the coefficient of static friction is less than that found in part (a) and the coefficient of kinetic friction between the box and the table is 0.3, find the time for the 2-kg mass to fall 2 m to the floor if the system starts from rest.

Figure 5-23 Problem 16.

17. A block of mass m_1 is attached to a cord of length L_1, which is fixed at one end. The mass moves in a horizontal circle supported by a frictionless table. A second block of mass m_2 is attached to the first by a cord of length L_2 and also moves in a circle, as shown in Figure 5-24. If the period of the motion is T, find the tension in each cord.

Figure 5-24 Problem 17.

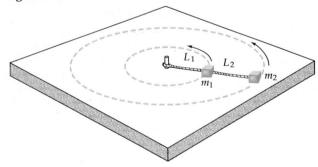

18. The apparatus shown in Figure 5-25, called *Atwood's machine*, is used to measure the acceleration due to gravity g by measuring the acceleration of the bodies. Assuming the string to be massless and the peg to be frictionless, show that the magnitude of the acceleration of either body and the tension in the string are

$$a = \frac{m_1 - m_2}{m_1 + m_2} g \quad \text{and} \quad T = \frac{2m_1m_2g}{m_1 + m_2}$$

Figure 5-25 Problems 18 and 50.

5-4 Pseudoforces

In Problems 19 through 23 the situations described take place in a boxcar that has an initial velocity $v = 0$ and an acceleration $\mathbf{a} = 5$ m/s^2 $\mathbf{i}$ (Figure 5-26). Work the exercises in the frame of the boxcar using pseudoforces and in an inertial frame using only real forces.

Figure 5-26 Boxcar initially at rest with a constant acceleration $\mathbf{a} = 5$ m/s^2 $\mathbf{i}$ to the right for Problems 19 through 23.

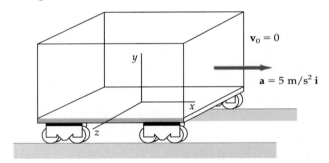

19. A 2-kg object slides along the frictionless floor of the boxcar with an initial velocity of 10 m/s $\mathbf{i}$ relative to the boxcar. (a) Describe the motion of the object. (b) When does the object reach its original position relative to the boxcar?

20. A 2-kg object slides along the frictionless floor of the boxcar with an initial transverse velocity of 10 m/s $\mathbf{k}$ relative to the boxcar. Describe the motion.

21. A 2-kg object slides along the floor of the boxcar, which is now assumed to be rough, with a coefficient of sliding friction of 0.3. The initial velocity of the object is 10 m/s $\mathbf{i}$ relative to the boxcar. Describe the motion of the object, assuming that the coefficient of static friction is 0.6.

22. A 2-kg object is suspended from the ceiling of the boxcar by a massless, unstretchable string. (a) What angle does the string make with the vertical? (b) Indicate all the forces acting on the object in each frame of reference.

23. A 6-kg object is attached to the ceiling of the boxcar by a massless spring of force constant 1000 N/m. By how much is the spring stretched?

5-5 Numerical Methods

24. An object falls under the influence of gravity and a drag force $F_d = -bv$. (a) Show that the acceleration of the object can be written

$$a = g\left(1 - \frac{v}{v_t}\right)$$

where $v_t = mg/b$ is the terminal speed. (b) Solve numerically to obtain a graph of v versus t and a graph of x versus t for $v_t = 60$ m/s.

Level II

25. An 800-kg car rolls down a very long, 6° grade. The drag force for motion of the car has the form $F_d = 100$ N + $(1.2$ N·s^2/m$^2)v^2$. What is the terminal velocity for the car rolling down this grade?

26. An 8-kg box and a 10-kg box that are attached by a rope that passes over a frictionless pulley slide on frictionless inclines, as shown in Figure 5-27. (a) Find the acceleration of the boxes and the tension in the rope. (b) The two boxes are replaced by two others of mass m_1 and mass m_2 such that there is no acceleration. Find whatever information you can about the mass of these two new boxes.

Figure 5-27 Problem 26.

27. The coefficient of static friction between the floor of a truck and a box resting on it is 0.30. The truck is traveling at 80 km/h. What is the least distance in which the truck can stop if the box is not to slide?

28. Two 100-kg boxes are dragged along a frictionless surface with a constant acceleration of 1 m/s^2, as shown in Figure 5-28. Each rope has a mass of 1 kg. Find the force **F** and the tension in the ropes at points A, B, and C.

Figure 5-28 Problem 28.

29. One end of a heavy rope of length 5 m and mass 4 kg is attached to the bottom of a 6-kg block on a frictionless horizontal table. At the other end of the rope, a constant force of 100 N is applied. (a) What is the acceleration of the system? (b) Give the tension in the rope as a function of position along the rope.

30. A block is on an incline whose angle can be varied. The angle is gradually increased from 0°. At 30°, the block starts to slide down the incline. It slides 3 m in 2 s. Calculate the coefficients of static and kinetic friction between the block and incline.

31. A platform scale calibrated in newtons is placed on the bed of a truck driven at a constant speed of 14 m/s. A box weighing 500 N is placed on the scale. Find the reading on the scale if (a) the truck passes over the crest of a hill with a radius of curvature 100 m, and (b) the truck passes through the bottom of a dip with a radius of curvature of 80 m.

32. A 100-g disk sits on a horizontally rotating turntable. The turntable makes one revolution each second. The disk is located 10 cm from the axis of rotation of the turntable. (a) What is the frictional force acting on the disk? (b) The disk will slide off the turntable if it is located at a radius larger than 16 cm from the axis of rotation. What is the coefficient of static friction?

33. In an amusement-park ride, riders stand against the wall of a spinning cylinder. The floor falls away and the riders are held up by friction. If the radius of the cylinder is 4 m, find the minimum number of revolutions per minute necessary when the coefficient of friction between a rider and the wall is 0.4.

34. Show with a force diagram how a motorcycle can travel in a circle on the inside vertical wall of a hollow cylinder. Assume reasonable parameters (coefficient of friction, radius of the circle, mass of the motorcycle, or whatever is required), and calculate the minimum speed needed.

35. The coefficient of friction between box A and the cart in Figure 5-29 is 0.6. The box has a mass of 2 kg. (a) Find the minimum acceleration a of the cart and box if the box is not to fall. (b) What is the magnitude of the frictional force in this case? (c) If the acceleration is greater than this minimum, will the frictional force be greater than in part (b)? Explain. (d) Show that for a box of any mass, the box will not fall if the acceleration is $a \geq g/\mu_s$, where μ_s is the coefficient of static friction.

Figure 5-29 Problem 35.

Cart A 2 kg   **a**

36. (a) A parachute creates enough air resistance to hold an 80-kg sky diver to a constant downward speed of

6.0 m/s after it has been open for a while. Assuming the force of air resistance is given by $f = bv^2$, calculate b for this case. (b) A sky diver drops in a free fall until his speed is 60 m/s before opening his parachute. If it opens instantaneously, calculate the initial upward force exerted by the parachute on the sky diver moving at 60 m/s. Explain why it is important that the parachute takes a few seconds to open.

37. A box of mass m rests on a horizontal table. The box is pulled by a massless rope with a force $\mathbf{F}$ at an angle θ as shown in Figure 5-30. The coefficient of static friction is 0.6. The minimum value of the force needed to move the box depends on the angle θ. (a) Discuss qualitatively how you would expect this force to depend on θ. (b) Compute the force for the angles $\theta = 0, 10, 20, 30, 40, 50$, and $60°$, and make a plot of F versus θ for $mg = 400$ N. From your plot, at what angle is it most efficient to apply the force to move the box?

Figure 5-30   Problems 37 and 49.

38. (a) Estimate the maximum acceleration of a car if 40 percent of its weight is on the tires that are powered by the engine, and if the coefficient of static friction is 0.7. (b) What is the shortest possible time in which this car can achieve a speed of 100 km/h? Assume the engine has unlimited power.

39. A road on a college campus has a sharp, level curve with a radius of 30 m that has a speed limit of 30 km/h. A serious accident occurs at this curve. Upon investigating the accident, a physics professor discovers that the curve has become covered with small, gravel-sized particles that act like miniature ball bearings and reduce the effective coefficient of static friction between the average tire and road from 0.7 to 0.2. (a) What is the maximum safe speed one can drive on this curve under normal conditions (with no gravel)? (b) What is the maximum safe speed one can drive on the curve when the road is covered with the gravel?

40. Consider the block of mass m_2 in Figure 5-31 to be a variable mass that can be adjusted until the block of mass m_1 is on the verge of sliding on the table. (a) If the critical mass m_2 is 5 kg and mass m_1 is 7 kg, what is the coefficient of static friction between the table and the block? (b) With a slight nudge, the system accelerates at 1 m/s². What is the coefficient of kinetic friction between the table and the block?

Figure 5-31   Problem 40.

41. A 100-kg mass is pulled along a frictionless surface by a force $\mathbf{F}$ such that its acceleration is 6 m/s² (see Figure 5-32). A 20-kg mass slides along the top of the 100-kg mass and has an acceleration of 4 m/s². (It thus slides back relative to the 100-kg mass.) (a) What is the frictional force exerted by the 100-kg mass on the 20-kg mass? (b) What is the net force acting on the 100-kg mass? What is the force F? (c) After the 20-kg mass falls off the 100-kg mass, what is the acceleration of the 100-kg mass?

Figure 5-32   Problem 41.

42. A 60-kg block slides along the top of a 100-kg block with an acceleration of 3 m/s² when a horizontal force F of 320 N is applied, as shown in Figure 5-33. The 100-kg block sits on a horizontal frictionless surface, but there is friction between the two blocks. (a) Find the coefficient of kinetic friction between the blocks. (b) Find the acceleration of the 100-kg block during the time that the 60-kg block maintains contact.

Figure 5-33   Problem 42.

43. Two bodies of mass m_1 and m_2 rest on a horizontal frictionless table, as shown in Figure 5-34. A force $\mathbf{F}$ is applied as shown. (a) If $m_1 = 2$ kg, $m_2 = 4$ kg, and $F = 3$ N, find the acceleration of the bodies and the contact force F_c exerted by one body on the other. (b) Find the contact force for general values of the masses of the bodies and show that if $m_2 = nm_1$, $F_c = nF/(n + 1)$.

Figure 5-34   Problem 43.

44. Two climbers on an icy (frictionless) slope, tied together by a 30-m rope, are in the predicament shown in

Figure 5-35 Problem 44.

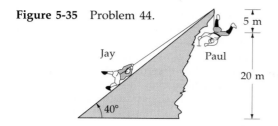

Jay

Paul

5 m

20 m

40°

Figure 5-35. At time $t = 0$ the speed of each is zero, but the top climber, Paul (mass 52 kg), has taken one step too many and his friend Jay (mass 74 kg) has dropped his pick. (a) Find the tension in the rope as Paul falls and his speed just before he hits the ground. (b) If Paul unties his rope after hitting the ground, find Jay's speed as he hits the ground. Assume the rope's length to be 30 m.

45. A 4-kg block resting on a 30° incline is attached to a second block of mass m by a cord that passes over a smooth peg, as shown in Figure 5-36. The coefficient of static friction between the block and the incline is 0.4. (a) Find the range of possible values for m for which the system will be in static equilibrium. (b) If $m = 1$ kg, the system will be in static equilibrium. What is the frictional force on the 4-kg block in this case?

Figure 5-36 Problem 45.

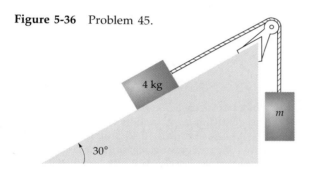

4 kg

m

30°

46. An automobile is going up a grade of 15° at a speed of 30 m/s. The coefficient of static friction between the tires and the road is 0.7. (a) What minimum distance does it take to stop the car? (b) What minimum distance would it take if the car were going down the grade?

Level III

47. A 2-kg block sits on a 4-kg block that is resting on a frictionless table (Figure 5-37). The coefficients of friction between the blocks are $\mu_s = 0.3$ and $\mu_k = 0.2$. (a) What is the maximum force F that can be applied if the 2-kg block is not to slide on the 4-kg block? (b) If F has half this value, find the acceleration of each block and the force of friction acting on each block. (c) If F is twice the value found in (a), find the acceleration of each block.

Figure 5-37 Problem 47.

2 kg

4 kg

F

48. In Figure 5-38, the mass $m_2 = 10$ kg slides on a frictionless table. The coefficients of static and kinetic friction between m_2 and $m_1 = 5$ kg are $\mu_s = 0.6$ and $\mu_k = 0.4$. (a) What is the maximum acceleration of m_1? (b) What is the maximum value of m_3 if m_1 moves with m_2 without slipping? (c) If $m_3 = 30$ kg, find the acceleration of each body and the tension in the string.

Figure 5-38 Problem 48.

m_1

m_2

m_3

49. A box of mass m rests on a horizontal table. The coefficient of static friction is μ_s. A force $\mathbf{F}$ is applied at an angle θ as shown in Figure 5-30. (a) Find the force F needed to move the box as a function of angle θ. (b) At the angle θ for which this force is minimum, the slope $dF/d\theta$ of the curve F versus θ is zero. Compute $dF/d\theta$ and show that this derivative is zero at the angle θ that obeys $\tan \theta = \mu_s$. Compare this general result with that obtained in Problem 37.

50. Using the Atwood's machine illustrated in Figure 5-25 (Problem 18), the acceleration of gravity g is determined by measuring the time t for the larger body of mass m_1 to fall a distance L. (a) Find an expression for g in terms of m_1, m_2, L, and t. (b) Show that if there is a small error in the time measurement dt, it will lead to an error in the determination of g by an amount dg given by $dg/g = -2\ dt/t$. If m_1 is 1 kg, find the value of m_2 such that g can be measured with an accuracy of ± 5 percent with a watch that is accurate to 0.1 s if $L = 3$ m. Assume that the only important uncertainty in the measurement is the time of fall.

51. A student rides a bicycle on a horizontal surface in a circle with a radius of 20 m. The bicycle and student are leaning inward at an angle of 15° with respect to the vertical. (a) What is the speed of the student? (b) If the frictional force is half its maximum value, what is the coefficient of static friction?

52. An object falls under the influence of gravity and a drag force $F_d = -bv$, as in Problem 24. (a) Show that the acceleration of the object can be written

$$a = \frac{dv}{dt} = g - \frac{b}{m} v$$

(b) Rearrange this equation to obtain

$$\frac{dv}{g - (b/m)v} = dt$$

(c) Integrate this equation to obtain the exact solution

$$v = \frac{mg}{b}(1 - e^{-bt/m}) = v_t(1 - e^{-gt/v_t})$$

(d) Plot v versus t for $v_t = 60$ m/s, and compare this with the numerical solution obtained in Problem 24.

53. Consider a bead of mass m that is free to move on a thin, circular wire of radius r. The bead is given an initial speed v_0, and there is a coefficient of sliding friction μ_k. The experiment is performed in a spacecraft drifting in space. Find the speed of the bead at any subsequent time t.

54. In Problem 53, (a) find the centripetal acceleration of the bead. (b) Find the tangential acceleration of the bead. (c) What is the magnitude of the resultant acceleration?

55. A curve of radius 30 m is banked so that a car traveling 40 km/h can round it even if the road is so icy that the coefficient of static friction is approximately zero. Find the range of speeds at which a car can travel around this curve without skidding if the coefficient of static friction between the road and the tires is 0.3.

56. Small spherical particles experience a viscous drag force given by Stokes' law, which is given by $F_d = 6\pi\eta rv$, where r is the radius of the particle, v is its speed, and η is the dynamic viscosity of the air or fluid medium. (a) Estimate the terminal speed of a spherical pollution particle of radius 10^{-5} m and density of 2000 kg/m³. Assume that the air is still and that η is 1.8×10^{-5} N·s/m². (b) Estimate the time it takes for such a particle to fall from a 100 m smokestack.

57. Figure 5-39 shows a 20-kg block sliding on a 10-kg block. All surfaces are frictionless. Find the acceleration of each block and the tension in the string that connects the blocks.

Figure 5-39   Problem 57.

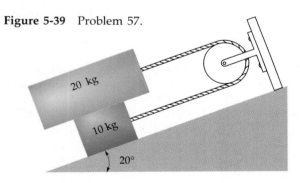

58. A 20-kg block with a pulley attached slides along a frictionless table. It is connected by a massless string to a 5-kg block, as shown in Figure 5-40. Find the acceleration of each block and the tension in the connecting string.

Figure 5-40   Problem 58.

59. A space station has two compartments, as shown in Figure 5-41. The station rotates at B rev/min. (a) A mass m rests on the floor of one of the compartments a distance r from the center of rotation as shown in the figure. What is the normal force exerted by the floor on the mass? (b) The mass is now dropped from the ceiling of the compartment. Describe its motion relative to the compartment. What forces (including pseudoforces) act on the mass while it is falling? (c) Explain qualitatively why the mass falls to the floor from the point of view of an inertial reference frame in which there are no forces acting on the mass.

Figure 5-41   Problem 59.

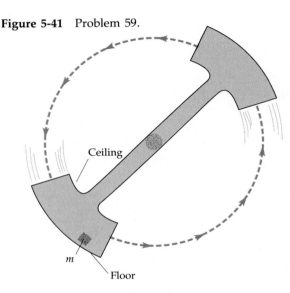

60. A 10-kg block rests on a 5-kg bracket as shown in Figure 5-42. The coefficients of friction between the 10-kg block and the bracket are $\mu_s = 0.40$ and $\mu_k = 0.30$. The bracket sits on a frictionless surface. (a) What is the maximum force F that can be applied if the 10-kg block is not to slide on the bracket? (b) What is the corresponding acceleration of the bracket?

Figure 5-42   Problem 60.

Chapter 6

Work and Energy

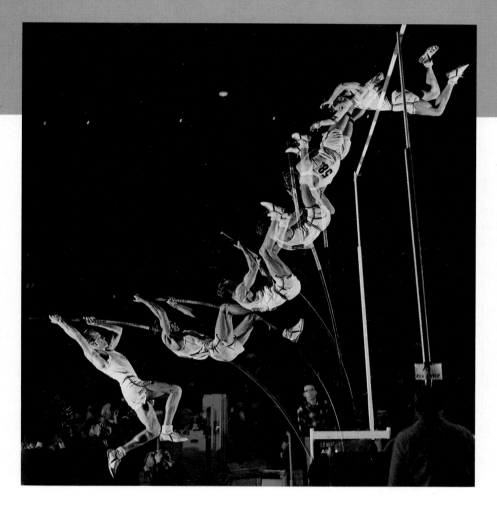

The action of a pole vaulter demonstrates several kinds of energy. First he transforms the internal chemical energy of his body into kinetic energy as he runs. Some of this kinetic energy is then converted into elastic potential energy, as shown in the deformation of the pole; and the rest is eventually converted into gravitational potential energy, which in turn is converted into kinetic energy as he drops. Mechanical energy is finally converted into heat energy when he reaches the ground.

Work and energy are among the most important concepts in physics, and they play important roles in our everyday life, as well. In physics, **work** is given a precise definition that differs from our everyday usage. Work is done on an object by a force only when the point of application of the force moves through a distance and there is a component of the force along the line of motion. Thus, when you exert a force on a sled and thereby pull it across the snow, you do work on the sled. However, if the sled were immobilized (for example, if it were chained to a tree) and you exert the same force on it, no work would be done on the sled because the point of application of the force does not move through a distance.

Closely associated with the concept of work is the concept of **energy**, which is the capacity to do work. When work is done by one system on another, energy is transferred between the two systems. For example, when you pull a sled, the work you do goes partly into the energy of motion of the sled, called its kinetic energy, and partly into the thermal energy that arises from the friction between the sled and the snow. At the same time, the internal chemical energy of your body decreases as you pull the sled. The net result is the transfer of the internal chemical energy of your body to the external kinetic energy of the sled plus thermal energy. One of the most important principles in science is the conservation of energy: The total energy of a system and its surroundings does not change. When the energy of a system decreases, there is always a corresponding increase in the energy of its surroundings or of another system.

There are many forms of energy. Kinetic energy is associated with the motion of an object. Potential energy is stored energy associated with the configuration of a system, such as the separation distance between some object and the earth. Thermal energy is associated with the motion of the molecules within a system and is closely connected with the temperature of the system.

In this chapter, we will explore the concepts of work, kinetic energy, and potential energy, and we will see how to use the law of conservation of energy to solve various problems. Later, when we study thermodynamics, we will look at heat energy, which is energy that is transferred due to differences in temperature, and we will explore the internal molecular energy of systems in more depth.

6-1   Work and Kinetic Energy: Motion in One Dimension with Constant Forces

We define the work done by a force on an object as the product of the force and the displacement of the point on which the force acts. If the force and the displacement are in different directions, then only the component of the force in the direction of the displacement does work. For simplicity, we will first consider the special case of constant forces and motion in one dimension. We will also assume that the object can be considered to be a particle. This eliminates such complications as the deformation of the object, in which case the displacement is different for different parts of the object. For a particle, the displacement of the point of application of the force is the same as the displacement of the particle.

If a force **F** makes an angle θ with the displacement Δx, as in Figure 6-1, the work done is

Figure 6-1 When a constant force **F** acts on a particle at an angle θ through a distance Δx, the work done on the particle is $F \cos \theta \, \Delta x = F_x \, \Delta x$.

Work by a constant force

$$W = F \cos \theta \, \Delta x = F_x \, \Delta x \qquad \text{6-1}$$

Work is a scalar quantity that is positive if Δx and F_x have the same signs and negative if they have opposite signs. The dimensions of work are those of force times distance. The SI unit of work and energy is the **joule** (J), which equals the product of a newton and a meter:

$$1 \text{ J} = 1 \text{ N·m} \qquad \text{6-2}$$

In the U.S. customary system, the unit of work is the **foot-pound.** The relation between these units of work is easily found using the relations between pounds and newtons and between meters and feet. The result is

$$1 \text{ J} = 0.738 \text{ ft·lb} \qquad \text{6-3}$$

Exercise

A force of 12 N is exerted on a box at an angle of $\theta = 20°$, as in Figure 6-1. How much work is done by the force as the box moves along the table a distance of 3 m?   (Answer: 33.8 J)

Figure 6-2 shows a boy pulling a sled across the snow by exerting a force **F** at an angle θ to the horizontal. For this discussion, we will assume that friction can be neglected. The other forces acting on the sled are then its weight $m\mathbf{g}$ and a vertical support force $\mathbf{F_n}$. The work done by the weight is zero because this force is perpendicular to the direction of motion. Similarly the support force $\mathbf{F_n}$ is perpendicular to the direction of motion and therefore does no work. The only force that does work on the sled is the horizontal force $F \cos \theta$. When there are several forces that do work on a particle, the total work is found by computing the work done by each force and summing. For a particle, the same result is obtained by summing all the forces to find the net force and then computing the work done by the net force. This is true because the displacement of the point of application is the same for each force and is equal to the displacement of the particle. In this case, the vertical upward force $F \sin \theta + F_n$ must equal the weight mg because the sled does not accelerate vertically. The net force acting on the sled is thus the horizontal force $F \cos \theta$.

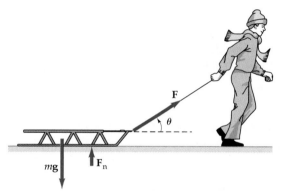

Figure 6-2   Boy pulling a sled. The work done by the boy as he pulls the sled a distance Δx is $F \cos \theta \, \Delta x$.

There is an important relation between the total work done on a particle and the initial and final speeds of the particle. If F_x is the net force acting on a particle, Newton's second law gives

$$F_x = ma_x$$

For a constant force, the acceleration is constant, and we can relate the distance the particle moves to its initial and final speed by using the constant-acceleration formula (Equation 2-13). If the initial speed is v_i and the final speed is v_f, we have

$$v_f^2 = v_i^2 + 2a_x \, \Delta x \qquad\qquad 6\text{-}4$$

Since the work done by the net force equals the total work done on the particle,

$$W_{total} = F_x \, \Delta x = ma_x \, \Delta x$$

Substituting $\frac{1}{2}(v_f^2 - v_i^2)$ for $a_x \, \Delta x$ we obtain

$$W_{total} = \tfrac{1}{2}mv_f^2 - \tfrac{1}{2}mv_i^2 \qquad\qquad 6\text{-}5$$

The quantity $\frac{1}{2}mv^2$ is called the **kinetic energy** K of the particle. It is a scalar quantity that depends on the particle's mass and speed:

$$K = \tfrac{1}{2}mv^2 \qquad\qquad 6\text{-}6 \qquad \textit{Kinetic energy defined}$$

The quantity on the right side of Equation 6-5 is the change in the kinetic energy of the particle, that is, the kinetic energy $\frac{1}{2}mv_f^2$ at the end of the interval minus the kinetic energy $\frac{1}{2}mv_i^2$ at the beginning of the interval. The total work done on the particle is therefore equal to the change in the kinetic energy of the particle:

Work–energy theorem

$$W_{total} = \Delta K = \tfrac{1}{2}mv_f^2 - \tfrac{1}{2}mv_i^2 \qquad 6\text{-}7$$

The total work done on a particle equals the change in the kinetic energy of the particle.

This result is known as the **work–energy theorem.** It holds whether the net force is constant or not, as we will see in the next section.

Exercise

A girl of mass 50 kg is running at 3.5 m/s. What is her kinetic energy? (Answer: 306 J)

Example 6-1

A 4-kg box is raised from rest a distance of 3 m by an upward applied force of 60 N. Find (*a*) the work done by the applied force, (*b*) the work done by gravity, and (*c*) the final speed of the box.

(*a*) The applied force is in the direction of motion ($\theta = 0°$), so the work done by it is positive:

$$W_{app} = F_{app} \cos 0° \, \Delta y = (60 \text{ N})(1)(3 \text{ m}) = 180 \text{ J}$$

(*b*) The force of gravity is opposite the direction of motion ($\theta = 180°$), so the work done by gravity is negative:

$$W_g = mg \cos 180° \, \Delta y = (4 \text{ kg})(9.81 \text{ N/kg})(-1)(3 \text{ m}) = -118 \text{ J}$$

The total work done on the box is thus $W_{total} = 180 \text{ J} - 118 \text{ J} = 62 \text{ J}$. Applying the work–energy theorem with $v_i = 0$, we obtain

$$W_{total} = 62 \text{ J} = \tfrac{1}{2}mv_f^2 - \tfrac{1}{2}mv_i^2 = \tfrac{1}{2}mv_f^2$$

(*c*) The final speed of the box is then

$$v_f = \sqrt{\frac{2W_{total}}{m}} = \sqrt{\frac{2(62 \text{ J})}{4 \text{ kg}}} = 5.57 \text{ m/s}$$

Example 6-2

If the mass of the sled in Figure 6-2 is 5 kg and the boy exerts a force of 12 N at 30°, find the work done by the boy and the final speed of the sled after it moves 3 m, assuming that it starts from rest and there is no friction.

The forces are shown in the figure. The vertical forces are the force due to gravity,

$$mg = (5 \text{ kg})(9.81 \text{ N/kg}) = 49.0 \text{ N}$$

the upward component of the force exerted by the rope,

$$F_y = (12 \text{ N}) \sin 30° = 6 \text{ N}$$

and the vertical support force exerted by the ground, which equals $mg - F_y$ because there is no vertical acceleration,

$$F_n = mg - F_y = 49.0 \text{ N} - 6 \text{ N} = 43.0 \text{ N}$$

The only horizontal force is

$$F_x = (12 \text{ N}) \cos 30° = 10.4 \text{ N}$$

The total work done on the sled is the product of the component of **F** in the direction of motion (10.4 N) and the distance traveled (3 m):

$$W_{total} = F_x \, \Delta x = (10.4 \text{ N})(3 \text{ m}) = 31.2 \text{ J}$$

The total work done on the sled equals the change in its kinetic energy. If the sled starts from rest, its kinetic energy after traveling 3 m is therefore 31.2 J. Its speed can be found from

$$W_{total} = \Delta K = \tfrac{1}{2}mv_f^2 - \tfrac{1}{2}mv_i^2 = \tfrac{1}{2}mv_f^2 - 0$$

$$\tfrac{1}{2}mv_f^2 = K = 31.2 \text{ J}$$

$$v_f = \sqrt{\frac{2K}{m}} = \sqrt{\frac{2(31.2 \text{ J})}{5 \text{ kg}}} = 3.53 \text{ m/s}$$

Atlas does no work holding up the earth because there is no motion.

We could have found the final speed in Examples 6-1 and 6-2 by first finding the acceleration of the box or sled and then using the constant-acceleration formulas. Setting the total work equal to the change in kinetic energy provides an alternative method to using Newton's laws of motion to solve mechanics problems. When the forces are not constant, this method is often much easier than using Newton's laws.

We have seen that for work to be done on a particle, the particle must move through a distance. But what if we are holding a weight at a fixed position a distance h off the floor, as the man in Figure 6-3 is doing? In everyday usage, we might say that it takes work to do this, but using our scientific definition, no work is done by a force acting on a stationary object. We could eliminate the effort of holding the weight merely by tying the rope to some other object, in which case the weight would be supported with no help from us. Although no external work is done when we hold a weight in a fixed position, we do become tired from this activity. While we hold the weight, nerve impulses continually trigger contractions of muscle fibers inside our body. As the muscle fibers contract and relax, work is done on the molecular level inside our body. In this process, some of our internal chemical energy is converted into thermal energy.

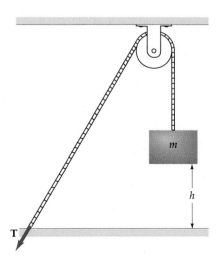

Figure 6-3 No external work is done by the woman holding the weight at a fixed position. The same task could be accomplished by tying the rope to a fixed point.

Questions

1. A heavy box is to be moved from the top of one table to the top of another table of the same height on the other side of the room. Is work required to do this?

2. To get out of bed in the morning, do you have to do work?

3. By what factor does the kinetic energy of a car change when its speed is doubled?

4. An object moves in a circle at constant speed. Does the force that accounts for its acceleration do work on it? Explain.

5. Is it possible to exert a force that does work on an object without increasing the kinetic energy of the object? If so, give an example.

Figure 6-4 The work done by a constant force is represented graphically as the area under the F_x-versus-x curve.

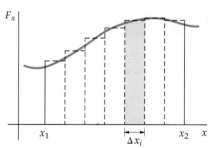

Figure 6-5 A variable force can be approximated by a series of constant forces acting over small intervals. The work done by one of these constant forces is the rectangular area beneath that force. The total work done by the variable force from x_1 to x_2 is the total area under the curve for this interval.

Work by a variable force

6-2   Work Done by a Variable Force

In the previous section, we defined the work done by a constant force F_x as $W = F_x \, \Delta x$. In Figure 6-4, we have plotted a constant force F_x as a function of position x. The work done on a particle whose displacement is Δx is represented by the area under the force-versus-position curve, which is indicated by the shading in the figure. However, many forces encountered in nature are not constant but vary with position. For example, when you stretch a spring, the force exerted by the spring is proportional to the amount the spring is stretched. Similarly, the gravitational force exerted by the earth on a spaceship varies inversely with the square of the distance from the spaceship to the center of the earth. We can use the graphical representation of work as the area under a force-versus-position curve to extend our definition of work to cases in which the force varies with position.

Figure 6-5 shows a variable force F_x as a function of position x. In the figure, we have divided the interval from x_1 to x_2 into a set of smaller intervals Δx_i. If each interval is small enough, we can approximate the variable force by a series of constant forces, as shown in the figure. For each interval, the work done by the constant force is the area of the rectangle beneath the force. The sum of these rectangular areas is the sum of the work done by the set of constant forces that approximates the varying force. As can be seen in the figure, this area is approximately equal to the area under the curve. In the limit of infinitesimally small Δx_i, the sum of the areas of the rectangles equals the area under the curve. We can therefore define the work done by a variable force to be the area under the F_x-versus-x curve for that force. (Note that this is the same procedure we used in Chapter 2 when we found the displacement Δx to be the area under the v-versus-t curve.)

$$W = \lim_{\Delta x_i \to 0} \sum_i F_x \, \Delta x_i = \text{area under the } F_x\text{-versus-}x \text{ curve} \qquad 6\text{-}8$$

This limit is the integral of F_x over x. Thus, the work done by a variable force F_x acting on a particle as it moves from point x_1 to point x_2 is

$$W = \int_{x_1}^{x_2} F_x \, dx = \text{area under the } F_x\text{-versus-}x \text{ curve} \qquad 6\text{-}9$$

We now show that the total work done on a particle equals the change in kinetic energy of the particle even when the net force varies with position. Consider Figure 6-5 again. For each rectangular area, the force is constant, so the work done by the constant force for that interval equals the change in the kinetic energy over that interval. When we sum over all the intervals, we sum the changes in kinetic energy for all the intervals, which gives us the total or net change in kinetic energy for the entire interval from x_1 to x_2. The total area under the curve therefore equals the sum of the changes in the kinetic energy for each interval Δx_i, which in turn equals the net change in the kinetic energy for the complete interval, $\Delta K = \frac{1}{2}mv_2^2 - \frac{1}{2}mv_1^2$. This result is the same as Equation 6-7, provided that W_{total} is interpreted to be the integral given by Equation 6-9.

Example 6-3

A 4-kg block on a frictionless table (Figure 6-6a) is attached to a horizontal spring that obeys Hooke's law and exerts a force $F_x = -kx$, where x is measured from the equilibrium position of the block and the force constant $k = 400$ N/m. The spring is compressed to $x_1 = -5$ cm. Find (a) the work done by the spring on the block as the block moves from $x_1 = -5$ cm to its equilibrium position $x_2 = 0$ and (b) the speed of the block at $x_2 = 0$.

(a) Figure 6-6b is a graph of this force versus distance. The work done on the block as it moves from x_1 to $x_2 = 0$ equals the triangular area under the F_x-versus-x curve between these limits, which is shaded in the figure. The area of this triangle is one-half its base times its height. The base is 5 cm = 0.05 m, and the height is the value of the force at x_1, which is

$$F_x = -kx = -(400 \text{ N/m})(-0.05 \text{ m}) = +20 \text{ N}$$

The work done is thus

$$W = \int_{-0.05 \text{ m}}^{0} F_x \, dx = \frac{1}{2}(0.05 \text{ m})(20 \text{ N}) = 0.500 \text{ N·m} = 0.500 \text{ J}$$

The work is positive because the force is in the direction of motion. This is indicated in the figure by the fact that the area is above the x axis. We can also compute this work using integral calculus:

$$W = \int_{x_1}^{0} F_x \, dx = \int_{x_1}^{0} -kx \, dx = -k\int_{x_1}^{0} x \, dx$$

$$= -\frac{1}{2}kx^2 \Big|_{x_1}^{0} = \frac{1}{2}kx_1^2 = \frac{1}{2}(400 \text{ N/m})(0.05 \text{ m})^2 = 0.500 \text{ J}$$

(b) We find the speed of the block from the fact that its kinetic energy is 0.500 J:

$$K = \frac{1}{2}mv^2 = 0.500 \text{ J}$$

The speed is then

$$v = \sqrt{\frac{2K}{m}} = \sqrt{\frac{2(0.500 \text{ J})}{4 \text{ kg}}} = 0.50 \text{ m/s}$$

Figure 6-6 (a) Block attached to a spring for Example 6-3. The spring is compressed from equilibrium and released. (b) Plot of the force F_x exerted by the spring versus x. The work done by the spring as the block moves from x_1 to x_2 is the shaded area.

(a)

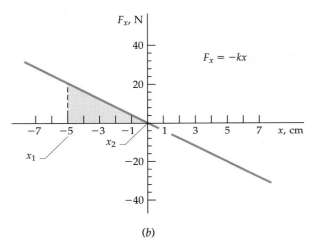

(b)

Note that we could not have found this result by first finding the acceleration and then using the constant-acceleration equations. The force varies with position, so the acceleration also varies.

Since the force exerted on the block by the spring is $-kx$ whether x is positive or negative, the work done by a spring on a block as the block moves from some initial compression or extension x to its equilibrium position $x = 0$ is given by

$$W = \int_x^0 -kx\ dx = \tfrac{1}{2}kx^2 \qquad\qquad 6\text{-}10$$

Example 6-4

A force F_x varies with x as shown in Figure 6-7. Find the work done by the force on a particle as the particle moves from $x = 0$ to $x = 6$ m.

We find the work done by calculating the area under the F_x-versus-x curve. From 0 to 4 m, the force is constant and the area is that of a rectangle, (5 N)(4 m) = 20 J. From 4 m to 6 m, the force is decreasing at a constant rate and the area is that of the triangle shown in the figure: $\tfrac{1}{2}$(5 N)(2 m) = 5 J. The total area is 20 J + 5 J = 25 J, which equals the work done by the force.

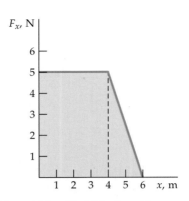

Figure 6-7 Variable force for Example 6-4. The work done equals the shaded area.

Exercise

The force shown in Figure 6-7 is the only force that acts on a particle of mass 3 kg. If the particle starts from rest at $x = 0$, how fast is it moving when it reaches $x = 6$ m?   (Answer: 4.08 m/s)

Question

6. How does the work required to stretch a spring 2 cm compare with that required to stretch it 1 cm?

Wind farm at Altamount pass in California uses hundreds of windmills to convert wind energy into electrical energy.

6-3   Work and Energy in Three Dimensions and the Dot Product

The work done by a force is defined to be the product of the displacement of the point of application of the force and the component of the force in the direction of the displacement. The component of the force in the direction of the displacement is important because it changes the speed of the particle. Figure 6-8 shows a particle of mass m that is acted on by a single force $\mathbf{F}$ as it moves along an arbitrary curve in space. We consider a small displacement along the curve. The force $\mathbf{F}$ makes an angle ϕ with the displacement. By Newton's second law, the tangential component of this force F_s is related to the rate of change of speed of the particle, $F_s = m\, dv/dt$, and the perpendicular component is related to the centripetal acceleration, $F_\perp = mv^2/r$, where r is the radius of curvature of the path at this point. The perpendicular component does not affect the speed of the particle. Instead, it changes the direction of the particle's velocity. The work done by the force on the particle for a small displacement Δs is

$$\Delta W = F_s \, \Delta s$$

To find the work done by the force as the particle moves along the curve from point 1 to point 2, we compute the product $F_s \, \Delta s$ for each element of the path and sum. In the limit of smaller and smaller displacement elements, this sum becomes an integral:

$$W = \int_{s_1}^{s_2} F_s \, ds$$

As in the case of one-dimensional motion, we can show that the total work done on the particle equals the change in its kinetic energy. From Newton's second law, we have

$$F_s = m \frac{dv}{dt}$$

If we think of the speed as a function of the distance s measured along the curve, we can apply the chain rule for derivatives (rule 3 in Table A-4 of Appendix A):

$$\frac{dv}{dt} = \frac{dv}{ds}\frac{ds}{dt} = v\frac{dv}{ds}$$

where we have used the fact that ds/dt is the speed v. The work done by the net force, which for a particle is the total work, is thus

$$W_{\text{total}} = \int_{s_1}^{s_2} F_s \, ds = \int_{s_1}^{s_2} m \frac{dv}{dt} \, ds = \int_{s_1}^{s_2} mv \frac{dv}{ds} \, ds = \int_{v_1}^{v_2} mv \, dv$$

or

$$W_{\text{total}} = \int_{s_1}^{s_2} F_s \, ds = \tfrac{1}{2}mv_2^2 - \tfrac{1}{2}mv_1^2 \qquad \text{6-11}$$

Work–energy theorem in three dimensions

Equation 6-11, along with its one-dimensional counterpart, Equation 6-7, follows directly from the definition of work and from Newton's second law of motion.

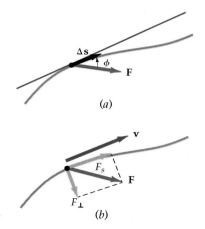

Figure 6-8   (*a*) A particle moving along an arbitrary curve in space. (*b*) The perpendicular component of the force $F_\perp$ changes the direction of the particle's motion but not its speed. The tangential component F_s changes the particle's speed but not its direction. F_s equals the mass m times the tangential acceleration dv/dt. Only this component does work.

The Dot Product

If ϕ is the angle between the force **F** acting on a particle traveling along a curve in space and a small displacement $d\mathbf{s}$ along that curve, the component of the force parallel to $d\mathbf{s}$ is $F_s = F \cos \phi$. The work done by the force over the small displacement is then

$$dW = F_s \, ds = (F \cos \phi) \, ds$$

This type of scalar combination of two vectors and the cosine of the angle between them occurs often in physics, and is called the **scalar product** of the vectors. The scalar product of two general vectors **A** and **B** is written **A·B** and defined as

Dot product defined

$$\mathbf{A \cdot B} = AB \cos \phi \qquad\qquad 6\text{-}12$$

where ϕ is the angle between **A** and **B**. Because of the way it is written, the scalar product is also called the **dot product**. The dot product **A·B** can be thought of as the product of A and $B \cos \phi$, which is the component of **B** in the direction of **A**, or alternately as the product of B and $A \cos \phi$, which is the component of **A** in the direction of **B** (Figure 6-9).

If **A·B** = 0, then **A** = 0 or **B** = 0 or **A** and **B** are perpendicular. If **A** and **B** are perpendicular, their dot product is zero because $\phi = 90°$ and $\cos 90° = 0$. If **A** and **B** are parallel, the dot product is just the product of their magnitudes. The dot product of a vector with itself is the square of the magnitude of the vector:

$$\mathbf{A \cdot A} = A^2$$

It follows from the definition of the dot product that it is independent of the order of multiplication; that is, **A·B** = **B·A**. This is known as the *commutative rule of multiplication*. The dot product also obeys the *distributive rule of multiplication*:

$$(\mathbf{A + B}) \cdot \mathbf{C} = \mathbf{A \cdot C} + \mathbf{B \cdot C}$$

Figure 6-9 Geometric representation of the dot product **A·B**. We can think of this as the product of A times $B \cos \phi$ or B times $A \cos \phi$.

We sometimes wish to write the dot product in terms of the rectangular components of the two vectors. Consider the dot product of the vectors $\mathbf{A} = A_x\mathbf{i} + A_y\mathbf{j} + A_z\mathbf{k}$ and $\mathbf{B} = B_x\mathbf{i} + B_y\mathbf{j} + B_z\mathbf{k}$. Since the unit vectors **i**, **j**, and **k** are mutually perpendicular, the dot product of two different unit vectors is zero; that is, $\mathbf{i \cdot j} = \mathbf{i \cdot k} = \mathbf{j \cdot k} = 0$. In addition, the dot product of a unit vector with itself is 1; that is, $\mathbf{i \cdot i} = \mathbf{j \cdot j} = \mathbf{k \cdot k} = 1$. Thus, for the dot product **A·B**, we have

$$\mathbf{A \cdot B} = (A_x\mathbf{i} + A_y\mathbf{j} + A_z\mathbf{k}) \cdot (B_x\mathbf{i} + B_y\mathbf{j} + B_z\mathbf{k})$$
$$= A_xB_x + A_yB_y + A_zB_z \qquad\qquad 6\text{-}13$$

In terms of the dot-product notation, the work dW done by a force **F** during a small displacement $d\mathbf{s}$ is written

$$dW = F \cos \phi \, ds = \mathbf{F} \cdot d\mathbf{s} \qquad\qquad 6\text{-}14$$

and the work done on the particle as it moves from point 1 to point 2 is written

Work—general definition

$$W = \int_{s_1}^{s_2} \mathbf{F} \cdot d\mathbf{s} \qquad\qquad 6\text{-}15$$

When several forces $\mathbf{F}_i$ act on a particle whose displacement is $d\mathbf{s}$, the total work is

$$\Delta W_{total} = \mathbf{F}_1 \cdot d\mathbf{s} + \mathbf{F}_2 \cdot d\mathbf{s} + \cdots = \sum_i \mathbf{F}_i \cdot d\mathbf{s} = \left(\sum_i \mathbf{F}_i \right) \cdot d\mathbf{s}$$

or

$$\Delta W_{total} = \mathbf{F}_{net} \cdot d\mathbf{s} \qquad\qquad 6\text{-}16$$

Example 6-5

A particle is given a displacement $\Delta\mathbf{s} = 2\text{ m i} - 5\text{ m j}$ along a straight line. During the displacement, a constant force $\mathbf{F} = 3\text{ N i} + 4\text{ N j}$ acts on the particle. Find the work done by the force and the component of the force in the direction of the displacement.

The work done by the force is

$$W = \mathbf{F} \cdot \Delta\mathbf{s} = (3\text{ N i} + 4\text{ N j}) \cdot (2\text{ m i} - 5\text{ m j})$$

$$= 6\text{ N·m} - 20\text{ N·m} = -14\text{ N·m}$$

In terms of the magnitudes of the force and displacement, the work is

$$W = \mathbf{F} \cdot \Delta\mathbf{s} = F \cos\phi \, \Delta s$$

The component of the force in the direction of the displacement is $F\cos\phi$, which equals the work divided by the magnitude of the displacement. We find Δs from

$$\Delta\mathbf{s} \cdot \Delta\mathbf{s} = (\Delta s)^2 = (2\text{ m i} - 5\text{ m j}) \cdot (2\text{ m i} - 5\text{ m j})$$

$$= 4\text{ m}^2 + 25\text{ m}^2 = 29\text{ m}^2$$

so $\Delta s = \sqrt{29}$ m. The component of $\mathbf{F}$ in the direction of the displacement is therefore

$$F \cos\phi = \frac{W}{\Delta s} = \frac{-14\text{ N·m}}{\sqrt{29}\text{ m}} = -2.60\text{ N}$$

The component of the force in the direction of the displacement is negative so the work done is negative.

Example 6-6

Find the angle between the vectors $\mathbf{A} = 3\text{ m i} + 2\text{ m j}$ and $\mathbf{B} = 4\text{ m i} - 3\text{ m j}$.

We find the angle ϕ between the vectors from the definition of the dot product:

$$\mathbf{A} \cdot \mathbf{B} = AB \cos\phi$$

$$\cos\phi = \frac{\mathbf{A} \cdot \mathbf{B}}{AB}$$

The dot product of these vectors is

$$\mathbf{A} \cdot \mathbf{B} = A_x B_x + A_y B_y = (3\text{ m})(4\text{ m}) + (2\text{ m})(-3\text{ m})$$

$$= 12\text{ m}^2 - 6\text{ m}^2 = 6\text{ m}^2$$

The magnitudes of the vectors are obtained from

$$\mathbf{A} \cdot \mathbf{A} = A^2 = A_x^2 + A_y^2$$

$$= (2\text{ m})^2 + (3\text{ m})^2 = 13\text{ m}^2$$

$$A = \sqrt{13}\text{ m}$$

and

$$\mathbf{B} \cdot \mathbf{B} = B^2 = B_x^2 + B_y^2$$
$$= (4 \text{ m})^2 + (-3 \text{ m})^2 = 25 \text{ m}^2$$
$$B = 5 \text{ m}$$

Then, substituting these values into the equation for $\cos \phi$, we get

$$\cos \phi = \frac{\mathbf{A} \cdot \mathbf{B}}{AB} = \frac{6 \text{ m}^2}{(\sqrt{13} \text{ m})(5 \text{ m})} = 0.333$$
$$\phi = 70°$$

Exercise

Find $\mathbf{A} \cdot \mathbf{B}$ for (a) $\mathbf{A} = 3 \text{ m } \mathbf{i} + 4 \text{ m } \mathbf{j}$ and $\mathbf{B} = 5 \text{ m } \mathbf{i} + 8 \text{ m } \mathbf{j}$ and (b) $\mathbf{A} = 2 \text{ m/s } \mathbf{i} + 6 \text{ m/s } \mathbf{j}$ and $\mathbf{B} = 5 \text{ m/s } \mathbf{i} - 3 \text{ m/s } \mathbf{j}$.
[Answers: (a) 47 m^2, (b) $-8 \text{ m}^2/\text{s}^2$]

Example 6-7

A skier of mass m skis down a frictionless hill that has a constant angle of inclination θ as shown in Figure 6-10. The skier starts from rest at a height h. Considering the skier to be a particle, find the work done on the skier by all the forces and the speed of the skier at the bottom of the hill.

The forces acting on the skier are the force of gravity $m\mathbf{g}$ and the contact force $\mathbf{F_n}$ exerted by the snow, as shown in Figure 6-10a. Since the force $\mathbf{F_n}$ is perpendicular to the hill and to the motion of the skier, it has no component in the direction of motion of the skier and therefore does no work on the skier. The only force that does work is the weight $m\mathbf{g}$. The angle between this force and the displacement is $\phi = 90° - \theta$ (Figure 6-10b). The component of the weight in the direction of motion is therefore $mg \cos \phi = mg \cos (90° - \theta) = mg \sin \theta$. For a displacement $\Delta \mathbf{s}$ of the skier down the incline, the force of gravity does work

$$\Delta W = m\mathbf{g} \cdot \Delta \mathbf{s} = (mg \cos \phi) \Delta s = (mg \sin \theta) \Delta s$$

The total work done by gravity when the skier moves a distance s down the incline is

$$W = \sum \Delta W = \sum (mg \sin \theta) \Delta s$$

Since the force exerted by the earth is constant, $\sum (mg \sin \theta) \Delta s = (mg \sin \theta) \sum \Delta s = (mg \sin \theta)s$. We can see from Figure 6-10a that the

Figure 6-10   (a) Skier skiing down a hill of constant slope for Example 6-7. (b) Free-body diagram for the skier. The net force is $mg \sin \theta$, which is the component of the weight in the direction of the displacement Δs.

(a)

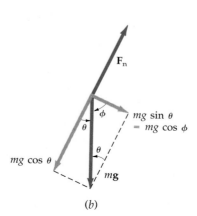

(b)

total distance s measured along the incline is related to the initial height h by $\sin \theta = h/s$ or $h = s \sin \theta$, so the work done by gravity is

$$W = (mg \sin \theta)s = mgh$$

Since this is the total work done by all the forces acting on the skier, it equals the change in the kinetic energy of the skier:

$$W_{\text{total}} = mgh = \Delta K = \tfrac{1}{2}mv^2 - 0$$

The speed of the skier at the bottom of the hill is thus given by

$$v = \sqrt{2gh}$$

This result is the same as if the skier had dropped the total distance h in free fall. If the angle θ were changed to θ', the skier would travel a different distance $s' = h/\sin \theta'$ to drop the same vertical distance h, but the component of $m\mathbf{g}$ parallel to the motion, $mg \sin \theta'$, would also change, making the work done, $(mg \sin \theta')s' = mgh$, the same. Thus, the work done by gravity on the skier is mgh, independent of the angle of incline of the hill.

The results of Example 6-7 can be generalized. Consider a skier who is skiing down a hill of arbitrary shape as shown in Figure 6-11. The small displacement $\Delta \mathbf{s}$ is tangent to the hill. The work done by gravity for this displacement is $(mg \cos \phi) \Delta s$, where ϕ is the angle between the displacement and the downward force of gravity. The quantity $\Delta s \cos \phi$ is just Δh, the vertical distance dropped over Δs. As the skier skis down the hill, the angle ϕ varies, but for each displacement $\Delta \mathbf{s}$, the work done is $\Delta W = m\mathbf{g} \cdot \Delta \mathbf{s} = mg \cos \phi \, \Delta s = mg \, \Delta h$. Thus, the total work done by gravity is mgh, where h is the total vertical distance the skier descends. If the hill is frictionless, the weight of the skier is the only force that does work. In this case, the speed of the skier after descending a vertical distance h is obtained from

$$\tfrac{1}{2}mv^2 - \tfrac{1}{2}mv_0^2 = mgh$$

where v_0 is the initial speed. This result applies only if the hill is frictionless. If there is friction between the skis and the snow, the frictional force will do work on the skier, which will be negative because the frictional force is in the direction opposite the motion. The work done on the skier by the frictional force depends on the length and shape of the hill and the coefficient of kinetic friction.

Questions

7. Suppose there is a net force acting on a particle but it does no work. Can the particle be moving in a straight line?

8. How are the vectors $\mathbf{A}$ and $\mathbf{B}$ related if $\mathbf{A} \cdot \mathbf{B} = -AB$?

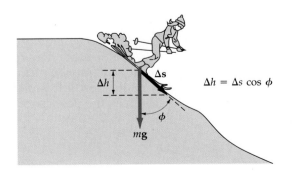

$$\Delta h = \Delta s \cos \phi$$

Figure 6-11 Skier skiing down a hill of arbitrary shape. The work done by the earth for the displacement $\Delta \mathbf{s}$ is $(mg \cos \phi) \Delta s = mg \, \Delta h$, where Δh is the vertical component of the displacement.

The ski lift does work on the skiers, which is stored as potential energy.

6-4   Work and Energy for Particle Systems: Potential Energy

When we have a system consisting of two or more particles, we must be careful when we calculate the work done by a force acting on a part of the system because the displacement of the point of application of the force is not always equal to the displacement of the entire system. In fact, different parts of a system may have different displacements.

In many cases, the work done *on a system* does not result in a change in the kinetic energy of the system but is instead stored as **potential energy.** Consider a skier of mass m going up a hill on a ski lift that raises the skier at a negligible speed to a height h. The lift does work on the skier of the amount mgh, independent of the angle of incline of the lift. The kinetic energy of the skier does not change because the gravitational force of the earth on the skier does work of the amount $-mgh$, so the total work done on the skier is zero. Let us now consider the earth and the skier (excluding the lift) to be a system consisting of two particles. The work done by gravity on the skier is internal to this system. The total work done *on this system* is the work mgh done by the lift, which we are considering to be external to our system. This work is stored as potential energy mgh of the earth–skier system. When the skier skis down a frictionless slope, this potential energy is converted into kinetic energy of the system, which is just the kinetic energy of the skier since the motion of the earth is negligible. Note that when the skier is lifted to the top of the hill, the work done by gravity is negative and the potential energy of the system increases, whereas when the skier skis down the hill, the work done by gravity is positive and the potential energy of the system decreases. The total work done by gravity when the skier moves to the top of the hill and back down to her original position is zero. The force of gravity is called a **conservative force.** In general

> A force is conservative if the total work it does on a particle is zero when the particle moves around any closed path returning to its initial position.

It follows that

> The work done by a conservative force on a particle is independent of how the particle moves from one point to another.

This is illustrated in Figure 6-12 on facing page, which shows three possible paths connecting point 1 and point 2. If the work done by a conservative force as the particle moves from 1 to 2 along one of the paths is W, it must be $-W$ when the particle returns along any of the paths because the total round-trip work must be zero. Since the work done by a conservative force on a particle does not depend on the path, it can depend only on the endpoints 1 and 2. (For example, we saw that the work done by gravity on a skier when the skier skies down a hill of height h is mgh, independent of the angle of incline of the hill.) We can use this property to define the potential-energy function U that is associated with a conservative force. It is defined such that the work done by a conservative force equals the decrease in the potential-energy function:

$$W = \int \mathbf{F} \cdot d\mathbf{s} = -\Delta U$$

or

$$\Delta U = U_2 - U_1 = -W = -\int_{s_1}^{s_2} \mathbf{F} \cdot d\mathbf{s}$$ 6-17a

For infinitesimal displacement, we have

Potential-energy function defined

$$dU = -\mathbf{F} \cdot d\mathbf{s}$$ 6-17b

The potential-energy function of a system depends on the configuration of the system. For the earth–skier system, it depends on the distance of the skier from the center of the earth. Since only a change in the potential energy is defined, the value of the function U for a system having any particular configuration is not specified by the definition. We are free to choose U to be zero at some convenient reference point. The potential energy at any other point is then the difference between the potential energy at that point and that at the reference point. For example, if the gravitational potential energy of the earth–skier system is chosen to be zero when the skier is at the bottom of the hill, its value when the skier is at a height h above that level is mgh. Or we could choose the potential energy to be zero when the skier is at sea level, in which case its value at any other point would be mgy, where y is measured from sea level.

Figure 6-12 (a) Three paths in space connecting the points 1 and 2. If the work done by a conservative force along path A from 1 to 2 is W, the work done on the return trip along path B must be $-W$ because the round-trip work is zero. It follows that the work done as a particle goes from point 1 to 2 is the same along any path connecting the two points. (b) Gold mine in Brazil. The work done by the earth on a miner as he descends to the bottom is the same no matter which ladder he uses.

(a)

(b)

We can calculate the potential-energy function associated with the gravitational force near the surface of the earth from Equation 6-17b. For the force $\mathbf{F} = -mg\mathbf{j}$, we have

$$dU = -\mathbf{F} \cdot d\mathbf{s} = -(-mg\mathbf{j}) \cdot (dx\,\mathbf{i} + dy\,\mathbf{j} + dz\,\mathbf{k}) = +mg\,dy$$

Integrating, we obtain

$$U = U_0 + mgy$$ 6-18

Gravitational potential energy near the earth's surface

where U_0 is the value of the potential energy at $y = 0$.

Exercise

A 55-kg girl is on the third floor of a building, which is 8 m above the ground floor. What is the potential energy U of the girl–earth system if (a) U is chosen to be zero on the ground floor and (b) U is chosen to be zero on the second floor, which is 4 m above the ground floor? [Answer: (a) 4.32 kJ, (b) 2.16 kJ]

Example 6-8

A bottle of mass 0.350 kg falls from rest from a shelf that is 1.75 m above the floor. Find the original potential energy of the bottle–earth system relative to the floor and its kinetic energy just before it hits the floor.

If we choose the potential energy of the bottle–earth system to be zero when the bottle is on the floor, the potential energy when the bottle is at $y = 1.75$ m is

$$U = mgy$$

$$= (0.350 \text{ kg})(9.81 \text{ N/kg})(1.75 \text{ m})$$

$$= 6.01 \text{ J}$$

As the bottle falls, the earth does work on the bottle. The total work done by the earth as the bottle falls a distance $y = 1.75$ m is $W = mgy = 6.01$ J. The kinetic energy of the bottle just before it hits the floor is thus 6.01 J. In this example, the original potential energy of the bottle–earth system is converted into kinetic energy of the bottle.

Figure 6-13   A mass attached to a spring. To stretch the spring, a force $F_{app} = +kx$ must be applied to the block.

Another example of a conservative force is that exerted by a spring. Figure 6-13 shows a system consisting of a spring and a block. The block is at rest on a frictionless table at $x = 0$ with the spring unstretched. Let us now slowly push the block with a force equal and opposite to that exerted by the spring so that the spring is extended. Since the spring exerts a force $F_{sb} = -kx$ on the block, we must exert an equal and opposite force $F_{app} = +kx$ to push the block without acceleration. The work done by the applied force is

$$W_{app} = \int_0^x kx \, dx = \tfrac{1}{2}kx^2$$

The spring exerts a force on the block opposite the direction of the displacement, so it does negative work on the block:

$$W_{sb} = \int_0^x -kx \, dx = -\tfrac{1}{2}kx^2$$

Since the total work done on the block is zero, its kinetic energy does not change. The block exerts a force on the spring F_{bs}, which is the reaction force to the force F_{sb} exerted by the spring on the block:

$$F_{bs} = -F_{sb} = +kx$$

The displacement of the point of application of force F_{bs} is the same as the displacement of the block, so the block does work on the spring equal in magnitude to W_{sb}:

$$W_{bs} = +\tfrac{1}{2}kx^2$$

We cannot assign a value for the displacement of the spring because different parts of the spring have different displacements. The other force exerted on the spring is that of the wall to which it is fixed. The total work done on the *block–spring system* is W_{app}. This work is stored as potential energy of the block–spring system. When the block is released, the spring does positive work on the block as the potential energy of the system is converted into kinetic energy. The round-trip work done on the block by the spring when the block moves from its original position out to some position x and then back is zero. The force of the spring is thus a conservative force. We can calculate the potential-energy function associated with this force from Equation 6-17b. We have

$$dU = -\mathbf{F} \cdot d\mathbf{s} = -F_x \, dx = -(-kx) \, dx = +kx \, dx$$

Then

$$U = \tfrac{1}{2}kx^2 + U_0$$

where U_0 is the potential energy when $x = 0$, that is, when the spring is unstretched. Choosing U_0 to be zero, we have

$$U = \tfrac{1}{2}kx^2$$

6-19   *Potential energy of a spring*

Exercise

A 3-kg block is hung vertically from a spring with a force constant of 600 N/m. (*a*) By how much is the spring stretched when the block is in equilibrium? (*b*) How much potential energy is stored in the spring–block system?   [Answer: (*a*) 4.9 cm, (*b*) 0.72 J]

Although potential energy is associated with a system of particles, we sometimes have systems in which only one particle moves. For example, in the earth–skier system, the earth's motion is negligible. Since the configuration of the system is expressed in terms of the position of the skier relative to some chosen reference position, we sometimes speak of the potential energy of this system as the potential energy of the skier. Similarly, we may refer to the potential energy of the block–spring system as the potential energy of the block.

Not all forces are conservative. An example of a nonconservative force is a force, such as a push or a pull, applied by a human agent. Consider a box on a rough horizontal table. Suppose you push the box around some closed path so that the box ends up at its original position. The work that you do is not generally zero. It depends on how great a force you decide to exert on the box. Since the work done by such an applied force is not zero when the box completes a round trip, no potential-energy function can be defined for this force. The work done on the box by the force of kinetic friction is also not zero. Since the force of kinetic friction is always opposite the direction of motion, the total round-trip work done by kinetic friction will be negative. The force of kinetic friction is therefore not a conservative force, and no potential-energy function can be defined for it.

Question

9. When you climb a mountain, is the work done on you by gravity different if you take a short, steep trail instead of a long, gentle trail? If not, why do you find one trail easier than the other?

6-5   Potential Energy and Equilibrium in One Dimension

The potential-energy function U associated with a conservative force $\mathbf{F} = F_x \mathbf{i}$ is related to the force by Equation 6-17b:

$$dU = -\mathbf{F} \cdot d\mathbf{s} = -F_x\, dx$$

The force is therefore the negative derivative of the potential-energy function:

$$F_x = -\frac{dU}{dx} \qquad\qquad 6\text{-}20$$

We can illustrate this general relation for the block–spring system by differentiating the function $U = \frac{1}{2}kx^2$. We obtain

$$F_x = -\frac{dU}{dx} = -\frac{d}{dx}\left(\tfrac{1}{2}kx^2\right) = -kx$$

Figure 6-14 shows a plot of the potential-energy function $U = \frac{1}{2}kx^2$ versus the displacement x for a block–spring system. The derivative of this function is represented graphically as the slope of the tangent line to the curve. The force is thus equal to the negative of the slope of the curve. When x is positive, the slope is positive and the force F_x is negative. When x is negative the slope is negative and the force F_x is positive. In either case, the force is in the direction such as to accelerate the block toward lower potential energy.

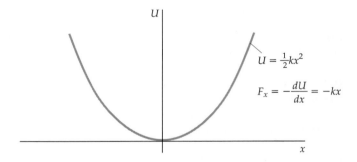

Figure 6-14 Plot of the potential-energy function U versus displacement x for the object on a spring. The force $F_x = -dU/dx$ is the negative slope of this curve. It is zero at the minimum in the curve, which occurs at $x = 0$. For positive values of x the force is negative, and for negative values of x the force is positive. A minimum in a potential-energy curve is a point of stable equilibrium because a displacement to either side of this point results in a force directed toward the equilibrium position.

$$U = \tfrac{1}{2}kx^2$$

$$F_x = -\frac{dU}{dx} = -kx$$

At the position on the potential-energy curve where the slope is zero, the force is zero and, if F_x is the only force acting on it, the block is in equilibrium; that is, the block will remain at rest in this position if it is placed there at rest.

Condition for equilibrium   A particle is in equilibrium if the net force acting on it is zero.

This occurs at $x = 0$, where the spring is unstretched. For x slightly greater than zero, the force $F_x = -dU/dx$ is negative. Thus, if the block is displaced slightly in the positive x direction, the spring force will accelerate the block back toward its equilibrium position at $x = 0$. Similarly, if x is negative, F_x is positive. If the block is displaced in the negative x direction, the spring force accelerates the block in the positive x direction back toward its equilibrium position. This equilibrium is called **stable equilibrium.**

In stable equilibrium a small displacement results in a restoring force that accelerates the particle back toward its equilibrium position.

Figure 6-15 shows a potential-energy curve with a maximum rather than a minimum at $x = 0$. Such a curve could represent the potential energy of a skier at the top of a hill between two valleys. For this curve, when x is positive, the slope is negative and the force F_x is positive, and when x is negative, the slope is positive and the force F_x is negative. Again, the force is in the direction such as to accelerate the particle toward lower potential energy.

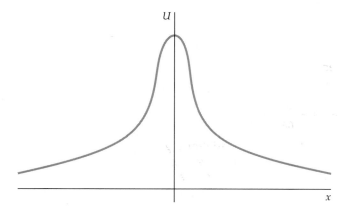

Figure 6-15 For this potential-energy curve, $F_x = -dU/dx$ is zero at $x = 0$, but this equilibrium is unstable because a displacement in either direction results in a force directed away from the equilibrium position.

At $x = 0$, $dU/dx = -F_x = 0$, so a particle placed at this point is in equilibrium. However, this equilibrium is **unstable equilibrium**. For x slightly greater than 0, the force F_x is positive. Thus, if a particle in equilibrium at $x = 0$ is given a small positive displacement x, it will experience a positive force F_x that will accelerate it to the right away from its equilibrium position. Similarly, for x slightly less than 0, F_x is negative. If the particle is displaced slightly to the left of its equilibrium position, it will accelerate to the left, away from its equilibrium position.

In unstable equilibrium a small displacement results in a force that accelerates the particle away from its equilibrium position.

Both maximum and minimum points on the potential-energy curve are positions of equilibrium. At minimum points the equilibrium is stable, and at maximum points it is unstable.

A conservative force always tends to accelerate a particle toward lower potential energy.

Figure 6-16 shows a potential-energy curve that is flat in the region near $x = 0$. If a particle is placed at the point $x = 0$, it will experience no force so the particle will be in equilibrium at that point. If the particle is displaced slightly in either direction, the force will again be zero. This is an example of **neutral equilibrium.**

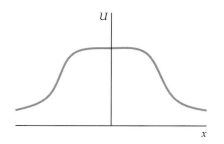

Figure 6-16 Plot of a potential-energy function U for which $x = 0$ is a position of neutral equilibrium. The force $F_x = -dU/dx$ is zero at $x = 0$ and at neighboring points. If a particle is displaced in either direction from $x = 0$, it experiences no force, and so remains in equilibrium.

In neutral equilibrium a small displacement results in zero force so the particle is again in equilibrium.

Example 6-9

The force between two atoms in a diatomic molecule can be represented approximately by the potential-energy function

$$U = U_0\left[\left(\frac{a}{x}\right)^{12} - 2\left(\frac{a}{x}\right)^6\right] \qquad 6\text{-}21$$

where U_0 and a are constants. (a) At what value of x is the potential energy zero? (b) Find the force F_x. (c) At what value of x is the potential energy a minimum? What is the value of this minimum potential energy?

(a) Setting $U = 0$ and solving for x, we obtain

$$U_0\left[\left(\frac{a}{x}\right)^{12} - 2\left(\frac{a}{x}\right)^6\right] = 0$$

$$\left(\frac{a}{x}\right)^{12} = 2\left(\frac{a}{x}\right)^6$$

$$\left(\frac{a}{x}\right)^6 = 2$$

$$x = \frac{a}{2^{1/6}}$$

(b) We find the force from Equation 6-20. We have

$$U = U_0\left[\left(\frac{a}{x}\right)^{12} - 2\left(\frac{a}{x}\right)^6\right] = U_0(a^{12}x^{-12} - 2a^6x^{-6})$$

$$F_x = -\frac{dU}{dx} = -U_0[a^{12}(-12x^{-13}) - 2a^6(-6x^{-7})]$$

$$= \frac{12U_0}{a}\left[\left(\frac{a}{x}\right)^{13} - \left(\frac{a}{x}\right)^7\right]$$

Figure 6-17 Plot of potential energy U versus separation x for a diatomic molecule. The minimum potential energy is $-U_0$, where U_0 is approximately equal to the magnitude of the energy needed to separate the atoms. The total energy of the atoms in a diatomic molecule in its lowest energy state is somewhat greater than $-U_0$.)

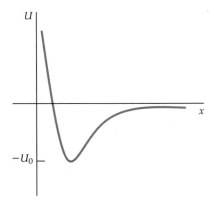

(c) When the potential energy is minimum, the force is zero. We see from the above equation that $F_x = 0$ when $x = a$. Setting $x = a$ in Equation 6-21, we obtain $U = -U_0$ for the minimum value of the potential energy. Figure 6-17 shows a plot of U versus x. The energy U_0 is approximately equal to the energy needed to separate the atoms, which is called the dissociation energy.

6-6   The Conservation of Mechanical Energy

The negative sign in the definition of the potential-energy function in Equation 6-17 is introduced so that the work done by a conservative force on a particle equals the decrease in the potential energy of the system. Let us consider a system in which work is done on only one of the particles such as a skier–earth system. If a conservative force is the only force that does work on the particle, the work done by the force equals the decrease in the potential energy of the system and also equals the increase in the kinetic energy of the particle (which is the increase in the kinetic energy of the system):

$$W_{\text{total}} = \int \mathbf{F} \cdot d\mathbf{s} = -\Delta U = +\Delta K \qquad \qquad 6\text{-}22$$

Hence,

$$\Delta K + \Delta U = \Delta(K + U) = 0 \qquad \qquad 6\text{-}23$$

The sum of the kinetic energy and the potential energy of the system is called the **total mechanical energy** E:

$$E = K + U \qquad \qquad 6\text{-}24$$

If only conservative forces do work, Equation 6-23 states that the change in the total mechanical energy is zero. Thus, the total mechanical energy remains constant during the motion of the particle.

$$E = K + U = \text{constant} \qquad \qquad 6\text{-}25 \qquad \textit{Conservation of mechanical energy}$$

This is known as the **law of conservation of mechanical energy** and is the origin of the expression "conservative force."

   We can extend our discussion to systems in which work is done on more than one particle. Figure 6-18 shows a system consisting of two blocks of mass m_1 and m_2 resting on a frictionless table connected by a spring of negligible mass. The spring is stretched by applying a force to each mass as shown in Figure 6-18b. The work done by the applied forces to stretch the spring by an amount x is stored as potential energy $\Delta U = \frac{1}{2}kx^2$ in the system.

(a)

(b)

Figure 6-18   (a) A system consisting of two blocks connected by a spring. (b) The work done by the forces applied to stretch the spring equals the change in the potential energy of the system.

When the blocks are released, the only force that does work on each block is the conservative force exerted by the spring. The work done on each block equals the change in the kinetic energy of that block. The total work done by the spring therefore equals the change in the total kinetic energy of the two blocks. But this total work also equals the (negative) change in the potential energy of the system. Thus, Equations 6-22 through 6-25 also apply to this system, where K is the total kinetic energy of the system, U is the total potential energy of the system, and $E = K + U$ is the total mechanical energy of the system.

Applications of the Conservation of Mechanical Energy

Consider a problem in which a single object of mass m moves in one dimension under the influence of a conservative force. Writing $\frac{1}{2}mv^2$ for the kinetic energy and $U(x)$ for the potential energy in Equation 6-25, we obtain

$$\frac{1}{2}mv^2 + U(x) = E \qquad\qquad 6\text{-}26a$$

If the total (constant) energy E is known, Equation 6-26a allows the determination of the speed v as a function of x. We have

$$v = \sqrt{\frac{2[E - U(x)]}{m}} \qquad\qquad 6\text{-}26b$$

In problems in which we are not concerned with the time variable t, Equation 6-26 is often much easier to use than is Newton's second law. For example, consider the simple problem of a skier skiing down a hill starting from rest at an initial height h above the bottom of a hill. If we choose the potential energy of the skier–earth system to be zero at the bottom of the hill, the original potential energy is mgh. This equals the total energy because the initial kinetic energy is zero. At any other height y above the bottom of the hill, the potential energy is mgy. The speed v at any height y is then given by the conservation of mechanical energy:

$$\frac{1}{2}mv^2 + mgy = E = mgh$$
$$v = \sqrt{2g(h - y)}$$

Example 6-10

A pendulum consists of a bob of mass m attached to a string of length L. It is pulled aside so that the string makes an angle θ_0 with the vertical and is released from rest. Find the speed v at the bottom of the swing and the tension in the string at that time.

The two forces acting on the bob (neglecting air resistance) are the force of gravity $m\mathbf{g}$, which is conservative, and the tension $\mathbf{T}$, which is perpendicular to the motion and therefore does no work. The mechanical energy of the bob–earth system is therefore conserved in this problem.

Let us choose the gravitational potential energy to be zero at the bottom of the swing. Initially, the bob is at a height h above the bottom and at rest. Its kinetic energy is zero and the potential energy of the system is mgh. The initial total energy of the system is thus

$$E_i = K_i + U_i = 0 + mgh$$

As the pendulum swings down, potential energy is converted into kinetic energy. The final energy at the bottom of the swing is therefore

$$E_f = K_f + U_f = \tfrac{1}{2}mv^2 + 0 = \tfrac{1}{2}mv^2$$

Conservation of energy then gives

$$E_f = E_i$$
$$\tfrac{1}{2}mv^2 = mgh$$

To find the speed in terms of the initial angle θ_0, we must relate h to θ_0. According to Figure 6-19, the distance h is related to θ_0 and the length of the pendulum L by

$$h = L - L\cos\theta_0 = L(1 - \cos\theta_0)$$

Thus, the speed at the bottom is found from

$$v^2 = 2gh = 2gL(1 - \cos\theta_0)$$

The waterfall in this 1961 lithograph by the Swiss artist M. C. Escher violates the law of conservation of energy. As the water falls, part of its potential energy is converted into the kinetic energy of the waterwheel. How does the water get back to the top of the waterfall?

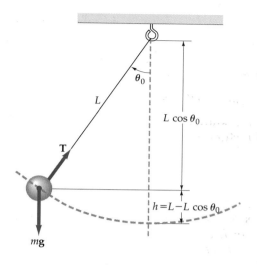

Figure 6-19   Simple pendulum for Example 6-10. The tension is perpendicular to the motion and does no work. The speed of the bob at the bottom is found from the conservation of energy $\frac{1}{2}mv^2 = mgh$, where the initial height h above the bottom is related to the initial angle θ_0 by $h = L - L \cos \theta_0$.

To find the tension at the bottom, we use Newton's second law. (One of the things that makes physics challenging is that we must apply everything we have learned previously as well as the material studied in this chapter.) At the bottom of the swing, the forces on the bob are the weight mg acting down and the tension $\mathbf{T}$ acting up. Since the bob is moving in a circle of radius L with a speed v, it has a centripetal acceleration v^2/L. This acceleration is directed toward the center of the circle, which is upward at this point. Newton's second law thus gives

$$T - mg = \frac{mv^2}{L} = 2mg(1 - \cos \theta_0)$$

$$T = mg + 2mg(1 - \cos \theta_0)$$

Note that if the bob is released from $\theta_0 = 90°$, the tension at the bottom is 3 times the force of gravity mg.

The speed of the bob at the bottom could also be found using Newton's laws, but the solution is difficult and requires calculus because the acceleration tangential to the curve varies with the angle θ and therefore with time, so the constant-acceleration formulas do not apply. (See Problem 88.)

Multiflash photograph of a simple pendulum. As the bob descends, gravitational potential energy is converted into kinetic energy, and the speed increases as indicated by the increased spacing of the recorded positions. The speed decreases as the bob moves up, and the kinetic energy is changed into potential energy.

Figure 6-20 Block pushed against a spring for Example 6-11. In this example, the initial potential energy in the spring is converted first into kinetic energy and then into gravitational potential energy.

Example 6-11

A 2-kg block is pushed against a spring that has a force constant of 500 N/m, compressing it 20 cm. The block is then released, and the spring projects it along a frictionless horizontal surface and then up a frictionless incline of angle 45° as shown in Figure 6-20. How far up the incline does the block travel?

After the spring is released, the only forces that do work are the force exerted by the spring and the force of gravity. Since both these forces are conservative, the total mechanical energy of the block–spring–earth system is conserved. In this case, the total mechanical energy consists of kinetic energy of the block $\frac{1}{2}mv^2$, the potential energy of the spring $\frac{1}{2}kx^2$, and the gravitational potential energy mgh. We choose $h = 0$ at the horizontal surface. The initial energy is then just the potential energy of the spring:

$$E_i = \tfrac{1}{2}kx^2 = \tfrac{1}{2}(500 \text{ N/m})(0.20 \text{ m})^2 = 10 \text{ J}$$

When the block leaves the spring, its kinetic energy is 10 J, the initial potential energy of the compressed spring. As the block slides up the incline, its speed decreases until it is zero at its maximum height h. At this height, the block is at rest so its energy is entirely gravitational potential energy:

$$E_f = mgh$$

Setting the final energy equal to the initial energy, we obtain

$$mgh = 10 \text{ J}$$

$$h = \frac{10 \text{ J}}{mg} = \frac{10 \text{ J}}{(2 \text{ kg})(9.81 \text{ N/kg})} = 0.51 \text{ m}$$

The distance s the block travels up the incline is found from

$$\frac{h}{s} = \sin 45° = 0.707$$

$$s = \frac{h}{\sin 45°} = \frac{0.51 \text{ m}}{0.707} = 0.721 \text{ m}$$

Exercise

Find the speed of the block in Example 6-11 after it leaves the spring. (Answer: 3.16 m/s)

Example 6-12

A spring with a force constant of k hangs vertically. A block of mass m is attached to the unstretched spring and allowed to fall from rest. Find the maximum distance the block falls before it begins moving upward.

Again, we have two kinds of potential energy, gravitational potential energy and the potential energy of the stretched spring. The initial and final positions of the block are shown in Figure 6-21. We let y be the distance the block falls, and we choose the gravitational potential energy of the block to be zero at the original position $y = 0$. The initial potential energy of the spring is zero because the spring is unstretched at this position. Since the block is initially at rest, its kinetic energy is initially zero. The total initial energy is thus zero:

$$E_i = K_i + U_i = 0$$

After the block has fallen some distance y, it has kinetic energy $\frac{1}{2}mv^2$. The gravitational potential energy is $-mgy$, and the potential energy of the spring is $\frac{1}{2}ky^2$. The potential energy of the system is then

$$U = -mgy + \tfrac{1}{2}ky^2$$

and the total energy is

$$E = K + U = \tfrac{1}{2}mv^2 - mgy + \tfrac{1}{2}ky^2$$

Since mechanical energy is conserved, the total energy always equals the initial energy, which is zero. Setting $E = E_i = 0$, we obtain

$$\tfrac{1}{2}mv^2 - mgy + \tfrac{1}{2}ky^2 = 0 \qquad 6\text{-}27$$

Equation 6-27 relates the speed v to the distance dropped y. As the block drops, its speed first increases, then reaches some maximum value, and then decreases until it is again zero when the block is at its lowest point, which we will call y_m. We can find y_m by setting $v = 0$ in Equation 6-27:

$$-mgy_m + \tfrac{1}{2}ky_m^2 = 0$$

The two solutions to this equation are $y_m = 0$, which corresponds to the initial position, and

$$y_m = \frac{2mg}{k}$$

Figure 6-21   Block attached to an unstretched spring and dropped for Example 6-12. Gravitational potential energy is converted into the kinetic energy of the block plus the potential energy of the spring. At the lowest point, where the block is momentarily at rest, the loss in gravitational potential energy of the block equals the gain in potential energy of the spring.

Figure 6-22 shows the plot of the potential energy $U = -mgy + \frac{1}{2}ky^2$ versus the displacement y for Example 6-12. The potential energy is initially zero, then decreases to some minimum value, at which time the kinetic energy of the block is maximum, and then increases back to zero, at which time the block is at rest at its lowest point. It is instructive to find the value of y for which the potential energy is minimum. At this point, the slope dU/dy is zero, and the net force acting on the block $F_y = -dU/dy$ is zero. We can find the value of y for which the potential energy is minimum by computing the derivative dU/dy and setting it equal to zero. Computing dU/dy gives

$$\frac{dU}{dy} = \frac{d}{dy}(-mgy + \tfrac{1}{2}ky^2) = -mg + ky$$

The net force acting on the block is thus

$$F_y = -\frac{dU}{dy} = mg - ky$$

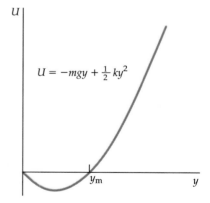

Figure 6-22   Potential-energy function $U = -mgy + \frac{1}{2}ky^2$ versus y for Example 6-12. The initial potential energy is chosen to be zero, and since the block is initially at rest, the total energy is zero. At the minimum in the potential-energy curve, the kinetic energy of the block is maximum. The value of y at this minimum point is $y = mg/k$, which is found by setting $dU/dy = 0$.

This force is zero when $y = mg/k$, which is the value of y for which the upward force exerted by the spring on the block equals the weight of the block. This is the equilibrium position of the block hanging on the spring.

Example 6-13

Two blocks of mass m_1 and m_2 are initially at rest attached to a light string that passes over a frictionless peg as shown in Figure 6-23. Find the speed of either block when the heavier one falls a distance h.

Figure 6-23 Atwood's machine for Example 6-13.

This device, called *Atwood's machine*, was developed in the eighteenth century to measure the acceleration of gravity g. As we will see, if the values of m_1 and m_2 are not too different, the acceleration of either object is a small fraction of g. It could be easily measured with the rather crude timing devices available in the eighteenth century, whereas a direct measurement of g was difficult if not impossible. (In his experiment, Atwood used a pulley rather than a peg. If the mass of the pulley can be neglected, the analysis is the same as for a frictionless peg.)

Since the peg is frictionless, the tension in the string is uniform. Provided that the string does not stretch, the total work done by the tension is zero because when the lighter object moves upward in the direction of the tension, the heavier object moves downward the same distance in the opposite direction. The only force that does work on the two-object system is the force of gravity. Therefore, the total mechanical energy of the system is conserved.

Let us assume that m_2 is greater than m_1 and choose the potential energy of the system to be zero at the initial position of the blocks. Since they are at rest at this position, the total energy is zero. Calling this initial energy E_i, we have

$$E_i = 0$$

Let v be the speed of m_1 after it has moved up a distance h. Then its kinetic energy is $\frac{1}{2}m_1v^2$. Since the connecting string does not stretch, m_2 must move down the same distance h and acquire the same speed v. Its kinetic energy is $\frac{1}{2}m_2v^2$. Since m_1 moves up a distance h and m_2 moves down a distance h, the potential energy of the system is $m_1gh - m_2gh$. The energy at this time E_f is thus

$$E_f = \tfrac{1}{2}m_1v^2 + \tfrac{1}{2}m_2v^2 + m_1gh - m_2gh$$

Setting the final energy equal to the initial energy, we obtain

$$\tfrac{1}{2}m_1v^2 + \tfrac{1}{2}m_2v^2 + m_1gh - m_2gh = 0$$

or

$$\tfrac{1}{2}(m_1 + m_2)v^2 = (m_2 - m_1)gh \qquad 6\text{-}28$$

Note that the left side of Equation 6-28 is the kinetic energy gained by the system and the right side is the net potential energy lost. Solving for v^2 gives

$$v^2 = \frac{2(m_2 - m_1)}{(m_1 + m_2)}\,gh \qquad 6\text{-}29$$

which relates the speed of either mass to the distance it moves. Comparing Equation 6-29 with the constant-acceleration equation $v^2 = 2ah$, we see that the acceleration is given by

$$a = \frac{(m_2 - m_1)}{(m_1 + m_2)}\,g$$

In the laboratory experiment, the acceleration a is measured, and the unknown value of g is then obtained from

$$g = \frac{(m_1 + m_2)}{(m_2 - m_1)} a$$

This problem could also be solved by applying Newton's second law to each of the bodies and eliminating the tension T from the two equations obtained.

Exercise

What is the magnitude of the acceleration of either block in Example 6-13 if the masses are $m_1 = 3$ kg and $m_2 = 5$ kg? (Answer: $a = 0.25$ g $= 2.45$ m/s^2)

We have seen that the law of conservation of mechanical energy as expressed in Equation 6-25 can be used as an alternative to Newton's laws for solving certain problems in mechanics. Since this equation was derived from Newton's laws, any problem that can be solved using the conservation of mechanical energy can also be solved directly from Newton's laws. In spite of this, energy conservation is an extremely useful tool for the analysis of problems.

6-7   The Generalized Work–Energy Theorem

When both conservative and nonconservative forces do work, the total mechanical energy of a system does not remain constant. Let us again consider a system in which work is done on only one of the particles. Consider the particle to be acted on by a nonconservative force $\mathbf{F}_{nc}$ and two conservative forces $\mathbf{F}_1$ and $\mathbf{F}_2$ so that the net force is

$$\mathbf{F}_{net} = \mathbf{F}_{nc} + \mathbf{F}_1 + \mathbf{F}_2$$

(Our discussion can be easily extended to include any number of forces.) According to the work–energy theorem, the total work done by these forces equals the change in the kinetic energy of the particle (and therefore of the system):

$$W_{total} = \int \mathbf{F}_{nc}\cdot d\mathbf{s} + \int \mathbf{F}_1\cdot d\mathbf{s} + \int \mathbf{F}_2\cdot d\mathbf{s} = W_{nc} + W_1 + W_2 = \Delta K \quad \text{6-30}$$

where W_{nc} is the work done by the nonconservative force, W_1 is that done by force $\mathbf{F}_1$, and W_2 is that done by force $\mathbf{F}_2$. For each conservative force we define a potential-energy function U_i in the usual way (Equation 6-17):

$$W_i = -\Delta U_i$$

Equation 6-30 can then be written

$$W_{nc} - \Delta U_1 - \Delta U_2 = \Delta K$$

or

$$W_{nc} = \Delta U_1 + \Delta U_2 + \Delta K = \Delta E \qquad \text{6-31} \qquad \textit{Generalized work–energy theorem}$$

where

$$E = U_1 + U_2 + K$$

is the total mechanical energy of the system. Equation 6-31 is the **generalized work–energy theorem:**

> The work done by a nonconservative force acting on a particle equals the change in the total mechanical energy of the system.

This modified form of the work–energy theorem is the most convenient in many applications since explicit calculation of the work done is required only for the nonconservative forces. When there is no work done by nonconservative forces, the total mechanical energy is conserved, that is, it remains constant.

Work done by kinetic friction   We have derived the generalized work–energy theorem for a particle, but the most common nonconservative force is that of kinetic friction, which is a force exerted by one surface on another. As we have stated, we must be careful when we calculate the work done by a force acting on an extended object because the displacement of the point of application of the force is not in general equal to the displacement of the object. Consider a block on a horizontal table for which the coefficient of kinetic friction is μ_k. Suppose the block has an initial horizontal velocity v_i, and it slides a distance Δx before coming to rest. The initial mechanical energy of the block is its initial kinetic energy

$$K_i = \tfrac{1}{2}mv_i^2$$

The final mechanical energy of the block is zero. The work done on the block by the force of friction is not $-f\,\Delta x$, where f is the force of friction, because the displacement of the point of application of the force is not the same as the displacement of the block Δx. However, the total mechanical energy change *is* equal to $-f\,\Delta x$. We can see this by applying Newton's second law to the block. We have

$$F_{net} = ma$$

$$-f = ma$$

Multiplying both sides by Δx, we obtain

$$-f\,\Delta x = ma\,\Delta x = m(\tfrac{1}{2}v_f^2 - \tfrac{1}{2}v_i^2) = -\tfrac{1}{2}mv_i^2$$

where we have used the constant-acceleration formula $2a\,\Delta x = v_f^2 - v_i^2$, and $v_f = 0$. It can be shown that $-f\,\Delta x$ does equal the total mechanical energy that is converted to thermal energy because of the frictional forces.[*] This is the same as would be obtained from the generalized work–energy theorem if $-f\,\Delta x$ were the work done by friction on the block. We may thus use the generalized work–energy equation in problems involving kinetic friction acting on extended objects if we use $-f\,\Delta x$ for the work done by the nonconservative force. (The product of a force acting on an object times the displacement of the *object*, such as $-f\,\Delta x$, is called pseudowork as discussed in Section 7-5.)

Example 6-14

A 5-kg sled is sliding with an initial speed of 4 m/s. If the coefficient of friction between the sled and the snow is 0.14, how far will the sled slide before coming to rest?

The initial mechanical energy of the system is the original kinetic energy of the sled:

$$E_i = K_i = \tfrac{1}{2}mv^2 = \tfrac{1}{2}(5\text{ kg})(4\text{ m/s})^2 = 40\text{ J}$$

[*]The work done by kinetic friction is examined in detail in "Work and Heat Transfer in the Presence of Sliding Friction" by B. A. Shewood and W. H. Bernard, *American Journal of Physics*, **52**, 1001 (1984)

The final mechanical energy is zero. The change in the mechanical energy of the sled–snow system is thus

$$\Delta E = E_f - E_i = -E_i = -40 \text{ J}$$

The only nonconservative force is that of kinetic friction. The normal force exerted by the snow on the sled equals the weight of the sled,

$$F_n = mg = (5 \text{ kg})(9.81 \text{ N/kg}) = 49.0 \text{ N}$$

The magnitude of the frictional force on the sled is then

$$f = \mu_k F_n = (0.14)(49.0 \text{ N}) = 6.86 \text{ N}$$

When the sled moves a distance Δx, the "work" done by friction is

$$W_{nc} = f_x \, \Delta x = (-6.86 \text{ N}) \, \Delta x$$

Setting the total work done by the nonconservative force equal to the change in the mechanical energy of the system we obtain

$$W_{nc} = \Delta E = -40 \text{ J}$$

$$(-6.86 \text{ N}) \, \Delta x = -40 \text{ J}$$

$$\Delta x = \frac{-40 \text{ J}}{-6.86 \text{ N}} = 5.83 \text{ m}$$

Example 6-15

A horizontal force of 25 N is applied to a 4-kg box, which is initially at rest on a rough horizontal table. The coefficient of kinetic friction between the box and table is 0.35. Find the speed of the box after it has been pushed 3 m.

In this problem, we have two nonconservative forces acting—the applied force and the force of kinetic friction. The applied force does work on the box given by

$$W_{app} = F_{app} \, \Delta x = (25 \text{ N})(3 \text{ m}) = 75 \text{ J}$$

Since the normal force exerted by the table on the box equals the weight of the box, the force of kinetic friction exerted on the box is

$$f = \mu_k F_n = \mu_k mg = (0.35)(4 \text{ kg})(9.81 \text{ N/kg}) = 13.7 \text{ N}$$

The work done by the force of kinetic friction is given by

$$W_f = -f \, \Delta x = -(13.7 \text{ N})(3 \text{ m}) = -41.1 \text{ J}$$

The total work done by nonconservative forces is thus

$$W_{nc} = W_{app} + W_f = 75 \text{ J} - 41.1 \text{ J} = 33.9 \text{ J}$$

If we take the potential energy of the box to be zero on the table, the total mechanical energy of this system is just the kinetic energy of the box. The generalized work–energy theorem then gives

$$W_{nc} = \Delta E = \Delta K = \tfrac{1}{2}mv^2 = 33.9 \text{ J}$$

$$v = \sqrt{\frac{2(33.9 \text{ J})}{m}} = \sqrt{\frac{2(33.9 \text{ J})}{4 \text{ kg}}} = 4.12 \text{ m/s}$$

Figure 6-24 Child on a slide for Example 6-16. In this example, mechanical energy is not conserved because of friction.

Example 6-16

A child of mass 40 kg slides down a rough slide inclined at 30° (Figure 6-24). The coefficient of kinetic friction between the child and the slide is $\mu_k = 0.2$. If the child starts from rest at the top of the slide, a height 4 m above the bottom, how fast is she traveling when she reaches the bottom?

The forces acting on the child are gravity, the normal force F_n, and the frictional force $f = \mu_k F_n$ exerted by the slide. The normal force exerted by the slide is $F_n = mg \cos 30°$. Since the normal force is perpendicular to the motion, it does no work. The work done by the nonconservative force of friction is the negative of the force f times the distance s traveled. Since the original height h is 4 m and $\sin 30° = h/s = 0.5$, the distance s is 8 m. The work done by friction is thus

$$W_f = -fs = -(\mu_k mg \cos 30°)s$$
$$= -(0.2)(40 \text{ kg})(9.81 \text{ N/kg})(0.866)(8 \text{ m}) = -544 \text{ J}$$

According to the work–energy theorem, the −544 J of work done by friction equals the change in the total mechanical energy of the child–slide–earth system. Since the kinetic energy K_i of the child at the top of the slide is zero, the original energy of the system is its potential energy U_i. If we choose the potential energy to be zero at the bottom of the slide, the original energy is

$$E_i = U_i = mgh = (40 \text{ kg})(9.81 \text{ N/kg})(4 \text{ m}) = 1570 \text{ J}$$

The work–energy theorem then gives

$$W_{nc} = \Delta E = E_f - E_i = E_f - 1570 \text{ J} = -544 \text{ J}$$

The final energy is therefore

$$E_f = 1570 \text{ J} - 544 \text{ J} = 1026 \text{ J}$$

When the child reaches the bottom of the slide the final energy is the kinetic energy of the child $\frac{1}{2}mv^2$. Then

$$E_f = K_f = \tfrac{1}{2}mv^2 = 1026 \text{ J}$$

Solving for v, we obtain

$$v = \sqrt{\frac{2K_f}{m}} = \sqrt{\frac{2(1026 \text{ J})}{40 \text{ kg}}} = 7.16 \text{ m/s}$$

The total mechanical energy of the child–slide–earth system is thus decreased from its original value of 1570 J to its final value of 1570 J − 544 J = 1026 J, which equals the kinetic energy of the child at the bottom, $\frac{1}{2}mv^2$.

Example 6-17

A 4-kg block hangs by a light string that passes over a smooth peg and is connected to a 6-kg block that rests on a rough table (Figure 6-25). The coefficient of kinetic friction is $\mu_k = 0.2$. The 6-kg block is pushed against a spring, which has a force constant of 180 N/m, compressing it 30 cm. It is then released. Find the speed of the blocks when the 4-kg block has fallen a distance of 40 cm.

Figure 6-25 Two blocks connected by a light string for Example 6-17. The 6-kg block is pushed against the spring and released. As the two blocks move, the potential energy of the spring decreases, the gravitational potential energy of the 4-kg block decreases, and mechanical energy is lost because of friction.

We consider the system as consisting of the earth, the table, the spring, and the two blocks $m_1 = 6$ kg and $m_2 = 4$ kg, and we choose the initial gravitational potential energy to be zero. The initial energy is then just the potential energy of the spring:

$$E_i = \tfrac{1}{2}kx^2 = \tfrac{1}{2}(180 \text{ N/m})(0.30 \text{ m})^2 = 8.1 \text{ J}$$

The work done by friction as the 6-kg block moves a distance $x_1 = 0.40$ m is

$$W_{nc} = -\mu_k m_1 g x_1$$

$$= -(0.2)(6 \text{ kg})(9.81 \text{ N/kg})(0.40 \text{ m}) = -4.7 \text{ J}$$

The work–energy theorem then gives

$$W_{nc} = E_f - E_i = -4.7 \text{ J}$$

$$E_f = E_i - 4.7 \text{ J} = 8.1 \text{ J} - 4.7 \text{ J} = 3.4 \text{ J}$$

After the 4-kg block has fallen a distance y, the two blocks are moving with the same speed v, and the total energy of the system is

$$E_f = \tfrac{1}{2}(m_1 + m_2)v^2 - m_2 g y$$

For $y = 40$ cm $= 0.4$ m, we have

$$E_f = \tfrac{1}{2}(6 \text{ kg} + 4 \text{ kg})v^2 - (4 \text{ kg})(9.81 \text{ N/kg})(0.4 \text{ m})$$

$$= (5 \text{ kg})v^2 - 15.7 \text{ J}$$

Setting this equal to 3.4 J and solving for v, we obtain

$$E_f = (5 \text{ kg})v^2 - 15.7 \text{ J}$$

$$= 3.4 \text{ J}$$

$$v = \sqrt{\frac{19.1 \text{ J}}{5 \text{ kg}}} = 1.95 \text{ m/s}$$

Like the law of conservation of mechanical energy, the work–energy theorem of Equation 6-31 can be used as an alternative to Newton's laws for solving certain problems in mechanics. Since Equation 6-31 was derived from Newton's laws, any problem that can be solved using the work–energy theorem can also be solved directly from Newton's laws. Like energy conservation, the work–energy theorem is a useful tool for the analysis of many problems.

Questions

10. If a pulley with nonnegligible mass is used rather than a frictionless peg in Atwood's machine, the pulley has kinetic energy of rotation (which will be studied in Chapter 8). Would you expect the inclusion of this kinetic energy to increase or decrease the speed v for a given distance h?

11. What are the advantages and disadvantages of solving mechanics problems using energy methods compared with using Newton's laws?

6-8   The Conservation of Energy

In the macroscopic world, nonconservative forces are always present to some extent, the most common being frictional forces. Another type of nonconservative force is that involved in the deformations of objects. For example, if a spring is stretched beyond its elastic limit, it becomes permanently deformed, and the work done in stretching the spring is not recovered when the spring is released. Some of the work done in deforming the spring is dissipated into thermal energy, so the spring becomes warmer. You may have observed that when you bend a coat hanger back and forth until it breaks, the coat hanger becomes warm.

Since mechanical energy is often not conserved, the importance of energy was not realized until the nineteenth century, when it was discovered that the disappearance of macroscopic mechanical energy is accompanied by the appearance of thermal energy, usually indicated by an increase in temperature. We now know that, on the microscopic scale, this thermal energy consists of the kinetic and potential energies of the molecules in the system. When we carefully define a system—for example, several objects and their local surroundings—we find that even when thermal energy and other types of energy, such as chemical energy, are included, the total energy of the system does not always remain constant. The energy of a system often decreases because of some form of radiation, such as sound waves from a collision between two objects, water waves produced by a ship, or electromagnetic waves produced by accelerated charges in a radio antenna. The energy of a system can also increase through the absorption of radiant energy. For example, the earth absorbs radiant energy from the sun. However, *the increase or decrease in the energy of a system can always be accounted for by the appearance or disappearance of some kind of energy somewhere else.* This experimental result is known as the **law of conservation of energy.** It is one of the most important laws in all science. Let E_{sys} be the total energy of a given system, E_{in} be the energy put into the system, and E_{out} be the energy that leaves the system. The law of conservation of energy then states:

Law of conservation of energy

$$E_{in} - E_{out} = \Delta E_{sys}$$

6-32

A common method of putting energy into a system is to do work on it. When work is done on a system, energy is transferred to the system from the agent or system doing the work. Energy put into a system may appear as an increase in the mechanical energy of the system, such as when a nonconservative force acts on a single particle within the system (the work–energy theorem), or it may appear as internal energy of the system. Energy can also be transferred into or out of a system in the form of heat. We will discuss internal energy and heat in more detail in Chapter 16 when we study the first law of thermodynamics, which is essentially a statement of the law of conservation of energy.

Example 6-18

A ball of putty of mass m is released from rest from a height h above the floor. It falls to the floor. (Plop!) Find the work done on the ball and discuss the application of the law of conservation of energy to this situation.

The two forces acting on the ball are the force of gravity and the contact force exerted by the floor. The work done by gravity is $+mgh$.

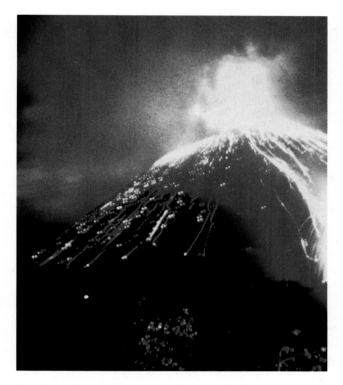

Geothermal energy is released dramatically in the eruption of Cerro Negro in Nicaragua.

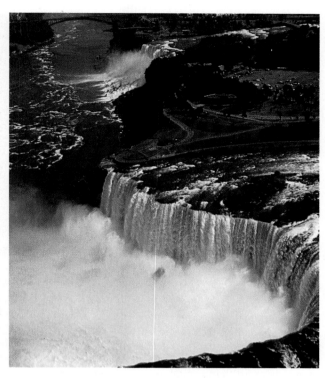

Niagara Falls. The potential energy of the water at the top of the falls is used to produce electrical energy.

The work done by the floor is zero because the point of application of the force does not move. Thus the total work done on the ball is mgh. Energy is transferred to the ball by the work done on it by gravity. This energy appears as kinetic energy of the ball before it hits the floor and as thermal energy within the ball after. The ball warms slightly and the energy is eventually transferred to the ball's surroundings as heat.

If we consider a system consisting of the ball and earth, there is no work done on the system because the work done by gravity is internal to the system. The original potential energy of the system is converted into internal thermal energy of the ball. Note that we *cannot* treat the ball of putty as a particle and apply the work–energy theorem. Although the only work done is by a conservative force, the total mechanical energy of the ball is not conserved.

There are situations in which some of a system's internal energy is converted into mechanical energy with no work being done by any outside agent. Consider a car that starts from rest and accelerates along a horizontal road such that the wheels do not slip. The net force on the car that accelerates it is the force of static friction exerted by the road. This force does no work. According to our definition of work, the point of application of the force must move through a distance. Since the road and the tire are always instantaneously at rest, no energy is transferred to the car by the road. The increase in the kinetic energy of the car comes from the chemical energy in the fuel that is being burned in the car engine. According to the work–energy theorem, the total energy of the car does not change. Its mechanical kinetic energy increases, so its internal chemical energy must decrease by

(a)

(b)

(a) This entire pizza loaded with meat and cheese contains only about 16 megajoules of energy. (b) The Quasar 3C 273 is shown imaged via x-ray energy. The x-ray energy emitted by this quasar is more than a million times that emitted by the entire Milky Way.

the same amount. Walking provides a similar example. To walk forward, you push back on the floor and the floor pushes forward on you with the force of static friction. This force accelerates you forward, but it does no work. There is no displacement of the point of application of the force and no energy is transferred from the floor to your body. The kinetic energy of your body comes from the conversion of chemical energy in your body derived from the food you eat.

Example 6-19

A man of mass m walks with a small constant speed up a flight of stairs to a height h. Find the work done on the man, and discuss the application of energy conservation to this situation.

There are two forces exerted on the man, the force of gravity exerted by the earth, and the contact force of the stairs on the man's feet. The work done by gravity is $-mgh$. This work is negative because the force is in the opposite direction of the displacement. The work done by the stairs is zero because the point of application of this force does not undergo any displacement. Thus the total work done on the man is $-mgh$. Since this is less than zero, we conclude from the law of conservation of energy that (neglecting any heat losses) the internal energy of the man decreases by mgh. (Actually, heat losses are not negligible for this problem. Because the body is relatively inefficient, the amount of chemical energy converted in the man's body will be considerably greater than mgh. The excess energy is eventually transferred from the man to his surroundings as heat.)

If we consider a system consisting of the man and the earth, there is no work done on the system because the work done by gravity is internal to this system. The mechanical energy of this system is increased by the increase in potential energy mgh. This increase in mechanical potential energy comes from a decrease in the internal chemical energy of the man. Again we note that we cannot treat the man as a particle. In this case the total mechanical energy of the man–earth system is increased even though there is no work done by any nonconservative force acting on the system.

6-9   Power

Power is the rate at which energy is transferred from one system to another. Consider a particle with instantaneous velocity **v**. In a short time interval dt, the particle has a displacement $d\mathbf{s} = \mathbf{v}\,dt$. The work done by a force **F** acting on the particle during this time interval is

$$dW = \mathbf{F}\cdot d\mathbf{s} = \mathbf{F}\cdot\mathbf{v}\,dt$$

The rate at which the force does work is the power input P of the force:

$$P = \frac{dW}{dt} = \mathbf{F}\cdot\mathbf{v} \qquad\qquad 6\text{-}33 \qquad \textit{Power defined}$$

The SI unit of power, one joule per second, is called a **watt** (W):

$$1\ \text{J/s} = 1\ \text{W} \qquad\qquad 6\text{-}34$$

Power should not be confused with work or energy. A car is said to be powerful if it can convert the chemical energy in its fuel into kinetic energy (or potential energy if the car is climbing a hill) in a short period of time. You can increase the chemical energy of a car by increasing the amount of fuel it carries. To increase its power, you must increase the *rate* at which the fuel is burned by increasing the number or size of the cylinders. When you pay your electric or gas bill to the power company you are paying for energy, not power. You are usually charged by the kilowatt-hour (kW·h). A **kilowatt-hour** of energy is

$$1\ \text{kW·h} = (10^{3}\ \text{W}) \times (3600\ \text{s}) = 3.6 \times 10^{6}\ \text{W·s} = 3.6\ \text{MJ} \qquad 6\text{-}35$$

In the U.S. customary system, the unit of energy is the foot-pound and the unit of power is the foot-pound per second. A common multiple of this unit is called a **horsepower** (hp):

$$1\ \text{hp} = 550\ \text{ft·lb/s} = 746\ \text{W}$$

In this power plant in Kansas, energy stored in the fossil fuel coal (black area at the right) is released by burning the coal to produce steam; the steam is then used to drive turbines to produce electricity. The excess heat is dissipated by the cooling towers.

Example 6-20

A small motor is used to power a lift that raises a load of bricks weighing 800 N to a height of 10 m in 20 s. What is the minimum power the motor must provide?

If the bricks are lifted without acceleration, the upward force equals the force of gravity, which is 800 N. The speed of the bricks is (10 m)/(20 s) = 0.5 m/s. Since the applied force is in the direction of motion, the power input of the force is

$$P = Fv = (800 \text{ N})(0.5 \text{ m/s}) = 400 \text{ N·m/s} = 400 \text{ J/s} = 400 \text{ W}$$

If there are no mechanical energy losses—to frictional forces, for example—the motor must have a power output of 400 W, which is a little more than $\frac{1}{2}$ horsepower.

Example 6-21

A 1000-kg car travels at a constant speed of 100 km/h = 28 m/s = 62 mi/h up a 10 percent grade. (A 10 percent grade means that the road is inclined at an angle θ given by tan θ = 0.1; that is, for each 10 m traveled horizontally, the gain in elevation is 1 m.) The total frictional force (rolling friction plus air drag) acting on the car is 700 N. Neglecting internal energy losses, what is the minimum power that must be delivered by the car's engine?

Some of the power delivered by the car's engine goes into increasing the potential energy of the car as it climbs the hill, and some goes into work against friction. From Figure 6-26 we can see that when the car travels a distance s along the road, it climbs a height h, which is related to s by $h = s \sin \theta \approx s \tan \theta = s/10$. We can use the approximation tan $\theta \approx \sin \theta$ because the angle is small. The potential energy of the car is thus

$$U = mgh = (0.1)mgs$$

The rate of change of the potential energy is

$$\frac{dU}{dt} = mg \frac{dh}{dt} = (0.1)mg \frac{ds}{dt} = (0.1)mgv$$

where $v = ds/dt$ is the speed of the car. Substituting the numerical values for m and g, we obtain

$$\frac{dU}{dt} = (0.1)(1000 \text{ kg})(9.81 \text{ N/kg})(28 \text{ m/s}) = 27.5 \text{ kW}$$

The power input of the force of friction is

$$P_f = \mathbf{f} \cdot \mathbf{v} = -fv = -(700 \text{ N})(28 \text{ m/s}) = -19.6 \text{ kW}$$

The total power that the engine must deliver is thus 27.5 kW + 19.6 kW = 47.1 kW = 63.1 hp. Actually, the power needed is considerably greater than this because a car is only about 15 percent efficient; that is, only about 15 percent of the power produced by the engine is delivered to power the car. The other 85 percent is lost to the cooling system, heat exhaust, and internal friction.

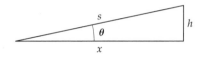

$$\tan \theta = h/x \sim \sin \theta = h/s$$

Figure 6-26 Relation between height climbed and distance traveled for car driving up a 10 percent grade. Here tan θ = 0.10 is given, and sin $\theta \approx$ tan θ because the angle is small.

Summary

1. The work done by a constant force is the product of the component of the force in the direction of motion and the displacement of the point of application of the force:

$$W = F \cos \theta \, \Delta x = F_x \, \Delta x$$

2. The work done by a variable force equals the area under the force-versus-distance curve:

$$W = \int_{x_1}^{x_2} F_x \, dx = \text{area under the } F_x\text{-versus-}x \text{ curve}$$

3. Kinetic energy is energy associated with the motion of an object and is related to its mass and speed by

$$K = \tfrac{1}{2}mv^2$$

4. The total work done on a particle equals the change in the kinetic energy of the particle. This is the work–energy theorem:

$$W_{\text{total}} = \Delta K = \tfrac{1}{2}mv_f^2 - \tfrac{1}{2}mv_i^2$$

5. The SI unit of work and energy is the joule (J):

$$1 \, \text{J} = 1 \, \text{N·m}$$

6. The dot product or scalar product of two vectors is defined by

$$\mathbf{A} \cdot \mathbf{B} = AB \cos \phi$$

where ϕ is the angle between the vectors. In terms of the components of the vectors, the dot product is

$$\mathbf{A} \cdot \mathbf{B} = A_x B_x + A_y B_y + A_z B_z$$

The work done on a particle by a force $\mathbf{F}$ for a small displacement $d\mathbf{s}$ is written

$$dW = \mathbf{F} \cdot d\mathbf{s}$$

and the work done on a particle that moves from point 1 to point 2 is

$$W = \int_{s_1}^{s_2} \mathbf{F} \cdot d\mathbf{s}$$

7. A force is conservative if the total work it does on a particle is zero when the particle moves along any closed path that returns it to its initial position. The work done by a conservative force on a particle is independent of how the particle moves from one point to another.

8. The potential energy of a system is the energy associated with the configuration of the system. The change in the potential energy of a system is defined as the negative of the work done by a conservative force acting on the system:

$$dU = -\mathbf{F} \cdot d\mathbf{s}$$

$$\Delta U = U_2 - U_1 = -W = -\int_{s_1}^{s_2} \mathbf{F} \cdot d\mathbf{s}$$

The work done on a system by a conservative force equals the decrease in the potential energy of the system. The absolute value of potential energy is unimportant. Only changes in potential energy are important.

9. The gravitational potential energy of an object of mass m at a height y above some reference point is

$$U = mgy$$

The potential energy of a spring with a force constant k when it is stretched or compressed a distance x from equilibrium is given by

$$U = \tfrac{1}{2}kx^2$$

10. In one dimension, a conservative force equals the negative derivative of the potential-energy function associated with it:

$$F_x = -\frac{dU}{dx}$$

At a minimum on the curve of the potential-energy function versus the displacement, the force is zero and the system is in stable equilibrium. At a maximum, the force is zero and the system is in unstable equilibrium. A conservative force always tends to accelerate a particle toward a position of lower potential energy.

11. If only conservative forces do work on an object, the sum of the kinetic and potential energies of the object remains constant:

$$E = K + U = \tfrac{1}{2}mv^2 + U = \text{constant}$$

This is the law of conservation of mechanical energy.

12. The work done by a nonconservative force acting on a particle equals the change in the total mechanical energy of the system:

$$W_{nc} = \Delta(U + K) = \Delta E$$

This is the generalized work–energy theorem.

The conservation of mechanical energy and the generalized work–energy theorem can be used as an alternative to Newton's laws to solve mechanics problems that require the determination of the speed of a particle as a function of its position.

13. The total energy of a system can include other types of energy such as thermal energy or internal chemical energy as well as mechanical energy. The energy of a system can be changed by various means such as emission or absorption of radiation, work done on the system, or heat transfer. The increase or decrease in the energy of the system can always be accounted for by the appearance or disappearance of some kind of energy somewhere else, an experimental result known as the law of conservation of energy:

$$E_{in} - E_{out} = \Delta E_{sys}$$

14. Power is the rate at which energy is transferred from one system to another. If a force **F** acts on a particle that moves with velocity **v** the power input of the force is

$$P = \frac{dW}{dt} = \mathbf{F} \cdot \mathbf{v}$$

The SI unit of power is the watt (W), which equals one joule per second. A common energy unit is the kilowatt-hour, which equals 3.6 megajoules.

Suggestions for Further Reading

"Energy", Chapter 7 of *World Resources 1988–89: A Report by the World Resources Institute and the International Institute for Environment and Development in Collaboration with the United Nations Environment Programme*, Basic Books, Inc., New York, 1988.

Charts, tables, and text present trends in energy use and conservation around the world. This book also contains similar information on population, food, forests, wildlife, the atmosphere, and the oceans.

Dyson, Freeman J.: "Energy in the Universe," *Scientific American*, September 1971, p. 50.

This article introduces a Scientific American *special issue on energy by examining the different forms in which energy manifests itself and how it is converted from one form to another.*

Gosz, James R., Richard T. Holms, Gene E. Likens, and F. Herbert Bermann: "The Flow of Energy in a Forest Ecosystem," *Scientific American*, March 1978, p. 92.

The various ways in which energy from the sun is transferred among living things in a forest and how efficiently they use it are examined.

Heinrich, Bernd: "The Energetics of the Bumblebee," *Scientific American*, April 1973, p. 96.

Kingsolver, Joel G.: "Butterfly Engineering," *Scientific American*, August 1985, p. 106.

These two articles describe how bumblebees and butterflies metabolize nectar and absorb thermal energy from the sun to achieve body temperatures high enough to enable them to fly.

Walker, Jearl: "The Amateur Scientist: Fly Casting Illuminates the Physics of Fishing," *Scientific American*, July 1985, p. 122.

The conservation of kinetic energy helps to explain how it is possible to cast a fly so far; a consideration of drag makes the model more realistic; and the forces involved in fighting a fish once it is on the line are analyzed.

Review

A. Objectives: After studying this chapter, you should:

1. Know the definitions of work, kinetic energy, potential energy, and power.

2. Be able to distinguish between conservative and nonconservative forces, and know the criterion for a force to be conservative.

3. Be able to find the potential-energy function from a given conservative force as a function of position.

4. Be able to state the law of conservation of mechanical energy and use it in solving problems.

5. Be able to state the work–energy theorem and use it in solving problems.

B. Define, explain, or otherwise identify:

Work
Energy
Joule
Foot-pound
Kinetic energy
Work–energy theorem
Scalar product
Dot product
Potential energy
Conservative force
Gravitational potential energy
Potential energy of a spring

Stable equilibrium
Unstable equilibrium
Total mechanical energy
Law of conservation of mechanical energy
Generalized work–energy theorem
Law of conservation of energy
Power
Watt
Kilowatt-hour
Horsepower

C. True or false: If the statement is true, explain why it is true. If it is false, give a counterexample.

1. Only the net force acting on an object can do work.

2. No work is done on a particle that remains at rest.

3. Work is the area under the force-versus-time curve.

4. A force that is always perpendicular to the velocity of a particle does no work on the particle.

5. A kilowatt-hour is a unit of power.

6. Only conservative forces can do work.

7. If only conservative forces act, the kinetic energy of a particle does not change.

8. The work done by a conservative force decreases the potential energy associated with that force.

Problems

Level I

6-1 Work and Kinetic Energy: Motion in One Dimension with Constant Forces

1. A 10-g bullet has a speed of 1.2 km/s. (*a*) What is its kinetic energy in joules? (*b*) What is the bullet's kinetic energy if its speed is halved? (*c*) If its speed is doubled?

2. Find the kinetic energy in joules of (*a*) a 0.145-kg baseball moving with a speed of 40 m/s and (*b*) a 60-kg jogger running at a steady pace of 9 minutes per mile.

3. From the conversion factors for converting newtons to pounds and meters to feet, derive the relation between the joule and the foot-pound given in Equation 6-3.

4. A 5-kg box is raised from rest a distance of 4 m by a vertical force of 80 N. Find (*a*) the work done by the force, (*b*) the work done by gravity, and (*c*) the final kinetic energy of the box.

6-2 Work Done by a Variable Force

5. A 2-kg particle is moving with a speed of 3 m/s when it is at $x = 0$. It is subjected to a single force F_x that varies with position as shown in Figure 6-27. (*a*) What is the kinetic energy of the particle when it is at $x = 0$? (*b*) How much work is done by the force as the particle moves from $x = 0$ to $x = 4$ m? (*c*) What is the speed of the particle when it is at $x = 4$ m?

Figure 6-27   Problem 5.

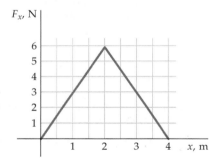

6. A 4-kg particle is initially at rest at $x = 0$. It is subjected to a single force F_x that varies with position as shown in Figure 6-28. Find the work done by the force as the particle moves (*a*) from $x = 0$ to $x = 3$ m and (*b*) from $x = 3$ m to $x = 6$ m. Find the kinetic energy of the particle when it is at (*c*) $x = 3$ m and (*d*) $x = 6$ m.

7. A force F_x is related to the position of a particle by the formula $F_x = Cx^3$, where C is a constant. Find the work done by this force on the particle when the particle moves from $x = 0$ to $x = 3$ m.

6-3 Work and Energy in Three Dimensions and the Dot Product

8. A block of mass 6 kg slides down a frictionless incline. The angle of the incline is 60°. (*a*) List all the forces acting on the block, and find the work done by each force when the block slides 2 m (measured along the incline). (*b*) What is the total work done on the block? What is the speed of the block after it has slid 2 m if (*c*) it starts from rest and (*d*) it starts with an initial speed of 3 m/s?

9. A 100-kg cart is raised up a 1-m step by being rolled up an incline formed by a plank of length L that has been laid from the lower level to the top of the step. (Assume the rolling is equivalent to sliding without friction.) (*a*) Find the force parallel to the incline needed to push the cart up without acceleration for $L = 3$, 4, and 5 m. (*b*) Calculate directly from Equation 6-15 the work needed to push the cart up the incline for each value of L. (*c*) Since the work found in (*b*) is the same for each value of L, what advantage, if any, is there in choosing one length over another?

10. A 2-kg object attached to a string moves with a speed of 1.5 m/s in a circle of radius 3 m on a frictionless horizontal surface. (*a*) Find the tension in the string. (*b*) List the forces acting on the object, and find the work done by each force during 1 revolution.

11. Two vectors **A** and **B** have magnitudes of 6 m and make an angle of 60° with each other. Find **A·B**.

12. Find **A·B** for the following vectors: (*a*) **A** = 2i − 6j, **B** = −3i + 2j; (*b*) **A** = 4i + 4j, **B** = 2i − 3j; and (*c*) **A** = 3i + 4j, **B** = 4i − 3j.

13. Find the angles between the vectors **A** and **B** in Problem 12.

14. A 2-kg object is given a displacement **Δs** = 3 mi + 3 mj − 2 mk along a straight line. During the displacement, a constant force **F** = 2 Ni − 1 Nj + 1 Nk acts on the object. (*a*) Find the work done by **F** for this displacement. (*b*) Find the component of **F** in the direction of the displacement.

15. (*a*) Given any vector **A** = A_xi + A_yj + A_zk, prove that the x component of **A** is given by **A·i**. (*b*) Find the unit vector that is parallel to **A**. (*c*) Find the component of the vector 2i + j + k in the direction of the vector 3i + 4j.

Figure 6-28   Problem 6.

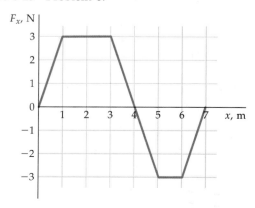

6-4 Work and Energy for Particle Systems: Potential Energy

16. A man of mass 80 kg climbs up stairs to a height of 5 m. What is the increase in the potential energy of the man?

17. State which of the following forces are conservative and which are nonconservative: (*a*) the frictional force exerted on a sliding box, (*b*) the force exerted by a linear spring that obeys Hooke's law, (*c*) the force of gravity, and (*d*) the wind resistance on a moving car.

18. A 2-kg book is held at a height of 20 m above the ground and is released at $t = 0$. (*a*) What is the original potential energy of the book relative to the ground? (*b*) From Newton's laws, find the distance the book falls in 1 s and its speed at $t = 1$ s. (*c*) Find the potential energy and the kinetic energy of the book at $t = 1$ s. (*d*) Find the kinetic energy and the speed of the book just before it hits the ground.

19. A 2-kg box slides down a long, frictionless incline of angle 30°. It starts from rest at time $t = 0$ at the top of the incline at a height of 20 m above the ground. (*a*) What is the original potential energy of the box relative to the ground? (*b*) From Newton's laws, find the distance the box travels in 1 s and its speed at $t = 1$ s. (*c*) Find the potential energy and the kinetic energy of the box at $t = 1$ s. (*d*) Find the kinetic energy and the speed of the box just as it reaches the bottom of the incline.

20. How high must an object be lifted to gain an amount of potential energy equal to the kinetic energy it has when it is moving at a speed of 20 m/s?

21. A force $F_x = 4$ N is constant. (*a*) Find the potential-energy function U associated with this force for an arbitrary choice of where $U = 0$. (*b*) Find U such that $U = 0$ at $x = 6$ m. (*c*) Find U such that $U = 12$ J at $x = 6$ m.

22. A spring has a force constant of $k = 10^4$ N/m. How far must it be stretched for its potential energy to be (*a*) 50 J and (*b*) 100 J?

6-5 Potential Energy and Equilibrium in One Dimension

23. Figure 6-29 shows the plot of a potential-energy function U versus x. (*a*) At each point indicated, state whether the force F_x is positive, negative, or zero. (*b*) At which point does the force have the greatest magnitude? (*c*) Identify any equilibrium points, and state whether the equilibrium is stable, unstable, or neutral.

Figure 6-29 Problem 23.

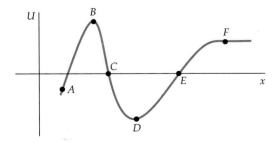

24. (*a*) Find the force F_x associated with the potential-energy function $U = Ax^4$, where A is a constant. (*b*) At what point or points is the force zero?

25. A potential-energy function is given by $U = C/x$, where C is a positive constant. (*a*) Find the force F_x as a function of x. (*b*) Is this force directed toward the origin or away from it? (*c*) Does the potential energy increase or decrease as x increases? (*d*) Answer parts (*b*) and (*c*) if C is a negative constant.

26. On the potential-energy curve for U versus y shown in Figure 6-30, the segments AB and CD are straight lines. Sketch a plot of the force F_y versus y.

Figure 6-30 Problem 26.

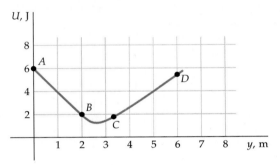

6-6 The Conservation of Mechanical Energy

27. A woman places a 2-kg block against a horizontal spring of force constant $k = 300$ N/m and compresses it 9 cm. (*a*) Find the work done by the woman and the work done by the spring. (*b*) The block is released and leaves the spring when the spring is at its original position. Find the speed of the block when it leaves the spring.

28. A 3-kg block slides along a frictionless horizontal surface with speed of 7 m/s. After sliding a distance of 2 m, it encounters a frictionless ramp inclined at an angle of 40° to the horizontal. How far up the ramp does the block slide before coming momentarily to rest?

29. The 3-kg object in Figure 6-31 is released from rest at a height of 5 m on a curved frictionless ramp. At the foot of the ramp is a spring of force constant $k = 400$ N/m. The object slides down the ramp and into the spring, compressing it a distance x before coming momentarily to rest. (*a*) Find x. (*b*) What happens to the object after it comes to rest?

Figure 6-31 Problem 29.

30. The system shown in Figure 6-32 is at rest when the lower string is cut. Find the speed of the objects when they are at the same height.

31. An object slides along the frictionless track shown in Figure 6-33. Initially, it is at point P, headed downhill with speed v_0. Describe the motion of the object in as much detail as you can for (a) $v_0 = 7$ m/s and (b) $v_0 = 12$ m/s. (c) What is the least speed needed by the object to get past point Q?

32. In Figure 6-34, the blocks are initially at rest. Choose $U = 0$ at this initial position. (a) Write an expression for the total mechanical energy of the system after the 2-kg block has fallen a distance y. (b) Find the speed of the 2-kg mass after it has fallen from rest a distance of 2 m, assuming no friction.

33. A 15-g ball is shot from a spring gun whose spring has a force constant of 600 N/m. The spring can be compressed 5 cm. How high can the ball be shot if the gun is aimed vertically?

6-7 The Generalized Work–Energy Theorem

34. A 2000-kg car moving at an initial speed of 25 m/s along a horizontal road skids to a stop in 60 m. (a) Find the work done by the force of kinetic friction. (b) Find the coefficient of kinetic friction between the tires and the road.

35. An 8-kg sled is initially at rest on a horizontal road. The coefficient of kinetic friction between the sled and the road is 0.4. The sled is pulled a distance of 3 m by a force of 40 N applied to the sled at an angle of 30° to the horizontal. (a) Find the work done by the applied force. (b) Find the work done by friction. (c) Find the change in the kinetic energy of the sled. (d) Find the speed of the sled after it has traveled 3 m.

36. Suppose that the surfaces in Problem 28 are not frictionless and that the coefficient of kinetic friction between the block and the surfaces is 0.30. Find (a) the speed of the block when it reaches the ramp and (b) the distance up the ramp the block slides before coming momentarily to rest.

37. The 2-kg block in Figure 6-35 slides down a frictionless curved ramp from rest at a height of 3 m. It then slides 9 m on a rough horizontal surface before coming to rest. (a) What is the speed of the block at the bottom of the ramp? (b) How much work is done by friction? (c) What is the coefficient of friction between the block and the horizontal surface?

38. The coefficient of kinetic friction between the 4-kg block and the table in Figure 6-34 is 0.35. (a) Find the work done by friction when the 2-kg block falls a distance y. (b) Find the total mechanical energy E of the system after the 2-kg block falls a distance y, assuming that $E = 0$ initially. (c) Use your result for (b) to find the speed of either block after the 2-kg block falls 2 m.

6-8 The Conservation of Energy

39. In a volcanic eruption, 4 km³ of mountain with a density of 1600 kg/m³ was lifted an average height of 500 m. (a) How much energy in joules was released in this eruption? (b) The energy released by bombs is measured

in megatons of TNT, where 1 megaton of TNT = 4.2×10^{15} J. Convert your answer for (a) to megatons of TNT.

Figure 6-32   Problem 30.

Figure 6-33   Problem 31.

Figure 6-34   Problems 32 and 38.

Figure 6-35   Problem 37.

40. An 80-kg physics student climbs a 120-m hill. (*a*) What is the increase in the gravitational potential energy of the student? (*b*) Where does this energy come from? (*c*) The student's body is 20 percent efficient; that is, for every 20 J that are converted to mechanical energy, 100 J of internal energy are expended, with 80 J going into heat. How much internal energy is expended by the student during the climb?

41. A 70-kg skater pushes off the wall of a skating rink, acquiring a speed of 4 m/s. (*a*) How much work is done on the skater? (*b*) What is the change in the mechanical energy of the skater? (*c*) Discuss the conservation of energy as applied to the skater.

6-9 Power

42. A 4-kg box is lifted by a force equal to the weight of the box. The box moves upward at a constant velocity of 2 m/s. (*a*) What is the power input of the force? (*b*) How much work is done by the force in 3 s?

43. A constant horizontal force $F = 3$ N drags a box along a rough horizontal surface at a constant speed v. The force does work at the rate of 5 W. (*a*) What is the speed v? (*b*) How much work is done by F in 3 s?

44. A single force of 5 N acts in the x direction on a 12-kg object. (*a*) If the object starts from rest at $x = 0$ at time $t = 0$, find its velocity v as a function of time. (*b*) Write an expression for the power input as a function of time. (*c*) What is the power input of the force at time $t = 3$ s?

45. Water flows over Victoria Falls, which is 100 m high, at an average rate of 1.4×10^6 kg/s. If all the potential energy of the water were converted into electric energy, how much power could be produced by these falls?

46. On a clear day, solar energy falls on a certain house at an average rate of 400 W/m^2 for 8 h. How much energy is collected by a large glass window of area 40 m^2?

47. Our bodies convert internal chemical energy into work and heat at the rate of about 100 W, which is called our metabolic rate. (*a*) How much internal chemical energy do we use in 24 h? (*b*) The energy comes from the food that we eat and is usually measured in kilocalories, where 1 kcal = 4.184 kJ. How many kilocalories of food energy must we ingest per day if our metabolic rate is 100 W?

48. Find the power input of a force **F** acting on a particle that moves with a velocity **v** for (*a*) **F** = 3 N **i** + 4 N **k**, **v** = 2 m/s **i**; (*b*) **F** = 5 N **i** − 6 N **j**, **v** = −5 m/s **i** + 4 m/s **j**; and (*c*) **F** = 2 N **i** + 4 N **j**, **v** = 6 m/s **i** + 3 m/s **j**.

Level II

49. A child is swinging from a suspended rope 4.0 m long that will break if the tension equals twice the weight of the child. (*a*) What is the greatest angle θ_0 the rope can make with the vertical during a swing if the rope is not to break? (*b*) If the angle slightly exceeds that found in (*a*), what is the speed of the child when the rope breaks?

50. A pendulum of length L has a bob of mass m. It is released from some angle θ_1. The string hits a peg at a distance x directly below the pivot as shown in Figure 6-36, effectively shortening the length of the pendulum.

Figure 6-36   Problem 50.

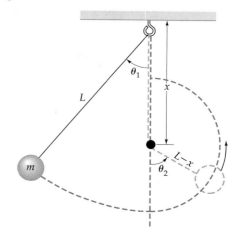

Find the maximum angle θ_2 between the string and the vertical when the bob is to the right of the peg.

51. A force F_x acting on a particle is shown as a function of x in Figure 6-37. (*a*) From the graph, calculate the work done by the force when the particle moves from $x = 0$ to the following values of x: −4, −3, −2, −1, 0, 1, 2, 3, and 4 m. (*b*) Plot the potential energy U versus x for the range of values of x from −4 m to +4 m, assuming that $U = 0$ at $x = 0$.

Figure 6-37   Problem 51.

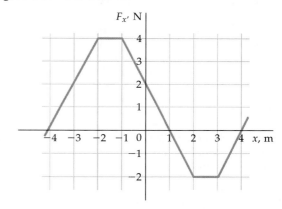

52. Repeat Problem 51 for the force F_x shown in Figure 6-38.

Figure 6-38   Problem 52.

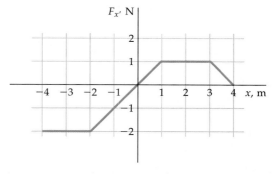

53. A 3-kg object is moving with a speed of 1.50 m/s in the x direction. When it passes the origin, it is acted on by a single force F_x, which varies with x as shown in Figure 6-39. (a) Find the work done by the force from $x = 0$ to $x = 2$ m. (b) What is the kinetic energy of the object at $x = 2$ m? (c) What is the speed of the object at $x = 2$ m? (d) Find the work done on the object from $x = 0$ to $x = 4$ m. (e) What is the speed of the object at $x = 4$ m?

Figure 6-39 Problem 53.

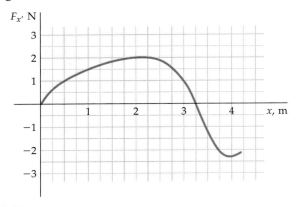

54. A T-bar tow is required to pull 80 skiers up a 600-m slope inclined at 15° to the horizontal at a speed of 2.5 m/s. The coefficient of kinetic friction is 0.06. Find the motor power required if the mass of the average skier is 75 kg.

55. A 0.2-kg ball is thrown upward with an initial speed of 24 m/s at an angle of 30° with the horizontal from the edge of a 60-m cliff. Neglecting air resistance, find (a) the initial kinetic energy of the ball, (b) the potential energy and kinetic energy of the ball when it is at its highest point, (c) the kinetic energy of the ball when it reaches the ground 60 m below the cliff, and (d) the speed of the ball just before it strikes the ground. Assume that $U = 0$ at the top of the cliff.

56. A small block is attached to a rubberlike material that exerts a force $F_x = -kx - ax^2$ when it is stretched a distance x ($x > 0$), where k and a are constants. Find the work done on the block by the material when it is stretched from $x = 0$ to $x = A$.

57. The force acting on a particle moving along the x axis is given by $F_x = -ax^2$, where a is a constant. Calculate the potential-energy function U relative to $U = 0$ at $x = 0$, and sketch a graph of U versus x.

58. A ski jumper of mass 70 kg starts from rest at A in Figure 6-40. His speed is 30.0 m/s at B and 23.0 m/s at C, where the distance BC is 30 m. (a) How much work is done by friction on the skier as he moves from B to C? (b) Find the greatest height above the level of C attained by the jumper.

59. A 2-kg block is released 4 m from a massless spring with a force constant $k = 100$ N/m that is fixed along a frictionless plane inclined at 30° as shown in Figure 6-41. (a) Find the maximum compression of the spring. (b) If the plane is not frictionless and the coefficient of kinetic friction between it and the block is 0.2, find the maximum compression. (c) For the rough incline, how far up the incline will the block travel after leaving the spring?

60. A train with a total mass of 2×10^9 g rises 707 m in a travel distance of 62 km at an average speed of 15.0 km/h. If the frictional force is 0.8 percent of the weight, find (a) the kinetic energy of the train, (b) the total change in its potential energy, (c) the work done on the train by friction, and (d) the power output of the train's engines.

61. A woman carries a large bucket of water to the top of a 40-m tower at a constant upward speed. The bucket has a mass of 10 kg and initially contains 30 kg of water, but it leaks at a constant rate and only 10 kg of water are in the bucket at the top of the tower. (a) Write an expression for the mass of the bucket plus water as a function of the height y climbed. (b) Find the work done by the woman on the bucket.

62. Two men stand at the edge of a 10-m cliff. Simultaneously, they throw balls at an initial speed of 10 m/s, one straight up and the other straight down. (a) Neglecting air resistance, find the speed of each ball when it reaches the bottom of the cliff. (b) Discuss qualitatively how the speeds would compare if air resistance were not neglected.

63. A 2-kg box is projected with an initial speed of 3 m/s up a rough plane inclined at 60° to the horizontal. The coefficient of kinetic friction is 0.3. (a) List all the forces acting on the box, and find the work done by each as the box slides up the plane. (b) How far up the plane does the box slide before it stops momentarily? (c) Find the work done by each force acting on the box as it slides back down the plane. (d) Find its speed when it reaches its initial position.

64. One expects accelerating to consume more energy than driving at a constant speed. (a) Calculate the energy required to give a 1200-kg car a speed of 50 km/h neglect-

Figure 6-40 Problem 58.

Figure 6-41 Problem 59.

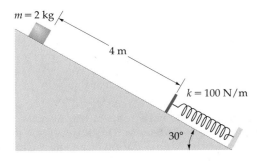

ing friction. (*b*) If friction of all types results in a retarding force of 300 N at a speed of 50 km/h, what is the energy needed to move the car a distance of 300 m at a constant speed of 50 km/h? (*c*) Assuming the energy losses due to friction in part (*a*) are 75 percent of those found in part (*b*), estimate the ratio of the energy consumption for the two cases considered.

65. A particle of mass m moves in a horizontal circle of radius r on a rough table. It is attached to a string fixed at the center of the circle. The speed of the particle is initially v_0. After completing one full trip around the circle, the speed of the particle is $\frac{1}{2}v_0$. (*a*) Find the work done by friction during that one revolution in terms of m, v_0, and r. (*b*) What is the coefficient of kinetic friction? (*c*) How many more revolutions will the particle make before coming to rest?

66. Water from behind a dam flows through a large turbine at a rate of 1.5×10^6 kg/min. The turbine is located 50 m below the surface of the reservoir, and the water leaves the turbine with a speed of 5 m/s. (*a*) Neglecting any energy dissipation, what is the power output of the turbine? (*b*) How many U.S. citizens would be supplied with energy by this dam if each citizen uses 3×10^{11} J of energy per year?

67. An elastic string has a natural length L_0 and a spring constant k. When an object of mass m is hanging from it vertically, the string stretches by y_0. One end of the string is attached to the top of a frictionless plane inclined at 30° to the horizontal. With the string lying along the incline, the object of mass m is attached to it and is released with the string in its unstretched condition. How far does the object slide down the incline before coming to rest for the first time?

68. Vectors **A**, **B**, and **C** form a triangle as shown in Figure 6-42. The angle between **A** and **B** is θ, and the vectors are related by $\mathbf{C} = \mathbf{A} - \mathbf{B}$. Compute $\mathbf{C}\cdot\mathbf{C}$ in terms of A, B, and θ, and derive the law of cosines, $C^2 = A^2 + B^2 - 2AB\cos\theta$.

Figure 6-42   Problem 68.

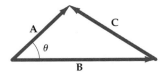

69. A car of mass 1500 kg traveling at 24 m/s is at the foot of a hill that rises 120 m in 2.0 km. At the top of the hill, the speed of the car is 10 m/s. Find the average power delivered by the car's engine, neglecting any frictional losses.

70. A straight rod of negligible mass is mounted on a frictionless pivot as shown in Figure 6-43. Masses m_1 and m_2 are suspended at distances l_1 and l_2 as shown. (*a*) Write an expression for the gravitational potential energy of the masses as a function of the angle θ made by the rod and the horizontal. (*b*) For what angle θ is the potential energy a minimum? Is the statement "systems tend to move toward minimum potential energy" consistent with your

result? (*c*) Show that if $m_1l_1 = m_2l_2$, the potential energy is the same for all values of θ. (When this holds, the system will balance at any angle θ. This result is known as *Archimedes' law of the lever*.)

Figure 6-43   Problem 70.

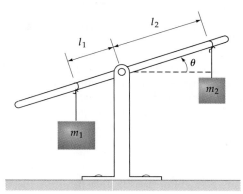

71. A force $\mathbf{F} = 2x^2\ \text{N/m}^2\ \mathbf{i}$ is applied to a particle. Find the work done on the particle as it moves a total distance of 5 m (*a*) parallel to the y axis from point $x = 2$ m, $y = 2$ m to point $x = 2$ m, $y = 7$ m and (*b*) in a straight line from $x = 2$ m, $y = 2$ m to $x = 5$ m, $y = 6$ m.

72. In a volcanic eruption, a 200-g piece of porous volcanic rock is thrown vertically upward with an initial speed of 40 m/s. It travels upward a distance of 50 m before it begins to fall back to the earth. (*a*) What is the initial kinetic energy of the rock? (*b*) How much work does air friction do on the rock? (*c*) If air friction does seven-tenths as much work on the rock on the way down as on the way up, what is the velocity of the rock when it returns to its initial position?

73. Two blocks of equal mass M are attached to the ends of a very light string hung over two frictionless pegs as shown in Figure 6-44. A third block of mass m is attached to the middle of the string between the pegs. (*a*) Find the potential energy of the system as a function of the distance y shown in the figure. (*b*) Find the equilibrium distance y_0 using the potential-energy function. Check your answer by analyzing the forces.

Figure 6-44   Problem 73.

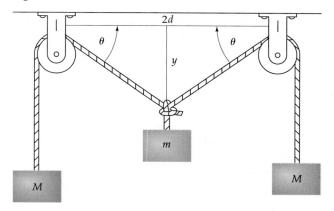

74. A large nuclear power plant produces 3000 MW of power by nuclear fission, which converts mass m into energy according to Einstein's equation $E = mc^2$, where $c = 3 \times 10^8$ m/s is the speed of light. (a) How much mass is converted into energy in one year? (b) In a coal-burning power plant, each kilogram of coal releases 31 MJ of energy when burned. How many kilograms of coal are needed each year for a 3000-MW plant?

75. In one model of running, the energy expended is assumed to go into accelerating and decelerating the legs. If the mass of the leg is m and the running speed is v, the energy needed to accelerate the leg from rest to v is $\frac{1}{2}mv^2$, and the same energy is needed to decelerate the leg back to rest for the next stride. Thus, the energy required for each stride is mv^2. Assume that the mass of a man's leg is 10 kg and that he runs at a speed of 3 m/s with 1 m between one footfall and the next. Therefore, the energy he must provide to his legs in each second is $3 \times mv^2$. Calculate the rate of the man's energy expenditure using this model and assuming that his muscles have an efficiency of 25 percent.

76. A rope with a mass per unit length of u lies coiled on the floor. (a) If a length y of the rope is picked up, what force F is required to hold it above the floor as shown in Figure 6-45? (b) Integrate $F\,dy$ from $y = 0$ to $y = L$ to find the total work needed to pick up a length L of the rope. (c) What is the potential energy of the rope when a length L has been picked up?

Level III

77. A small block of mass m slides without friction along the loop-the-loop track shown in Figure 6-46. The block starts from point P a distance h above the bottom of the loop. (a) What is the kinetic energy of the block when it reaches the top of the loop? (b) What is its acceleration at the top of the loop, assuming that it stays on the track? (c) What is the least value of h for which the block will reach the top of the loop without leaving the track?

78. A pendulum consists of a bob of mass m attached to a light rod of length L and negligible mass. The other end of the rod is fixed to a frictionless pivot. The pendulum is released when the bob is directly over the pivot. Find the force exerted by the rod on the bob when (a) the bob is at its lowest point, (b) the rod is at an angle of 30° below the horizontal, and (c) the rod is at an angle of 30° above the horizontal.

79. A pendulum bob is released from rest at an initial angle of θ_0 measured from $\theta = 0$ at the lowest point. Prove that the tension in the string at the bottom is greater than the tension when the bob is at rest by the amount $3K/L$, where L is the length of the string and K is the maximum kinetic energy of the bob.

80. A skier starts from rest at height H above the center of a rounded hummock of radius 4.0 m as shown in Figure 6-47. There is negligible friction. Find the maximum value of H for which the skier remains in contact with the snow at the peak of the hummock.

81. The skier of Problem 80 starts with a small initial speed from the top of the rounded hummock (Figure

6-47). If friction can be neglected, find (a) his speed v as a function of the angle θ, and (b) the angle θ at which he loses contact with the slope.

82. A 5-kg block is held against a spring of force constant 20 N/cm, compressing it 3 cm. The block is released and the spring extends, pushing the block along a rough horizontal surface. The coefficient of friction between the surface and the block is 0.2. (a) Find the work done on the block by the spring as it extends from its compressed position to its equilibrium position. (b) Find the work done by friction on the block while it moves the 3 cm to the equilibrium position of the spring. (c) What is the speed of the block when the spring is at its equilibrium position? (d) If the block is not attached to the spring, how far will it slide along the rough surface before coming to rest?

Figure 6-45   Problem 76.

Figure 6-46   Problem 77.

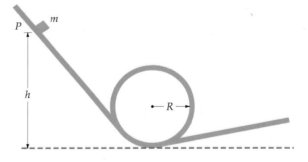

Figure 6-47   Problems 80 and 81.

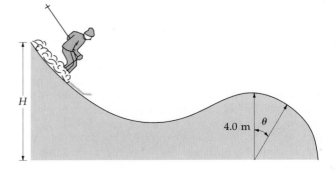

83. The potential-energy function of a particle of mass 4 kg in a force field is described by

$$U = 3x^2 - x^3 \qquad x \le 3 \text{ m}$$
$$U = 0 \qquad x \ge 3 \text{ m}$$

where U is in joules and x is in meters. (a) For what values of x is the force F_x zero? (b) Sketch a plot of U versus x. (c) Discuss the stability of the equilibrium for the values of x found in (a). (d) If the total energy of the particle is 12 J, what is its speed at $x = 2$ m?

84. A force is given by $F_x = Ax^{-3}$, where $A = 8$ N·m³. (a) For positive values of x, does the potential energy associated with this force increase or decrease with increasing x? You can determine the answer to this question by imagining what happens to a particle that is placed at rest at some point x and is then released. (b) Find the potential-energy function U associated with this force such that U approaches zero as x approaches infinity. (c) Sketch U versus x.

85. A force in the xy plane is given by $\mathbf{F} = A(10a\mathbf{i} + 3x\mathbf{j})$, where A and a are constants, F is in newtons, and x is in meters. Suppose that the force acts on a particle as it moves from an initial position $x = 4$ m, $y = 1$ m to a final position $x = 4$ m, $y = 4$ m. Show that this force is not conservative by computing the work done by the force for at least two different paths.

86. A force in the xy plane is given by $\mathbf{F} = (F_0/r)(y\mathbf{i} - x\mathbf{j})$, where F_0 is a constant and $r = \sqrt{x^2 + y^2}$. (a) Show that the magnitude of this force is F_0 and that its direction is perpendicular to $\mathbf{r} = x\mathbf{i} + y\mathbf{j}$. (b) Find the work done by this force on a particle that moves in a circle of radius 5 m centered at the origin.

87. A theoretical formula for the potential energy associated with the nuclear force between two protons, two neutrons, or a neutron and a proton is the *Yukawa potential*:

$$U = -U_0\left(\frac{a}{x}\right)e^{-x/a}$$

where U_0 and a are constants. (a) Sketch U versus x using $U_0 = 4$ pJ and $a = 2.5$ fm. (b) Find the force F_x. (c) Compare the magnitude of the force at the separation $x = 2a$ to that at $x = a$. (d) Compare the magnitude of the force at the separation $x = 5a$ to that at $x = a$.

88. The bob of a pendulum of length L is pulled aside so the string makes an angle θ_0 with the vertical and released. In Example 6-10, the conservation of energy was used to obtain the speed of the bob at the bottom of its swing. In this problem, you are to obtain the same result using Newton's second law. (a) Show that the tangential component of Newton's second law gives $dv/dt = -g \sin \theta$, where v is the speed and θ is the angle made by the string and the vertical. (b) Show that v can be written

$$v = L\frac{d\theta}{dt}$$

(c) Use this result and the chain rule for derivatives to obtain

$$\frac{dv}{dt} = \frac{dv}{d\theta}\frac{d\theta}{dt} = \frac{dv}{d\theta}\frac{v}{L}$$

(d) Combine the results of (a) and (c) to obtain

$$v \, dv = -gL \sin \theta \, d\theta$$

(e) Integrate the left side of the equation in part (d) from $v = 0$ to the final speed v and the right side from $\theta = \theta_0$ to $\theta = 0$, and show that the result is equivalent to $v = \sqrt{2gh}$, where h is the original height of the bob above the bottom.

89. (a) Using $v = dx/dt$, show that Equation 6-26b can be written

$$\frac{dx}{\sqrt{E - U}} = \sqrt{\frac{2}{m}} \, dt$$

For one-dimensional motion, U is a function of x, so the left side of this equation depends only on x (and not on t) and the right side depends only on t. (b) Apply this result to a particle of mass m on a spring with a force constant k oscillating with an amplitude A so that $U = \frac{1}{2}kx^2$ and $E = \frac{1}{2}kA^2$ to obtain the equation

$$\frac{dx}{\sqrt{A^2 - x^2}} = \sqrt{\frac{k}{m}} \, dt$$

(c) Integrate both sides of this equation to obtain an expression relating the position of the mass x to the time t, assuming that $x = A$ at time $t = 0$.

Chapter 7

Systems of Particles and the Conservation of Momentum

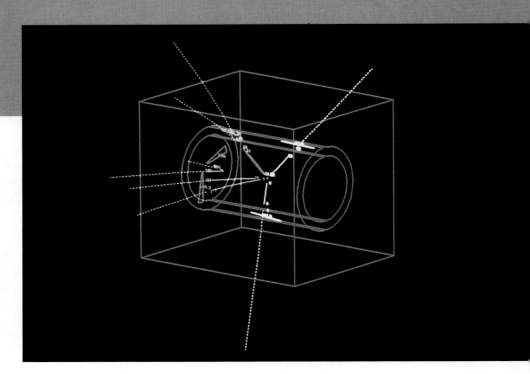

The collisions of elementary particles provide information about the basic interactions and the structure of matter. In this photograph, a proton and an antiproton collide producing a shower of other particles including the rare Z particle. The Z particle, which is electrically neutral, leaves no track, but it quickly decays into other particles, which are charged and do leave tracks. In this case it decays into an electron (pink track) and a positron (white track). The energy of the electron and positron can be measured by the curvature of their tracks in a magnetic field. The total energy of these decay particles equals $m_Z c^2$, where m_Z is the predicted mass of the Z particle.

Although our discussion of Newton's laws has been in terms of the motion of point particles, many of our applications have been to extended objects, such as blocks, balls, and even automobiles. We will justify these applications in this chapter, where we consider an extended object to be a system of point particles and assume that Newton's laws apply to each particle. We will show that there is one point, called the **center of mass,** of a system that moves as if the mass of the system were concentrated at that point and the external forces acting on the system were acting exclusively on that point. The motion of any object or system of particles, no matter how complex, can be described in terms of the motion of the center of mass (which may be thought of as the bulk motion of the system) plus the motion of the individual particles in the system relative to the center of mass.

We will also introduce an important new quantity, momentum, which is the product of the mass of a particle and its velocity. Momentum is important because the total momentum of a system of particles remains constant if the net external force acting on the system is zero. Like energy, momentum is a conserved quantity in an isolated system. Momentum conservation is very useful in analyzing collisions between billiard balls, automobiles, and subatomic particles in a nuclear reaction. It is also useful in analyzing the motion of a jet plane or rocket, and the recoil of a rifle.

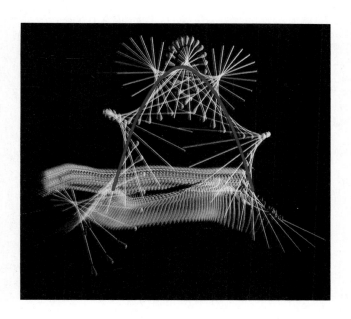

Figure 7-1 Multiflash photograph of a baton thrown into the air. The center of mass follows the same simple parabolic path it would if it were a single point particle.

Figure 7-1 is a multiflash photograph of a baton thrown into the air. Although the motion of the baton is complicated, the motion of one point, the center of mass, is simple. While the baton is in the air, the center of mass follows a parabolic path, the same path that would be followed by a point particle. We will show in general that the acceleration of the center of mass of a system of particles equals the net external force acting on the system divided by the total mass of the system. For the baton thrown into the air, the acceleration of the center of mass is **g** downward.

7-1   The Center of Mass

Let us first consider a simple system of two particles in one dimension. Let x_1 and x_2 be the coordinates of the particles relative to some arbitrarily chosen origin. The center-of-mass coordinate X_{cm} is then defined by

$$MX_{cm} = m_1x_1 + m_2x_2 \qquad 7\text{-}1$$

where $M = m_1 + m_2$ is the total mass of the system. For this case of just two particles, the center of mass lies at some point on the line joining the particles (Figure 7-2a). This is easily seen if we choose our origin to coincide with one of the particles, say m_1 (Figure 7-2b). Then x_2 is the distance d between the particles. The center-of-mass coordinate for this choice of origin is then obtained from Equation 7-1:

$$MX_{cm} = m_1x_1 + m_2x_2 = m_1(0) + m_2d$$

$$X_{cm} = \frac{m_2}{M}d = \frac{m_2}{m_1 + m_2}d \qquad 7\text{-}2$$

For particles of equal mass, the center of mass is halfway between the particles. Otherwise, it is closer to the particle of greater mass (Figure 7-2c).

Figure 7-2 The center of mass of a two-particle system. (a) If the particles have equal masses, the center of mass is midway between them. (b) If the origin is chosen at m_1, the position of the center of mass is given by $X_{cm} = m_2d/(m_1 + m_2)$. (c) If the particles have unequal masses, the center of mass is closer to the more massive particle.

(a)   (b)

(c)

Exercise
A 4-kg mass is at the origin and a 2-kg mass is at $x = 6$ cm. Find X_{cm}. (Answer: $X_{cm} = 2$ cm)

We can generalize from our special case of two particles in one dimension to many particles in three dimensions. If we have N particles, the x coordinate of the center of mass X_{cm} is defined by

$$MX_{cm} = m_1x_1 + m_2x_2 + m_3x_3 + \cdots + m_Nx_N = \sum_i m_ix_i \qquad 7\text{-}3a$$

where again $M = \Sigma\, m_i$ is the total mass of the system. Similar equations define the y and z coordinates of the center of mass:

$$MY_{cm} = \sum_i m_iy_i \qquad 7\text{-}3b$$

$$MZ_{cm} = \sum_i m_iz_i \qquad 7\text{-}3c$$

In vector notation, if $\mathbf{r}_i = x_i\mathbf{i} + y_i\mathbf{j} + z_i\mathbf{k}$ is the position vector of the ith particle, the position vector of the center of mass $\mathbf{R}_{cm}$ is given by

Center of mass defined, discrete systems

$$M\mathbf{R}_{cm} = \sum_i m_i\mathbf{r}_i \qquad 7\text{-}4$$

where $\mathbf{R}_{cm} = X_{cm}\mathbf{i} + Y_{cm}\mathbf{j} + Z_{cm}\mathbf{k}$.

For a continuous body, the sum in Equation 7-4 is replaced by an integral:

Center of mass defined, continuous systems

$$M\mathbf{R}_{cm} = \int \mathbf{r}\, dm \qquad 7\text{-}5$$

where dm is an element of mass that is located at position $\mathbf{r}$ as shown in Figure 7-3.

If two point masses are connected by a light rod of negligible mass, the system will balance on a pivot at the center of mass (Figure 7-4a). This follows from the fact that the gravitational potential energy of a system of particles is the same as it would be if the total mass were concentrated at the center of mass as we will now show. Let y_i be the height above the earth of the ith particle in a general system. The potential energy of the system is

$$U = \sum_i m_igy_i = g\sum_i m_iy_i$$

But

$$\sum_i m_iy_i = MY_{cm}$$

so

$$U = MgY_{cm} \qquad 7\text{-}6$$

If we try to balance our two particles on a light rod at some point other than the center of mass, the system will rotate until the potential energy is a minimum, which occurs when the center of mass is at its lowest possible point directly below the pivot (Figure 7-4b). (Recall from Section 6-5 that a conservative force tends to accelerate a particle toward lower potential energy.)

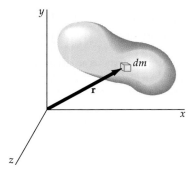

Figure 7-3 Mass element dm located at $\mathbf{r}$ for finding the center of mass by integration.

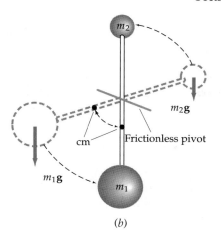

Figure 7-4 (*a*) Two masses connected by a light rod balance if a pivot is placed at the center of mass. (*b*) If the pivot is not placed at the center of mass, the potential energy of the system is least when the center of mass is directly below the pivot.

(*a*)

(*b*)

Figure 7-5 illustrates a simple experimental method for finding the center of mass of any plane object, such as a board or metal sheet. We suspend the object from a frictionless pivot at any point on the object. It will hang in equilibrium with its center of mass directly below the pivot. Thus, if we draw a vertical line downward from the pivot, the center of mass must lie somewhere on the line. We can then suspend the object from another point and draw a second vertical line. The center of mass will lie at the intersection of the two lines.

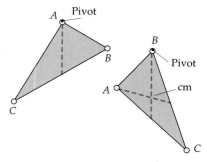

Example 7-1

Find the center of mass of a system consisting of three particles: $m_1 =$ 2 kg at the origin, $m_2 = 4$ kg on the y axis at $y = 3$ m, and $m_3 = 6$ kg on the x axis at $x = 4$ m (Figure 7-6).

Using Equation 7-3a, we have

$$MX_{cm} = \sum_i m_i x_i = m_1 x_1 + m_2 x_2 + m_3 x_3$$

where $M = m_1 + m_2 + m_3 = 12$ kg. Then

$$X_{cm} = \frac{\sum_i m_i x_i}{M} = \frac{(2 \text{ kg})(0) + (4 \text{ kg})(0) + (6 \text{ kg})(4 \text{ m})}{12 \text{ kg}} = \frac{24 \text{ kg·m}}{12 \text{ kg}} = 2 \text{ m}$$

Similarly, from Equation 7-3b we have

$$Y_{cm} = \frac{\sum_i m_i y_i}{M} = \frac{(2 \text{ kg})(0) + (4 \text{ kg})(3 \text{ m}) + (6 \text{ kg})(0)}{12 \text{ kg}} = 1 \text{ m}$$

Figure 7-5 The center of mass of an irregularly shaped plane object lies directly below any point about which the object is pivoted. This property can be used to locate the center of mass of the object.

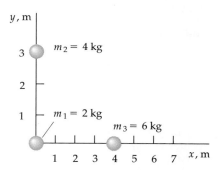

Figure 7-6 Three-particle system for Example 7-1.

There is a useful technique for solving Example 7-1 and other more complex problems. For instance, Equation 7-4 for this problem is

$$M\mathbf{R}_{cm} = m_1 \mathbf{r}_1 + m_2 \mathbf{r}_2 + m_3 \mathbf{r}_3$$

The first two terms on the right side of this equation are related to the center of mass of the first two particles:

$$m_1 \mathbf{r}_1 + m_2 \mathbf{r}_2 = (m_1 + m_2)\mathbf{r}'_{cm}$$

where $\mathbf{r}'_{cm}$ is the center of mass of the first two particles alone. The center of mass of the three-particle system can thus be written

$$M\mathbf{R}_{cm} = (m_1 + m_2)\mathbf{r}'_{cm} + m_3 \mathbf{r}_3$$

Figure 7-7   Figure 7-6 with the particles m_1 and m_2 replaced by a single particle of mass $m_1 + m_2$ located at the center of mass of the original two particles. The three-particle system has now been reduced to a two-particle system.

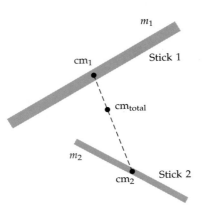

Figure 7-8   The center of mass of a system consisting of two uniform sticks can be found by treating each stick as a point particle at its individual center of mass.

Figure 7-9   Calculation of the center of mass of a uniform stick by integration. The mass element dm at a distance x from the origin is treated as a point particle.

We can thus solve this problem by first finding the center of mass for two of the particles and then replacing them with a single particle of total mass $m_1 + m_2$ at that center of mass. This is illustrated in Figure 7-7, where the first two particles have been replaced by a single particle of mass 6 kg on the y axis at $y = 2$ m. The center of mass of this particle and the 6-kg particle on the x axis is then halfway between the two equal mass particles, as shown in the figure. Figure 7-8 shows how this technique can be used to find the center of mass of two sticks whose mass is uniformly distributed. The center of mass of each stick separately is at the center of the stick, which can be found by balancing the stick. We can then replace the sticks with point masses and find their center of mass in the usual way.

For regularly shaped objects, we can use symmetry to find the center of mass. For example, the center of mass of a uniform cylinder or sphere is at the geometric center of the object.

We now illustrate the use of integration to find the center of mass.

Example 7-2

Find the center of mass of a uniform stick of mass M and length L.

This simple example, in which we already know the result from symmetry, illustrates the technique for setting up the integration indicated in Equation 7-5. We first set up a coordinate system with the x axis along the stick and one end of the stick at the origin as in Figure 7-9. Let us call the mass per unit length of the stick λ. Since the stick is uniform, $\lambda = M/L$. In the figure, we have indicated a mass element dm of length dx at a distance x from the origin. Since M is the total mass of the stick, the mass of an element of length dx is

$$dm = M\,\frac{dx}{L} = \frac{M}{L}\,dx = \lambda\,dx$$

Equation 7-5 thus gives

$$MX_{cm} = \int x\,dm = \int_0^L x\lambda\,dx = \frac{\lambda x^2}{2}\Big|_0^L = \frac{\lambda L^2}{2}$$

Using $\lambda = M/L$, we obtain the expected result

$$X_{cm} = \frac{\lambda L^2}{2M} = \frac{M}{L}\left(\frac{L^2}{2M}\right) = \frac{1}{2}L$$

Example 7-3

Find the center of mass of a semicircular hoop.

This calculation is easiest if we choose the origin to be on the line of symmetry of the hoop (the y axis) at the center of curvature as shown in Figure 7-10. Then $X_{cm} = 0$ because of symmetry. (For every mass element at $+x$, there is an equal mass element at $-x$.) However, Y_{cm} is not zero because all of the mass is at positive values of y. In the figure, we have indicated a mass element of length $ds = R\,d\theta$. Since the total length

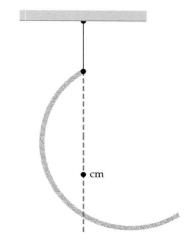

Figure 7-10 Geometry for the calculation of the center of mass of a semicircular hoop by integration. The center of mass lies on the y axis.

of the hoop is πR, the mass per unit length is $\lambda = M/\pi R$, where M is the total mass. The mass of the element chosen is thus

$$dm = \lambda \ ds = \lambda R \ d\theta$$

The y coordinate of the mass element is related to the angle θ by $y = R \sin \theta$. The angle θ varies from 0 to π. We thus have

$$MY_{\text{cm}} = \int y \ dm = \int y\lambda \ ds = \int y\lambda R \ d\theta$$

$$= \int_0^\pi (R \sin \theta)\lambda R \ d\theta = R^2\lambda \int_0^\pi \sin \theta \ d\theta = 2R^2\lambda$$

since

$$\int_0^\pi \sin \theta \ d\theta = -\cos \theta \ \Big|_0^\pi = 2$$

Using $\lambda = M/\pi R$, we have

$$MY_{\text{cm}} = 2R^2\frac{M}{\pi R}$$

$$Y_{\text{cm}} = \frac{2R}{\pi}$$

The center of mass is not within the body in this case.

Figure 7-11 shows how the location of the center of mass can be found by suspending a semicircular hoop from one end and then from some other point.

Figure 7-11 The center of mass of a semicircular hoop can be found by suspending it from two points.

Question

1. Give an example of a three-dimensional object that has no mass at its center of mass.

7-2 Motion of the Center of Mass of a System

The motion of an extended object such as the baton in Figure 7-1 is complicated. To describe it, we first describe the motion of the center of mass. We find the velocity of the center of mass by differentiating Equation 7-4 with respect to time. We have

$$M\mathbf{R}_{\text{cm}} = \sum_i m_i\mathbf{r}_i$$

$$M\frac{d\mathbf{R}_{\text{cm}}}{dt} = m_1\frac{d\mathbf{r}_1}{dt} + m_2\frac{d\mathbf{r}_2}{dt} + \cdots = \sum_i m_i\frac{d\mathbf{r}_i}{dt}$$

or

$$M\mathbf{V}_{\text{cm}} = m_1\mathbf{v}_1 + m_2\mathbf{v}_2 + \cdots = \sum_i m_i\mathbf{v}_i \qquad 7\text{-}7$$

We differentiate again to obtain the acceleration of the center of mass:

$$M\mathbf{A}_{cm} = m_1\mathbf{a}_1 + m_2\mathbf{a}_2 + \cdots = \sum_i m_i\mathbf{a}_i \qquad 7\text{-}8$$

According to Newton's second law, the mass of each particle times its acceleration equals the net force acting on the particle. We can therefore replace the quantity $m_i\mathbf{a}_i$ with $\mathbf{F}_i$, the net force acting on the ith particle.

The forces acting on a particle can be separated into two categories: internal forces due to interactions with other particles within the system and external forces due to agents outside the system:

$$\mathbf{F}_i = m_i\mathbf{a}_i = \mathbf{F}_{i,\text{int}} + \mathbf{F}_{i,\text{ext}}$$

Substituting this into Equation 7-8, we obtain

$$M\mathbf{A}_{cm} = \sum_i \mathbf{F}_{i,\text{int}} + \sum_i \mathbf{F}_{i,\text{ext}} \qquad 7\text{-}9$$

According to Newton's third law, for each internal force acting on one particle, there is an equal but opposite force acting on another particle. For example, if particle m_1 exerts a force on particle m_2, particle m_2 exerts an equal and opposite force on m_1. The internal forces thus occur in pairs of equal and opposite forces. When we sum over all the particles in the system, the internal forces cancel out and we are left with only the external forces. Equation 7-9 then becomes

Newton's second law for a system

$$\mathbf{F}_{net,ext} = \sum_i \mathbf{F}_{i,\text{ext}} = M\mathbf{A}_{cm} \qquad 7\text{-}10$$

This equation states that the net external force acting on the system equals the total mass M of the system times the acceleration of the center of mass $\mathbf{A}_{cm}$. This has the same form as Newton's second law for a single particle of mass M located at the center of mass and under the influence of the net external force. So

The center of mass of a system moves like a particle of mass $M = \Sigma\, m_i$ under the influence of the net external force acting on the system.

This theorem is important because it shows us how to describe the motion of one point, the center of mass, for any system of particles no matter how extensive the system may be. The center of mass of the system behaves just like a single particle that is subject only to the external forces. The individual motions of the particles of the system are usually much more complex. For example, the motion of a pair of masses connected by a spring and thrown into the air is quite involved. The masses tumble and turn as they move and oscillate along the line joining them. But the center of mass moves as if it were a single particle—it follows a simple parabolic trajectory.

This theorem for the motion of the center of mass is the justification for our earlier treatment of large objects as point particles. All large objects can be considered to be made up of many small particles whose motions are governed by Newton's laws. No matter how complicated the motion of the object, the center of mass moves like a simple particle. The baton thrown into the air in Figure 7-1 is an example. The only external force acting is that of gravity so the center of mass of the baton moves in a simple parabolic path. We note that Equation 7-10 does not give a complete description of the complicated motion of the baton. It applies only to the motion of one point, the center of mass.

Figure 7-12 shows two particles of equal mass m connected by a massless spring on a frictionless table. An external force $\mathbf{F}_{app}$ is applied to one of the particles. The vertical force of gravity acting on each particle is balanced by the normal force exerted by the table. The net external force acting on this system is thus the applied force $\mathbf{F}_{app}$. Since the total mass of the system is $2m$, the center of mass of this system, which lies midway between the particles, has an acceleration given by

$$\mathbf{A}_{cm} = \frac{\mathbf{F}_{app}}{2m}$$

If we have the same two masses but no spring, that is, if we have two unconnected masses on a table, and an external force $\mathbf{F}_{app}$ is applied to one of them (Figure 7-13), the acceleration of the center of mass will be the same as if the masses were connected by the spring. However, the internal motions of the masses of the two systems will be quite different. Thus, Equation 7-10 applies to any system even if the parts of the system are not interacting, and even if the external forces are applied to only parts of the system.

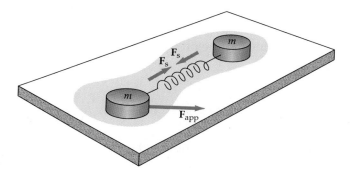

Figure 7-12   Two particles of equal mass m connected by a spring. The forces exerted by the spring are internal forces. These forces cancel when all the forces acting on the system are added up. The net force acting on the two-particle system is therefore the applied force $\mathbf{F}_{app}$.

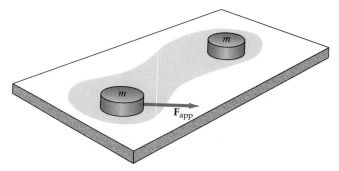

Figure 7-13   The same two particles as Figure 7-12 without the spring. The net force acting on the two-particle system is the applied force $\mathbf{F}_{app}$, even though this force acts on only one of the particles. The acceleration of the center of mass of the system is $\mathbf{F}_{app}/2m$.

Figure 7-14 shows a projectile that explodes into two equal parts at the top of its flight so that one piece falls straight down. Since the only external force acting on the projectile whether it is in one piece or two is that of gravity, the center of mass continues on its parabolic path just as if there had been no explosion.

Figure 7-14   A projectile that explodes into two equal pieces at the top of its trajectory. The center of mass, which is midway between the fragments, continues on its original parabolic path.

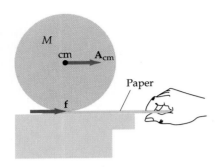

Figure 7-15 A cylinder resting on a piece of paper. When the paper is pulled to the right, the net external force acting on the cylinder is that of friction to the right, so the center of mass of the cylinder must accelerate toward the right. The cylinder rolls backward relative to the paper because the acceleration of the paper is greater than that of the cylinder.

Figure 7-15 shows a cylinder resting on a piece of paper on a table. When the paper is pulled to the right, the cylinder rolls backward on the paper. However, if you mark the original position of the cylinder on the table, you will notice that the cylinder moves to the right relative to the table. (That is, the paper moves further to the right than the cylinder does. So, relative to the paper, the cylinder moves to the left.) This must be because the net external force acting on the cylinder is that of friction between it and the paper, and this force is to the right. Thus, the center of mass of the cylinder must accelerate to the right in the direction of the net force.

Questions

2. A self-made man is said to have pulled himself up by his bootstraps. Discuss this act from the point of view of internal and external forces and the motion of the center of mass of the system.

3. If only external forces can cause the center of mass of a system to accelerate, how can an automobile be accelerated by its motor?

7-3   The Conservation of Momentum

The **momentum** of a particle is defined as the product of its mass and velocity:

Momentum of a particle defined

$$\mathbf{p} = m\mathbf{v} \qquad\qquad 7\text{-}11$$

Momentum is a vector quantity. The momentum of a particle may be thought of as measuring the difficulty of bringing the particle to rest. For example, a heavy truck has more momentum than a light car traveling at the same speed. It takes a greater force to stop the truck than it does the car in a given time. (The quantity $m\mathbf{v}$ is sometimes referred to as the *linear* momentum of a particle to distinguish it from the *angular* momentum, which will be discussed in Chapter 8.) Newton's second law can be written in terms of the momentum of a particle. Differentiating Equation 7-11, we obtain

$$\frac{d\mathbf{p}}{dt} = \frac{d(m\mathbf{v})}{dt} = m\frac{d\mathbf{v}}{dt} = m\mathbf{a}$$

Substituting the force $\mathbf{F}_{net}$ for $m\mathbf{a}$, we obtain

$$\mathbf{F}_{net} = \frac{d\mathbf{p}}{dt} \qquad\qquad 7\text{-}12$$

Thus, the net force acting on a particle equals the time rate of change of the particle's linear momentum. Newton's original statement of his second law was in fact in this form.

The concept of momentum is important because, if no net external force acts on a system of particles, the total momentum of the system is conserved; that is, it remains constant over time. Consider two particles that exert equal and opposite forces on each other but have no other forces acting on them. If $\mathbf{F}_{12}$ is the force exerted by particle 1 on particle 2 and $\mathbf{F}_{21}$ is that exerted by particle 2 on particle 1, we have

$$\mathbf{F}_{21} = \frac{d\mathbf{p}_1}{dt}$$

and

$$\mathbf{F}_{12} = \frac{d\mathbf{p}_2}{dt}$$

Adding these equations and using the fact that by Newton's third law $\mathbf{F}_{12} = -\mathbf{F}_{21}$, we obtain

$$0 = \frac{d\mathbf{p}_1}{dt} + \frac{d\mathbf{p}_2}{dt}$$

$$\frac{d(\mathbf{p}_1 + \mathbf{p}_2)}{dt} = 0$$

or

$$\mathbf{p}_1 + \mathbf{p}_2 = \text{constant}$$

This result can be extended to systems of many particles. The total momentum $\mathbf{P}$ of the system is the sum of the momenta of the individual particles:

$$\mathbf{P} = \sum_i m_i\mathbf{v}_i = \sum_i \mathbf{p}_i$$

According to Equation 7-7, the total momentum of the system equals the total mass M times the velocity of the center of mass:

$$\mathbf{P} = \sum_i m_i\mathbf{v}_i = M\mathbf{V}_{cm} \qquad\qquad 7\text{-}13 \qquad \textit{Total momentum of a system}$$

Differentiating this equation, we obtain

$$\frac{d\mathbf{P}}{dt} = M\frac{d\mathbf{V}_{cm}}{dt} = M\mathbf{A}_{cm}$$

But according to Equation 7-10, the mass times the acceleration of the center of mass equals the net external force acting on the system. Thus,

$$\sum_i \mathbf{F}_{i,\text{ext}} = \mathbf{F}_{\text{net,ext}} = \frac{d\mathbf{P}}{dt} \qquad\qquad 7\text{-}14$$

When the net external force acting on a system of particles is zero, the rate of change of the total momentum is zero, and the total momentum of the system remains constant:

$$\mathbf{P} = M\mathbf{V}_{cm} = \sum_i m_i\mathbf{v}_i = \text{constant} \qquad\qquad 7\text{-}15 \qquad \textit{Conservation of momentum}$$

This result is known as the **law of conservation of momentum:**

If the net external force on a system is zero, the velocity of the center of mass of the system is constant and the total momentum of the system is conserved; that is, it remains constant.

This law is one of the most important in physics. It applies, for example, to any system isolated from its surroundings so that there are no external forces acting on it. It is more widely applicable than the law of conservation of mechanical energy because internal forces exerted by one particle in the system on another are often not conservative. Thus, these internal forces can change the total mechanical energy of the system, but since they always occur in pairs, they cannot change the total momentum of the system. The conservation of momentum is particularly useful when considering collisions, which we will do in Section 7-6.

Example 7-4

A man of mass 70 kg and a boy of mass 35 kg are standing together on a smooth ice surface for which friction is negligible. If they push each other apart and the man moves away with speed of 0.3 m/s relative to the ice, how far apart are they after 5 s (Figure 7-16)?

We take the man and boy together to be the system. The force exerted by the man on the boy is equal and opposite to that exerted by the boy on the man. The force of gravity on each is balanced by a corresponding normal force exerted by the ice. Since there is no friction, the net external force on the system is zero. The acceleration of the center of mass is therefore zero, and since it is initially at rest, it remains at rest at its initial position. Since the man and boy are standing at rest initially, the total momentum of the system is zero, and since the net external force is zero, the total momentum remains zero. Therefore, after they push each other, they must have equal and opposite momenta. Since the motion takes place in one dimension, we can take care of the vector nature of the momenta by choosing the momentum to the right to be positive and that to the left to be negative. If the man is moving to the right, his momentum is

$$p_m = m_m v_m = (70 \text{ kg})(0.3 \text{ m/s}) = 21 \text{ kg·m/s}$$

The momentum of the boy is

$$p_b = m_b v_b = (35 \text{ kg})v_b$$

Setting the total momentum equal to zero, we obtain

$$p_m + p_b = 21 \text{ kg·m/s} + (35 \text{ kg})v_b = 0$$

$$v_b = -\frac{21 \text{ kg·m/s}}{35 \text{ kg}} = -0.6 \text{ m/s}$$

Since the man has twice the mass of the boy and moves in one direction with a speed of 0.3 m/s, the boy moves in the opposite direction with a speed of 0.6 m/s. After 5 s, the man has moved 1.5 m and the boy has moved 3 m, so they are 4.5 m apart.

Note that the mechanical energy of this system is not conserved. The forces the man and boy exert on each other are not conservative. In this case, the mechanical energy of the system is increased as they push each other apart since the kinetic energy was initially zero and the potential energy does not change. This energy comes from a decrease in internal chemical energy of the man and boy.

Figure 7-16 The net external force acting on the man–boy system is zero, so the total momentum of the system is conserved and remains zero. The momentum of the man to the right is equal in magnitude to that of the boy to the left. Since the mass of the man is twice that of the boy, his speed must be half that of the boy.

Exercise

Find the final kinetic energy of the man–boy system in Example 7-4.
(Answer: 9.45 J)

Example 7-5

A bullet of mass 0.01 kg moves horizontally with a speed of 400 m/s and embeds itself in a block of mass 0.39 kg that is initially at rest on a frictionless table (Figure 7-17). Find (*a*) the final velocity of the bullet and block and (*b*) the initial and final mechanical energy of the bullet–block system.

(*a*) Since there are no external horizontal forces on the bullet–block system, the horizontal component of the momentum of the system is conserved. The total initial horizontal momentum P_i before the bullet strikes the block is just that of the bullet

$$P_i = m_1 v_1 = (0.01 \text{ kg})(400 \text{ m/s}) = 4 \text{ kg·m/s}$$

Afterward, the bullet and block move together with a common velocity V_f. The total final momentum P_f is

$$P_f = (m_1 + m_2)V_f = (0.4 \text{ kg})V_f$$

Since the total momentum is conserved, the final momentum equals the initial momentum:

$$(0.4 \text{ kg})V_f = 4 \text{ kg·m/s}$$

$$V_f = 10 \text{ m/s}$$

Since the bullet and block both move together with this final velocity, the center of mass must also move with this velocity.

We could have found the velocity of the center of mass from Equation 7-13:

$$P = MV_{cm} = m_1 v_1 + m_2 v_2$$

$$= (0.01 \text{ kg})(400 \text{ m/s}) + (0.39 \text{ kg})(0)$$

$$= 4 \text{ kg·m/s}$$

$$V_{cm} = \frac{4 \text{ kg·m/s}}{0.01 \text{ kg} + 0.39 \text{ kg}} = 10 \text{ m/s}$$

(*b*) The initial mechanical energy of the bullet–block system is the kinetic energy of the bullet:

$$E_i = K_i = \tfrac{1}{2}m_1 v_1^2 = \tfrac{1}{2}(0.01 \text{ kg})(400 \text{ m/s})^2 = 800 \text{ J}$$

Since the potential energy of the system does not change, the final mechanical energy is the kinetic energy of the bullet and block moving together:

$$E_f = K_f = \tfrac{1}{2}(0.4 \text{ kg})(10 \text{ m/s})^2 = 20 \text{ J}$$

In this case, most of the initial mechanical energy (780 J out of 800 J) is lost because of the large nonconservative forces between the bullet and the block. Some of the energy goes into thermal energy, some goes into producing permanent deformations of the block and bullet, and a small amount goes into sound waves produced when the bullet strikes the block. A bullet embedding itself in a block is an example of an inelastic collision, which we will study in Section 7-6.

Figure 7-17   Bullet striking a block for Example 7-5. Since there is no net external force acting on the system, the momentum of the system and the velocity of the center of mass are constant.

Questions

4. As a pendulum bob swings back and forth, is the momentum of the bob conserved? Explain why or why not.

5. How is the recoil of a rifle or a cannon related to momentum conservation?

6. A man is stranded in the middle of an ice rink that is perfectly frictionless. How can he get to the edge?

7. A girl jumps from a boat to a dock. Why does she have to jump with more energy than she would need if she were jumping the same distance from one dock to another?

7-4   The Center-of-Mass Reference Frame

As we have seen, when the net external force on a system is zero, the velocity of the center of mass is constant. It is often convenient to choose a coordinate system with the origin at the center of mass. Then, relative to the original coordinate system, this coordinate system moves with a constant velocity V_{cm}. The reference frame attached to the center of mass is called the **center-of-mass reference frame**. Relative to this frame, the velocity of the center of mass is zero. Since the total momentum of a system equals the total mass times the velocity of the center of mass, the total momentum is also zero in the center-of-mass frame. This frame is also called the **zero-momentum reference frame**.

In the analysis of problems, it is sometimes convenient to transform from a given frame to the center-of-mass frame and back. This is straightforward. We transform from some original frame to the center-of-mass frame by subtracting the velocity of the center of mass V_{cm} in the original frame from the velocity of each particle in that frame. Consider, for example, a simple two-particle system in a reference frame in which one particle of mass m_1 is moving with a velocity v_1 and a second particle of mass m_2 is moving with a velocity v_2 (Figure 7-18). The velocities in the center-of-mass frame are u_1 and u_2, given by

$$u_1 = v_1 - V_{cm} \qquad \qquad 7\text{-}16a$$

and

$$u_2 = v_2 - V_{cm} \qquad \qquad 7\text{-}16b$$

where, as usual, the velocity of the center of mass in the original frame is

$$V_{cm} = \frac{m_1 v_1 + m_2 v_2}{m_1 + m_2}$$

Since the total momentum is zero in the center-of-mass frame, the particles have equal and opposite momenta in this frame.

Original reference frame

(a)

Center-of-mass reference frame

(b)

Figure 7-18 (a) Two particles moving in a general reference frame in which the center of mass has a velocity V_{cm}. (b) In the center-of-mass reference frame, the center of mass is at rest and the particles have equal and opposite momenta. The velocities in the two frames are related by $u_1 = v_1 - V_{cm}$ and $u_2 = v_2 - V_{cm}$.

Example 7-6

A 2.5-kg box moves with velocity $v_1 = 10$ m/s i and a 3.5-kg box moves with velocity $v_2 = -2$ m/s i. Find (a) the total momentum, (b) the velocity of the center of mass, and (c) the velocity of each box in the center-of-mass reference frame.

(a) The total momentum is

$$P = m_1 v_1 + m_2 v_2 = (2.5 \text{ kg})(10 \text{ m/s } i) + (3.5 \text{ kg})(-2 \text{ m/s } i)$$

$$= (25 \text{ kg·m/s} - 7 \text{ kg·m/s}) \, i = 18 \text{ kg·m/s } i$$

(b) The velocity of the center of mass is then

$$\mathbf{V}_{cm} = \frac{\mathbf{P}}{M} = \frac{m_1\mathbf{v}_1 + m_2\mathbf{v}_2}{m_1 + m_2} = \frac{18 \text{ kg·m/s } \mathbf{i}}{(2.5 \text{ kg} + 3.5 \text{ kg})} = 3 \text{ m/s } \mathbf{i}$$

(c) The velocities of the boxes in the center-of-mass reference frame are

$$\mathbf{u}_1 = \mathbf{v}_1 - \mathbf{V}_{cm} = 10 \text{ m/s } \mathbf{i} - 3 \text{ m/s } \mathbf{i} = 7 \text{ m/s } \mathbf{i}$$

and

$$\mathbf{u}_2 = \mathbf{v}_2 - \mathbf{V}_{cm} = -2 \text{ m/s } \mathbf{i} - 3 \text{ m/s } \mathbf{i} = -5 \text{ m/s } \mathbf{i}$$

As a check, we compute the total momentum in the center-of-mass frame

$$\mathbf{P}_{cm} = m_1\mathbf{u}_1 + m_2\mathbf{u}_2 = (2.5 \text{ kg})(7 \text{ m/s } \mathbf{i}) + (3.5 \text{ kg})(-5 \text{ m/s } \mathbf{i})$$

$$= 17.5 \text{ kg·m/s } \mathbf{i} - 17.5 \text{ kg·m/s } \mathbf{i} = 0$$

In the center-of-mass reference frame, the total momentum is zero.

Example 7-7

Analyze the collision in Example 7-5 in the center-of-mass frame.

The velocity of the center of mass in Example 7-5 was found to be $V_{cm} = 10$ m/s. The velocity of the bullet in the center-of-mass frame is thus

$$u_1 = v_1 - V_{cm} = 400 \text{ m/s} - 10 \text{ m/s} = 390 \text{ m/s}$$

and that of the block is

$$u_2 = v_2 - V_{cm} = 0 - 10 \text{ m/s} = -10 \text{ m/s}$$

In the center-of-mass frame, the initial momentum is zero:

$$P_i = m_1u_1 + m_2u_2 = (0.01 \text{ kg})(390 \text{ m/s}) + (0.39 \text{ kg})(-10 \text{ m/s})$$

$$= 3.90 \text{ kg·m/s} - 3.90 \text{ kg·m/s} = 0$$

After the collision, the two particles stick together and the total momentum is zero, so the velocity of both particles is zero.

The initial kinetic energy of the system in the center-of-mass frame is

$$K_i = \tfrac{1}{2}m_1u_1^2 + \tfrac{1}{2}m_2u_2^2$$

$$= \tfrac{1}{2}(0.01 \text{ kg})(390 \text{ m/s})^2 + \tfrac{1}{2}(0.39 \text{ kg})(-10 \text{ m/s})^2$$

$$= 760.5 \text{ J} + 19.5 \text{ J} = 780 \text{ J}$$

The final kinetic energy in the center-of-mass frame is zero because both particles are at rest in this frame. Thus, all the initial mechanical energy is lost in this frame. Note that the mechanical energy lost is 780 J, the same as was calculated in the original reference frame in which the block was initially at rest.

7-5   Kinetic Energy of a System of Particles

Although the total linear momentum of a system of particles must be constant if the net external force on the system is zero, the total mechanical energy of the system may change. As we saw in Examples 7-4 and 7-5, internal forces that cannot change the total momentum may be nonconservative and thus change the total mechanical energy of the system. In Example

7-4 the mechanical energy increased, whereas in Example 7-5 the mechanical energy decreased. There is an important theorem concerning the kinetic energy of a system of particles that allows us to treat the energy of complex systems more easily and gives us insight into energy changes within a system:

> The kinetic energy of a system of particles can be written as the sum of two terms: (1) the kinetic energy associated with the motion of the center of mass, $\frac{1}{2}MV_{cm}^2$, where M is the total mass of the system; and (2) the kinetic energy associated with the motion of the particles of the system relative to the center of mass, $\sum \frac{1}{2}m_i u_i^2$, where u_i is the velocity of the ith particle relative to the center of mass.

The kinetic energy of motion relative to the center of mass is the kinetic energy as observed in the center-of-mass reference frame.

We will now prove this theorem. The kinetic energy of a system of particles is the sum of the kinetic energies of the individual particles:

$$K = \sum_i \tfrac{1}{2}m_i v_i^2 = \sum_i \tfrac{1}{2}m_i(\mathbf{v}_i \cdot \mathbf{v}_i)$$

where $\mathbf{v}_i$ is the velocity of the ith particle. From Equations 7-16a, and b, we know that the velocity of the ith particle can be written as the sum of the velocity of the center of mass $\mathbf{V}_{cm}$ and the velocity of the particle relative to the center of mass $\mathbf{u}_i$, which is the velocity of the particle in the center-of-mass frame:

$$\mathbf{v}_i = \mathbf{V}_{cm} + \mathbf{u}_i$$

The kinetic energy of the system is then

$$K = \sum_i \tfrac{1}{2}m_i(\mathbf{v}_i \cdot \mathbf{v}_i) = \sum_i \tfrac{1}{2}m_i(\mathbf{V}_{cm} + \mathbf{u}_i) \cdot (\mathbf{V}_{cm} + \mathbf{u}_i)$$

$$= \sum_i \tfrac{1}{2}m_i V_{cm}^2 + \sum_i \tfrac{1}{2}m_i u_i^2 + \mathbf{V}_{cm} \cdot \sum_i m_i \mathbf{u}_i$$

where in the last term we have removed $\mathbf{V}_{cm}$ from the sum because it is the same for each particle, that is, it refers to the system and not any particular particle. The quantity $\sum m_i \mathbf{u}_i$ is the momentum of the system as seen in the center-of-mass reference frame, which is zero. Setting this term equal to zero, we obtain our desired result:

Kinetic energy of a system

$$K = \tfrac{1}{2}MV_{cm}^2 + K_{rel} \qquad\qquad 7\text{-}17$$

where $M = \sum m_i$ is the total mass of the system and

$$K_{rel} = \sum_i \tfrac{1}{2}m_i u_i^2 \qquad\qquad 7\text{-}18$$

is the kinetic energy of motion relative to the center of mass; that is, it is the kinetic energy as seen in the center-of-mass reference frame. This energy is the same in any reference frame because it depends only on the speeds of the particles relative to the center of mass. However, the energy associated with the motion of the center of mass depends on the reference frame. It is $\frac{1}{2}MV_{cm}^2$ in any frame in which the center of mass is moving with a velocity $\mathbf{V}_{cm}$. In the center-of-mass frame, it is zero. In that case, all the kinetic energy is the energy of relative motion.

We will have many occasions to use Equation 7-17 in the following chapters. For example, the kinetic energy of a rolling ball will be written as the sum of $\frac{1}{2}MV_{cm}^2$ and the energy of relative motion, which in this case is the energy of rotation.

If there are no external forces acting on the system, the velocity of the center of mass is constant and the first term in Equation 7-17 does not change. Only the relative energy can increase or decrease in an isolated system. We will have occasion to use this result when we study collisions in Section 7-6. In collisions, the external forces are negligible, so the kinetic energy of the center of mass remains constant. We saw an example of this in Examples 7-5 and 7-7, in which a bullet embeds itself in a block. In the frame of Example 7-5, in which the block is originally at rest, the initial energy was found to be 800 J. In the center-of-mass frame of Example 7-7, the initial energy was found to be 780 J. All of this mechanical energy was lost in the collision. The energy of the center of mass (in the original frame), $\frac{1}{2}MV_{cm}^2 = \frac{1}{2}(0.4 \text{ kg})(10 \text{ m/s})^2 = 20$ J, was not changed by the collision.

Pseudowork

The kinetic energy associated with motion of the center of mass can be related to the net force acting on a system and the displacement of the center of mass. From Newton's second law for a system of particles, we have (Equation 7-10)

$$\mathbf{F}_{net} = M\mathbf{A}_{cm} = M\frac{d\mathbf{V}_{cm}}{dt}$$

If we take the dot product of the net force and the displacement of the center of mass $d\mathbf{R}_{cm}$, we have

$$\mathbf{F}_{net} \cdot d\mathbf{R}_{cm} = M\frac{d\mathbf{V}_{cm}}{dt} \cdot d\mathbf{R}_{cm} = M\frac{d\mathbf{R}_{cm}}{dt} \cdot d\mathbf{V}_{cm} = M\mathbf{V}_{cm} \cdot d\mathbf{V}_{cm}$$

Integrating over some finite displacement, we obtain

$$\int \mathbf{F}_{net} \cdot d\mathbf{R}_{cm} = \int_{V_{cm,1}}^{V_{cm,2}} MV_{cm}\, dV_{cm} = \frac{1}{2}MV_{cm,2}^2 - \frac{1}{2}MV_{cm,1}^2 \qquad \text{7-19}$$

Equation 7-19 looks like the work–energy theorem, but the left side does *not* represent work done on the system because the displacement $d\mathbf{R}_{cm}$ is the displacement of the center of mass, which is not generally equal to the displacement of the point of application of any of the forces acting on the system. In addition, the right side of Equation 7-19 is not the change in the total energy of the system, but merely the change in the energy associated with the motion of the center of mass. The quantity $\int \mathbf{F}_{net} \cdot d\mathbf{R}_{cm}$ is called **pseudowork**. We can understand Equation 7-19 from a simple example of a car accelerating from rest along a road, such that its wheels do not slip. The net force acting on the car is the force of static friction $\mathbf{f}$ acting in the direction of motion of the car. If V_{cm} is the speed of the center of mass of the car after it has traveled a distance Δx, Equation 7-19 gives

$$f\,\Delta x = \Delta K_{cm} = \frac{1}{2}MV_{cm}^2$$

This is the same result as is obtained using Newton's second law for a system of particles (Equation 7-10). The pseudowork done by the force of static friction exerted by the road on the car has nothing to do with any energy transfer between the road and the car. In fact, the road does not do any work on the car. The distance Δx is the displacement of the center of mass of the car and not of the point of application of the frictional force which is static. The increase in the mechanical energy of the car results from the conversion of internal chemical energy of the fuel in the car.

7-6   Collisions in One Dimension

In a collision, two objects approach each other, interact strongly, and move apart. Before the collision, when they are far apart, the objects move with constant velocities. After the collision, they move with constant but altered velocities. Usually, we wish to find the final velocities of the objects when their initial velocities and the characteristics of the collision are given.

A collision can be brief, like the click of two billiard balls, or it can take centuries, as when two stars collide in space. However, in all collisions, the objects interact strongly with each other only during this collision time. When there are external forces present, these forces are much smaller than the forces of interaction during the collision and can be neglected. For example, billiard balls are gradually slowed down by friction with the table, but the forces of friction are negligible compared to the forces of interaction during their brief collision.

When the total kinetic energy of the two objects is the same after the collision as before, the collision is said to be an **elastic collision.** When the total kinetic energy is not the same after the collision, the collision is said to be an **inelastic collision.** Inelastic collisions occur between macroscopic systems when nonconservative forces act that change the mechanical energy of the system. An example is a ball of putty dropped onto the floor. Inelastic collisions occur between microscopic systems when one of the systems is left in a different state of internal energy. An example is the scattering of an electron by an atom. If the internal energy of the atom is not changed, the total kinetic energy of the atom and the electron is the same after the collision as before and the collision is elastic. However, sometimes the internal energy of the atom is changed (that is, the atom is excited to a higher internal energy state), and the total kinetic energy of the atom and the electron is less after the collision, in which case the collision is inelastic.

In an inelastic collision the kinetic energy relative to the center of mass is changed, but the kinetic energy of the center of mass $\frac{1}{2}MV_{cm}^2$ remains constant because the velocity of the center of mass is unchanged when the external forces acting on the system are negligible. In one special case, all the relative energy is lost and the objects move together with the velocity of the center of mass. Such a collision is called a **perfectly inelastic collision.** The bullet embedding itself in a block in Example 7-5 is a typical example of a perfectly inelastic collision. In most collisions, only part of K_{rel} is lost; these collisions are neither elastic nor perfectly inelastic.

In this section, we consider collisions in one dimension only. Figure 7-19 shows an object of mass m_1 moving with an initial velocity v_{1i} toward a second object of mass m_2 moving with an initial velocity v_{2i} (which we assume to be less than v_{1i} so that the bodies collide). Let v_{1f} and v_{2f} be the final velocities of the objects after the collision. The velocities can be positive or negative, depending on whether the objects are moving to the right or to the left. Conservation of momentum gives

$$m_1v_{1f} + m_2v_{2f} = m_1v_{1i} + m_2v_{2i} \qquad 7\text{-}20$$

Equation 7-20 provides one relation between the two unknown velocities v_{1f} and v_{2f} (if the initial velocities v_{1i} and v_{2i} are given). To find these unknowns, we must have a second relation. The second relation comes from energy considerations.

Collision of baseball with bat. Note the deformation of the ball due to the large force exerted by the bat during the brief time of contact. As the ball leaves the bat, it springs back to its original shape, converting the elastic potential energy of deformation into kinetic energy.

Figure 7-19   A general collision of two objects in one dimension.

Before collision

After collision

Elastic Collisions

For elastic collisions, the initial and final energies are equal. If there is no change in the internal potential energy of the system, the final kinetic energy equals the initial kinetic energy:

$$\tfrac{1}{2}m_1v_{1f}^2 + \tfrac{1}{2}m_2v_{2f}^2 = \tfrac{1}{2}m_1v_{1i}^2 + \tfrac{1}{2}m_2v_{2i}^2 \qquad 7\text{-}21$$

Equations 7-20 and 7-21 are sufficient to determine the final velocities of the two objects, from their initial velocities. However, the quadratic nature of Equation 7-21 presents algebraic difficulties in solving elastic collision problems. These difficulties can be avoided by combining the equations to obtain a third equation that is linear in the velocities. Rearranging Equation 7-21, we have

$$m_2(v_{2f}^2 - v_{2i}^2) = m_1(v_{1i}^2 - v_{1f}^2)$$

or

$$m_2(v_{2f} - v_{2i})(v_{2f} + v_{2i}) = m_1(v_{1i} - v_{1f})(v_{1i} + v_{1f}) \qquad 7\text{-}22$$

Similarly, Equation 7-20 can be rearranged to yield

$$m_2(v_{2f} - v_{2i}) = m_1(v_{1i} - v_{1f}) \qquad 7\text{-}23$$

Dividing Equation 7-22 by Equation 7-23, we obtain

$$v_{2f} + v_{2i} = v_{1i} + v_{1f}$$

which can be written

$$v_{2f} - v_{1f} = -(v_{2i} - v_{1i}) \qquad 7\text{-}24$$

The relative velocity $v_2 - v_1$ is the velocity of object 2 as seen from object 1. If the objects are to collide, $v_{2i} - v_{1i}$ must be negative (provided that object 1 is on the left). Then $-(v_{2i} - v_{1i})$ is the speed of approach of the two objects. After the collision, the objects move away from each other, so $v_{2f} - v_{1f}$ must be positive. It is the speed of recession after the collision. Equation 7-24 states the following important result:

> For elastic collisions, the relative speed of recession after the collision equals the relative speed of approach before the collision.

In solving problems involving elastic collisions, it is nearly always easiest to use Equations 7-20 and 7-24 to find the final velocities, thereby avoiding the quadratic terms in the conservation-of-energy equation 7-21. It should be remembered that Equation 7-24 was derived using the conservation of mechanical energy, so it applies only to elastic collisions.

Example 7-8

A 4-kg block moving to the right at 6 m/s makes an elastic collision with a 2-kg block moving to the right at 3 m/s (Figure 7-20). Find the velocities v_{1f} and v_{2f}.

The conservation-of-momentum Equation 7-20 gives

$$(4 \text{ kg})v_{1f} + (2 \text{ kg})v_{2f} = (4 \text{ kg})(6 \text{ m/s}) + (2 \text{ kg})(3 \text{ m/s}) = 30 \text{ kg·m/s}$$

or

$$4v_{1f} + 2v_{2f} = 30 \text{ m/s}$$

Figure 7-20   Collision of two blocks for Example 7-8.

The velocity of block 2 relative to block 1 before the collision is

$$v_{2i} - v_{1i} = 3 \text{ m/s} - 6 \text{ m/s} = -3 \text{ m/s}$$

The speed of approach is thus 3 m/s, which must equal the speed of recession. Equation 7-24 then gives

$$v_{2f} - v_{1f} = -(-3 \text{ m/s}) = +3 \text{ m/s}$$

Combining this equation with $4v_{1f} + 2v_{2f} = 30$ m/s and solving for the two final velocities, we obtain

$$v_{2f} = 7 \text{ m/s} \quad \text{and} \quad v_{1f} = 4 \text{ m/s}$$

To check our results, we can compute the initial and final kinetic energies of the blocks:

$$K_i = \tfrac{1}{2}(4 \text{ kg})(6 \text{ m/s})^2 + \tfrac{1}{2}(2 \text{ kg})(3 \text{ m/s})^2 = 72 \text{ J} + 9 \text{ J} = 81 \text{ J}$$

$$K_f = \tfrac{1}{2}(4 \text{ kg})(4 \text{ m/s})^2 + \tfrac{1}{2}(2 \text{ kg})(7 \text{ m/s})^2 = 32 \text{ J} + 49 \text{ J} = 81 \text{ J}$$

Our results are consistent with conservation of energy.

In the center-of-mass reference frame, an elastic collision is particularly simple. The velocity of each object is merely reversed by the collision. We can prove this by writing the initial and final kinetic energies in terms of the momenta of the bodies. Let u_{1i} and u_{2i} be the initial velocities relative to the center of mass. The initial momentum of the first object is

$$p_{1i} = m_1 u_{1i}$$

and that of the second object is

$$p_{2i} = m_2 u_{2i}$$

Since the total momentum in the center-of-mass reference frame is zero, we have

$$p_{2i} = -p_{1i} \qquad\qquad 7\text{-}25$$

The kinetic energy of an object can be written in terms of its momentum:

$$K = \tfrac{1}{2}mu^2 = \frac{(mu)^2}{2m}$$

or

$$K = \frac{p^2}{2m} \qquad\qquad 7\text{-}26$$

The initial kinetic energy of the two bodies is thus

$$K_i = \frac{p_{1i}^2}{2m_1} + \frac{p_{2i}^2}{2m_2} = p_{1i}^2 \left(\frac{1}{2m_1} + \frac{1}{2m_2} \right)$$

where we have used $p_{2i}^2 = p_{1i}^2$. Similarly, the final kinetic energy written in terms of the final momenta of the bodies is

$$K_f = \frac{p_{1f}^2}{2m_1} + \frac{p_{2f}^2}{2m_2} = p_{1f}^2 \left(\frac{1}{2m_1} + \frac{1}{2m_2} \right)$$

where again $p_{2f}^2 = p_{1f}^2$. Setting the final kinetic energy equal to the initial kinetic energy gives

$$p_{1f}^2 = p_{1i}^2$$

or

$$p_{1f} = \pm p_{1i} \qquad\qquad 7\text{-}27a$$

This also implies that

$$p_{2f} = \pm p_{2i} \qquad\qquad 7\text{-}27b$$

The plus signs in Equations 7-27a and b correspond to no collision at all, which would be the case if the bodies were initially heading away from each other. Since only conservation of momentum and conservation of energy were used to derive these equations, it is not surprising that the results include the case of no collision. If there is a collision, we have

$$p_{1f} = -p_{1i} \quad \text{and} \quad p_{2f} = -p_{2i} \qquad 7\text{-}28$$

Since the velocity of each object is its momentum divided by its mass, Equation 7-28 implies

$$u_{1f} = -u_{1i} \quad \text{and} \quad u_{2f} = -u_{2i} \qquad 7\text{-}29$$

which is the result we desired to prove. Thus, in an elastic collision in one dimension, each object is merely turned around in the center-of-mass reference frame and leaves the collision with the same speed and energy it had before the collision.

Example 7-9

Find the final velocities for the elastic collision in Example 7-8 by transforming to the center-of-mass reference frame.

We first find the velocity of the center of mass in the original reference frame (Figure 7-21a):

$$MV_{cm} = m_1 v_{1i} + m_2 v_{2i}$$

$$(6\ \text{kg})V_{cm} = (4\ \text{kg})(6\ \text{m/s}) + (2\ \text{kg})(3\ \text{m/s}) = 30\ \text{kg·m/s}$$

$$V_{cm} = 5\ \text{m/s}$$

We now transform to the center-of-mass reference frame (Figure 7-21b) by subtracting the velocity of the center of mass from the initial velocities of the blocks in the original frame. We have

$$u_{1i} = v_{1i} - V_{cm} = 6\ \text{m/s} - 5\ \text{m/s} = 1\ \text{m/s}$$

and

$$u_{2i} = v_{2i} - V_{cm} = 3\ \text{m/s} - 5\ \text{m/s} = -2\ \text{m/s}$$

Figures 7-21b and c show the collision in the center-of-mass frame. Since each object is reversed by the collision, the final velocities in this frame are

$$u_{1f} = -u_{1i} = -(1\ \text{m/s}) = -1\ \text{m/s}$$

and

$$u_{2f} = -u_{2i} = -(-2\ \text{m/s}) = +2\ \text{m/s}$$

We now transform back into the original frame by adding $V_{cm} = 5$ m/s to each velocity (Figure 7-21d):

$$v_{1f} = u_{1f} + V_{cm} = -1\ \text{m/s} + 5\ \text{m/s} = 4\ \text{m/s}$$

and

$$v_{2f} = u_{2f} + V_{cm} = 2\ \text{m/s} + 5\ \text{m/s} = 7\ \text{m/s}$$

which are the same as the results we found in Example 7-8.

(*a*) Initial conditions

(*b*) Step 1; Transform to the center-of-mass frame by subtracting V_{cm}

(*c*) Step 2; Solve collision

(*d*) Step 3; Transform back to the original frame by adding V_{cm}

Figure 7-21 Finding the final velocities after the collision in Example 7-8 by transferring to the center-of-mass reference frame. (*a*) In the original frame, the center of mass is moving to the right at 5 m/s. (*b*) We transfer to the center-of-mass frame by subtracting 5 m/s from the initial velocity of each block. The center of mass is at rest in this frame, and the blocks move toward each other with equal and opposite momenta. (*c*) Each block has its velocity reversed in an elastic collision. (*d*) We transfer back to the original frame by adding 5 m/s to the final velocity of each block.

Example 7-10

An object of mass m_1 moving with speed v_{1i} makes an elastic collision with a second object of mass m_2 that is initially at rest (Figure 7-22). What is the velocity of the second object after the collision?

This problem is most easily worked in the center-of-mass frame. We first find the velocity of the center of mass in the original frame by setting the total momentum equal to $m_1 v_{1i}$:

$$(m_1 + m_2)V_{cm} = m_1 v_{1i}$$

$$V_{cm} = \frac{m_1 v_{1i}}{m_1 + m_2}$$

To transform to the center-of-mass frame, we subtract V_{cm} from the velocity of each object in the original frame. Since the second object was originally at rest, its velocity in the center-of-mass frame is

$$u_{2i} = -V_{cm}$$

After the collision, the velocity of the second object is

$$u_{2f} = +V_{cm}$$

We transform back to the original frame by adding V_{cm} to each velocity in the center-of-mass frame. The final velocity of the object that was originally stationary is therefore

$$v_{2f} = +2V_{cm} = \frac{2m_1 v_{1i}}{m_1 + m_2} \qquad\qquad 7\text{-}30$$

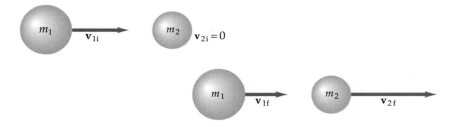

Figure 7-22   Elastic collision of an object of mass m_1 moving with a velocity $\mathbf{v}_{1i}$ and a second object that is initially at rest for Example 7-10.

If the first object in Example 7-10 is much more massive than the second one that is originally at rest, we can neglect m_2 compared to m_1 in the denominator of Equation 7-30. Then the final velocity of the second object is approximately

$$v_{2f} \approx 2v_{1i}$$

We can understand this as follows. The massive object will not be affected very much by a collision with a much less massive stationary object. A cannonball, for example, will hardly be slowed down if it collides with a stationary beach ball. Before the collision, the relative velocity of approach is v_{1i}. Then after the collision, the relative velocity of recession must be v_{1i}. For a first approximation, we can neglect any change in the velocity of the first object. Since it continues to move with a velocity v_{1i}, the velocity of the smaller object must be $2v_{1i}$.

An example of such a collision is that of a golf club with a golf ball. Careful measurement of the distances the club and the ball travel between successive flashes in Figure 7-23 shows that the speed of the ball is indeed twice that of the club.

Although energy is conserved in an elastic collision, energy is transferred from one object to the other. In cases like Example 7-10, in which one object is initially stationary, the energy transferred to the initially stationary object is its final kinetic energy:

$$K_{2f} = \frac{1}{2} m_2 v_{2f}^2 = \frac{2 m_2 m_1^2 v_{1i}^2}{(m_1 + m_2)^2} = \frac{4 m_1 m_2}{(m_1 + m_2)^2} \left(\frac{1}{2} m_1 v_{1i}^2 \right) \qquad 7\text{-}31$$

where we have used Equation 7-30 for v_{2f}.

Inelastic Collisions

For perfectly inelastic collisions, the second relation between the final velocities is that they are equal to each other and to the velocity of the center of mass:

$$v_{1f} = v_{2f} = V_{cm}$$

This result combined with conservation of momentum gives

$$(m_1 + m_2)V_{cm} = m_1 v_{1i} + m_2 v_{2i} \qquad 7\text{-}32$$

This is the same as Equation 7-13. Example 7-5 is a typical example of a perfectly inelastic collision.

For the special case of collisions in which one of the objects is initially at rest, we can relate the initial and final energies in a simple way by writing the kinetic energies in terms of the momenta as we did in Equation 7-26. Let the incoming object have mass m_1 and velocity v_{1i} and the second object, which is initially at rest, have mass m_2. The total momentum of the system is the momentum of the incoming object:

$$p = m_1 v_{1i}$$

The initial kinetic energy is

$$K_i = \frac{p^2}{2 m_1} \qquad 7\text{-}33$$

After the collision, the two objects move together as a single particle with a mass $m_1 + m_2$ and a velocity V_{cm}. Since momentum is conserved, the final momentum is also equal to p. The final kinetic energy can similarly be written as the square of the momentum divided by twice the mass, where the mass is now $m_1 + m_2$. The final kinetic energy is thus

$$K_f = \frac{p^2}{2(m_1 + m_2)} \qquad \text{7-34}$$

From a comparison of Equations 7-33 and 7-34, it is clear that the final energy is less than the initial energy. The ratio of the final and initial kinetic energies is

$$\frac{K_f}{K_i} = \frac{m_1}{m_1 + m_2} \qquad \text{7-35}$$

This result holds only if the collision is perfectly inelastic and the object with mass m_2 is initially at rest.

Exercise

A car of mass 2000 kg that is moving at 25 m/s runs head-on into a car of mass 1500 kg that is initially at rest. If the collision is perfectly inelastic, (a) find the speed of the two cars after the collision and (b) find the ratio of the final kinetic energy to the initial kinetic energy.
[Answers: (a) 14.3 m/s, (b) 0.57]

Figure 7-24 shows a bullet of mass m_1 moving with initial speed v_{1i} colliding inelastically with a block of mass m_2 that is suspended as a pendulum. After the collision the block–bullet system rises to some height h. This device is called a **ballistic pendulum.** Measurement of the height h can be used to determine the initial speed of the bullet. During the collision, mechanical energy is not conserved but momentum is. The kinetic energy just after the collision is related to the initial kinetic energy of the bullet by Equation 7-35, which we obtained from conservation of momentum:

$$K_f = \frac{m_1}{m_1 + m_2} K_i = \left(\frac{m_1}{m_1 + m_2}\right)\frac{1}{2}m_1 v_{1i}^2$$

After the collision, mechanical energy is conserved, and the kinetic energy is converted to potential energy $(m_1 + m_2)gh$:

$$(m_1 + m_2)gh = K_f = \left(\frac{m_1}{m_1 + m_2}\right)\frac{1}{2}m_1 v_{1i}^2$$

Figure 7-24 (a) A ballistic pendulum. The height h is related to the speed of the bullet–block system v_f by the conservation of energy. The speed v_f can be found from the conservation of momentum during the inelastic collision. (b) Inelastic collision of bullet and apple.

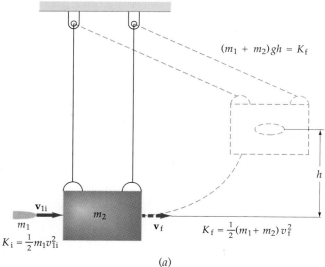

$(m_1 + m_2)gh = K_f$

$K_f = \frac{1}{2}(m_1 + m_2)v_f^2$

$K_i = \frac{1}{2}m_1 v_{1i}^2$

(a)

(b)

so

$$v_{1i} = \left(\frac{m_1 + m_2}{m_1}\right)\sqrt{2gh} \qquad\qquad 7\text{-}36$$

Exercise

For a ballistic pendulum, if the mass of the bullet is 12 g and its initial speed is 240 m/s and the mass of the block is 2 kg, find (a) the original kinetic energy of the bullet, (b) the kinetic energy of the bullet–block system just after the collision, and (c) the height h to which the bullet–block system rises. [Answers: (a) 346 J, (b) 2.06 J, (c) 10.4 cm]

In general, collisions lie somewhere between the extreme cases of elastic, in which the relative velocities are reversed, and perfectly inelastic, in which there is no relative velocity after the collision. The **coefficient of restitution** e, which is a measure of the elasticity of a collision, is defined as the ratio of the relative speed of recession to the relative speed of approach. For any collision, then, the relative velocity of recession is

$$v_{2f} - v_{1f} = -e(v_{2i} - v_{1i}) \qquad\qquad 7\text{-}37$$

For an elastic collision, $e = 1$. For a perfectly inelastic collision, $e = 0$.

Inelastic collision of two cars in a crash test simulation.

Example 7-11

The coefficient of restitution for steel on steel is measured by dropping a steel ball onto a steel plate that is rigidly attached to the earth. If the ball is dropped from a height of 3 m and rebounds to a height of 2.5 m, what is the coefficient of restitution?

The speed of the ball just before it strikes the plate is

$$v_i = \sqrt{2gh_i} = \sqrt{2(9.81 \text{ m/s}^2)(3 \text{ m})} = 7.67 \text{ m/s}$$

For the ball to rebound to a height of 2.5 m, it must leave the plate with an upward speed v_f given by

$$v_f = \sqrt{2gh_f} = \sqrt{2(9.81 \text{ m/s}^2)(2.5 \text{ m})} = 7.00 \text{ m/s}$$

Since the plate is rigidly attached to the earth, we can neglect its rebound velocity. Thus, the relative velocity of approach of the ball and plate is 7.67 m/s and the velocity of recession is 7.00 m/s. The coefficient of restitution is therefore

$$e = \frac{7.00 \text{ m/s}}{7.67 \text{ m/s}} = 0.913$$

Example 7-12

A 2-kg box moving to the right with speed 5 m/s collides with a 3-kg box that is moving in the same direction at 2 m/s, as shown in Figure 7-25. After the collision the 3-kg box moves at 4.2 m/s. Find (a) the velocity of the 2-kg box after the collision and (b) the coefficient of restitution for this collision.

Figure 7-25 Inelastic collision for Example 7-12.

Before collision   After collision

(a) The initial momentum of the 2-box system is

$$P_i = (2 \text{ kg})(5 \text{ m/s}) + (3 \text{ kg})(2 \text{ m/s})$$

$$= 16 \text{ kg·m/s}$$

If v_1' is the velocity of the 2-kg box after the collision, the final momentum is

$$P_f = (2 \text{ kg})v_1' + (3 \text{ kg})(4.2 \text{ m/s})$$

Conservation of momentum gives

$$(2 \text{ kg})v_1' + (3 \text{ kg})(4.2 \text{ m/s}) = 16 \text{ kg·m/s}$$

$$2\,v_1' + 12.6 \text{ m/s} = 16 \text{ m/s}$$

$$v_1' = \frac{16 \text{ m/s} - 12.6 \text{ m/s}}{2}$$

$$= 1.7 \text{ m/s}$$

(b) The fact that v_1' is positive means that the 2-kg box continues to move to the right. The relative speed of approach before the collision is 5 m/s − 2 m/s = 3 m/s. After the collision, the relative speed of recession is 4.2 m/s − 1.7 m/s = 2.5 m/s. The coefficient of restitution is thus

$$e = \frac{2.5 \text{ m/s}}{3.0 \text{ m/s}} = 0.833$$

Exercise

Find (a) the velocity of the center of mass and (b) the velocity of each box relative to the center of mass in Example 7-12, and (c) work that example in the center of mass reference frame.   [Answers: (a) $V_{cm} = 3.2$ m/s, (b) $u_1 = 1.8$ m/s, $u_2 = -1.2$ m/s, $u_1' = -1.5$ m/s, $u_2' = 1.0$ m/s, (c) $e = 0.833$]

Questions

8. Under what conditions can all the initial kinetic energy be lost in a collision?

9. Describe a perfectly inelastic collision as viewed in the center-of-mass reference frame.

7-7   Collisions in Three Dimensions

In three dimensions, the vector nature of conservation of linear momentum is important. Perfectly inelastic collisions present no special difficulty. The total initial momentum is found by adding the initial momentum vectors of the two bodies. Since the objects stick together and their final momentum equals the initial momentum, they move off in the direction of the resultant total momentum with velocity $\mathbf{V}_{cm}$ given by

$$\mathbf{V}_{cm} = \frac{\mathbf{P}}{m_1 + m_2} \qquad \text{7-38}$$

where $\mathbf{P}$ is the total momentum of the system.

Example 7-13

A small car of mass 1.2 Mg (1.2×10^3 kg) traveling east at 60 km/h collides at an intersection with a truck of mass 3 Mg traveling north at 40 km/h, as shown in Figure 7-26. The car and truck stick together. Find the velocity of the wreckage just after the collision.

Figure 7-26 A perfectly inelastic collision of a car and truck. After the collision, the wreckage moves off with a velocity whose magnitude and direction are determined by the conservation of momentum.

We choose our coordinate system so that initially the car is traveling in the x direction and the truck is traveling in the y direction. The initial momentum of the car is

$$\mathbf{p}_c = (1.2 \text{ Mg})(60 \text{ km/h})\mathbf{i} = 72 \text{ Mg·km/h } \mathbf{i}$$

and that of the truck is

$$\mathbf{p}_t = (3 \text{ Mg})(40 \text{ km/h})\mathbf{j} = 120 \text{ Mg·km/h } \mathbf{j}$$

These momentum vectors are shown in Figure 7-27. The total momentum is

$$\mathbf{P} = \mathbf{p}_c + \mathbf{p}_t = 72 \text{ Mg·km/h } \mathbf{i} + 120 \text{ Mg·km/h } \mathbf{j}$$

The wreckage has a total mass of 4.2 Mg, and moves with the velocity of the center of mass, which is given by

$$\mathbf{V}_{cm} = \frac{\mathbf{P}}{M} = \frac{72 \text{ Mg·km/h } \mathbf{i} + 120 \text{ Mg·km/h } \mathbf{j}}{4.2 \text{ Mg}}$$

$$= 17.1 \text{ km/h } \mathbf{i} + 28.6 \text{ km/h } \mathbf{j}$$

The magnitude of the final velocity is found using the pythagorean theorem

$$V_{cm}^2 = (17.1 \text{ km/h})^2 + (28.6 \text{ km/h})^2 = 1.11 \times 10^3 \text{ (km/h)}^2$$

$$V_{cm} = 33.3 \text{ km/h}$$

$\mathbf{p}_c = 72$ Mg·km/h $\mathbf{i}$
$\mathbf{p}_t = 120$ Mg·km/h $\mathbf{j}$
$\mathbf{P} = 72$ Mg·km/h $\mathbf{i}$
      $+ 120$ Mg·km/h $\mathbf{j}$
$\tan \theta = \dfrac{120}{72} = 1.67$

The direction of the final velocity is the same as that of the momentum vector. The angle θ in the figure is found from

$$\tan \theta = \frac{P_y}{P_x} = \frac{120 \text{ Mg·km/h}}{72 \text{ Mg·km/h}} = 1.67$$

$$\theta = 59°$$

Figure 7-27 Momentum vector diagram for the collision in Figure 7-26. The original momentum of the system, which equals the final momentum, is found by the vector addition of the initial momenta of the car and truck.

Collisions that are not perfectly inelastic are more complicated in three dimensions. Figure 7-28 shows an object of mass m_1 moving with velocity $\mathbf{v}_{1i}$ along the x axis toward an object of mass m_2 that is initially at rest. The "closeness" of the collision is measured by the distance between the centers measured perpendicular to the direction of $\mathbf{v}_{1i}$. This distance b, called the impact parameter, is indicated in the figure. After the collision, object 1 moves off with velocity $\mathbf{v}_{1f}$, making an angle θ_1 with its initial velocity, and object 2 moves with velocity $\mathbf{v}_{2f}$, making an angle θ_2 with $\mathbf{v}_{1i}$. Conservation of momentum gives

$$\mathbf{P} = m_1\mathbf{v}_{1i} = m_1\mathbf{v}_{1f} + m_2\mathbf{v}_{2f} \qquad 7\text{-}39$$

Before collision

After collision

Figure 7-28 Off-center collision of two bodies. The distance b is the impact parameter.

We can see from this equation that the vector $\mathbf{v}_{2f}$ must lie in the plane formed by $\mathbf{v}_{1i}$ and $\mathbf{v}_{1f}$, which we will take to be the xy plane. Assuming that we know the initial velocity $\mathbf{v}_{1i}$, we have four unknowns to find: the x and y components of each final velocity; or alternatively, the two final speeds v_{1f} and v_{2f}, the angle of deflection θ_1 of body 1, and the angle of recoil θ_2 of body 2. The x and y components of the conservation-of-momentum equation give us two of the needed relations among these quantities:

$$P_x = m_1 v_{1i} = m_1 v_{1f} \cos\theta_1 + m_2 v_{2f} \cos\theta_2 \qquad 7\text{-}40$$

and

$$P_y = 0 = m_1 v_{1f} \sin\theta_1 - m_2 v_{2f} \sin\theta_2 \qquad 7\text{-}41$$

A third relation comes from energy considerations. If the collision is elastic, we have

$$\tfrac{1}{2}m_1 v_{1i}^2 = \tfrac{1}{2}m_1 v_{1f}^2 + \tfrac{1}{2}m_2 v_{2f}^2 \qquad 7\text{-}42$$

To be able to solve for the four unknowns, we need another relation. The fourth relation depends on the impact parameter b and on the type of interacting force exerted by the bodies on each other. If the impact parameter is zero, for example, the collision is head-on and can be treated as a one-dimensional collision. (The angle θ_2 is then 0, and θ_1 is either 0 or $-180°$, depending on the relative masses of the particles.) In practice, the fourth relation is often found experimentally, by measuring the angle of deflection or the angle of recoil. Such a measurement can then give us information about the type of interacting force between the bodies.

We omit further discussion of elastic collisions in three dimensions except for the interesting special case of the off-center elastic collision of two objects of equal mass when one is initially at rest. Figure 7-29 shows the geometry of such a collision. If $\mathbf{v}_{1i}$ and $\mathbf{v}_{1f}$ are the initial and final velocities of object 1 and $\mathbf{v}_{2f}$ is the final velocity of object 2, conservation of momentum gives

$$m\mathbf{v}_{1i} = m\mathbf{v}_{1f} + m\mathbf{v}_{2f}$$

or

$$\mathbf{v}_{1i} = \mathbf{v}_{1f} + \mathbf{v}_{2f}$$

These vectors form the triangle shown in Figure 7-29b. Conservation of energy for this collision gives

$$\tfrac{1}{2}m v_{1i}^2 = \tfrac{1}{2}m v_{1f}^2 + \tfrac{1}{2}m v_{2f}^2$$

or

$$v_{1f}^2 + v_{2f}^2 = v_{1i}^2 \qquad 7\text{-}43$$

Equation 7-43 is the pythagorean theorem for a right triangle formed by the vectors $\mathbf{v}_{1f}$, $\mathbf{v}_{2f}$, and $\mathbf{v}_{1i}$, with the hypotenuse of the triangle being $\mathbf{v}_{1i}$. Thus, for this special case, the final velocity vectors $\mathbf{v}_{1f}$ and $\mathbf{v}_{2f}$ are perpendicular to each other, as shown in Figure 7-29b.

Figure 7-29   (a) Off-center elastic collision of two spheres of equal mass when one sphere is initially at rest. After the collision, the spheres move off at right angles to each other. (b) The velocity vectors for this collision form a right triangle.

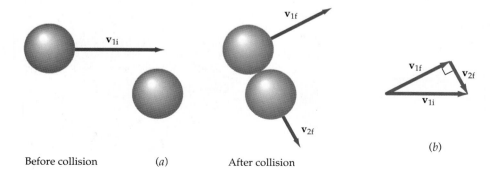

Before collision   (a)   After collision

(b)

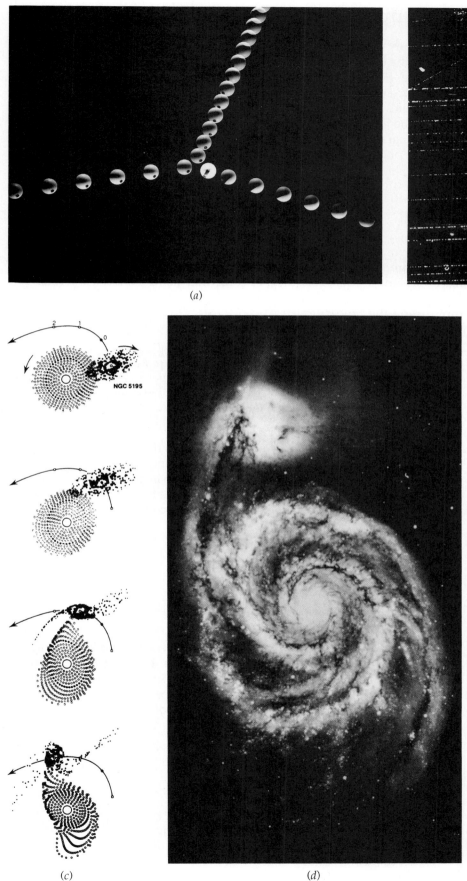

(a)

(b)

(c)

(d)

(a) Multiflash photograph of an elastic, off-center collision of two balls of equal mass. The dotted ball is incident from the left and strikes the striped ball, which is initially at rest. The final velocities of the two balls are perpendicular to each other. (b) Proton-proton collision in a liquid hydrogen bubble chamber. The incident proton enters from the left and interacts with a stationary proton in the chamber. The two particles then move off at right angles after the collision. The slight curvature of the tracks is due to a magnetic field. (c) Computer simulation of the collision of two galaxies [shown in (d)] at four different times. The time scale is in units of 100 million years. (d) Photograph from the Kitt Peak National Observatory of the collision of the large spiral galaxy NGC 5194 (bottom), which has a mass of about 10^{11} solar masses, and a much younger galaxy NGC 5195 (top), which is about three times less massive.

7-8   Impulse and Time Average of a Force

In our study of collisions, we have said little about the forces of interaction between the bodies except that they are usually very large and act for a very short time. Figure 7-30 shows the variation over time in the magnitude of a typical force exerted by one body on another during a collision. Before time t_i the bodies are apart and the force is zero. When the bodies come into contact, the force rises steeply and then falls back to zero at time t_f, when the bodies separate. The time of contact $\Delta t = t_f - t_i$ is usually very small, perhaps only about a millisecond. The **impulse I** of the force is a vector defined by

Impulse defined

$$I = \int_{t_i}^{t_f} F\, dt \qquad\qquad 7\text{-}44$$

The area under the F-versus-t curve is the magnitude of the impulse of the force. Assuming that **F** is the net force and using Newton's second law $F = dp/dt$, we can see that the impulse equals the total change in momentum during the time interval:

$$I = \int_{t_i}^{t_f} F\, dt = \int_{t_i}^{t_f} \frac{dp}{dt}\, dt = p_f - p_i = \Delta p \qquad\qquad 7\text{-}45$$

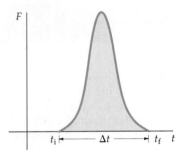

Figure 7-30 Typical time variation of the force exerted by one object on another during a collision. The force becomes very large, but it acts for only a very short time. The area under the F-versus-t curve is the magnitude of the impulse of the force. The impulse exerted on an object equals the change in its momentum.

From Equation 7-45, we can see that the units of impulse are newton-seconds or kilogram-meters per second.

For a general force **F**, the impulse depends on the times t_i and t_f, but the forces that occur in collisions are zero except during a very small time interval as shown in Figure 7-30. For these forces, the impulse does not depend on the time interval as long as t_i is any time before the collision and t_f is any time afterward. It is for this type of force that the concept of impulse is most useful.

The **time average of a force** for the interval $\Delta t = t_f - t_i$ is defined as

Time average of a force defined

$$F_{av} = \frac{1}{\Delta t} \int_{t_i}^{t_f} F\, dt = \frac{I}{\Delta t} \qquad\qquad 7\text{-}46$$

The average force is the constant force that gives the same impulse as the actual force in the time interval Δt. F_{av} is shown in Figure 7-31. It is often useful to compute the average force for a collision in order to compare it with other forces, such as frictional forces or the force of gravity. The average force can often be estimated by finding the time of the collision from a reasonable estimate of the distance traveled by one of the bodies during the collision.

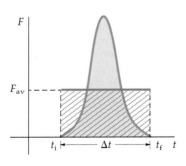

Figure 7-31 The average force F_{av} for some time interval is the constant force that gives the same impulse as the actual force during that interval. The rectangular area $F_{av}\,\Delta t$ is the same as the area under the F-versus-t curve.

Example 7-14

An egg of mass 50 g rolls off a table 1 m high and splatters on the floor. (*a*) Find the impulse exerted on the egg by the floor. (*b*) Assuming that the egg travels 2 cm (about half the smaller diameter of a typical egg) after it first comes into contact with the floor, estimate the collision time and the average force exerted by the floor on the egg.

(*a*) We can find the speed of the egg just before it hits the floor from the constant acceleration formula $v^2 = 2gy$, where g is the acceleration of gravity, and y is the distance the egg falls. For $y = 1$ m,

$$v^2 = 2gy = 2(9.81 \text{ m/s}^2)(1 \text{ m}) = 19.6 \text{ m}^2/\text{s}^2$$

The speed is then

$$v = \sqrt{19.6} \text{ m/s} \approx 4.4 \text{ m/s}$$

The momentum of the egg just before it hits the floor is downward and has the magnitude

$$p_i = mv = (0.05 \text{ kg})(4.4 \text{ m/s})$$
$$= 0.22 \text{ kg·m/s}$$

Since the final momentum is zero, the change in momentum is 0.22 kg·m/s in the upward direction. Thus, the impulse exerted by the floor on the egg is upward and has a magnitude of 0.22 kg·m/s = 0.22 N·s.

(*b*) Taking $\frac{1}{2}(4.4 \text{ m/s}) = 2.2$ m/s for the average speed of the egg during the collision, the time of the collision is estimated to be

$$\Delta t = \frac{\Delta y}{v_{av}} = \frac{0.02 \text{ m}}{2.2 \text{ m/s}} = 0.009 \text{ s} = 9 \text{ ms}$$

The average force is then

$$F_{av} = \frac{\Delta(mv)}{\Delta t} = \frac{(0.05 \text{ kg})(4.4 \text{ m/s})}{0.009 \text{ s}} \approx 24 \text{ N}$$

This is about 50 times the weight of the egg.

The hammer delivers a large impulsive force to drive the nail into the board.

Example 7-15

Estimate the force exerted by the seat belt on an 80-kg car driver when the car, originally moving at 25 m/s (about 56 mi/h), crashes into a fixed object.

We assume that the car travels about 1 m as the front end of the car crumples during the collision. This is also the distance traveled by the driver during the collision if he is wearing a seat belt. We also assume that the acceleration of the car is uniform as it crashes. Therefore, the average speed of the car during the collision is one-half of the initial speed, or 12.5 m/s. The time of the collision is then

$$\Delta t = \frac{1 \text{ m}}{12.5 \text{ m/s}} = 0.08 \text{ s}$$

The average acceleration is

$$a_{av} = \frac{\Delta v}{\Delta t} = \frac{25 \text{ m/s}}{0.08 \text{ s}} = 312 \text{ m/s}^2$$

This acceleration is about $32g$, that is, about 32 times the free-fall acceleration of gravity. The average force exerted by the seat belt on the driver is then

$$F_{av} = ma_{av}$$
$$= (80 \text{ kg})(312 \text{ m/s}^2) = 25,000 \text{ N}$$

This force is great enough to break the driver's ribs and cause other chest injuries, but he may survive the crash. Were he not wearing a seat belt, he would continue to move at 25 m/s until he hit the dashboard or windshield. His stopping distance would then be considerably less than 1 m, with the acceleration and force being correspondingly greater.

Example 7-16

What are reasonable magnitudes for the impulse I, average force F_{av}, and collision time Δt for a golf club hitting a golf ball? A typical golf ball has a mass $m = 45$ g and a radius $r = 2$ cm. For a typical drive, a reasonable range is $R = 160$ m (about 175 yd).

If air resistance is neglected, the speed v_0 of the ball when it leaves the club is related to the range R by

$$R = \frac{v_0^2}{g} \sin 2\theta_0$$

where θ_0 is the angle of the projection. Taking $\theta_0 = 45°$ (which corresponds to the maximum range), we find

$$v_0^2 = \frac{Rg}{\sin 2\theta_0} = \frac{(160 \text{ m})(9.81 \text{ m/s}^2)}{1} \approx 1600 \text{ m}^2/\text{s}^2$$

$$v_0 = 40 \text{ m/s}$$

The magnitude of the impulse is thus

$$I = \int F \, dt = \Delta p = mv_0 = (0.045 \text{ kg})(40 \text{ m/s})$$
$$= 1.8 \text{ kg·m/s} = 1.8 \text{ N·s}$$

A reasonable estimate for the distance traveled by the ball while it is in contact with the golf club is the radius of the golf ball, so $\Delta x = 2$ cm. Since the ball starts from rest and its final speed is 40 m/s, its average speed is 20 m/s (assuming constant acceleration). The collision time, that is, the time needed for the ball to travel 2 cm at an average speed of 20 m/s, is

$$\Delta t = \frac{\Delta x}{v_{av}} = \frac{0.02 \text{ m}}{20 \text{ m/s}} = 0.001 \text{ s}$$

The magnitude of the average force is therefore

$$F_{av} = \frac{I}{\Delta t} = \frac{1.8 \text{ N·s}}{0.001 \text{ s}} = 1800 \text{ N}$$

We can see that the average force exerted by the club, 1800 N, is much larger than any of the other forces acting on the ball. For example, the weight of the ball is only about 0.44 N, and the frictional forces exerted on the ball by the grass or tee are even less, assuming the coefficient of friction is less than 1.

Questions

10. Why can friction and the force of gravity usually be neglected in collision problems?

11. Explain why a safety net can save the life of a circus performer.

12. How might you estimate the collision time of a baseball and bat?

13. Why might a wine glass survive a fall onto a carpet but not onto a concrete floor?

7-9 Jet Propulsion

Jet propulsion is an interesting application of Newton's third law and the conservation of momentum. It is, for instance, the means by which the squid and octopus propel themselves. They squirt water from their bodies with great force, and the expelled water exerts an equal and opposite force on the squid or octopus, propelling it forward. A rocket gets its thrust by burning fuel and exhausting the gas created out the back. The rocket exerts a force on the exhaust gas, and by Newton's third law, the gas exerts an equal and opposite force on the rocket, propelling it forward. The momentum lost by the expelled gas is equal to the momentum gained by the rocket. A misconception prevalent before rockets in space were commonplace was that a rocket needs air to push against. This is not true. The rocket pushes against its own exhaust gas, which pushes back against the rocket. Jet propulsion is in fact more efficient in empty space where there is no air resistance.

We will develop an equation that describes the motion of a rocket. The description is complicated because the mass of the rocket changes continuously as it burns fuel and expels the exhaust gas. The easiest approach is to compute the change in the momentum of the total system (including the exhaust gas) for some time interval and set this change equal to the impulse exerted on the system by the external forces acting on it. Let F_{ext} be the net external force acting on the rocket, m be the mass of the rocket (plus the unburned fuel inside), and v be the speed of the rocket relative to the earth at time t (Figure 7-32a). At a later time $t + \Delta t$, the rocket has expelled gas of mass $|\Delta m|$ (Figure 7-32b). We use the absolute-value sign because the mass of the expelled gas is equal in magnitude to the change in the mass of the rocket Δm, which is negative. Thus, at time $t + \Delta t$, the rocket has a mass $m - |\Delta m|$ and is moving at a speed $v + \Delta v$. If the gas is exhausted at a speed u_{ex} *relative to the rocket*, its velocity at time $t + \Delta t$ relative to the earth is $v - u_{ex}$. The initial momentum of the system at time t is

$$P_i = mv$$

The momentum of the system at time $t + \Delta t$ is

$$P_f = (m - |\Delta m|)(v + \Delta v) + |\Delta m| (v - u_{ex})$$

$$= mv + m\,\Delta v - v\,|\Delta m| - |\Delta m|\,\Delta v + v\,|\Delta m| - u_{ex}\,|\Delta m|$$

$$\approx mv + m\,\Delta v - u_{ex}\,|\Delta m|$$

where we have dropped the term $|\Delta m|\,\Delta v$, which is the product of two very small quantities and therefore negligible compared with the others if the time interval Δt is very small. Computing the change in momentum and setting it equal to the impulse, we have

$$\Delta P = P_f - P_i = m\,\Delta v - u_{ex}\,|\Delta m| = F_{ext}\,\Delta t$$

We now divide by the time interval and take the limit as Δt approaches zero. The term $\Delta v/\Delta t$ approaches the derivative dv/dt, which is the acceleration, and the term $|\Delta m|/\Delta t$ approaches $|dm/dt|$, the absolute value of the rate of change of the mass of the rocket. This gives us the **rocket equation:**

$$m\,\frac{dv}{dt} = u_{ex}\left|\frac{dm}{dt}\right| + F_{ext} \qquad\qquad 7\text{-}47 \qquad \textit{Rocket equation}$$

The quantity $u_{ex}\,|dm/dt|$ is called the **thrust** of the rocket:

$$F_{th} = u_{ex}\left|\frac{dm}{dt}\right| \qquad\qquad 7\text{-}48$$

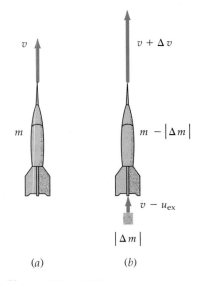

Figure 7-32 (a) Rocket moving with an initial speed v. (b) After a time interval Δt, the rocket has a mass $m - |\Delta m|$ and moves with a speed $v + \Delta v$. Gas exhausted from the rocket at a speed u_{ex} relative to the rocket is moving with speed $v - u_{ex}$. The change in the momentum of the rocket–exhaust system equals the impulse $F_{ext}\,\Delta t$.

When the rocket moves near the surface of the earth, the external force F_{ext} is the weight of the rocket. It is negative in Equation 7-47 because it is directly opposite the direction of the velocity, provided that the rocket is moving upward. Thus, the thrust must be greater than the weight of the rocket if the rocket is to accelerate upward. After we substitute $F_{ext} = -mg$ and divide by m, Equation 7-47 becomes

$$\frac{dv}{dt} = -g + \frac{u_{ex}}{m}\left|\frac{dm}{dt}\right| \qquad \text{7-49}$$

To solve Equation 7-49 for the velocity v, we must know the exhaust speed relative to the rocket u_{ex} and the rate the rocket burns fuel $|dm/dt|$. The solution of this equation is complicated by the fact that m is not constant but a function of time. For example, if the rocket burns fuel at a constant rate R, the mass of the rocket at any time is $m = m_i - Rt$, where m_i is the initial mass. Since dm/dt is negative, we write $|dm/dt| = -dm/dt$. Equation 7-49 is then

$$\frac{dv}{dt} = -g - \frac{u_{ex}}{m}\frac{dm}{dt}$$

or

$$dv = -g\,dt - u_{ex}\frac{dm}{m}$$

Assuming that g is constant and integrating from $t = 0$ to $t = t_b$ when the fuel is completely burned, we obtain

$$\int_{v_i}^{v_f} dv = -\int_0^{t_b} g\,dt - u_{ex}\int_{m_i}^{m_f}\frac{dm}{m}$$

$$v_f - v_i = -gt_b - u_{ex}\ln\frac{m_f}{m_i}$$

where we have used $\int(dm/m) = \ln m$ (Appendix A). Using $-\ln m_f/m_i = \ln(m_i/m_f)$, we obtain

$$v_f - v_i = +u_{ex}\ln\frac{m_i}{m_f} - gt_b \qquad \text{7-50}$$

Equation 7-50 gives the change in velocity for a rocket moving in a constant gravitational field in terms of the exhaust speed, the time for the fuel to burn t_b, and the ratio of the initial mass to final mass. For a rocket moving in free space with no external force, the change in velocity is given by

$$v_f - v_i = +u_{ex}\ln\frac{m_i}{m_f} \qquad \text{(no external force)} \qquad \text{7-51}$$

The mass of the rocket without any fuel is called the **payload.** If the payload is just 10 percent of the total initial mass, that is, if 90 percent of the initial mass is fuel, the ratio m_i/m_f when the fuel is spent will be 10. For a rocket moving with $v_i = 0$ and no external force, the final speed will then be

$$v_f = u_{ex}\ln 10 = 2.3 u_{ex}$$

The logarithmic nature of Equations 7-50 and 7-51 severely restricts the final speeds attainable. For example, if we increase the fuel so that the payload is only 1 percent of the initial mass, the final speed with no external force is $4.6u_{ex}$, just twice that for a 10-percent payload.

(a)

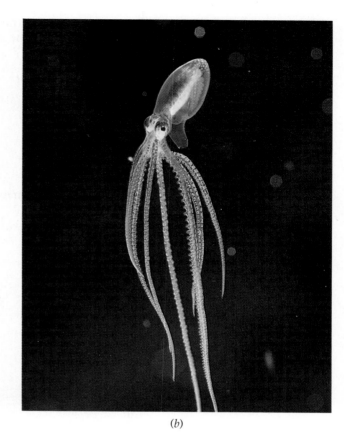

(b)

(a) Launch of a satellite, using a Thor booster rocket. (b) The octopus also moves by jet propulsion. However, instead of burning fuel like a rocket, the octopus takes in water and expels it at a high velocity.

Example 7-17

The Saturn V rocket used in the Apollo moon-landing program had an initial mass m_i of 2.85×10^6 kg, a payload of 27 percent, a burn rate $|dm/dt|$ of 13.84×10^3 kg/s, and a thrust F_{th} of 34×10^6 N. Find (a) the exhaust speed, (b) the burn time t_b, (c) the initial acceleration at lift-off, (d) the acceleration at burnout t_b, and (e) the final speed of the rocket.

(a) The exhaust speed is obtained from Equation 7-48:

$$u_{ex} = \frac{F_{th}}{|dm/dt|} = \frac{34 \times 10^6 \text{ N}}{13.84 \times 10^3 \text{ kg/s}} = 2.46 \text{ km/s}$$

(b) Since the payload is 27 percent, the mass of fuel burned is 73 percent of the initial mass or $m_{fuel} = (0.73)(2.85 \times 10^6 \text{ kg}) = 2.08 \times 10^6$ kg. The time to burn this much fuel at a rate of 13.84×10^3 kg/s is

$$t_b = \frac{2.08 \times 10^6 \text{ kg}}{13.84 \times 10^3 \text{ kg/s}} = 150 \text{ s}$$

(c) The initial acceleration is

$$\frac{dv}{dt} = -g + \frac{u_{ex}}{m_i} \left| \frac{dm}{dt} \right|$$

$$= -9.81 \text{ m/s}^2 + \frac{2.46 \text{ km/s}}{2.85 \times 10^6 \text{ kg}} (13.84 \times 10^3 \text{ kg/s})$$

$$= -9.81 \text{ m/s}^2 + 11.95 \text{ m/s}^2 = 2.14 \text{ m/s}^2 \approx 0.21g$$

(d) At burnout, the final mass is $m_f = (0.27)(2.85 \times 10^6 \text{ kg}) = 7.70 \times 10^5$ kg, so the acceleration is

$$\frac{dv}{dt} = -g + \frac{u_{ex}}{m_f}\left|\frac{dm}{dt}\right|$$

$$= -9.81 \text{ m/s}^2 + \frac{2.46 \text{ km/s}}{7.70 \times 10^5 \text{ kg}}(13.84 \times 10^3 \text{ kg})$$

$$= -9.81 \text{ m/s}^2 + 44.26 \text{ m/s}^2$$

$$= 34.4 \text{ m/s}^2 \approx 3.5 \text{ g}$$

(e) The final speed of the rocket is

$$v_f = +u_{ex} \ln \frac{m_i}{m_f} - gt_b$$

$$= (2.46 \text{ km/s}) \ln \frac{m_i}{0.27m_i} - (9.81 \text{ m/s}^2)(150 \text{ s})$$

$$= 3.22 \text{ km/s} - 1.47 \text{ km/s}$$

$$= 1.75 \text{ km/s}$$

Summary

1. The position of the center of mass of a system of particles is defined by

$$M\mathbf{R}_{cm} = m_1\mathbf{r}_1 + m_2\mathbf{r}_2 + \cdots = \sum_i m_i\mathbf{r}_i$$

where M is the total mass of the system and $\mathbf{R}_{cm}$ is the position vector from the chosen origin to the center of mass.

2. The center of mass of a system moves like a particle having a mass equal to the total mass of the system under the influence of the net external force acting on the system:

$$\mathbf{F}_{net,ext} = M\mathbf{A}_{cm}$$

3. The momentum of a particle is defined as its mass times its velocity:

$$\mathbf{p} = m\mathbf{v}$$

Newton's second law states that the net force acting on a particle equals the time rate of change of its momentum:

$$\mathbf{F}_{net} = \frac{d\mathbf{p}}{dt}$$

The kinetic energy of a particle can be written in terms of its momentum:

$$K = \frac{p^2}{2m}$$

4. The total mass of a system times the velocity of the center of mass equals the total momentum of the system:

$$\mathbf{P} = \sum_i m_i\mathbf{v}_i = M\mathbf{V}_{cm}$$

5. If the net external force acting on a system is zero, the total momentum of the system is conserved. This is the law of conservation of momentum.

6. The center-of-mass reference frame is the frame that moves with the center of mass. In this frame, the total momentum is zero.

7. The kinetic energy of a system of particles can be written as the sum of the kinetic energy associated with the motion of the center of mass, $\frac{1}{2}MV^2_{cm}$, and the energy of motion relative to the center of mass, $K_{rel} = \Sigma \frac{1}{2}m_i u_i^2$, which is the kinetic energy as observed in the center-of-mass frame:

$$K = \tfrac{1}{2}MV^2_{cm} + K_{rel}$$

8. An elastic collision is one in which the total kinetic energy of the two objects is the same before and after the collision. An inelastic collision is one in which the total kinetic energy is changed by the collision.

9. In a perfectly inelastic collision, the bodies stick together and move with the velocity of the center of mass.

10. For an elastic collision, the relative speed of recession of the bodies after the collision equals the relative speed of approach before the collision.

11. The coefficient of restitution e is a measure of the elasticity of a collision and is defined as the ratio of the relative speed of recession to the relative speed of approach. For an elastic collision, $e = 1$; for a perfectly inelastic collision, $e = 0$.

12. The impulse of a force is defined as the integral of the force over the time interval during which the force acts. The impulse of the net force equals the total change in the momentum of a particle:

$$\mathbf{I} = \int_{t_i}^{t_f} \mathbf{F}\, dt = \Delta\mathbf{p}$$

The time average of a force for the interval $\Delta t = t_f - t_i$ is

$$\mathbf{F}_{av} = \frac{1}{\Delta t}\int_{t_i}^{t_f} \mathbf{F}\, dt = \frac{\mathbf{I}}{\Delta t}$$

13. A rocket gets its thrust by burning fuel and exhausting the resulting gases. The force exerted by the exhaust gases on the rocket propels the rocket forward. The equation for motion of a rocket obtained from Newton's laws is

$$m\frac{dv}{dt} = u_{ex}\left|\frac{dm}{dt}\right| + F_{ext}$$

In this equation, u_{ex} is the speed of the exhaust gas relative to the rocket. If the external force of gravity is constant, the final velocity of the rocket is related to the exhaust speed of the gas and the initial and final mass of the rocket:

$$v_f = +u_{ex}\ln\frac{m_i}{m_f} - gt_b$$

where t_b is the burn time.

Discovery of the Neutrino

Ralph A. Llewellyn
University of Central Florida

The concept of energy conservation was first set forth by the great biologist and physicist Herman von Helmholtz in 1847 when he generalized the results of a series of painstaking experiments that had been conducted some years previously by James Joule. The latter had shown that when energy is converted from one form into another no energy was either lost or created. Joule demonstrated, for example, that the kinetic energy of a paddle wheel turning in water is converted into thermal energy of the water. Similarly, the mechanical energy of a piston is converted into potential energy of a gas when the gas is compressed by the piston. That the total amount of energy of a closed system remains unchanged was tested and confirmed by innumerable experiments in the years following. Even Albert Einstein's discovery of the equivalence of mass and energy, embodied in the famous equation $E = mc^2$, did not imply a violation of the law of conservation of energy. It simply added a new kind of conversion, that of energy to mass and mass to energy, to those already known.

The law of conservation of energy was based entirely on experimental observation. There existed no fundamental physical theory that *predicted* the conservation of total energy. Nor, in fact, does such a theory or equation exist now. It was in this context that, in 1896, Henri Becquerel discovered radioactivity. What Becquerel observed was the emission of previously unknown "rays" from certain materials (initially, salts containing uranium). This strange radiation was subsequently shown—by Becquerel, Ernest Rutherford, and others—to consist of three distinct types, called alpha, beta, and gamma rays, which differed in their ability to penetrate matter. They had other significant differences, too, but shared the common characteristic that the energy of each individual ray was very large.

Following Rutherford's discovery of the atomic nucleus and Niels Bohr's use of the nuclear model in his explanation of the structure of atoms (both in 1913), it became clear that the large energies of the radioactive rays meant that they were associated with events occurring inside the nucleus itself. The study of radioactivity therefore afforded scientists the exciting opportunity to probe the internal structure of the nucleus. Precise measurements of the energy balances in the nuclear events involving alpha, beta, and gamma rays* were high priority experiments, since data from corresponding experiments on the energy of electromagnetic radiation (light) emitted by atoms had formed the foundation of Bohr's successful theory of atomic structure.

For any given nuclear decay involving only the emission of a gamma ray, the gamma rays were observed to all have the same energy (that is, to be "monoenergetic"). In particular, the energy of the gamma ray (E_γ) was exactly equal to the energy of the radioactive nucleus before emission (E_i) minus the nucleus after emission (E_f):

$$E_\gamma = E_i - E_f$$

Thus, the emission of gamma rays by nuclei conserved energy.

The same result held true for any given nuclear decay involving only the emission of an alpha ray. For a given nuclear decay, the alpha rays were monoenergetic, with energy E_α given by

$$E_\alpha = E_i - E_f$$

This again indicated that energy was conserved in the nuclear decay.

However, those nuclear events resulting in the emission of a beta ray appeared different: When a nucleus decayed from some specific initial state of energy E_i to a specific final state of energy E_f, the

Ralph A. Llewellyn has been Professor of Physics at the University of Central Florida since 1980. After earning a Ph.D. in experimental nuclear physics at Purdue University in 1962, he taught and chaired physics departments at Rose–Hulman Institute of Technology and Indiana State University. He has also been a member of the scientific staff of the National Academy of Sciences/National Research Council, where he directed studies on alternative energy sources, energy conservation, and world energy resources. When not doing physics, his recreational interests include photography and bicycle racing.

---

*It is now known that alpha rays are helium nuclei, beta rays are electrons or positrons (antielectrons), and gamma rays are high-energy electromagnetic radiation.

beta rays were not monoenergetic but were emitted in a continuous spectrum of energies from 0 to $E_{max} = E_i - E_f$. (Also, the beta rays were subsequently discovered to be electrons.) The energy balance equation for beta decay was, therefore,

$$E_\beta \leq E_{max} = E_i - E_f$$

The difference between the initial and final states of the nucleus ($E_i - E_f$) was generally *not* the energy present in the beta ray (E_β), which varied. This seemed to indicate that energy was not being conserved during beta decay.

This presented physics with a monumental problem, one capable of severely damaging the theoretical basis of the science. After all of these years it is difficult for us to realize how serious this matter was. Many physicists were prepared to abandon the law of conservation of energy, at least for nuclear events, on the logical grounds that a general law founded on experimental results must be discarded if further experiment fails to confirm it. Notable among these physicists was Neils Bohr himself. The proposal was made that the energy principle be rewritten as

$$E_{\substack{final \\ system}} \leq E_{\substack{initial \\ system}}$$

This form would preserve energy conservation in all experiments, including the disappearance of energy in beta decay (with the "less than" inequality), and still disallow the *creation* of energy.

That this was not a good solution to the problem was pointed out by Wolfgang Pauli. In a letter to Lise Meitner and Hans Geiger in December 1930, Pauli noted that beta decay seemed to imply not only a failure of the conservation of energy but also the failure of the conservation of linear momentum (Equation 7-15) and angular momentum (to be discussed in Chapter 8). These failures were no less serious: Furthermore, inserting the inequality sign into the energy conservation law did not save the others.

In the same letter, Pauli went on to suggest a way out of the dilemma. He made the proposal that a new fundamental particle existed and that one (or more) of these new particles was emitted along with the electron in beta decay. Regarding the new particle, he made several predictions. It would have no electric charge, since electric charge was already conserved in beta decay without it. It would have an extremely high ability to penetrate matter, since no one had ever seen one. (This high penetrability would imply a weak interaction with matter, which in turn would imply that the particle would be hard to detect.) Its mass would be zero, or nearly so, since beta rays with energy nearly equal to the maxi-

mum possible value (E_{max}) were occasionally seen in beta decays.

The proposed new particle would carry energy equal to the difference $E_{max} - E_\beta$ in each individual beta decay, thus preserving the law of conservation of energy. Its physical behavior would also be consistent with the preservation of linear momentum and angular momentum.

To appreciate how innovative Pauli's proposal was, one must remember that—at the time he made it—only two fundamental particles were known, the electron and the proton. The existence of both had been deduced from their appearance and behavior in many, many experiments. No one had ever before *invented* a new particle. Pauli acknowledged that his solution might appear an unlikely one, but he went on to say, "Nothing ventured, nothing gained" and to urge the "dear radioactive folks [to] put it to the test and judge."

Pauli went public with his proposal for this strange new particle at a meeting of the American Physical Society in June 1931. Despite his substantial reputation as a theoretical physicist, the response was decidedly underwhelming. Most physicists were highly skeptical of the proposal for a new, apparently undetectable particle. A consensus emerged that Bohr was correct. For some reason, energy was not conserved in nuclear processes.

In 1933, Pauli was encouraged by James Chadwick's discovery a few months earlier of the neutron (a new uncharged particle with mass nearly the same as the proton). Pauli then argued forcefully against accepting a breakdown of the conservation laws in beta decay:

> The electric charge is conserved in the process, and I don't see why conservation of charge should be more fundamental than conservation of energy and momentum.—If the conservation laws were not valid, one would have to conclude from these relations that a beta disintegration occurs always with a loss of energy and never a gain; this conclusion implies an irreversibility of these processes with respect to time, which doesn't seem to me at all acceptable.*

Enrico Fermi was much attracted by Pauli's proposed new particle, which he called the *neutrino* (meaning "little neutral one" in Italian) in order to distinguish it from Chadwick's massive neutron. Fermi developed an entirely new theory of beta decay incorporating the neutrino. Fermi's theory was enormously successful in accounting accurately for experimental observations involving beta decay.

*Wolfgang Pauli, "Structures et Propriétés des Noyaux Atomiques," *Proceedings of the Solvay Congress 1933*, Paris, Gauthier-Villars, 1934, p. 324.

Continued

This theory all but guaranteed the existence of the neutrino with precisely the properties that Pauli had suggested, despite—as was claimed at the time—the utter impossibility of its observation in any experiment.

Time passed. Our knowledge of the fundamental particles that make up the atom, the forces that act between them, and the conservation laws that govern their interactions, all expanded and improved. Physicists began to realize that the neutrino played a far more central role in particle physics than even Pauli had at first suspected. As a consequence, it became a matter of high priority to verify that the neutrino existed and that it behaved according to the assumptions that were rapidly being incorporated into nuclear and particle theories. In 1949, Chalmers Sherwin measured the linear momentum of beta rays and simultaneously measured the linear momentum of the recoiling nuclei. He showed that a single massless particle, corresponding to Pauli's neutrino, emitted along with the beta ray could precisely satisfy both the conservation of linear momentum and the conservation of energy.

That neutrinos interact exceedingly weakly with matter means, for example, that the average distance a typical neutrino will travel from where it is created to where it first interacts with matter is of the order of 1000 light-years. This makes neutrinos very hard to detect in *any* experiment. In order to have a reasonable chance of detecting even a few neutrinos, an enormous number of them must pass through whatever experimental apparatus is set up to detect them.

It was not until 25 years after Pauli first proposed the neutrino that the advent of large nuclear reactors made available sources of neutrinos in the numbers necessary to make detection experiments conceivable. Finally, in 1956, Clyde Cowan and Frederick Reines, using the very high neutrino flux (about 10^{15} neutrinos per square centimeter per second) from a reactor at Savannah River, were able to detect the unmistakable "signature" of a nuclear event that could only have been initiated by a neutrino (actually, by an antineutrino, but that's another story!). Thus was the last element of doubt regarding Pauli's proposal effectively removed.

Since that time the complex, fundamental properties of the neutrino have continued to be the focus of active theoretical and experimental study. We believe that the neutrinos (we're now sure there are at least six different kinds) are among the very few, truly fundamental particles. By "fundamental," we mean that they do not themselves decay. Neutrino experiments, once considered impossible, are now performed at all the world's major particle accelerators and are enabling important advances in our search for deeper understanding of the physical universe. Neutrino astronomy is an area of active research because neutrinos, due to the weakness of their interaction with matter, are the only particles that can easily escape the interior of the densest stars. They bring information about the genesis of stars and the earliest moments of the universe.

And it all began with a letter written by Wolfgang Pauli. "Nothing ventured nothing gained" indeed!

This photograph shows the tracks of a complicated spray of particles that are emitted when a neutrino (entering from the left) collides with a proton in the Big European Bubble Chamber at CERN.

Suggestions for Further Reading

Damask, Arthur C. "Forensic Physics of Vehicle Accidents," *Physics Today*, March 1987, p. 36.

The law of conservation of momentum and the concept of the center of mass are extremely useful in reconstructing collisions between vehicles or between a vehicle and a pedestrian.

Walker, Jearl: "The Amateur Scientist: Success in Racquetball Is Enhanced by Knowing the Physics of the Collision of Ball with Wall," *Scientific American*, September 1984, p. 215.

The laws of conservation of momentum and energy are applied to the collisions of a racquetball.

Review

A. Objectives: After studying this chapter, you should:

1. Be able to find the center of mass of a system of discrete particles and of a continuous object.

2. Be able to state Newton's second law for a system of particles.

3. Be able to use the conservation of momentum to solve problems.

4. Know the definition of the center-of-mass reference frame and be able to transform into it and out of it to work collision problems.

5. Know that the kinetic energy of a system can be written as the sum of the kinetic energy of the motion of the center of mass and the kinetic energy of the motion of the system relative to the center of mass.

6. Be able to solve elastic and inelastic collision problems in one dimension.

7. Know the relation between impulse and momentum, and be able to use it to estimate the magnitude of the large forces that occur in collisions.

8. Be able to apply Newton's laws to rocket motion and to derive the equation for the motion of a rocket.

B. Define, explain, or otherwise identify:

Center of mass
Newton's second law for a system
Momentum
Law of conservation of momentum
Center-of-mass reference frame
Zero-momentum reference frame
Elastic collision
Inelastic collision
Perfectly inelastic collision

Ballistic pendulum
Coefficient of restitution
Impulse
Time-average force
Rocket equation
Thrust
Payload

C. True or false: If the statement is true, explain why. If it is false, give a counterexample.

1. The momentum of a heavy object is greater than that of a light object moving at the same speed.

2. In a perfectly inelastic collision, all the kinetic energy of the bodies is lost.

3. The momentum of a system may be conserved even when mechanical energy is not.

4. Kinetic energy is conserved in an elastic collision.

5. In an elastic collision, the relative speed of recession after the collision equals the relative speed of approach before the collision.

6. The velocity of the center of mass of a system equals the total momentum of the system divided by its total mass.

Problems

Level I

7-1 The Center of Mass

1. Three point masses of 2 kg each are located on the x axis at the origin, $x = 0.20$ m and $x = 0.50$ m. Find the center of mass of the system.

2. A 24-kg child is 20 m from an 86-kg man. Where is the center of mass of this system?

3. Three objects of 2 kg each are located as follows: object 1 is at $x = 10$ cm, $y = 0$; object 2 is at $x = 0$, $y = 10$ cm; and object 3 is at $x = 10$ cm, $y = 10$ cm. Find the location of the center of mass.

4. Find the center of mass x_{cm} of the three masses shown in Figure 7-33.

5. The uniform sheet of plywood in Figure 7-34 has a mass of 20 kg. Find the x and y coordinates of the center of mass.

6. Alley Oop's club-axe consists of a symmetrical 8-kg stone attached to the end of a uniform 2.5-kg stick. The dimensions of the club-axe are shown in Figure 7-35. How far is the center of mass from the handle end of the club-axe?

7. Three small balls A, B, and C, with masses of 300, 100, and 100 g, respectively, are connected by massless rods. The balls are located as shown in Figure 7-36. What are the coordinates of the center of mass?

7-2 Motion of the Center of Mass of a System

8. Two 3-kg masses have velocities $\mathbf{v}_1 = 2$ m/s $\mathbf{i} +$ 3 m/s $\mathbf{j}$ and $\mathbf{v}_2 = 4$ m/s $\mathbf{i} - 6$ m/s $\mathbf{j}$. Find the velocity of the center of mass.

9. A 1500-kg car is moving to the west with a speed of 20 m/s, and a 3000-kg truck is moving east with a speed of 16 m/s. Find the velocity of the center of mass of the system.

10. A force $\mathbf{F} = 12$ N $\mathbf{i}$ is applied to the 300-g ball in Figure 7-36. What is the acceleration of the center of mass?

11. A constant force $\mathbf{F} = 26$ N $\mathbf{i}$ is applied to the 8-kg mass of Problem 4 at $t = 0$. (*a*) Find the velocity of the center of mass at $t = 5$ s. (*b*) What is the location of the center of mass at $t = 5$ s?

7-3 The Conservation of Momentum

12. An open railroad car of mass 20,000 kg is rolling without friction along a track with a speed of 5 m/s when it starts to rain. After the car has collected 2000 kg of water, what is its new speed?

13. A girl of mass 50 kg jumps out of the bow of a 250-kg canoe that is initially at rest. If her velocity is 7.5 m/s to the right, what is the velocity of the canoe after she jumps?

14. Two masses of 5 kg and 10 kg resting on a frictionless table are connected by a compressed spring. When the spring is released, the smaller mass has a velocity of 8 m/s to the left. What is the velocity of the larger mass?

15. A model-train car of mass 250 g traveling with a speed of 0.50 m/s couples to another car of mass 400 g that is initially at rest. What is the speed of the cars after they have coupled together?

7-4 The Center-of-Mass Reference Frame and

7-5 Kinetic Energy of a System of Particles

16. (*a*) Find the total kinetic energy of the two model-train cars of Problem 15 before they couple. (*b*) Find the initial velocities of the two cars relative to the center of mass of the system, and use them to calculate the initial kinetic energy of the system relative to the center of mass.

Figure 7-33 Problems 4 and 11.

Figure 7-34 Problem 5.

Figure 7-35 Problem 6.

Figure 7-36 Problem 7 and 10.

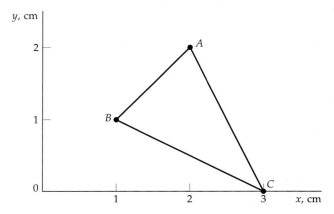

(*c*) Find the kinetic energy of the center of mass. (*d*) Compare your answers for parts (*b*) and (*c*) with that for part (*a*).

17. Describe how a solid ball can move such that (*a*) its total kinetic energy is just the energy of motion of its center of mass and (*b*) its total kinetic energy is the energy of its motion relative to its center of mass.

18. Two bowling balls are moving with the same velocity, but one just slides down the alley, whereas the other rolls down the alley. Which ball has more energy?

19. A 3-kg block is traveling to the right at 5 m/s, and a second 3-kg block is traveling to the left at 2 m/s. (*a*) Find the total kinetic energy of the two blocks in this reference frame. (*b*) Find the velocity of the center of mass of the two-body system. (*c*) Find the velocities of the two blocks relative to the center of mass. (*d*) Find the kinetic energy of the motion of the blocks relative to the center of mass. (*e*) Show that your answer for part (*a*) is greater than your answer for part (*d*) by the kinetic energy of the center of mass.

20. Repeat Problem 19 for a second block having a mass of 5 kg and moving to the right at 3 m/s.

7-6 Collisions in One Dimension

21. A 150-g glob of clay is thrown horizontally with a speed of 5 m/s at a 1-kg block that is initially at rest on a frictionless surface. If the clay sticks to the block, what is the speed of the combined system?

22. A 2000-kg car traveling to the right at 30 m/s is chasing a second car of the same mass that is traveling to the right at 10 m/s. (*a*) If the two cars collide and stick together, what is their speed just after the collision? (*b*) What fraction of the initial kinetic energy of the cars is lost during this collision? Where does it go?

23. An 85-kg running back moving at 7 m/s makes a perfectly inelastic collision with a 105-kg linebacker who is initially at rest. What is the speed of the players just after their collision?

24. Find the final velocity of each block in Problem 19 if they make (*a*) a perfectly inelastic collision and (*b*) an elastic collision.

25. Repeat Problem 24 using the blocks in Problem 20.

26. A 5.0-kg object with a speed of 4.0 m/s collides head-on with a 10-kg object moving toward it with a speed of 3.0 m/s. (*a*) If the 10-kg object stops dead after the collision, what is the final speed of the 5-kg object? (*b*) Is the collision elastic?

27. A heavy ball of mass 5 kg hits an 85-kg man in the chest and bounces directly back at a speed of 2 m/s. (*a*) If the initial speed of the ball is 8 m/s and the man is initially at rest, find the speed imparted to the man by the collision. (*b*) Is this an elastic or inelastic collision?

28. According to the official rules of racquetball, a ball acceptable for tournament play must bounce to a height of between 173 and 183 cm when dropped from a height of 254 cm at room temperature. What is the acceptable range of values for the coefficient of restitution for the racquetball–floor system?

29. A ball bounces to 80 percent of its original height. (*a*) What fraction of its mechanical energy is lost each time it bounces? (*b*) What is the coefficient of restitution of the ball–floor system?

7-7 Collisions in Three Dimensions

30. A 1500-kg car traveling north at 70 km/h collides at an intersection with a 2000-kg car traveling west at 55 km/h. The two cars stick together. (*a*) What is the total momentum of the system before the collision? (*b*) Find the magnitude and direction of the velocity of the wreckage just after the collision.

31. In a pool game, the cue ball, which has an initial speed of 5 m/s, makes an elastic collision with the 8 ball, which is initially at rest. After the collision, the 8 ball moves at an angle of 30° with the original direction of the cue ball. (*a*) Find the direction of motion of the cue ball after the collision. (*b*) Find the speed of each ball. Assume that the balls have equal mass.

32. Figure 7-37 shows the results of a collision of two objects of unequal mass. (*a*) Find the speed v_2 of the larger mass after the collision and the angle θ_2. (*b*) Show that the collision is elastic.

Figure 7-37   Problem 32.

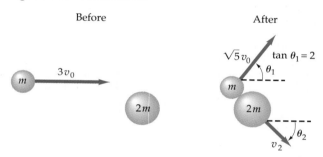

7-8 Impulse and the Time Average of a Force

33. A soccer ball of mass 0.43 kg leaves the foot of the kicker with an initial speed of 25 m/s. (*a*) What is the impulse imparted to the ball by the kicker? (*b*) If the foot of the kicker is in contact with the ball for 0.008 s, what is the average force exerted by the foot on the ball?

34. A 0.3-kg brick is dropped from a height of 8 m. It hits the ground and comes to rest. (*a*) What is the impulse exerted by the ground on the brick? (*b*) If it takes 1.3 ms from the time the brick first touches the ground until it comes to rest, what is the average force exerted by the ground on the brick?

35. When a 0.15-kg baseball is hit, its velocity changes from +20 m/s to −20 m/s. (*a*) What is the magnitude of the impulse delivered by the bat to the ball? (*b*) If the baseball is in contact with the bat for 1.3 ms, what is the average force exerted by the bat on the ball?

36. A 300-g handball moving with a speed of 5.0 m/s strikes the wall at an angle of 40° and then bounces off with the same speed at the same angle. It is in contact with the wall for 2 ms. What is the average force exerted by the ball on the wall?

37. In Problem 23, if the time of the collision is 0.1 s, what is the average force each player exerts on the other?

38. In Problem 27, if the time of the collision is 0.05 s, what is the average force exerted by the ball on the man?

7-9 Jet Propulsion

39. A rocket burns fuel at a rate of 200 kg/s and exhausts the gas at a relative speed of 6 km/s. Find the thrust of the rocket.

40. The payload of a rocket is 5 percent of its total mass, the rest being fuel. If the rocket starts from rest and moves with no external forces acting on it, what is its final velocity if the exhaust velocity of its gas is 5 km/s?

41. A rocket moves in free space with no external forces acting on it. It starts from rest and has an exhaust speed of 3 km/s. Find the final velocity if the payload is (*a*) 20 percent, (*b*) 10 percent, (*c*) 1 percent.

Level II

42. A small car of mass 800 kg is parked behind a small truck of mass 1600 kg on a level road. The brakes of both the car and the truck are off so that they are free to roll with negligible friction. A man sitting on the tailgate of the truck exerts a constant force on the car with his feet as shown in Figure 7-38. The car accelerates at 1.2 m/s^2. (*a*) What is the acceleration of the truck? (*b*) What is the magnitude of the force exerted on either the truck or the car?

Figure 7-38 Problem 42.

43. A 16-g bullet is fired into the bob of a ballistic pendulum of mass 1.5 kg. When the bob is at its maximum height, the strings make an angle of 60° with the vertical. The length of the pendulum is 2.3 m. Find the speed of the bullet.

44. A 3-kg body moving at 4 m/s makes an elastic collision with a stationary body of mass 2 kg. Use conservation of momentum and the fact that the relative velocity of recession equals the relative velocity of approach to find the velocity of each body after the collision. Check your answer by calculating the initial and final kinetic energies of each body.

45. A 6-kg projectile is launched at an angle of 30° with the horizontal and an initial speed of 40 m/s. At the top of its flight, it explodes into two parts with masses of 2 and 4 kg. The fragments move horizontally just after the explosion, and the 2-kg piece lands back at the launch site. (*a*) Where does the 4-kg piece land? (*b*) Compute the ki-

netic energy of the projectile just before the explosion and the total kinetic energy of the fragments just after the explosion to find the energy of the explosion.

46. A 3-kg bomb slides along a frictionless horizontal plane in the *x* direction at 6 m/s. It explodes into two pieces, one of mass 2 kg and the other of mass 1 kg. The 1-kg piece moves along the horizontal plane in the *y* direction at 4 m/s. (*a*) Find the velocity of the 2-kg piece. (*b*) What is the velocity of the center of mass after the explosion?

47. A 2-kg body moving at 3 m/s to the right collides with a 3-kg body moving at 2 m/s to the left. The coefficient of restitution is 0.4. Find the velocity of each body after the collision.

48. A 2-kg body moving at 6 m/s collides with a 4-kg body that is initially at rest. After the collision, the 2-kg body moves backward at 1 m/s. (*a*) Find the velocity of the 4-kg body after the collision. (*b*) Find the energy lost in the collision. (*c*) What is the coefficient of restitution for this collision?

49. The ratio of the mass of the earth to the mass of the moon is $M_E/m_m = 81.3$. The radius of the earth is about 6370 km, and the distance from the earth to the moon is about 384,000 km. (*a*) Locate the center of mass of the earth–moon system relative to the surface of the earth. (*b*) What external forces act on the earth–moon system? (*c*) In what direction is the acceleration of the center of mass of this system? (*d*) Assume that the center of mass of this system moves in a circular orbit around the sun. How far must the center of the earth move in the radial direction (toward or away from the sun) during the 14 days between the time the moon is farthest from the sun (full moon) and the time it is closest to the sun (new moon)?

50. A circular plate of radius *r* has a circular hole cut out of radius *r*/2 (Figure 7-39). Find the center of mass of the plate. *Hint:* The hole can be represented by two disks superimposed, one of mass *m* and the other of mass −*m*.

51. Using the hint from Problem 50, find the center of mass of a solid sphere of radius *r* that has a spherical cavity of radius *r*/2 as shown in Figure 7-40.

52. A peculiar baseball bat of length *L* has a linear density (mass per unit length) given by $\lambda = \lambda_0(1 + x^2/L^2)$. Find the *x* coordinate of the center of mass in terms of *L*. See Figure 7-41.

53. A 40-kg girl gets on her 10-kg wagon on level ground with two 5-kg bricks. She throws the bricks horizontally off the back of the wagon one at a time at a speed of 7 m/s relative to herself. (*a*) How fast does she go after throwing the second brick? (*b*) How fast would she go if she threw both bricks at the same time at 7 m/s relative to herself?

54. A hammer of mass 0.8 kg is used to drive nails of mass 30 g into wood. When the hammer has an impact speed of 0.5 m/s, the nail penetrates 2.0 cm at a single blow. Find (*a*) the common speed of the hammer and nail immediately after impact, assuming the collision is perfectly inelastic; (*b*) the time the nail is in motion, assuming that its initial speed is acquired in a negligible time, after

Figure 7-39   Problem 50.

Figure 7-40   Problem 51.

Figure 7-41   Problem 52.

which there is uniform deceleration; and (*c*) the average resisting force of the wood against the nail as it penetrates.

55. You throw a 150-g ball to a height of 40 m. (*a*) Use a reasonable value for the distance the ball moves while it is in your hand to calculate the average force exerted by your hand and the time the ball is in your hand while you throw it. (*b*) Is it reasonable to neglect the weight of the ball while it is being thrown?

56. A 2000-kg car traveling at 90 km/h crashes into a concrete wall that does not give at all. (*a*) Estimate the time of the collision, assuming that the center of the car travels halfway to the wall with constant deceleration. (Use any reasonable length for the car.) (*b*) Estimate the average force exerted by the wall on the car.

57. A ball moving at 10 m/s makes an off-center elastic collision with another ball of equal mass that is initially at rest. The incoming ball is deflected at an angle of 30° from its original direction of motion. Find the velocity of each ball after the collision.

58. A proton of mass m makes a head-on elastic collision with a stationary carbon nucleus of mass $12m$. The speed of the proton is 300 m/s. (*a*) Find the velocity of the center

of mass of the system. (*b*) Find the velocity of the proton after the collision in the center-of-mass reference frame. (*c*) Find the velocity of the proton after the collision in the lab reference frame.

59. A handball of mass 300 g is thrown straight against a wall with a speed of 8 m/s. It rebounds with the same speed. (*a*) What impulse is delivered to the wall? (*b*) If the ball is in contact with the wall for 0.003 s, what average force is exerted on the wall by the ball? (*c*) The ball is caught by a player who brings it to rest. In the process, her hand moves back 0.5 m. What is the impulse received by the player? (*d*) What average force was exerted on the player by the ball?

60. Wistful Willie is skating along on very smooth ice at 3 m/s, daydreaming of nothing in particular, when he collides head-on with Dennis the Menace, who is moving in the opposite direction at 5 m/s. Willie's mass is 80 kg and Dennis's mass is 30 kg. After the collision, Willie and Dennis stick together. (*a*) Find the common speed of the boys after the collision. (*b*) Find the velocity of their center of mass before the collision. (*c*) Find the velocity of their center of mass after the collision. (*d*) If the time of impact is 0.1 s, what is the average force exerted on Willie by Dennis? *Note:* After 0.1 s, the boys move along together but they no longer exert forces on one another.

61. A 13-kg block is at rest on a level floor. A 400-g glob of putty is thrown at the block such that it travels horizontally, hits the block, and sticks to it. The block and putty slide 15 cm along the floor. If the coefficient of sliding friction is 0.4, what is the initial speed of the putty?

62. A bullet of mass m_1 is fired with a speed v into the bob of a ballistic pendulum of mass m_2. The bob is attached to a very light rod of length L that is pivoted at the other end. The bullet is stopped in the bob. Find the minimum v such that the bob will swing through a complete circle.

63. A bullet of mass m_1 is fired with a speed v into the bob of a ballistic pendulum of mass m_2. Find the maximum height h attained by the bob if the bullet passes through the bob and emerges with a speed $v/2$.

64. The force acting on a particle of mass 5.0 kg is given by the function $F(t) = 3t^2$ N. If the particle starts from rest, what will its velocity be after 5 s?

65. A rocket has initial mass of 30,000 kg of which 20 percent is the payload. It burns fuel at a rate of 200 kg/s and exhausts its gas at a relative speed of 1.8 km/s. Find (*a*) the thrust of the rocket, (*b*) the time until burnout, and (*c*) its final speed assuming it moves upward near the surface of the earth where the gravitational field g is constant.

66. Find the initial acceleration and the acceleration at burnout for the rocket in Problem 65.

Level III

67. The great limestone caverns were formed by dripping water. (*a*) If water droplets of 10 mL fall from a height of 5 m at a rate of 10 per minute, what is the average force exerted on the limestone floor by the droplets of water? (*b*) Compare this force to the weight of a water droplet.

68. A favorite game at picnics is the egg toss. Two people toss a raw egg back and forth as they move further apart. If the force required to break the egg's shell is about 5 N and the mass of the egg is 50 g, estimate the maximum separation distance for the egg throwers. Make whatever assumptions seem reasonable.

69. A rocket for mining the asteroid belt between Mars and Jupiter is designed like a large scoop. Its mass m is 10^5 kg, and it is traveling at an initial velocity of 10^4 m/s. It overtakes asteroids traveling at 100 m/s. If it scoops up asteroids at the rate of 100 kg/s, what will its speed be after 1 h of mining?

70. What thrust must the mining rocket in Problem 69 have for it to continue travel at a constant speed? Ignore any decrease in the mass of the rocket due to the exhaust of gases.

71. In the billiards shot shown in Figure 7-42, the initial direction of the cue ball is perpendicular to the line joining the centers of the other two balls, which are touching. The cue ball strikes both balls simultaneously. Use the symmetry of the situation along with the conservation of energy and momentum to find the final velocities of all three balls.

Figure 7-42 Problem 71.

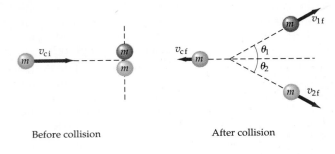

Before collision   After collision

72. Show that in a one-dimensional collision between two particles, the fractional relative-energy loss is related to the coefficient of restitution by

$$-\frac{\Delta K_{rel}}{K_{rel}} = 1 - e^2$$

73. Show that for an elastic collision between two particles of equal mass when one of the particles is initially at rest, the energy transferred to the originally stationary particle is $(\sin^2 \theta)E_0$, where E_0 is the initial energy of the incoming particle, and θ is the angle of its deflection.

74. Use integration to find the center of mass of a uniform plate cut in the shape of a half circle. Choose the origin to be at the center of curvature of the plate, with the y axis on the bisector.

75. Use integration to find the center of mass of the right isosceles triangle shown in Figure 7-43.

Figure 7-43 Problem 75.

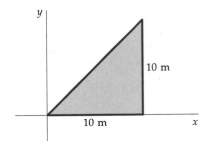

76. A stream of glass beads, each with a mass of 0.5 g, comes out of a horizontal tube at 100 per second. The beads fall a distance of 0.5 m to a balance pan and bounce back to their original height as shown in Figure 7-44. How much mass must be placed in the other pan of the balance to keep the pointer at zero?

Figure 7-44 Problem 76.

77. A rocket has a payload of 5000 kg and a fuel supply of 20,000 kg. It starts from rest, and it burns fuel at a rate of 200 kg/s and exhausts the gases at $u_{ex} = 6$ km/s. (a) Find the final velocity of the rocket if it is in free space where there is no gravity. (b) Find the velocity after the fuel burns up if the rocket moves against a uniform gravitational field g. (c) In part (b), if the rocket starts from the earth's surface, is it reasonable to neglect the variation in g with height above the earth's surface?

78. A particle has an initial speed v_0. It collides with a second particle that is at rest and is deflected through an angle ϕ. Its speed after the collision is v. The second particle recoils. Its velocity makes an angle θ with the initial direction of the first particle. (a) Show that

$$\tan \theta = \frac{v \sin \phi}{v_0 - v \cos \phi}$$

(b) Do you have to assume that the collision is either elastic or inelastic to get the result in part (a)?

Chapter 8

Rotation

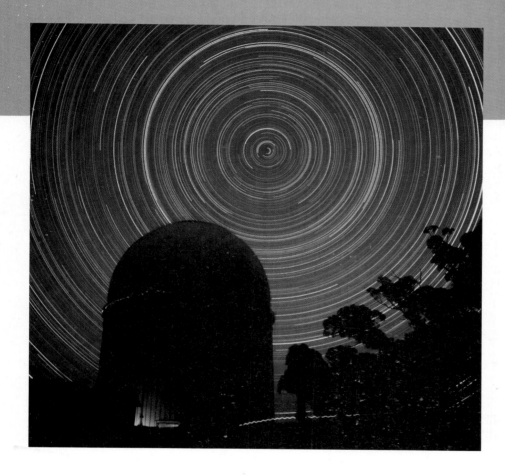

Star tracks in a time exposure of the night sky.

There are many examples of rotational motion in the world around us. Merry-go-rounds rotate, as do high divers as they somersault and ice skaters as they pirouette. The earth rotates about its axis. Balls and cylinders rotate as they roll. Molecules, atoms, and even electrons rotate.

In this chapter, we will learn how to describe rotational motion. First we will define the kinematic quantities angular displacement, angular velocity, and angular acceleration, quantities that are the rotational analogs of displacement, velocity, and acceleration, which we use to describe linear motion. We will see that the rotation of a rigid body with constant angular acceleration is described by equations that are the rotational analogs of the equations we have already dealt with for linear motion with constant acceleration. We will then study the dynamics of rotational motion and the rotational analog of Newton's second law of motion. Again, we will find that dynamic quantities we encountered in our study of linear motion—force, mass, and momentum—have corresponding rotational analogs—torque, moment of inertia, and angular momentum.

8-1   Angular Velocity and Angular Acceleration

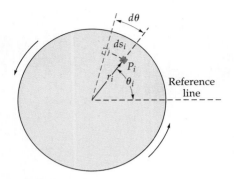

Figure 8-1   A disk rotating about a fixed axis through its center. The distance ds_i moved by the ith particle in some time interval depends on r_i, but the angular displacement $d\theta$ is the same for all the particles on the disk.

Figure 8-1 shows a disk that is free to rotate about its axis, which is through its center and perpendicular to the disk. When the disk rotates, different parts of it move at different speeds. For example, a point near the rim moves faster than a point near the axis. It therefore makes no sense to speak of the speed or velocity of the disk as it rotates. However, when a point near the rim moves through a complete circle, so does any other point on the disk. In a given time, the angle swept out by a line from the axis of rotation to any point on the disk is the same whether the point is on the rim, near the axis, or at any other place on the disk. The angle swept out by a radial line on the disk is thus a characteristic of the disk as a whole, as is the rate at which the angle changes.

To develop the concepts of angular velocity and angular acceleration, we consider the disk to be made up of many small point particles. As the disk turns, the distance between any two particles remains fixed. Such a system is called a **rigid body.**

Let us focus our attention on a typical particle of mass m_i on the disk. We can specify the position of the particle P_i by its distance r_i from the center of the disk and the angle θ_i between a line from the center to the particle and a reference line that is fixed in space, as shown in Figure 8-1. During a small time interval dt, the particle moves along the arc of a circle a distance ds_i given by

$$ds_i = v_i \, dt \qquad\qquad 8\text{-}1$$

where v_i is the speed of the particle. During this time, the radial line to the particle sweeps out an angle $d\theta$ relative to a line fixed in space. The measure of this angle in radians is the arc length ds_i divided by r_i:

$$d\theta = \frac{ds_i}{r_i} \qquad\qquad 8\text{-}2$$

(If you are unfamiliar with the radian measure of angles, you should read the trigonometry review section in Appendix A.) Although the distance ds_i varies from particle to particle, the angle $d\theta$ swept out in a given time is the same for all the particles of the disk. It is called the **angular displacement** of the disk. When the disk makes one complete revolution, the arc length Δs_i is $2\pi r_i$ and the angular displacement is

$$\Delta\theta = \frac{2\pi r_i}{r_i} = 2\pi \text{ rad} = 360° = 1 \text{ rev}$$

The rate of change of the angle with respect to time, $d\theta/dt$, is the same for all the particles of the disk. It is called the **angular velocity** ω of the disk:

Angular velocity defined

$$\omega = \frac{d\theta}{dt} \qquad\qquad 8\text{-}3$$

The angular velocity ω is positive for counterclockwise rotation, for which θ is increasing, and negative for clockwise rotation, for which θ is decreasing. (Later in this chapter, we will see that, for general rotations, the angular velocity is a vector quantity that points along the axis of rotation. For rotation about a fixed axis, there are but two possible directions for the angular velocity, corresponding to clockwise or counterclockwise rotation. This is analogous to linear motion in one dimension in which the direction of the velocity is either positive or negative.) The units of angular velocity are radians per second. Since a radian is a dimensionless unit, the dimensions of angular velocity are those of reciprocal time (T^{-1}). The magnitude of the angular velocity is called the **angular speed.** Although the angular motion of a disk is often described using other units, such as revolutions per minute, it is important to remember that many of the equations we will use involving the angular velocity ω are valid only when the angles are expressed in radians and the angular velocity, in radians per second. To convert from revolutions to radians, we use

$$1 \text{ rev} = 2\pi \text{ rad}$$

Exercise

A record rotates at 33.3 revolutions per minute. What is its angular speed?   (Answer: 3.49 rad/s)

The rate of change of angular velocity with respect to time is called the **angular acceleration** α:

$$\alpha = \frac{d\omega}{dt} = \frac{d^2\theta}{dt^2}$$

8-4   *Angular acceleration defined*

The units of angular acceleration are radians per second per second (rad/s^2). The angular acceleration is positive if the angular velocity ω is increasing and negative if ω is decreasing.

We can relate the tangential velocity of a particle on the disk to the angular velocity of the disk using Equations 8-2 and 8-3:

$$v_{it} = \frac{ds_i}{dt} = \frac{r_i \, d\theta}{dt} = r_i\omega$$

8-5

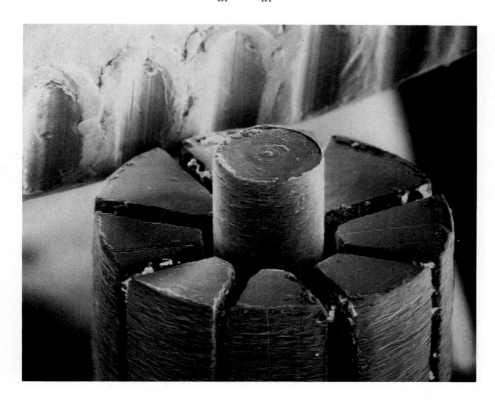

A tiny motor called a "wobble motor." This motor, which has a diameter of the order of a millimeter, has achieved angular speeds in excess of 120,000 rev/min. The edge of a dime is visible in the background.

Similarly, the tangential acceleration of a particle on the disk is related to the angular acceleration of the disk:

$$a_{it} = \frac{dv_i}{dt} = r_i \frac{d\omega}{dt}$$

Thus

$$a_{it} = r_i \alpha \qquad \text{8-6}$$

Each particle of the disk also has a radial acceleration, the centripetal acceleration, which points inward along the radial line and has the magnitude

$$a_{ic} = \frac{v_i^2}{r_i} = \frac{(r_i \, \omega)^2}{r_i} = r_i \, \omega^2 \qquad \text{8-7}$$

Exercise

A point on the rim of a record is 15 cm from the axis of rotation. Find the tangential velocity v_t, the tangential acceleration a_t, and the centripetal acceleration a_c of the point if the record is rotating at a constant angular velocity of 3.49 rad/s. (Answers: $v_t = 52.4$ cm/s, $a_t = 0$, $a_c = 183$ cm/s^2)

The angular displacement θ, angular velocity ω, and angular acceleration α are analogous to the linear displacement x, linear velocity v, and linear acceleration a in one-dimensional motion. Because of the similarity of the definitions of the rotational and linear quantities, much of what we learned in Chapter 2 will be useful in treating problems involving the rotation of a rigid body. For example, the equations for constant angular acceleration are the same as Equations 2-9 through 2-13, with θ replacing x, ω replacing v, and α replacing a. Thus,

$$\omega = \omega_0 + \alpha t \qquad \text{8-8}$$

is the rotational analog of

$$v = v_0 + at$$

and

$$\theta = \theta_0 + \omega_0 t + \tfrac{1}{2}\alpha t^2 \qquad \text{8-9}$$

is the rotational analog of

$$x = x_0 + v_0 t + \tfrac{1}{2}at^2$$

where ω_0 and θ_0 are the initial values of the angular velocity and angular position, respectively. As with the constant-linear-acceleration equations, we can eliminate time from these equations to obtain an equation relating the angular displacement, angular velocity, and angular acceleration:

$$\omega^2 = \omega_0^2 + 2\alpha(\theta - \theta_0) \qquad \text{8-10}$$

which is the rotational analog of

$$v^2 = v_0^2 + 2a(x - x_0)$$

W = ANGULAR VELOCITY

θ = ANGULAR DISPLACEMENT

α = ANGULAR ACCELERATION

Example 8-1

A disk rotates with constant angular acceleration $\alpha = 2$ rad/s². If the disk starts from rest, how many revolutions does it make in 10 s?

This is analogous to the linear problem of finding the distance traveled by a particle in a given time if it starts from rest with constant acceleration. The number of revolutions is related to the angular displacement by the fact that each revolution is an angular displacement of 2π rad. Thus, we need to find the angular displacement $\theta - \theta_0$ in radians for a time of 10 s and multiply by the conversion factor (1 rev)/(2π rad).

Equation 8-9 relates the angular displacement to the time. We are given $\omega_0 = 0$ (the disk starts from rest). Thus,

$$\theta - \theta_0 = \omega_0 t + \tfrac{1}{2}\alpha t^2 = 0 + \tfrac{1}{2}(2 \text{ rad/s}^2)(10 \text{ s})^2 = 100 \text{ rad}$$

The number of revolutions is therefore

$$100 \text{ rad} \times \frac{1 \text{ rev}}{2\pi \text{ rad}} = 15.9 \text{ rev}$$

Example 8-2

Find the angular speed of the disk in Example 8-1 after 10 s.

Using Equation 8-8 we have

$$\omega = \omega_0 + \alpha t = 0 + (2 \text{ rad/s}^2)(10 \text{ s}) = 20 \text{ rad/s}$$

As a check of this result and that of the previous example, we can also find the angular speed from Equation 8-10:

$$\omega^2 = 2\alpha(\theta - \theta_0) = 2(2 \text{ rad/s}^2)(100 \text{ rad}) = 400 \text{ rad}^2/\text{s}^2$$

or

$$\omega = \sqrt{400 \text{ rad}^2/\text{s}^2} = 20 \text{ rad/s}$$

Question

1. Two points are on a disk turning at constant angular velocity, one point on the rim and the other halfway between the rim and the axis. Which point moves the greater distance in a given time? Which turns through the greater angle? Which has the greater speed? The greater angular velocity? The greater tangential acceleration? The greater angular acceleration? The greater centripetal acceleration?

8-2 Torque and Moment of Inertia

When you wish to set a top spinning you twist it. In Figure 8-2a, a disk resting flat on a horizontal surface is set spinning by the forces F_1 and F_2 exerted at the edge of the disk. Note that the locations of the points of application of these forces are important. The same two forces applied so that their lines of action pass through the center of the disk, as in Figure 8-2b, will not set the disk spinning. (The line of action of a force is the line along which the force acts.) The perpendicular distance between the line of action of a force and the axis of rotation is called the **lever arm** ℓ of the force. The product of a force times its lever arm is called a **torque** τ. The torque exerted on an object by a force is the quantity that affects the angular velocity of the object.

(a)

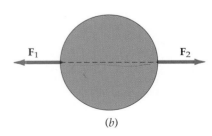

(b)

Figure 8-2 (a) The forces F_1 and F_2 will set the disk spinning about its axis. (b) If the same two forces are applied such that their lines of action pass through the center of the disk, they will not set the disk spinning.

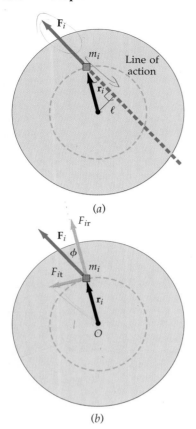

(a)

(b)

Figure 8-3 A force $\mathbf{F}_i$ acting on the *i*th particle of a disk pivoted about its center. (a) The lever arm ℓ is the perpendicular distance from the line of action of the force to the axis of rotation. The axis of rotation here is directly into the page. (b) The force is resolved into a radial component $F_{ir} = F_i \cos \phi$ and a perpendicular component $F_{it} = F_i \sin \phi$. The radial component does not affect the rotation of the disk.

Moment of inertia defined

Figure 8-3a shows a single force $\mathbf{F}_i$ acting on the *i*th particle of a disk. The lever arm of this force is $\ell = r_i \sin \phi$, where ϕ is the angle between the force $\mathbf{F}_i$ and the position vector $\mathbf{r}_i$ to the point of application of the force. The torque exerted by this force is thus

$$\tau_i = F_i \ell = F_i r_i \sin \phi \qquad 8\text{-}11$$

In Figure 8-3b, we have resolved the force $\mathbf{F}_i$ into two components, $F_{ir} = F_i \cos \phi$ along the radial line $\mathbf{r}_i$ and $F_{it} = F_i \sin \phi$ perpendicular to the radial line. The radial component F_{ir} has no effect on the rotation of the disk. The torque exerted by the force $\mathbf{F}_i$ can be written in terms of F_{it}. From Equation 8-11, we have

$$\tau_i = F_i \ell = F_i r_i \sin \phi = F_{it} r_i \qquad 8\text{-}12$$

We will now show that the angular acceleration of a rigid body is proportional to the net torque acting on it. Let us assume that $\mathbf{F}_i$ is the net external force acting on the *i*th particle. The tangential acceleration of the *i*th particle is, by Newton's second law,

$$F_{it} = m_i a_{it} = m_i r_i \alpha \qquad 8\text{-}13$$

where we have used the relation $a_{it} = r_i \alpha$ between the tangential acceleration of the *i*th particle and the angular acceleration of the object (Equation 8-6). If we multiply each side by r_i, we obtain

$$r_i F_{it} = m_i r_i^2 \alpha \qquad 8\text{-}14$$

The left side of Equation 8-15 is the torque $\tau_i = r_i F_{it}$ exerted by the force $\mathbf{F}_i$ about the pivot O. We thus have

$$\tau_i = m_i r_i^2 \alpha \qquad 8\text{-}15$$

If we now sum over all the particles in the object, we obtain

$$\sum_i \tau_i = \sum_i m_i r_i^2 \alpha \qquad 8\text{-}16$$

The quantity $\sum \tau_i$ is the net torque acting on the object, which we will denote by τ_{net}. For a rigid body, the angular acceleration is the same for all the particles of the object and can therefore be taken out of the sum. The quantity $\sum m_i r_i^2$ is a property of the object and the axis of rotation called the **moment of inertia** I:

$$I = \sum_i m_i r_i^2 \qquad 8\text{-}17$$

The torque exerted by a wrench on a nut is proportional to the force and to the lever arm. Charlie Chaplin could exert a greater torque with the same force if he held the wrenches nearer their ends.

In this equation, the distance r_i is the distance from the ith particle *to the axis of rotation*. Usually, this distance is not the same as the distance from the ith particle to the origin, though for a disk with the origin at the center on the axis, these distances are the same. The moment of inertia is a measure of the resistance of an object to changes in its rotational motion. It depends on the distribution of the mass of the object relative to the axis of rotation of the object. It is a property of the object (and the axis of rotation), just as the mass m is a property of the object that measures its resistance to changes in translational motion. For systems consisting of a small number of discrete particles, we can compute the moment of inertia about a given axis directly from Equation 8-17. For the more common case of a continuous object, such as a wheel, the moment of inertia about a given axis is computed using calculus. We will illustrate such calculations in Section 8-4. Table 8-1 lists the moments of inertia of various uniform objects.

In terms of the moment of inertia, Equation 8-16 becomes

$$\tau_{\text{net}} = I\alpha \qquad \text{8-18} \qquad \textit{Newton's second law for rotation}$$

Equation 8-18 is the rotational analog of Newton's second law for linear motion:

$$\mathbf{F}_{\text{net}} = m\mathbf{a}$$

The net torque is analogous to the net force, the moment of inertia is analogous to the mass, and the angular acceleration is analogous to the linear acceleration.

In Chapter 7, we saw that the net force acting on a system of particles is equal to the net *external* force acting on the system because, by Newton's

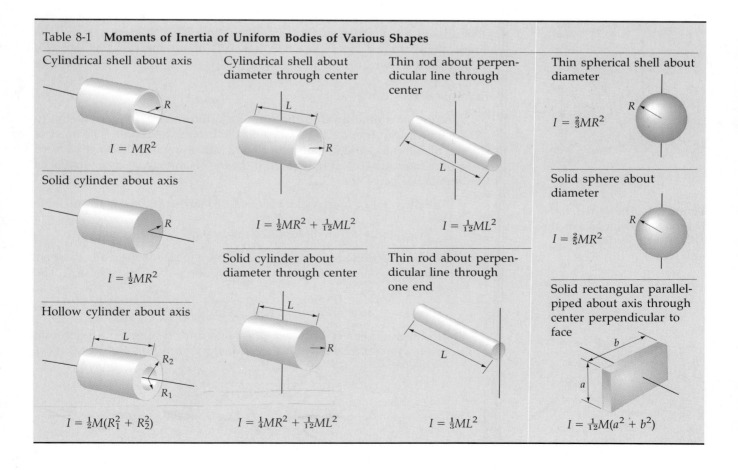

Table 8-1   **Moments of Inertia of Uniform Bodies of Various Shapes**

Cylindrical shell about axis
$$I = MR^2$$

Cylindrical shell about diameter through center
$$I = \tfrac{1}{2}MR^2 + \tfrac{1}{12}ML^2$$

Thin rod about perpendicular line through center
$$I = \tfrac{1}{12}ML^2$$

Thin spherical shell about diameter
$$I = \tfrac{2}{3}MR^2$$

Solid cylinder about axis
$$I = \tfrac{1}{2}MR^2$$

Solid cylinder about diameter through center
$$I = \tfrac{1}{4}MR^2 + \tfrac{1}{12}ML^2$$

Thin rod about perpendicular line through one end
$$I = \tfrac{1}{3}ML^2$$

Solid sphere about diameter
$$I = \tfrac{2}{5}MR^2$$

Hollow cylinder about axis
$$I = \tfrac{1}{2}M(R_1^2 + R_2^2)$$

Solid rectangular parallelpiped about axis through center perpendicular to face
$$I = \tfrac{1}{12}M(a^2 + b^2)$$

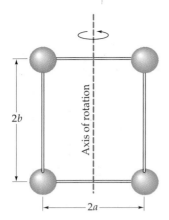

Figure 8-4 Four particles of equal mass connected by massless rods and rotating about an axis through the plane of the particles and through their center of mass for Example 8-3.

Figure 8-5 The system of Figure 8-4 rotating about an axis through two of the particles.

Figure 8-6 A string wrapped around a disk for Example 8-4.

third law, the internal forces exerted by the particles within the system on one another cancel in pairs. The treatment of internal torques exerted by the particles within a system on one another is somewhat complicated and will be deferred until Section 8-7. There we will show that the internal torques sum to zero if the force exerted by one particle on another acts along the line joining the particles. Assuming that this holds for all the internal forces, we find that the net torque acting on a system is also equal to the net *external* torque acting on the system.

Example 8-3

Four particles of mass m are connected by massless rods to form a rectangle of sides $2a$ and $2b$ as shown in Figure 8-4. The system rotates about an axis in the plane of the figure through the center. Find the moment of inertia about this axis.

From the figure, we can see that the distance from each particle to the axis of rotation is a. The moment of inertia of each particle about this axis is therefore ma^2, and since there are four particles, the total moment of inertia of the body is

$$I = 4ma^2$$

The distance b plays no role at all because it is not related to the distance from any mass to the axis of rotation.

Exercise

Find the moment of inertia of the system of Example 8-3 for rotation about an axis parallel to the first axis but passing through two of the masses as shown in Figure 8-5. (Answer: $I = 8ma^2$)

Example 8-3 and the exercise illustrate the fact that the moment of inertia depends on the location of the axis of rotation. The moment of inertia is larger about the axis in Figure 8-5 than it is about the axis parallel to it through the center of mass in Figure 8-4.

Example 8-4

A string is wound around the rim of a uniform disk that is pivoted to rotate without friction about a fixed axis through its center. The mass of the disk is 3 kg, and its radius is 25 cm. The string is pulled with a force **F** of magnitude 10 N (Figure 8-6). If the disk is initially at rest, what is its angular velocity after 5 s?

The moment of inertia of a uniform disk about its axis is (from Table 8-1)

$$I = \tfrac{1}{2}MR^2 = \tfrac{1}{2}(3 \text{ kg})(0.25 \text{ m})^2 = 9.38 \times 10^{-2} \text{ kg·m}^2$$

Since the direction of the string as it leaves the rim of the disk is always tangent to the disk, the lever arm of the force it exerts is just R. The applied torque is thus

$$\tau = FR = (10 \text{ N})(0.25 \text{ m}) = 2.5 \text{ N·m}$$

To find the angular velocity, we must first find the angular acceleration from Newton's second law for rotational motion (Equation 8-18):

$$\alpha = \frac{\tau_{\text{net}}}{I} = \frac{2.5 \text{ N·m}}{0.0938 \text{ kg·m}^2} = 26.7 \text{ rad/s}^2$$

Since α is constant, we find ω from Equation 8-8, setting $\omega_0 = 0$:

$$\omega = \omega_0 + \alpha t = 0 + (26.7 \text{ rad/s}^2)(5 \text{ s}) = 133 \text{ rad/s}$$

Example 8-5

An object of mass m is tied to a light string wound around a wheel of moment of inertia I and radius R (Figure 8-7). The wheel bearing is frictionless, and the string does not slip on the rim. Find the tension in the string and the acceleration of the object.

The only force exerting a torque on the wheel is the tension **T** in the string, which has a lever arm R and produces a clockwise rotation. Taking the clockwise direction to be positive, we have

$$TR = I\alpha \qquad \text{8-19}$$

Two forces act on the suspended object, the upward tension **T**, and the downward force of gravity $m\mathbf{g}$. Taking the downward direction to be positive, so that a and α both have the same sign, we have from Newton's second law

$$mg - T = ma \qquad \text{8-20}$$

There are three unknowns, T, a, and α, in these two equations. The string provides a constraint by which we can relate a and α. Since the string does not slip, its speed is equal to the speed of a point on the rim, and its acceleration is the same as the tangential acceleration of a point on the rim. Its acceleration is thus

$$a = R\alpha \qquad \text{8-21}$$

Substituting a/R for α in Equation 8-19 gives

$$TR = I\frac{a}{R}$$

or

$$a = \frac{TR^2}{I} \qquad \text{8-22}$$

Substituting this result for a into Equation 8-20 gives

$$mg - T = m\frac{TR^2}{I}$$

or

$$T\left(1 + \frac{mR^2}{I}\right) = mg$$

$$T = \frac{I}{I + mR^2}mg$$

We can use this value for T in Equation 8-22 to find a:

$$a = \frac{mR^2}{I + mR^2}g$$

Figure 8-7 Object of mass m tied to a string wrapped around a rotating wheel for Example 8-5. Since the string does not slip, the acceleration of the object equals the tangential acceleration of a point on the rim, $a = R\alpha$.

Questions

2. Can an object rotate if there is no torque acting?

3. Can a given rigid body have more than one moment of inertia?

4. Does an applied net torque always increase the angular speed of an object?

5. If the angular velocity of an object is zero at some instant, does this mean that the net torque on the object must be zero?

8-3   Rotational Kinetic Energy

When a rotating object turns through a small angular displacement $d\theta$, the ith particle moves through a distance $ds_i = r_i\,d\theta$. If a force $\mathbf{F}_i$ acts on the ith particle, it does work

$$dW_i = F_{it}\,ds_i = F_{it}r_i\,d\theta = \tau_i\,d\theta$$

In general, the work done by a torque τ when an object turns through a small angle $d\theta$ is

$$dW = \tau\,d\theta \qquad\qquad 8\text{-}23$$

Equation 8-23 is analogous to the similar result for linear motion in one dimension, $dW = F_s\,ds$. The rate at which the torque does work is the power input of the torque:

$$P = \frac{dW}{dt} = \tau\,\frac{d\theta}{dt}$$

or

Power
$$P = \tau\omega \qquad\qquad 8\text{-}24$$

Equation 8-24 is the rotational analog of $P = F_s v_s$.

The total work done on a system equals the change in the kinetic energy of the system (if the system has no change in potential energy and there are no energy losses). For an object rotating about an axis through its center of mass, the kinetic energy of the object is the kinetic energy relative to the center of mass, K_{rel}, as discussed in Chapter 7. This kinetic energy is the sum of the kinetic energies of the individual particles in the body:

$$K = \sum_i \tfrac{1}{2}m_i v_i^2 = \sum_i \tfrac{1}{2}m_i(r_i\omega)^2 = \tfrac{1}{2}\sum_i m_i r_i^2\omega^2$$

or

Kinetic energy of rotation
$$K = \tfrac{1}{2}I\omega^2 \qquad\qquad 8\text{-}25$$

Equation 8-25 is the rotational analog of $K = \tfrac{1}{2}mv^2$ for linear motion.

The Archimedes screw is a device for lifting water. The rotational work done by the torque exerted at the handle is converted into increased potential energy of the water.

Example 8-6

A uniform disk of mass 3 kg and radius 12 cm rotates at 480 rev/min. Find its kinetic energy.

From Table 8-1, the moment of inertia of a uniform disk is given by

$$I = \tfrac{1}{2}mR^2 = \tfrac{1}{2}(3 \text{ kg})(0.12 \text{ m})^2 = 0.0216 \text{ kg·m}^2$$

The angular velocity is

$$\omega = \left(\frac{480 \text{ rev}}{60 \text{ s}}\right)\left(\frac{2\pi \text{ rad}}{1 \text{ rev}}\right) = 50.3 \text{ rad/s}$$

The kinetic energy is then

$$K = \tfrac{1}{2}I\omega^2 = \tfrac{1}{2}(0.0216 \text{ kg·m}^2)(50.3 \text{ rad/s})^2 = 27.3 \text{ J}$$

Note that we drop the dimensionless unit radian and use $1 \text{ kg·m}^2/\text{s}^2 = 1 \text{ J}$.

The Crab Pulsar is one of the fastest rotating neutron stars known, but it is slowing down. It appears to blink on (left) and off (right) like a rotating spotlight about 30 times per second, but the period is increasing by about 10^{-5} seconds per year. The loss in rotational energy, which is equivalent to the power output of 100,000 suns, appears as light emitted by electrons accelerated in the magnetic field of the pulsar.

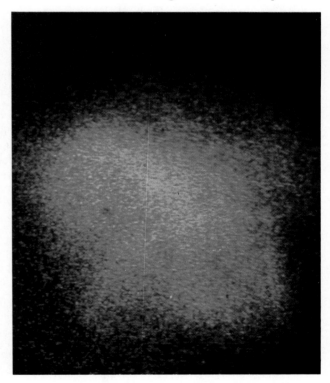

Example 8-7

A car engine delivers 380 N·m of torque at 3200 rev/min. Find the power output of the engine.

The angular speed corresponding to 3200 rev/min is

$$\omega = \left(\frac{3200 \text{ rev}}{\text{min}}\right)\left(\frac{2\pi \text{ rad}}{1 \text{ rev}}\right)\left(\frac{1 \text{ min}}{60 \text{ s}}\right) = 335 \text{ rad/s}$$

The power output of the engine is given by Equation 8-25:

$$P = \tau\omega = (380 \text{ N·m})(335 \text{ rad/s}) = 127 \text{ kW}$$

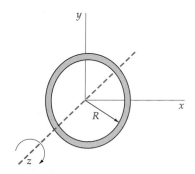

Figure 8-8   A hoop rotating about an axis perpendicular to the plane of the hoop and through its center. Since all the mass of the hoop is at a distance R from the axis, the moment of inertia is MR^2.

8-4   Calculating the Moment of Inertia

For continuous bodies, the sum in Equation 8-17 for the moment of inertia is replaced by an integral:

$$I = \int r^2 \, dm \qquad\qquad 8\text{-}26$$

where r is the distance of the mass element dm from the axis of rotation.

Example 8-8

Find the moment of inertia of a hoop of mass M and radius R about an axis through the center and perpendicular to the plane of the hoop (Figure 8-8).

In this case, all the mass is at a distance $r = R$, so the moment of inertia is simply

$$I = \int r^2 \, dm = R^2 \int dm = MR^2$$

Example 8-9

Find the moment of inertia of a stick of uniform density about an axis perpendicular to the stick through one end.

The mass element dm is shown in Figure 8-9. It is at a distance x from the axis of rotation. Since the total mass M is uniformly distributed along the length L, the linear mass density is $\lambda = M/L$. Thus, $dm = \lambda \, dx = (M/L) \, dx$. The moment of inertia about the y axis is

$$I_y = \int_0^L x^2 \, dm = \int_0^L x^2 \frac{M}{L} \, dx = \frac{M}{L} \int_0^L x^2 \, dx$$

$$= \frac{M}{L} \frac{1}{3} x^3 \Big|_0^L = \frac{ML^3}{3L} = \frac{1}{3} ML^2$$

The moment of inertia about the z axis is also $\frac{1}{3}ML^2$, and that about the x axis is zero, if all the mass is right on the x axis.

Figure 8-9   Geometry for setting up the integral to calculate the moment of inertia of a uniform stick rotating about an axis perpendicular to the stick and through one end.

Example 8-10

Find the moment of inertia of a uniform disk about the axis through its center and perpendicular to the plane of the disk.

We expect that I will be smaller than MR^2 since the mass of a disk is not concentrated at $r = R$ like that of a hoop but, rather, is uniformly distributed from $r = 0$ to $r = R$. We calculate I by taking mass elements dm as shown in Figure 8-10. Each mass element is a hoop of radius r and thickness dr. The moment of inertia of each element is $r^2 \, dm$. Since the area of each element is $dA = 2\pi r \, dr$, the mass of each element is

$$dm = \frac{M}{A} \, dA = \frac{M}{A} 2\pi r \, dr$$

where $A = \pi R^2$ is the area of the disk. We thus have

$$I = \int r^2 \, dm = \int_0^R r^2 \frac{M}{A} 2\pi r \, dr = \frac{2\pi M}{\pi R^2} \int_0^R r^3 \, dr = \frac{2M}{R^2} \frac{R^4}{4} = \frac{1}{2} MR^2$$

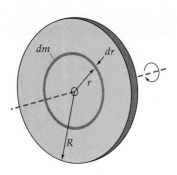

Figure 8-10   Geometry for setting up the integral to calculate the moment of inertia of a uniform disk rotating about an axis through its center and perpendicular to the plane of the disk.

Example 8-11

Find the moment of inertia of a cylinder of uniform density about its axis (Figure 8-11).

We can consider the cylinder to be a set of disks, each with mass m_i and moment of inertia $\frac{1}{2}m_iR^2$. Then the moment of inertia of the complete cylinder is

$$I = \sum_i \tfrac{1}{2}m_iR^2 = \tfrac{1}{2}R^2 \sum_i m_i = \tfrac{1}{2}MR^2$$

where $M = \Sigma\ m_i$ is the total mass of the cylinder.

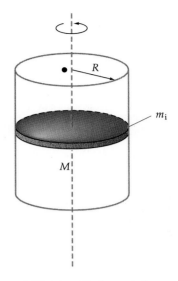

Example 8-12

Find the moment of inertia of a sphere of uniform density about an axis through its center.

We calculate this moment of inertia by treating the sphere as a set of disks as shown in Figure 8-12. We consider a disk element at height x above the center. The radius of the disk element is

$$r = \sqrt{R^2 - x^2}$$

The volume of the disk is the area times the height, $dV = \pi r^2\ dx$. If M is the total mass of the sphere and V is its volume, the mass per unit volume is M/V. The mass of each disk is then

$$dm = \frac{M}{V}\ dV = \frac{M}{V}\ \pi r^2\ dx = \frac{M}{V}\ \pi(R^2 - x^2)\ dx$$

The moment of inertia of each disk element is

$$dI = \frac{1}{2}r^2\ dm = \frac{1}{2}(R^2 - x^2)\left[\frac{M}{V}\ \pi(R^2 - x^2)\ dx\right]$$

$$= \frac{1}{2}\frac{M}{V}\ \pi(R^2 - x^2)^2\ dx$$

Figure 8-11 A cylinder rotating about its axis can be considered to be a stack of disks of mass m_i. Since each disk has a moment of inertia $\frac{1}{2}m_iR^2$, the moment of inertia of the cylinder is $\frac{1}{2}MR^2$.

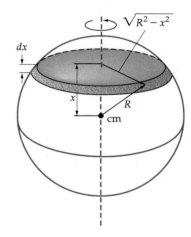

When x varies from 0 to R, the top half of the sphere is covered. The total moment of inertia of the sphere is therefore twice the integral of dI from $x = 0$ to $x = R$:

$$I = 2\int_0^R \frac{1}{2}\frac{M}{V}\ \pi(R^2 - x^2)^2\ dx$$

The integral in this equation can be evaluated by writing $(R^2 - x^2)^2 = R^4 - 2R^2x^2 + x^4$ and integrating term by term. The result is

$$\int_0^R (R^2 - x^2)^2\ dx = \frac{8R^5}{15} \qquad\qquad 8\text{-}27$$

The moment of inertia of the sphere is thus

$$I = \frac{\pi M}{V}\ \frac{8R^5}{15} = \frac{2}{5}MR^2$$

where we have used $V = \frac{4}{3}\pi R^3$.

Figure 8-12 Geometry for the calculation of the moment of inertia of a uniform sphere rotating about a diameter (Example 8-12). The sphere can be thought of as a set of disks of varying radii.

Exercise

Derive Equation 8-27 by expanding the integrand and integrating term by term.

The Parallel-Axis Theorem

We can often simplify the calculation of the moments of inertia for various bodies by using general theorems that relate the moment of inertia about one axis of the body to that about another axis. The parallel-axis theorem relates the moment of inertia about an axis through the center of mass of a body to that about a second, parallel axis. Let I_{cm} be the moment of inertia about an axis through the center of mass of a body and I be that about a parallel axis a distance h away. The parallel-axis theorem states that

$$I = I_{cm} + Mh^2$$

where M is the total mass of the body. Example 8-3 and the exercise following it illustrated a special case of this theorem with $h = a$, $M = 4m$, and $I_{cm} = 4ma^2$.

We can prove this theorem using the result developed in Chapter 7 that the kinetic energy of a system of particles is the sum of the kinetic energy of the motion of the center of mass plus the kinetic energy of the motion relative to the center of mass:

$$K = \tfrac{1}{2}MV_{cm}^2 + K_{rel} \qquad\qquad 8\text{-}28$$

Consider a rigid object rotating with an angular velocity ω about an axis a distance h away from a parallel axis through the center of mass as shown in Figure 8-13. When the body rotates through an angle $\Delta\theta$ measured about the axis of rotation, it rotates through the same angle $\Delta\theta$ measured about any other parallel axis. The motion of the object relative to the center of mass is thus a rotation about the center-of-mass axis with the same angular velocity ω. The kinetic energy of the motion relative to the center of mass is thus

$$K_{rel} = \tfrac{1}{2}I_{cm}\omega^2$$

The velocity of the center of mass relative to any point on the axis of rotation is

$$V_{cm} = h\omega$$

The kinetic energy of the motion of the center of mass is thus

$$\tfrac{1}{2}MV_{cm}^2 = \tfrac{1}{2}M(h\omega)^2 = \tfrac{1}{2}M\omega^2h^2$$

When the total kinetic energy of the body is written as $\tfrac{1}{2}I\omega^2$, Equation 8-28 becomes

$$K = \tfrac{1}{2}I\omega^2 = \tfrac{1}{2}M\omega^2h^2 + \tfrac{1}{2}I_{cm}\omega^2 = \tfrac{1}{2}(Mh^2 + I_{cm})\omega^2$$

Thus

Parallel-axis theorem $$I = Mh^2 + I_{cm} \qquad\qquad 8\text{-}29$$

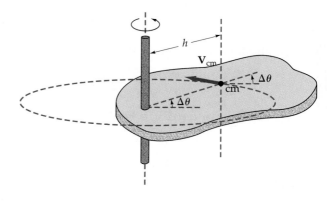

Figure 8-13 An object rotating about an axis parallel to an axis through the center of mass and a distance h from it.

Example 8-13

Find the moment of inertia of the hoop of Example 8-8 about an axis perpendicular to the hoop but through its edge as shown in Figure 8-14.

This calculation is easily done using the parallel-axis theorem with $h = R$ and the result from Example 8-8 that $I_{cm} = MR^2$:

$$I = I_{cm} + Mh^2 = MR^2 + MR^2 = 2MR^2$$

Example 8-14

Find the moment of inertia of a stick of uniform density about the y' axis through the center of mass (Figure 8-15).

In Example 8-9, we found the moment of inertia about an axis through the end of a uniform stick to be $\frac{1}{3}ML^2$. Since this axis is at a distance $h = \frac{1}{2}L$ from the center of mass of the stick, the parallel-axis theorem gives

$$I = I_{cm} + M(\tfrac{1}{2}L)^2 = \tfrac{1}{3}ML^2$$

or

$$I_{cm} = \tfrac{1}{3}ML^2 - \tfrac{1}{4}ML^2 = \tfrac{1}{12}ML^2$$

Figure 8-14 A hoop rotating about an axis that is perpendicular to the plane of the hoop and passes through the edge of the hoop for Example 8-13. The moment of inertia is found to be $2MR^2$ using the parallel-axis theorem.

The Perpendicular-Axis Theorem

The perpendicular-axis theorem applies only to plane figures. It relates the moments of inertia about two perpendicular axes in the plane of the figure to the moment of inertia about a third axis that is perpendicular to the plane of the figure. If x, y, and z are perpendicular axes for a figure that lies in the xy plane, the perpendicular-axis theorem states that the moment of inertia about the z axis equals the sum of the moments of inertia about the x and y axes.

Figure 8-16 shows a figure in the xy plane. The distance of some mass element dm from the x axis is y. The moment of inertia about the x axis is thus $I_x = \int y^2 \, dm$. Similarly, the moment of inertia about the y axis is $I_y = \int x^2 \, dm$. The moment of inertia about the z axis perpendicular to the figure is

$$I_z = \int r^2 \, dm$$

But for each element, $r^2 = x^2 + y^2$. Thus,

$$I_z = \int r^2 \, dm = \int (x^2 + y^2) \, dm = \int x^2 \, dm + \int y^2 \, dm$$

or

$$I_z = I_y + I_x \qquad\qquad 8\text{-}30$$

Figure 8-15 A uniform stick rotating about an axis perpendicular to the stick and passing through its center of mass for Example 8-14. The moment of inertia is found to be $\frac{1}{12}ML^2$ using the parallel-axis theorem.

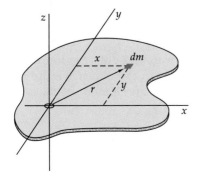

Figure 8-16 A plane figure with the z axis perpendicular to the plane. The moment of inertia about the z axis equals the sum of the moments of inertia about the x and y axes.

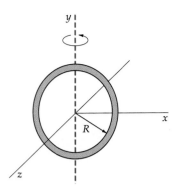

Figure 8-17   A hoop rotating about an axis through its center and in the plane of the hoop for Example 8-15. The moment of inertia is $\frac{1}{2}MR^2$, which is found using the perpendicular-axis theorem.

Example 8-15

Find the moment of inertia of the hoop in Example 8-8 about an axis that is a diameter of the hoop (Figure 8-17).

We take the hoop to be in the xy plane with the origin at the center of the hoop. By symmetry $I_x = I_y$. Since we have already found I_z in Example 8-8 to be MR^2, we have

$$I_z = I_y + I_x$$
$$= 2I_x = MR^2$$

Therefore,

$$I_x = I_y = \tfrac{1}{2}MR^2$$

Question

6. Through what point in a body must the axis of rotation pass if the moment of inertia is to be a minimum?

8-5   Angular Momentum

In Chapter 7 we defined the linear momentum **p** of a particle and showed that Newton's second law can be written

$$\mathbf{F}_{net} = \frac{d\mathbf{p}}{dt} \qquad\qquad 8\text{-}31$$

The **angular momentum** of a particle is defined as follows. For a particle moving in a circle of radius r with an angular velocity ω (Figure 8-18), the angular momentum L relative to the center of the circle is defined as the product of the magnitude of the linear momentum mv and the radius r:

$$L = mvr = m(r\omega)r$$
$$= mr^2\omega = I\omega \qquad\qquad 8\text{-}32$$

where $I = mr^2$ is the moment of inertia of the particle about an axis through the center of the circle and perpendicular to the plane of motion. The direction of L is the same as that of ω. For counterclockwise motion, ω and L are usually taken to be positive; then for clockwise motion, they are negative.

For general motion, the angular momentum of a particle relative to the origin O is defined as

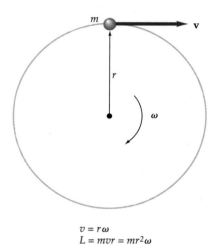

$v = r\omega$
$L = mvr = mr^2\omega$

Figure 8-18   The angular momentum of a particle moving in a circle is $L = mvr = mr^2\omega$.

Angular momentum defined

$$L = mvr_\perp = mvr \sin \theta \qquad\qquad 8\text{-}33$$

where v is the velocity of the particle and $r_\perp = r \sin \theta$ is the perpendicular distance from the line of motion to O as shown in Figure 8-19. Note that the particle has angular momentum about point O even though the particle is not moving in a circle.

The total angular momentum of a rotating object is found by summing the angular momenta of all the elements in the object. Figure 8-20 shows a rotating disk. The angular momentum of an element of mass m_i is

$$L_i = m_i r_i^2 \omega$$

When we sum over all the elements of the disk, we obtain

$$L = \sum_i L_i = \sum_i m_i r_i^2 \omega$$

Figure 8-19   A particle of mass m moving with a velocity **v** along a line a distance $r_\perp$ from the origin O. The angular momentum of the particle about the point O is $L = mvr_\perp$.

or

$$L = I\omega \qquad \text{8-34}$$

Angular momentum of a rotating object

Equation 8-34 is the rotational analog of the equation for linear momentum, $p = mv$. It holds for objects rotating about a fixed axis and also for objects rotating about an axis that is moving such that it remains parallel to itself, such as when a ball or cylinder rolls along a line.

Example 8-16

A 2.4-kg particle moves in a circle of radius 1.5 m with a speed of 3 m/s. (*a*) Find the angular momentum of the particle relative to the center of the circle. (*b*) Find the angular momentum relative to the origin of the same particle moving at 3 m/s along the line $y = 1.5$ m.

(*a*) From Equation 8-32, we obtain

$$L = mvr = (2.4 \text{ kg})(3 \text{ m/s})(1.5 \text{ m}) = 10.8 \text{ kg·m}^2/\text{s}$$

We could also compute the angular momentum using Equation 8-34. The moment of inertia of the particle relative to an axis through the center of the circle and perpendicular to the plane of the circle is $I = mr^2 = (2.4 \text{ kg})(1.5 \text{ m})^2 = 5.40 \text{ kg·m}^2$, and the angular speed is $\omega = v/r = (3 \text{ m/s})/(1.5 \text{ m}) = 2 \text{ rad/s}$. The angular momentum is then

$$L = I\omega = (5.40 \text{ kg·m}^2)(2 \text{ rad/s}) = 10.8 \text{ kg·m}^2/\text{s}$$

(*b*) For the same particle moving along the line $y = 1.5$ m, $r_\perp = 1.5$ m. The angular momentum of the particle is then (Equation 8-33)

$$L = mvr_\perp = (2.4 \text{ kg})(3 \text{ m/s})(1.5 \text{ m}) = 10.8 \text{ kg·m}^2/\text{s}$$

The angular momentum is the same whether the particle is moving in a circle of radius 1.5 m or is moving in a straight line that is a perpendicular distance of 1.5 m from the origin.

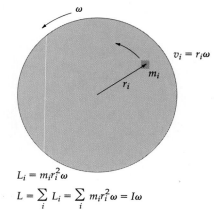

$$L_i = m_i r_i^2 \omega$$
$$L = \sum_i L_i = \sum_i m_i r_i^2 \omega = I\omega$$

Figure 8-20 The angular momentum of a disk is calculated by summing the angular momenta of all the mass elements m_i: $L = \sum L_i = \sum m_i r_i^2 \omega = I\omega$.

Newton's second law for rotation can be stated in a form analogous to Equation 8-31:

$$\tau_{\text{net}} = \frac{dL}{dt} = \frac{d(I\omega)}{dt} \qquad \text{8-35}$$

where τ_{net} is the net external torque exerted on the system.*

The net external torque acting on a system equals the rate of change of the angular momentum of the system.

For a rigid body, the moment of inertia is constant and Equation 8-35 becomes

$$\tau_{\text{net}} = \frac{d(I\omega)}{dt} = I\frac{d\omega}{dt} = I\alpha$$

which is the same as Equation 8-18. However, for a general system of particles, *the moment of inertia need not remain constant.* Equation 8-35 holds whether or not the moment of inertia is constant.

*The net torque is also the net external torque because the internal torques sum to zero, as will be discussed in Section 8-7.

Example 8-17

An object of mass m hangs from a string wrapped around a wheel of radius R and moment of inertia I as in Example 8-5 (Figure 8-7, page 235). Use Equation 8-35 to find the angular acceleration of the wheel.

Let v be the speed of the object and ω be the angular speed of the wheel at some time. Since the string does not slip, these quantities are related by $v = R\omega$. Relative to the center of the wheel, the angular momentum of the object is

$$L_{\mathrm{o}} = mvR = mR^2\omega$$

and the angular momentum of the wheel is

$$L_{\mathrm{w}} = I\omega$$

The total angular momentum of the system is thus

$$L = L_{\mathrm{o}} + L_{\mathrm{w}} = (I + mR^2)\omega$$

The only torque acting on the system relative to the center of the wheel is that due to the force of gravity mg acting on the object. Since the lever arm of this force relative to the center of the wheel is R, the torque is

$$\tau = mgR$$

Setting the torque equal to the rate of change of the angular momentum, we obtain

$$mgR = \frac{dL}{dt} = \frac{d}{dt}[(I + mR^2)\omega]$$

$$= (I + mR^2)\frac{d\omega}{dt}$$

$$= (I + mR^2)\alpha$$

The angular acceleration of the wheel is therefore

$$\alpha = \frac{mgR}{I + mR^2}$$

This agrees with the result found in Example 8-5 for the linear acceleration of the object $a = R\alpha$.

When the net external torque acting on a system is zero, we have

$$\frac{dL}{dt} = 0$$

or

$$L = \text{constant} \qquad \qquad 8\text{-}36$$

Equation 8-36 is a statement of the **law of conservation of angular momentum.**

Conservation of angular momentum

If the net external torque acting on a system is zero, the total angular momentum of the system is constant.

This is analogous to the law of conservation of linear momentum, which states that when the net external *force* acting on a system is zero, the total *linear* momentum is constant. Table 8-2 lists the rotational equations developed in this chapter along with the analogous equations for linear motion.

Table 8-2 Comparison of Linear Motion and Rotational Motion

Linear motion		Rotational motion	
Displacement	Δx	Angular displacement	$\Delta \theta$
Velocity	$v = \dfrac{dx}{dt}$	Angular velocity	$\omega = \dfrac{d\theta}{dt}$
Acceleration	$a = \dfrac{dv}{dt} = \dfrac{d^2x}{dt^2}$	Angular acceleration	$\alpha = \dfrac{d\omega}{dt} = \dfrac{d^2\theta}{dt^2}$
Constant-acceleration equations	$v = v_0 + at$ $\Delta x = v_{av}\,\Delta t$ $v_{av} = \frac{1}{2}(v_0 + v)$ $x = x_0 + v_0 t + \frac{1}{2}at^2$ $v^2 = v_0^2 + 2a\,\Delta x$	Constant–angular–acceleration equations	$\omega = \omega_0 + \alpha t$ $\Delta \theta = \omega_{av}\,\Delta t$ $\omega_{av} = \frac{1}{2}(\omega_0 + \omega)$ $\theta = \theta_0 + \omega_0 t + \frac{1}{2}\alpha t^2$ $\omega^2 = \omega_0^2 + 2\alpha\,\Delta \theta$
Mass	m	Moment of inertia	I
Momentum	$p = mv$	Angular momentum	$L = I\omega$
Force	F	Torque	τ
Kinetic energy	$K = \frac{1}{2}mv^2$	Kinetic energy	$K = \frac{1}{2}I\omega^2$
Power	$P = Fv$	Power	$P = \tau\omega$
Newton's second law	$F_{net} = \dfrac{dp}{dt} = ma$	Newton's second law	$\tau_{net} = \dfrac{dL}{dt} = I\alpha$

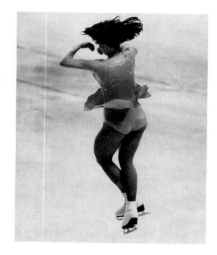

Figure 8-21 Because the torque exerted by the ice is small, the angular momentum of the skater is approximately constant. When she reduces her moment of inertia by drawing in her arms, her angular velocity increases.

If a system is isolated from its surroundings, so that there are no external forces or torques acting on it, three quantities are conserved: energy, linear momentum, and angular momentum. The law of conservation of angular momentum is a fundamental law of nature. Even on the microscopic scale of atomic and nuclear physics, where newtonian mechanics does not hold, the angular momentum of an isolated system is constant over time. There are many examples of the conservation of angular momentum in everyday life. Consider, for example, the ice skater spinning on the tips of her skates in Figure 8-21. Because the torque exerted by the ice is small, the angular momentum of the skater is approximately constant. When she draws her arms in toward her body, the moment of inertia of her body about a vertical axis through her body is reduced. Since her angular momentum $L = I\omega$ must remain constant, as I decreases her angular velocity ω increases; that is, she spins at a faster rate. Figure 8-22 is a multiflash photograph of a diver. His center of mass moves along a parabolic path after he leaves the board. The angular momentum is provided by the initial torque due to the force of the board. (This force does not pass through the diver's center of mass if he leans forward as he jumps.) If the diver wants to somersault in the air, he draws in his arms and legs, thereby decreasing his moment of inertia and increasing his angular velocity.

Figure 8-22 Multiflash photograph of a diver. The diver's center of mass moves along a parabolic path after he leaves the board. The angular momentum is provided by the initial torque due to the force of the board. This force does not pass through the diver's center of mass if he leans forward as he jumps. If the diver wanted to undergo $1\frac{1}{2}$ revolutions in the air, he would draw in his arms and legs, decreasing his moment of inertia to increase his angular velocity.

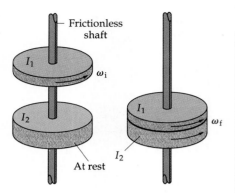

Figure 8-23 Inelastic rotational collision for Example 8-18. Since the only torques acting are internal to the system, angular momentum is conserved.

Example 8-18

A disk with a moment of inertia I_1 is rotating with an initial angular velocity ω_i about a frictionless shaft. It drops onto another disk with a moment of inertia I_2 that is initially at rest on the same shaft (see Figure 8-23). Because of surface friction, the two disks eventually attain a common angular velocity ω_f. Find this final common angular velocity.

Each disk exerts a torque on the other, but there is no torque external to the two-disk system, and so angular momentum of the system is conserved. Initially, the total angular momentum of the system L_i is the angular momentum of the first disk

$$L_i = I_1\omega_i$$

When both disks are rotating together, the total angular momentum is

$$L_f = I_1\omega_f + I_2\omega_f$$
$$= (I_1 + I_2)\omega_f$$

Setting the final angular momentum equal to the initial angular momentum, we obtain

$$(I_1 + I_2)\omega_f = I_1\omega_i$$

The final angular velocity is thus

$$\omega_f = \frac{I_1}{I_1 + I_2}\omega_i$$

In the collision of the two disks in Example 8-18, mechanical energy is not conserved. We can see this by writing the energy in terms of the angular momentum. In general, if a system with a moment of inertia I is rotating with an angular velocity ω, its kinetic energy is

$$K = \frac{1}{2}I\omega^2 = \frac{(I\omega)^2}{2I}$$

But $I\omega$ is the angular momentum L. The kinetic energy is therefore related to the angular momentum by

$$K = \frac{L^2}{2I} \qquad\qquad 8\text{-}37$$

The initial kinetic energy in Example 8-18 is

$$K_i = \frac{L_i^2}{2I_1}$$

and the final kinetic energy is

$$K_f = \frac{L_f^2}{2(I_1 + I_2)}$$

Since the final angular momentum L_f equals the initial angular momentum L_i, the final kinetic energy is less than the initial kinetic energy by the factor $I_1/(I_1 + I_2)$. This interaction of the disks is analogous to a one-dimensional inelastic collision of two objects.

The rotating plates in the transmission of a truck make inelastic collisions when engaged.

Example 8-19

A merry-go-round of radius 2 m and moment of inertia 500 kg·m² is rotating about a frictionless pivot. It makes one revolution every 5 s. A child of mass 25 kg originally standing at the center walks out to the rim. Find the new angular speed of the merry-go-round.

Since there is no external torque acting on the child–merry-go-round system, the angular momentum of the system remains constant. Originally, the child is at the center, where he has no appreciable moment of inertia about the axis and therefore no angular momentum. When he is at the rim, he has an angular momentum $I_c\omega_f$, where $I_c = mr^2$ is the moment of inertia of the child relative to the axis of the merry-go-round, and ω_f is the final angular velocity of the child and the merry-go-round. Since the mass of the child is 25 kg and r is 2 m, the moment of inertia of the child when he is at the rim is

$$I_c = (25 \text{ kg})(2 \text{ m})^2 = 100 \text{ kg·m}^2$$

If I_m is the moment of inertia of the merry-go-round and ω_i is its initial angular velocity, we have from the conservation of angular momentum

$$L_i = L_f$$

$$I_m\omega_i = I_m\omega_f + I_c\omega_f = (I_m + I_c)\omega_f$$

$$\omega_f = \frac{I_m}{I_m + I_c}\omega_i = \frac{500 \text{ kg·m}^2}{(500 + 100) \text{ kg·m}^2}\omega_i = \frac{5}{6}\omega_i$$

Since the merry-go-round originally makes one revolution every 5 s, its initial angular velocity is $\frac{1}{5}$ rev/s or 0.4π rad/s. The final angular velocity is therefore

$$\omega_f = (\tfrac{5}{6})\tfrac{1}{5} \text{ rev/s} = \tfrac{1}{6} \text{ rev/s}$$

(There is no particular need here to convert the given units of angular velocity to radians per second.) Thus, after the child reaches the rim, the merry-go-round rotates at one revolution every 6 s.

When the child is at the center of the merry-go-round, he is at rest. As he walks out, he begins moving in a circle. The force that accelerates the child is the force of friction exerted on him by the merry-go-round. This force has a component that is tangential to the circle that produces a torque that increases the angular momentum of the child. The child exerts an equal and opposite frictional force on the merry-go-round. The torque associated with this force decreases the angular momentum of the merry-go-round.

(a)

(b)

(a) The conservation of angular momentum has a dramatic impact on astronauts in space. Here, astronaut Dale Gardner prepares to capture a spinning Westar VI communications satellite. (b) Gardner used a large tool called the apogee kick motor, a capture device, to stabilize the satellite sufficiently to capture it for return to earth in the cargo bay of the Discovery.

As in Example 8-18, we can calculate the initial and final mechanical energy of the child and the merry-go-round. Since the angular momentum does not change and the final moment of inertia of the system is greater than the initial momentum of inertia, we can see from Equation 8-37 that the total kinetic energy decreases. The detailed analysis of the energy transfers in this system is complicated. As the child moves outward, his tangential velocity increases. With each step, he makes an inelastic collision with a part of the merry-go-round that is further out and therefore moving faster than he is. Mechanical energy is lost in each of these inelastic collisions. Conversely, if the child walks inward, the moment of inertia of the child–merry-go-round system decreases, so from Equation 8-37, the total kinetic energy of the system must increase. This energy comes from the internal chemical energy of the child.

Example 8-20

The same child as in Example 8-19 runs along a path tangential to the rim of the merry-go-round, which is initially at rest, with an initial speed v = 2.5 m/s and jumps on the merry-go-round (Figure 8-24). What is the final angular velocity of the child and the merry-go-round together?

We do not expect mechanical energy to be conserved because the child makes an inelastic collision with the rim of the merry-go-round. Linear momentum is not conserved either. The pivot of the merry-go-round exerts an impulse during the collision, but because it is frictionless, it cannot exert any torque. Thus, angular momentum about the pivot is conserved. (This example demonstrates that we need not have circular motion to have angular momentum.) The initial angular momentum of the child relative to the pivot of the merry-go-round is

$$L_i = mvR = (25 \text{ kg})(2.5 \text{ m/s})(2 \text{ m})$$
$$= 125 \text{ kg·m}^2/\text{s}$$

Figure 8-24 A child runs tangentially to the rim of a merry-go-round and jumps on for Example 8-20. If the pivot is frictionless, angular momentum is conserved.

When the child is on the merry-go-round, the angular momentum of the child plus the merry-go-round is

$$L_f = (I_c + I_m)\omega_f$$

$$= (100 \text{ kg·m}^2 + 500 \text{ kg·m}^2)\omega_f$$

where we have used $I_c = 100$ kg·m^2 and $I_m = 500$ kg·m^2 from Example 8-19. Setting the final angular momentum equal to the initial angular momentum, we have

$$(600 \text{ kg·m}^2)\omega_f = 125 \text{ kg·m}^2/\text{s}$$

$$\omega_f = \frac{125 \text{ kg·m}^2/\text{s}}{600 \text{ kg·m}^2}$$

$$= 0.208 \text{ rad/s}$$

Note that the unit of ω_f comes out to be s^{-1}, which we have written as rad/s to conform with the usual unit for angular velocity.

Exercise

Calculate the initial and final kinetic energies of the child–merry-go-round system in Example 8-20. (Answer: $K_i = 78.2$ J, $K_f = 13.0$ J)

Questions

7. A woman sits on a spinning piano stool with her arms folded. If she extends her arms, what happens to her angular velocity?

8. In Example 8-19, does the merry-go-round do work on the child?

9. Is it easier to crawl radially outward or radially inward on a rotating merry-go-round? Why?

10. Folk wisdom says that a cat always lands on its feet. If a cat starts falling with its feet up, how can it land on its feet without violating the law of conservation of angular momentum?

8-6   Rolling Objects

Figure 8-25 shows a ball of radius R rolling along a plane surface. If the ball rolls without slipping, its rotational and translational motions are simply related. As the ball turns through the angle ϕ shown in the figure, the point of contact between the ball and the plane moves a distance s that is related to ϕ by

$$s = R\phi \qquad \qquad 8\text{-}38$$

Since the center of the ball lies directly over the point of contact, the center of mass of the ball also moves through the distance s. The velocity of the center of mass and the angular velocity of rotation are therefore related by

$$V_{cm} = \frac{ds}{dt} = R\frac{d\phi}{dt}$$

or

$$V_{cm} = R\omega \qquad \qquad 8\text{-}39 \qquad \textit{Rolling condition}$$

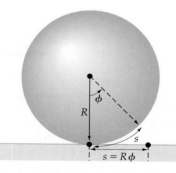

Figure 8-25   When a ball rolls without slipping, the center of mass moves a distance $s = R\phi$ as the ball rotates through an angle ϕ.

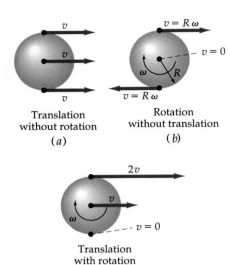

Translation
without rotation
(a)

Rotation
without translation
(b)

Translation
with rotation
(c)

Figure 8-26 (a) Translation without rotation. The top and bottom of the ball move with the same velocity. (b) Rotation without translation. The top of the ball moves to the right with a speed $v = R\omega$ relative to the center, which is at rest. The bottom moves to the left with the same speed relative to the center. (c) Rolling without slipping is a combination of translation and rotation. If the center moves with speed v, the top moves with speed $2v$ and the bottom of the ball is momentarily at rest.

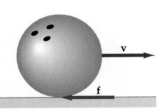

Figure 8-27 A bowling ball moving with no initial rotation. The frictional force **f** exerted by the floor reduces the speed of the center of mass and increases the angular speed until the rolling condition $v = R\omega$ is reached.

Equation 8-39 (or the equivalent Equation 8-38) is called the **rolling condition**. It holds whenever a ball or cylinder rolls without slipping. We can also write the rolling condition in terms of the linear and angular accelerations:

$$\frac{dV_{cm}}{dt} = R\frac{d\omega}{dt}$$

or

$$A_{cm} = R\alpha \qquad 8\text{-}40$$

When a ball rotates with an angular speed ω, a point on the edge of the ball has a speed $R\omega$ relative to the center of the ball. Since the center of the ball is moving with a speed $R\omega$ relative to the surface, and since the point of the ball in contact with the surface is moving backward with this same speed relative to the center of the ball, the point in contact is instantaneously at rest relative to the surface (see Figure 8-26). If a frictional force is exerted by the surface on the ball, it is static friction and no energy is dissipated.

Example 8-21

A homogeneous sphere of radius 12 cm and mass 30 kg is rolling, without slipping, on a horizontal floor at 2 m/s. How much work is required to stop the sphere?

The work needed to stop the sphere equals the initial kinetic energy of the sphere, which is the kinetic energy of translation of the center of mass, $\frac{1}{2}mv^2$, plus the kinetic energy of rotation relative to the center of mass, $\frac{1}{2}I_{cm}\omega^2$. We use $I_{cm} = \frac{2}{5}mR^2$ for the moment of inertia of the sphere. The rolling condition is $v = R\omega$. Thus, we obtain for the total kinetic energy

$$K = \tfrac{1}{2}mv^2 + \tfrac{1}{2}I_{cm}\omega^2 = \tfrac{1}{2}mv^2 + \tfrac{1}{2}(\tfrac{2}{5}mR^2)\left(\frac{v}{R}\right)^2$$

$$= \tfrac{1}{2}mv^2 + \tfrac{1}{5}mv^2 = \tfrac{7}{10}mv^2 = \tfrac{7}{10}(30\text{ kg})(2\text{ m/s})^2 = 84\text{ J}$$

We can apply these ideas to several interesting problems. We will first consider what happens when a bowling ball is thrown such that initially it has no rotation. As the ball slides along the floor, a force of sliding friction opposes its motion (Figure 8-27). This frictional force causes the ball to rotate, and it also reduces the linear velocity of the center of mass of the ball. The angular velocity increases and the linear velocity decreases until the rolling condition $V_{cm} = R\omega$ is met. Then the ball rolls without slipping, and there is no sliding friction between the surfaces.

Example 8-22

A bowling ball of mass M and radius R is thrown such that the instant it touches the floor it is moving horizontally with speed $v_0 = 5$ m/s and is not rotating. The coefficient of kinetic friction between the ball and the floor is $\mu_k = 0.3$. Find (a) the time the ball slides before the rolling condition is met and (b) the distance the ball slides before it rolls without slipping.

(a) While the ball is sliding, the net force on the ball is the force of kinetic friction $\mu_k Mg$ in the direction opposite its velocity. The acceleration of the center of mass of the ball has the magnitude $a = F_{net}/M = \mu_k g$. At any time t while the ball is sliding, the velocity of the ball is given by

$$v = v_0 - at = v_0 - \mu_k gt$$

The net torque about the center of mass of the ball, while the ball is sliding, is that due to kinetic friction. Its magnitude is

$$\tau = \mu_k MgR$$

From Table 8-1, the moment of inertia of the ball is $I = \frac{2}{5}MR^2$. The angular acceleration of the ball is then

$$\alpha = \frac{\tau}{I} = \frac{\mu_k MgR}{\frac{2}{5}MR^2} = \frac{5}{2}\left(\frac{\mu_k g}{R}\right)$$

While the ball is sliding, the angular velocity of the ball at any time t is

$$\omega = \alpha t = \frac{5}{2}\left(\frac{\mu_k g}{R}\right)t$$

At some time $t = t_1$, the rolling condition $v = R\omega$ is met and the ball stops sliding. Setting v equal to $R\omega$ at time $t = t_1$ gives

$$v = v_0 - \mu_k g t_1 = R\omega = \tfrac{5}{2}\mu_k g t_1$$

$$v_0 = \tfrac{5}{2}\mu_k g t_1 + \mu_k g t_1 = \tfrac{7}{2}\mu_k g t_1$$

Solving for t_1, we obtain

$$t_1 = \frac{2v_0}{7\mu_k g} = \frac{2(5\ m/s)}{7(0.3)(9.81\ m/s^2)} = 0.485\ s$$

(b) Since the acceleration of the center of mass of the ball is constant, the average velocity while it is sliding is the average of the initial and final velocities. From the equations above, we see that, at $t = t_1$, $v = \frac{5}{7}v_0$. The distance traveled during this time is thus

$$s = v_{av}t_1 = \tfrac{1}{2}(v_0 + v)t_1 = \tfrac{1}{2}(v_0 + \tfrac{5}{7}v_0)$$

$$= \tfrac{1}{2}(\tfrac{12}{7})(5\ m/s)(0.485\ s) = 2.08\ m$$

We will next consider the problem of where to strike the cue ball when playing pool or billiards. If the cue ball is struck with a horizontal force **F** from the cue stick such that the line of action of **F** passes through the ball's center of mass, the ball will begin to move initially with no rotation like the bowling ball we just considered. If the ball is struck such that the line of action is below the center of mass, it will start moving with back spin. The force of sliding friction will reduce the back spin and will eventually produce forward spin that increases until the rolling condition of Equation 8-39 is met. Where should the ball be struck so that it will roll without slipping from the beginning? We can find this by requiring the initial linear acceleration and angular acceleration of the ball produced by the cue stick to obey the rolling condition. Let **F** be the force exerted by the cue stick at a height x above the center of the ball (Figure 8-28). Since the cue stick imparts a very large impulsive force for a very short time, we can neglect friction during this time. Newton's second law for linear motion gives

$$F = ma = mR\alpha \qquad\qquad 8\text{-}41$$

where we have substituted the rolling condition $a = R\alpha$. Newton's second law for rotation gives

$$\tau = Fx = I\alpha \qquad\qquad 8\text{-}42$$

Dividing Equation 8-42 by Equation 8-41 to eliminate F and α gives

$$x = \frac{I}{mR}$$

Figure 8-28 Cue stick hitting the cue ball such that the line of action is a distance x above the center. If x is chosen properly, the cue ball will roll without slipping from the beginning.

Figure 8-29 Cue ball with top spin. The force of friction increases the speed of the center of mass and reduces the angular speed until the rolling condition is met.

Using $I = \frac{2}{5}mR^2$ for the moment of inertia for a ball, we obtain

$$x = \frac{2}{5}R$$

Thus, if the ball is struck such that the line of action of the force is $\frac{2}{5}R$ above the center, it will begin to roll without slipping. If the line of action is higher than this, the ball will have top spin. There will then be a force of friction in the direction of motion (Figure 8-29) that increases the linear velocity of the ball and decreases its rotational velocity until the rolling condition is met.

The final problem we will consider is that of spheres, hoops, or cylinders rolling down an incline. Our analysis of this type of problem is simplified by an important theorem concerning the angular momentum of a system relative to the center of mass:

> The net torque about the center of mass equals the rate of change of the angular momentum relative to the center of mass no matter how the center of mass is moving.

$$\tau_{\text{net,cm}} = \frac{dL_{\text{cm}}}{dt} \qquad\qquad 8\text{-}43$$

This is the same as Equation 8-35 except that the torque and the angular momentum are computed relative to the center of mass rather than relative to some fixed point. This theorem is important because, for example, when a ball rolls down an incline, the center of mass is accelerating, and we would not necessarily expect Newton's second law for rotation to hold in a noninertial reference frame. Equation 8-43 is true because of the special properties of the center of mass of a system. A proof of this theorem is given in most intermediate-level mechanics books.*

Figure 8-30 shows a sphere of mass m and radius R rolling down an incline. (Our analysis applies equally to a cylinder or hoop.) The forces acting are the weight $m\mathbf{g}$ downward, the normal force $\mathbf{F}_n$ that balances the normal component of the weight, and the force of friction $\mathbf{f}$ acting up the incline. If the object rolls without slipping, the friction is static, as previously discussed. As the object accelerates down the incline, the angular velocity of rotation must increase if it is to roll without slipping. Thus, the angular momentum relative to the center of mass must increase. This increase is due to the torque exerted by the frictional force. (The weight and normal force both act through the center of mass and therefore exert no torque about it.) The angular momentum about the center of mass is

$$L_{\text{cm}} = I_{\text{cm}}\omega$$

where I_{cm} is the moment of inertia about the center of mass, which is constant. The torque exerted by friction is fR. Setting the torque equal to the rate of change of the angular momentum, we have

$$\tau_{\text{net}} = fR = \frac{dL_{\text{cm}}}{dt} = I_{\text{cm}}\frac{d\omega}{dt}$$

or

$$fR = I_{\text{cm}}\alpha \qquad\qquad 8\text{-}44$$

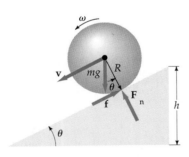

Figure 8-30 Forces acting on a sphere rolling down an incline. The frictional force acting up the incline exerts a torque about the center of mass of the sphere that increases its angular velocity so that the rolling condition is maintained as the sphere accelerates down the incline.

The linear acceleration of the center of mass of the object is down the incline. The net force down the incline is $mg\sin\theta - f$. Newton's second law therefore gives

$$mg\sin\theta - f = mA_{\text{cm}} \qquad\qquad 8\text{-}45$$

*See, for example, G. R. Fowles, *Analytical Mechanics*, pp. 201–203, Holt, New York, 1977.

We can solve Equations 8-44 and 8-45 for the unknowns α, A_{cm}, and f using Equation 8-40 for the rolling condition. Substituting A_{cm}/R for α in Equation 8-44, we have

$$fR = I_{cm} \frac{A_{cm}}{R}$$

$$f = \frac{I_{cm}}{R^2} A_{cm} \qquad\qquad 8\text{-}46$$

Substituting this result into Equation 8-45 gives

$$mg \sin \theta - \frac{I_{cm}}{R^2} A_{cm} = mA_{cm}$$

$$A_{cm} = \frac{mg \sin \theta}{m + I_{cm}/R^2} \qquad\qquad 8\text{-}47$$

For a sphere, $I_{cm} = \frac{2}{5}mR^2$, so the acceleration is

$$A_{cm} = \frac{mg \sin \theta}{m + \frac{2}{5}m} = \frac{5}{7}g \sin \theta \qquad \text{(sphere)} \qquad 8\text{-}48$$

For a cylinder, $I_{cm} = \frac{1}{2}mR^2$, so the acceleration is

$$A_{cm} = \frac{mg \sin \theta}{m + \frac{1}{2}m} = \frac{2}{3}g \sin \theta \qquad \text{(cylinder)} \qquad 8\text{-}49$$

For a hoop, $I_{cm} = mR^2$, so the acceleration is

$$A_{cm} = \frac{1}{2}g \sin \theta \qquad \text{(hoop)} \qquad 8\text{-}50$$

Note that the accelerations given by Equations 8-48, 8-49, and 8-50 are independent of the radius of the object. If we release a sphere, a cylinder, and a hoop at the top of an incline and if they all roll without slipping, the sphere will reach the bottom first because it has the greatest acceleration. The cylinder will be second and the hoop third (Figure 8-31). If an object could slide down the incline without friction, it would reach the bottom before any of the rolling objects.

We can use these results and Equation 8-46 to find the force of friction. For a cylinder with $I_{cm} = \frac{1}{2}mR^2$, for example, we have

$$f = \frac{I_{cm}}{R^2} A_{cm} = \frac{\frac{1}{2}mR^2}{R^2} (\frac{2}{3}g \sin \theta)$$

or

$$f = \frac{1}{3}mg \sin \theta \qquad\qquad 8\text{-}51$$

Figure 8-31   A sphere, a cylinder, and a hoop are started together from rest at the top of an incline. The sphere reaches the bottom first, followed by the cylinder and then the hoop.

Note that we have found the force of friction without considering the coefficient of friction. Since the cylinder is rolling without slipping, the friction is static friction, which is usually not equal to its maximum value $\mu_s F_n$, where μ_s is the coefficient of static friction. For the general case,

$$f \le \mu_s F_n = \mu_s mg \cos \theta$$

Using Equation 8-51 for f for a cylinder that is rolling without slipping, we have

$$f = \frac{1}{3}mg \sin \theta \le \mu_s mg \cos \theta$$

or

$$\tan \theta \le 3\mu_s \qquad\qquad 8\text{-}52$$

Thus, if the tangent of the angle of incline is greater than $3\mu_s$, the cylinder will slip as it moves down the incline.

Exercise

A cylinder rolls down a plane inclined at $\theta = 50°$. What is the minimum value of the coefficient of static friction for which the cylinder will roll without slipping?   (Answer: 0.40)

Exercise

Find (a) the force of friction for a hoop rolling down an incline and (b) the maximum value of tan θ for which the hoop will roll without slipping. [Answer: (a) $f = \frac{1}{2}mg \sin \theta$, (b) $\tan \theta \leq 2\mu_s$]

The linear acceleration of any object rolling down an incline is less than $g \sin \theta$ because of the frictional force directed up the incline. We can find the speed of the object at the bottom of the incline by using the constant-acceleration formulas or from the conservation of mechanical energy. (Since the friction is static, there is no dissipation of mechanical energy.) We will use the conservation of mechanical energy here. At the top of the incline, the total energy is potential energy mgh. At the bottom, the total energy is kinetic energy, which is the translational kinetic energy of the center of mass, $\frac{1}{2}mv^2$, plus the kinetic energy of rotation relative to the center of mass, $\frac{1}{2}I_{cm}\omega^2$. The conservation of energy therefore gives

$$\frac{1}{2}mv^2 + \frac{1}{2}I_{cm}\omega^2 = mgh$$

We can use the rolling condition to eliminate either v or ω. Substituting $\omega = v/R$, we obtain

$$\frac{1}{2}mv^2 + \frac{1}{2}I_{cm}\left(\frac{v}{R}\right)^2 = mgh$$

$$v^2 = \frac{2mgh}{m + I_{cm}/R^2} \qquad\qquad 8\text{-}53$$

For a cylinder with $I_{cm} = \frac{1}{2}mR^2$, for example, this gives us

$$v^2 = \frac{2mgh}{m + \frac{1}{2}m} = \frac{4}{3}gh$$

Note that this result is independent of the radius of the cylinder and that it is less than $2gh$, which would be the result for an object sliding down the incline with no friction.

Water spouts off the Grand Bahama Islands offer a stunning visualization of rotational motion.

8-7 The Vector Nature of Rotation and the Cross Product

For general rotational motion, torque, angular velocity, and angular momentum are vector quantities. We have not had to worry about the vector nature of these quantities in our discussion so far because we have considered only rotation about a fixed axis (or in the case of rolling objects, about an axis that remains parallel to a fixed direction in space). This rotation was therefore analogous to linear motion in one dimension, where we could use a plus or minus sign to indicate the direction of the velocity or acceleration. When the axis of rotation is not fixed in space, the vector nature of rotational quantities becomes important.

Figure 8-32 A disk rotating about an axis through its center and perpendicular to its plane. All directions in the plane of the disk are equivalent. The direction of the angular velocity $\boldsymbol{\omega}$ is chosen to be along the axis of rotation according to the right-hand rule.

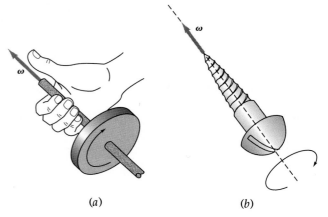

(a)　　　　(b)

Figure 8-33 (a) The right-hand rule for determining the direction of the angular velocity $\boldsymbol{\omega}$. When the fingers of the right hand curl in the direction of rotation, the thumb points in the direction of $\boldsymbol{\omega}$. (b) The direction of $\boldsymbol{\omega}$ is that of the advance of a rotating right-hand screw.

Consider the rotating disk in Figure 8-32. Since all directions in the plane of the disk are equivalent by symmetry, we describe the "direction" of rotation by giving the direction of the axis of rotation. We therefore choose the angular velocity vector $\boldsymbol{\omega}$ to be along the axis of rotation. We arbitrarily choose the direction of $\boldsymbol{\omega}$ to be given by the **right-hand rule,** which is illustrated in Figure 8-33a. If the axis of rotation is grasped by the right hand with the fingers following the rotation, the extended thumb points in the direction of $\boldsymbol{\omega}$. The direction of $\boldsymbol{\omega}$ is also that of the advance of a rotating right-hand screw, as shown in Figure 8-33b. Thus, if the rotation is counterclockwise, as in Figure 8-32, $\boldsymbol{\omega}$ is outward; if it is clockwise, $\boldsymbol{\omega}$ is inward.

Figure 8-34 shows a force $\mathbf{F}$ acting on a particle at some position $\mathbf{r}$ relative to the origin O. The torque exerted by this force relative to the origin O is defined as a vector of magnitude $Fr \sin \phi$, where ϕ is the angle between $\mathbf{F}$ and $\mathbf{r}$, that has a direction perpendicular to the plane formed by $\mathbf{F}$ and $\mathbf{r}$. This is the same definition of torque we have been using, except that we have now assigned the torque a direction. If $\mathbf{F}$ and $\mathbf{r}$ are in the xy plane, as in the figure, the torque is along the z axis. The torque is written as the **vector product** or **cross product** of $\mathbf{r}$ and $\mathbf{F}$:

$$\boldsymbol{\tau} = \mathbf{r} \times \mathbf{F} \qquad\qquad 8\text{-}54$$

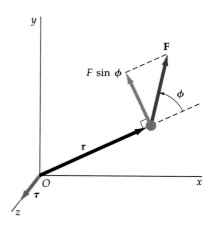

Figure 8-34 The torque $\boldsymbol{\tau}$ exerted by a force $\mathbf{F}$ acting on a particle at position $\mathbf{r}$ is perpendicular to both $\mathbf{F}$ and $\mathbf{r}$ and has a magnitude $Fr \sin \phi$.

The cross product of two vectors **A** and **B** is defined to be a vector **C** whose magnitude equals the area of the parallelogram formed by the two vectors (Figure 8-35*a*). The direction of **C** is perpendicular to the plane containing **A** and **B** in the sense given by the right-hand rule as **A** is rotated into **B** through the smaller angle between these vectors (Figure 8-35*b*). If ϕ is the angle between the two vectors and $\hat{n}$ is a unit vector that is perpendicular to each in the sense described, the cross product of **A** and **B** is

Cross product defined

$$\mathbf{A} \times \mathbf{B} = (AB \sin \phi)\hat{n} \qquad \text{8-55}$$

Figure 8-35 (*a*) The vector product **A** × **B** is a vector **C** that is perpendicular to both **A** and **B** and has a magnitude $AB \sin \phi$, which equals the area of the parallelogram shown. (*b*) The direction of **A** × **B** is given by the right-hand rule when **A** is rotated into **B** through the angle ϕ.

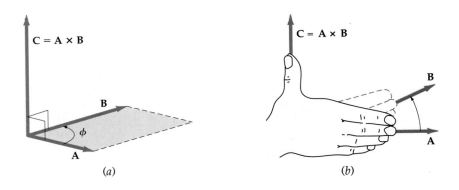

(*a*)   (*b*)

If **A** and **B** are parallel, **A** × **B** is zero. It follows from the definition of the cross product (Equation 8-55) that

$$\mathbf{A} \times \mathbf{A} = 0 \qquad \text{8-56}$$

and

$$\mathbf{A} \times \mathbf{B} = -\mathbf{B} \times \mathbf{A} \qquad \text{8-57}$$

It should be noted that the order in which two vectors are multiplied when determining the cross product is significant. Unlike the multiplication of ordinary numbers, changing the order of multiplication of the two vectors in a cross product changes the sign of the result. Some properties of the cross product of two vectors are

1. The cross product obeys a distributive law under addition:

$$\mathbf{A} \times (\mathbf{B} + \mathbf{C})$$
$$= \mathbf{A} \times \mathbf{B} + \mathbf{A} \times \mathbf{C} \qquad \text{8-58}$$

2. If **A** and **B** are functions of some variable such as t, the derivative of **A** × **B** follows the usual product rule for derivatives:

$$\frac{d}{dt}(\mathbf{A} \times \mathbf{B}) = \mathbf{A} \times \frac{d\mathbf{B}}{dt} + \frac{d\mathbf{A}}{dt} \times \mathbf{B} \qquad \text{8-59}$$

Note that we must keep the order straight here since, for example, **B** × $d\mathbf{A}/dt = -(d\mathbf{A}/dt)$ × **B**.

Figure 8-36 The unit vectors **i**, **j**, and **k** are mutually perpendicular and have magnitude 1. The cross product of **i** and **j** is **i** × **j** = (1)(1) sin 90°**k** = **k**. Similarly **j** × **k** = **i**, and **k** × **i** = **j**.

3. The unit vectors **i**, **j**, and **k**, which are mutually perpendicular, have cross products given by:

$$\mathbf{i} \times \mathbf{j} = \mathbf{k} \qquad \mathbf{j} \times \mathbf{k} = \mathbf{i} \qquad \mathbf{k} \times \mathbf{i} = \mathbf{j}$$

(See Figure 8-36.)

And,

$$\mathbf{i} \times \mathbf{i} = \mathbf{j} \times \mathbf{j} = \mathbf{k} \times \mathbf{k} = 0$$

We can use these results to express the cross product of two vectors in terms of the vectors' rectangular components:

$$\mathbf{A} \times \mathbf{B} = (A_x\mathbf{i} + A_y\mathbf{j} + A_z\mathbf{k}) \times (B_x\mathbf{i} + B_y\mathbf{j} + B_z\mathbf{k})$$

$$= A_xB_x\mathbf{i} \times \mathbf{i} + A_xB_y\mathbf{i} \times \mathbf{j} + A_xB_z\mathbf{i} \times \mathbf{k} + A_yB_x\mathbf{j} \times \mathbf{i} + A_yB_y\mathbf{j} \times \mathbf{j}$$
$$+ A_yB_z\mathbf{j} \times \mathbf{k} + A_zB_x\mathbf{k} \times \mathbf{i} + A_zB_y\mathbf{k} \times \mathbf{j} + A_zB_z\mathbf{k} \times \mathbf{k}$$

$$= 0 + A_xB_y\mathbf{k} + A_xB_z(-\mathbf{j}) + A_yB_x(-\mathbf{k}) + 0 + A_yB_z\mathbf{i}$$
$$+ A_zB_x\mathbf{j} + A_zB_y(-\mathbf{i}) + 0$$

or,

$$\mathbf{A} \times \mathbf{B} = (A_yB_z - A_zB_y)\mathbf{i} + (A_zB_x - A_xB_z)\mathbf{j} + (A_xB_y - A_yB_x)\mathbf{k} \quad 8\text{-}60a$$

This result can also be written in terms of a determinant:

$$\mathbf{A} \times \mathbf{B} = \begin{vmatrix} \mathbf{i} & \mathbf{j} & \mathbf{k} \\ A_x & A_y & A_z \\ B_x & B_y & B_z \end{vmatrix} \quad 8\text{-}60b$$

4. For the special case in which **A** and **B** are in the xy plane, $A_z = B_z = 0$ and **A** × **B** has only a z component:

$$\mathbf{A} \times \mathbf{B} = (A_xB_y - A_yB_x)\mathbf{k} \qquad \text{(for \textbf{A} and \textbf{B} in the } xy \text{ plane)} \quad 8\text{-}61$$

The direction of rotation is changed by this bevel gear in a diesel engine.

Example 8-23

Find **A** × **B** for **A** = 5**i** + 6**j** and **B** = 3**i** − 2**j**.

Computing the cross product directly, we have

$$\mathbf{A} \times \mathbf{B} = (5\mathbf{i} + 6\mathbf{j}) \times (3\mathbf{i} - 2\mathbf{j})$$

$$= 15(\mathbf{i} \times \mathbf{i}) - 10(\mathbf{i} \times \mathbf{j}) + 18(\mathbf{j} \times \mathbf{i}) - 12(\mathbf{j} \times \mathbf{j})$$

$$= 15(0) - 10(\mathbf{k}) + 18(-\mathbf{k}) - 12(0)$$

$$= -10\mathbf{k} - 18\mathbf{k}$$

$$= -28\mathbf{k}$$

Alternatively, since both **A** and **B** are in the xy plane, we may use Equation 8-61

$$\mathbf{A} \times \mathbf{B} = (A_xB_y - A_yB_x)\mathbf{k}$$

$$= [(5)(-2) - (6)(3)]\mathbf{k}$$

$$= -28\mathbf{k}$$

The angular momentum of a particle can be written as a vector cross product. Figure 8-37 shows a particle of mass m moving with a velocity **v** at a position **r** relative to the origin O. The linear momentum of the particle is **p** = m**v**. The angular momentum **L** of the particle relative to the origin O is defined to be the cross product of **r** and **p**:

$$\mathbf{L} = \mathbf{r} \times \mathbf{p} \qquad 8\text{-}62$$

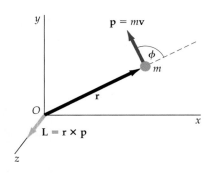

Figure 8-37 A particle with a momentum **p** at position **r** has angular momentum relative to the origin O given by **L** = **r** × **p**. For **r** and **p** in the xy plane, as shown here, **L** is along the z axis.

Angular momentum of a particle defined

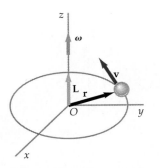

Figure 8-38 A particle moving in a circle has an angular momentum relative to the center of the circle that is parallel to the angular velocity and is given by $L = I\omega$.

Like torque, angular momentum is defined relative to a point in space. Figure 8-38 shows a particle moving in a circle in the xy plane with the center of the circle at the origin. The speed v of the particle and the magnitude of its angular velocity ω are related by $v = r\omega$. The angular momentum of the particle relative to the center of the circle is

$$\mathbf{L} = \mathbf{r} \times \mathbf{p} = \mathbf{r} \times m\mathbf{v} = (rmv \sin 90°)\mathbf{k} = rmv\mathbf{k} = mr^2\omega\mathbf{k}$$

The angular momentum is in the same direction as the angular velocity. Since mr^2 is the moment of inertia for a single particle about the z axis, we have

$$\mathbf{L} = mr^2\boldsymbol{\omega} = I\boldsymbol{\omega}$$

Figure 8-39 shows the angular-momentum vector $\mathbf{L}'$ for the same particle moving in the same circle but with $\mathbf{L}'$ computed relative to a point on the z axis that is not at the center of the circle. In this case, the angular momentum is not parallel to the angular velocity $\boldsymbol{\omega}$, which is along the z axis. In Figure 8-40, we have added a second particle of equal mass moving in the same circle. The angular-momentum vectors $\mathbf{L}_1'$ and $\mathbf{L}_2'$ are shown relative to the same point as in Figure 8-39. The total angular momentum $\mathbf{L}_1'$ and $\mathbf{L}_2'$ of the two-particle system is again parallel to the angular velocity $\boldsymbol{\omega}$. In this case the axis of rotation, the z axis, passes through the center of mass of the two-particle system, and the mass distribution is symmetric about this axis. Such an axis is called a **symmetry axis**. For any system of particles that rotates about a symmetry axis, the total angular momentum (which is the sum of the angular momenta of the individual particles) is parallel to the angular velocity and is given by

Angular momentum of a system

$$\mathbf{L} = I\boldsymbol{\omega} \qquad \text{8-63}$$

We will now show that Newton's second law implies that the rate of change of the angular momentum of a particle equals the net torque acting on the particle. If we have a number of forces acting on a particle, the net torque relative to the origin O is the sum of the torques due to each force:

$$\boldsymbol{\tau}_{\text{net}} = (\mathbf{r} \times \mathbf{F}_1) + (\mathbf{r} \times \mathbf{F}_2) + \cdots = \mathbf{r} \times \mathbf{F}_{\text{net}}$$

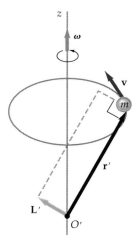

Figure 8-39 The angular momentum $\mathbf{L}'$ of a particle moving in a circle calculated relative to a point not at the center of the circle. In this case, $\mathbf{L}'$ is not parallel to the angular velocity $\boldsymbol{\omega}$.

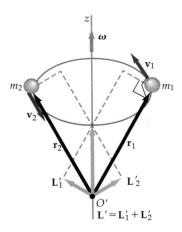

Figure 8-40 Angular-momentum vectors for two particles moving in a circle calculated relative to a point on an axis about which the mass distribution is symmetrical. The total angular momentum $\mathbf{L}_1' + \mathbf{L}_2'$ is parallel to the angular velocity $\boldsymbol{\omega}$.

According to Newton's second law, the net force equals the rate of change of the linear momentum $d\mathbf{p}/dt$. Thus,

$$\boldsymbol{\tau}_{net} = \mathbf{r} \times \mathbf{F}_{net} = \mathbf{r} \times \frac{d\mathbf{p}}{dt} \qquad 8\text{-}64$$

We can compute the rate of change of the angular momentum using the product rule for derivatives:

$$\frac{d\mathbf{L}}{dt} = \frac{d}{dt}(\mathbf{r} \times \mathbf{p}) = \left(\frac{d\mathbf{r}}{dt} \times \mathbf{p}\right) + \mathbf{r} \times \frac{d\mathbf{p}}{dt}$$

The first term on the right of this equation is zero because

$$\frac{d\mathbf{r}}{dt} \times \mathbf{p} = \mathbf{v} \times m\mathbf{v} = 0$$

Thus,

$$\frac{d\mathbf{L}}{dt} = \mathbf{r} \times \frac{d\mathbf{p}}{dt} \qquad 8\text{-}65$$

Comparing Equations 8-64 and 8-65 we have

$$\boldsymbol{\tau}_{net} = \frac{d\mathbf{L}}{dt} \qquad 8\text{-}66$$

The net torque acting on a system of particles is the sum of the individual torques acting on the system. The generalization of Equation 8-66 to a system of particles is thus

$$\sum_i \boldsymbol{\tau}_i = \sum_i \frac{d\mathbf{L}_i}{dt} = \frac{d}{dt}\sum_i \mathbf{L}_i = \frac{d\mathbf{L}}{dt}$$

In this equation, the sum of the torques may include internal torques as well as those due to forces external to the system. However, it is found experimentally that, if there is no net external torque, the angular momentum of a system is conserved. Thus, the sum of the internal torques must be zero. Therefore,

$$\boldsymbol{\tau}_{net,ext} = \frac{d\mathbf{L}}{dt} \qquad 8\text{-}67$$

Newton's second law for rotation

The net external torque acting on a system equals the rate of change of the angular momentum of a system.

When the net external torque is zero, the angular momentum of the system is conserved. Equation 8-67 is the rotational analog of $\mathbf{F}_{net,ext} = d\mathbf{p}/dt$ for linear motion.

The law of conservation of angular momentum is an experimental law that is independent of Newton's laws of motion. However, the fact that the internal torques of a system cancel is suggested by Newton's third law. Consider the two particles shown in Figure 8-41. Let $\mathbf{F}_1$ be the force exerted by particle 2 on particle 1 and $\mathbf{F}_2$ be that exerted by particle 1 on particle 2. By Newton's third law, $\mathbf{F}_2 = -\mathbf{F}_1$. The sum of the torques exerted by these forces about the origin O is

$$\boldsymbol{\tau}_1 + \boldsymbol{\tau}_2 = (\mathbf{r}_1 \times \mathbf{F}_1) + (\mathbf{r}_2 \times \mathbf{F}_2) = (\mathbf{r}_1 \times \mathbf{F}_1) + (\mathbf{r}_2 \times -\mathbf{F}_1)$$

$$= (\mathbf{r}_1 - \mathbf{r}_2) \times \mathbf{F}_1$$

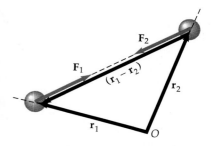

Figure 8-41   The internal forces $\mathbf{F}_1$ and $\mathbf{F}_2$ produce no torque about O if they act along the line joining the particles.

The vector $\mathbf{r}_1 - \mathbf{r}_2$ is along the line joining the two particles. If $\mathbf{F}_1$ acts parallel to the line joining m_1 and m_2, $\mathbf{F}_1$ and $\mathbf{r}_1 - \mathbf{r}_2$ are either parallel or antiparallel, and

$$(\mathbf{r}_1 - \mathbf{r}_2) \times \mathbf{F}_1 = 0$$

If this is true for all the internal forces, the internal torques cancel in pairs.

Questions

11. What is the angle between a particle's linear momentum $\mathbf{p}$ and its angular momentum $\mathbf{L}$?

12. A particle moves along a straight line at constant speed. How does its angular momentum about any point vary over time?

13. A particle moving at constant velocity has zero angular momentum about a particular point. Show that the particle either has passed through that point or will pass through it.

8-8   Motion of a Gyroscope

We will now consider the motion of a gyroscope, in which the axis of rotation changes direction. Such motions are usually very complicated. A large demonstration gyroscope can be made from a bicycle wheel by weighting the rim to increase its moment of inertia. If you have access to such a device (most physics-department stockrooms have one), you should play with it to experience its somewhat surprising properties. For example, if it is set spinning with its axis horizontal and you push horizontally on one end of the axle, the axle tends to move up or down rather than in the direction it is pushed. To understand this behavior, we need to take into account the vector nature of the rotational quantities discussed in the previous section.

Figure 8-42 shows a gyroscope consisting of a bicycle wheel that is free to turn on an axle. The axle is pivoted at a point a distance D from the center of the wheel, but it is free to turn in any direction. We will try to achieve a qualitative understanding of the complicated motion of such a system by using Newton's second law for rotation as expressed in Equation 8-67:

$$\boldsymbol{\tau}_{\text{net,ext}} = \frac{d\mathbf{L}}{dt}$$

All that we really need to remember is that the *change* in the angular momentum of the wheel must be in the direction of the net torque acting on the wheel.

When the axle is held horizontal and is then released, the wheel simply falls if it is not initially spinning. The torque about point O is MgD in the horizontal direction, perpendicular to the axle as shown in Figure 8-42. As the wheel falls, it has an angular momentum in the direction of the torque. This angular momentum is due to motion of the wheel itself; that is, it is associated with the motion of the center of mass of the wheel. (Since the center of mass accelerates downward, the upward force F exerted by the support at O is evidently less than Mg.)

We will now consider what happens when the wheel is initially set spinning, and the axle is held horizontal and is then released. The wheel now has a large angular momentum along its axle (to the right in the figure) due to its spin. It is quite easy in practice to make this spin angular momentum very large so that, to a first approximation, we can neglect the contribution due to the motion of the center of mass of the wheel. If the wheel were to fall as before, the axle would point downward, resulting in a large compo-

Figure 8-42 A bicycle-wheel gyroscope. The force of gravity $M\mathbf{g}$ produces a torque about the pivot that is perpendicular to the axle. This torque causes a change in the angular momentum in the direction of the torque. If the wheel is not spinning and is released, it just falls down. In this case, the change in angular momentum produces angular momentum due to the motion of the center of mass of the wheel.

nent of angular momentum in the downward direction. But there is no torque in the downward direction. The torque is horizontal. In Figure 8-43, a large component of the angular momentum is shown along the axis of the wheel. The change in the angular momentum $d\mathbf{L}$ in the direction of the torque is also indicated. What actually happens when the axle of the spinning wheel is held horizontal and is then released is that the axle moves in a horizontal plane in the direction into the paper. The wheel must move this way when it is spinning so that the *change* in the angular momentum is in the direction of the net torque. This motion, which is often quite surprising when first encountered, is called **precession.**

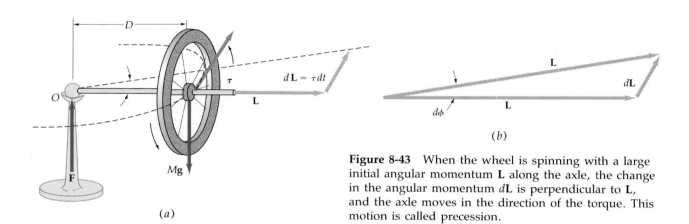

(a)

(b)

Figure 8-43   When the wheel is spinning with a large initial angular momentum $\mathbf{L}$ along the axle, the change in the angular momentum $d\mathbf{L}$ is perpendicular to $\mathbf{L}$, and the axle moves in the direction of the torque. This motion is called precession.

We can calculate the angular velocity of precession as follows. In a small time interval dt, the change in the angular momentum has a magnitude dL:

$$dL = \tau\, dt = MgD\, dt$$

From Figure 8-43, the angle $d\phi$ through which the axle moves is

$$d\phi = \frac{dL}{L} = \frac{MgD\, dt}{L}$$

The angular velocity of the precession is thus

$$\omega_p = \frac{d\phi}{dt} = \frac{MgD}{L} \qquad\qquad 8\text{-}68$$

(*Left*) A rotating bicycle wheel supported by a string is used by Professor Carpenter to demonstrate the precession of a gyroscope. (*Right*) A simple demonstration gyroscope.

If the angular momentum due to the spin of the wheel is very large, the precession will be very slow; that is, the angular velocity of the precession associated with the axle of the wheel moving in a horizontal plane will be very small.

If you perform this experiment, you will note a small up and down oscillation of the axle. This motion is called **nutation.** It can be eliminated by giving the axle a slight horizontal push when it is released. We can understand nutation qualitatively from a consideration of the directions of the angular momentum of the wheel and the torques exerted on it. Even when the precession is slow, there is a small amount of angular momentum associated with the motion of the center of mass of the wheel. The direction of this small angular momentum is upward. If the axle is simply released with no push, there is no component of torque in the upward or downward direction. Thus, unless the wheel is given an initial push (and thus an initial torque in the upward direction), the axle of the wheel will dip down slightly to provide a small component of angular momentum in the downward direction associated with the spin to cancel out the small upward angular momentum associated with the motion of the wheel itself. A detailed analysis of the motion of the gyroscope shows that if the wheel is not given an initial push, the axle will initially dip down, then overshoot, and then oscillate up and down.

8-9   Static and Dynamic Imbalance

When the axis of rotation is not a symmetry axis, the angular momentum is not necessarily parallel to the angular velocity. Figure 8-44 shows a simple dumbbell consisting of equal masses connected by a light rod that is rotating about an axis which passes through the center of mass but is not perpendicular to the rod. Each mass is moving in a circle of radius $r \sin \theta$ with a speed $(r \sin \theta)\omega$, where θ is the angle between the rod and the axis of rotation and ω is the angular velocity. The angular momentum of mass 1 is $\mathbf{L}_1 = \mathbf{r}_1 \times m\mathbf{v}_1$. Its direction is in the plane of the paper, perpendicular to the rod, and its magnitude is $L_1 = (mr^2 \sin \theta)\omega$. The angular momentum of mass 2 is $\mathbf{L}_2 = \mathbf{r}_2 \times m\mathbf{v}_2$. Since $\mathbf{r}_2 = -\mathbf{r}_1$ and $\mathbf{v}_2 = -\mathbf{v}_1$, $\mathbf{L}_2 = \mathbf{L}_1$. The total angular momentum of the system is thus $(2mr^2 \sin \theta)\omega$ in the direction perpendicular to the rod as shown in the figure. It is not parallel to $\boldsymbol{\omega}$, which is along the axis of rotation. As the dumbbell rotates, the angular-momentum vector rotates with its tip moving in a circle as shown in the figure. If the angular velocity is constant, the angular momentum has a constant magnitude, but its direction changes. Thus, even if the angular velocity is constant, the angular-momentum vector changes, so there must be a net torque on the system. This torque is exerted by the bearings supporting the system. The rotating system exerts an equal and opposite torque on the support bearings, causing wear. Such a system is said to be **dynamically imbalanced.** A calculation of the rate of change of the angular momentum shows that it is proportional to ω^2. Thus, the greater the angular velocity, the greater the wear on the bearings. Note that there is no static imbalance. If we pivot the dumbbell at its center of mass, it will be in static equilibrium at any orientation. If we have a system of unknown mass distribution (such as an automobile wheel), we cannot detect or correct dynamic imbalance by static-imbalance methods.

We can correct the dynamic imbalance of our two-mass system by making the angle θ equal to 90°, as in Figure 8-45, or by adding masses, as in Figure 8-46. Then $\mathbf{L}$ and $\boldsymbol{\omega}$ will be parallel, and no torque will be required to maintain a constant angular velocity.

Figure 8-47 shows a wheel rotating about an axis that is off center but parallel to the symmetry axis through the center of mass and a distance h

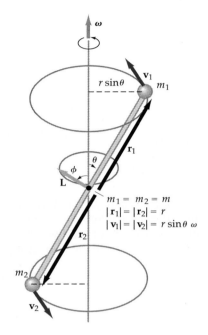

Figure 8-44   Particles of equal mass rotating about an axis through their center of mass that is not a symmetry axis. The angular momentum $\mathbf{L}$ is perpendicular to the line joining the particles and makes an angle ϕ with the angular velocity $\boldsymbol{\omega}$. Since $\mathbf{L}$ changes direction as the system rotates, a torque must be exerted by the bearings.

In the figure: $r \sin\theta$, $\mathbf{v}_1$, m_1, $\mathbf{r}_1$, θ, ϕ, $\mathbf{L}$, $\mathbf{r}_2$, m_2, $\mathbf{v}_2$

$m_1 = m_2 = m$
$|\mathbf{r}_1| = |\mathbf{r}_2| = r$
$|\mathbf{v}_1| = |\mathbf{v}_2| = r \sin\theta\, \omega$

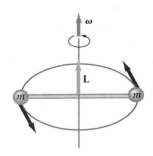

Figure 8-45   This arrangement of two rotating particles of equal mass is dynamically balanced. The axis of rotation is now a symmetry axis and $\mathbf{L}$ is parallel to $\boldsymbol{\omega}$. As the system rotates, $\mathbf{L}$ does not change.

---

from it. The angular momentum about a point on the axis of rotation in the plane of the wheel is the sum of the angular momentum about the center of mass and the angular momentum associated with the motion of the center of mass, both of which are parallel to the axis of rotation, which is the direction of $\boldsymbol{\omega}$. If the wheel rotates with constant angular velocity, the angular momentum is constant and no torque is required. However, since the center of mass moves in a circle of radius h, there must be a net force on the wheel of magnitude $MV_{cm}^2/h = Mh\omega^2$ directed toward the axis of rotation from the center of mass. This force is exerted by the bearings. Again, the equal but opposite force exerted on the bearings causes wear on them, particularly if ω is large. This **static imbalance** can be detected and corrected by pivoting the wheel on a horizontal axle and balancing it so that the axis of rotation passes through the center of mass.

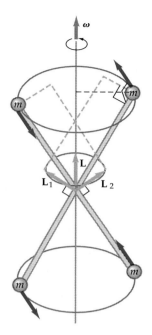

Figure 8-46 This arrangement is also dynamically balanced. Here, too, **L** is parallel to $\boldsymbol{\omega}$ and does not change as the system rotates.

Summary

1. The angular velocity of an object is the rate of change of the angle between a line fixed on the object and one fixed in space:

$$\omega = \frac{d\theta}{dt}$$

The angular acceleration is the rate of change of the angular velocity:

$$\alpha = \frac{d\omega}{dt} = \frac{d^2\theta}{dt^2}$$

2. The linear speed of a particle a distance r from the axis of rotation is related to the angular velocity of the particle by

$$v = r\omega$$

Similarly, the tangential acceleration of such a particle is related to the angular acceleration of the particle by

$$a_t = r\alpha$$

The centripetal acceleration of such a particle is given by

$$a_c = \frac{v^2}{r} = r\omega^2$$

Figure 8-47 This system is dynamically balanced because **L** is parallel to $\boldsymbol{\omega}$, but it is statically imbalanced. As the system rotates, the bearings exert no torque, but they must exert a force to accelerate the center of mass.

3. The equations for rotation of a rigid body with constant angular acceleration are

$$\omega = \omega_0 + \alpha t$$

$$\theta = \theta_0 + \omega_0 t + \tfrac{1}{2}\alpha t^2$$

and

$$\omega^2 = \omega_0^2 + 2\alpha(\theta - \theta_0)$$

These equations are analogous to those for motion with constant acceleration in one dimension.

4. The torque exerted by a force on a particle is defined as the product of the force and the lever arm. Newton's second law for the rotation of a rigid body about a fixed axis is

$$\tau_{net} = I\alpha$$

where I is the moment of inertia of a system of particles, which is defined as

$$I = \sum_i m_i r_i^2$$

The moment of inertia for rotational motion is analogous to the mass in translational motion.

5. The kinetic energy of a rotating body is given by

$$K = \tfrac{1}{2} I \omega^2$$

and the power input of a torque is given by

$$P = \tau \omega$$

6. The angular momentum of a system of particles with a common angular speed ω is

$$L = I \omega$$

7. The generalized form of Newton's second law for rotational motion is

$$\tau_{\text{net}} = \frac{dL}{dt} = \frac{d(I\omega)}{dt}$$

8. In an isolated system, the net torque is zero and the angular momentum of the system is conserved.

9. When a ball or cylinder of radius R rolls without slipping, the velocity of the center of mass of the body is related to the angular velocity by the rolling condition:

$$V_{\text{cm}} = R\omega$$

The acceleration is similarly related to the angular acceleration by

$$A_{\text{cm}} = R\alpha$$

10. The vector product or cross product of two vectors is defined by

$$\mathbf{A} \times \mathbf{B} = (AB \sin \phi)\hat{\mathbf{n}}$$

where ϕ is the angle between the vectors and $\hat{\mathbf{n}}$ is a unit vector that is perpendicular to the plane of $\mathbf{A}$ and $\mathbf{B}$ in the sense given by the right-hand rule as $\mathbf{A}$ is rotated into $\mathbf{B}$.

11. When the axis of rotation changes direction in space, the vector nature of rotational quantities is important. The direction of the angular velocity $\boldsymbol{\omega}$ is along the axis of rotation in the sense given by the right-hand rule. The torque exerted by a force $\mathbf{F}$ about some point P is

$$\boldsymbol{\tau} = \mathbf{r} \times \mathbf{F}$$

If a particle has a linear momentum $\mathbf{p} = m\mathbf{v}$, its angular momentum relative to some point P is

$$\mathbf{L} = \mathbf{r} \times \mathbf{p}$$

The angular momentum of a system of particles rotating about a symmetry axis through the center of mass is

$$\mathbf{L} = I\boldsymbol{\omega}$$

12. The complicated motion of a gyroscope can be understood in terms of the directional properties of torque and angular momentum. When a system such as the bicycle-wheel gyroscope has a large initial angular momentum and the net torque acting is perpendicular to the angular momentum, the system moves such that the change in the angular momentum is in the direction of the torque. This motion is called precession.

13. When a body rotates about an axis that is not a symmetry axis, the angular momentum may not be parallel to the angular velocity, in which case the body is dynamically imbalanced and the direction of the angular momentum changes even when the body rotates at constant angular velocity.

Suggestions for Further Reading

Frohlich, Cliff: "The Physics of Somersaulting and Twisting," *Scientific American*, March 1980, p. 154.

Though they may appear to, the midair rotations of divers and cats do not violate the law of conservation of angular momentum. With the aid of excellent illustrations, this article explains why.

Jones, David E. H.: "The Stability of the Bicycle," *Physics Today*, vol. 23, no. 4, 1970.

An amusing account of the author's attempt to test experimentally the idea, among others, that the gyroscopic effect of the spinning front wheel of a bicycle is responsible for its stability.

Laws, Kenneth: "The Physics of Dance," *Physics Today*, February 1985, p. 25.

Walker, Jearl: "The Amateur Scientist: The Essence of Ballet Maneuvers Is Physics," *Scientific American*, June 1982, p. 146.

Both articles present analyses of various dance movements from a physical point of view, showing that dance may be appreciated on more than just a purely esthetic level. Various jumps and turns are discussed, as well as the dancer's problem of maintaining balance.

Walker, Jearl: "The Amateur Scientist: The Physics of Spinning Tops, Including Some Far-Out Ones," *Scientific American*, March 1981, p. 182.

Precession, nutation, and other, more unusual aspects of top mechanics are explained in a nonmathematical fashion.

Walker, Jearl: "The Amateur Scientist: Thinking About Physics While Scared to Death (on a Falling Roller Coaster)," *Scientific American*, October 1983, p. 162.

Momentum and its contribution to a thrilling roller-coaster ride, the interconversion of potential and kinetic energy, and the rotational dynamics of several amusement-park rides are discussed.

Walker, Jearl: "The Amateur Scientist: Why Sidespin Helps the Bowler—and How to Keep Scoring Strikes," *Scientific American*, March 1988, p. 110.

A look at the theory behind good bowling technique.

Review

A. Objectives: After studying this chapter, you should:

1. Be able to state the definitions of angular velocity, angular acceleration, and moment of inertia.

2. Be able to state the equations for rotation with constant angular acceleration and use these equations to work problems.

3. Be able to state Newton's second law for rotational motion and apply it to problems.

4. Be able to calculate the moment of inertia for discrete or continuous mass distributions.

5. Be able to apply the conservation of angular momentum to problems in which the moment of inertia changes.

6. Be able to state the condition for rolling without slipping and apply it to problems involving translation and rotation.

7. Be able to describe qualitatively the motion of a gyroscope.

B. Define, explain, or otherwise identify:

Rigid body
Angular displacement
Angular velocity
Angular speed
Angular acceleration
Lever arm
Torque
Moment of inertia
Parallel-axis theorem
Perpendicular-axis theorem
Angular momentum
Law of conservation of angular momentum
Rolling condition
Right-hand rule
Vector product
Cross product
Symmetry axis
Precession
Nutation
Dynamic imbalance
Static imbalance

C. True or false: If the statement is true, explain why it is true. If it is false, give a counterexample.

1. Angular velocity and linear velocity have the same dimensions.

2. All parts of a rotating wheel have the same angular velocity.

3. All parts of a rotating wheel have the same angular acceleration.

4. The moment of inertia of a body depends on the location of the axis of rotation.

5. The moment of inertia of a body depends on the angular velocity of the body.

6. If the net torque on a body is zero, the angular momentum must be zero.

7. If the net torque on a rotating system is zero, the angular velocity of the system cannot change.

Problems

Level I

8-1 Angular Velocity and Angular Acceleration

1. A particle moves in a circle with a radius of 100 m with a constant speed of 20 m/s. (*a*) What is its angular velocity in radians per second about the center of the circle? (*b*) How many revolutions does it make in 30 s?

2. A wheel starts from rest with a constant angular acceleration of 2 rad/s^2. (*a*) What is its angular velocity after 5 s? (*b*) Through what angle has the wheel turned after 5 s? (*c*) How many revolutions has it made in 5 s? (*d*) After 5 s, what is the speed and acceleration of a point 0.3 m from the axis of rotation?

3. A turntable rotating at $33\frac{1}{3}$ rev/min is shut off. It brakes with constant angular acceleration and comes to rest in 2 min. (*a*) Find the angular acceleration. (*b*) What is the average angular velocity of the turntable? (*c*) How many revolutions does it make before stopping?

4. A disk of radius 10 cm begins rotating about its axis from rest with a constant angular acceleration of 10 rad/s^2. At $t = 5$ s, what are (*a*) the angular velocity of the disk and (*b*) the tangential acceleration a_t and the centripetal acceleration a_c of a point on the edge of the disk?

5. A spot on a record is on the radial line $\theta = 0°$ relative to a point fixed in space. At $t = \frac{1}{4}$ s after the turntable is turned on, the spot has advanced to $\theta = 10°$. Assuming constant angular acceleration, how long will it be before the record is rotating at 33.3 rev/min?

6. A Ferris wheel of radius 12 m rotates once in 25 s. (*a*) What is its angular velocity in radians per second? (*b*) What is the linear speed of a passenger? (*c*) What is the centripetal acceleration of a passenger?

7. A cyclist starts from rest and pedals such that the wheels of his bike have a constant angular acceleration. After 10 s, the wheels have made 5 rev. (*a*) What is the angular acceleration of the wheels? (*b*) What is the angular velocity of the wheels after 10 s? (*c*) If the radius of the wheel is 36 cm, and the wheel rolls without slipping, how far has the cyclist traveled in 10 s?

8-2 Torque and Moment of Inertia

8. A 0.5-kg particle falls freely under the influence of gravity. (*a*) It is at $y = 10$ m and $x = 2$ m at time t_1. What is the torque about the origin exerted by gravity on the particle at this time? (*b*) At some later time, the particle is at $y = 0$ and $x = 2$ m. What is the torque about the origin at this time?

9. A disk-shaped grindstone of mass 2 kg and radius 7 cm is spinning at 700 rev/min. After the power is shut off, a woman continues to sharpen her axe by holding it against the grindstone for 10 s until the grindstone stops rotating. (*a*) Find the angular acceleration of the grindstone, assuming it to be constant. (*b*) What is the torque exerted by the axe on the grindstone? (Assume there are no other frictional torques.)

10. To start a playground merry-go-round rotating, a rope is wrapped around it and pulled. A force of 200 N is exerted on the rope for 10 s. During this time, the merry-go-round, which has a radius of 2 m, makes one complete rotation. (*a*) Find the angular acceleration of the merry-go-round, assuming it to be constant. (*b*) What torque is exerted by the rope on the merry-go-round? (*c*) What is the moment of inertia of the merry-go-round?

11. A rope is wrapped around a 3-kg cylinder of radius 10 cm that is free to turn about its axis. The rope is pulled with a force of 15 N. The cylinder is initially at rest at $t = 0$. (*a*) Find the torque exerted by the rope, and the angular acceleration of the cylinder. (*b*) Find the angular velocity of the cylinder at time $t = 4$ s.

8-3 Rotational Kinetic Energy

12. The particles in Figure 8-48 are connected by a very light rod whose moment of inertia may be neglected. They rotate about the y axis with an angular velocity of 2 rad/s. (*a*) Find the speed of each particle, and use it to calculate the kinetic energy of this system directly from $\Sigma\frac{1}{2}m_iv_i^2$. (*b*) Find the moment of inertia about the y axis, and calculate the kinetic energy from $E_K = \frac{1}{2}I\omega^2$.

Figure 8-48   Problem 12.

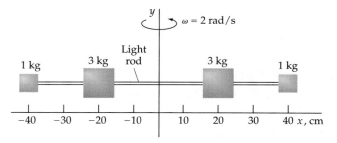

13. The four particles at the corners of the square in Figure 8-49 are connected by massless rods. The masses of the particles are $m_1 = m_3 = 3$ kg and $m_2 = m_4 = 4$ kg. The length of a side of the square is $L = 2$ m. (*a*) Find the moment of inertia of the system about an axis that is perpendicular to the plane of the particles and passes through m_4. (*b*) How much work is required to produce a rotation of 2 rad/s about this axis?

Figure 8-49 Problems 13, 17, and 19.

14. Four 2-kg particles are located at the corners of a rectangle of sides 3 and 2 m (Figure 8-50). (*a*) Find the moment of inertia of this system about an axis that is perpendicular to the plane of the masses and passes through one of the masses. (*b*) The system is set rotating about this axis with a kinetic energy of 184 J. Find the number of revolutions the system makes per minute.

Figure 8-50 Problems 14 and 20.

15. A solid ball of mass 1.2 kg and diameter 16 cm is rotating about its diameter at 90 rev/min. (*a*) What is the kinetic energy of rotation? (*b*) If an additional 2 J of energy are supplied to the rotational energy, what is the new angular speed of the ball?

16. An engine develops 500 N·m of torque at 3500 rev/min. Find the power developed by the engine.

8-4 Calculating the Moment of Inertia

17. Use the parallel-axis theorem and your results for Problem 13 to find the moment of inertia of the four-particle system in Figure 8-49 about an axis that is perpendicular to the plane of the masses and passes through the center of mass of the system. Check your result by direct computation.

18. A tennis ball has a mass of 57 g and a diameter of 7 cm. Find the moment of inertia about its diameter. Assume the ball is a thin spherical shell.

19. (*a*) Find the moment of inertia I_x for the four-particle system of Figure 8-49 about the x axis, which passes through m_3 and m_4. (*b*) Find I_y for the system about the y axis, which passes through m_1 and m_4. (*c*) Use Equation 8-30 to calculate the moment of inertia I_z about the z axis, which passes through m_4 and is perpendicular to the plane of the figure.

20. (*a*) Use the parallel-axis theorem to find the moment of inertia about an axis that is parallel to the z axis and passes through the center of mass of the system in Figure 8-50. (*b*) Let x' and y' be axes in the plane of the figure that pass through the center of mass and are parallel to the sides of the rectangle. Compute $I_{x'}$ and $I_{y'}$, and use your results and Equation 8-30 to check your result for part (*a*).

21. Use Equation 8-30 to find the moment of inertia of a disk of radius R and mass M about an axis in the plane of the disk that passes through its center (Figure 8-51).

22. Use the parallel-axis theorem to find the moment of inertia of a solid sphere of mass M and radius R about an axis that is tangent to the sphere (Figure 8-52).

Figure 8-51 Problem 21.   **Figure 8-52** Problem 22.

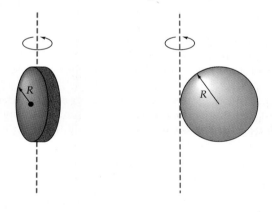

8-5 Angular Momentum

23. A 3-kg particle moves at a constant speed of 4 m/s around a circle of radius 5 m. (*a*) What is its angular momentum about the center of the circle? (*b*) What is its moment of inertia about an axis through the center of the circle and perpendicular to the plane of the motion? (*c*) What is the angular velocity of the particle?

24. A 3-kg particle moves at constant speed of 4 m/s along a straight line. (*a*) What is its angular momentum about a point 5 m from the line? (*b*) Describe qualitatively how its angular velocity about that point varies with time.

25. A particle travels in a circular path. (*a*) If its linear momentum p is doubled, how is its angular momentum affected? (*b*) If the radius of the circle is doubled but the speed is unchanged, how is the angular momentum of the particle affected?

26. A 2-kg particle moves in a circle of radius 3 m. Its angular momentum relative to the center of the circle depends on time according to $L = (4 \text{ N·m})t$. (a) Find the torque acting on the particle. (b) Find the angular velocity of the particle as a function of time.

27. A planet moves in an elliptical orbit about the sun with the sun at one focus of the ellipse (Figure 8-53). (a) What is the torque produced by the gravitational force of attraction of the sun for the planet? (b) At position A, the planet is a distance r_1 from the sun and is moving with a speed v_1 perpendicular to the line from the sun to the planet. At position B, it is at distance r_2 and is moving with speed v_2, again perpendicular to the line from the sun to the planet. What is the ratio of v_1 to v_2 in terms of r_1 and r_2?

Figure 8-53   Problem 27.

8-6 Rolling Objects

28. A homogeneous cylinder of radius 15 cm and mass 50 kg is rolling without slipping along a horizontal floor at 6 m/s. How much work is needed to stop the cylinder?

29. Find the percentages of the total kinetic energy associated with rotation and with translation for an object that is rolling without slipping if the object is (a) a uniform sphere, (b) a uniform cylinder, and (c) a hoop.

30. A hoop of radius 0.50 m and mass 0.8 kg is rolling without slipping at a speed of 20 m/s toward an incline of slope 30°. How far up the incline will the hoop roll, assuming it rolls without slipping?

31. A ball rolls without slipping down an incline of angle θ. Find, in terms of the coefficient of friction μ_s, (a) the acceleration of the ball, (b) the force of friction, and (c) the maximum angle of the incline for which the ball will roll without slipping.

32. A ball rolls without slipping along a horizontal plane. Show that the frictional force acting on the ball must be zero. *Hint:* Consider a possible direction for the action of the frictional force and what effects such a force would have on the velocity of the center of mass and on the angular velocity.

8-7 The Vector Nature of Rotation and the Cross Product

33. A force of magnitude F is applied horizontally in the negative x direction to the rim of a disk of radius R as shown in Figure 8-54. Write **F** and **r** in terms of the unit vectors **i**, **j**, and **k**, and compute the torque produced by the force about the origin at the center of the disk.

34. Compute the torque about the origin for the force **F** = $-mg$**j** acting on a particle at **r** = x**i** + y**j**, and show that this torque is independent of the y coordinate.

35. Find **A** × **B** for (a) **A** = 6**i** and **B** = 6**i** + 6**j**, (b) **A** = 6**i** and **B** = 6**i** + 6**k**, and (c) **A** = 3**i** + 3**j** and **B** = −2**i** + 2**j**.

36. Under what conditions is the magnitude of **A** × **B** equal to **A** · **B**?

37. A particle of mass 3 kg moves with velocity **v** = 2 m/s **i** along the line $z = 0$, $y = 4.3$ m. (a) Find the angular momentum **L** relative to the origin when the particle is at $x = 12$ m, $y = 4.3$ m. (b) A force **F** = −3 N **i** is applied to the particle. Find the torque relative to the origin due to this force.

8-8 Motion of a Gyroscope

38. The angular momentum of the propeller of a small airplane points forward. (a) As the plane takes off, the nose lifts up and the airplane tends to veer to one side. To which side does it veer and why? (b) If the plane is flying horizontally and suddenly turns to the right, does the nose of the plane tend to move up or down? Why?

39. A bicycle wheel of radius 30 cm is mounted at the middle of an axle 60 cm long. The tire and rim weigh 36 N. The wheel is spun at 10 rev/s, and the axle is then placed in a horizontal position with one end resting on a pivot. (a) What is the angular momentum due to the spinning of the wheel? (Treat the wheel as a hoop.) (b) What is the angular velocity of precession? (c) How long does it take for the axle to swing through 360° around the pivot? (d) What is the angular momentum associated with the motion of the center of mass, that is, due to the precession? In what direction is this angular momentum?

40. A uniform disk of mass 2 kg and radius 6 cm is mounted in the center of a 10-cm axle and spun at 900 rev/min. The axle is then placed in a horizontal position with one end resting on a pivot. The other end is given an initial horizontal velocity such that the precession is smooth with no nutation. (a) Find the angular velocity of precession. (b) What is the speed of the center of mass during the precession? (c) What are the magnitude and direction of the acceleration of the center of mass? (d) What are the vertical and horizontal components of the force exerted by the pivot?

8-9 Static and Dynamic Imbalance

41. A uniform disk of radius 30 cm, thickness 3 cm, and mass 5 kg rotates at $\omega = 10$ rad/s about an axis parallel to the symmetry axis but 0.5 cm from that axis. (a) Find the net force on the bearings due to this imbalance. (b) Where should a 100-g mass be placed on the disk to correct this problem?

Figure 8-54   Problem 33.

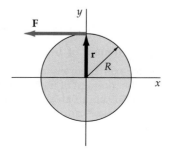

Level II

42. A flywheel is a uniform disk of mass 100 kg and radius of 1.2 m. It rotates with an angular velocity of 1200 rev/min. (*a*) A constant tangential force is applied at a radial distance of 0.5 m. What work must this force do to stop the wheel? (*b*) If the wheel is brought to rest in 2 min, what torque does the force produce? What is the magnitude of the force? (*c*) How many revolutions does the wheel make in these 2 min?

43. A uniform disk of radius 0.12 m and mass 5 kg is pivoted such that it rotates freely about its axis. A string wrapped around the disk is pulled with a force of 20 N (Figure 8-55). (*a*) What is the torque exerted on the disk? (*b*) What is the angular acceleration of the disk? (*c*) If the

Figure 8-55   Problem 43.

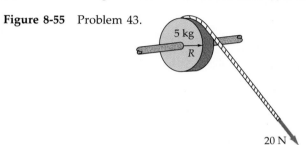

disk starts from rest, what is its angular velocity after 3 s? (*d*) What is its kinetic energy after 3 s? (*e*) What is its angular momentum after 3 s? (*f*) Find the total angle θ the disk turns through in 3 s, and (*g*) show that the work done by the torque $\tau\theta$ equals the kinetic energy.

44. A 2000-kg block is lifted by a steel cable that passes over a pulley to a motor-driven winch (Figure 8-56). The radius of the winch drum is 30 cm, and the moment of inertia of the pulley is negligible. (*a*) What force must be exerted by the cable to lift the block at a constant velocity of 8 cm/s? (*b*) What torque does the cable exert on the winch drum? (*c*) What is the angular velocity of the winch drum? (*d*) What power must be developed by the motor to drive the winch drum?

Figure 8-56   Problem 44.

r = 30 cm

(2000 kg)*g*

45. A homogeneous cylinder of mass 100 kg and radius 0.3 m is mounted such that it turns without friction on its fixed symmetry axis. It is rotated by a drive belt that wraps around its perimeter and exerts a constant torque. At time $t = 0$, its angular velocity is zero. At time $t = 30$ s, its angular velocity is 600 rev/min. (*a*) What is its angular momentum at $t = 30$ s? (*b*) At what rate is the angular

momentum increasing? (*c*) What is the torque acting on the cylinder? (*d*) What is the magnitude of the force acting on the rim of the cylinder?

46. A particle is traveling with a constant velocity **v** along a line that is a distance *b* from the origin *O*. Let *dA* be the area swept out by the position vector from *O* to the particle in time *dt* (Figure 8-57). Show that *dA/dt* is constant in time and equal to $\frac{1}{2}L/m$, where *L* is the angular momentum of the particle about the origin.

Figure 8-57   Problem 46.

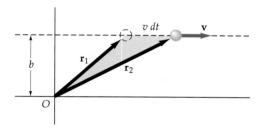

47. A 15-g coin of diameter 1.5 cm is spinning at 10 rev/s about a vertical diameter at a fixed point on a tabletop. (*a*) What is the angular momentum of the coin about its center of mass? (*b*) What is its angular momentum about a point on the table 10 cm from the coin? If the coin spins about a vertical diameter at 10 rev/s, but it also travels in a straight line across the tabletop at 5 cm/s, (*c*) what is the angular momentum of the coin about a point on the line of motion? (*d*) What is the angular momentum of the coin about a point 10 cm from the line of motion? (There are two answers to this question. Explain why and give both.)

48. (*a*) Assuming that the incline in Figure 8-58 is frictionless and that the string passes through the center of mass of m_2, find the net torque acting on the system (the two masses and the pulley) about the center of the pulley. (*b*) Write an expression for the total angular momentum of the system about the center of the pulley when the masses are moving with a speed *v*. Assume the pulley has a moment of inertia *I* and a radius *r*. (*c*) Find the acceleration of the masses from your results for parts (*a*) and (*b*) by setting the net torque equal to the rate of change of the angular momentum of the system.

Figure 8-58   Problem 48.

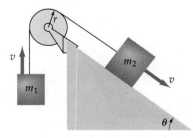

49. Calculate the kinetic energy of rotation of the earth, and compare it with the kinetic energy of motion of the earth's center of mass. Assume the earth to be a homogeneous sphere of mass 6.0×10^{24} kg and radius 6.4×10^6 m. The radius of the earth's orbit is 1.5×10^{11} m.

50. A 2-kg mass attached to a string of length 1.5 m moves in a horizontal circle as a conical pendulum (Figure 8-59). The string makes an angle $\theta = 30°$ with the vertical. (a) Show that the angular momentum of the mass about the point of support P has a horizontal component toward the center of the circle as well as a vertical component, and find these components. (b) Find the magnitude of $d\mathbf{L}/dt$, and show that it equals the magnitude of the torque exerted by gravity about the point of support.

Figure 8-59 Problem 50.

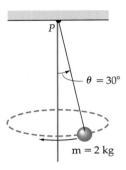

51. A wheel mounted on an axis that is not frictionless is initially at rest. A constant external torque of 50 N·m is applied to the wheel for 20 s. At the end of the 20 s, the wheel has an angular velocity of 600 rev/min. The external torque is then removed, and the wheel comes to rest after 120 s more. (a) What is the moment of inertia of the wheel? (b) What is the frictional torque, which is assumed to be constant?

52. A 1-Mg car is being unloaded by a winch. At the moment shown in Figure 8-60, the gearbox shaft of the winch breaks, and the car falls from rest. During the car's fall there is no slipping between the rope, the pulley, and the winch drum. The moment of inertia of the winch drum is 320 kg·m² and that of the pulley is 4 kg·m². The radius of the winch drum is 0.80 m and that of the pulley is 0.30 m. Find the speed of the car as it hits the water.

Figure 8-60 Problem 52.

53. A typical car engine delivers about 2 MJ of mechanical energy per kilometer on the average. A car is designed to use the energy stored in a flywheel in a vacuum container. If the mass of the flywheel is not to exceed 100 kg and its angular velocity is not to exceed 400 rev/s, find the smallest radius of the flywheel for which the car can travel 300 km without the flywheel having to be recharged. (Assume the flywheel to be a uniform cylinder.)

54. The system in Figure 8-61 is released from rest. The 30-kg block is 2 m above the floor. The pulley is a uniform disk with a radius of 10 cm and mass of 5 kg. Find (a) the speed of the 30-kg block just before it hits the floor, (b) the angular speed of the pulley at that time, (c) the tensions in the strings, and (d) the time it takes for the 30-kg block to reach the floor. Assume the string does not slip on the pulley.

Figure 8-61 Problem 54.

55. A meterstick is pivoted at one end such that it can swing freely in a vertical plane. It is released from rest in a horizontal position. (a) What is the angular velocity of the stick when it is vertical? (b) What force is exerted by the pivot when the stick is vertical? (Answer in terms of the mass m of the stick.)

56. A uniform disk of mass M and radius R is pivoted such that it can rotate freely about a horizontal axis through its center and perpendicular to the plane of the disk. A small particle of mass m is attached to the rim of the disk at the top, directly above the pivot. The system is given a gentle start, and the disk begins to rotate. (a) What is the angular velocity of the disk when the particle is at its lowest point? (b) At this point, what force must be exerted on the particle by the disk to keep it on the disk?

57. A uniform, 100-kg cylinder of radius 0.60 m is placed flat on some smooth ice. Two skaters wind ropes around the cylinder in the same sense. The skaters then pull on their ropes as they skate away, exerting constant forces of 40 N and 60 N, respectively, for 5 s (Figure 8-62). Describe the motion of the cylinder. That is, what are its acceleration and velocity and the position of its center of mass as functions of time, and what are its angular acceleration and angular velocity as functions of time?

Figure 8-62   Problem 57.

58. A vertical grinding wheel is a uniform disk of mass 50 kg and radius 40 cm. It has a handle of radius 60 cm of negligible mass. A 20-kg load is attached to the handle when it is in the horizontal position. Find (*a*) the initial angular acceleration of the wheel and (*b*) the maximum angular velocity of the wheel.

59. A uniform sphere of mass M and radius R is free to rotate about a horizontal axis through its center. A string is wrapped around the sphere and is attached to a body of mass m as shown in Figure 8-63. Find (*a*) the acceleration of the body and (*b*) the tension in the string.

Figure 8-63   Problem 59.

60. A uniform rectangular plate has mass m and sides a and b. (*a*) Show by integration that the moment of inertia of the plate about an axis that is perpendicular to the plate and passes through one corner is $\frac{1}{3}m(a^2 + b^2)$. (*b*) What is the moment of inertia about an axis that is perpendicular to the plate and passes through its center of mass?

61. A uniform ball of radius r rolls without slipping along the loop-the-loop track in Figure 8-64. The ball starts at rest at a height h above the bottom of the loop. (*a*) If it is not to leave the track at the top of the loop, what is the least value h can have (in terms of the radius R of the loop)? (*b*) What would h have to be if, instead of rolling, the ball slides without friction?

Figure 8-64   Problem 61.

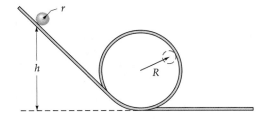

62. A particle moves in a circle of radius r with an angular velocity $\boldsymbol{\omega}$. (*a*) Show that its velocity is $\mathbf{v} = \boldsymbol{\omega} \times \mathbf{r}$. (*b*) Show that its centripetal acceleration is $\mathbf{a}_c = \boldsymbol{\omega} \times \mathbf{v} = \boldsymbol{\omega} \times (\boldsymbol{\omega} \times \mathbf{r})$.

63. A uniform cylinder of mass M and radius R has a string wrapped around it. The string is held fixed, and the cylinder falls vertically as shown in Figure 8-65. (*a*) Show that the acceleration of the cylinder is downward with a magnitude $a = 2g/3$. (*b*) Find the tension in the string.

Figure 8-65   Problem 63.

64. A car is powered by the energy stored in a single flywheel with an angular momentum $\mathbf{L}$. Discuss the problems that would arise for various orientations of $\mathbf{L}$ and various maneuvers of the car. For example, what would happen if $\mathbf{L}$ points vertically upward and the car travels over a hilltop or through a valley, or what would happen if $\mathbf{L}$ points forward or to one side and the car attempts to turn to the left or right. In each case, consider the direction of the torque exerted on the car by the road.

65. A man stands on a frictionless platform that is rotating with an angular speed of 2.0 rev/s. His arms are outstretched, and he holds a heavy weight in each hand. The moment of inertia of the man, the extended weights, and the platform is 5 kg·m². When the man pulls the weights inward toward his body the moment of inertia decreases to 2 kg·m². (*a*) What is the resulting angular speed of the platform? (*b*) What is the change in kinetic energy of the system? (*c*) Where did this increase in energy come from?

66. A man of mass m stands at the edge of a rotating circular turntable. The turntable has a radius R and a moment of inertia I, and it rotates without friction. The angular velocity about the vertical axis through the center of the turntable is 8 rad/s. The man walks radially inward. What is the final angular velocity of the system? Assume that $mR^2 = 3I$ and that the man has a moment of inertia of $I/10$ when he is standing at the center.

67. A professor stands at the center of a turntable that can rotate without friction. She begins to rotate a heavy ball on the end of a 0.8-m chain about her head. The ball has a mass of 2 kg, and it makes one revolution every 3 s. The professor and platform have a moment of inertia of 0.5 kg·m². (*a*) What is the angular speed of the professor? (*b*) What is the total kinetic energy of the ball, professor, and platform?

68. A 0.1-kg yo-yo consists of two solid disks of radius 10 cm joined together by a massless rod of radius 1 cm and a string wrapped around the rod. One end of the string is held fixed and is under constant tension T as the yo-yo is released. Find the acceleration of the yo-yo and the tension T.

69. A uniform boom 5.0 m long and having a total mass of 150 kg is connected to the ground by a hinge at the bottom and is supported by a horizontal cable as shown in Figure 8-66. (a) What is the tension in the cable? (b) If the cable is cut, what is the angular acceleration of the boom the instant the cable is cut? (c) What is the angular velocity of the boom when it is horizontal?

Figure 8-66   Problem 69.

70. A cue ball of radius r is initially at rest on a horizontal pool table. It is struck by a horizontal cue stick that delivers an impulse of magnitude $P_0 = F_{av} \Delta t$. (We use P_0 for the impulse rather than I_0 to avoid confusion with the moment of inertia I.) The stick strikes the ball at a point h above the ball's point of contact with the table (Figure 8-67). Show that the ball's initial angular velocity ω_0 is related to the initial linear velocity of its center of mass v_0 by $\omega_0 = 5v_0(h - r)/2r^2$.

Figure 8-67   Problem 70.

71. The sun's radius is 6.96×10^8 m and it rotates with a period of 25.3 d. Estimate its new period of rotation if it collapses with no loss of mass to become a neutron star of radius 5 km.

72. A uniform cylinder of mass m_1 and radius R is pivoted on frictionless bearings. A string wrapped around the cylinder connects to a mass m_2, which is on a frictionless incline of angle θ as shown in Figure 8-68. The system is released from rest with m_2 a height h above the bottom of the incline. (a) What is the acceleration of m_2? (b) What is the tension in the string? (c) What is the total energy of the system when m_2 is at height h? (d) What is the total energy when m_2 is at the bottom of the incline and has a speed v? (e) What is the speed v? (f) Evaluate your answers for the extreme cases of $\theta = 0°$, $\theta = 90°$, and $m_1 = 0$.

Figure 8-68   Problem 72.

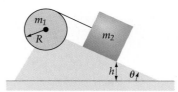

73. An Atwood's machine has two objects of mass $m_1 = 500$ g and $m_2 = 510$ g, connected by a string of negligible mass that passes over a frictionless pulley (Figure 8-69). The pulley is a uniform disk with a mass of 50 g and a radius of 4 cm. The string does not slip on the pulley. (a) Find the acceleration of the objects. (b) What is the tension in the string supporting m_1? In the string supporting m_2? By how much do they differ? (c) What would your answers have been if you had neglected the motion of the pulley?

Figure 8-69   Problem 73.

74. Two objects are attached to ropes that are attached to wheels on a common axle as shown in Figure 8-70. The total moment of inertia of the two wheels is 40 kg·m². The radii of the wheels are $R_1 = 1.2$ m and $R_2 = 0.4$ m. (a) If $m_1 = 24$ kg, find m_2 such that the system is in equilibrium. (b) If 12 kg is gently added to the top of m_1, find the angular acceleration of the wheels and the tensions in the ropes.

Figure 8-70   Problem 74.

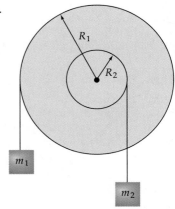

Level III

75. A mass m is attached to a light string that passes through a small hole in a frictionless tabletop. Initially, the mass is sliding with a speed v_0 in a circle of radius r_0 about the hole. A woman under the table now begins to pull the string slowly in. (*a*) Show that when the mass is moving in a circle of radius r, the tension in the string is $T = L_0^2/mr^3$, where $L_0 = mv_0r_0$ is the initial angular momentum. (*b*) The string is pulled in until the radius of the circular orbit is r_f. Using the result of part (*a*), calculate the work done by integrating $\mathbf{T}\cdot d\mathbf{r}$, and show that the work done is $(L_0^2/2m)(r_f^{-2} - r_0^{-2})$. (*c*) What is the velocity v_f when the radius of the circle is r_f? Show that the work done as calculated in part (*b*) equals the change in kinetic energy $\frac{1}{2}mv_f^2 - \frac{1}{2}mv_0^2$. (*d*) If $m = 0.5$ kg, $r_0 = 1$ m, $v_0 = 3$ m/s, and the maximum tension the string can withstand without breaking is 200 N, find the minimum of r_f before the string breaks.

76. (*a*) Assuming the earth to be a homogeneous sphere of radius r and mass m, show that the period T of the earth's rotation about its axis is related to its radius by $T = (4\pi m/5L)r^2$, where L is the angular momentum of the earth due to its rotation. (*b*) Suppose that the radius r changes by a very small amount Δr due to some internal effect such as thermal expansion. Show that the fractional change in the period ΔT is given approximately by $\Delta T/T = 2\,\Delta r/r$. *Hint*: Use the differentials dr and dT to approximate the changes in these quantities. (*c*) By how many kilometers would the earth need to expand for the period to change by $\frac{1}{4}$ d/y so that leap years would no longer be necessary?

77. The polar ice caps contain about 2.3×10^{19} kg of ice. This mass contributes essentially nothing to the moment of inertia of the earth because it is located at the poles, close to the axis of rotation. Estimate the change in the length of the day that would be expected if the polar ice caps were to melt and the water were distributed uniformly over the surface of the earth. (The moment of inertia of a spherical shell of mass m and radius r is $\frac{2}{3}mr^2$.)

78. Show that the moment of inertia of a spherical shell of radius R and mass m is $\frac{2}{3}mR^2$. This can be done by direct integration or, more easily, by finding the increase in the moment of inertia of a solid sphere when its radius changes. To do this, first show that the moment of inertia of a solid sphere of density ρ is $I = \frac{8}{15}\pi\rho R^5$. Then compute the change dI in I for a change dR, and use the fact that the mass of the shell is $m = 4\pi R^2\rho\,dR$.

79. Figure 8-71 shows a pair of uniform spheres, each of mass 500 g and radius 5 cm. They are mounted on a uniform rod that has a length $L = 30$ cm and a mass $m = 60$ g. (*a*) Calculate the moment of inertia of this system about an axis perpendicular to the rod through the center of the rod using the approximation that the two spheres can be treated as point particles that are 20 cm from the axis of rotation and that the mass of the rod is negligible. (*b*) Calculate the moment of inertia exactly and compare your result with your answer for part (*a*).

80. A hollow cylinder has mass m, an outside radius R_2, and an inside radius R_1. Show that its moment of inertia about its symmetry axis is given by $I = \frac{1}{2}m(R_2^2 + R_1^2)$.

81. A bowling ball of mass M and radius R is thrown such that at the instant it touches the floor it is moving horizontally with a speed v_0 and is not rotating. It slides for a time t_1 a distance s_1 before it begins to roll without slipping. (*a*) If μ_k is the coefficient of sliding friction between the ball and the floor, find s_1, t_1, and the final speed v_1 of the ball. (*b*) Evaluate these quantities for $v_0 = 8$ m/s and $\mu_k = 0.4$.

82. A uniform spherical ball is set rotating about a horizontal axis with an angular speed ω_0 and is placed on the floor. If the coefficient of sliding friction between the ball and the floor is μ_k, find the speed of the center of mass of the ball when it begins to roll without slipping.

83. A uniform rod of mass M and length L is pivoted at one end and hangs as shown in Figure 8-72 such that it is free to rotate about its pivot without friction. It is struck by a horizontal force that delivers an impulse $P_0 = F_{av}\,\Delta t$ at a distance x below the pivot as shown. (*a*) Show that the initial speed of the center of mass of the rod is given by $V_0 = 3P_0x/2ML$. (*b*) Find the impulse delivered by the pivot, and show that this impulse is zero if $x = 2L/3$. This point is called the *center of percussion* of the rod.

Figure 8-71 Problem 79.

Figure 8-72 Problem 83.

84. The rod in Problem 83 is struck by a piece of putty of mass $m = M/6$ that is moving horizontally with an initial speed v_0. The putty strikes the end of the rod and sticks to it. (*a*) Find the initial angular velocity of the rod with the putty attached just after the collision. (*b*) What fraction of the initial energy of the putty is lost in this inelastic collision?

85. Two heavy, uniform disks are connected by a short rod. The system is placed on an inclined plane such that the disks hang over the sides. Each disk has a mass of 20 kg and a radius of 30 cm, and the solid connecting rod has a radius of 2 cm and a mass of 1 kg. The plane is inclined at 30° and the rod rolls on the plane without slipping. (a) What is the acceleration of the system down the plane? (b) What is the angular acceleration of the system? (c) If the system starts from rest, what is its kinetic energy of translation when it has rolled 2 m down the incline? (d) What is the kinetic energy of rotation for the 2-m displacement?

86. A heavy, homogeneous cylinder has a mass m and a radius R. It is accelerated by a force T, which is applied through a rope wound around a light drum of radius r that is attached to the cylinder (Figure 8-73). The coefficient of static friction is sufficient for the cylinder to roll without slipping. (a) Find the frictional force. (b) Find the acceleration a of the center of the cylinder. (c) Is it possible to choose r so that a is greater than T/m? How? (d) What is the direction of the frictional force in the circumstances of part (c)?

Figure 8-73   Problem 86.

87. A uniform solid ball has a mass of 20 g and a radius of 5 cm. It rests on a horizontal surface. A sharp force is applied to the ball in the horizontal direction 9 cm above the horizontal surface. The force rises linearly from 0 to a peak value of 40,000 N in 10^{-4} s and then decreases linearly to 0 in 10^{-4} s. (a) What is the velocity of the ball after impact? (b) What is the angular velocity of the ball after impact? (c) What is the velocity of the ball when it begins to roll without sliding? (d) For how long does the ball slide on the surface? Assume $\mu_k = 0.5$.

88. A 0.3-kg billiard ball of radius 3 cm is given a sharp blow by a cue stick. The applied impulse is horizontal and passes through the center of the ball. The initial velocity of the ball is 4 m/s. The coefficient of kinetic friction is 0.6. (a) For how many seconds does the ball slide before it begins to roll without slipping? (b) How far does it slide? (c) What is its velocity once it begins rolling without slipping?

89. A uniform stick of length L and mass M is hinged at one end. It is released from rest at an angle θ_0 with the vertical. Show that when the angle with the vertical is θ, the hinge exerts a force F_r along the stick and a force F_t perpendicular to the stick given by

$$F_r = \tfrac{1}{2}Mg(5 \cos \theta - 3 \cos \theta_0)$$

and

$$F_t = \tfrac{1}{4}Mg \sin \theta$$

90. A wheel of radius R rolls without slipping at a speed V. (a) Show that the x and y coordinates of point P in Figure 8-74 are $r_0 \cos \theta$ and $R + r_0 \sin \theta$, respectively.

(b) Show that the total velocity $\mathbf{v}$ of point P has the components $v_x = V + (r_0 V \sin \theta)/R$ and $v_y = -(r_0 V \cos \theta)/R$. (c) Show that $\mathbf{v}$ and $\mathbf{r}$ are perpendicular to each other by calculating their scalar product. (d) Show that $v = r\omega$, where $\omega = V/R$ is the angular velocity of the wheel. (e) These results demonstrate that, in the case of rolling without slipping, the motion is the same as if the rolling object were instantaneously rotating about the point of contact with an angular speed $\omega = V/R$. Calculate the kinetic energy of the wheel, assuming that it is in pure rotation about point O, and show that the result is the same as that calculated by summing the translational kinetic energy of the center of mass and the rotational kinetic energy of the wheel about its center of mass.

Figure 8-74   Problem 90.

91. A bowling ball of radius R is given an initial velocity v_0 down the lane and a forward spin $\omega_0 = 3v_0/R$. The coefficient of kinetic friction is μ_k. (a) What is the speed of the ball when it begins to roll without slipping? (b) For how long does the ball slide before it begins to roll without slipping? (c) What distance does the ball slide down the lane before it begins rolling without slipping?

92. A billiard ball initially at rest is given a sharp impulse by a cue stick. The impulse is horizontal and is applied at a distance $2R/3$ below the centerline (Figure 8-75). The initial speed of the ball is v_0. (a) What is the initial angular speed ω_0? (b) What is the speed of the ball once it begins to roll without slipping? (c) What is the initial kinetic energy of the ball? (d) What is the frictional work done on the ball as it slides on the table?

Figure 8-75   Problem 92.

93. A marble rolls from rest without sliding from the top of a large sphere, which is held fixed. Find the angle from the top of the sphere of the point where the marble breaks contact with the sphere. The radius of the marble is 1 cm and that of the sphere is 80 cm.

Chapter 9

Static Equilibrium of a Rigid Body

The Oakland–San Francisco Bay Bridge. The part shown here is a suspension bridge supported by steel cables from one anchor post to another.

If an object is stationary and remains stationary, it is said to be in static equilibrium. Being able to determine the forces acting on an object in static equilibrium has many important applications, particularly in engineering. For example, the forces exerted by the cables of a suspension bridge must be known so that the cables can be designed to be strong enough to support the bridge. Similarly, cranes must be designed to not topple over when lifting a weight.

In this chapter, we will first examine the conditions necessary for static equilibrium to exist. We will then define the center of gravity and look at some examples of how to calculate the forces necessary for a body to be in static equilibrium. Finally, we will discuss the stability of equilibrium.

9-1   Conditions for Equilibrium

In Chapter 4, we found that a necessary condition for a particle to remain stationary is that the net force acting on the particle be zero. Under this condition, the particle does not accelerate, and if its velocity is initially zero, the particle remains at rest. Since the acceleration of the center of mass of an object equals the net force acting on the object divided by the total mass of the object, this condition is also necessary for a rigid body to be in equilibrium. However, even when the center of mass of an object is at rest, the object may rotate. Thus, it is also necessary that the net torque about the center of mass be zero. If the center of mass of an object is at rest and there is no rotation about it, there can be no rotation about any point. Thus, for static equilibrium to exist, the net torque acting on an object must be zero about any point. This condition is often useful in solving problems because it allows us to choose any convenient point for computing torques.

Conditions for equilibrium

The two necessary conditions for a rigid body to be in static equilibrium are therefore

1.   The net external force acting on the body must be zero:

$$\mathbf{F}_{net} = 0 \qquad\qquad 9\text{-}1$$

2.   The net external torque about any point must be zero:

$$\tau_{net} = 0 \qquad\qquad 9\text{-}2$$

As we have seen, we can describe the vector nature of rotation about a fixed axis as being positive or negative. We choose either clockwise rotation or counterclockwise rotation to be positive and the other to be negative. An alternative statement of our second condition is then

> For static equilibrium to exist, the sum of the torques that tend to produce clockwise rotation about any point must equal the sum of the torques that tend to produce counterclockwise rotation about that point.

We will refer to a torque that tends to produce clockwise rotation as a clockwise torque and one that tends to produce counterclockwise rotation as a counterclockwise torque.

Two massive, counter-weighted cranes lift a refinery tower to its vertical position.

Example 9-1

A 3-m board of negligible weight rests with its ends on scales as shown in Figure 9-1a. A small, 60-N weight rests on the board 2.5 m from the left end and 0.5 m from the right end. Find the readings on the scales.

Figure 9-1b shows the free-body diagram for the board. The force $\mathbf{F}_L$ is that exerted by the scale at the left end of the board. Since the board exerts an equal but opposite force on the scale, the magnitude of $\mathbf{F}_L$ is the reading on the left scale. Similarly, the magnitude of $\mathbf{F}_R$ is the reading on the right scale. Taking up to be positive, we have from our first condition for equilibrium (the net force must be zero)

$$F_L + F_R - 60 \text{ N} = 0$$

We can get a second relation between F_L and F_R by considering torques. If we consider the point at the weight to be the "pivot" point, we have two torques, a clockwise torque of magnitude $F_L(2.5 \text{ m})$ and a counterclockwise torque of magnitude $F_R(0.5 \text{ m})$. Choosing counterclockwise to be positive, we have

$$0.5F_R - 2.5F_L = 0$$

or

$$F_R = 5F_L$$

Substituting this result into the previous equation, we obtain

$$F_L + 5F_L = 60 \text{ N}$$

or

$$F_L = 10 \text{ N}$$

and

$$F_R = 60 \text{ N} - F_L = 50 \text{ N}$$

The scale readings are thus 10 N for the left scale and 50 N for the right one. The scale on the right supports the greater weight, as expected.

(a)

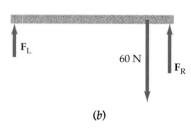

(b)

Figure 9-1 (a) Board of negligible weight on scales for Example 9-1. (b) Free-body diagram for the board.

This nearly completed globe sports complex in Stockholm achieves stability and support from its spherical shape.

Although there is nothing incorrect in this solution, there is an easier way to solve the problem without having to solve two equations for two unknowns. If we compute the torques about a point on the line of action of one of the unknown forces, that force will not enter into the equation because its lever arm will be zero. For instance, let us first consider the torques about the left scale. The weight produces a clockwise torque of magnitude (60 N)(2.5 m) = 150 N·m, and $\mathbf{F_R}$ produces a counterclockwise torque F_R(3 m). Setting the net torque equal to zero, we get

$$F_R(3 \text{ m}) - 150 \text{ N·m} = 0$$

or

$$F_R = 50 \text{ N}$$

We can then find F_L immediately from F_L = 60 N − 50 N = 10 N. Alternatively, we can consider the torques about the right scale. We then have (60 N)(0.5 m) − F_L(3 m) = 0 or F_L = 10 N. Whenever there are two ways of solving a problem, it is a good idea to use one method to check the results of the other.

In general, if we have several unknown forces, we can reduce the work involved in solving the problem by computing the torques about a point on the line of action of one of the unknown forces so that that force does not enter into the equation.

Question

1. If the net torque about some point is zero, must it be zero about any other point? Explain.

9-2   The Center of Gravity

When two or more parallel forces act on an object, they can be replaced by a single equivalent force that is equal to the sum of the forces and is applied at a point such that the torque produced by the single equivalent force equals the net torque produced by the original forces. Figure 9-2 shows a force F_1 acting on a rod at point x_1 and a second force F_2 acting at point x_2. The net force $\Sigma F = F_1 + F_2$ will produce the same torque about O if it is applied at a distance x_r, which is given by

$$x_r \sum F = F_1 x_1 + F_2 x_2 \qquad\qquad 9\text{-}3$$

We can use this result to show that the force of gravity exerted on the various parts of an object can be replaced by a single force, the total weight, acting at a point we call the center of gravity. In Figure 9-3 we have divided an object into many smaller objects. If we make the divisions small enough, we can consider the smaller objects to be particles. The weight of each particle is w_i and the total weight of the object is $W = \Sigma w_i$. Generalizing Equation 9-3 to this case of several parallel forces and using $\Sigma F = W$, we have for the point of application of the net force X_{cg}

$$X_{cg}W = \sum_i w_i x_i \qquad\qquad 9\text{-}4$$

Center of gravity defined

Equation 9-4 defines the x coordinate of the center of gravity. The **center of gravity** is that point at which the total weight of an object acts so that the torque it produces about any point is equal to that produced by the weights of the individual particles making up the object.

If the acceleration of gravity does not vary over the object (as is nearly always the case), we can write $w_i = m_i g$ and $W = Mg$ and cancel the common factor g. Then

$$X_{cg}Mg = \sum_i m_i g x_i$$

or

$$MX_{cg} = \sum_i m_i x_i \qquad\qquad 9\text{-}5$$

This is the same as Equation 7-3a, which gives the x coordinate for the center of mass. Thus, when the acceleration of gravity does not vary over an object, the center of gravity and the center of mass coincide.

If we choose our origin to be at the center of gravity, $X_{cg} = 0$. Then, according to Equation 9-4,

$$\sum_i w_i x_i = 0$$

Thus, we can think of the center of gravity of an object as being that point about which the forces of gravity acting on all the particles of the object produce zero torque. We can use the methods discussed in Chapter 7 for locating the center of mass to locate the center of gravity. For example, we can locate the center of gravity of a stick by balancing it on a pivot. The point at which the stick balances is the center of gravity of the stick. For a uniform object, such as a uniform sphere or stick, the center of gravity is at the object's geometric center.

Figure 9-2 The two parallel forces F_1 and F_2 can be replaced by a single net force ΣF that has the same effect. The point of application of ΣF is chosen such that it produces the same torque about any point as is produced by the original forces.

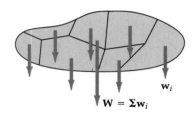

Figure 9-3 The weights of all the particles of an object can be replaced by the total weight W of the object acting at the center of gravity.

Questions

2. Must there be any material at the center of gravity of an object?

3. If the acceleration of gravity is not constant over an object, which point, the center of mass or the center of gravity, is at the pivot point when the object is balanced?

9-3   Some Examples of Static Equilibrium

In this section, we will look at several examples of problems involving the static equilibrium of extended objects.

Example 9-2

Work Example 9-1 for the case in which the board is uniform and weighs 20 N.

Figure 9-4   Free-body diagram for the uniform board of weight 20 N on scales for Example 9-2.

Since the board is uniform, its center of gravity is at the center of the board. The forces involved are shown in the free-body diagram of Figure 9-4. We first consider the torques about the left end of the board. The weight of the board produces a clockwise torque of magnitude (20 N)(1.5 m) = 30 N·m about this point. This adds to the clockwise torque (60 N)(2.5 m) = 150 N·m produced by the 60-N weight. Setting the net torque equal to zero gives

$$F_R(3 \text{ m}) - 30 \text{ N·m} - 150 \text{ N·m} = 0$$

$$F_R = \frac{180 \text{ N·m}}{3 \text{ m}}$$

$$= 60 \text{ N}$$

Since the total downward force is now 60 N + 20 N = 80 N, the upward force exerted by the left scale must be F_L = 80 N − 60 N = 20 N. The two scales share the additional weight of 20 N equally, as would be expected because this weight is midway between the scales.

Example 9-3

A 60-N weight is held in the hand with the forearm making a 90° angle with the upper arm as shown in Figure 9-5. The biceps muscle exerts a force $\mathbf{F_m}$ that is 3.4 cm from the pivot point O at the elbow joint. Neglecting the weight of the arm and hand, what is the magnitude of $\mathbf{F_m}$ if the distance from the weight to the pivot point is 30 cm?

Figure 9-5   Hand holding a weight for Example 9-3.

The counterclockwise torque about O exerted by the muscle must balance the clockwise torque exerted by the weight. Setting these torques equal, we have

$$F_m(3.4 \text{ cm}) = (60 \text{ N})(30 \text{ cm})$$

$$F_m = \frac{(60 \text{ N})(30 \text{ cm})}{3.4 \text{ cm}}$$

$$= 529 \text{ N}$$

The force exerted by the muscle is much greater than the 60-N weight because of the very short lever arm of the muscle about the pivot.

In the figure, we have isolated the forearm, hand, and weight. Since this system is not translating (as well as not rotating), the net vertical force on the system must be zero. We have calculated that the upward force of the muscle must be 529 N to ensure rotational equilibrium about O. Since the downward force of the weight is only 60 N, there must be an additional downward force of 529 N − 60 N = 469 N. The line of action of this force must pass through O, or there would be an additional torque about O. This force is exerted by the upper arm at the elbow joint.

Example 9-4

A sign of mass 20 kg hangs from the end of a rod of length 2 m and mass 4 kg (Figure 9-6a). A wire is attached to the end of the rod and to a point 1 m above point O. Find the tension in the wire and the force exerted by the wall on the rod at point O.

(a)

(b)

Figure 9-6   (a) A sign hanging from a rod for Example 9-4. (b) Free-body diagram showing the forces acting on the rod.

Figure 9-6b shows the forces acting on the rod. Since $g = 9.81$ N/kg, the weight of the sign is 196 N and that of the rod is 39.2 N. The force exerted by the wall has the components F_x and F_y. The tension **T** has been resolved into its x and y components. Since we do not know the force exerted by the wall, we choose point O for computing torques. The weight of the sign and that of the rod produce clockwise torques about O, and the y component of the tension in the wire produces a counterclockwise torque about O. Setting these torques equal, we have

$$T_y(2 \text{ m}) = (196 \text{ N})(2 \text{ m}) + (39.2 \text{ N})(1 \text{ m}) = 431 \text{ N·m}$$

$$T_y = 215.5 \text{ N}$$

The x component of the tension can be related to T_y and the angle θ. From Figure 9-6b, we have

$$\frac{T_y}{-T_x} = \tan \theta$$

And, from Figure 9-6a, we have

$$\tan \theta = \frac{1 \text{ m}}{2 \text{ m}} = \frac{1}{2}$$

Thus,

$$T_x = -2T_y = -2(215.5 \text{ N}) = -431 \text{ N}$$

(T_x is negative because it points to the left.) The magnitude of the tension is

$$T = \sqrt{T_x^2 + T_y^2} = \sqrt{(-431 \text{ N})^2 + (215.5 \text{ N})^2} = 482 \text{ N}$$

The force exerted by the wall on the rod at O is found from the first condition for equilibrium. That is, the horizontal component F_x must equal 431 N to balance the horizontal component of the tension. Setting the upward forces equal to the downward forces, we have

$$F_y + T_y = 196 \text{ N} + 39.2 \text{ N} = 235 \text{ N}$$

$$F_y = 235 \text{ N} - T_y = 235 \text{ N} - 215.5 \text{ N} = 19.5 \text{ N}$$

This mobile by Alexander Calder hangs suspended over the courtyard of the East Wing of the National Gallery in Washington, D.C.

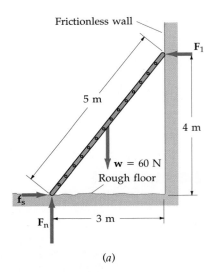

Frictionless wall

F_1

5 m

4 m

$w = 60$ N

Rough floor

f_s

F_n

3 m

(a)

P

F_1

4 m

F_2

w

F_n

θ'

f_s

1.5 m

(b)

Figure 9-7 (a) A ladder on a rough floor leaning against a frictionless wall for Example 9-5. (b) Free-body diagram for the ladder. Since the ladder is in static equilibrium under the influence of three nonparallel forces, F_2, w, and F_1, the lines of action of the three forces must all intersect at a common point P.

Example 9-5

A uniform, 5-m ladder weighing 60 N leans against a frictionless vertical wall (Figure 9-7a). The foot of the ladder is 3 m from the wall. What is the minimum coefficient of static friction necessary between the ladder and the floor if the ladder is not to slip?

The forces acting on the ladder are the force due to gravity **w** acting downward at the ladder's center of gravity, the force F_1 exerted horizontally by the wall (since the wall is frictionless, it exerts only a normal force), and the force exerted by the floor, which consists of a normal force F_n and a horizontal force of static friction f_s. From the first condition for static equilibrium, we have

$$F_n = w = 60 \text{ N} \quad \text{and} \quad F_1 = f_s$$

Since we know neither f_s nor F_1, we must use the second condition for static equilibrium and compute the torques about some convenient point. We choose the point of contact between the ladder and the floor because both F_n and f_s act at this point and will therefore not appear in our torque equation. The torque exerted by the force of gravity about this point is clockwise, with a magnitude of 60 N times the lever arm 1.5 m. The torque exerted by F_1 about this point is counterclockwise, with a magnitude of F_1 times the lever arm 4 m. The second condition for static equilibrium thus gives

$$F_1(4 \text{ m}) - (60 \text{ N})(1.5 \text{ m}) = 0$$

$$F_1 = 22.5 \text{ N}$$

This equals the magnitude of the frictional force. Since the frictional force f_s is related to the normal force F_n by

$$f_s \leq \mu_s F_n$$

we have

$$\mu_s \geq \frac{f_s}{F_n} = \frac{22.5 \text{ N}}{60 \text{ N}} = 0.375$$

where μ_s is the coefficient of static friction.

There is another way to solve the problem in Example 9-5. Let $F_2 = f_s + F_n$ be the force exerted on the ladder by the floor. The ratio f_s/F_n is the cotangent of the angle θ' between the force F_2 and the horizontal. We can find this angle as follows. If we extend the lines of action of the forces **w** and F_1, they meet at point P in Figure 9-7b. The torques exerted by these two forces about this point are zero. Since F_2 is the only other force acting, it must also exert zero torque about point P, so its line of action must also pass through this point. Note that the force F_2 is *not* along the line of the ladder. The force exerted by the ladder on the floor, which equals $-F_2$, is not generally along the line of the ladder. This differs from a tension force in a string, which is always along the string. From the figure, we can see that if the line of action of F_2 passes through point P, the cotangent of the angle made by F_2 with the horizontal is

$$\cot \theta' = \frac{1.5 \text{ m}}{4 \text{ m}} = 0.375 = \frac{f_s}{F_n} \leq \mu_s$$

We note that

If an object is in static equilibrium under the influence of three non-parallel forces, the lines of action of the forces must intersect at one point.

Example 9-6

A ladder of length L and negligible weight leans against a frictionless vertical wall making an angle θ with the horizontal. The coefficient of friction between the ladder and the floor is μ_s. A man of weight W climbs up the ladder. Show that the maximum distance s he can climb before the ladder slips is given by $s = \mu_s L \tan \theta$.

Figure 9-8 shows the man on the ladder and the forces involved. When the man is on the ladder, the clockwise torque exerted by his weight about the support point at the floor is balanced by the counterclockwise torque due to the force $\mathbf{F}_1$ exerted by the wall. As he climbs the ladder, both these torques increase, so the force $\mathbf{F}_1$ must increase. But the horizontal force $\mathbf{F}_1$ is balanced by the force of friction $\mathbf{f}_s$, which is limited to the value $\mu_s F_n$. Thus, there is a maximum distance s up the ladder for which equilibrium can be maintained. In the figure, the lines of action of the weight and the force $\mathbf{F}_1$ have been extended so that they meet at point P. The line of action of the force exerted by the floor on the ladder $\mathbf{F}_2 = \mathbf{F}_n + \mathbf{f}_s$ must also pass through point P. In the figure, θ is the angle made by the ladder with the horizontal, and θ' is that made by $\mathbf{F}_2$ with the horizontal. Since the friction is limiting, we have

$$f_s = \mu_s F_n$$

or

$$\mu_s = \frac{f_s}{F_n} = \cot \theta' \qquad 9\text{-}6$$

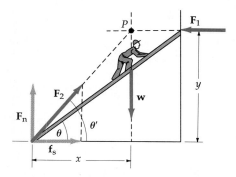

Figure 9-8 A man climbing on a ladder of negligible weight. The lines of action of the three forces acting on the ladder must intersect at a common point P.

If y is the height of the top of the ladder above the floor and x is the horizontal distance from the point of contact between the floor and the ladder to the man, we have from the geometry of the figure

$$\tan \theta' = \frac{y}{x} \qquad 9\text{-}7$$

$$\sin \theta = \frac{y}{L} \qquad 9\text{-}8$$

and

$$\cos \theta = \frac{x}{s} \qquad 9\text{-}9$$

Combining Equations 9-6 and 9-7, we have

$$x = \frac{y}{\tan \theta'} = y \cot \theta' = \mu_s y$$

Using this result in Equation 9-9, we have for the distance s

$$s = \frac{x}{\cos \theta} = \frac{\mu_s y}{\cos \theta} = \frac{\mu_s L \sin \theta}{\cos \theta} = \mu_s L \tan \theta$$

where we have used $y = L \sin \theta$ from Equation 9-8.

Questions

4. Can a ladder be in equilibrium standing on a frictionless floor and leaning against a wall?

5. If an object is in equilibrium under three forces, must the forces all be in the same plane?

9-4   Couples

We have seen that we can replace a set of parallel forces with a single force, equal to the sum of the parallel forces, that acts at such a point so as to produce the same rotational torque about any point as the parallel forces. This idea was illustrated in Figure 9-2 for two parallel forces. Further, we used this idea when we replaced the forces of gravity acting on the various parts of an object with a single force, the weight of the object, acting at the center of gravity. However, two forces that are equal in magnitude but opposite in direction and have different lines of action cannot be replaced by a single force. Such a pair of forces, called a **couple,** tends to produce rotation, but their net force is zero. Consider the couple shown in Figure 9-9. The torque produced by these forces about point O is

$$\tau = Fx_2 - Fx_1 = F(x_2 - x_1) = FD \qquad\qquad 9\text{-}10$$

where F is the magnitude of either force and $D = x_2 - x_1$ is the distance between them. This result does not depend on the choice of the point O.

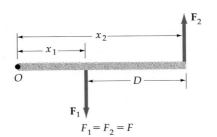

Figure 9-9 Two equal and opposite forces constitute a couple. The torque exerted by a couple has the same value FD about any point in space.

> The torque produced by a couple is the same about all points in space.

Two unequal, antiparallel forces, such as those shown in Figure 9-10, can be replaced by a single force that is equal to the net force acting at the center of mass plus a couple whose torque equals the torque about the center of mass of the original forces. In general, any number of forces can be replaced by a single force equal to the net force and a couple. Since the net force exerted by a couple is zero, it can be balanced only by a second couple that exerts an equal but opposite torque. For example, the forces $\mathbf{F_n}$ and $\mathbf{w}$ in Figure 9-7b for Example 9-5 form a couple with a torque of $(60\ \text{N})(1.5\ \text{m}) = 90\ \text{N·m}$. It is balanced by the forces $\mathbf{f_s}$ and $\mathbf{F_1}$, which also form a couple. Since the separation of these forces is $D = 4\ \text{m}$, their magnitude must be $(90\ \text{N·m})/(4\ \text{m}) = 22.5\ \text{N}$, as found in that example.

Figure 9-10 Two antiparallel forces of unequal magnitude can be replaced by a single force $\Sigma\mathbf{F}$ acting at the center of gravity plus a couple whose torque is equal to the torque exerted by the original forces about the center of gravity.

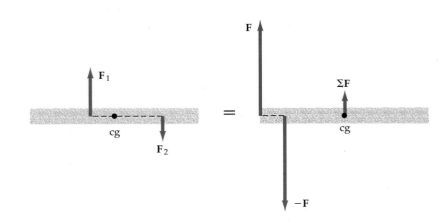

9-5   Stability of Equilibrium

The equilibrium of an object can be classified according to three categories: stable, unstable, or neutral. **Stable equilibrium** occurs when the torques or forces that arise from a small displacement of the object urge the object back toward its equilibrium position. Stable equilibrium is illustrated in Figures 9-11 and 9-12. When the box in Figure 9-11 is rotated slightly about one end, the resulting torque about the pivot point tends to restore the box to its original position. Similarly, if the marble resting at the bottom of the hemispherical bowl in Figure 9-12 is displaced slightly, the tangential component of its weight urges it back toward its initial position.

Unstable equilibrium, illustrated in Figures 9-13 and 9-14, occurs when the forces or torques that arise from a small displacement of the object urge the object away from its equilibrium position. A slight rotation of the narrow stick in Figure 9-13 causes it to fall over because the torque due to its weight urges it away from its original position. Similarly, if the marble on the overturned bowl in Figure 9-14 is displaced slightly, the tangential component of the weight urges the marble away from its original position.

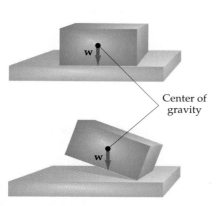

Figure 9-11   An example of stable equilibrium. When the box is rotated slightly, the torque exerted by the weight about the pivot point tends to restore the box to its original position.

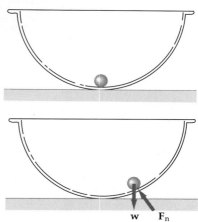

Figure 9-12   Another example of stable equilibrium. If the marble is displaced slightly in any direction, there is a net force that pushes the marble back toward its original position.

Figure 9-13   An example of unstable equilibrium. If the stick is rotated slightly, the torque exerted by the weight about the pivot point tends to rotate the stick away from its original position.

Figure 9-14   Another example of unstable equilibrium. If the marble is displaced slightly, there is a net force that pushes the marble further from its original position.

The gymnast improves her balance by bringing her center of gravity as close to the base of support as possible.

Figure 9-15   An example of neutral equilibrium. If the cylinder is rotated slightly, it is again in equilibrium. There are no torques or forces that tend to rotate the cylinder either toward or away from its original position.

The cylinder resting on a horizontal surface in Figure 9-15 illustrates **neutral equilibrium.** If the cylinder is rotated slightly, there is no torque or force that urges it either back toward its original position or away from it.

In summary, if a system is disturbed slightly from its equilibrium position, the equilibrium is stable if the system returns to its original position, unstable if it moves farther away, and neutral if there are no torques or forces urging it in either direction.

Since "disturbed slightly" is a relative term, stability is also relative. One example of equilibrium may be more or less stable than another. Figure 9-16a shows a stick that is not as narrow as that in Figure 9-13 balanced on end. Here, if the disturbance is very small (Figure 9-16b), the stick will move back toward its original position, but if the disturbance is great enough that the center of gravity no longer lies over the base of support (Figure 9-16c), the stick will fall over. We can improve the stability of a system by either lowering its center of gravity or widening its base of support. Figure 9-17 shows a nonuniform stick that has its center of gravity near one end. If the

(a)   (b)   (c)

Figure 9-16   Stability of equilibrium is relative. If the stick in (a) is rotated slightly, as in (b), it returns to its original equilibrium position as long as the center of gravity lies over the base of support. (c) If the rotation is too great, the center of gravity is no longer over the base of support, and the stick falls over.

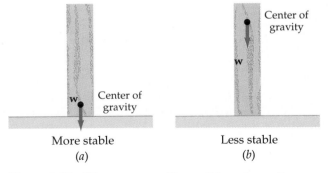

More stable   Less stable
(a)   (b)

Figure 9-17   When a nonuniform stick rests on its heavy end with its center of gravity low, as in (a), the equilibrium is more stable than when its center of gravity is high, as in (b).

stick is stood on its heavy end so that its center of gravity is low (Figure 9-17a), it is much more stable than if it is stood on the other end so that its center of gravity is high (Figure 9-17b). In Figure 9-18, the center of gravity lies below the point of support of the system. This system is stable for any displacement because the resulting torque always rotates the system back toward its equilibrium position.

A human being standing or walking upright has difficulty maintaining balance because the center of gravity must be kept over the base of support, the feet, which is relatively small. An animal walking on four legs has a much easier time because its base of support is larger and its center of gravity is lower.

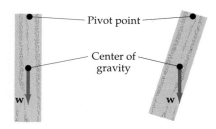

Figure 9-18 When a stick is pivoted such that its center of gravity is below the pivot point, the equilibrium is stable no matter how far the stick is displaced from equilibrium.

Questions

6. How should a hiker with a heavy backpack stand so as to maintain balance?

7. How do linemen in football improve their balance?

8. Why is it so much more difficult to stand on one foot than on two?

(a)

(b)

(a) The tightrope walker uses a long, heavy pole to help maintain balance. (b) In this fascinating demonstration of stable equilibrium, the center of gravity of the two-fork system lies under the point of support of the toothpick.

Summary

1. The two conditions that are necessary for a rigid body to be in static equilibrium are
    1. The net external force acting on the body must be zero:

        $$\mathbf{F}_{net} = 0$$

    2. The net external torque about any point must be zero:

        $$\boldsymbol{\tau}_{net} = 0$$

    An alternative statement of the second condition is that the sum of the torques that tend to produce clockwise rotation about any point must equal the sum of the torques that tend to produce counterclockwise rotation about that point.

2. When two or more parallel forces act on an object, they can be replaced by a single equivalent force that is equal to the sum of the forces and is applied at a point such that the torque produced by the single equivalent force equals the net torque produced by the original forces.

3. The force of gravity exerted on the various parts of an object can be replaced by a single force, the total weight of the object W, acting at the center of gravity. The x coordinate of the center of gravity X_{cg} relative to some origin is given by

$$X_{cg}W = \sum_i w_i x_i$$

When the acceleration of gravity is the same at all points of an object, the center of gravity coincides with the center of mass.

4. If an object is in static equilibrium under the influence of three nonparallel forces, the lines of action of the forces must intersect at one point.

5. Two equal and opposite forces that have different lines of action constitute a couple. The torque produced by a couple is the same about all points in space. Its magnitude is the product of the magnitude of either force times the distance between their lines of action.

6. The equilibrium of an object can be classified according to three categories: stable, unstable, or neutral. An object resting on some surface will be in equilibrium if its center of gravity lies over its base of support. The stability of the equilibrium of an object can be improved by lowering its center of gravity or by increasing the size of its base of support.

Suggestions for Further Reading

Mark, Robert: "The Structural Analysis of Gothic Cathedrals," *Scientific American*, November 1972, p. 90.

This article discusses the use of the technique of optical stress analysis of scale models in an analysis of the static and wind-loaded behaviors of Gothic cathedrals.

Steinman, David B.: "Bridges," *Scientific American*, November 1954, p. 60.

This article presents a history of bridge building, an empirical discipline until the eighteenth century, when physical principles began to be studied and applied to it. The aerodynamics of the Tacoma Narrows Bridge and the reasons for its collapse are also considered.

Review

A. Objectives: After studying this chapter, you should:

1. Be able to state the conditions necessary for static equilibrium to exist and apply them to problems.

2. Be able to discuss stable, unstable, and neutral equilibrium.

B. Define, explain, or otherwise identify:

Center of gravity
Couple
Stable equilibrium
Unstable equilibrium
Neutral equilibrium

C. True or false: If the statement is true, explain why it is true. If it is false, give a counterexample.

1. $\Sigma F = 0$ is sufficient for static equilibrium to exist.

2. $\Sigma F = 0$ is necessary for static equilibrium to exist.

3. In static equilibrium, the net torque about any point is zero.

4. The center of gravity is always at the geometric center of a body.

5. An object is in equilibrium only when there are no forces acting on it.

Problems

Level I

9-1 Conditions for Equilibrium

1. A seesaw consists of a 4-m board pivoted at the center. A 28-kg child sits on one end of the board. Where should a 40-kg child sit to balance the seesaw?

2. In Figure 9-19, Misako is about to do a push-up. Her center of gravity lies directly above point P on the floor, which is 0.9 m from her feet and 0.6 m from her hands. If her mass is 54 kg, what is the force exerted by the floor on her hands?

Figure 9-19   Problem 2.

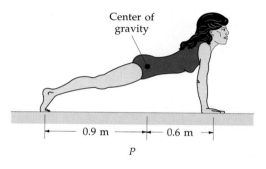

3. Juan and Betina are carrying a 60-kg weight on a 4-m board as shown in Figure 9-20. The mass of the board is 10 kg. Since Juan spends most of his time reading cookbooks, whereas Betina regularly does push-ups, they place the weight 2.5 m from Juan and 1.5 m from Betina. Find the force in newtons exerted by each to carry the weight.

Figure 9-20   Problem 3.

4. Misako wishes to measure the strength of her biceps muscle by exerting a force on a test strap as shown in Figure 9-21. The strap is 28 cm from the pivot point at the elbow, and her biceps muscle is attached at a point 5 cm from the pivot point. If the scale reads 18 N when she exerts her maximum force, what force is exerted by the biceps muscle?

Figure 9-21   Problem 4.

5. A 90-N board 12 m long rests on two supports, each 1 m from the end of the board. A 360-N block is placed on the board 3 m from one end as shown in Figure 9-22. Find the force exerted by each support on the board.

Figure 9-22   Problem 5.

6. Figure 9-23 shows a mobile consisting of four weights hanging on three rods of negligible mass. Find the value of each of the unknown weights if the mobile is to balance. *Hint:* Find weight w_1 first.

Figure 9-23   Problem 6.

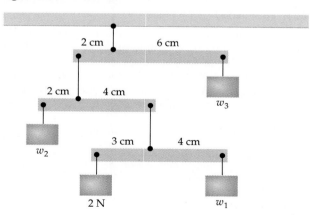

7. A crutch is pressed against the sidewalk with a force $\mathbf{F}_c$ along its own direction as shown in Figure 9-24. This force is balanced by a normal force $\mathbf{F}_n$ and a frictional force $\mathbf{f}_s$, as shown in the figure. (*a*) Show that when the force of friction is at its maximum value, the coefficient of friction is related to the angle θ by $\mu_s = \tan \theta$. (*b*) Explain how this result applies to the forces on your foot when you are not using a crutch. (*c*) Why is it advantageous to take short steps when walking on ice?

Figure 9-24 Problem 7.

9-2 The Center of Gravity

8. Two spheres of radius R rest on a horizontal table with their centers a distance $4R$ apart. One sphere has twice the weight of the other sphere. Where is the center of gravity of this system?

9. Uniform wire is bent into the shapes shown in Figure 9-25. Find the center of gravity for each shape.

Figure 9-25 Problem 9.

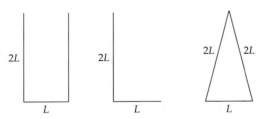

10. An automobile has 58 percent of its weight on the front wheels. The front and back wheels are separated by 8 m. Where is the center of gravity located with respect to the front wheels?

11. The height of the center of gravity of a man standing erect is determined by weighing the man as he lies on a board of negligible weight supported by two scales as shown in Figure 9-26. If the man's height is 188 cm and the left scale reads 445 N and the right scale reads 400 N, where is his center of gravity relative to his feet?

Figure 9-26 Problem 11.

Center of gravity

445 N   400 N

12. Each of the objects shown in Figure 9-27 is suspended from the ceiling by a thread attached to the point marked × on the object. Describe the orientation of each suspended object with a diagram.

Figure 9-27 Problem 12.

9-3 Some Examples of Static Equilibrium

13. A uniform 10-m log of mass 100 kg rests in a horizontal position on two supports. One support is 2 m from the left end of the log and the other is 4 m from the right end. Find the forces exerted on the log by the supports.

14. A 10-m beam of mass of 300 kg extends over a ledge as shown in Figure 9-28. The beam is not held down but simply rests on the surface. A 60-kg student has positioned the beam so that he can walk to the end of it. How far from the edge of the ledge can the beam extend?

Figure 9-28 Problem 14.

15. A man uses a crowbar that is 1 m long to lift a heavy crate off the ground. The crowbar has a rigid fulcrum 10 cm from one end as shown in Figure 9-29. (*a*) If the man exerts a downward force of 600 N on one end of the crowbar, what is the upward force exerted on the crate by the other end? (*b*) The ratio of the forces at the ends of the crowbar is called the mechanical advantage of the crowbar. What is the mechanical advantage?

Figure 9-29 Problem 15.

16. A gravity board for locating the center of gravity of a person consists of a horizontal board supported by a fulcrum at one end and by a scale at the other end. A physics student lies horizontally on the board with the top of his head above the fulcrum point as shown in Figure 9-30. The scale is 2 m from the fulcrum. The student has a mass of 70 kg, and when he is on the gravity board, the scale advances 250 N. Where is the center of gravity of the student?

Figure 9-30   Problem 16.

250 N

2 m

9-4 Couples

17. Two 80-N forces are applied to opposite corners of a rectangular plate as shown in Figure 9-31. Find the torque produced by this couple.

18. Resolve each force in Problem 17 into its horizontal and vertical components, producing two couples. The algebraic sum of the two component couples equals the resultant couple. Use this result to find the perpendicular distance between the lines of action of the two forces.

Figure 9-31   Problems 17 and 18.

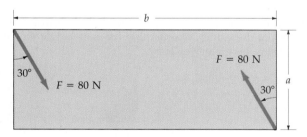

b

30°

F = 80 N

F = 80 N

30°

a

19. A uniform cube of side a and mass M rests on a horizontal surface. A horizontal force **F** is applied to the top of the cube as shown in Figure 9-32. This force is not sufficient to move or tip the cube. (*a*) Show that the force of static friction exerted by the surface and the applied force constitute a couple, and find the torque exerted by the couple. (*b*) This couple is balanced by the couple consisting of the normal force exerted by the surface and the weight of the cube. Use this fact to find the effective point of application of the normal force when $F = Mg/3$. (*c*) What is the greatest magnitude of **F** for which the cube will not tip?

Figure 9-32   Problem 19.

F

a

a   M

f_s

9-5 Stability of Equilibrium

There are no problems for this section.

Level II

20. A 3-m board of mass 5 kg is hinged at one end. A force **F** is applied vertically at the other end to lift a 60-kg box, which rests on the board 80 cm from the hinge as shown in Figure 9-33. (*a*) Find the magnitude of the force needed to hold the board stationary at $\theta = 30°$. (*b*) Find the force exerted by the hinge at this angle. (*c*) Find the force **F** and the force exerted by the hinge if **F** is exerted perpendicular to the board when $\theta = 30°$.

Figure 9-33   Problem 20.

F

80 cm

30°

3 m

21. A square plate is produced by welding together four smaller square plates, each of side a as shown in Figure 9-34. Plate 1 weighs 40 N; plate 2, 60 N; plate 3, 30 N; and plate 4, 50 N. Find the center of gravity (x_{cg}, y_{cg}).

Figure 9-34   Problem 21.

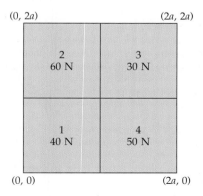

(0, 2a)   (2a, 2a)

2 60 N	3 30 N
1 40 N	4 50 N

(0, 0)   (2a, 0)

22. A uniform rectangular plate has a circular section of radius R cut out as shown in Figure 9-35. Find the center of gravity of the system. *Hint:* Do not integrate. Use superposition of a rectangular plate minus a circular plate.

Figure 9-35   Problem 22.

23. A cylinder of weight **W** is supported by a frictionless trough formed by a plane inclined at 30° to the horizontal on the left and one inclined at 60° on the right as shown in Figure 9-36. Find the force exerted by each plane on the cylinder.

Figure 9-36   Problem 23.

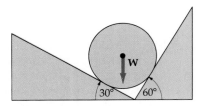

24. An 80-N weight is supported by a cable attached to a strut hinged at point A (Figure 9-37). The strut is supported by a second cable under tension T_2. The mass of the strut is negligible. (*a*) What are the three forces acting on the strut? (*b*) Show that the vertical component of the tension T_2 must equal 80 N. (*c*) Find the force exerted on the strut by the hinge.

25. Find the force exerted on the strut by the hinge at A for the arrangement in Figure 9-38 if (*a*) the strut is weightless and (*b*) the strut weighs 20 N.

Figure 9-37   Problem 24.   **Figure 9-38**   Problem 25.

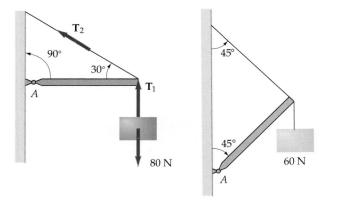

26. A horizontal plank 8.0 m long is used by pirates to make their victims walk the plank. A pirate of mass 105 kg stands on the shipboard end of the plank to prevent it

from tipping. Find the maximum distance the plank can overhang for a 63-kg victim to be able to walk to the end if (*a*) the mass of the plank is negligible and (*b*) the mass of the plank is 25 kg.

27. A box that is 2 by 1 by 1 m of uniform mass is placed on end on a hinged rough plank as shown in Figure 9-39. The plank is inclined at an angle θ, which is slowly increased. The coefficient of friction is large enough to prevent the box from sliding before it tips over. Find the greatest angle that can be attained without tipping the box over.

Figure 9-39   Problem 27.

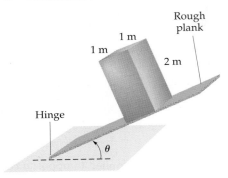

28. A uniform 18-kg door, 2.0 m high by 0.8 m wide, is hung from two hinges that are 20 cm from the top and 20 cm from the bottom. If each hinge supports half the weight of the door, find the magnitude and direction of the horizontal components of the forces exerted by the two hinges on the door.

29. A 900-N boy sits on top of a ladder of negligible weight that rests on a frictionless floor (Figure 9-40). There is a cross brace halfway up the ladder. The angle at the apex is $\theta = 30°$. (*a*) What is the force exerted by the floor on each leg of the ladder? (*b*) Find the tension in the cross brace. (*c*) If the cross brace is moved down toward the bottom of the ladder (maintaining the same angle θ), will its tension be greater or less?

Figure 9-40   Problem 29.

30. Two painters are working from a plank 5.0 m long that is suspended from the top of a building by two ropes attached to the ends of the plank. Either rope will break when its tension exceeds 1 kN. Painter A (mass 80 kg) is working at a distance of 1.0 m from one end. Find the range of positions available to painter B if her mass is 60 kg and the plank has mass of 20 kg.

31. A wheel of mass M and radius R rests on a horizontal surface against a step of height h ($h < R$). The wheel is to be raised over the step by a horizontal force F applied to the axle of the wheel (Figure 9-41). Find the force F necessary to raise the wheel over the step.

Figure 9-41   Problem 31.

32. One end of a uniform, 100-kg beam 10 m long is hinged at a vertical wall. It is held horizontal by a cable that connects to the beam 6 m from the wall as shown in Figure 9-42. A 400-kg weight is suspended from the free end of the beam. (*a*) What is the tension in the cable? (*b*) What is the horizontal force on the hinge? (*c*) What is the vertical force on the beam at the hinge?

Figure 9-42   Problems 32 and 33.

33. The cable in Figure 9-42 must remain attached to the wall 8 m above the hinge, but its length can vary so that it can be connected to the beam at various distances x from the wall. How far from the wall should it be attached so that the force on the hinge has no vertical component?

34. A tall, uniform, rectangular block sits on an inclined plane as shown in Figure 9-43. If $\mu_s = 0.4$, does the block slide or fall over as the angle θ is slowly increased?

Figure 9-43   Problem 34.

35. A cylinder of mass M and radius R rolls against a step of height h as shown in Figure 9-44. When a horizontal force F is applied to the top of cylinder, the cylinder remains at rest. (*a*) What is the normal force exerted by the floor on the cylinder? (*b*) What is the horizontal force exerted by the corner of the step on the cylinder? (*c*) What is the vertical component of the force exerted by the corner on the cylinder?

Figure 9-44   Problems 35 and 36.

36. For the cylinder in Problem 35, find the minimum horizontal force F that will roll the cylinder over the step if the cylinder does not slide on the corner.

37. A strong man holds one end of a 3-m rod of mass 5 kg at rest in a horizontal position. (*a*) What total force does the man exert on the rod? (*b*) What total torque does the man exert on the rod? (*c*) If you approximate the effort of the man with two forces that act in opposite directions and are separated by the width of the man's hand, which is taken to be 10 cm, what are the magnitudes and directions of the two forces?

38. A large gate weighing 200 N is supported by hinges at the top and bottom and is further supported by a wire as shown in Figure 9-45. (*a*) What must the tension in the wire be for the force on the upper hinge to have no horizontal component? (*b*) What is the horizontal force on the lower hinge? (*c*) What are the vertical forces on the hinges?

Figure 9-45   Problem 38.

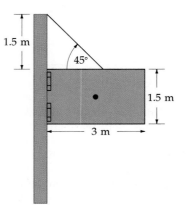

39. The diving board shown in Figure 9-46 has a mass of 30 kg. Find the force on the supports when a 70-kg diver stands at the end of the diving board. Give the direction of each support force as a tension or a compression.

Figure 9-46   Problem 39.

40. A ladder rests against a frictionless vertical wall. The coefficient of static friction between the ladder and the floor is 0.3. What is the smallest angle at which the ladder will remain stationary?

41. A uniform rectangular box 3 m high sits on its base, which is 1 m by 1 m, on a horizontal platform in a truck that is moving along a level road. What deceleration will cause the box to tip over?

Level III

42. A uniform log with a mass of 100 kg, a length of 4 m, and a radius of 12 cm is held in an inclined position as shown in Figure 9-47. The coefficient of static friction between the log and the horizontal surface is 0.6. The log is on the verge of slipping to the right. Find the tension in the support wire and the angle the wire makes with the vertical wall.

Figure 9-47   Problem 42.

43. Find the center of gravity of a uniform wire that has been bent into a semicircle as shown in Figure 9-48. Use the coordinate system given in the figure.

Figure 9-48   Problem 43.

44. A uniform ladder of length L and weight 200 N leans against a wall. The coefficients of static friction are 0.4 between the ladder and the vertical wall and 0.7 between the ladder and the floor. An 80-kg firefighter is four-fifths of the way up the ladder when it begins to slip. Find the angle θ between the ladder and the ground.

45. The legs of a tripod make equal angles of 90° with each other at the apex, where they join together. A 100-kg block hangs from the apex. What are the compressional forces in the three legs?

46. Six identical bricks are stacked one on top of the other lengthwise to produce a leaning tower with the maximum offset that will still allow the tower to stand. (a) Starting from the top, give the maximum possible offset for each successive brick. (b) What is the total protrusion or offset of the six bricks?

47. A tall, uniform, rectangular block sits on an inclined plane as shown in Figure 9-49. A cord is attached to the top of the block to prevent it from falling down the incline. What is the maximum angle θ for which the block will not slide on the incline? Let b/a be 4 and $\mu_s = 0.8$.

Figure 9-49   Problem 47.

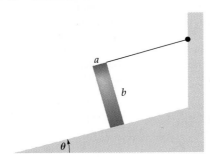

48. A thin rail of length 10 m and mass 20 kg is supported at a 30° incline. One support is 2 m and the other is 6 m from the lower end of the rail. Friction prevents the rail from sliding off the supports. Find the force (magnitude and direction) exerted on the rail by each support.

49. A ladder leans against a large smooth sphere of radius R that is fixed in place on a horizontal surface. The ladder makes an angle of 60° with the horizontal surface and has a length $5R/2$. (a) What is the force that the sphere exerts on the ladder? (b) What is the frictional force that prevents the ladder from slipping? (c) What is the normal force that the horizontal surface exerts on the ladder?

50. A uniform sphere of radius R and mass M is held at rest on an inclined plane of angle θ by a horizontal string as shown in Figure 9-50. Let $R = 20$ cm, $M = 3$ kg, and $\theta = 30°$. (a) Find the tension in the string. (b) What is the normal force exerted on the sphere by the inclined plane? (c) What is the frictional force acting on the sphere?

Figure 9-50   Problem 50.

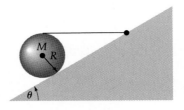

Chapter 10

Gravity

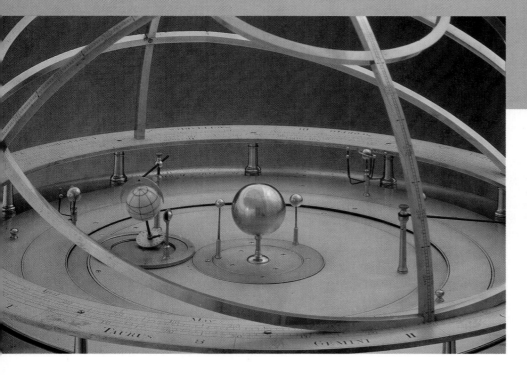

A mechanical model of the solar system, called an orrery, in the collection of Historical Scientific Instruments at Harvard University.

The gravitational force is the weakest of the four basic interactions occurring among elementary particles discussed in Chapter 4. It is negligibly small in the interactions of elementary particles. It is also difficult to observe the gravitational force between objects of everyday life even though their masses may be thousands of kilograms. Yet gravity is of primary importance when we consider interactions that involve very large objects such as planets, moons, and stars. It is gravity that binds us to the earth and keeps the earth and the other planets within the solar system. The gravitational force plays an important role in the evolution of stars and in the behavior of galaxies. In a sense, it is gravity that holds the universe together. In this chapter, we will study the force of gravity in some detail. We begin by stating Kepler's empirical laws of planetary motion and then discuss how these laws are related to Newton's law of gravity.

10-1 Kepler's Laws

The nighttime sky with its myriad stars and shining planets has always fascinated humankind. The apparent movements of the stars and planets relative to the earth have been observed and charted by astronomers for many centuries.

The apparent motions of the planets as seen from the earth. This photograph was taken in the Munich planetarium, where the motions of the planets over a period of seventeen years are being simulated.

A model of the universe, published by Ptolemy in about 140 A.D., had the earth at the center, with the sun and stars moving around the earth in simple circles and the planets moving around the earth in more complicated paths consisting of small circles, called epicycles, superimposed on larger circles (Figure 10-1). This somewhat complicated model was in basic agreement with the naked-eye observations of that time and was universally accepted for over 14 centuries. It was replaced in 1543 by the simpler but highly controversial model of Copernicus in which the sun and other stars were fixed and the planets, including the earth, revolved in circles around the sun. (This model was controversial because it did not place the earth at the center of the universe, contrary to the religious doctrine of the day.)

Figure 10-1   Ptolemy's model of the universe. The primary motion of the planets and stars is clockwise with a period of 1 day. To account for the motion of the planets relative to the stars, the planets are given a smaller counterclockwise angular velocity. This model is called a geocentric model because it places the earth at the center of the universe.

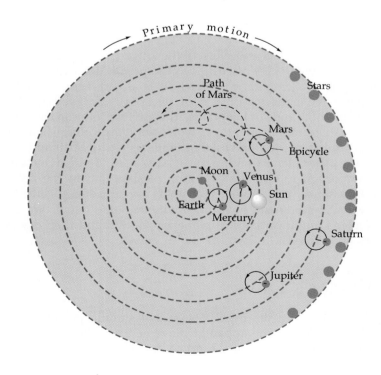

Toward the end of the sixteenth century, the astronomer Tycho Brahe studied the motions of the planets and made observations that were considerably more accurate than those previously available. Using Brahe's data, Johannes Kepler, after much trial and error, discovered that the actual paths of the planets about the sun were ellipses (Figure 10-2). He also showed that the planets did not move with constant speed but moved faster when close to the sun than when farther away. Finally, Kepler developed a precise mathematical relation between the period of a planet and its average distance from the sun. Kepler stated his results in three empirical laws of planetary motion. Ultimately, these laws provided the basis for Newton's discovery of the law of gravity.

Kepler's three laws are

> Law 1. All planets move in elliptical orbits with the sun at one focus.
>
> Law 2. A line joining any planet to the sun sweeps out equal areas in equal times.
>
> Law 3. The square of the period of any planet is proportional to the cube of the planet's mean distance from the sun.

Kepler's laws

Figure 10-3 shows an **ellipse.** The points labeled F are called the foci. The distance a is called the semimajor axis, and b is called the semiminor axis. You can draw an ellipse by taking a piece of string, fixing each end at a focus (for example, with thumb tacks), and using it to guide your pencil as is shown in the figure. As the foci are moved closer together, the ellipse begins to resemble a circle. Indeed, a circle is a special case of an ellipse in which the

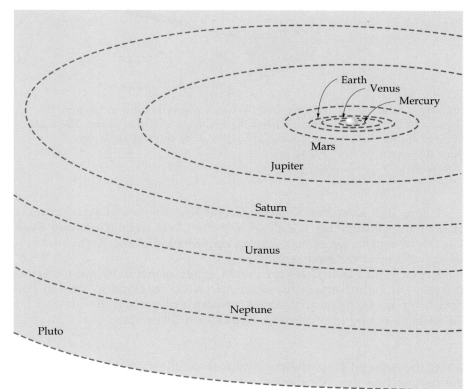

Figure 10-2 Orbits of the planets around the sun.

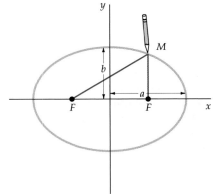

Figure 10-3 An ellipse is the locus of points for which the sum of the distances from two foci F is constant. Such a figure can be drawn by fastening a string at the foci and using it to guide a pencil (shown at M) to trace out the ellipse. The distance a is called the semimajor axis, and b is called the semiminor axis of the ellipse. If the foci coincide, a and b are equal and the ellipse is a circle.

(a)

(b)

(a) The tightly woven orbits about the sun of many of the known comets, including Halley's comet, are depicted in this drawing.
(b) Halley's comet as photographed over Duxbury Beach, Massachusetts, in March 1986.

two foci coincide. Figure 10-4 shows an elliptical path of a planet with the sun at one focus. The point P, at which the planet is closest to the sun, is called the perihelion, while point A, the farthest distance, is called the aphelion. The earth's orbit is nearly circular, with the distance to the sun at perihelion being 91.5 million miles and that at aphelion being 94.5 million miles. The semimajor axis equals half the sum of these distances, which is 93 million miles for the earth's orbit. This turns out to be the mean distance from the earth to the sun during its orbit.

Figure 10-4 Elliptical path of a planet with the sun at one focus. Point P, where the planet is closest to the sun, is called the perihelion, and point A, where it is farthest, is called the aphelion. The average distance between the planet and the sun equals the semimajor axis.

Figure 10-5 When a planet is close to the sun, it moves faster than when it is farther away. The areas swept out in a given time interval are equal.

Figure 10-5 illustrates Kepler's second law, the law of equal areas. A planet moves faster when it is closer to the sun than when it is farther away. The law of equal areas is related to the conservation of angular momentum, as we will see in the next section.

Kepler's third law relates the period of any planet to its mean distance from the sun, which equals the semimajor axis of its elliptical path. In algebraic form, if r is the mean distance between a planet and the sun and T is the planet's period of revolution, Kepler's third law states that

$$T^2 = Cr^3 \qquad\qquad \text{10-1}$$

where the constant C has the same value for all the planets. We will show in the next section that (for the special case of a circular orbit) this law is a simple consequence of the fact that the force exerted by the sun on a planet varies inversely with the square of the distance from the sun to the planet.

Example 10-1

The mean distance of Jupiter from the sun is 5.20 astronomical units (AU), where

$$1 \text{ AU} = 1.50 \times 10^{11} \text{ m} = 93.0 \times 10^6 \text{ mi}$$

is the mean distance between the earth and the sun. What is the period of Jupiter?

From Kepler's third law, the square of the period is proportional to the cube of the mean distance from the sun. Taking the square root of both sides of Equation 10-1, we have

$$T = \sqrt{C} r^{3/2}$$

If T_E and r_E are the period and mean distance for the earth, and T_J and r_J are the period and mean distance for Jupiter, we have

$$\frac{T_J}{T_E} = \frac{\sqrt{C} \, r_J^{3/2}}{\sqrt{C} \, r_E^{3/2}}$$

$$= \left(\frac{r_J}{r_E}\right)^{3/2} = \left(\frac{5.20 \text{ AU}}{1 \text{ AU}}\right)^{3/2}$$

$$T_J = (5.20)^{3/2} T_E = 11.9(1 \text{ y}) = 11.9 \text{ y}$$

Exercise

The period of Neptune is 164.8 y. What is its mean distance from the sun?   (Answer: 30.1 AU)

10-2   Newton's Law of Gravity

Although Kepler's laws were an important step in understanding the motion of the planets, they were still merely empirical rules obtained from the astronomical observations of Brahe. It remained for Newton to take the giant step forward and attribute the acceleration of a planet in its orbit to a force exerted by the sun on the planet that varied inversely with the square of the distance between the sun and the planet. Others besides Newton had proposed that such a force existed, but Newton was able to prove that a force that varied inversely with the square of the separation distance would result in the elliptical orbits observed by Kepler. He then made the bold assumption that such a force existed between any two objects in the universe. (Before Newton, it was not even generally accepted that the laws of physics observed on earth were applicable to the heavenly bodies.)

Newton's law of gravity postulates that every object exerts an attractive force on every other object that is proportional to the masses of the two objects and inversely proportional to the square of the distance separating them. Newton's law of gravity can be written as a simple vector equation. Let m_1 and m_2 be two point masses separated by a distance r_{12}, which is the magnitude of the vector $\mathbf{r}_{12}$ pointing from mass m_1 to m_2 (Figure 10-6). The force $\mathbf{F}_{12}$ exerted by mass m_1 on m_2 is then

Figure 10-6   (a) A particle of mass m_1 at position $\mathbf{r}_1$ and a second particle of mass m_2 at $\mathbf{r}_2$. The vector $\mathbf{r}_{12}$ points from m_1 to m_2. (b) The force $\mathbf{F}_{12}$ exerted by m_1 on m_2 is toward m_1 and varies inversely as the square of the distance between the two particles.

$$\mathbf{F}_{12} = -\frac{Gm_1m_2}{r_{12}^2}\,\hat{\mathbf{r}}_{12}$$

10-2   *Newton's law of gravity*

where $\hat{\mathbf{r}}_{12} = \mathbf{r}_{12}/r_{12}$ is a unit vector pointing from m_1 to m_2 and G is the **universal gravitational constant** and has the value

$$G = 6.67 \times 10^{-11} \text{ N·m}^2/\text{kg}^2$$

10-3

The force $\mathbf{F}_{21}$ exerted by m_2 on m_1 is the negative of $\mathbf{F}_{12}$, by Newton's third law. That is, $\mathbf{F}_{21}$ is equal in magnitude to $\mathbf{F}_{12}$ but opposite in direction. The magnitude of the gravitational force exerted by a particle of mass m_1 on another particle of mass m_2 a distance r away is thus given by

$$F = \frac{Gm_1m_2}{r^2} \qquad\qquad 10\text{-}4$$

Newton published his theory of gravitation in 1686, but it was not until a century later that an accurate experimental determination of G was made by Cavendish, as will be discussed in Section 10-3.

We can use the known value of G to compute the gravitational attraction between two ordinary objects.

Example 10-2

Find the force of attraction between two balls, each of mass 1 kg, when their centers are 10 cm apart.

We can treat each ball as a point mass. The magnitude of the force on either ball exerted by the other is then

$$F = \frac{(6.67 \times 10^{-11}\ \text{N·m}^2/\text{kg}^2)(1\ \text{kg})(1\ \text{kg})}{(0.1\ \text{m})^2}$$

$$= 6.67 \times 10^{-9}\ \text{N}$$

This example demonstrates that the gravitational force exerted by an object of ordinary size on another such object is extremely small. For example, the weight of a 1-kg object is 9.81 N, more than a billion times the force calculated in Example 10-2. We can usually neglect the gravitational force between objects compared with the other forces acting on them. The gravitational attraction can be noticed only if at least one of the objects is extremely massive, as with an ordinary object and the earth, or if great care is taken to eliminate other forces on the objects, as Cavendish did in determining G.

Exercise

Find the gravitational force that attracts a 65-kg boy to a 50-kg girl when they are 0.5 m apart, assuming that they are point masses. (Answer: 8.67×10^{-7} N)

Newton showed that, in general, when an object (such as a planet or comet) moves around a force center (such as the sun) to which it is attracted by a force that varies as $1/r^2$, the path of the object is either an ellipse, a parabola, or a hyperbola. The parabolic and hyperbolic paths apply to objects (if there are any) that make one pass by the sun and never return. Such orbits are not closed orbits. The only possible closed orbits in an inverse-square force field are ellipses. (A circle is a special case of an ellipse.) Thus, Kepler's first law is a direct consequence of Newton's law of gravity.

Kepler's second law, the law of equal areas, follows from the fact that the force exerted by the sun on a planet is directed toward the sun. Such a force is called a **central force.** Since the force on a planet is along the line from the planet to the sun, it has no torque about the sun. We know from our study of angular momentum that, if the net torque on an object is zero, the angular momentum of the object is conserved. Figure 10-7 shows a planet moving in an elliptical orbit around the sun. In time dt, the planet moves a distance $v\,dt$ and sweeps out the area indicated in the figure. This is half the area of the parallelogram formed by the vectors $\mathbf{r}$ and $\mathbf{v}\,dt$, which is

Figure 10-7   The area swept out by a planet in time dt is half the area of the parallelogram indicated, which is the magnitude of the cross product $\mathbf{r} \times \mathbf{v}\, dt$. The area is thus proportional to the angular momentum of the planet about the sun.

$\mathbf{r} \times \mathbf{v}\, dt$. Thus, the area dA swept out in time dt by the radius vector $\mathbf{r}$ is

$$dA = \frac{1}{2}|\mathbf{r} \times \mathbf{v}\, dt| = \frac{1}{2m}\,|\mathbf{r} \times m\mathbf{v}\, dt|$$

or

$$dA = \frac{1}{2m}\, L\, dt \qquad\qquad 10\text{-}5$$

where $\mathbf{L} = \mathbf{r} \times m\mathbf{v}$ is the angular momentum of the planet relative to the sun. The area swept out in a given time interval dt is therefore proportional to the angular momentum L. Since L is constant during the motion of the planet, the area swept out in a given time interval dt is the same for all parts of the orbit, which is Kepler's second law.

We will now show that Newton's law of gravity implies Kepler's third law for the special case of a circular orbit. We consider a planet moving about the sun with speed v in a circular orbit of radius r. Since the planet is moving in a circle, we know it has centripetal acceleration v^2/r. This acceleration is provided by the force of attraction between the sun and the planet, which is given by Newton's law of gravity. From Newton's second law of motion, we have

$$F = m_{\mathrm{p}}a$$
$$\frac{GM_{\mathrm{s}}m_{\mathrm{p}}}{r^2} = m_{\mathrm{p}}\frac{v^2}{r} \qquad\qquad 10\text{-}6$$

where M_{s} is the mass of the sun and m_{p} is that of the planet. Solving for v^2, we obtain

$$v^2 = \frac{GM_{\mathrm{s}}}{r} \qquad\qquad 10\text{-}7$$

We can now relate the speed v of the planet to its period T. Since the planet moves a distance $2\pi r$ in time T, its speed is

$$v = \frac{2\pi r}{T} \qquad\qquad 10\text{-}8$$

Substituting this expression for v in Equation 10-7, we obtain

$$v^2 = \frac{4\pi^2 r^2}{T^2} = \frac{GM_{\mathrm{s}}}{r}$$

or

$$T^2 = \frac{4\pi^2}{GM_{\mathrm{s}}}\, r^3 \qquad\qquad 10\text{-}9 \qquad \textit{Kepler's third law}$$

Equation 10-9 is Kepler's third law. For a general elliptical path, the radius r is replaced by the mean distance between the planet and the sun, which equals the semimajor axis of the ellipse. This equation also applies to the orbits of the satellites of any planet if we replace the mass of the sun M_{s} with the mass of the planet. For example, it applies to the earth's moon and to all the artificial satellites orbiting the earth if the sun's mass M_{s} is replaced by

Johannes Kepler (1571–1630).

the earth's mass M_E. We also note that, since G is known, we can determine the mass of a planet by measuring the period T and the mean orbital radius r of a moon orbiting it.

Example 10-3

Mars has a moon with a period of 460 min and a mean orbital radius of 9.4 Mm. What is the mass of Mars?

Replacing M_s in Equation 10-9 with the mass of Mars M and using $r = 9.4 \times 10^6$ m, $T = 460(60)$ s, and $G = 6.67 \times 10^{-11}$ N·m²/kg², we obtain

$$M = \frac{4\pi^2 r^3}{GT^2} = \frac{4\pi^2(9.4 \times 10^6 \text{ m})^3}{(6.67 \times 10^{-11} \text{ N·m}^2/\text{kg}^2)[460(60) \text{ s}]^2} = 6.45 \times 10^{23} \text{ kg}$$

As a further check on the validity of the inverse-square nature of the gravitational force, Newton compared the acceleration of the moon in its orbit with the acceleration of objects near the surface of the earth (such as the legendary apple). He made the bold assumption that the force that causes the moon to circle the earth has the same origin as the force that causes objects near the earth's surface to fall toward the earth, namely the gravitational attraction due to the earth. He first assumed that the earth and moon could be treated as point particles with their total masses concentrated at their centers. Since the distance to the moon is about 60 times the radius of the earth, the acceleration of objects near the surface of the earth ($g = 9.81$ m/s²) should be $60^2 = 3600$ times the acceleration of the moon. The moon's centripetal acceleration can be calculated from its known distance from the center of the earth and its period:

$$a_m = \frac{v^2}{r} = \frac{(2\pi r/T)^2}{r} = \frac{4\pi^2 r}{T^2}$$

For $r = 3.84 \times 10^8$ m and $T = 27.3$ d, the result is $a_m = 2.72 \times 10^{-3}$ m/s². Comparing this result with g, we have

$$\frac{g}{a_m} = \frac{9.81 \text{ m/s}^2}{2.72 \times 10^{-3} \text{ m/s}^2} = 3607 \approx 60^2$$

The calculations agreed "pretty nearly," according to Newton: "I thereby compared the force requisite to keep the Moon in her orb with the force of gravity at the surface of the Earth, and found them answer pretty nearly."

The assumption that the earth and moon can be treated as point particles in the calculation of the force on the moon is reasonable because the moon is far from the earth compared with the radius of either the earth or the moon, but such an assumption is certainly questionable when applied to the force exerted by the earth on an object near its surface. After considerable effort, Newton was able to prove that the force exerted by any spherically symmetric object on a point mass either on or outside its surface is the same as if all the mass of the object were concentrated at its center. The proof involves integral calculus, which Newton developed to solve this problem. This proof is given in Section 10-7.

The force exerted by the earth on any mass m at a distance r from the center of the earth is directed toward the earth and has the magnitude given by Equation 10-4 with m_1 equal to the mass of the earth M_E and m_2 replaced by m:

$$F = \frac{GM_E m}{r^2} \qquad\qquad 10\text{-}10$$

As mentioned in Chapter 4, the gravitational force on a mass divided by the mass is called the gravitational field. The gravitational field of the earth at a distance r (with r greater than the radius of the earth) points toward the earth and has the magnitude $g(r)$ given by

$$g(r) = \frac{F}{m} = \frac{GM_E}{r^2}$$

10-11   *Gravitational field of the earth*

The force exerted by the earth on an object at some height h above the surface of the earth is given by Equation 10-10 with $r = R_E + h$. The force exerted by the earth on an object of mass m near the earth's surface is also given by Equation 10-10 but with $r = R_E$:

$$F = \frac{GM_E m}{R_E^2}$$

When this is the only force acting on the object, the object is in free fall and has an acceleration

$$a = \frac{F}{m} = \frac{GM_E}{R_E^2} = g = 9.81 \text{ m/s}^2$$

10-12

Since $g = 9.81 \text{ m/s}^2$ is easily measured and the radius of the earth is known, Equation 10-12 can be used to determine either the constant G or the mass of the earth M_E if one of these quantities is known. Newton estimated the value of G from an estimate of the mass of the earth. When Cavendish determined G some 100 years later by measuring the force between small spheres of known mass and separation, he called his experiment "weighing the earth."

Example 10-4

What is the free-fall acceleration of an object 200 km above the earth's surface?

The force on the object is given by Equation 10-10 with $r = R_E + 200$ km. The acceleration is then

$$a = \frac{F}{m} = \frac{GM_E}{r^2}$$

Rather than substituting the values of G and M_E into this equation, we can use Equation 10-12 to write the product GM_E in terms of the radius of the earth and the free-fall acceleration of gravity g near the surface of the earth:

$$GM_E = gR_E^2$$

10-13

The acceleration at distance r is then

$$a = \frac{GM_E}{r^2} = \frac{gR_E^2}{r^2} = g\frac{R_E^2}{r^2}$$

10-14

Using $R_E = 6.37$ Mm $= 6370$ km and $r = R_E + 200$ km $= 6570$ km, we obtain

$$a = (9.81 \text{ m/s}^2)\left(\frac{6370 \text{ km}}{6570 \text{ km}}\right)^2$$

$$= 9.22 \text{ m/s}^2$$

Exercise

At what distance h above the surface of the earth is the acceleration of gravity half its value at sea level?   (Answer: $h = 2640$ km)

Example 10-5

A satellite orbits the earth in a circular orbit. Find its period (*a*) if the satellite is just above the surface of the earth and (*b*) if the satellite is at an altitude of 300 km. (Assume that air resistance can be neglected.)

(*a*) We can apply Equation 10-9 (Kepler's third law) to satellites orbiting the earth if we replace the mass of the sun M_s with the mass of the earth M_E. We then have

$$T^2 = \frac{4\pi^2}{GM_E} r^3$$

Again, it is convenient to replace GM_E with gR_E^2:

$$T^2 = \frac{4\pi^2}{gR_E^2} r^3$$

If the satellite is just above the surface of the earth, $r = R_E$ and

$$T^2 = \frac{4\pi^2}{gR_E^2} R_E^3 = \frac{4\pi^2 R_E}{g}$$

Therefore

$$T = 2\pi\sqrt{\frac{R_E}{g}} = 2\pi\sqrt{\frac{6.37 \times 10^6 \text{ m}}{9.81 \text{ m/s}^2}} = 5.06 \times 10^3 \text{ s} = 84.4 \text{ min}$$

(*b*) At an altitude of 300 km above the earth's surface, $r = 6370$ km $+ 300$ km $= 6670$ km $= 6.67 \times 10^6$ m. Since T is proportional to $r^{3/2}$, we can find T at this distance from

$$T = (84.4 \text{ min})\left(\frac{r}{R_E}\right)^{3/2}$$

$$= (84.4 \text{ min})\left(\frac{6.67 \times 10^6 \text{ m}}{6.37 \times 10^6 \text{ m}}\right)^{3/2}$$

$$= 90.4 \text{ min}$$

Exercise

Find the radius of a circular orbit of a satellite that orbits the earth with a period of one day. (If such a satellite is in orbit over the equator and moves in the same direction as the rotation of the earth, it appears stationary relative to the earth. Most satellites are parked in such an orbit, which is called a geosynchronous orbit.)   (Answer: $r = 6.63R_E = 4.22 \times 10^7$ m)

Questions

1. Why don't you feel the gravitational attraction of a large building when you walk near it?

2. Astronauts orbiting the earth in a satellite 300 km above the surface of the earth feel weightless. Why? Is the force of gravity exerted by the earth on them negligible at this height?

10-3   Measurement of the Universal Gravitational Constant

A knowledge of the universal gravitational constant G is not only of intrinsic interest, but it also has practical applications in the determination of the density distributions of the interiors of the earth, the moon, other planets, and distant stars. The first measurement of G was made by Henry Cavendish in 1798. Figure 10-8a is a schematic sketch of the apparatus Cavendish used to measure the gravitational force between two small objects. Two small spheres, each of mass m_2, are at the ends of a light rod suspended by a fine fiber. A torque is required to turn the two spheres through the angle θ from their equilibrium position because the fiber must be twisted. Careful measurement shows that the torque required to turn the fiber through a given angle is proportional to the angle. The constant of proportionality, called the torsion constant, can be determined, and the fiber and the suspended masses can be used to measure very small torques. This arrangement, called a torsion balance, was invented in the eighteenth century by John Michell. Charles Augustin de Coulomb used a similar device in 1785 to determine the electric force between charged particles, which is now referred to as Coulomb's law (Chapter 18). Cavendish used a refined and especially sensitive torsion balance in his determination of G.

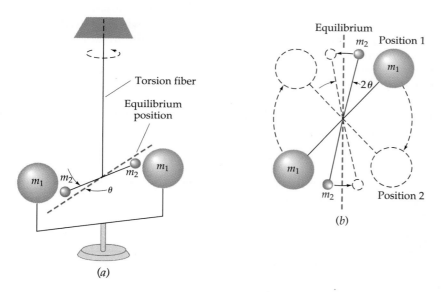

(a)

(b)

Figure 10-8   (a) Schematic drawing of Cavendish's apparatus for measuring G. Because of the gravitational attraction of the large spheres of mass m_1 for the nearby small spheres of mass m_2, the torsion fiber is turned through a very small angle θ from its equilibrium position. (b) The apparatus as seen from above. The large spheres are reversed so that they are at the same distance from the equilibrium position of the balance but on the other side. The fiber then turns through the angle 2θ. Once the torsion constant of the fiber has been determined, measurement of this angle makes it possible to determine the force exerted by m_1 on m_2, which in turn allows G to be determined.

In Cavendish's experiment two large spheres, each of mass m_1, are placed near the small spheres as shown in Figure 10-8a. The apparatus is allowed to come to equilibrium. Since the apparatus is so sensitive and the gravitational force is so small, this takes hours. Instead of measuring the deflection angle directly, Cavendish reversed the positions of the large spheres as shown by the dashed lines in Figure 10-8b. If the torsion balance is again allowed to come to equilibrium, it will turn through angle 2θ in response to the reversal of the torque. Once the torsion constant has been determined, the forces between the masses m_1 and m_2 can be determined from the measurement of this angle. When the masses and their separations are known, G can be calculated. Cavendish obtained a value for G within about 1 percent of the presently accepted value given by Equation 10-3.

Cavendish's measurement of G has been repeated by other experimenters with various improvements and refinements. In one variation, the torsion balance is used as an oscillator, and the period of oscillation, which

Original drawing of the
Cavendish apparatus.

depends on the attraction of the nearby masses, is measured. (We will study oscillations in Chapter 12.) In another, the attracting masses are moved back and forth around the torsion balance in resonance with the oscillations and the amplitude of the vibration is measured. In a different kind of procedure, G is determined by measuring the effect of a large attracting mass on one of the masses of a very sensitive, equal-arm chemical balance. All measurements of G are difficult because of the extreme weakness of the gravitational attraction. Consequently, the value of G is known today only to about 1 part in 10,000. Although G was one of the first physical constants ever measured, it remains one of the least accurately known. Table 10-1 lists some of the results of the various measurements of G.

Gravitational Torsion Balance used in student laboratories for the measurement of G. A tiny angular deflection of the balance results in a large angular deflection of the laser beam that reflects from a mirror on the balance.

Table 10-1   **Measurements of G**

Experimenter	Year	Method	G (10^{-11} N·m²/kg²)
Cavendish	1798	Torsion balance, deflection	6.754
Poynting	1891	Common balance	6.698
Boys	1895	Torsion balance, deflection	6.658
Von Eötvös	1896	Torsion balance, deflection	6.65
Heyl	1930	Torsion balance, period	
		Gold	6.678
		Platinum	6.664
		Glass	6.674
Zahradniček	1933	Torsion balance, resonance	6.659
Heyl and Chrzanowski	1942	Torsion balance, period	6.673
Luther and Towler	1982	Torsion balance, period	6.6726

The very small magnitude of G means that the gravitational force exerted by one object of ordinary size on another such object is extremely small, as we saw in Example 10-2. Such forces can be observed only if extreme care is taken to balance all other forces acting on the objects, as must be done to measure G.

10-4   Gravitational and Inertial Mass

The property of an object responsible for the gravitational force it exerts on another object is called its **gravitational mass.** On the other hand, the property of an object that measures its resistance to acceleration is called its **inertial mass.** We have used the same symbol m for these two properties because, experimentally, the gravitational and inertial masses of an object are equal. The fact that the gravitational force exerted by an object is proportional to its inertial mass is a characteristic unique to the force of gravity among all the forces we know, and it is a matter of considerable interest. One consequence is that all objects near the surface of the earth fall with the same acceleration if air resistance is neglected. This fact has seemed surprising to all since it was discovered. The well-known story of how Galileo demonstrated it by dropping objects from the Leaning Tower of Pisa is just one example of the excitement this discovery aroused in the sixteenth century.

We could easily imagine that the gravitational and inertial masses of an object were not the same. Suppose we write m_G for the gravitational mass and m for the inertial mass. The force exerted by the earth on an object near its surface would then be

$$F = \frac{GM_E m_G}{R_E^2} \qquad \text{10-15}$$

where M_E is the gravitational mass of the earth. The free-fall acceleration of the object near the earth's surface would then be

$$a = \frac{F}{m} = \left(\frac{GM_E}{R_E^2} \right) \frac{m_G}{m} \qquad \text{10-16}$$

If gravity were just another property of matter like color or hardness, it might be reasonable to expect that the ratio m_G/m would depend on such things as the chemical composition of the object, its temperature, or some other physical characteristics of the object. The free-fall acceleration would then be different for different objects. The experimental fact, however, is that a is the same for all bodies. This means that the ratio m_G/m is the same for every object. Since this is the case, we need not maintain the distinction between m_G and m and can set $m_G = m$. (This amounts to choosing the ratio to be 1, which in turn determines the magnitude and units of G in the law of gravity.) We must keep in mind, however, that the equivalence of gravitational and inertial mass is an experimental law, that is limited by the accuracy of experiment. Experiments testing this equivalence were carried out by Simon Stevin in the 1580s. Galileo publicized this law widely, and his contemporaries made considerable improvements in the experimental accuracy with which the law was established.

The most precise early comparisons of gravitational and inertial mass were made by Newton. Through experiments using simple pendulums rather than falling bodies, Newton was able to establish the equivalence between gravitational and inertial mass to an accuracy of about 1 part in 1000. Experiments comparing gravitational and inertial mass have improved steadily over the years. Their equivalence is now established to about 1 part in 10^{12}. Thus, the equivalence of gravitational and inertial mass is one of the best established of all physical laws.

The equivalence of gravitational and inertial mass is the basis for the principle of equivalence, which is the foundation of Einstein's general theory of relativity. We will study this theory in Chapter 34.

10-5   Escaping the Earth

In the last three decades, the idea of escaping from the earth's gravitational field has changed from fantasy to reality. Space probes have been sent out to the far reaches of the solar system. Some of these probes are expected to orbit the sun, whereas others will leave the solar system and drift on into outer space. In this section, we will look at the problem of escaping from the gravitational field of the earth (or of the sun). We will see that there is a minimum initial speed, called the escape speed, that is required of an object to escape from the earth, and we will calculate this escape speed using Newton's law of gravity.

Near the surface of the earth, the force of attraction between the earth and some object of mass m is a constant mg, which is independent of the height of the object above the earth's surface. The gravitational field near the surface of the earth is said to be uniform. If we project an object vertically upward with initial speed v in a uniform gravitational field, it will rise to a maximum height h that we can easily calculate from the conservation of energy. (We will neglect air resistance, which is important for practical calculations but does not affect the essential ideas of this discussion.) If the gravitational potential energy is zero at the earth's surface, the initial potential energy of the object is zero and the initial kinetic energy is $\frac{1}{2}mv^2$. At the maximum height h, the kinetic energy is zero and the potential energy is mgh. The conservation of energy gives

$$\tfrac{1}{2}mv^2 = mgh \qquad\qquad\qquad 10\text{-}17$$

or

$$v^2 = 2gh \qquad\qquad\qquad 10\text{-}18$$

According to Equation 10-18, if we increase the initial speed of the object, the maximum height attained by the object increases. But for any initial speed, no matter how great, there is some maximum height h. That is, there is no initial speed v great enough to allow an object to escape from a uniform gravitational field.

However, we know from Newton's law of gravity that the gravitational field of the earth is not uniform but decreases as $1/r^2$, where r is the distance from the center of the earth. If we project an object upward with a very large initial speed so that the object moves a distance comparable to the radius of the earth, we must take into account the decrease in the gravitational force on the object to calculate correctly the maximum height the object attains.

(a) Proposed orbit of the Galileo satellite launched in 1989 and due to arrive at Jupiter in 1995. (b) The Galileo satellite, which will orbit Jupiter and send back high quality images and instrument data about the planet and its natural moons.

(a)

(b)

Thus, in using conservation of energy, we need to find the correct expression for the potential energy of an object in the gravitational field of the earth, which varies as $1/r^2$.

We first show that any central force whose magnitude depends only on the radial coordinate r is conservative. We do this by showing that the work done by such a force on a particle that moves from some initial position P_1 to some final position P_2 is independent of the path followed and therefore depends only on the initial and final positions. We consider the general central force

$$\mathbf{F} = F_r \hat{\mathbf{r}}$$

where $\hat{\mathbf{r}}$ is a unit vector in the radial direction and F_r depends only on r. The work done by this force on a particle whose displacement is $d\mathbf{s}$ is

$$dW = \mathbf{F} \cdot d\mathbf{s} = (F_r \hat{\mathbf{r}}) \cdot d\mathbf{s} = F_r \, dr$$

(In general, the displacement $d\mathbf{s}$ has both a radial component $dr\,\hat{\mathbf{r}}$ and a component perpendicular to $\hat{\mathbf{r}}$. However, since $\mathbf{F}$ has only a radial component, the dot product $\mathbf{F} \cdot d\mathbf{s} = F_r \, dr$.) Figure 10-9 shows two paths connecting the initial and final positions of the particle. The work done by $\mathbf{F}$ along path 1, which is a simple radial path, is the integral of $F_r \, dr$ from r_1 to r_2. Path 2 is made up of radial segments and arc segments that are perpendicular to the radial lines. No work is done along these perpendicular segments because $\mathbf{F}$ and $d\mathbf{s}$ are perpendicular and therefore $\mathbf{F} \cdot d\mathbf{s} = 0$. The total work done along the radial segments of path 2 is the same as that done along path 1 because F_r depends only on r. Thus, the work done is independent of the path, so the force $\mathbf{F}$ is conservative.

To find the potential-energy function $U(r)$ for the system consisting of a particle of mass m and the earth, we use the general definition of potential energy (Equation 6-17):

$$dU = -\mathbf{F} \cdot d\mathbf{s} = -\left(-\frac{GM_E m}{r^2}\,\hat{\mathbf{r}}\right) \cdot d\mathbf{s}$$

where $\mathbf{F}$ is the force exerted by the earth on the particle and $d\mathbf{s}$ is a general displacement of the particle. The change in the potential energy of the object–earth system for a displacement $d\mathbf{s}$ is

$$dU = -\mathbf{F} \cdot d\mathbf{s} = -F_r \, dr = -\left(-\frac{GM_E m}{r^2}\right) dr = +\frac{GM_E m}{r^2} \, dr \qquad 10\text{-}19$$

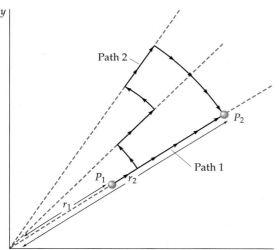

Figure 10-9 As a particle moves from point P_1 to point P_2, the work done on it by a central force $\mathbf{F}$ that depends only on r is the same along the radial path 1 as it is along path 2, which is made up of radial segments and arc segments. The work done along the arc segments is zero because $\mathbf{F}$ is perpendicular to $d\mathbf{s}$.

When the particle is displaced from some initial position r_1 to some final position r_2, the change in the potential energy is

$$U(r_2) - U(r_1) = \int_{r_1}^{r_2} dU = \int_{r_1}^{r_2} \frac{GM_E m}{r^2}\, dr$$

$$= -\left.\frac{GM_E m}{r}\right|_{r_1}^{r_2} = \frac{GM_E m}{r_1} - \frac{GM_E m}{r_2} \qquad \text{10-20}$$

If we choose the potential energy to be zero when the particle is at the earth's surface ($r_1 = R_E$), we have

$$U(r_2) - 0 = \frac{GM_E m}{R_E} - \frac{GM_E m}{r_2}$$

At some general position r ($r > R_E$), the potential-energy function $U(r)$ is thus

Gravitational potential energy with $U = 0$ at earth's surface

$$U(r) = \frac{GM_E m}{R_E} - \frac{GM_E m}{r} \qquad U = 0 \text{ at } r = R_E \qquad \text{10-21}$$

If we write this potential-energy function in terms of the distance above the surface of the earth, $y = r - R_E$, we can compare this function with mgy, which we obtained previously for an object near the surface of the earth at a height y above the surface. Putting the two terms in Equation 10-21 over a common denominator $R_E r$, we obtain

$$U(r) = \frac{GM_E m}{R_E} - \frac{GM_E m}{r} = \frac{GM_E m}{R_E r}(r - R_E)$$

Using $y = r - R_E$, and writing $GM_E = gR_E^2$ from Equation 10-13, we have

$$U(r) = \frac{GM_E m}{R_E r} y = m\left(\frac{GM_E}{R_E^2}\right) y \frac{R_E}{r} = mgy \frac{R_E}{r} \qquad \text{10-22}$$

The potential energy is thus mgy times the quantity R_E/r. Near the surface of the earth, r and R_E are approximately equal, and the potential energy is approximately mgy, as obtained by assuming the gravitational force to be constant. Figure 10-10 shows this potential energy as a function of r. The blue solid line is the potential-energy function $mgy = mg(r - R_E)$ for a constant gravitational force. We note that the actual potential-energy function for the varying gravitational force does not become infinitely large as r in-

Figure 10-10 Graph of the gravitational potential energy $U(r)$ versus r. For r just slightly greater than R_E, $U(r)$ is approximately equal to $mgy = mg(r - R_E)$, which is indicated by the blue solid line. At large values of r, $U(r)$ approaches U_{max}.

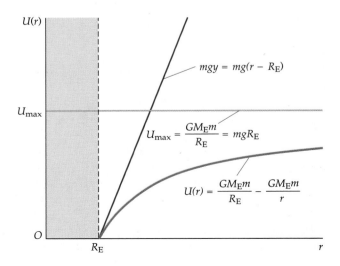

creases. Instead, as r becomes larger and larger, the second term in Equation 10-21 gets smaller and smaller, and the potential energy approaches a maximum value U_{max} given by

$$U_{max} = \frac{GM_E m}{R_E} = mgR_E \qquad \text{10-23}$$

where we have again used $GM_E = gR_E^2$. Note that this result can also be obtained from Equation 10-22 using $y \approx r$ for very large r.

Example 10-6

A projectile is fired straight up from the surface of the earth with an initial speed $v_1 = 8$ km/s. Find the maximum height it reaches, neglecting air resistance.

We use conservation of energy. Initially, at the surface of the earth, the potential energy of the projectile is zero and its kinetic energy is $\frac{1}{2}mv_1^2$. When it is at a height $h = r - R_E$, its potential energy is given by Equation 10-21. We wish to find the height for which its final speed $v_2 = 0$. Conservation of energy gives

$$\tfrac{1}{2}mv_1^2 + U_1 = \tfrac{1}{2}mv_2^2 + U_2$$

$$\tfrac{1}{2}mv_1^2 + 0 = 0 + \left(\frac{GM_E m}{R_E} - \frac{GM_E m}{r} \right)$$

Simplifying this equation, we obtain

$$v_1^2 = \frac{2GM_E}{R_E} - \frac{2GM_E}{r} = \frac{2GM_E}{R_E}\left(1 - \frac{R_E}{r} \right)$$

We can simplify this further by using $GM_E = gR_E^2$. Then

$$v_1^2 = 2gR_E\left(1 - \frac{R_E}{r} \right)$$

It is sufficient to solve for R_E/r:

$$1 - \frac{R_E}{r} = \frac{v_1^2}{2gR_E}$$

$$\frac{R_E}{r} = 1 - \frac{v_1^2}{2gR_E}$$

$$= 1 - \frac{(8000 \text{ m/s})^2}{2(9.81 \text{ m/s}^2)(6.37 \times 10^6 \text{ m})}$$

$$= 1 - 0.512 = 0.488$$

Then

$$r = \frac{R_E}{0.488} = 2.05R_E$$

The maximum height reached is therefore $h = r - R_E = 1.05R_E$.

We are now ready to find the initial speed an object must have to escape from the earth's gravitational field. If we project an object upward from the earth with some initial kinetic energy, the kinetic energy decreases and the potential energy increases as the object rises. But the potential energy cannot increase by more than U_{max}, as given by Equation 10-23. Therefore, this is the most that the kinetic energy can decrease. If the initial kinetic energy is greater than U_{max}, the object will still have some kinetic energy when r is very great (or even when r is infinite). Thus, the object will escape from the earth if the initial kinetic energy is greater than U_{max}. The critical speed v_e associated with this critical energy is called the **escape speed.** It is found from

$$\tfrac{1}{2}mv_e^2 = U_{max} = \frac{GM_E m}{R_E} = mgR_E$$

or

Escape speed

$$v_e = \sqrt{\frac{2GM_E}{R_E}} = \sqrt{2gR_E} \qquad\qquad 10\text{-}24$$

Using $g = 9.81 \text{ m/s}^2$ and $R_E = 6.37$ Mm, we obtain

$$v_e = \sqrt{2(9.81 \text{ m/s}^2)(6.37 \times 10^6 \text{ m})} = 11.2 \text{ km/s}$$

This is about 6.95 mi/s or about 25,000 mi/h.

The energy $U_{max} = mgR_E$ is called the **binding energy.** If the kinetic energy of an object at the earth's surface is less than the binding energy, the object will not leave the earth but will rise to some maximum height h and then fall back to earth. If its initial kinetic energy is greater than the binding energy, the object will continue moving away from the earth forever. Thus, the escape speed is just that speed corresponding to a kinetic energy that is equal to the binding energy. An earth–mass system is said to be bound or unbound depending on whether the kinetic energy of the mass at the earth's surface is less or greater than the binding energy.

The magnitude of the escape speed for a planet or moon relative to the thermal speeds of gas molecules determines the kind of atmosphere a planet or moon can have. The average kinetic energy of gas molecules, $(\tfrac{1}{2}mv^2)_{av}$, is proportional to the absolute temperature T (Chapter 15). The average speed of a gas molecule then depends on the temperature and varies inversely with the mass of the molecule. At the surface of the earth, the speeds of oxygen and nitrogen molecules are much lower than the escape speed, so these gases can exist in our atmosphere. For the lighter molecules hydrogen and helium, however, a considerable fraction of the molecules have speeds greater than the escape speed. Hydrogen and helium gases are therefore not found in our atmosphere. The escape speed at the surface of the moon can be calculated from Equation 10-24, with the acceleration of gravity on the moon and the radius of the moon replacing g and R_E. The escape speed for the moon is 2.3 km/s, considerably smaller than that for the earth and too small for any atmosphere to exist.

Example 10-7

Find the escape speed at the surface of Mercury, which has a mass $M = 3.31 \times 10^{23}$ kg and a radius $R = 2.44$ Mm.

Equation 10-24 can be used for the escape speed at the surface of Mercury if we replace the mass of the earth M_E with the mass of Mercury, and replace the radius of the earth R_E with the radius of Mercury. Then

$$v_e = \sqrt{\frac{2GM}{R}} = \sqrt{\frac{2(6.67 \times 10^{-11} \text{ N·m}^2/\text{kg}^2)(3.31 \times 10^{23} \text{ kg})}{2.44 \times 10^6 \text{ m}}}$$

$$= 4.25 \times 10^3 \text{ m/s} = 4.25 \text{ km/s}$$

Questions

3. What is the effect of air resistance on the escape speed near the earth's surface?

4. Would it be possible in principle for the earth to escape from the solar system?

10-6   Potential Energy, Total Energy, and Orbits

For the gravitational potential-energy function of Equation 10-21, we chose the potential energy to be zero at the surface of the earth. Since only changes in potential energy are important, we could have chosen the potential energy to be zero at any arbitrary position. Although it seems natural to choose the potential energy to be zero at the earth's surface when we are working problems involving objects near the earth's surface, there are many situations in which this is not at all the most convenient choice. For example, when considering the potential energy associated with a planet and the sun, there is no reason to want the potential energy to be zero at the surface of the sun. In fact, it is nearly always more convenient to choose the gravitational potential energy of a two-object system to be zero when the separation of the two objects is infinite. For a general two-object system consisting of one object of mass m and another of mass M, the change in the potential energy when the system moves from a separation of r_1 to a separation of r_2 is given by Equation 10-20:

Earth as seen from Apollo 11 orbiting the moon on July 31, 1969.

$$U(r_2) - U(r_1) = \frac{GMm}{r_1} - \frac{GMm}{r_2}$$

Setting r_2 equal to some general separation r and $r_1 = \infty$, we have

$$U(r) - U(\infty) = -\frac{GMm}{r}$$

Choosing U to be zero at $r = \infty$ then gives

$$U(r) = -\frac{GMm}{r} \qquad U = 0 \text{ at } r = \infty \qquad \qquad 10\text{-}25$$

Gravitational potential energy with $U = 0$ at infinite separation

(This choice for zero potential energy seems to have the disadvantage of making the potential energy always negative, but this is not really a disadvantage. It is like choosing the potential energy to be zero at the ceiling of a room rather than at the floor when you are doing a problem involving an object in a room.)

Figure 10-11 is a plot of $U(r)$ versus r for this choice of $U = 0$ at $r = \infty$ for an object of mass m and the earth of mass M_E. This function begins at the negative value $U = -(GM_Em)/R_E = -mgR_E$ at the earth's surface and increases as r increases, approaching zero at infinite r. The maximum increase in potential energy is still mgR_E, as it was when U was chosen to be zero at the earth's surface. The criterion for an object to escape is also the same: the kinetic energy at the earth's surface must be greater than or equal to mgR_E. Since the potential energy at the earth's surface is $-mgR_E$, the total energy $E = K + U$ must be greater than or equal to zero for the object to escape.

In Figure 10-11, two possible values for the total energy E are shown on the graph: E_1, which is negative, and E_2, which is positive. A negative total

Figure 10-11 The same potential-energy curve as in Figure 10-10, except that $U(r)$ is chosen to be zero at $r = \infty$. For this choice, a particle is bound if its total energy E is less than zero and unbound if E is greater than zero.

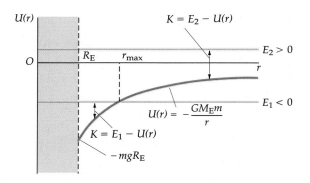

energy merely means that the kinetic energy at the earth's surface is less than mgR_E, so the magnitude of the kinetic energy is never greater than the magnitude of the potential energy. From the figure, we can see that when the total energy is negative, the total-energy line intersects the potential-energy curve at some maximum separation r_{max} and the system is bound. On the other hand, when the total energy is positive, there is no intersection and the system is unbound. The criteria for bound and unbound systems can be simply stated:

If $E < 0$, the system is bound.

If $E > 0$, the system is unbound.

The binding energy is, of course, the same regardless of the choice of where the potential energy is zero.

When we choose the gravitational potential energy to be zero at infinite separation, we can discuss the kind of orbit taken by an object that moves in the gravitational field of the sun. The potential energy of an object of mass m at a distance r from the sun is

$$U(r) = -\frac{GM_s m}{r} \qquad \text{10-26}$$

where M_s is the mass of the sun. The kinetic energy of the object is $\frac{1}{2}mv^2$. If the total energy, kinetic plus potential, is less than zero, the orbit will be an ellipse (or a circle), and the object will be bound to the sun. That is, it cannot escape from the gravitational field of the sun. On the other hand, if the total energy is positive, the orbit will be a hyperbola, which means that the object will make one swing around the sun and leave, never to return again. If the total energy is exactly zero, the orbit will be a parabola, and again the object will escape. That is, when the total energy is zero or positive, the object is not bound to the gravitational field of the sun.

Example 10-8

Show that the total energy of a satellite in a circular orbit equals half its potential energy.

The total energy of an orbiting satellite is the sum of its potential energy and its kinetic energy. If it is a distance r from the center of the earth, its potential energy is

$$U = -\frac{GM_E m}{r}$$

Since it is moving in a circle of radius r, it has a centripetal acceleration of magnitude v^2/r. From Newton's second law, $\mathbf{F} = m\mathbf{a}$, we have

$$\frac{GM_Em}{r^2} = \frac{mv^2}{r}$$

The kinetic energy of the satellite is thus

$$K = \frac{1}{2}mv^2 = \frac{1}{2}\frac{GM_Em}{r}$$

Thus, the kinetic energy has a magnitude half that of the potential energy. This result holds for any circular orbit in an inverse-square force field. The total energy of the satellite is thus

$$E = K + U = \frac{1}{2}\frac{GM_Em}{r} - \frac{GM_Em}{r} = -\frac{1}{2}\frac{GM_Em}{r} = \frac{1}{2}U(r)$$

On the energy diagram in Figure 10-11, the total energy of a satellite orbiting in a circle would be halfway between the potential energy $U(r)$ and zero. The minimum additional energy such a satellite must be given to escape from the earth is thus $(\frac{1}{2}GM_Em)/r$. That is, the kinetic energy of the orbiting satellite must be doubled.

Example 10-9

A satellite of mass 450 kg orbits the earth in a circular orbit at 6.83 Mm above the earth's surface. Find (*a*) the potential energy, (*b*) the kinetic energy, and (*c*) total energy of the satellite.

(*a*) The distance between the satellite and the center of the earth is

$$r = R_E + h = 6.37 \text{ Mm} + 6.83 \text{ Mm} = 13.2 \text{ Mm}$$

From Equation 10-26, the potential energy of the satellite is

$$U = -\frac{GM_Em}{r}$$
$$= -\frac{(6.67 \times 10^{-11} \text{ N·m}^2/\text{kg}^2)(5.98 \times 10^{24} \text{ kg})(450 \text{ kg})}{1.32 \times 10^7 \text{ m}} = -13.6 \text{ GJ}$$

(*b*) In the previous example, we found that in a circular orbit the kinetic energy is half the magnitude of the potential energy. The kinetic energy is therefore

$$K = \frac{1}{2}\frac{GM_Em}{r} = -\frac{1}{2}U = -\frac{1}{2}(-13.6 \text{ GJ}) = 6.80 \text{ GJ}$$

(*c*) The total energy is the sum of the potential and kinetic energies.

$$E = K + U = 6.80 \text{ GJ} + (-13.6 \text{ GJ}) = -6.80 \text{ GJ}$$

The total energy equals the negative of the kinetic energy.

Question

5. An object (say, a newly discovered comet) enters the solar system and makes a pass around the sun. How can we tell if the object will return many years later, or if it will never return?

Figure 10-12 A uniform spherical shell of mass M and radius R. The force exerted by the shell on a mass outside the shell is the same as if the shell were a point mass at the center. The force exerted by the shell on a mass inside the shell is zero.

10-7   The Gravitational Field of a Spherical Shell and a Solid Sphere

One of Newton's motivations for developing calculus was to be able to prove that the gravitational field of a solid sphere at a point outside the sphere is the same as if all the mass of the sphere were concentrated at its center. This proof is not difficult, but it appears so because the geometry is somewhat involved, and the use of spherical coordinates may be unfamiliar. Before we give this proof, we will discuss some of the important results derived from it.

Figure 10-12 shows a uniform spherical shell of mass M and radius R. We will show below that the force exerted by this shell on a point mass m_0 located a distance r from the center of the shell is given by

$$\mathbf{F} = -\frac{GMm_0}{r^2}\hat{\mathbf{r}} \qquad r > R$$

and

$$\mathbf{F} = 0 \qquad\qquad r < R$$

That is, if m_0 is outside the shell, the force exerted on it by the shell is the same as if all the mass of the shell were concentrated at its center, whereas if m_0 is inside the shell, the shell exerts no force on it. This surprising result can be understood from Figure 10-13, which shows a point mass m_0 inside a spherical shell. In this figure, the masses of the shell segments m_1 and m_2 are related by

$$m_2 = m_1\left(\frac{r_2^2}{r_1^2}\right)$$

Since the force due to each mass is proportional to $1/r^2$, the force due to the smaller mass on the left is exactly balanced by that due to the larger mass on the right.

The gravitational field $\mathbf{g}$ of the shell is the force exerted on m_0 divided by m_0:

$$\mathbf{g} = \frac{\mathbf{F}}{m_0}$$

The gravitational field of a spherical shell of mass M and radius R is thus

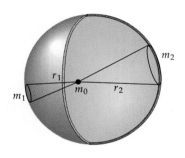

Figure 10-13 A point mass m_0 inside a uniform spherical shell feels no net force. The attraction of the part of the shell of mass m_1 is balanced by that of mass m_2, which is farther away but larger.

Gravitational field of a spherical shell

$$\mathbf{g} = -\frac{GM}{r^2}\hat{\mathbf{r}} \qquad r > R \qquad\qquad 10\text{-}27a$$

$$\mathbf{g} = 0 \qquad\qquad r < R \qquad\qquad 10\text{-}27b$$

The gravitational field of a uniform spherical shell of mass M at a point outside the shell is the same as that due to a point mass M at the center of the shell. Inside a uniform spherical shell, the gravitational field is zero.

We will now use this result to find the gravitational field of a solid sphere. For a point some distance r, outside the sphere, the problem is very simple. We merely consider the solid sphere to consist of a continuous set of spherical shells. Since the field due to each shell is the same as if its mass were concentrated at the center of the shell, the field due to the entire sphere is the same as if the entire mass of the sphere were concentrated at its center:

$$g_r = -\frac{GM}{r^2} \qquad r > R \qquad\qquad 10\text{-}28$$

M = total mass

This result holds whether or not the sphere has a constant density as long as the density depends only on r so that spherical symmetry is maintained.

We can also find the gravitational field of a solid sphere of constant density at a point inside the sphere at a distance r from the center, where r is less than the radius R of the sphere. This would apply, for example, to finding the weight of an object at the bottom of a deep mine shaft. As we have seen, the field inside a spherical shell is zero. Thus, in Figure 10-14, the mass of the sphere outside r exerts no force at or inside r. Therefore, only the mass M' within the radius r contributes to the gravitational field at r. This mass produces a field equal to that of a point mass M' at the center of the sphere. The fraction of the total mass of the sphere within r is equal to the ratio of the volume of a sphere of radius r to that of a sphere of radius R. Thus, for a uniform mass distribution, if M is the total mass of the sphere, M' is given by

Figure 10-14   A uniform solid sphere of mass M. The gravitational field outside the sphere is the same as if the sphere were a point mass at the origin. Inside the sphere at some distance $r < R$, only the mass M' that is within r contributes to the gravitational field.

$$M' = \frac{\frac{4}{3}\pi r^3}{\frac{4}{3}\pi R^3}M = \frac{r^3}{R^3}M \qquad\qquad 10\text{-}29$$

The gravitational field at the distance r is thus

$$g_r = -\frac{GM'}{r^2} = -\frac{(GMr^3)/R^3}{r^2}$$

or

$$g_r = -\frac{GM}{R^3}r \qquad r < R \qquad\qquad 10\text{-}30$$

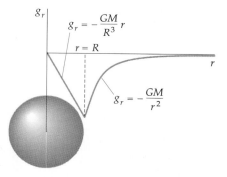

Figure 10-15   Plot of g_r versus r for a uniform solid sphere of mass M. The magnitude of the field increases with r inside the sphere and decreases as $1/r^2$ outside.

The magnitude of the field increases with distance r inside the sphere. Figure 10-15 shows a plot of the field g_r as a function of r for a solid sphere of constant mass density.

The rings and three moons of Saturn. Because of the $1/r^2$ dependence, the gravitational force on the near side of a satellite is much greater than that on the far side if the orbit is very near the planet. This force difference, called a **tidal force,** is great enough to break up large objects that are close to the planet, hence the rings of small particles orbiting close to Saturn. The three moons shown orbit at much greater distances where the tidal forces are much smaller.

Derivation of the Equation for the Gravitational Field of a Spherical Shell (Optional)

We will derive the equation for the gravitational field of a spherical shell in two steps. We will first find the gravitational field on the axis of a ring of uniform mass. We then apply our result to a spherical shell, which we can consider to be a set of coaxial rings.

Figure 10-16 shows a ring of total mass m and radius a and a point mass m_0 on the axis of the ring a distance x from its center. We have chosen an element of mass dm on the ring that is small enough to be considered a point particle. The element is at a distance s from the point mass on the axis, and the line joining the element and the point mass makes an angle α with the axis of the ring.

The force exerted by the element on m_0 is toward the element and has the magnitude dF given by

$$dF = \frac{G(dm)m_0}{s^2}$$

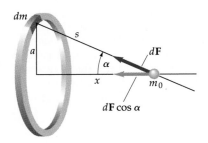

Figure 10-16 A uniform ring of total mass m and a point mass m_0 on its axis. The force on m_0 exerted by the element dm is toward the element. The total force exerted on m_0 by the ring is along the axis of the ring.

From the symmetry of the figure, we can see that when we sum over all the elements of the ring, the net force will be along the axis of the ring; that is, the perpendicular components will sum to zero. For example, the perpendicular component of the force on m_0 shown in the figure will be canceled by that due to another element of the ring directly opposite the one shown. The net force on m_0 will therefore be in the negative x direction. The x component of the force due to the element dm is

$$dF_x = -dF \cos \alpha = -\frac{G(dm)m_0}{s^2} \cos \alpha$$

The gravitational field dg_x due to the ring at the position of the point mass is this force divided by m_0:

$$dg_x = \frac{dF_x}{m_0} = -\frac{G\,dm}{s^2} \cos \alpha$$

We obtain the total field by summing over all the elements of the ring:

$$g_x = -\int \frac{G\,dm}{s^2} \cos \alpha$$

Since s and α are the same for all points on the ring, they are constants as far as the integration is concerned. Thus,

$$g_x = -\frac{G}{s^2} \cos \alpha \int dm = -\frac{Gm}{s^2} \cos \alpha \qquad \text{10-31}$$

where $m = \int dm$ is the total mass of the ring. We can write this result in terms of the distance x by using

$$s^2 = x^2 + a^2$$

and

$$\cos \alpha = \frac{x}{s} = \frac{x}{\sqrt{x^2 + a^2}}$$

Then,

$$g_x = -\frac{Gmx}{(x^2 + a^2)^{3/2}} \qquad \text{10-32}$$

We will now use this result to calculate the gravitational field of a spherical shell of mass M and radius R at a point a distance r from the center of the shell. We first consider the case in which the point is outside the shell. By

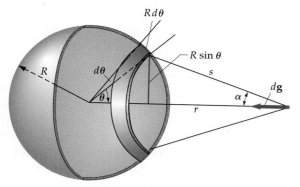

Figure 10-17 A uniform spherical shell of radius R and total mass M. The strip shown can be considered to be a ring of width $R\,d\theta$ and circumference $2\pi R \sin\theta\,d\theta$. The gravitational field of the shell is found by summing over all such strips on the shell, which is done by integrating from $\theta = 0$ to $\theta = 180°$.

symmetry, the field must be radial. The geometry is shown in Figure 10-17. We choose for our element of mass the strip shown, which can be considered to be a ring. The radius of the strip is $R \sin\theta$ and the width is $R\,d\theta$. If the total area of the shell is A, the area dA of the strip is the circumference, $2\pi R \sin\theta$, times the width, $R\,d\theta$. If M is the total mass of the shell, the mass of the strip of area dA is

$$dM = M\,\frac{dA}{A} = M\,\frac{2\pi R^2 \sin\theta\,d\theta}{4\pi R^2}$$

or

$$dM = \frac{M}{2}\sin\theta\,d\theta \qquad\qquad 10\text{-}33$$

where we have used $A = 4\pi R^2$ for the total area of the shell. Using Equation 10-31 for the field of a ring, we obtain

$$dg_r = -\frac{G\,dM}{s^2}\cos\alpha$$

Using Equation 10-33 for dM, we have

$$dg_r = -\frac{GM \sin\theta\,d\theta}{2s^2}\cos\alpha \qquad\qquad 10\text{-}34$$

Before integrating over the entire shell, we must eliminate two of the three related variables s, θ, and α. It turns out to be easiest to write everything in terms of s, which varies from $s = r - R$ at $\theta = 0$ to $s = r + R$ at $\theta = 180°$. By the law of cosines, we have

$$s^2 = r^2 + R^2 - 2rR\cos\theta$$

Differentiating gives

$$2s\,ds = +2rR\sin\theta\,d\theta$$

or

$$\sin\theta\,d\theta = \frac{s\,ds}{rR}$$

An expression for $\cos\alpha$ can be obtained by applying the law of cosines to the same triangle. We have

$$R^2 = s^2 + r^2 - 2sr\cos\alpha$$

or

$$\cos\alpha = \frac{s^2 + r^2 - R^2}{2sr}$$

Substituting these results into Equation 10-34 gives

$$dg_r = -\frac{GM}{2s^2}\frac{s\,ds}{rR}\frac{s^2 + r^2 - R^2}{2sr} = -\frac{GM}{4r^2R}\left(1 + \frac{r^2 - R^2}{s^2}\right)ds \qquad 10\text{-}35$$

The field due to the entire shell is found by integrating from $s = r - R$ ($\theta = 0$) to $s = r + R$ ($\theta = 180°$).

$$g_r = -\frac{GM}{4r^2R}\int_{r-R}^{r+R}\left(1 + \frac{r^2 - R^2}{s^2}\right) ds$$

$$= -\frac{GM}{4r^2R}\left[s - \frac{r^2 - R^2}{s}\right]_{r-R}^{r+R}$$

Substitution of the upper and lower limits yields $4R$ for the quantity in brackets. Thus,

$$g_r = -\frac{GM}{r^2}$$

which is Equation 10-27*a*.

We will now calculate the gravitational field of the shell at a point located inside the shell. This calculation is identical to that for a point outside the shell except that s now varies from $R - r$ to $R + r$. Thus,

$$g_r = -\frac{GM}{4r^2R}\left[s - \frac{r^2 - R^2}{s}\right]_{R-r}^{R+r}$$

Substitution of these upper and lower limits yields 0. Therefore,

$$g_r = 0$$

which is Equation 10-27*b*.

Question

6. Explain why the gravitational field increases with r rather than decreasing as $1/r^2$ as one moves out from the center inside a solid sphere of uniform mass.

Summary

1. Kepler's three laws are

    Law 1. All planets move in elliptical orbits with the sun at one focus.
    Law 2. A line joining any planet to the sun sweeps out equal areas in equal times.
    Law 3. The square of the period of any planet is proportional to the cube of the planet's mean distance from the sun.

    Kepler's laws can be derived from Newton's law of gravity. The first and third laws follow from the fact that the force exerted by the sun on the planets varies inversely as the square of the separation distance. The second law follows from the fact that the force exerted by the sun on a planet is along the line joining them and so the angular momentum of the planet is conserved. Kepler's laws also hold for any object orbiting another in an inverse-square field, such as a satellite orbiting a planet.

2. Newton's law of gravity postulates that every object exerts an attractive force on every other object that is proportional to the masses of the two objects and inversely proportional to the square of the distance separating them. The gravitational force exerted by a particle of mass m_1 on

another particle of mass m_2 a distance r away is given by

$$\mathbf{F}_{12} = -\frac{Gm_1m_2}{r_{12}^2}\hat{\mathbf{r}}_{12}$$

where G, the universal gravitational constant, has the value

$$G = 6.67 \times 10^{-11} \text{ N·m}^2/\text{kg}^2$$

3. The gravitational force exerted by the earth on a mass m a distance r from the center of the earth is directed toward the earth and has magnitude

$$F = \frac{GM_Em}{r^2}$$

The gravitational force on a mass divided by the mass is called the gravitational field. The gravitational field of the earth points toward the earth and has magnitude

$$g(r) = \frac{GM_E}{r^2}$$

4. The fact that all bodies have the same free-fall acceleration near the surface of the earth implies that the gravitational mass and the inertial mass of an object are equal. This equivalence of gravitational and inertial mass has been established experimentally to a very high degree of accuracy.

5. The gravitational potential energy of an object of mass m at a distance r from the center of the earth is given by

$$U(r) = U_{\max} - \frac{GM_Em}{r} \qquad U = 0 \text{ at } r = R_E$$

where $U_{\max} = GM_Em/R_E = mgR_E$ and the potential energy is chosen to be zero at the earth's surface. If an object is given an initial kinetic energy at the earth's surface greater than or equal to $U_{\max}$, the object will escape from the earth. In the absence of air resistance, the speed needed for escape is approximately 11.2 km/s.

6. When the gravitational potential energy is chosen to be zero for an infinite separation of the bodies, $U(r)$ is given by

$$U(r) = -\frac{GMm}{r} \qquad U = 0 \text{ at } r = \infty$$

and the orbits of objects in a gravitational field can be simply classified. If the total energy of an orbiting object is less than zero, the object is bound and the orbit is an ellipse. If the total energy is greater than or equal to zero, the object is not bound and the orbit is a parabola (for zero energy) or a hyperbola (for positive energy).

7. The gravitational field of a uniform spherical shell outside the shell is the same as if all the mass of the shell were concentrated at the center. The field inside the shell is zero. These results are obtained directly from Newton's law of gravity by integration. The gravitational field of a uniform solid sphere of mass M and radius R at a point a distance r from the center of the sphere is given by

$$g_r = -\frac{GM}{r^2} \qquad r > R$$

$$g_r = -\frac{GM}{R^3}r \qquad r < R$$

Isaac Newton (1642–1727)

A.P. French
Massachusetts Institute of Technology

When one considers all that Newton achieved, and the cultural and scientific environment in which he achieved it, there is reason to regard him as the greatest scientist—and perhaps the greatest genius—who has ever lived.

Despite the godlike status that some have tried to ascribe to him, he was of course a fallible human being. He groped and fumbled, as all scientists do, in approaching a new problem; he even fudged things a bit sometimes in trying to reconcile theory with observation. But the extent and profundity of his discoveries remain without parallel.

This would not have been predictable from his beginnings as a somewhat unloved child growing up in the English countryside, though he did demonstrate practical ingenuity and a probing curiosity, and was a good enough student to be admitted to Cambridge University. It was there that the real Newton emerged, with astonishing suddenness, as he approached the end of his undergraduate career. This young man, still an undergraduate, devoured most of what was then known in mathematics and began making his own original contributions. His

A. P. French was educated at Cambridge University. He was a member of the British Mission at Los Alamos in the Manhattan Project during World War II. From 1948 to 1955 he taught physics and did nuclear research at the Cavendish Laboratory, Cambridge. He then emigrated to the U.S.A. and joined the faculty of the University of South Carolina. In 1962 he moved to M.I.T., where he has been a member of the permanent faculty since 1964. His chief interests for many years have been in curriculum development, teaching, and the history of physics. He was President of the American Association of Physics Teachers in 1985–1986.

physics began at about the same time with investigations into optics, and through a succession of controlled experiments he built up a picture of what light is and how it behaves. Here his powers as an experimentalist revealed themselves—the brilliantly analytical mind behind the experiments that it devised.

His first concern seems to have been with the way in which the eye perceives color. The eye as an image-forming device was already well understood, but color perception was much more mysterious. (It still is.) One of the 21-year-old Newton's first experiments, in 1664, took the form of pressing one side of his eyeball with a fingertip and observing the colored rings that appeared to be formed around the pressure point. A little later he did the truly horrifying experiment of thrusting a bodkin into his eye socket, between the bone and the eyeball, to study the same phenomenon closer to the center of the retina (Figure 1). He also stared directly at the Sun and studied the subsequent color impressions he obtained by looking at light or dark objects. It is a miracle that he did not blind himself or do himself serious injury.

Figure 1 Newton's experiment on colors produced by pressing his eyeball.

He then concentrated on investigating the refraction of light through prisms—in particular the white light from the Sun (Figure 2). It had been the belief that colors were the result of the modification of pure white light by materials. What Newton con-

Figure 2 One of Newton's first experiments in forming an extended spectrum (from Voltaire, *The Elements of Sir Isaac Newton's Philosophy* 97).

Figure 3 Newton's crossed-prism experiment (from Newton, *Opticks* 36).

cluded was that the various colors were already present in the white light, that this was itself complex, and that the prism merely resolved it into its components. In all kinds of ways he sought to demonstrate the correctness of his analysis. One very beautiful experiment was to pass the light emerging from a first prism through a second prism oriented at right angles to it (Figure 3). This showed that no further modification of the light occurred; once the separation of the white light into its various components had been achieved with the first prism, the second prism merely refracted the individual colors by differing amounts and produced a final spectrum along a diagonal line, rather than a pattern filling a rectangle.

Newton's sustained and many-faceted attack on a single problem was something for which there was essentially no precedent or parallel in the scientific literature of his time. It was something qualitatively different from the mere collection of precise observations. Even in our own day, his approach could be taken as a model for the exploration of a new phenomenon.

In 1669, at the age of twenty-six, Newton was appointed Professor of Mathematics at Cambridge University. He chose to give his inaugural course of lectures on the subject of optics, and their final section—another Newtonian *tour de force*—was a quantitative explanation of the rainbow. A year or two later he designed and constructed the first reflecting telescope, the metal mirror of which was not only ground and polished by him, but was made of an alloy evolved during the chemical (and alchemical) researches that he had been conducting at the same time. All of this work was described in his *Opticks*, which was completed in the 1670s but not published until 1704. It is essentially a blow-by-blow account of his work in this field, written in an easy and conversational style, and makes excellent reading.

His most glorious achievements, however, were in mechanics, and were published in detail in his masterpiece, *Mathematical Principles of Natural Philosophy*, universally known as the *Principia*. Its style is in total contrast to the *Opticks*. Quite apart from the fact that its first editions were in classical Latin, the whole presentation is forbidding. Using pure geometry with a virtuosity that few mathematicians today could emulate, Newton embodied his arguments in a Euclidean framework of axioms, theorems, lemmas, and corollaries. At least, this is the general character of the first two parts. In the third part, "The System of the World," Newton proceeded to apply his theory to all the celestial phenomena that were known or knowable in his time. It was a dazzling performance.

Although the *Principia* was not published until 1687, Newton's conjectures about the motions of the Moon and the planets had begun about twenty years earlier, soon after his graduation. There seems to be no reason to doubt the basic truth of the story of Newton and the apple: that in 1666, having left Cambridge for a while on account of the Great Plague, he was moved by the fall of an apple to speculate if the Moon itself was falling toward the earth in a similar way. He was at that time already aware of Kepler's discovery (1619) that the squares of the periodic times of the planets are in proportion to the cubes of their mean distances from the Sun. Newton deduced, to quote his own words (written before 1669), that "the endeavour of receding from the Sun will be reciprocally as the squares of the distances from the Sun"—that is, an inverse-square law of "centrifugal force." The germ of Newton's theory of universal gravitation was undoubtedly present here, but his full understanding and exploitation of the theory did not take place until, in the space of about two years (starting in 1684), he wrote it all out in the *Principia* (chiefly at the urging of his devoted admirer Edmond Halley, who financed the publication of the work out of his own pocket).

Continued

Newton had special reason for dividing the *Principia* into three separate parts (''Books''). Book I, ''The Motion of Bodies,'' is a development of the basic dynamics of particles, and embodies the theoretical substructure for the celestial mechanics of Book III. Book II, ''The Motion of Bodies in Resisting Mediums,'' had as its main result, and its main goal, a demonstration of the invalidity of Descartes' hypothesis that the circular motions of heavenly bodies resulted from their being swept around in huge fluid vortices. So Book I is concerned with the general laws of motion, and Book II with fluid mechanics (in which Newton solved a number of important problems, such as aerodynamic drag and a calculated value for the speed of sound).

In Book I Newton began with his ideas about space and time. Although he recognized that the only measurements we can make of positions and velocities are relative ones, he believed that space itself—and acceleration, the central quantity in his laws of motion—must be absolute. The proof? Take a bucket of water hanging at the end of a twisted rope, and let it spin (he did the experiment). Initially, the bucket turns, but the water remains at rest and its surface is flat. Later, the water picks up the rotation and its surface becomes concave. If the bucket is then suddenly stopped, the water continues to rotate and its surface remains concave. Clearly, the important feature is not the relative rotation of bucket and water, and Newton concluded that what matters is the rotation of the water (and the associated radial acceleration) within a fundamental reference frame, closely related to the space defined by the fixed stars. (The bulging of the Earth in its own rotation points to the same conclusion.) The deep implications of this have continued to be debated into our own day.

Newton's laws of motion themselves (Chapter 4) were based on some very simple experiments: observations of collisions between different kinds and sizes of objects. From them, Newton extracted his quantitative concept of momentum and the role of force in changing the momentum of a given body. He also demonstrated one of the most important and fundamental relationships in physics, the proportionality of weight to mass. He did this by showing that pendulums of the same length all had the same period of swing, regardless of the mass or composition of the pendulum bob. This result is, in effect, a more refined and precise test of the fact that different objects have equal accelerations under gravity down any given slope (including free vertical fall). Coupled with Newton's second law, this result implied that gravitational force is proportional to inertial mass—to an accuracy, Newton claimed, of about one part in a thousand.

His work in optics and in basic mechanics (not to mention pure mathematics and the invention of calculus) would already have been enough to mark Newton as one of the greatest scientists of all time. But the crowning glory of his work was surely the third book of the *Principia*, in which he analyzed the workings of the universe as a magnificent machine.

Newton began, perhaps surprisingly, with the satellites of Jupiter. The discovery of the four principal satellites by Galileo in 1610, using one of the earliest telescopes, was one of the great events in astronomical history—the first ''additions'' to the solar system since antiquity. Galileo himself had recorded their positions, night after night, for many months. As a result he was able to state their orbital periods with considerable accuracy. The data showed clearly that Jupiter's satellites, like the planets going around the sun, conformed to Kepler's third law (Section 10-1). For Newton, this was powerful evidence that a truly universal inverse-square law of gravitation was at work. He asked John Flamsteed, the first Astronomer Royal (and a man whom Newton treated very shabbily) to provide up-to-date observations. In fact, Flamsteed's results were not greatly different from those of Galileo, who must have been a magnificent observer. In 1687 Giovanni Cassini provided similar data for the five recently discovered satellites of Saturn.

From all this information, plus his fundamental law of gravitation, Newton was able to compare the masses of Earth, Sun, Jupiter, and Saturn, because the motion of any satellite involves a force proportional to the central mass that attracts it, and this fact is reflected in the orbital period. Newton, armed with his theories, had thus gained access to hitherto unknowable facts about the universe. This was exploration on a grand scale.

Another of Newton's great achievements was his application of universal gravitation to the motions of comets. Newton had first undertaken detailed observations of a comet in 1664, while he was still an undergraduate, but his interest in such matters subsided until December 1680, when a spectacular comet became visible, moving away from the Sun toward the outer reaches of the solar system. Newton himself charted its course. At the same time John Flamsteed was following it from the Royal Observatory at Greenwich. Flamsteed was also familiar with a less brilliant comet that had been seen approaching the sun only a month earlier. He proposed to Newton that these were one and the same object. Newton at first strongly resisted this suggestion; such a complete reversal of direction seemed inexplicable to him at the time. But the idea grew on him, and when, in the process of writing the *Principia* in 1685, he made a full assault on the problem of comets, he took the comet of 1680/81 as his chief case. Meanwhile, in 1682, his friend Halley had observed the comet that came to bear his name.

Comets, in general, are such small objects that they are invisible until they enter the inner regions of the solar system, where their trajectories can be excellently fitted as parabolas. In fact, as Newton recognized, the orbit of a comet is either hyperbolic, in which case it makes only one close encounter with the sun, or a very elongated ellipse, in which case it returns periodically (like Halley's comet, with its period of 76 years).

Another of Newton's great calculations involved the Earth itself. Using once again the inverse-square law of gravitational attraction, he showed how the Moon and the Sun would combine to create a double bulge in the ocean, with the maxima diametrically opposite one other. A given place on the spinning Earth would then encounter two maxima and two minima of ocean level every day. Newton knew that the actual behavior of the tides is much more complicated than this, but he was able to explain the way in which the effects of Sun and Moon combine differently at different times to produce the monthly cycle of tides of different heights. A much simpler problem that Newton solved was that of the Earth's shape—an oblate spheroid whose equatorial radius, because of "centrifugal force," is about one part in 300 bigger than the polar radius.

In all of the problems described up to this point, Newton felt himself to be on pretty solid ground; but no account of his role as an explorer of the universe would be complete without some mention of his more tentative speculations. They reveal a mind reaching out in all directions to understand how the universe is put together and how it works. Newton's curiosity in these matters knew no limits, and his brilliant scientific imagination is always in evidence.

The *Principia* itself contains some fascinating examples. Several of them have a bearing on the ages of the Earth and the solar system. In his discussion of the heating of comets by the sun (which creates their tails through vaporization of material, combined with the solar wind) Newton offers an estimate of how long it would have taken for the Earth to cool down if it had started out as a red-hot ball. His result was at least 50,000 years—far shorter than Baron William Kelvin calculated on the same basis two centuries later, but disturbingly long for any of Newton's contemporaries who might have been tempted to believe Bishop Ussher's date of 4004 B.C. for the Creation.

A more definite conjecture concerned the persistence of planetary motions. Astronomical records made it certain that these motions had continued undiminished for thousands of years; clearly they experienced no detectable resistance. Was space simply a void? Newton was not sure, but certainly there could be no "gross matter." This fitted in well with what Newton had calculated about the Earth's atmosphere. He had discovered the theoretical law of an exponential decrease of atmospheric pressure and density with height. Both would fall by a factor of a million for each 75 miles or so of altitude, and hence the space between the planets must be an essentially perfect vacuum. And yet . . . If there was no intervening medium, how was the force of gravity transmitted? He expressed his thoughts on this in a famous letter to Richard Bentley (see p. 89).

At the end of his *Opticks*, in a series of what he called Queries, Newton set down a number of his final speculations about the physical world. Many, but not all, are indeed concerned with optics. The next to last Query seems to foreshadow Einstein: "Are not gross Bodies and Light convertible into one another, and may not Bodies receive much of their Activity from the Particles of Light which enter their Composition?" It sounds just like a qualitative statement of $E = mc^2$! Of course it was not, and could not have been; Newton's concern here was to compare the absorption and emission of light with various other chemical and biological transformations. But it is yet one more example of his boundless capacity for speculation.

The last Query introduced a lengthy discussion in which Newton surveyed the whole range of the forces of nature. He began: "Have not the small Particles of Bodies certain Powers, Virtues, or Forces, by which they act at a distance, not only upon the Rays of Light for reflecting, refracting, and inflecting them, but also upon one another for producing a great Part of the Phenomena of Nature?" He then went on to discuss chemical reactions, surface tension phenomena, and the cohesion of solids, and speculated on the possible role of electric and magnetic forces in producing them. He expressed a belief in a fundamental uniformity of physical phenomena, saying: "And thus Nature will be very comfortable to her self and very simple, performing all the great Motions of the Heavenly Bodies by the Attraction of Gravity which intercedes those Bodies, and almost all the small ones of their Particles by some other attractive and repelling Powers which intercede the Particles." It was a great vision, which he had done more than any other person to make possible. It is still not complete, more than three centuries later, but this unification of the forces of nature remains the physicist's ultimate dream.

Suggestions for Further Reading

Cohen, I. Bernard: "Newton's Discovery of Gravity," *Scientific American*, March 1981, p. 166.

A historical article in which the development of Newton's ideas leading to his formulation of the law of gravity is put into the context of the ideas of other natural philosophers of the time.

Shamos, Morris H.: "The Law of Gravitation—Henry Cavendish," in *Great Experiments in Physics*, Henry Holt and Company, New York, 1959. Reprinted by Dover, 1987.

Cavendish's description of his experiments, with editorial annotations for clarity and a biographical sketch.

Whipple, Fred L., and J. Allen Hynek: "Observations of Satellite I," *Scientific American*, December 1957, p. 37.

An account of the American effort to determine precisely the orbit of the first artificial earth satellite, launched by the U.S.S.R. in 1957.

Review

A. Objectives: After studying this chapter, you should:

1. Be able to state Kepler's three empirical laws of planetary motion.

2. Be able to derive Kepler's third law of planetary motion, $T^2 \propto r^3$, for circular orbits.

3. Be able to discuss gravitational and inertial mass.

4. Be able to give the value for the speed needed to escape from the earth.

5. Be able to sketch a graph of gravitational potential energy as a function of separation distance and to discuss the calculation of the escape speed.

6. Be able to discuss the relationship between the total energy of an orbiting body and the type of orbit.

7. Be able to describe the gravitational field of a spherical shell and of a solid sphere.

B. Define, explain, or otherwise identify:

Kepler's laws

Ellipse
Newton's law of gravity
Universal gravitational constant
Central force
Gravitational mass
Inertial mass
Escape speed
Binding energy

C. True or false: If the statement is true, explain why it is true. If it is false, give a counterexample.

1. Kepler's law of equal areas implies that gravity varies inversely with the square of the distance.

2. The planet closest to the sun, on the average, has the shortest period of revolution about the sun.

3. The force that causes an apple to fall has the same origin as the force that causes the moon to move in a circle about the earth.

Problems

Level I

10-1 Kepler's Laws

1. Suppose a small planet with a period of 5 y were discovered. What would be its mean distance from the sun?

2. Halley's comet has a period of about 76 y. What is its mean distance from the sun?

3. The comet Kohoutek has a period estimated to be at least 10^6 y. What is its mean distance from the sun?

4. The radius of the earth's orbit is 1.496×10^{11} m and that of Uranus is 2.87×10^{12} m. What is the period of Uranus?

5. The mean distance of Pluto from the sun is 39.5 AU. Find the period of Pluto.

6. A planet moves around a massive sun with constant angular momentum. When the planet is at perihelion, it has a speed of 5×10^4 m/s and is 1.0×10^{15} m from the sun.

The orbital radius increases to 2.2×10^{15} m at aphelion. What is the planet's speed at aphelion?

7. A comet orbits the sun with constant angular momentum. It has a maximum radius of 150 AU, where its speed is 7×10^3 m/s. The comet's closest approach to the sun is 0.4 AU. What is its speed at perihelion?

10-2 Newton's Law of Gravity

8. One of Jupiter's moons, Io, has a mean orbital radius of 422 Mm and a period of 1.53×10^5 s. (*a*) Find the mean orbital radius of another of Jupiter's moons, Callisto, whose period is 1.44×10^6 s. (*b*) Use the known value of G to compute the mass of Jupiter.

9. Uranus has a moon, Umbriel, whose mean orbital radius is 267 Mm and whose period is 3.58×10^5 s. (*a*) Find the period of another of Uranus's moons, Oberon, whose mean orbital radius is 586 Mm. (*b*) Use the known value of G to find the mass of Uranus.

10. The mass of Saturn is 5.69×10^{26} kg. (*a*) Find the period of its moon Mimas, whose mean orbital radius is 186 Mm. (*b*) Find the mean orbital radius of its moon Titan, whose period is 1.38×10^6 s.

11. Calculate the mass of the earth from the period of the moon $T = 27.3$ d, its mean orbital radius $r_m = 384$ Mm, and the known value of G.

12. Use the period of the earth (1 y), its mean orbital radius (1.496×10^{11} m), and the value of G to calculate the mass of the sun.

13. Calculate the mass of the earth using the known values of G, g, and R_E.

14. A body is dropped from a height of 6.37 Mm above the surface of the earth. What is its initial acceleration?

15. Suppose you land on a planet of another solar system that has the same mass per unit volume as the earth but has 10 times the earth's radius. What would you weigh on this planet compared with what you weigh on earth?

10-3 Measurement of the Universal Gravitational Constant

16. The masses in a Cavendish apparatus are $m_1 = 10$ kg and $m_2 = 10$ g, the separation of their centers is 5 cm, and the rod separating the two small masses is 20 cm long. (*a*) What is the force of attraction between the large and small masses? (*b*) What torque must be exerted by the suspension to balance these forces?

17. The masses in a Cavendish apparatus are $m_1 = 12$ kg and $m_2 = 15$ g, and the separation of their centers is 5 cm. (*a*) What is the force of attraction between these two masses? (*b*) If the rod separating the two small masses is 18 cm long, what torque must be exerted by the suspension to balance the torque exerted by gravity?

10-4 Gravitational and Inertial Mass

18. A standard object, which is defined as having a mass of exactly 1 kg, is given an acceleration of 2.6587 m/s^2 when a certain force is applied to it. A second object of unknown mass acquires an acceleration of 1.1705 m/s^2 when the same force is applied to it. (*a*) What is the mass of the second object? (*b*) Is the mass that you determined in part (*a*) gravitational or inertial mass?

19. The weight of a standard object, which is defined as having a mass of exactly 1 kg, is measured to be 9.81 N. In the same laboratory, a second object weighs 56.6 N. (*a*) What is the mass of the second object? (*b*) Is the mass you determined in part (*a*) gravitational or inertial mass?

10-5 Escaping the Earth

20. The planet Saturn has a mass 95.2 times that of the earth and a radius 9.47 times that of the earth. Find the escape speed for objects near the surface of Saturn.

21. Find the escape speed for a rocket leaving the moon. The acceleration of gravity on the moon is 0.166 times that on earth, and the moon's radius is $0.273R_E$.

22. Suppose that in space there is a duplicate earth, except that it has no atmosphere, is not rotating, and is not in motion around any sun. What initial velocity must a spacecraft have to travel vertically upward a distance above the surface of the planet equal to one earth radius?

10-6 Potential Energy, Total Energy, and Orbits

23. (*a*) Taking the potential energy to be zero at infinite separation, find the potential energy of a 100-kg object at the surface of the earth. (Use 6.37 Mm for the earth's radius.) (*b*) Find the potential energy of the same object at a height above the earth's surface equal to the earth's radius. (*c*) What would be the escape speed for a body projected from this height?

24. A satellite with a mass of 300 kg moves in a circular orbit 5×10^7 m above the earth's surface. (*a*) What is the gravitational force on the satellite? (*b*) What is the speed of the satellite? (*c*) What is the period of the satellite?

"Castle in the Pyrenees" by René Magritte.

10-7 The Gravitational Field of a Spherical Shell and a Solid Sphere

25. A spherical shell has a radius of 2 m and a mass of 300 kg. What is the gravitational field (*a*) 0.5 m, (*b*) 1.9 m, and (*c*) 2.5 m from the center of the shell?

26. A spherical shell has a radius of 2 m and a mass of 300 kg, and its center is located at the origin of a coordinate system. Another spherical shell with a radius of 1 m and mass 150 kg is inside the larger shell with its center at 0.6 m on the x axis. What is the gravitational force of attraction between the two shells?

Level II

27. A point mass m_0 is initially at the surface of a large sphere of mass M and radius R. How much work is needed to remove it to a very large distance away from the large sphere?

28. Two concentric uniform spherical shells have masses M_1 and M_2 and radii a and $2a$ as shown in Figure 10-18. (*a*) What is the gravitational force on a point mass m located a distance $3a$ from the center of the shells? (*b*) What is the force on the point mass if it is located a distance $1.9a$ from the center of the shells? (*c*) What is the force on the point mass if it is located a distance $0.9a$ from the center of the shells?

Figure 10-18   Problem 28.

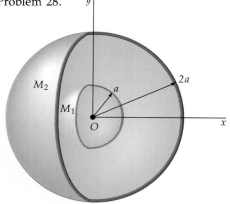

29. The inner spherical shell in Problem 28 is shifted such that its center is now at $x = 0.8a$. The points $3a$, $1.9a$, and $0.9a$ lie along the same radial line from the center of the larger spherical shell. (*a*) What is the force on m at $3a$? (*b*) What is the force on m at $1.9a$? (*c*) What is the force on m at $0.9a$?

30. A uniform sphere of radius 100 m and density 2000 kg/m^3 is in free space far from other massive objects. (*a*) Find the gravitational field outside of the sphere as a function of r. (*b*) Find the gravitational field inside the sphere as a function of r.

31. A particle is projected from the surface of the earth with a speed equal to twice the escape speed. When it is very far from the earth, what is its speed? (Neglect air resistance.)

32. What initial speed should a particle be given if it is to have a final speed when it is very far from the earth equal to the escape speed?

33. A space probe sent from the earth is to have a speed of 50 km/s when it is very far from the earth. What speed is needed for the probe at the surface of the earth? (Neglect air resistance.)

34. An object is dropped from rest from a height of 4×10^6 m above the surface of the earth. If there were no air resistance, what would its speed be when it strikes the earth?

35. An object is projected upward from the surface of the earth with an initial speed of 4 km/s. Find the maximum height it reaches. (Neglect air resistance.)

36. (*a*) Calculate the energy in joules necessary to launch a 1-kg mass from the earth with the escape speed. (*b*) Convert this energy to kilowatt-hours. (*c*) If energy can be obtained at 10 cents per kilowatt-hour, what is the minimum cost of giving an 80-kg astronaut enough energy to escape the earth's gravitational field?

37. Two planets of equal mass orbit a much more massive star (Figure 10-19). Planet m_1 moves in a circular orbit of radius 100 Gm with a period of 2 y. Planet m_2 moves in an elliptical orbit, with its closest distance $r_1 = 100$ Gm and its farthest distance $r_2 = 180$ Gm. (*a*) Using the fact that the mean radius of an elliptical orbit is the length of the semimajor axis, find the period of m_2's orbit. (*b*) What is the mass of the star? (*c*) Which planet has the greater speed at point P? Which has the greater total energy? (*d*) How does the speed of m_2 at point P compare with its speed at point A?

Figure 10-19   Problem 37.

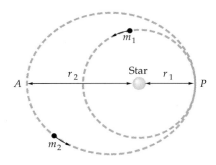

38. A thick spherical shell of mass M and uniform density has an inner radius R_1 and an outer radius R_2. Find the gravitational field g_r as a function of r for all possible values of r.

39. Two point particles of mass m are on the y axis at $y = +a$ and $y = -a$ as shown in Figure 10-20. (*a*) Show that the force exerted by these two particles on a third particle of mass m_0 located on the x axis at a distance x from the origin is given by

$$\mathbf{F} = -\frac{2Gmm_0x}{(x^2 + a^2)^{3/2}}\,\mathbf{i}$$

(*b*) What is the gravitational field $\mathbf{g}$ on the x axis due to the two particles on the y axis? (*c*) Show that g_x due to the two particles on the y axis is approximately $-2Gm/x^2$ when x is much greater than a. (*d*) Show that the maximum value of $|g_x|$ occurs at the points $x = \pm a/\sqrt{2}$.

Figure 10-20   Problem 39.

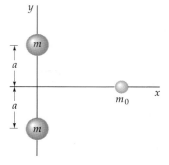

40. (*a*) Sketch a plot of the gravitational field g_x versus x due to a uniform ring of mass M and radius R whose axis is the x axis. (*b*) At what points is the magnitude of g_x maximum?

41. Five equal masses M are equally spaced on the arc of a semicircle of radius R as shown in Figure 10-21. A mass m is located at the center of curvature of the arc. If M is 3 kg, m is 2 kg, and R is 10 cm, what is the force on m due to the five masses?

Figure 10-21   Problem 41.

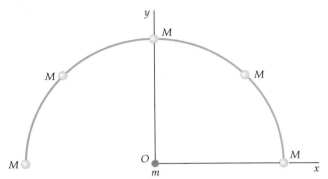

42. (a) Show that the gravitational field of a ring of uniform mass is zero at the center of the ring. (b) Figure 10-22 shows a point P in the plane of the ring but not at its center. Consider two elements of the ring of length s_1 and s_2 at distances of r_1 and r_2, respectively. What is the ratio of the masses of these elements? Which produces the greater gravitational field at point P? What is the direction of the field at point P due to these elements? (c) What is the direction of the gravitational field at point P due to the entire ring? (d) Suppose that the gravitational field varied as $1/r$ rather than $1/r^2$. What would be the net gravitational field at point P due to the two elements? (e) How would your answers to parts (b) and (c) differ if point P were inside a spherical shell of uniform mass rather than inside a plane circular ring?

Figure 10-22   Problem 42.

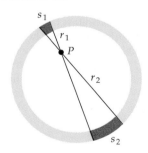

43. A superconducting gravity meter can measure changes in gravity of the order $\Delta g/g = 10^{-11}$. (a) Estimate the maximum range at which an 80-kg person can be detected by this gravity meter. Assume that the gravity meter is stationary, and that the person's mass can be considered to be concentrated at his or her center of gravity. (b) What vertical change in the position of the gravity meter in the earth's gravitational field is detectable?

44. A spaceship traveling on a line from the earth to the moon passes through a point where the force of gravity of the earth and that of the moon exactly cancel. How far from the center of the earth is this special point?

45. During a solar eclipse, when the moon is between the earth and the sun, a student's weight is affected by the gravitational pull of the moon and the sun. (a) If the pull of the earth on the student is 800 N, what is the force of the moon on the student? (b) What is the force of the sun on the student? (c) What percentage correction due to the pull of the sun and moon should be applied to the reading of a very accurate scale to obtain the student's weight?

46. At the airport, a physics student weighs 800 N. The student boards a jet plane that rises to an altitude of 15,000 m. What is the student's loss in weight?

47. Suppose the earth were a sphere of uniform mass. If there were a deep elevator shaft going 15,000 m into the earth, what would be the loss in weight at the bottom of this deep shaft for a student who weighs 800 N at the surface of the earth?

Level III

48. Assume that the earth is a sphere of uniform mass and that a small-diameter hole has been drilled from the surface to the center. (a) How much work would be needed to bring a small object of mass m from the center of the earth to the surface? (b) If the object were dropped into the hole at the surface of the earth, what would its speed be when it reaches the earth's center?

49. A uniform sphere of mass M is located near a thin, uniform rod of mass m and length L as shown in Figure 10-23. Find the gravitational force of attraction exerted by the sphere on the rod.

Figure 10-23   Problem 49.

50. A uniform rod of mass $M = 20$ kg and length $L = 5$ m is bent into a semicircle. What is the gravitational force exerted by the rod on a point mass $m = 0.1$ kg located at the center of the circular arc?

51. A spherical shell has a radius R and a mass M. (a) Write expressions for the force exerted by the shell on a point mass m_0 when m_0 is outside the shell and when it is inside the shell. (b) What is the potential-energy function $U(r)$ for this system when the mass m_0 is at a distance r ($r \geq R$) if $U = 0$ at $r = \infty$? Evaluate this function at $r = R$. (c) Using the general relation for $dU = -\mathbf{F}\cdot d\mathbf{r} = -F_r\,dr$, show that U is constant everywhere inside the shell. Using the fact that U is continuous everywhere, including at $r = R$, find the value of the constant U inside the shell. (d) Sketch $U(r)$ versus r for all possible values of r.

52. A uniform rod of mass M and length L lies along the x axis with its center at the origin. Consider an element of length dx at a distance x from the origin. (a) Show that this element produces a gravitational field at a point x_0 on the x axis ($x_0 > \frac{1}{2}L$) given by

$$dg_x = -\frac{GM}{L(x_0 - x)^2}\, dx$$

(b) Integrate this result over the length of the rod to find the total gravitational field at the point x_0 due to the rod. (c) What is the force on an object of mass m_0 at x_0?

53. In this problem, you are to find the gravitational potential energy of the rod in Problem 52 and a point mass m_0 that is on the x axis at x_0. (a) Show that the potential energy of an element of the rod dm and m_0 is given by

$$dU = -\frac{GM_0\, dm}{x_0 - x} = -\frac{GMm_0}{L(x_0 - x)}\, dx$$

where $U = 0$ at $x_0 = \infty$. (b) Integrate your result for part (a) over the length of the rod to find the total potential energy for the system. Write your result as a general function $U(x)$ by setting x_0 equal to a general point x. (c) Compute the force on m_0 at a general point x from $F_x = -dU/dx$, and compare your result with that for part (c) of Problem 52.

54. Our galaxy can be considered to be a large disk of radius R and mass M of approximately uniform mass density. (a) Find the gravitational potential energy of a 1-kg mass on the axis of the disk at a distance x from it due to a ring element of radius r and thickness dr. (b) Integrate your result for part (a) to find the total gravitational potential energy of a 1-kg mass at a distance x due to the disk. (c) From $F_x = -dU/dx$ and your result for part (b), find the gravitational field g_x on the axis of the disk.

55. A sphere of radius R has its center at the origin. It has a uniform mass density ρ_0, except that there is a spherical cavity in it of radius $r = \frac{1}{2}R$ centered at $x = \frac{1}{2}R$ as shown in Figure 10-24. Find the gravitational field at points on the x axis for $|x| > R$. (Hint: The cavity may be thought of as a sphere of mass $m = \frac{4}{3}\pi r^3 \rho_0$ plus a sphere of mass $-m$.)

Figure 10-24   Problem 55.

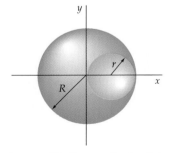

56. For the sphere with the cavity in it in Problem 55, show that the gravitational field inside the cavity is uniform, and find its magnitude and direction.

57. A straight, smooth tunnel is dug through a spherical planet whose mass density ρ_0 is constant. The tunnel passes through the center of the planet and is perpendicular to the planet's axis of rotation, which is fixed in space. The planet rotates with an angular velocity ω such that objects in the tunnel have no acceleration relative to the tunnel. Find ω.

58. A plumb bob near a large mountain is slightly deflected from the vertical by the gravitational attraction of the mountain. Estimate the order of magnitude of the angle of deflection using any assumptions you like.

59. Both the sun and the moon exert gravitational forces on the oceans of the earth, causing tides. (a) Show that the ratio of the force exerted by the sun to that exerted by the moon is $M_s r_m^2 / M_m r_s^2$, where M_s and M_m are the masses of the sun and moon and r_s and r_m are the distances from the earth to the sun and to the moon. Evaluate this ratio. (b) Even though the sun exerts a much greater force on the oceans than the moon does, the moon has a greater effect on the tides because it is the difference in the force from one side of the earth to the other that is important. Differentiate the expression

$$F = \frac{Gm_1 m_2}{r^2}$$

to calculate the change in F due to a small change in r. Show that $dF/F = (-2\, dr)/r$. (c) During one full day, the distance from the sun or moon to an ocean changes (because of the earth's rotation) by at most the diameter of the earth. Show that for a small change in distance, the change in the force exerted by the sun is related to the change in the force exerted by the moon by

$$\frac{\Delta F_s}{\Delta F_m} \approx \frac{M_s r_m^3}{M_m r_s^3}$$

and calculate this ratio.

60. The solid surface of the earth has a density of about 3000 kg/m^3. A spherical deposit of heavy metals with a density of 8000 kg/m^3 and a radius of 1000 m is centered 2000 m below the surface. Find $\Delta g/g$ directly above this deposit, where Δg is the increase in the gravitational field due to the deposit.

61. Two identical spherical hollows are made in a lead sphere of radius R. The hollows have a radius $R/2$. They touch the outside surface of the sphere and its center as shown in Figure 10-25. The mass of the lead sphere before hollowing was M. (a) Find the force of attraction of a small sphere of mass m to the lead sphere at the position shown in the figure. (b) What is the attractive force if m is located right at the surface of the lead sphere?

Figure 10-25   Problem 61.

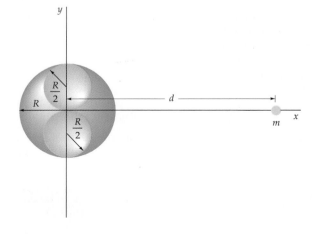

Chapter 11

Mechanics of Solids and Fluids

Scanning-electron micrograph of a fiber-reinforced ceramic composite that was loaded to near the point of complete failure of the ceramic material. The fibers bridge the cracks allowing the composite to carry a high load despite localized damage. Such reinforced materials are used in turbine motors that rotate at over 100,000 rev/min.

The states of matter in bulk can be conveniently divided into solids and fluids. Solids tend to be rigid and to maintain their shape, whereas fluids do not maintain their shape but flow. Fluids include both liquids, which flow under gravity until they occupy the lowest possible regions of their containers, and gases, which expand to fill their containers regardless of their shapes. The distinction between solids and liquids is not sharp. Although ice is considered a solid, the flow of glaciers is well known. Similarly, glasses, and even rocks under great pressure, tend to flow slightly over long periods of time.

In this chapter, we will look at some of the mechanical properties of solids and of fluids at rest and in motion.

11-1 Density

An important property of a substance is the ratio of its mass to its volume, which is called its **density:**

$$\text{Density} = \frac{\text{mass}}{\text{volume}}$$

The Greek letter ρ (rho) is usually used to denote density:

Density defined

$$\rho = \frac{m}{V}$$ 11-1

Because the gram was originally defined as the mass of one cubic centimeter of water, the density of water in cgs units is 1 g/cm^3. Converting to SI units of kilograms per cubic meter, we obtain for the density of water

$$\rho = \frac{1 \text{ g}}{\text{cm}^3} \times \frac{\text{kg}}{10^3 \text{ g}} \times \left(\frac{100 \text{ cm}}{\text{m}}\right)^3 = 10^3 \text{ kg/m}^3$$ 11-2

The density of water varies with temperature. Equation 11-2 gives its maximum value, which occurs at 4°C. A convenient unit of volume is the **liter** (L):

$$1 \text{ L} = 10^3 \text{ cm}^3 = 10^{-3} \text{ m}^3$$

In terms of this unit, the density of water is 1.00 kg/L.

When an object's density is greater than that of water, it will sink in water. When its density is less, it will float. In fact, we will show in Section 11-4 that, for floating objects, the fraction of the volume of an object that is submerged in any liquid equals the ratio of the density of the object to that of the liquid. For example, ice has a density of approximately 0.92 g/cm^3, and it floats in water with about 92 percent of its volume submerged. Table 11-1 lists the densities of some common materials. Since the density of water in cgs units is 1 g/cm^3, these units are slightly more convenient to use than SI units. The ratio of the density of a substance to that of water is called the **specific gravity** of the substance. The specific gravity is a dimensionless number that equals the magnitude of the density when it is expressed in grams per cubic centimeter (or in kilograms per liter). The specific gravity of a substance can be found by dividing its density by 10^3 kg/m^3. For example, the specific gravity of aluminum is 2.7, and that of ice is 0.92. The specific gravities of objects that sink in water range from 1 to about 22.5 (for the densest element, osmium).

Although most solids and liquids expand slightly when heated and contract slightly when subjected to an increase in external pressure, these changes in volume are relatively small, so we can say that the densities of most solids and liquids are approximately independent of temperature and pressure. On the other hand, the density of a gas depends strongly on the pressure and temperature, so the temperature and pressure must be specified when giving the densities of gases. In Table 11-1, the densities are given at standard conditions (atmospheric pressure at sea level and a temperature of 0°C). Note that the densities of gases are considerably less than those of liquids or solids. For example, the density of water is about 800 times that of air under standard conditions.

A glacier flows into Glacier Bay, Alaska.

Exercise

A solid metal cube 8 cm on a side has a mass of 4.08 kg. (*a*) What is the density of the cube? (*b*) If the cube is made from a single element listed in Table 11-1, what is the element?   [Answers: (*a*) 7.97 kg/L, (*b*) iron]

In the U.S. customary system, the term **weight density,** which is defined as the ratio of the weight of an object to its volume, is often used. The weight density is the product of the density ρ and the acceleration due to gravity g:

$$\rho g = \frac{w}{V} = \frac{mg}{V}$$ 11-3

The weight density of water is

$$\rho_w g = 62.4 \ \text{lb/ft}^3 \qquad\qquad 11\text{-}4$$

The weight density of any other material can be found by multiplying its specific gravity by 62.4 lb/ft³.

Exercise

Find the weight density of copper.   (Answer: 559 lb/ft³)

Example 11-1

A lead brick is 5 by 10 by 20 cm. How much does it weigh?

The volume of the brick is

$$V = (5 \ \text{cm})(10 \ \text{cm})(20 \ \text{cm}) = 1000 \ \text{cm}^3 = 10^{-3} \ \text{m}^3$$

From Table 11-1, the density of lead is $11.3 \times 10^3 \ \text{kg/m}^3$. Its mass is therefore

$$m = \rho V = (11.3 \times 10^3 \ \text{kg/m}^3)(10^{-3} \ \text{m}^3) = 11.3 \ \text{kg}$$

and its weight is

$$w = mg = (11.3 \ \text{kg})(9.81 \ \text{N/kg}) = 111 \ \text{N} \approx 25 \ \text{lb}$$

Example 11-2

A 200-mL flask is filled with water at 4°C. When the flask is heated to 80°C, 6 g of water spill out. What is the density of water at 80°C? (Assume that the expansion of the flask is negligible.)

Since the density of water at 4°C is 1 g/cm³ and 200 mL = 200 cm³, the mass of the water originally in the flask is

$$m = \rho V = (1 \ \text{g/cm}^3)(200 \ \text{cm}^3) = 200 \ \text{g}$$

Since 6 g spill out of the flask at 80°C, the mass of the remaining water is 200 g − 6 g = 194 g. This mass of water now occupies 200 mL, so the density of water at 80°C is

$$\rho = \frac{m}{V} = \frac{194 \ \text{g}}{200 \ \text{cm}^3} = 0.97 \ \text{g/cm}^3$$

Question

1. What is the approximate specific gravity of your body?

11-2   Stress and Strain

If a solid object is in equilibrium but is subjected to forces that tend to stretch, shear, or compress it, the shape of the object changes. If the object returns to its original shape when the forces are removed, it is said to be elastic. Most objects are elastic for forces up to a certain limit called the **elastic limit.** If the forces are too great and the elastic limit is exceeded, the object does not return to its original shape but is permanently deformed.

Figure 11-1a shows a solid bar subjected to a stretching or tensile force **F** to the right and an equal but opposite force to the left. In Figure 11-1b, we concentrate on a small element of the bar of length L. Since this element is in equilibrium, the forces exerted on it by adjacent elements to the right must

Table 11-1   Densities of Selected Substances*

Substance	Density, kg/m³
Aluminum	2.70×10^3
Bone	$1.7\text{–}2.0 \times 10^3$
Brick	$1.4\text{–}2.2 \times 10^3$
Cement	$2.7\text{–}3.0 \times 10^3$
Copper	8.93×10^3
Earth (average)	5.52×10^3
Glass (common)	$2.4\text{–}2.8 \times 10^3$
Gold	19.3×10^3
Ice	0.92×10^3
Iron	7.96×10^3
Lead	11.3×10^3
Wood (oak)	$0.6\text{–}0.9 \times 10^3$
Alcohol (ethanol)	0.806×10^3
Gasoline	0.68×10^3
Mercury	13.6×10^3
Seawater	1.025×10^3
Water	1.00×10^3
Air	1.293
Helium	0.1786
Hydrogen	0.08994
Steam (100°C)	0.6

*$t = 0°C$ and $P = 1$ atm unless otherwise indicated.

Figure 11-1   (a) Solid bar subjected to a stretching force **F**. (b) A small length of the bar. The force per unit area is the stress S.

equal those exerted by adjacent elements to the left. If the element is not too near the end of the bar, these forces will be distributed uniformly over the cross-sectional area of the bar. The ratio of the force F to the cross-sectional area A is called the **tensile stress:**

$$\text{Stress} = \frac{F}{A} \qquad\qquad 11\text{-}5$$

The forces exerted on the bar tend to stretch the bar. The fractional change in the length of the bar $\Delta L/L$ is called the **strain:**

$$\text{Strain} = \frac{\Delta L}{L} \qquad\qquad 11\text{-}6$$

Figure 11-2 shows a graph of the strain versus the stress for a typical solid bar. The graph is linear up to point A. The result that strain varies linearly with stress is known as Hooke's law. (This is the same behavior as that of a coiled spring for small stretching.) Point B in Figure 11-2 is the elastic limit of the material. If the bar is stretched beyond this point, it will not return to its original length but is permanently deformed. If an even greater stress is applied, the material eventually breaks, as indicated by point C. The ratio of the stress to the strain in the linear region of the graph is a constant called **Young's modulus** Y:

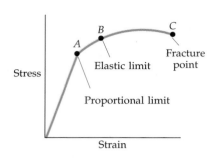

Figure 11-2 A graph of stress versus strain. Up to point A, the strain is proportional to the stress. Beyond the elastic limit at point B, the bar will not return to its original length when the stress is removed. At point C, the bar fractures.

Young's modulus defined

$$Y = \frac{\text{stress}}{\text{strain}} = \frac{F/A}{\Delta L/L} \qquad\qquad 11\text{-}7$$

The units of Young's modulus are newtons per square meter (or pounds per square inch). Approximate values of Young's modulus for various materials are listed in Table 11-2.

Table 11-2 Young's Modulus Y and Strengths of Various Materials*			
Material	Y, GN/m²†	Tensile strength, MN/m²	Compressive strength, MN/m²
Aluminum	70	90	
Bone			
Tensile	16	200	
Compressive	9	—	270
Brass	90	370	
Concrete	23	2	17
Copper	110	230	
Iron (wrought)	190	390	
Lead	16	12	
Steel	200	520	520

*These values are representative. Actual values for particular samples may differ.
†1 GN = 10^3 MN = 10^9 N.

Example 11-3

A certain man's biceps muscle has a maximum cross-sectional area of $12 \text{ cm}^2 = 1.2 \times 10^{-3} \text{ m}^2$. What is the stress in the muscle if it exerts a force of 300 N?

From the definition of tensile stress, we have

$$\text{Stress} = \frac{F}{A} = \frac{300 \text{ N}}{1.2 \times 10^{-3} \text{ m}^2} = 2.5 \times 10^5 \text{ N/m}^2$$

The maximum stress that can be exerted is approximately the same for all muscles. Greater forces can be exerted by muscles with greater cross-sectional areas.

Example 11-4

A 500-kg mass is hung from a 3-m steel wire with a cross-sectional area of 0.15 cm². By how much does the wire stretch?

The weight of a 500-kg mass is

$$mg = (500 \text{ kg})(9.81 \text{ N/kg}) = 4.90 \times 10^3 \text{ N}$$

The stress of the wire is

$$S = \frac{F}{A} = \frac{4.9 \times 10^3 \text{ N}}{0.15 \text{ cm}^2} = 3.27 \times 10^4 \text{ N/cm}^2 = 3.27 \times 10^8 \text{ N/m}^2$$

From Table 11-2, we find Young's modulus for steel to be about $2.0 \times 10^{11} \text{ N/m}^2$. The strain is therefore

$$\frac{\Delta L}{L} = \frac{S}{Y} = \frac{3.27 \times 10^8 \text{ N/m}^2}{2.0 \times 10^{11} \text{ N/m}^2} = 1.63 \times 10^{-3}$$

Since the wire is 300 cm long, the amount it stretches is

$$\Delta L = (1.63 \times 10^{-3})L = (1.63 \times 10^{-3})(300 \text{ cm}) = 0.49 \text{ cm}$$

Exercise

A wire 1.5 m long has a cross-sectional area of 2.4 mm². It is hung vertically and stretches 0.32 mm when a 10-kg block is attached to it. Find (*a*) the stress, (*b*) the strain, and (*c*) Young's modulus for the wire. [Answers: (*a*) $4.09 \times 10^7 \text{ N/m}^2$, (*b*) 2.13×10^{-4}, (*c*) 192 GN/m²]

If a bar is subject to forces that tend to compress it rather than stretch it, the stress is called **compressive stress.** For many (but not all) materials, Young's modulus for compressive stress is the same as that for tensile stress if ΔL in Equation 11-7 is taken to be the decrease in the length of the bar.

(*a*) These flying buttresses make use of the compressive strength and weight of stone to support Notre Dame Cathedral in Paris. (*b*) A sharp karate chop to the cinderblocks is enough to shatter them indicating that their resistance to a "bending load" is much less than their compressive strength.

(*a*)

(*b*)

(Bone is an important exception that has different Young's moduli for compression and extension.) If the tensile or compressive stress is too great, the bar breaks. The stress at which breakage occurs is called the **tensile strength,** or in the case of compression, the **compressive strength.** Approximate values of the tensile and compressive strengths of various materials are also listed in Table 11-2.

In Figure 11-3, a force $\mathbf{F}_s$ is applied tangentially to the top of a book. Such a force is called a **shear force.** The ratio of the shear force F_s to the area A is called the **shear stress:**

$$\text{Shear stress} = \frac{F_s}{A}$$

Figure 11-3 The application of the horizontal force $\mathbf{F}_s$ to the book causes a shearing stress defined as the force per unit area. The ratio $\Delta X/L = \tan \theta$ is the shear strain.

A shear stress tends to deform the book as shown in Figure 11-3. The ratio $\Delta X/L$ is called the **shear strain:**

$$\text{Shear strain} = \frac{\Delta X}{L} = \tan \theta$$

where θ is the shear angle shown in the figure. The ratio of the shear stress to the shear strain is called the **shear modulus** M_s:

Shear modulus defined

$$M_s = \frac{\text{shear stress}}{\text{shear strain}} = \frac{F_s/A}{\Delta X/L} = \frac{F_s/A}{\tan \theta} \qquad 11\text{-}8$$

The shear modulus is also known as the **torsion modulus.** The fact that this modulus is approximately constant for small stresses, which implies that the shear strain varies linearly with the shear stress, is Hooke's law for torsional stress. In a torsion balance, such as that used in Cavendish's apparatus for measuring the universal gravitational constant G, the torque (which is related to the stress) is proportional to the angle of twist (which equals the strain for small angles). Approximate values of the shear modulus for various materials are listed in Table 11-3.

Table 11-3 Approximate Values of the Shear Modulus M_s of Various Materials

Material	M_s, GN/m^2
Aluminum	30
Brass	36
Copper	42
Iron	70
Lead	5.6
Steel	84
Tungsten	150

11-3 Pressure in a Fluid

Fluids differ from solids in that they are unable to support a shear stress. Thus, they deform to fill a container of any shape. When a body is submerged in a fluid such as water, the fluid exerts a force perpendicular to the surface of the body at each point on the surface. If the body is small enough so that we can neglect any differences in the depth of the fluid, the force per unit area exerted by the fluid is the same at each point on the surface of the body. This force per unit area is called the **pressure** P of the fluid:

Pressure defined

$$P = \frac{F}{A} \qquad 11\text{-}9$$

The SI unit of pressure is the newton per square meter (N/m^2), which is called the **pascal** (Pa):

$$1 \text{ Pa} = 1 \text{ N/m}^2 \qquad 11\text{-}10$$

In the U.S. customary system, pressure is usually given in pounds per square inch (lb/in^2). Another common unit of pressure is the atmosphere (atm), which is approximately the air pressure at sea level. The atmosphere is now defined to be 101.325 kilopascals, which is approximately 14.70 lb/in^2:

$$1 \text{ atm} = 101.325 \text{ kPa} = 14.70 \text{ lb/in}^2 \qquad 11\text{-}11$$

Other units of pressure in common use will be discussed later.

The pressure due to a fluid pressing against a body tends to compress the body. The ratio of the pressure to the fractional decrease in volume $(-\Delta V/V)$ is called the **bulk modulus** B:

$$B = -\frac{P}{\Delta V/V} \qquad 11\text{-}12 \qquad \textit{Bulk modulus defined}$$

Since all materials decrease in volume when they are subjected to external pressure, a minus sign is introduced in Equation 11-12 to make B positive. The pressure exerted by a fluid is equivalent to a compressive stress, and the fractional decrease in the volume $(-\Delta V/V)$ is the compressive strain. The inverse of the bulk modulus is called **compressibility** k:

$$k = \frac{1}{B} = -\frac{\Delta V/V}{P} \qquad 11\text{-}13$$

The more difficult material is to compress, the smaller the fractional change $\Delta V/V$ for a given pressure and, hence, the smaller the compressibility k. The concepts of the bulk modulus and compressibility apply to liquids and gases as well as to solids. Solids and liquids are relatively incompressible; that is, they have small values of compressibility and large values for the bulk modulus. These values are also relatively independent of temperature and pressure. Gases, on the other hand, are easily compressed, and the values of B and k depend strongly on the pressure and temperature. Table 11-4 lists values of the bulk modulus for various materials.

As any scuba diver knows, the pressure in a lake or ocean increases as depth increases. Similarly, the pressure of the atmosphere decreases as altitude increases. (This is why aircraft cabins must be pressurized.) For a liquid such as water whose density is constant throughout, the pressure increases linearly with depth. We can see this easily by considering the column of liquid of height h and cross-sectional area A shown in Figure 11-4. The pressure at the bottom of the column must be greater than the pressure at the top of the column to support the weight of the column. The mass of this liquid column is

$$m = \rho V = \rho A h$$

and its weight is

$$w = mg = \rho A h g$$

If P_0 is the pressure at the top and P is the pressure at the bottom, the net upward force exerted by this pressure difference is $PA - P_0 A$. Setting this net upward force equal to the weight of the column, we obtain

$$PA - P_0 A = \rho A h g$$

or

$$P = P_0 + \rho g h \qquad (\rho \text{ constant}) \qquad 11\text{-}14$$

Table 11-4   **Approximate Values of the Bulk Modulus B of Various Materials**

Material	B, GN/m^2
Aluminum	70
Brass	61
Copper	140
Iron	100
Lead	7.7
Mercury	27
Steel	160
Tungsten	200
Water	2.0

Figure 11-4 Column of water of height h and cross-sectional area A. The pressure P at the bottom must be greater than the pressure P_0 at the top to balance the weight of the water.

Example 11-5

Find the pressure at a depth of 10 m below the surface of a lake if the pressure at the surface is 1 atm.

Using Equation 11-14 with $P_0 = 1$ atm $= 101$ kPa, $\rho = 10^3$ kg/m^3, and $g = 9.81$ N/kg, we have

$$P = 101 \text{ kPa} + (10^3 \text{ kg/m}^3)(9.81 \text{ N/kg})(10 \text{ m})$$

$$= 101 \text{ kPa} + (9.81 \times 10^4 \text{ N/m}^2)$$

$$= 101 \text{ kPa} + 98.1 \text{ kPa} = 199 \text{ kPa} = 1.97 \text{ atm}$$

The pressure at a depth of 10 m is nearly twice that at the surface.

The result that the pressure at a depth h is greater than that at the top by the amount $\rho g h$ holds for a liquid in any container, independent of the shape of the container. Furthermore, the pressure is the same at all points at the same depth. Thus, if we increase P_0, say, by pressing down on the top surface with a piston, the increase in pressure is the same throughout the liquid. This is known as **Pascal's principle,** named after Blaise Pascal (1623–1662):

Pascal's principle

> Pressure applied to an enclosed liquid is transmitted undiminished to every point in the fluid and to the walls of the container.

A common application of Pascal's principle is the hydraulic lift shown in Figure 11-5. When force $\mathbf{F}_1$ is applied to the smaller piston, the pressure in the liquid is increased by F_1/A_1. The upward force exerted by the liquid on the larger piston is this increase in pressure times the area A_2. Calling this force $\mathbf{F}_2$, we have

$$F_2 = \frac{F_1}{A_1} A_2 = \frac{A_2}{A_1} F_1$$

If A_2 is much greater than A_1, a small force $\mathbf{F}_1$ can be used to exert a much larger force $\mathbf{F}_2$ to lift a weight placed on the larger piston.

Figure 11-5 Hydraulic lift. A small force $\mathbf{F}_1$ on the small piston produces a change in pressure that is transmitted by the liquid to the large piston. Since the area of the large piston is much greater than that of the small piston, the force $\mathbf{F}_2$ on the large piston is much greater than $\mathbf{F}_1$.

Example 11-6

The large piston in a hydraulic lift has a radius of 20 cm. What force must be applied to the small piston of radius 2 cm to raise a car of mass 1500 kg?

The weight of the car is

$$mg = (1500 \text{ kg})(9.81 \text{ N/kg}) = 1.47 \times 10^4 \text{ N}$$

The force that must be applied is therefore

$$F_1 = \frac{A_1}{A_2} F_2 = \frac{\pi r_1^2}{\pi r_2^2} mg$$

$$= \frac{(2 \text{ cm})^2}{(20 \text{ cm})^2} (1.47 \times 10^4 \text{ N}) = 147 \text{ N} \approx 33 \text{ lb}$$

Figure 11-6 The hydrostatic paradox. The water level is the same regardless of the shape of the vessel. The shaded portion of the water is supported by the sides of the container.

Figure 11-6 shows water in a container with sections of different shapes. At first glance, it might seem that the pressure in the largest section of the container would be greatest and that the water would therefore be forced into the smallest section of the container to a greater height. That this does not happen is known as the hydrostatic paradox. The pressure depends only on the depth of the water, not on the shape of the container, so at the same height the pressure is the same in all parts of the container, as can be shown experimentally. Although the water in the largest section of the container weighs more than that in the smaller sections, some of this weight is supported by the normal force exerted by the sides of the largest section of the container, which in this case has a component upward. In fact, the shaded portion of the water is completely supported by the sides of the container.

We can use the result that the pressure difference is proportional to the depth of the fluid to measure unknown pressures. Figure 11-7 shows the simplest pressure gauge, the open-tube manometer. The top of the tube is open to the atmosphere at pressure P_{at}. The other end of the tube is at pressure P, which is to be measured. The difference $P - P_{at}$ is equal to $\rho g h$, where ρ is the density of the liquid in the tube. The difference between the "absolute" pressure P and atmospheric pressure P_{at} is called the **gauge pressure**. The pressure you measure in your automobile tire is gauge pressure. When the tire is entirely flat, the gauge pressure is zero, and the absolute pressure in the tire is atmospheric pressure. The absolute pressure is obtained from gauge pressure by adding atmospheric pressure to it:

$$P = P_{gauge} + P_{at} \qquad 11\text{-}15$$

Figure 11-8 shows a mercury U-tube barometer, which is used to measure atmospheric pressure. The top end of the tube has been closed off and evacuated so that the pressure there is zero. The other end is open to the atmosphere at pressure P_{at}. The pressure P_{at} is given by $P_{at} = \rho g h$, where ρ is the density of mercury.

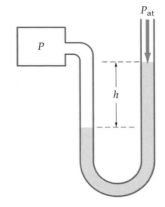

Figure 11-7   Open-tube manometer for measuring an unknown pressure P. The difference $P - P_{at}$ equals $\rho g h$.

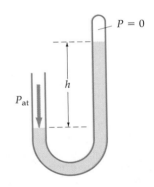

Figure 11-8   U-tube barometer for measuring atmospheric pressure P_{at}.

Example 11-7

At 0°C the density of mercury is 13.595×10^3 kg/m³. What is the height of the mercury column in a U-tube barometer if the pressure is 1 atm = 101.325 kPa?

We have

$$h = \frac{P}{\rho g} = \frac{1.01325 \times 10^5 \text{ N/m}^2}{(13.595 \times 10^3 \text{ kg/m}^3)(9.81 \text{ N/kg})}$$

$$= 0.7597 \text{ m} \approx 760 \text{ mm}$$

In practice, pressure is often measured in millimeters of mercury (commonly called **torr,** after the Italian physicist Torricelli), in inches of mercury (written inHg), and in inches or feet of water (written inH$_2$O or ftH$_2$O). The various units of pressure are related as follows:

$$1 \text{ atm} = 760 \text{ mmHg} = 760 \text{ torr} = 29.9 \text{ inHg}$$

$$= 33.9 \text{ ftH}_2\text{O} = 101.325 \text{ kPa} = 14.7 \text{ lb/in}^2 \qquad 11\text{-}16$$

$$1 \text{ mmHg} = 1 \text{ torr} = 1.316 \times 10^{-3} \text{ atm} = 133.3 \text{ Pa} \qquad 11\text{-}17$$

Other units commonly used on weather maps are the **bar** and the **millibar,** which are defined as

$$1 \text{ bar} = 10^3 \text{ millibars} = 100 \text{ kPa} \qquad 11\text{-}18$$

A pressure of 1 bar is just slightly less than 1 atm.

Example 11-8

The average (gauge) pressure in the aorta is about 100 mmHg. Convert this average blood pressure to pascals and pounds per square inch.

We can use the conversion factors implied in Equation 11-16. For example, we have 760 mmHg = 101.325 kPa. Thus,

$$P = 100 \text{ mmHg} \left(\frac{101.325 \text{ kPa}}{760 \text{ mmHg}} \right) = 13.3 \text{ kPa}$$

and

$$P = 100 \text{ mmHg} \left(\frac{14.7 \text{ lb/in}^2}{760 \text{ mmHg}} \right) = 1.93 \text{ lb/in}^2$$

Exercise

Convert a pressure of 45 kPa to (a) mmHg and (b) atmospheres. [Answers: (a) 338 mmHg, (b) 0.444 atm]

The relation between pressure and altitude for a gas such as air is more complicated than the relation between pressure and depth for a liquid because the density of a gas is not constant but depends on the pressure. In fact, to a good approximation, the density of a gas is proportional to the pressure. The pressure in a column of air decreases with height as you go up from the surface of the earth, like the pressure in a water column decreases as you go up from the bottom, but unlike the case with water pressure, the decrease in air pressure is not linear with distance. Instead, the air pressure decreases by a constant fraction for a given increase in height, as is shown in Figure 11-9. This type of decrease is called an exponential decrease. At a height of about 5.5 km (18,000 ft), the air pressure is half its value at sea level. If we go up another 5.5 km to an altitude of 11 km (a typical altitude for airliners), the pressure is again halved so that it is one-fourth its value at sea level, and so on. Since the density of air is proportional to the pressure, the density of air also decreases with altitude. Thus, for example, there is less oxygen available on a mountain than at normal elevations, which makes exercising difficult. At very high altitudes, such as those at which jet planes fly, the cabins must be pressurized.

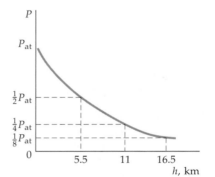

Figure 11-9 Variation in pressure with height above the earth's surface. For each increase in height of 5.5 km, the pressure decreases by half. This is an example of an exponential decrease.

11-4   Buoyancy and Archimedes' Principle

If a heavy object submerged in water is "weighed" by suspending it from a spring scale (Figure 11-10a), the scale reads less than when the object is weighed in air. This is because the water exerts an upward force that partially balances the force of gravity. This force is even more evident when we submerge a piece of cork. When completely submerged, the cork experiences an upward force from the water pressure that is greater than the force of gravity, so it accelerates up toward the surface, where it floats partially submerged. The force exerted by a fluid on a body submerged in it is called the **buoyant force.** It depends on the density of the fluid and the volume of the body but not on the composition or shape of the body and is equal in magnitude to the weight of the fluid displaced by the body.

Archimedes' principle

A body wholly or partially submerged in a fluid is buoyed up by a force equal to the weight of the displaced fluid.

This result is known as **Archimedes' principle.**

Archimedes (287–212 B.C.) had been given the task of determining whether a crown made for King Hieron II was of pure gold or whether it contained some cheaper metal, such as silver. The problem was to determine the density of the irregularly shaped crown without destroying it. As the story goes, Archimedes came upon the solution while bathing and immediately rushed naked through the streets of Syracuse shouting "Eureka" ("I have found it"). This flash of insight preceded Newton's laws, from which Archimedes' principle can be derived, by some 1900 years. What Archimedes found was a simple and accurate way to determine the specific gravity of the crown, which he could then compare with the specific gravity of gold. The specific gravity of an object is the weight of the object in air divided by the weight of an equal volume of water:

$$\text{Specific gravity} = \frac{\text{weight of object in air}}{\text{weight of equal volume of water}}$$

But, according to Archimedes' principle, the weight of an equal volume of water equals the buoyant force on the object when it is submerged. It is therefore equal to the *loss* in weight of the object when it is weighed while submerged in water. Thus,

$$\text{Specific gravity} = \frac{\text{weight of object in air}}{\text{weight loss when submerged in water}} \qquad 11\text{-}19$$

The specific gravity of the crown could be determined by weighing the crown in air and then again when it was submerged in water.

Example 11-9

The specific gravity of gold is 19.3. If a crown made of pure gold weighs 8 N in air, what should its weight be when it is submerged in water?

From Equation 11-19, we have for the weight loss in water

$$\text{Weight loss} = \frac{\text{weight in air}}{\text{specific gravity}} = \frac{8 \text{ N}}{19.3} = 0.415 \text{ N}$$

The crown should therefore weigh 8 N − 0.415 N = 7.59 N

Figure 11-10 (*a*) Weighing an object submerged in a fluid. (*b*) Free-body diagram showing the weight, the force of the spring $\mathbf{F}_s$, and the forces $\mathbf{F}_1$ and $\mathbf{F}_2$ exerted by the surrounding fluid. (*c*) The buoyant force $B = F_2 - F_1$ is the net force exerted on the object by the fluid. It is upward because the pressure at the bottom of the object is greater than that at the top.

(*a*) To determine the percentage of fat in this man's body, the density of his body is measured by weighing him while he is submerged under water. (*b*) Archimedes in his bath, at the legendary moment when he discovered how to measure the density of an object.

(*a*)

(*b*)

Figure 11-11 Figure 11-10 with the submerged body replaced by an equal volume of fluid. The forces F_1 and F_2 due to the pressure of the fluid are the same as in Figure 11-10. The buoyant force is thus equal to the weight w_f of the displaced fluid.

We can derive Archimedes' principle from Newton's laws by considering the forces acting on a portion of a fluid and noting that in static equilibrium the net force must be zero. Figure 11-10*b* shows the vertical forces acting on an object being weighed while submerged. These are the force of gravity **w** down, the force of the spring scale F_s acting up, a force F_1 acting down because of the fluid pressing on the top surface of the object, and a force F_2 acting up because of the fluid pressing on the bottom surface of the object. Since the spring scale reads a force less than the weight, the force F_2 must be greater in magnitude than the force F_1. The difference in magnitude of these two forces is the buoyant force $B = F_2 - F_1$. The buoyant force occurs because the pressure of the fluid at the bottom of the object is greater than that at the top.

In Figure 11-11 the spring scale has been eliminated and the submerged object has been replaced by an equal volume of fluid (indicated by the dotted lines). The buoyant force $B = F_2 - F_1$ acting on this volume of fluid is the same as the buoyant force that acted on our original object since the fluid surrounding the space is the same. Because this volume of fluid is in equilibrium, the net force acting on it must be zero. The upward buoyant force thus equals the downward weight of this volume of fluid:

$$B = w_f \qquad \text{11-20}$$

Note that this result does not depend on the shape of the submerged object. If we consider any irregularly shaped portion of fluid, there must be a buoyant force acting on it due to the surrounding fluid that is equal to the weight of that portion. Thus, we have derived Archimedes' principle.

We can see from Archimedes' principle that an object will float in fluid if the density of the object is less than that of the fluid. If ρ_f is the density of a fluid, a volume V of the fluid then has a mass $\rho_f V$ and a weight

$$w_f = \rho_f g V = B$$

The weight of the object can be written

$$w_o = \rho g V$$

where ρ is the density of the object. If the density of the object is greater than that of the fluid, the weight of the object will be greater than the buoyant force, and the object will sink unless supported. If ρ is less than ρ_f, the buoyant force will be greater than the weight of the object, and the object will accelerate up to the top of the fluid unless held down. At the top, it will float in equilibrium with a fraction of its volume submerged such that the weight of the fluid it displaces equals the weight of the object.

Hot-air balloons rising in the night sky over Albuquerque during a balloon festival.

Example 11-10

A cork has a density of 200 kg/m^3. Find the fraction of the volume of the cork that is submerged when the cork floats in water.

Let V be the volume of the cork and V' be the volume that is submerged when it floats in water. The weight of the cork is $\rho g V$, and the buoyant force is $\rho_w g V'$. Since the cork is in equilibrium, the buoyant force equals the weight:

$$B = w$$

$$\rho_w g V' = \rho g V$$

The fraction of the cork submerged is thus

$$\frac{V'}{V} = \frac{\rho}{\rho_w} = \frac{200 \text{ kg/m}^3}{1000 \text{ kg/m}^3} = \frac{1}{5} \qquad \text{11-21}$$

Thus, one-fifth of the cork is submerged.

For any object floating in a fluid, Equation 11-21 gives the fraction of the object that is submerged if we replace ρ_w with ρ_f, the density of the fluid. Since the density of ice is 920 kg/m³ and that of seawater is 1025 kg/m³, the fraction of an iceberg that is submerged in seawater is

$$\frac{V'}{V} = \frac{\rho}{\rho_f} = \frac{920 \text{ kg/m}^3}{1025 \text{ kg/m}^3} = 0.898$$

The great danger of icebergs to ships is related to the fact that only about 10 percent of an iceberg is visible above the water.

When an object submerged in a fluid is weighed, as in Figure 11-10a, the measured weight F_s, the reading on the spring scale, is less than the true weight w_o because of the upward buoyant force B:

$$F_s = w_o - B$$

Writing these forces in terms of the volume V of the object and the densities ρ of the object and ρ_f of the fluid, we have

$$F_s = \rho g V - \rho_f g V = \rho g V \left(1 - \frac{\rho_f}{\rho} \right)$$

or

$$F_s = w_o \left(1 - \frac{\rho_f}{\rho} \right) \qquad \qquad 11\text{-}22$$

Thus, the measured weight F_s of the submerged object is less than the true weight w_o by the factor $(1 - \rho_f/\rho)$.

Example 11-11

A block of an unknown material weighs 3 N in air and 1.89 N when submerged in water, as in Figure 11-10a. (a) What is the material? (b) What correction must be made for the buoyancy of air when the block is weighed in air? (The density of air is approximately 1.3 kg/m³.)

(a) We determine the material of the block by finding its density. According to Equation 11-19, the specific gravity of an object equals the weight of the object in air divided by the loss in weight when the object is submerged in water. Since for this example the loss in weight is 3 N − 1.89 N = 1.11 N, we have

$$\text{Specific gravity} = \frac{3 \text{ N}}{1.11 \text{ N}} = 2.70$$

The density of this material is thus 2.70 times that of water or 2.70 × 10³ kg/m³. Comparing this with the densities listed in Table 11-1, we see that the block is probably made of aluminum.

(b) The correction that must be made when an aluminum block is weighed in air is obtained from Equation 11-22. For aluminum in air, which has a density $\rho_f = 1.3$ kg/m³, we have

$$\frac{\rho_f}{\rho} = \frac{1.3 \text{ kg/m}^3}{2.7 \times 10^3 \text{ kg/m}^3} = 4.8 \times 10^{-4}$$

and

$$1 - \frac{\rho_f}{\rho} = 1 - 0.00048$$

This differs from 1 by only 0.048 percent, showing that we can usually neglect the buoyant force of air.

Exercise

A piece of lead (specific gravity = 11.3) weighs 80 N in air. What does it weigh when submerged in water?   (Answer: 72.9 N)

Questions

2. How can you estimate your average density at a swimming pool?

3. Why can you see "only the tip of the iceberg"?

4. Smoke usually rises from a smokestack, but it may sink on a very humid day. What can be concluded about the relative densities of humid air and dry air?

5. Why is it easier to float in salt water than in fresh water?

6. Fish can adjust their volume by varying the amount of oxygen and nitrogen gas (obtained from the blood) in a thin-walled sac under their spinal column called a swim bladder. Explain how this helps them swim.

7. A certain object has a density just slightly less than that of water so that it floats almost completely submerged. However, the object is more compressible than water. What happens if the floating object is given a slight push to submerge it?

Figure 11-12 A needle of length L being lifted off the surface of a liquid. Surface tension exerts a force on the needle toward the surface.

11-5 Surface Tension and Capillarity

A needle can be made to "float" on a water surface if it is placed there carefully. The forces that support the needle are not buoyant forces but are instead due to **surface tension.** In the interior of a liquid, a molecule is surrounded on all sides by other molecules, but at the surface, there are no molecules above the surface molecules. If a surface molecule is raised slightly, the molecular bonds between it and the adjacent molecules are stretched, and there is a restoring force that tends to pull the molecule back toward the surface. Similarly, when a needle is placed carefully on the surface, the surface molecules are depressed slightly and the adjacent molecules exert an upward restoring force on them, supporting the needle. Thus, the surface of a liquid is like a stretched elastic membrane. The force necessary to break the surface can be measured by lifting a needle off of the surface as shown in Figure 11-12. This force is found to be proportional to the length of the surface broken, which is twice the length of the needle since there is a surface film on both sides of the needle. If the needle has a mass m and length L, the force F needed to lift it off the surface is

$$F = \gamma 2L + mg \qquad \qquad 11\text{-}23$$

where γ is the **coefficient of surface tension,** which is the force per unit length exerted by the film. The value of γ for water is about 0.073 N/m. It is because of surface tension that small droplets of a liquid tend to be spherical.

(a) The formation of a water drop. The nearly spherical shape of the drop is the result of surface tension. (b) A spherical drop rises from a liquid as a result of another drop falling into it.

(a)

(b)

As the drop is formed, surface tension pulls the surface together, minimizing the surface area and making the drop spherical.

Exercise

A wire of length 12.0 cm is pulled off a water surface with its length parallel to the surface. What force in addition to the weight of the wire is needed? (Answer: 0.0175 N)

The attractive forces between a molecule in a liquid and other molecules in the liquid are called **cohesive forces.** The force between a molecule of liquid and another substance, such as the wall of a thin tube, is called an **adhesive force.** When the adhesive forces are large relative to the cohesive forces, as in the case of water and a glass surface, the liquid is said to wet the surface of the other substance. In this case, the surface of a column of liquid in a tube is concave upward as shown in Figure 11-13a. The contact angle θ_c between the wall and the surface indicates the relative strength of the cohesive and adhesive forces. For a liquid that wets the surface, the contact angle is less than 90°, as in Figure 11-13a. When the adhesive forces are small relative to the cohesive forces, as is the case for mercury and glass, the liquid does not wet the surface, and the surface is convex as shown in Figure 11-13b. In this case, the contact angle is greater than 90°. Cohesive and adhesive forces are difficult to calculate theoretically, but the angles of contact θ_c in Figures 11-13a and b can be measured. For water and glass, the contact angle is approximately 0°. For mercury and glass, it is about 140°.

When the surface of a liquid is concave upward, the surface tension at the wall of the tube has a component upward as shown in Figure 11-14. The liquid will rise in the tube until the net upward force on it due to the surface tension is balanced by the weight of the liquid. This rise is called **capillary action,** or just **capillarity,** and the tube is called a capillary tube. (The smallest blood vessels are also called capillaries.) In Figure 11-14, the liquid has risen to a height h in a thin capillary tube of radius r. The tube is open to atmospheric pressure at the top. The force holding the liquid up is the vertical component of the surface tension, $F \cos \theta_c$. Since the length of the contact surface is $2\pi r$, this vertical force is $\gamma 2\pi r \cos \theta_c$. If the slight curvature of the surface is neglected, the volume of the liquid in the tube is $\pi r^2 h$. Setting the net upward force equal to the weight, we get

$$\gamma 2\pi r \cos \theta_c = \rho(\pi r^2 h)g$$

or

$$h = \frac{2\gamma \cos \theta_c}{\rho r g} \qquad 11\text{-}24$$

Capillarity is responsible for the rise of a liquid in blotting paper and for the rise of oil in the wick of a lamp. An important effect of capillarity is the holding of water in soil in the small spaces between soil particles. Were it not for capillarity, all the rain water would trickle down to the water table, leaving the upper soil dry. Farming could then only be done in swampy areas, as is the case for rice farming.

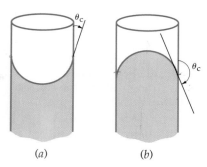

Figure 11-13 (*a*) Surface of a liquid in a narrow tube for the case in which the adhesive forces are greater than the cohesive forces. The surface is concave upwards, and the contact angle θ_c is less than 90°. (*b*) When the cohesive forces are greater than the adhesive forces, the surface of the liquid is convex, and the contact angle θ_c is greater than 90°.

COHESIVE — > 90°

ADHESIVE — < 90°

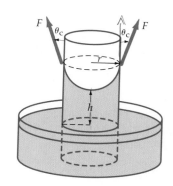

Figure 11-14 Liquid rising in a capillary tube. The upward force due to surface tension supports the weight of the column of liquid.

Example 11-12

How high will water rise in a tube of radius 0.1 mm if the contact angle is zero?

Using $\gamma = 0.073$ N/m for water, we have from Equation 11-24

$$h = \frac{2(0.073 \text{ N/m}) \cos 0°}{(1000 \text{ kg/m}^3)(0.0001 \text{ m})(9.81 \text{ N/kg})} = 0.149 \text{ m} = 14.9 \text{ cm}$$

(a) Water drops on the surface of a plant. The surface tension is large compared with the weight of the smaller drops, so they are nearly spherical in shape. (b) A water strider rests on the surface of a lake.

(a)

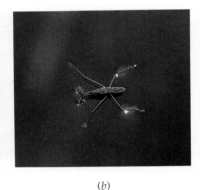

(b)

Example 11-13

The surface tension for mercury is 0.465 N/m and $\theta_c = 140°$. A glass capillary tube of radius 3 mm has been placed in a bowl of mercury. What is the height of the mercury in the tube relative to that in the bowl?

From Equation 11-24, we obtain

$$h = \frac{2(0.465 \text{ N/m}) \cos 140°}{(13.6 \times 10^3 \text{ kg/m}^3)(3 \times 10^{-3} \text{ m})(9.81 \text{ N/kg})}$$

$$= -1.78 \times 10^{-3} \text{ m} = -1.78 \text{ mm}$$

The mercury in the capillary tube is depressed 1.78 mm below the surface of the mercury in the bowl.

Question

8. A water bug walks on the surface of the lake. What keeps the bug from sinking?

11-6   Fluids in Motion and Bernoulli's Equation

The general flow of a fluid can be very complicated. Consider, for example, the rise of smoke from a burning cigarette (Figure 11-16 on facing page). At first the smoke rises in a regular stream, but soon turbulence sets in and the smoke begins to swirl irregularly. Turbulent flow is very difficult to treat even qualitatively. We will therefore consider only nonturbulent, steady-state flow as, for example, the smooth rising of the smoke before it becomes turbulent.

We will first consider a fluid that is flowing with no dissipation of mechanical energy. Such a fluid is called nonviscous. (We will treat viscosity in the next section.) We will also assume that the fluid is incompressible, which is a good approximation for most liquids. In an incompressible fluid, the density is constant throughout the fluid.

Figure 11-15 shows a fluid flowing in a tube of varying cross-sectional area. The shading on the left indicates the volume of fluid flowing into the tube at point 1 in some time Δt. If the speed of the fluid at this point is v_1 and the cross-sectional area of the tube is A_1, the volume flowing into the tube in time Δt is

$$\Delta V = A_1 v_1 \, \Delta t$$

Since we are assuming the fluid is incompressible, an equal volume of fluid must flow out of the tube at point 2 as indicated by the shading on the right.

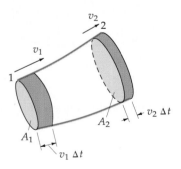

Figure 11-15   An incompressible fluid flowing in a tube of varying cross section. The shaded volumes are equal.

If the speed of the fluid at this point is v_2 and the cross-sectional area is A_2, the volume is $\Delta V = A_2 v_2 \, \Delta t$. Since these volumes are equal, we have

$$A_1 v_1 \, \Delta t = A_2 v_2 \, \Delta t$$

$$A_1 v_1 = A_2 v_2 \qquad \text{11-25}$$

The quantity Av is called the **volume flow rate** I_V. The dimensions of I_V are those of volume per time. In the steady flow of an incompressible fluid, the volume flow rate is the same at any point in the fluid:

$$I_V = vA = \text{constant} \qquad \text{11-26} \qquad \textit{Continuity equation}$$

Equation 11-26 is called the **continuity equation.**

Example 11-14

Blood flows in an aorta of radius 1.0 cm at 30 cm/s. What is the volume flow rate?

From Equation 11-26,

$$I_V = vA = (0.30 \text{ m/s})\pi(0.01 \text{ m})^2$$

$$= 9.42 \times 10^{-5} \text{ m}^3/\text{s}$$

It is customary to give the pumping rate of the heart in liters per minute. Using $1 \text{ L} = 10^{-3} \text{ m}^3$ and $1 \text{ min} = 60 \text{ s}$, we have

$$I_V = (9.42 \times 10^{-5} \text{ m}^3/\text{s}) \times \frac{1 \text{ L}}{10^{-3} \text{ m}^3} \times \frac{60 \text{ s}}{1 \text{ min}}$$

$$= 5.65 \text{ L/min}$$

(a)

(b)

Figure 11-16   (a) Smoke rising from a cigarette. The simple streamlined flow quickly becomes turbulent. (b) Vortex of turbulent flow formed at the tip of a fan blade by heated air above an alcohol lamp.

False-color view of turbulence of blood flowing into and out of the heart as seen by Magnetic Resonance Imaging. Systolic ejection from the left ventricle into the aorta is seen in red, and diastolic filling of the ventricles is in blue.

Example 11-15

Blood flows from a large artery of radius 0.3 cm, where its speed is 10 cm/s, into a region where the radius has been reduced to 0.2 cm because of thickening of the walls (arteriosclerosis). What is the speed of the blood in the narrower region?

If v_1 and v_2 are the initial and final speeds and A_1 and A_2 are the initial and final areas, Equation 11-25 gives

$$v_2 = \frac{A_1}{A_2}v_1 = \frac{\pi(0.3 \text{ cm})^2}{\pi(0.2 \text{ cm})^2}(10 \text{ cm/s}) = 22.5 \text{ cm/s}$$

Exercise

Water is flowing at 12 m/s in a horizontal pipe. If the pipe widens to twice its original diameter, what is the flow speed in the wide section? (Answer: 3 m/s)

We will now consider a fluid flowing in a tube that varies in elevation as well as in cross-sectional area as shown in Figure 11-17. We apply the work-energy theorem to the fluid that is contained initially between points 1 and 2 in Figure 11-17a. After some time Δt, this fluid will have moved along the tube and will be contained in the region between points 1' and 2' in Figure 11-17b. The only change between Figures 11-17a and b is for the dark green portions of the fluid mass. Let $\Delta m = \rho\, \Delta V$ be the mass of this fluid. The net effect on the fluid in time Δt is that the mass of fluid Δm is lifted from height y_1 to height y_2 and its speed is increased from v_1 to v_2. The change in the potential energy of this mass is

$$\Delta U = \Delta m\, gy_2 - \Delta m\, gy_1 = \rho\, \Delta V\, g(y_2 - y_1)$$

and the change in its kinetic energy is

$$\Delta K = \tfrac{1}{2}(\Delta m)v_2^2 - \tfrac{1}{2}(\Delta m)v_1^2 = \tfrac{1}{2}\rho\, \Delta V\,(v_2^2 - v_1^2)$$

The fluid following the mass of fluid Δm in the pipe (to the left of it) exerts a force on it to the right in the figure of magnitude $F_1 = P_1A_1$, where P_1 is the pressure at point 1. This force does work

$$W_1 = F_1\, \Delta x_1 = P_1A_1\, \Delta x_1 = P_1\, \Delta V$$

At the same time, the fluid preceding it (to the right) exerts a force $F_2 = P_2A_2$ on it to the left in the figure. This force does negative work because it opposes the motion:

$$W_2 = -F_2\, \Delta x_2 = -P_2A_2\, \Delta x_2 = -P_2\, \Delta V$$

Figure 11-17 Fluid moving in a pipe that varies in both height and cross-sectional area. The total work done by the forces $F_1 = P_1A_1$ and $F_2 = P_2A_2$ has the effect of raising the portion of the fluid shaded in dark green from height y_1 to y_2 and changing its speed from v_1 to v_2.

(a)

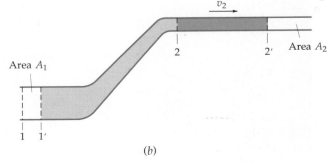

(b)

The total work done by these forces is

$$W_{total} = P_1 \, \Delta V - P_2 \, \Delta V = (P_1 - P_2) \, \Delta V$$

The work-energy theorem gives

$$W_{total} = \Delta U + \Delta K$$

so

$$(P_1 - P_2) \, \Delta V = \rho \, \Delta V \, g(y_2 - y_1) + \tfrac{1}{2}\rho \, \Delta V \, (v_2^2 - v_1^2)$$

If we divide by ΔV, we obtain

$$P_1 - P_2 = \rho g y_2 - \rho g y_1 + \tfrac{1}{2}\rho v_2^2 - \tfrac{1}{2}\rho v_1^2$$

When we collect all the quantities having a subscript 1 on one side and those having a subscript 2 on the other, this equation becomes

$$P_1 + \rho g y_1 + \tfrac{1}{2}\rho v_1^2 = P_2 + \rho g y_2 + \tfrac{1}{2}\rho v_2^2 \qquad \text{11-27a}$$

This result can be restated as

$$P + \rho g y + \tfrac{1}{2}\rho v^2 = \text{constant} \qquad \text{11-27b} \qquad \textit{Bernoulli's equation}$$

which means that this combination of quantities has the same value at any point along the pipe. Equation 11-27 is known as **Bernoulli's equation** for the steady, nonviscous flow of an incompressible fluid. To some extent, we can also apply Bernoulli's equation to compressible fluids such as gases.

A special application of Bernoulli's equation is for a fluid at rest. Then $v_1 = v_2 = 0$, and we obtain

$$P_1 - P_2 = \rho g(y_2 - y_1) = \rho g h$$

where $h = y_2 - y_1$ is the difference in height between points 2 and 1. This is the same as Equation 11-14.

We now give some examples using Bernoulli's equation in nonstatic situations.

Example 11-16

A large tank of water has a small hole a distance h below the water surface as shown in Figure 11-18. Find the speed of the water as it flows out the hole.

We apply Bernoulli's equation to points a and b in the figure. Since the diameter of the hole is much smaller than the diameter of the tank, we can neglect the velocity of the water at the top (point a). We then have

$$P_a + \rho g y_a = P_b + \tfrac{1}{2}\rho v_b^2 + \rho g y_b$$

Since both points a and b are open to the atmosphere, the pressures P_a and P_b are both equal to atmospheric pressure. Therefore,

$$v_b^2 = 2g(y_a - y_b) = 2gh$$

and

$$v_b = \sqrt{2gh}$$

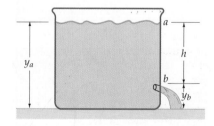

Figure 11-18   Water tank with small hole near the bottom. The speed of the water emerging from the hole is the same as if the water had dropped in free fall a distance $h = y_a - y_b$, a result known as Torricelli's law.

In Example 11-16, the water emerges from the hole with a speed equal to that it would have if it dropped in free fall a distance h. This is known as **Torricelli's law.**

Figure 11-19 Constriction in a pipe carrying a moving fluid. The pressure is smaller in the narrow section of the pipe where the fluid is moving faster.

In Figure 11-19, water is flowing through a horizontal pipe that has a constricted section. Since both sections of the pipe are at the same elevation, $y_1 = y_2$ in Equation 11-27a. Then Bernoulli's equation becomes

$$P + \tfrac{1}{2}\rho v^2 = \text{constant} \qquad \qquad 11\text{-}28$$

When the fluid moves into the constriction, the area A gets smaller, so the speed v must get larger since Av remains constant. But according to Equation 11-28, if the speed gets larger, the pressure must get smaller if $P + \tfrac{1}{2}\rho v^2$ is to remain constant. Thus, the pressure in the constriction is reduced. Equation 11-28 is an important result that applies to many situations in which we can ignore changes in height.

Venturi effect

> When the speed of a fluid increases, the pressure drops.

This result is often referred to as the **Venturi effect.**

Example 11-17

Water moves through a pipe at 4 m/s under a pressure of 200 kPa. The pipe narrows to half its original diameter. Find (*a*) the speed and (*b*) the pressure of the water in the narrower part of the pipe.

(*a*) Since the area of the pipe is proportional to the square of the diameter, the area of the narrow part of the pipe is one-fourth that of the original. Then, from the continuity equation $I_V = vA = \text{constant}$, the speed in the narrow part must be 4 times that in the wide part or 16 m/s.

(*b*) To find the pressure in the narrow part of the pipe, we use Equation 11-28:

$$P_1 + \tfrac{1}{2}\rho v_1^2 = P_2 + \tfrac{1}{2}\rho v_2^2$$

$$200 \text{ kPa} + \tfrac{1}{2}(1000 \text{ kg/m}^3)(4 \text{ m/s})^2 = P_2 + \tfrac{1}{2}(1000 \text{ kg/m}^3)(16 \text{ m/s})^2$$

$$200 \text{ kPa} + 8000 \text{ Pa} = P_2 + 128{,}000 \text{ Pa}$$

$$P_2 = 200 \text{ kPa} + 8 \text{ kPa} - 128 \text{ kPa} = 80 \text{ kPa}$$

Note that when we put ρ and v in their proper SI units, the units of $\tfrac{1}{2}\rho v^2$ are pascals. We can check this by writing out the units of ρv^2:

$$\frac{\text{kg}}{\text{m}^3} \times \frac{\text{m}^2}{\text{s}^2} = \frac{\text{kg·m/s}^2}{\text{m}^2} = \frac{\text{N}}{\text{m}^2} = \text{Pa}$$

where we have used 1 kg·m/s² = 1 N.

(*a*)

(*b*)

Air flow around a wing foil. (*a*) When the wing is horizontal, the flow is uniform and the pressure above the wing is the same as below it. (*b*) When the wing is tilted, the pressure is greater at the bottom of the wing than at the top of the wing, giving rise to the lift.

The Venturi effect can be used qualitatively to understand the lift of an airplane wing and the curve of the path of a baseball. An airplane wing is designed so that the air moves faster over the top of the wing than it does under the bottom of the wing, thus making the air pressure less on top of the wing than underneath. This difference in pressure results in a net force upward on the wing. Figure 11-20a shows a top view of the motion of a baseball thrown by a right-handed pitcher throwing a curve ball. As the ball spins, it tends to drag the air around with it. Figure 11-20b is drawn from the point of view of the ball being stationary (but spinning) with the air rushing past it. The air movement caused by the drag of the spinning ball adds to the velocity of the air rushing by on the left side of the ball and subtracts from it

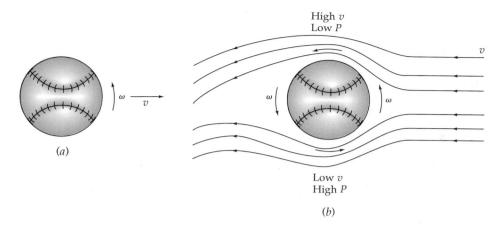

(a)

High v
Low P

Low v
High P

(b)

Figure 11-20 (a) Top view of baseball thrown with a counter-clockwise spin, as would be the case for a curve ball thrown by a right-handed pitcher. (b) In the frame of the ball, the ball is stationary (but spinning) and the air rushes past it. Because of its rough cover, the spinning ball drags the air around with it, making the air speed greater on the left side and less on the right. The pressure is therefore less on the left side, where the air speed is greater, so the ball curves to the left.

on the right. Thus, the speed of the air is greater on the left side of the ball than on the right, so according to Equation 11-28, the pressure on the left is less than that on the right. The ball therefore curves to the left.

Figure 11-21 shows an atomizer. When the bulb is squeezed, air is forced through the constriction in the horizontal tube, which reduces the pressure there below atmospheric pressure. Because of the resulting pressure difference, the liquid in the jar is pumped up through the vertical tube, enters the airstream, and emerges from the nozzle as a fine spray. A similar effect occurs in the carburetor of a gasoline engine.

Although Bernoulli's equation is very useful for qualitative descriptions of many features of fluid flow, such descriptions are often grossly inaccurate when compared with the quantitative results of experiments. Of course, gases like air are hardly incompressible, and liquids like water have viscosity, which invalidates the assumption of the conservation of mechanical energy. In addition, it is often difficult to maintain steady-state, streamlined flow without turbulence.

Figure 11-21 Atomizer. Squeezing the bulb at the left forces air through the narrow constriction. Because the pressure in the constriction is below atmospheric pressure, the liquid in the container below is forced into the airstream, resulting in a fine spray of droplets.

Questions

9. In Figure 11-17, the fluid is accelerated to a greater speed as it enters the narrow part of the pipe. What forces act on the fluid to produce this acceleration?

10. In a department store, a beach ball is supported by the airstream from a hose connected to the exhaust of a vacuum cleaner. Does the air blow under or over the ball to support it? Why?

11-7   Viscous Flow

According to Bernoulli's equation, when a fluid flows steadily through a long, narrow, horizontal pipe of constant cross section, the pressure along the pipe will be constant. In practice, however, we observe a pressure drop as we move along the direction of the flow. Looking at this in another way, a pressure difference is required to push a fluid through a horizontal pipe. This pressure difference is needed because of the drag force that is exerted by the pipe on the layer of fluid next to it and because of the drag force exerted by each layer of the fluid on an adjacent layer that is moving with a different velocity. These drag forces are called **viscous forces.** As a result of these viscous forces, the velocity of the fluid is not constant across the diameter of the pipe. Instead, it is greatest near the center of the pipe and least near the edge, where the fluid is in contact with the walls of the pipe (Figure

Figure 11-22 When a viscous fluid flows through a pipe, the speed is greatest at the center of the pipe. Near the walls of the pipe, the fluid tends to remain stationary.

Figure 11-23 Two plates of equal area with a viscous fluid between them. When the upper plate is moved relative to the lower one, each layer of fluid exerts a drag force on adjacent layers. The force needed to pull the upper plate is directly proportional to v and the area A and inversely proportional to z, the separation between the plates.

Table 11-5 Coefficients of Viscosity of Various Fluids		
Fluid	t, °C	η, mPa·s
Water	0	1.8
	20	1.00
	60	0.65
Blood (whole)	37	4.0
Engine oil		
(SAE 10)	30	200
Glycerin	0	10,000
	20	1,410
	60	81
Air	20	0.018

11-22). Let P_1 be the pressure at point 1 and P_2 be that at point 2, a distance L downstream from point 1. The pressure drop $\Delta P = P_1 - P_2$ is proportional to the volume flow rate:

$$\Delta P = P_1 - P_2 = I_V R \qquad 11\text{-}29$$

where $I_V = vA$ is the volume flow rate and the proportionality constant R is the resistance to flow, which depends on the length of the pipe L, the radius r, and the viscosity of the fluid.

Example 11-18

As blood flows from the aorta through the major arteries, the small arteries, the capillaries, and the veins to the right atrium, the (gauge) pressure drops from about 100 torr to zero. If the volume flow rate is 0.8 L/s, find the total resistance of the circulatory system.

In Example 11-8, we found that 100 torr = 13.3 kPa = 1.33×10^4 N/m². Using 1 L = 1000 cm³ = 10^{-3} m³, we have from Equation 11-29

$$R = \frac{\Delta P}{I_V} = \frac{1.33 \times 10^4 \text{ N/m}^2}{8 \times 10^{-4} \text{ m}^3/\text{s}} = 1.66 \times 10^7 \text{ N·s/m}^5$$

To define the coefficient of viscosity of a fluid, we consider a fluid that is confined between two parallel plates, each of area A, separated by a distance z as shown in Figure 11-23. The upper plate is pulled at a constant speed v by a force **F** while the bottom plate is held at rest. A force is needed to pull the upper plate because the fluid next to the plate exerts a viscous drag force opposing its motion. The speed of the fluid between the plates is essentially v near the upper plate and zero near the lower plate, and it varies linearly with separation between the plates. The force **F** is found to be directly proportional to v and A and inversely proportional to the plate separation z. The proportionality constant is the **coefficient of viscosity** η:

$$F = \eta \frac{vA}{z} \qquad 11\text{-}30$$

The SI unit of viscosity is the N·s/m² = Pa·s. An older cgs unit still in common use is the **poise** after the French physicist Poiseuille. These units are related by

$$1 \text{ Pa·s} = 10 \text{ poise} \qquad 11\text{-}31$$

Table 11-5 lists the coefficients of viscosity for various fluids at various temperatures. Generally, the viscosity of a liquid increases as the temperature decreases. Thus, in cold climates, a thinner grade of oil is used to lubricate automobile engines in the winter than in summer.

The resistance to flow R in Equation 11-29 for steady flow through a circular tube of radius r can be shown to be

$$R = \frac{8\eta L}{\pi r^4} \qquad 11\text{-}32$$

Equations 11-29 and 11-32 can be combined to give the pressure drop over a length L of a circular tube of radius r:

Poiseuille's law

$$\Delta P = \frac{8\eta L}{\pi r^4} I_V \qquad 11\text{-}33$$

Equation 11-33 is known as **Poiseuille's law.** Note the inverse r^4 dependence of the pressure drop. If the radius of the tube is halved, the pressure drop for a given volume flow rate is increased by a factor of 16; or a pressure 16 times as great is needed to pump the fluid through the tube at the original volume flow rate. Thus, if the diameter of the blood vessels or arteries is reduced for some reason, either the volume flow rate of the blood is greatly reduced, or the heart must work much harder to maintain the volume flow rate. For water flowing through a long garden hose, the pressure drop is fixed. It equals the difference in pressure between that at the water source and atmospheric pressure at the open end. The volume flow rate is then proportional to the fourth power of the radius. Thus, if the radius is halved, the volume flow rate drops by a factor of 16.

Poiseuille's law applies only to the laminar (nonturbulent) flow of a fluid of constant viscosity that is independent of the velocity of the fluid. Blood is a complex fluid consisting of solid particles of various shapes suspended in a liquid. The red blood cells, for example, are disk-shaped objects that are randomly oriented at low velocities but tend to become oriented at high velocities to facilitate the flow. Thus, the viscosity of blood decreases as the flow velocity increases, so Poiseuille's law is not strictly valid. Nevertheless, Poiseuille's law is a good approximation that is very useful for obtaining a qualitative understanding of blood flow.

When the flow velocity of a fluid becomes sufficiently great, laminar flow breaks down and turbulence sets in. The critical velocity above which the flow through a tube is turbulent depends on the density and viscosity of the fluid and on the radius of the tube. The flow of a fluid can be characterized by a dimensionless number called the **Reynolds number** N_R, which is defined by

$$N_R = \frac{2r\rho v}{\eta} \qquad 11\text{-}34$$

where v is the average velocity of the fluid. Experiments have shown that the flow will be laminar if the Reynolds number is less than about 2000 and turbulent if it is greater than 3000. Between these values, the flow is unstable and may change from one type to the other.

This turbulent motion of air around a vibrating cylinder in a wind tunnel is called **vortex shedding.** The cylinder was oscillating normal to the oncoming flow of air and the vortices were shed periodically in unison with the oscillations. The Reynolds number was 200. A sheet of minute aerosol particles was introduced into the wind tunnel upstream of the cylinder in order to make the flow pattern visible.

Example 11-19

Calculate the Reynolds number for blood flowing at 30 cm/s through an aorta of radius 1.0 cm. Assume that blood has a viscosity of 4 mPa·s and a density of 1060 kg/m³.

Since the Reynolds number is dimensionless, we can use any set of units as long as we are consistent. Putting each quantity in Equation 11-34 in SI units, we get

$$N_R = \frac{2r\rho v}{\eta} = \frac{2(0.01 \text{ m})(1060 \text{ kg/m}^3)(0.3 \text{ m/s})}{4 \times 10^{-3} \text{ Pa·s}} = 1590$$

Since the Reynolds number is less than 2000, this flow will be laminar rather than turbulent.

Summary

1. The density of a substance is the ratio of its mass to its volume:

$$\text{Density} = \frac{\text{mass}}{\text{volume}}$$

$$\rho = \frac{m}{V}$$

The specific gravity of a substance is the ratio of its density to that of water. An object sinks or floats in a given fluid depending on whether its density is greater than or less than that of the fluid. The densities of most solids and liquids are approximately independent of temperature and pressure, whereas those of gases depend strongly on these quantities. Weight density is the density times g. The weight density of water is 62.4 lb/ft³.

2. Tensile stress is the force per unit area applied to a body:

$$\text{Stress} = \frac{F}{A}$$

Strain is the fractional change in the length of the body:

$$\text{Strain} = \frac{\Delta L}{L}$$

Young's modulus is the ratio of stress to strain:

$$Y = \frac{\text{stress}}{\text{strain}} = \frac{F/A}{\Delta L/L}$$

The shear modulus is the ratio of the shear stress to the shear strain:

$$M_s = \frac{\text{shear stress}}{\text{shear strain}} = \frac{F_s/A}{\Delta X/L}$$

The (negative) ratio of the pressure to the fractional change in volume of an object is called its bulk modulus:

$$B = -\frac{P}{\Delta V/V}$$

The inverse of this ratio is the compressibility k.

3. Pressure of a fluid is the force per unit area exerted by the fluid:

$$P = \frac{F}{A}$$

The SI unit of pressure is the pascal (Pa), which is a newton per square meter:

$$1 \text{ Pa} = 1 \text{ N/m}^2$$

Many other units of pressure, such as atmosphere, bar, torr, pound per square inch, or millimeter of mercury, are often used. These units are related by

$$1 \text{ atm} = 101.325 \text{ kPa} = 760 \text{ mmHg} = 760 \text{ torr}$$

$$= 29.9 \text{ inHg} = 33.9 \text{ ftH}_2\text{O} = 14.71 \text{ lb/in}^2$$

Gauge pressure is the difference between the absolute pressure and atmospheric pressure.

4. Pascal's principle states that pressure applied to an enclosed liquid is transmitted undiminished to every point in the fluid and to the walls of the container.

5. In a liquid, such as water, the pressure increases linearly with depth:

$$P = P_0 + \rho g h$$

In a gas such as air, pressure decreases exponentially with altitude.

6. Archimedes' principle states that a body wholly or partially submerged in a fluid is buoyed up by a force equal to the weight of the displaced fluid.

7. Objects can be supported on the surface of a less dense fluid because of surface tension, which is the result of molecular forces at the surface of the fluid. These molecular forces are also responsible for the rise of a liquid in a thin tube, which is known as capillarity.

8. For the steady-state flow of an incompressible fluid, the volume flow rate is the same throughout the fluid:

$$I_V = vA = \text{constant}$$

This is called the continuity equation.

9. Bernoulli's equation

$$P + \rho g y + \tfrac{1}{2}\rho v^2 = \text{constant}$$

applies to steady-state, nonviscous flow without turbulence in which mechanical energy is conserved. For situations in which we can ignore changes in height, we have the important result that when the speed of the fluid increases, the pressure drops. This result, known as the Venturi effect, can be used to explain qualitatively the lift on an airplane wing and the curve of a baseball's path.

10. In viscous flow through a tube, the pressure drop is proportional to the volume flow rate and to the resistance, which in turn is inversely proportional to the fourth power of the radius of the tube:

$$\Delta P = I_V R = \frac{8\eta L}{\pi r^4} I_V$$

This is Poiseuille's law.

Aerodynamics of Bicycling

Robert G. Hunt
Johnson County Community College

Aerodynamics is the systematic study of forces exerted by air or other gases. In this chapter you have been introduced to some of the basic concepts of viscous flow of fluids; this essay will extend those ideas to illustrate some important technical aspects of bicycling.

One of the attractions of bicycling is the aerodynamic interaction, the feel of the "wind" as we ride. Another satisfying aspect is the ease with which a bicycle is set into motion and the effortless coasting of long distances because of the low levels of mechanical friction. These two factors oppose one another, with aerodynamic drag limiting speeds obtained.

As we start riding, we exert forces on the pedals that result in a forward-directed force that causes acceleration. This motion can be described by using the kinematics and dynamics developed in Chapters 2 through 6, if the net force can be determined. This, however, becomes a complicated problem. So far in this course, we have primarily considered constant or uniformly changing forces, but the forces exerted in this case are more complicated.

Robert G. Hunt is currently a physics instructor at Johnson County Community College, Overland Park, Kansas. He has taught 14 years at the high school and community college level. His undergraduate work was at the University of Kansas and his two Master's degrees are from Rensselaer Polytechnic Institute. He has worked as a chemical physicist at Midwest Research Institute and more recently as an environmental policy consultant and as a developer of environmental curricula at Franklin Associates. He is a lifelong bicycling enthusiast, and has developed a number of demonstrations and lab experiments using bicycles.

When we start riding on a flat terrain, we exert relatively large forces on the pedals, perhaps even standing on them to get started. The bicycle accelerates forward, described by Newton's second law $\mathbf{F} = m\mathbf{a}$. The bicycle continues to accelerate as we exert forces on the pedals, but eventually a level of physiological riding comfort is reached; for most people this is near one revolution of the pedals per second. As this point is approached, the force on the pedals is reduced gradually, until a constant speed is achieved. This terminal speed is typically 4 to 7 m/s (9 to 16 mi/h) for recreational riding on a ten- or three-speed bicycle.

Constant speed means that the acceleration is zero, so the net force is zero. Thus, the propelling force is balanced by the retarding forces. The two major retarding forces are mechanical friction and the aerodynamic drag on the bicycle and the rider. Mechanical friction includes that in the bicycle bearings, chains, and rolling tires. This typically accounts for about one-third or less of the retarding forces. The dominant forces arise from aerodynamic drag.

Earlier in this chapter we learned about fluid flow confined to tubes or pipes. For bicycles, the viscous fluid (air) is unconfined. It impinges directly onto the rider and the bicycle, and an analysis of the aerodynamic drag requires a different approach.

Aerodynamic drag occurs for two reasons. One is pressure that results from air molecules striking a surface and bouncing off, undergoing a momentum change and exerting a normal force on the surface. The other type of force arises from the sliding motion of air molecules along the surface. This is called air friction; it arises because air molecules collide with rough surfaces, and also because they have a weak chemical attraction for the surface. This causes drag forces to be exerted along the surface as the air rushes past.

Fluid flow can be either laminar or turbulent (see Section 11-7), depending on many factors such as speed, surface roughness, and the type of surface material. At sufficiently slow speeds, the flow will be laminar; this results in quite low drag forces. As the relative speed of the air and the surface increases, the laminar flow becomes unstable and layers of air begin to separate. The flow then becomes turbulent, characterized by whirling eddies of air (Figure 1). Turbulent boundary layers have much higher drag than laminar layers, but the highest drag is caused by instability at air velocities in the transition region between laminar and turbulent flow. In fact, aerodynamic drag can be 5 times greater in the transition speed ranges than aerodynamic drag for purely turbulent flow.

Figure 1 Air flow around a bicycle rider. The rider, moving left, is simulated here by a circular cylinder.

Laminar Flow

Turbulent Flow

It is this transition region that must be avoided to achieve low drag forces. The limits of the transition region are defined by the Reynolds number (Section 11-7). For analytical purposes, a bicycle rider can be simulated by a series of stacked cylinders of circular cross section. For this geometry, the transition region occurs at a Reynolds number of about 4×10^5. Unfortunately for bicycle riders, this occurs on very smooth surfaces at speeds of about 4 to 6 m/s, which is where most riding takes place. This means that very smooth surfaces, which encourage laminar flow, should be avoided except in some special cases, to be discussed later. This may seem surprising at first, but experienced riders report noticeable increases in drag when they don very smooth plastic rain gear. An analogy can be seen in the design of golf balls. They are intentionally dimpled to make them rough because a roughened surface ensures turbulent air flow; this allows the golf ball to travel longer distances than it would with a smooth surface.

Let's now look at some quantitative aspects of aerodynamic drag for recreational bicycling. The drag forces retarding forward motion can be characterized by

$$F_D = \tfrac{1}{2}C_D\rho Av^2 \qquad (1)$$

where F_D is the drag force, C_D is a dimensionless number known as the drag coefficient, ρ is the air density, A is the projected frontal area (usually the cross-sectional area perpendicular to the general direction of travel), and v is the relative velocity of the air and the surface over which the air is flowing. This equation includes both pressure and air friction effects.

Like many practical applications of physics, the equation for determining the drag force provides an estimate, not "the" answer. The primary reason for this is that aerodynamic drag is a phenomenon for which no simple model exists. For example, C_D cannot be calculated for most real objects, but is usually inferred from experimental data, often from measurements made in wind tunnels. The value obtained for C_D depends in a complicated way on shape, speed, type of material, temperature, and so on. However, for some applications, such as bicycle riding, C_D is relatively constant over the range of speeds of interest. This simplifies our approximation.

In Equation 1, ρ, as well as C_D, can be considered constant for a given bicycle and rider. The other two terms—A and v—are important variables. Speed is usually continuously changing while riding, and the square dependence makes it the dominant variable. The frontal area term includes both bicycle and rider. While the area of the bicycle will be constant during a ride, the rider may frequently change position. The rider's frontal area is several times greater than that of the bicycle and therefore is the most important component determining frontal area. The rider's clothes are also a factor; tight fitting clothes present a smaller frontal area than those that are loose fitting.

For a casual ride on a warm spring day, the following are typical:

$$C_D = 0.90 \qquad \rho = 1.3 \text{ kg/m}^3$$
$$A = 0.45 \text{ m}^2 \qquad v = 4.0 \text{ to } 7.0 \text{ m/s}$$

Figure 2 plots data points derived using Equation 1 and the above values for C_D, A, ρ, and v. Aerodynamic drag force increases from 4 N at 4 m/s to over 12 N at 7 m/s. This figure also shows the power necessary to sustain a given speed considering only

Figure 2 Force and power required to overcome aerodynamic drag at various speeds.

Continued

the drag force. Power can be found from $P = \mathbf{F} \cdot \mathbf{v}$ (Equation 6-33). For the situation we are considering here, the force and velocity are in the same direction, so Equation 6-33 reduces to $P = Fv$. From Equation 1 we have

$$P = \tfrac{1}{2}C_D \rho A v^3 \qquad (2)$$

That is, the power required to sustain motion depends on the cube of the speed and so increases rapidly as the speed increases.

Although aerodynamic drag is a dominant factor for bicycle riding, it is interesting to put it into perspective and to relate it to our own senses. We will now translate the results shown in Figure 2 into three physiological parameters to which we can all relate. These are (1) the average force exerted on the pedals, (2) the average rate at which we expend energy (i.e., our power output), and (3) food energy requirements.

From experimental data obtained with inexpensive mass-marketed ten- and three-speed bicycles (of the kind typically used by students), Table 1 was prepared. The mechanical bearing, chain, and rolling tire friction was added to the aerodynamic drag forces to find a total retarding force (second column). The drag forces (which can be obtained from Figure 2) range from $\tfrac{2}{3}$ of the total force at 4 m/s to $\tfrac{3}{4}$ of the total at 7 m/s. Provided that the bicycle is in its highest gear, measurements show that the average downward force on the pedals (third column) is 14.2 times the total retarding force. The average downward force is also reported as a percent of the rider's weight for a 68-kg rider (typical for cross-country riders). This translates to a body weight of 668 N (150 lb). To maintain the selected speeds requires 13 to 34 percent of body weight be exerted downward on the pedals. Compare this to the 20 percent body weight exertion that many people find comfortable for sustained riding.

Finally, the mechanical power output (calculated by $P = Fv$) is shown in the fifth column. We may estimate food requirements, assuming that the rider's body is 25 percent efficient at converting food energy into useful work. This illustrates the efficiency of locomotion by bicycle. Traveling at a sustained leisurely rate of 5 m/s, a rider could travel 18 km in 1 h on only 150 food calories, about that contained in two slices of plain bread.

Let's now examine how riders can (intentionally or unintentionally) affect aerodynamic drag. An easily altered factor is the frontal area of the rider. As a rider shifts position and moves about, the area changes. The values range from 0.3 to 0.6 m^2. The smaller value corresponds to a racing posture, crouching low and wearing tightly fitting clothes (Figure 3). A more comfortable posture in street clothes provides a frontal area of 0.45 m^2 (the value used in our sample calculation above). For an upright, touring, or standing posture the area becomes 0.6 m^2. Thus, the range of possible areas leads to a variation of a factor of two in the aerodynamic drag force.

Another important variable is the relative speed of the air and the surface. In our sample calculation, we assumed that the air was still (something it rarely is). If the wind blows from behind, then the drag will provide a forward force. In fact, if the tail wind equals the bicycle speed, the relative speed is zero and there is no aerodynamic drag on the rider. On the other hand, a head wind increases the drag force. Suppose the rider tries to maintain a ground speed of 7 m/s against a head wind of 7 m/s. The relative speed between the air and the surface will then be 14 m/s, resulting in a drag force of 49 N (from Equation 1). At a bicycle speed of 7 m/s the mechanical friction is about 4 N for a total retarding force of 53 N. Table 2 is a summary of the forces and other consequences to be expected. Much greater

Table 1   **Estimated Force and Power for Casual Recreational Bicycling**

| Speed, m/s | Total force to sustain motion, N | Average force on pedal (bike in high gear) | | Power output of rider, W |
		In newtons	As percentage of weight	
4	6	85	13	24
5	9	126	19	45
6	12	170	25	72
7	16	227	34	112

Table 2   **Consequences of Riding into a 7-m/s Head Wind at Two Bicycle Speeds**

| | Bicycle speed | |
	7 m/s (high gear)	3 m/s (medium gear)
Relative speed, m/s	14	10
Drag force, N	49	25
Total force, N	53	27
Force on pedal, N	753	189
Power output, W	371	81

Figure 3 A rider can intentionally lower aerodynamic drag by acquiring a racing posture (crouching low and wearing tightly fitting clothes).

forces are needed on the pedal than before—753 N are needed, compared to 227 N in still air. It is now necessary to stand with full weight on the pedals, and pull upward on the handles to achieve an even greater downward force on the pedals. The power requirement is now 371 W, well in excess of sustainable limits (of no more than 100 W).

In the above case, riders will normally decrease speed, and shift to a lower gear to reduce the force necessary to continue in the forward direction. Generally bicycle speed is reduced by about one-half of the wind speed by changing to a lower gear. The second column of Table 2 shows the results of shifting and slowing to 3 m/s. The forces on the pedal are reduced to a much more comfortable 189 N (28% of body weight), and to a sustainable power level of 81 W. Riders tend to adjust to changing aerodynamic conditions by changing speed and gears.

Still another factor that influences drag is the air density. While this will be relatively constant at a given location, it does change with altitude (Section 11-3) and temperature. For example, in the "mile-high" city of Denver, air density is 10 percent lower than at sea level, leading to a 10 percent reduction in aerodynamic drag. At even higher altitudes, drag may be reduced by another 10 percent. Temperature is less variable, but warm air has a lower density than cool air. A 5°C increase in air temperature produces a 1.5 percent reduction in drag.

Aerodynamic drag can also be reduced by lowering the drag coefficient. Some specially designed bicycles have a C_D of 0.1 or lower. This is accomplished by streamlining, carefully designing the shape of the bicycle so that laminar flow is preserved during the ride. This requires surrounding the bicycle and rider with a shaped surface, as shown in Figure 4. Frontal area is minimized by having the rider lie flat. However, these bicycles are quite expensive to build, are not comfortable to ride, and have never gained wide popularity.

There are many other interesting aspects of bicycling aerodynamics, which you are encouraged to pursue. Two good references for additional reading are listed with the Further Readings at the end of the chapter.

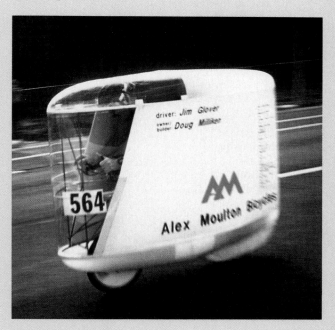

Figure 4 Jim Glover is shown here riding a bicycle designed by Alex Moulton, with which he set the world speed record of 82.53 km/h at Vancouver in 1986.

Suggestions for Further Reading

Hazen, David C., and Rudolf F. Lehnert: "Low-Speed Flight," *Scientific American*, April 1956, p. 46.

How an airplane wing generates lift and how viscosity of air helps to determine the minimum possible flying speed of an airplane.

Kyle, Chester R.: "How Wind Affects Cycling," *Bicycling*, May 1988, p. 194.

The effects of wind on the speed of bicycle riding.

Walker, Jearl: "The Amateur Scientist: Looking into the Ways of Water Striders, the Insects that Walk (and Run) on Water," *Scientific American*, November 1983, p. 188.

Walker explains how surface tension enables water striders to stay dry and how they are able to propel themselves.

Walker, Jearl: "The Amateur Scientist: What Forces Shape the Behavior of Water as a Drop Meanders Down a Windowpane?" *Scientific American*, September 1985, p. 138.

The effects of surface tension on both resting drops and small streams on glass or plexiglass are investigated.

Whitt, Frank Rowland, and David Gordon Wilson: *Bicycling Science*, second edition, MIT Press, Cambridge, 1982.

A comprehensive technical treatment of many aspects of bicycling.

Review

A. Objectives: After studying this chapter, you should:

1. Be able to state the definitions of density, specific gravity, weight density, Young's modulus, and the bulk modulus.

2. Be able to explain the buoyancy of objects, such as boats and hot-air balloons, using Archimedes' principle.

3. Be able to work problems involving buoyant forces on submerged or floating objects.

4. Be able to discuss the rise of a liquid in a capillary tube in terms of the surface tension of the liquid.

5. Be able to explain qualitatively the lift of an airplane wing and the curving of the path of a baseball's flight using the Venturi effect.

B. Define, explain, or otherwise identify:

Density
Liter
Specific gravity
Weight density
Elastic limit
Tensile stress
Strain
Young's modulus
Compressive stress
Tensile strength
Compressive strength
Shear force
Shear stress
Shear strain
Shear modulus
Torsion modulus
Pressure
Pascal
Bulk modulus
Compressibility

Pascal's principle
Gauge pressure
Torr
Bar
Millibar
Buoyant force
Archimedes' principle
Surface tension
Coefficient of surface tension
Cohesive forces
Adhesive forces
Capillarity
Volume flow rate
Continuity equation
Bernoulli's equation
Torricelli's law
Venturi effect
Viscous forces
Coefficient of viscosity
Poise
Poiseuille's law
Reynolds number

C. True or false: If the statement is true, explain why it is true. If it is false, give a counterexample.

1. The buoyant force on a submerged object depends on the shape of the object.

2. Young's modulus has the same dimensions as pressure.

3. If the density of an object is greater than that of water, it cannot float on the surface of the water.

4. In viscous flow, the pressure drop along a pipe is proportional to the flow rate.

Problems

Level I

11-1 Density

1. A copper cylinder is 6 cm long and has a radius of 2 cm. Find its mass.

2. Find the mass of a lead sphere of radius 2 cm.

3. Find the mass of air in a room 4 m by 5 m by 4 m.

4. A small flask used for measuring densities of liquids (called a pyconmeter) has a mass of 22.71 g. When it is filled with water, the total mass is 153.38 g, and when it is filled with milk, the total mass is 157.67 g. Find the density of milk.

5. A solid oak door is 200 cm high, 75 cm wide, and 4 cm thick. How much does it weigh?

6. A 60-mL flask is filled with mercury at 0°C. When the temperature rises to 80°C, 1.47 g of mercury spills out of the flask. Assuming the volume of the flask is constant, find the density of mercury at 80°C if its density at 0°C is 13,645 kg/m^3.

11-2 Stress and Strain

7. A 50-kg ball is suspended from a steel wire of length 5 m and radius 2 mm. By how much does the wire stretch?

8. Copper has a breaking stress of about 3×10^8 N/m^2. (a) What is the maximum load that can be hung from a copper wire of diameter 0.42 mm? (b) If half this maximum load is hung from the copper wire, by what percentage of its length will it stretch?

9. As a runner's foot touches the ground, the shearing force acting on an 8-mm-thick sole is as shown in Figure 11-24. If the force of 25 N is distributed over an area of 15 cm^2, find the angle of shear θ shown, given that the shear modulus of the sole is 1.9×10^5 N/m^2.

11-3 Pressure in a Fluid

10. Barometer readings are commonly given in inches of mercury. Find the pressure in inHg equal to 101 kPa.

11. The pressure on the surface of a lake is atmospheric pressure $P_{at} = 101$ kPa. (a) At what depth is the pressure twice atmospheric pressure? (b) If the pressure at the top of a deep pool of mercury is P_{at}, at what depth is the pressure $2P_{at}$?

12. Find the (a) absolute pressure and (b) the gauge pressure at the bottom of a diving pool of depth 5.0 m.

13. The top of a card table is 80 cm by 80 cm. What is the force exerted on it by the atmosphere? Why doesn't the table collapse?

14. When a woman in high heels takes a step, she momentarily places her entire weight on one heel of her shoe, which has a radius of 0.4 cm. If her mass is 56 kg, what is the pressure exerted on the floor by her heel?

15. A hydraulic lift is used to raise an automobile of mass 1500 kg. The radius of the shaft of the lift is 8 cm and that of the piston is 1 cm. How much force must be applied to the piston to raise the automobile?

16. Blood flows into the aorta through a circular opening of radius 0.9 cm. If the blood pressure is 120 torr, how much force must be exerted by the heart?

17. A car misses a turn and sinks into a shallow lake to a depth of 8 m. (a) If the area of the car door is 0.5 m^2, what is the force exerted on the outside of the door by the water? (b) What is the force exerted on the inside of the door by the air, assuming the inside of the car is at atmospheric pressure? (c) What should the occupant do to get the door open?

18. What pressure is required to reduce the volume of 1 kg of water from 1.00 L to 0.99 L?

19. Seawater has a bulk modulus of 2.3×10^9 N/m^2. Find the density of seawater at a depth where the pressure is 800 atm if the density at the surface is 1024 kg/m^3.

20. A 1500-kg car rests on four tires, each of which is inflated to a gauge pressure of 200 kPa. What is the area of contact of each tire with the road, assuming the four tires support the weight equally?

21. Blood plasma flows from a bag through a tube into a patient's vein, where the blood pressure is 12 mmHg. The specific gravity of blood plasma at 37°C is 1.03. What is the minimum elevation the bag must have so that the pressure of the plasma as it flows into the vein is at least 12 mmHg?

22. In the seventeenth century, Pascal performed the experiment shown in Figure 11-25. A wine barrel was filled with water and was then connected to a long tube. Water was added to the tube until the barrel burst. (a) If the radius of the lid was 20 cm and the height of the water in the tube was 12 m, calculate the force exerted on the lid. (b) If the tube had an inner radius of 3 mm, what mass of water in the tube caused the pressure that burst the barrel?

Figure 11-24 Problem 9.   **Figure 11-25** Problem 22.

11-4 Buoyancy and Archimedes' Principle

23. A 500-g piece of copper (specific gravity 9.0) is suspended from a spring scale and is submerged in water (Figure 11-26). What force does the spring scale read?

Figure 11-26 Problem 23.

Water

0.5 kg
Cu

24. When a 60-N stone is attached to a spring scale and is submerged in water, the scale reads 40 N. Find the specific gravity of the stone.

25. A block of an unknown material weighs 5 N in air and 4.55 N when submerged in water. (*a*) What is the density of the material? (*b*) Of what material is the block made?

26. A 5-kg iron block is attached to a spring scale and is submerged in a fluid of unknown density. The spring scale reads 6.16 N. What is the density of the fluid?

11-5 Surface Tension and Capillarity

27. When a capillary tube with a diameter of 0.8 mm is dipped into methanol, the methanol rises to a height of 15.0 mm. If the angle of contact is zero, find the surface tension of methanol (specific gravity 0.79).

28. Tiny tubes called xylem carry nutrients upward in a plant. Each tube has a radius of about 0.01 mm. How high will water rise by capillary action in such a tube, assuming the contact angle to be zero?

11-6 Fluids in Motion and Bernoulli's Equation

29. Blood flows in an aorta of radius 9 mm at 30 cm/s. (*a*) Calculate the volume flow rate in liters per minute. (*b*) Although the cross-sectional area of a capillary is much smaller than that of the aorta, there are many capillaries, so their total cross-sectional area is much larger. If all the blood from the aorta flows into the capillaries and the speed of flow through the capillaries is 1.0 mm/s, calculate the total cross-sectional area of the capillaries.

30. Water flows through a 3-cm-diameter hose at 0.65 m/s. The diameter of the nozzle is 0.30 cm. (*a*) At what speed does the water pass through the nozzle? (*b*) If the pump at one end of the hose and the nozzle at the other end are at the same height, and if the pressure at the nozzle is atmospheric pressure, what is the pressure at the pump?

31. Water is flowing at 3 m/s in a horizontal pipe under a pressure of 200 kPa. The pipe narrows to half its original diameter. (*a*) What is the speed of flow in the narrow section? (*b*) What is the pressure in the narrow section? (*c*) How do the volume flow rates in the two sections compare?

32. The pressure in a section of horizontal pipe with a diameter of 2 cm is 142 kPa. Water flows through the pipe at 2.80 L/s. What should the diameter of a constricted section of the pipe be for the pressure there to be 101 kPa?

33. During very high winds, the atmospheric pressure inside a house may blow off the roof because of the reduced pressure outside. Calculate the force on a square roof 15 m on a side if the wind speed over the roof is 30 m/s.

11-7 Viscous Flow

34. A horizontal tube with an inside diameter of 1.2 mm and a length of 25 cm has water flowing through it at 0.30 mL/s. Find the pressure difference required to drive this flow if the viscosity of water is 1.00 mPa·s.

35. Find the diameter of a tube that would give double the flow rate for the pressure difference in Problem 34.

36. Blood takes about 1.0 s to pass through a 1-mm-long capillary of the human circulatory system. If the diameter of the capillary is 7 μm and the pressure drop is 2.60 kPa, find the viscosity of blood.

Level II

37. Many people believe that if they float the top of a flexible snorkel tube out of the water (Figure 11-27), they will be able to breathe with it while walking under water. However, the water pressure opposes the expansion of the chest and the inflation of the lungs. Suppose you can just breath while lying on the floor with a 400-N weight on your chest. How far below the surface of the water could your chest be for you still to be able to breathe, assuming your chest has a frontal area of 0.09 m^2?

Figure 11-27 Problem 37.

38. When the ground becomes saturated with water during floods, pressure develops in the ground similar to that in water. This pressure forces water through the joints in concrete-block cellar walls. If this happens quickly enough to fill up the cellar with water, there may be no further damage. Otherwise, the upward pressure on the cellar floor may float the house like a ship. What upward force would be exerted on a 10 m by 10 m basement floor if the floor were 2 m below the surface of the water?

39. Under standard conditions, the density of air is 1.29 kg/m^3 and that of helium is 0.178 kg/m^3. A helium balloon lifts a basket and cargo of total weight 2000 N. What must be the volume of the balloon?

40. Figure 11-28 shows a small sphere that is supported on the surface of water by surface tension. The upward force due to surface tension is $2\pi r\gamma \cos \theta_c$. An insect of mass 0.002 g is supported on the surface of a lake by its six legs, each of which has an approximately spherical base of radius 0.02 mm. (a) Using

$$\tfrac{1}{6}w = \tfrac{1}{6}mg = 2\pi\gamma \cos \theta_c$$

where w is the weight of the insect and the factor $\tfrac{1}{6}$ is used because the six legs support the insect equally, calculate the angle θ_c. (b) If the mass of the insect increases with no increase in r, the angle θ_c decreases until the critical value $\theta_c = 0$ is reached. If the mass increases further, the insect cannot be supported by surface tension. Find the critical mass of the insect that can be supported by legs of this size.

Figure 11-28   Problems 40 and 59.

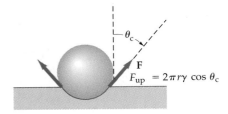

41. A large tank of water is tapped a distance h below the water surface by a small pipe as shown in Figure 11-29. Find the distance x reached by the water flowing out the pipe.

Figure 11-29   Problems 41 and 64.

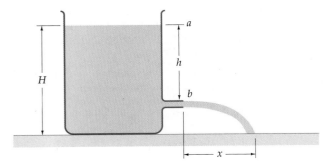

42. A force F is applied to a long wire of length L and cross-sectional area A. Show that if the wire is considered to be a spring, the force constant k is given by $k = AY/L$ and that the energy stored in the wire is $U = \tfrac{1}{2}F \, \Delta L$, where Y is Young's modulus and ΔL is the amount the wire has stretched.

43. A manometer is a U-tube that contains a liquid and is designed to measure small pressure differences between its two arms. If an oil ($\rho = 900$ kg/m^3) manometer can be read to ± 0.05 mm, what is the smallest pressure change that can be detected?

44. A 3-m by 3-m raft is 11 cm thick and is made of wood with a specific gravity of 0.6. How many 70-kg people can stand on the raft without getting their feet wet when the water is calm?

45. An object has neutral buoyancy when its density equals that of the liquid in which it is submerged, which means that it neither floats nor sinks. If the average density of an 85-kg diver is 0.96 kg/L, what mass of lead should be added to give him neutral buoyancy?

46. When you weigh yourself in air, the scale reading is less than the force exerted on you by gravity because of the buoyant force of air. Estimate the correction to the scale reading you should make to get your true weight.

47. A beaker of mass 1 kg containing 2 kg of water rests on a scale. A 2-kg block of aluminum (specific gravity 2.70) suspended from a spring scale is submerged in the water as shown in Figure 11-30. Find the readings of both scales.

Figure 11-30   Problem 47.

48. In Example 11-6, a force of 147 N is applied to a small piston to lift a car that weighs 14,700 N. Show that this does not violate the law of conservation of mechanical energy by showing that, when the car is lifted some distance h, the work done by the force acting on the small piston equals the work done by the large piston on the car.

49. A large piece of cork weighs 0.285 N in air. When held submerged under water by a spring scale as in Figure 11-31, the spring scale reads 0.855 N. Find the density of the cork.

Figure 11-31 Problem 49.

50. A building is to be demolished by a 400-kg steel ball swinging on the end of a 30-m steel wire of diameter 5.0 mm hanging from a tall crane. As the ball is swung through an arc from side to side, the wire makes an angle of 50° with the vertical at the top of the swing. Find the amount by which the wire is stretched at the bottom of the swing.

51. A firefighter holds a hose with a bend in it as shown in Figure 11-32. Water is expelled from the hose in a stream of radius 1.5 cm at a speed of 30 m/s. (a) What mass of water emerges from the hose in 1 s? (b) What is the horizontal momentum of this water? (c) Before reaching the bend, the water has momentum upward, whereas afterward, its momentum is horizontal. Draw a vector diagram of the initial and final momentum vectors, and find the change in the momentum of the water at the bend in 1 s. From this, find the force exerted on the water by the hose.

Figure 11-32 Problem 51.

52. A fountain designed to spray a column of water 12 m into the air has a 1-cm-diameter nozzle at ground level. The water pump is 3 m below the ground. The pipe to the nozzle has a diameter of 2 cm. Find the necessary pump pressure (neglecting the viscosity of the water).

53. A rectangular dam 30 m wide supports a body of water to a height of 25 m. (a) Neglecting atmospheric pressure, find the total force due to water pressure acting on a thin strip of the dam of height dy located at a depth y. (b) Integrate your result in part (a) to find the total horizontal force exerted by the water on the dam. (c) Why is it reasonable to neglect atmospheric pressure?

54. When a rubber strip with a 3 by 1.5 mm cross section is suspended vertically and various masses are attached to it, a student obtains the following data for length versus load:

Load, g	0	100	200	300	400	500
Length, cm	5.0	5.6	6.2	6.9	7.8	10.0

(a) Find Young's modulus for the rubber strip for small loads. (b) Find the energy stored in the strip when the load is 150 g. (See Problem 42.)

55. The steel E string of a violin is under a tension of 53 N. The diameter of the string is 0.20 mm, and its length under tension is 35.0 cm. Find (a) the unstretched length of this string, and (b) the work needed to stretch the string. (See Problem 42.)

Level III

56. A ship sails from seawater (specific gravity 1.03) into fresh water and therefore sinks slightly. When its load of 600 Mg is removed, it returns to its original level. Assuming the sides of the ship are vertical at the water line, find the mass of the ship before it was unloaded.

57. The hydrometer shown in Figure 11-33 is a device for measuring the density of liquids. The bulb contains lead shot, and the density can be read directly from the liquid level on the stem after the hydrometer has been calibrated. The bulb's volume is 20 mL, the stem is 15 cm long and has a diameter of 5.00 mm, and the mass of the glass is 6.0 g. (a) What mass of lead shot must be added so that the least density of liquid that can be measured is 0.9 kg/L? (b) What is the maximum density of liquid that can be measured?

Figure 11-33 Problem 57.

58. A hollow can with a small hole of radius 0.1 mm is pushed under water. At what depth will water start to flow into the can through the hole if the coefficient of surface tension of water is 0.073 N/m?

59. A small sphere rests on the surface of a liquid as shown in Figure 11-28. (a) Show that the angle θ_c is related to the radius r of the sphere, the density ρ of the sphere, and the coefficient of surface tension γ of the liquid by

$$\cos \theta_c = \frac{2}{3} \frac{r^2 \rho g}{\gamma}$$

(b) Find the radius of the largest copper sphere that can rest on water without sinking.

60. The volume of a cone of height h and base radius r is $V = \frac{1}{3}\pi r^2 h$. A conical vessel of height 25 cm resting on its base of radius 15 cm is filled with water. (a) Find the volume and weight of the water in the vessel. (b) Find the force exerted by the water on the base of the vessel. Explain how this force can be greater than the weight of the water.

61. The water in the cylindrical container in Figure 11-34 drains through a horizontal capillary of diameter 0.50 mm. Calculate the length of time needed for the water in the cylinder to drop from 10.0 cm to 5.0 cm if the viscosity of water is 1.00 mPa·s.

Figure 11-34   Problem 61.

5.0 cm

10.0 cm

20 cm

62. In this problem you are to prove that, for a spherical balloon of radius R with a surface tension γ, the pressure inside the balloon is given by $P = P_{at} + 2\gamma R$. First draw a line around the circumference of the balloon and consider the forces exerted on the right hemisphere of the balloon (Figure 11-35). The force to the right is due to the excess pressure against the hemispherical surface of the balloon, whereas the force to the left is due to the surface tension acting on the line.

63. A large beer keg of height H and cross-sectional area a_1 is filled with beer. The top is open to atmospheric pressure. At the bottom is a spigot opening of area A_2, which is much smaller than A_1. (a) Show that when the height of the beer is h, the speed of the beer leaving the spigot is approximately $\sqrt{2gh}$. (b) Show that for the approximation $A_2 \ll A_1$, the rate of change of the height h of the beer is given by

$$\frac{dh}{dt} = -\frac{A_2}{A_1}(2gh)^{1/2}$$

(c) Find h as a function of time if $h = H$ at $t = 0$. (d) Find the total time needed to drain the keg if $H = 2$ m, $A_1 = 0.8$ m^2, and $A_2 = (10^{-4})A_1$.

64. In Figure 11-29, (a) find the distance x at which the water strikes the ground as a function of h and H. (b) Show that there are two values of h that are equidistant from the point $h = \frac{1}{2}H$ that give the same distance x. (c) Show that x is a maximum when $h = \frac{1}{2}H$. What is the value of this maximum distance x?

Figure 11-35   Problem 62.

Part 2 Oscillations and Waves

Vibration patterns in a violin. The vibrations of the violin strings are transmitted via the bridge to the body of the violin. The vibrations of the body, made visible here by interference effects, resonate with the vibrations of the strings, thus giving the violin its tone.

Chapter 12

Oscillations

A vibrating tuning fork touching a water surface.

Oscillation occurs when a system is disturbed from a position of stable equilibrium. The most recognizable characteristic of oscillatory motion is that the motion is **periodic,** that is, it repeats itself. There are many familiar examples of oscillations: small boats bob up and down, clock pendulums swing back and forth, and the strings and reeds of musical instruments vibrate. Other, less familiar examples are the oscillations of air molecules in a sound wave and the oscillations of electric currents in radios or television sets.

Wave motion is closely related to oscillatory motion. Sound waves, for example, are produced by a vibrating string (such as a violin string), a vibrating oboe reed, a vibrating drum head, or the vibrations of your vocal cords when you speak. In each case, the vibrating system sets up oscillations in the nearby air molecules, and these oscillations are propagated through the air (or some other medium, such as water or a solid).

In this chapter, we will study some of the fundamentals of oscillating systems. In Chapters 13 and 14, we will study mechanical waves—waves on strings and sound waves.

12-1   Simple Harmonic Motion: A Mass on a Spring

A common and very important kind of oscillatory motion is simple harmonic motion. When an object is displaced from its equilibrium position, simple harmonic motion occurs if there is a restoring force that is proportional to the displacement. This condition is nearly always met, at least approximately, when the displacement from equilibrium is small.

A typical system that exhibits simple harmonic motion is an object attached to a spring, as illustrated in Figure 12-1. In equilibrium, the spring exerts no force on the object. When the object is displaced an amount x from its equilibrium position, the spring exerts a force $-kx$, as given by Hooke's law (Section 4-5):

$$F_x = -kx \qquad\qquad 12\text{-}1$$

The minus sign in Hooke's law arises because the force is opposite the direction of the displacement. If we choose x to be positive for displacements to the right, the force is negative (to the left) when x is positive and positive (to the right) when x is negative. Combining Equation 12-1 with Newton's second law, we have

$$F_x = -kx = ma = m\,\frac{d^2x}{dt^2}$$

or

$$a = \frac{d^2x}{dt^2} = -\left(\frac{k}{m}\right)x \qquad\qquad 12\text{-}2$$

The acceleration is proportional to the displacement and is oppositely directed. This is a general characteristic of simple harmonic motion and can, in fact, be used to identify systems that will exhibit simple harmonic motion:

> Whenever the acceleration of an object is proportional to its displacement and is oppositely directed, the object will move with simple harmonic motion.

If we displace an object from equilibrium and release it, it oscillates back and forth about its equilibrium position. The time for the object to make a complete oscillation is called the **period** T. The reciprocal of the period is the **frequency** f, which is the number of oscillations per second:

$$f = \frac{1}{T} \qquad\qquad 12\text{-}3$$

The unit of frequency is the reciprocal second (s^{-1}), which is called a **hertz** (Hz). For example, if the time for one complete oscillation is 0.25 s, the frequency is 4 Hz.

For an oscillating object, the displacement x as a function of time t can be obtained experimentally. For example, let us attach a marking pen to an object that is attached to a vertical spring and place the pen so it can write on a strip of paper that can be moved perpendicular to the direction of oscillation as shown in Figure 12-2. We then displace the object a distance A and begin to pull the paper to the left with constant speed when we release the object. The pen traces out the sinusoidal curve shown in the figure. The equation for this curve is

$$x = A \cos\left(\omega t + \delta\right) \qquad\qquad 12\text{-}4$$

Simple harmonic motion defined

Figure 12-1   An object attached to a spring resting on a friction-less table. The displacement x is measured from the equilibrium position. The displacement can be positive or negative depending on whether the spring is stretched or compressed from its natural length.

Conditions for simple harmonic motion

Figure 12-2   An object oscillating on a vertical spring. As the paper is moved with constant speed to the left, the marking pen attached to the object on the spring traces out the displacement x as a function of time t. (For this case, we have chosen x to be positive when the spring is compressed.)

where A, ω, and δ are constants. Motion in which the position varies with time according to Equation 12-4 is by definition **simple harmonic motion.** Note that $\cos(\omega t + \delta) = \sin(\omega t + \delta + \pi/2)$. Whether the equation is expressed as a cosine function or a sine function depends merely on when we choose $t = 0$. The maximum displacement from equilibrium is called the **amplitude** A. The argument of the cosine function, $\omega t + \delta$, is called the **phase** of the motion, and the constant δ is called the **phase constant.** During one complete cycle of the motion, the phase increases by 2π. At the end of the cycle, the object again has the same position and velocity as it had at the start of the cycle since $\cos(\omega t + \delta + 2\pi) = \cos(\omega t + \delta)$. We can determine the period T from the fact that the phase at time $t + T$ is just 2π plus the phase at time t:

$$\omega(t + T) + \delta = 2\pi + \omega t + \delta$$

or

$$\omega T = 2\pi$$

so

$$T = \frac{2\pi}{\omega} \qquad\qquad 12\text{-}5$$

From Equation 12-3, we obtain for the frequency

$$f = \frac{1}{T} = \frac{\omega}{2\pi} \qquad\qquad 12\text{-}6$$

The constant $\omega = 2\pi f$ is called the **angular frequency.** It has units of radians per second and dimensions of inverse time, the same as angular velocity, which is also designated by ω. In terms of the frequency or period, Equation 12-4 may be rewritten as

$$x = A \cos(2\pi ft + \delta) = A \cos\left(\frac{2\pi t}{T} + \delta\right) \qquad 12\text{-}7$$

The phase constant δ depends on when we choose $t = 0$. If we choose $t = 0$ when $x = A$, as in Figure 12-3a, the phase constant is zero and $x = A \cos \omega t$. On the other hand, if we choose $t = 0$ when $x = 0$, δ is either $\pi/2$ or $3\pi/2$, depending on whether x is increasing or decreasing at $t = 0$. For example, if x is increasing at $t = 0$ as in Figure 12-3b, $\delta = 3\pi/2$ and

$$x = A \cos\left(\omega t + \frac{3\pi}{2}\right) = A \sin \omega t$$

The general relation between the initial position x_0 and the constants A and δ is found by setting $t = 0$ in Equation 12-4. Then

$$x_0 = A \cos \delta \qquad\qquad 12\text{-}8$$

Figure 12-3 (a) Graph of x versus t for $\delta = 0$. At $t = 0$, the displacement has its maximum value A, and the object is at rest, which is indicated by the fact that the slope is zero. The time for one complete oscillation is the period T. (b) Graph of x versus t for $\delta = 3\pi/2$. At $t = 0$, the particle is at $x = 0$ and is moving toward positive x, which is indicated by the fact that the slope is positive.

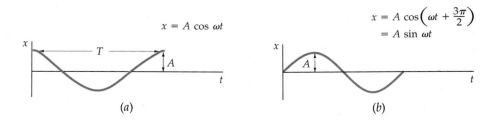

$x = A \cos \omega t$

A

T

(a)

$x = A \cos\left(\omega t + \frac{3\pi}{2}\right)$
$= A \sin \omega t$

A

(b)

We can show that x as given by Equation 12-4 is a solution to Equation 12-2 by differentiating x twice. The first derivative of x with respect to time gives the velocity v:

$$v = \frac{dx}{dt} = -A\omega \sin(\omega t + \delta) = A\omega \cos\left(\omega t + \delta + \frac{\pi}{2}\right) \qquad \text{12-9}$$

The phase of the velocity differs from that of the position by $\pi/2$ rad $= 90°$. When $\cos(\omega t + \delta)$ has the value $+1$ or -1, $\sin(\omega t + \delta) = 0$. Thus, when x is at its maximum or minimum value, the velocity is zero. Similarly, when $\sin(\omega t + \delta)$ has the value $+1$ or -1, $\cos(\omega t + \delta) = 0$. The speed is maximum when the object passes through its equilibrium position $x = 0$. We can relate the initial velocity v_0 to the constants A and δ by setting $t = 0$ in Equation 12-9. We obtain

$$v_0 = -A\omega \sin \delta \qquad \text{12-10}$$

Differentiating the velocity (Equation 12-9) with respect to time gives the acceleration of the object:

$$a = \frac{dv}{dt} = \frac{d^2x}{dt^2} = -\omega^2 A \cos(\omega t + \delta) \qquad \text{12-11}$$

or

$$a = -\omega^2 x \qquad \text{12-12}$$

Comparing this with $a = -(k/m)x$ for a mass on a spring (Equation 12-2), we see that $x = A \cos(\omega t + \delta)$ is a solution of $d^2x/dt^2 = -(k/m)x$ if the angular frequency ω is related to the force constant k and the mass m by

$$\omega^2 = \frac{k}{m} \qquad \text{12-13}$$

The frequency and period of a mass on a spring are thus related to the force constant k and the mass m by

$$f = \frac{\omega}{2\pi} = \frac{1}{2\pi}\sqrt{\frac{k}{m}} \qquad \text{12-14}$$

$$T = \frac{1}{f} = 2\pi\sqrt{\frac{m}{k}} \qquad \text{12-15}$$

Astronaut Alan L. Bean measures his body mass during the second Skylab mission. The total mass of the astronaut plus the apparatus is related to his frequency of vibration by Equation 12-14.

Period and frequency for an object on a spring

From this result, we can see that when k is large, as in the case of a stiff spring, the frequency is large. Similarly, if the mass is large, the frequency is small.

We have demonstrated by direct substitution that $x = A \cos(\omega t + \delta)$ is a solution of the differential equation $d^2x/dt^2 = -\omega^2 x$. Rather than solving the differential equation directly, we guessed the solution by using a physical picture of a pen attached to a mass on a spring so that it traces on paper. Although there are mathematical techniques for solving differential equations directly, they work only for some kinds of differential equations. There is no general method for solving an arbitrary differential equation. Solutions to differential equations are often based on guesswork aided by physical insight. A direct solution of the equation $d^2x/dt^2 = -\omega^2 x$ using the conservation of mechanical energy is outlined in Problem 89 of Chapter 6.

Figure 12-4 Plots of the displacement x, the velocity v, and the acceleration a as functions of time t for $\delta = 0$.

The variations with time of the displacement x, the velocity v, and the acceleration a for the special case of $\delta = 0$ are plotted in Figure 12-4. For $\delta = 0$, Equations 12-4, 12-9, and 12-11 become

$$x = A \cos \omega t \qquad\qquad 12\text{-}16a$$

$$v = -\omega A \sin \omega t \qquad\qquad 12\text{-}16b$$

and

$$a = -\omega^2 A \cos \omega t \qquad\qquad 12\text{-}16c$$

Initially, at time $t = 0$, the displacement is maximum, the velocity is zero, and the acceleration is negative and equal to $-\omega^2 A$. The velocity then becomes negative as the object moves back toward its equilibrium position. The object is at its equilibrium position $x = 0$ when $\cos \omega t = 0$. The acceleration at these times is also zero, and the speed has its maximum value of ωA because, when $\cos \omega t$ is zero, $\sin \omega t$ is either 1 or -1.

Example 12-1

A particle has displacement x given by

$$x = 0.3 \cos \left(2t + \frac{\pi}{6} \right)$$

where x is in meters and t is in seconds. (a) What are the frequency, period, amplitude, angular frequency, and phase constant of the motion? (b) Where is the particle at $t = 1$ s? (c) Find the velocity and acceleration at any time t. (d) Find the initial position and velocity of the particle.

(a) Comparing this equation with Equation 12-4, we see that the angular frequency $\omega = 2$ rad/s, the amplitude $A = 0.3$ m, and the phase constant $\delta = \pi/6$ rad. The frequency is thus $f = \omega/2\pi = 0.318$ Hz, and the period is $T = 1/f = 3.14$ s.

(b) At $t = 1$ s, the position of the particle is

$$x = 0.3 \cos \left[2(1) + \frac{\pi}{6} \right] = -0.245 \text{ m}$$

(c) The velocity is obtained from

$$v = \frac{dx}{dt} = -0.3 \sin \left(2t + \frac{\pi}{6} \right) \frac{d(2t)}{dt}$$

$$= -0.6 \sin \left(2t + \frac{\pi}{6} \right)$$

We differentiate again to obtain the acceleration

$$a = \frac{dv}{dt} = \frac{d}{dt} \left[-0.6 \sin \left(2t + \frac{\pi}{6} \right) \right]$$

$$= -0.6 \cos \left(2t + \frac{\pi}{6} \right) \frac{d(2t)}{dt} = -1.2 \cos \left(2t + \frac{\pi}{6} \right)$$

(d) We find the initial position and the initial velocity by substituting $t = 0$ into the expressions for x and v. We obtain

$$x_0 = 0.3 \cos \frac{\pi}{6} = 0.260 \text{ m}$$

and

$$v_0 = -0.6 \sin \frac{\pi}{6} = -0.300 \text{ m/s}$$

Exercise

A 0.8-kg object is attached to a spring of force constant $k = 400$ N/m. Find the frequency and period of the motion of the object when it is displaced from equilibrium.   (Answer: $f = 3.56$ Hz, $T = 0.281$ s)

Example 12-2

A 2-kg object stretches a spring 10 cm when it hangs vertically in equilibrium (Figure 12-5a). The object is then attached to the same spring while the object is resting on a frictionless table and the spring is fixed at one end as in Figure 12-1. The object is held a distance 5 cm from the equilibrium position and is released at $t = 0$. Find the amplitude A, the angular frequency ω, the frequency f, and the period T.

The force constant of the spring is determined from the first measurement. In equilibrium, the downward force mg must equal the upward force kx_0, where x_0 is the amount the spring is stretched from its natural length (Figure 12-5). Using $g = 9.81$ N/kg, $m = 2$ kg, and $x_0 = 10$ cm = 0.1 m, we have

$$k = \frac{mg}{x_0} = \frac{(2 \text{ kg})(9.81 \text{ N/kg})}{0.1 \text{ m}} = 196 \text{ N/m}$$

In the horizontal position, the spring is initially stretched 5 cm from equilibrium and released. Since the initial velocity v_0 is zero, the phase constant $\delta = 0$ (Equation 12-10), and the displacement x is given by Equation 12-16a: $x = A \cos \omega t$. The initial displacement is 5 cm so $A = 5$ cm.

The angular frequency is found from Equation 12-13:

$$\omega = \sqrt{\frac{k}{m}}$$

$$= \sqrt{\frac{196 \text{ N/m}}{2 \text{ kg}}} = 9.90 \text{ rad/s}$$

The frequency is then

$$f = \frac{\omega}{2\pi} = \frac{9.90 \text{ rad/s}}{2\pi} = 1.58 \text{ Hz}$$

The period is the reciprocal of the frequency:

$$T = \frac{1}{f} = \frac{1}{1.58 \text{ Hz}} = 0.63 \text{ s}$$

(a)   (b)

Figure 12-5 (a) An object hanging in equilibrium from a spring for Example 12-2. (b) Free-body diagram of the object. In equilibrium, the weight of the object mg is balanced by the upward force of magnitude kx_0 exerted by the spring.

Example 12-3

What is the maximum speed of the object on the spring in Example 12-2, and when does this maximum speed first occur?

The velocity of the object at any time is given by Equation 12-16b. The maximum value of the velocity occurs when $\sin \omega t = 1$ (or -1). The magnitude of the velocity at this time is

$$v_{max} = \omega A = (9.90 \text{ rad/s})(5 \text{ cm}) = 49.5 \text{ cm/s}$$

This maximum speed first occurs after one-fourth of a period, when the object first reaches equilibrium. This is $(0.63 \text{ s})/4 = 0.16$ s after the object is released. At this time, $\sin \omega t = +1$, and the velocity is negative, which means that the object is moving to the left.

Example 12-4

A second spring, identical to that in Example 12-2, is attached to a second object, also of mass 2 kg. This spring is stretched a distance 10 cm from equilibrium and is released at the same time as the first, which is again stretched to 5 cm. Which object reaches the equilibrium position first?

Figure 12-6 shows the initial positions of the objects, and Figure 12-7 shows a sketch of the position functions for the two objects. Both the springs have the same force constant and the masses of the two objects are equal. Only the amplitudes differ. But, according to Equation 12-15, the frequency and period depend only on k and m and not on the amplitude. Thus, the frequencies and periods of the motions are equal and both objects reach their equilibrium positions at the same time. The second object has twice as far to go to reach equilibrium as the first object, but it also begins with twice the initial acceleration.

Figure 12-6   Two objects on identical springs released simultaneously. They reach their equilibrium positions at the same time because the period does not depend on the amplitude. It depends only on the mass of the object and force constant of the spring, which are the same for both systems.

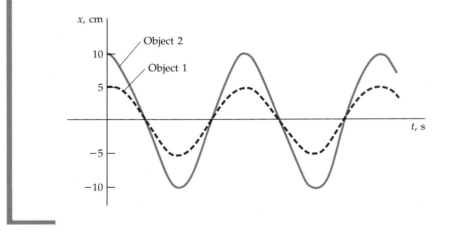

Figure 12-7   Plots of position versus time for the objects in Figure 12-6.

Example 12-4 illustrates a very important general property of simple harmonic motion:

In simple harmonic motion, the frequency and period are independent of the amplitude.

The fact that the frequency in simple harmonic motion is independent of the amplitude has important consequences in many fields. In music, for example, it means that the pitch (which corresponds to the frequency) of a note struck on the piano does not depend on how loudly the note is played (which corresponds to the amplitude). The pitch does not change as the amplitude of the note decreases. For many musical instruments, there is a slight dependence of the frequency on amplitude. The pitch of an oboe reed, for example, depends slightly on how hard it is blown because the vibration is not exactly simple harmonic. However, this effect can be corrected for by a skilled musician. If changes in amplitude had a large effect on the frequency, musical instruments would be unplayable.

Example 12-5

If the object in Example 12-2 is initially at $x_0 = 3$ cm and has an initial velocity $v_0 = -25$ cm/s, find the amplitude and phase constant for the motion.

Equations 12-8 and 12-10 relate the initial position and velocity to the amplitude and phase constant. We have

$$x_0 = A \cos \delta$$

and

$$v_0 = -\omega A \sin \delta$$

Then

$$\frac{v_0}{x_0} = \frac{-\omega A \sin \delta}{A \cos \delta} = -\omega \tan \delta$$

Using $\omega = 9.90$ rad/s from Example 12-2 and $x_0 = 3$ cm and $v_0 = -25$ cm/s, we obtain

$$\tan \delta = -\frac{v_0}{\omega x_0} = -\frac{-25 \text{ cm/s}}{(9.90 \text{ rad/s})(3 \text{ cm})} = 0.842$$

$$\delta = 0.70 \text{ rad}$$

The amplitude can be found from

$$A = \frac{x_0}{\cos \delta} = \frac{3 \text{ cm}}{\cos 0.70} = 3.9 \text{ cm}$$

Questions

1. How far does a particle oscillating with an amplitude A move in one full period? What is its displacement after one full period?

2. If you know that the speed of an oscillator of amplitude A is zero at certain times, can you say exactly what its displacement is at those times?

3. What is the magnitude of the acceleration of an oscillator of amplitude A and frequency f when its speed is maximum? When its displacement is maximum?

4. Can the acceleration and the displacement of a simple harmonic oscillator ever be in the same direction? The acceleration and the velocity? The velocity and the displacement? Explain.

5. The effect of the mass of a spring on the motion of an object attached to it is usually neglected. Describe qualitatively its effect when it is not neglected.

12-2   Simple Harmonic Motion and Circular Motion

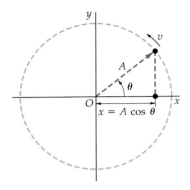

Figure 12-8   A particle moving with constant speed v in a circle of radius A. The angle θ increases with time, $\theta = \omega t + \delta$, where ω is the angular velocity of the circular motion. The x component of the motion is simple harmonic.

There is a simple but important mathematical relation between simple harmonic motion and circular motion with constant speed. Consider a particle moving with constant speed v in a circle of radius A as shown in Figure 12-8. Its angular velocity ω is constant and is related to its speed by $\omega = v/A$. The angular displacement of the particle relative to the x axis is given by

$$\theta = \omega t + \delta$$

where δ is the angular displacement at time $t = 0$. From the figure we see that the x component of the position of the particle is given by

$$x = A \cos \theta = A \cos (\omega t + \delta)$$

which is the same as Equation 12-4. Thus,

> The projection on a straight line of a particle moving with uniform circular motion is simple harmonic motion.

The frequency and period of the circular motion are the same as the corresponding frequency and period of the projected simple harmonic motion. The relation between circular motion and simple harmonic motion can be demonstrated with a turntable and an object hanging on a spring as shown in Figure 12-9. The shadow of the peg and that of the object are projected on a screen. If the period of the rotating turntable is adjusted to be equal to the period of the oscillating object, and the amplitude of the spring system equals the radius of the turntable, the shadows move together.

Figure 12-9   The shadow of a peg on a turntable and the shadow of an object on a spring are projected onto a screen. When the period of rotation of the turntable equals the period of oscillation of the object on the spring, the shadows move together.

Sinusoidal pattern produced by a rotating turbine blade as it moves through water.

The projection of the circular motion on the y axis is $y = A \sin \theta = A \sin (\omega t + \delta) = A \cos (\omega t + \delta - \pi/2)$. We can therefore consider simple harmonic motion to be one component of circular motion. Alternatively, the circular motion of a particle can be considered to be the combination of perpendicular simple harmonic motions having the same amplitude and frequency but a relative phase difference of $\pi/2$.

12-3   Energy in Simple Harmonic Motion

When an object oscillates on a spring, the kinetic energy of the object and potential energy of the object–spring system both vary with time, while their sum, the total energy, is constant (assuming no friction). The potential energy for a spring of force constant k stretched a distance x from equilibrium is given by Equation 6-19:

$$U = \tfrac{1}{2}kx^2 \qquad\qquad 12\text{-}17$$

The kinetic energy of an object of mass m moving with speed v is

$$K = \tfrac{1}{2}mv^2 \qquad\qquad 12\text{-}18$$

The total energy is the sum of the potential and kinetic energies:

$$E_{total} = U + K = \tfrac{1}{2}kx^2 + \tfrac{1}{2}mv^2 \qquad\qquad 12\text{-}19$$

When the displacement is maximum, $x = A$, the velocity is zero, and the total energy is

$$E_{total} = \tfrac{1}{2}kA^2 \qquad\qquad 12\text{-}20$$

Total energy in simple harmonic motion

This equation gives an important general property of simple harmonic motion:

> The total energy in simple harmonic motion is proportional to the square of the amplitude.

If we begin with the object at its maximum displacement, the total energy is originally just potential energy. As the object moves toward equilibrium, its kinetic energy increases and the potential energy of the system decreases. At the equilibrium point, the speed of the object is maximum, the potential energy of the system is zero, and the total energy is kinetic. As the object moves past the equilibrium point, its kinetic energy begins to decrease and the potential energy of the system increases until the object is again at its maximum displacement (in the other direction), at which time it stops momentarily, so its kinetic energy is again zero and the potential energy of the system is again maximum. At all times, the sum of the potential and kinetic energies is constant.

In Figure 12-10, the potential energy U is graphed as a function of x. This curve is a parabola with its minimum at the equilibrium position. (This is the same plot as Figure 6-14.) The total energy E_{total} is constant and is therefore

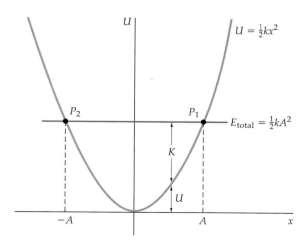

Figure 12-10   The potential-energy function $U = \tfrac{1}{2}kx^2$ for an object of mass m on a spring of force constant k. The horizontal line represents the total energy for an amplitude of A, $E_{total} = \tfrac{1}{2}kA^2$. The kinetic energy K is represented by the vertical distance $K = E_{total} - U$.

plotted as a straight line. This line intersects the potential-energy curve at $x = A$ and $x = -A$. The motion is restricted to $-A \le x \le +A$ since the total energy must be greater than or equal to the potential energy. The kinetic energy at any value of x is $K = E_{total} - U$.

We can demonstrate that the total energy for an object oscillating on a spring is constant by substituting the expressions for x and v in Equations 12-4 and 12-9 into Equations 12-17 and 12-18 for the potential and kinetic energies. We have

$$U = \tfrac{1}{2}k[A \cos(\omega t + \delta)]^2$$

or

$$U = \tfrac{1}{2}kA^2 \cos^2(\omega t + \delta) \qquad \text{12-21}$$

and

$$K = \tfrac{1}{2}m[-A\omega \sin(\omega t + \delta)]^2$$

Using $\omega^2 = k/m$ from Equation 12-13, the kinetic energy can be written

$$K = \tfrac{1}{2}kA^2 \sin^2(\omega t + \delta) \qquad \text{12-22}$$

Adding the potential and kinetic energies to get the total energy, we obtain

$$E_{total} = \tfrac{1}{2}kA^2 \cos^2(\omega t + \delta) + \tfrac{1}{2}kA^2 \sin^2(\omega t + \delta)$$

$$= \tfrac{1}{2}kA^2[\cos^2(\omega t + \delta) + \sin^2(\omega t + \delta)]$$

$$= \tfrac{1}{2}kA^2$$

where we have used the trigonometric identity $\cos^2\theta + \sin^2\theta = 1$.

In terms of the total energy, the potential and kinetic energies in simple harmonic motion are

$$U = E_{total} \cos^2\theta$$

and

$$K = E_{total} \sin^2\theta$$

where $\theta = \omega t + \delta$. Figure 12-11$a$ and b show plots of U and K as functions of t for $\delta = 0$. These curves have the same form, except that U is maximum when K is zero and U is zero when K is maximum. Their average values over one or more cycles are equal, and since $U + K = E_{total}$, these average values are given by

$$U_{av} = K_{av} = \tfrac{1}{2}E_{total} \qquad \text{12-23}$$

Figure 12-11   Plots of (a) the potential energy U and (b) the kinetic energy K as functions of t for $\delta = 0$. Both functions have the same shape and the same maximum value E_{total}. For each, the average value, indicated by the dashed horizontal line, is half the maximum value.

(a)

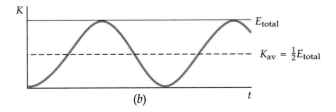

(b)

Example 12-6

A 3-kg object attached to a spring oscillates with an amplitude of 4 cm and a period of 2 s. (a) What is the total energy? (b) What is the maximum speed of the object?

(a) The total energy is $\frac{1}{2}kA^2$. The force constant of the spring, k, is related to the period by

$$T = 2\pi\sqrt{\frac{m}{k}}$$

so that

$$k = \frac{(2\pi)^2 m}{T^2} = \frac{(4\pi^2)(3\text{ kg})}{4\text{ s}^2} = 29.6\text{ N/m}$$

The total energy is therefore

$$E_{\text{total}} = \tfrac{1}{2}kA^2$$

$$= \tfrac{1}{2}(29.6\text{ N/m})(0.04\text{ m})^2$$

$$= 2.37 \times 10^{-2}\text{ J}$$

(b) We use the total energy to find the maximum speed. When the speed is maximum, the potential energy is zero and the total energy is the kinetic energy:

$$E_{\text{total}} = \tfrac{1}{2}mv_{\text{max}}^2 = 2.37 \times 10^{-2}\text{ J}$$

The maximum speed is therefore

$$v_{\text{max}} = \sqrt{\frac{2E_{\text{total}}}{m}}$$

$$= \sqrt{\frac{2(2.37 \times 10^{-2}\text{ J})}{3\text{ kg}}} = 0.126\text{ m/s}$$

We could have found the maximum speed from Equation 12-16b as we did in Example 12-3:

$$v_{\text{max}} = \omega A(\sin \omega t)_{\text{max}} = \omega A$$

Using $\omega = 2\pi/T$ (Equation 12-6), we have

$$v_{\text{max}} = \frac{2\pi A}{T} = \frac{2(3.14)(4\text{ cm})}{2\text{ s}}$$

$$= 12.6\text{ cm/s} = 0.126\text{ m/s}$$

Exercise

An object of mass 2 kg is attached to a spring of force constant 40 N/m. It is moving at 25 cm/s when it is at its equilibrium position. (a) What is the total energy of the object? (b) What is the amplitude of the motion? [Answers: (a) $E = 0.0625$ J, (b) $A = 5.59$ cm]

Question

6. If the amplitude of a simple harmonic oscillator is tripled, by what factor is the energy changed?

12-4   An Object on a Vertical Spring

When an object hangs from a vertical spring, as in Figure 12-12, there is a force mg downward in addition to the force of the spring $F_s = -ky$, where we are assuming that y is measured downward from the unstretched position of the spring. Newton's second law gives

$$m\frac{d^2y}{dt^2} = -ky + mg \qquad\qquad 12\text{-}24$$

Figure 12-12   (a) An unstretched vertical spring. (b) The spring stretches an amount $y_0 = mg/k$ when an object of mass m hangs from it in equilibrium. (c) The object oscillates about its equilibrium position $y = y_0$ with a displacement $y' = y - y_0$.

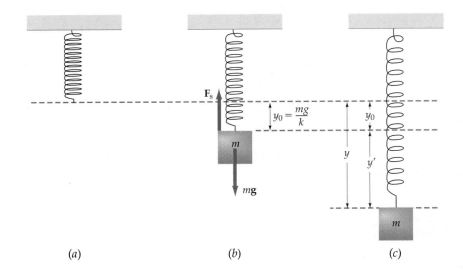

(a)   (b)   (c)

This differs from Equation 12-2 because of the constant term mg. We handle this extra term by changing to a new variable $y' = y - y_0$, where y_0 is the amount the spring is stretched when the object is in equilibrium. When the object is in equilibrium at $y = y_0$, Equation 12-24 becomes

$$0 = -ky_0 + mg$$

or

$$y_0 = \frac{mg}{k} \qquad\qquad 12\text{-}25$$

Since y' differs from y by only a constant, we have

$$\frac{dy'}{dt} = \frac{dy}{dt}$$

and

$$\frac{d^2y'}{dt^2} = \frac{d^2y}{dt^2}$$

Substituting d^2y'/dt^2 for d^2y/dt^2 and $y' + y_0$ for y in Equation 12-24, we obtain

$$m\frac{d^2y'}{dt^2} = -k(y' + y_0) + mg$$

$$= -ky' - ky_0 + mg$$

But $ky_0 = mg$, so

$$m \frac{d^2y'}{dt^2} = -ky' \qquad \text{12-26}$$

which has the familiar solution

$$y' = A \cos(\omega t + \delta)$$

Thus, the effect of the gravitational force mg is merely to shift the equilibrium position from $y = 0$ to $y' = 0$. When the object is displaced from this equilibrium position by the amount y', the unbalanced force is $-ky'$. The object oscillates about this equilibrium position with an angular frequency $\omega = \sqrt{k/m}$, the same as that for an object on a horizontal spring.

The potential energy of the spring relative to $U_s = 0$ at $y = 0$ is $\frac{1}{2}ky^2 = \frac{1}{2}k(y' + y_0)^2$. At the equilibrium point, it is $\frac{1}{2}ky_0^2$. If, instead, we choose the potential energy of the spring to be zero at the equilibrium position ($y = y_0$, $y' = 0$), the potential energy of the spring is

$$U_s = \frac{1}{2}k(y' + y_0)^2 - \frac{1}{2}ky_0^2$$

$$= \frac{1}{2}ky'^2 + ky_0y' + \frac{1}{2}ky_0^2 - \frac{1}{2}ky_0^2 = \frac{1}{2}ky'^2 + ky_0y'$$

But $ky_0 = mg$. Thus the potential energy *of the spring* relative to $U_s = 0$ at $y' = 0$ is

$$U_s = \frac{1}{2}ky'^2 + mgy' \qquad (U_s = 0 \text{ at } y' = 0)$$

When the spring is stretched by an amount y', the object is lowered by $\Delta h = -y'$, and the gravitational potential energy is changed by $mg(\Delta h) = -mgy'$. If the gravitational potential energy is also chosen to be zero at equilibrium ($y = y_0$, $y' = 0$), the gravitational potential energy at y' is

$$U_g = -mgy' \qquad (U_g = 0 \text{ at } y' = 0)$$

The total potential energy of the system relative to $U = 0$ at $y' = 0$, including both the potential energy of the spring and the gravitational potential energy, is therefore

$$U = U_s + U_g$$
$$= (\frac{1}{2}ky'^2 + mgy') - mgy'$$

or

$$U = \frac{1}{2}ky'^2 \qquad (U = 0 \text{ at } y' = 0) \qquad \text{12-27}$$

Total potential energy of object on a vertical spring

Thus, if we measure the displacement from the equilibrium position, we can forget about the effect of gravity.

Exercise

A 4-kg object is suspended from a spring of force constant $k = 200$ N/m.

(a) Find the amount y_0 the spring is stretched when in equilibrium.

(b) Find the total potential energy, including gravitational potential energy, when the spring is stretched an additional 12 cm from equilibrium. (Assume $U = 0$ at $y = y_0$.) (c) Find the period of oscillation. [Answers: (a) 19.6 cm (b) 1.44 J, (c) 0.889 s]

12-5   Pendulums

The Simple Pendulum

A familiar example of oscillatory motion is that of a pendulum. The motion of a pendulum is simple harmonic motion only if the amplitude of the motion is small. Figure 12-13 shows a simple pendulum consisting of a string of length L and a bob of mass m. The forces acting on the bob are its weight $m\mathbf{g}$ and the tension $\mathbf{T}$ in the string. When the string makes an angle ϕ with the vertical, the weight has the components $mg \cos \phi$ along the string and $mg \sin \phi$ perpendicular to the string in the direction of decreasing ϕ. Let s be the arc length measured from the bottom of the circle. The arc length is related to the angle ϕ by

$$s = L\phi \tag{12-28}$$

Figure 12-13   A simple pendulum. The forces on the bob are its weight $m\mathbf{g}$ and the tension $\mathbf{T}$. The tangential component of the net force is $-mg \sin \phi = -mg \sin (s/L)$. For small displacements, the motion of the pendulum is approximately simple harmonic motion.

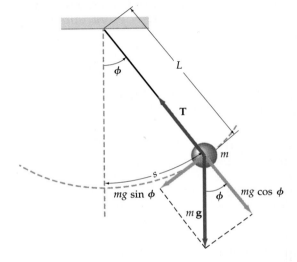

The tangential component of the acceleration of the bob is d^2s/dt^2. The tangential component of Newton's second law is

$$\Sigma F_t = -mg \sin \phi = m \frac{d^2s}{dt^2}$$

or

$$\frac{d^2s}{dt^2} = -g \sin \phi = -g \sin \frac{s}{L} \tag{12-29}$$

If s is much less than L, the angle $\phi = s/L$ is small, and we can approximate $\sin \phi$ by the angle ϕ. Using $\sin (s/L) \approx s/L$ in Equation 12-29, we have

$$\frac{d^2s}{dt^2} = -\frac{g}{L} s \tag{12-30a}$$

We can see that for angles small enough for $\sin \phi \approx \phi$ to be valid, the acceleration is proportional to the displacement. The motion of the pendulum is thus approximately simple harmonic motion for small displacements. Equation 12-30a can be written

$$\frac{d^2s}{dt^2} = -\omega^2 s \tag{12-30b}$$

where

$$\omega^2 = \frac{g}{L} \tag{12-31}$$

The solution of Equation 12-30b is $s = s_0 \cos (\omega t + \delta)$, where s_0 is the maximum displacement measured along the arc of the circle. The period of the motion is

$$T = \frac{2\pi}{\omega} = 2\pi\sqrt{\frac{L}{g}}$$

12-32   *Period of a simple pendulum*

According to Equation 12-32 the greater the length of a pendulum, the greater the period, which is consistent with experimental observation. Note that the period does not depend on the mass. This is because the restoring force is proportional to the mass. The acceleration $a = F/m$ is therefore independent of the mass. Note also that the frequency and period are independent of the amplitude of oscillation, a general feature of simple harmonic motion.

It is often more convenient to describe the motion of a simple pendulum in terms of its angular displacement ϕ. Using $s = L\phi$ in Equation 12-29, we have

$$\frac{d^2(L\phi)}{dt^2} = -g \sin \phi$$

or

$$\frac{d^2\phi}{dt^2} = -\frac{g}{L} \sin \phi$$

12-33

which for small ϕ becomes

$$\frac{d^2\phi}{dt^2} = -\frac{g}{L}\phi = -\omega^2\phi$$

12-34

The solution of Equation 12-34 is

$$\phi = \phi_0 \cos (\omega t + \delta)$$

12-35

where $\phi_0 = s_0/L$ is the maximum angular displacement. The criterion for simple harmonic motion stated in terms of these angular quantities is that the angular acceleration must be proportional to the angular displacement and oppositely directed, as in Equation 12-34.

Example 12-7

Find the period of a pendulum of length 1 m.

From Equation 12-32, we have

$$T = 2\pi\sqrt{\frac{1 \text{ m}}{9.81 \text{ m/s}^2}} = 2.01 \text{ s}$$

The validity of this result can be easily demonstrated experimentally.

The acceleration due to gravity can be easily measured using a simple pendulum. You need only measure the length L with a meterstick and the period T by determining the time for one oscillation. (One usually measures the time for n oscillations and then divides by n to reduce the error in the time measurement.) The acceleration due to gravity is then determined by solving Equation 12-32 for g:

$$g = \frac{4\pi^2 L}{T^2}$$

12-36

When the amplitude of vibration is not small, the motion of a pendulum is periodic, but it is not simple harmonic. In particular, the period does have

Motion of a simple pendulum. The bob is shown at equal time intervals. It moves faster at the bottom, as shown by the greater spacing of the images.

a slight dependence on the amplitude. This dependence is usually expressed in terms of the angular amplitude ϕ_0. For amplitudes that are not necessarily small, the period can be shown to be given by

$$T = T_0\left[1 + \frac{1}{2^2}\sin^2\frac{1}{2}\phi_0 + \frac{1}{2^2}\left(\frac{3}{4}\right)^2\sin^4\frac{1}{2}\phi_0 + \cdots\right] \qquad 12\text{-}37$$

where $T_0 = 2\pi\sqrt{L/g}$ is the period for very small amplitudes.

Example 12-8

A simple pendulum clock is calibrated to keep accurate time at an angular amplitude of $\phi_0 = 10°$. When the amplitude has decreased to the point that it is very small, how much time does the clock gain in one day?

From Equation 12-37, the original period is approximately

$$T \approx T_0(1 + \tfrac{1}{4}\sin^2 5°)$$

since for $\phi_0 = 10°$, $\sin^4\frac{1}{2}\phi_0$ in Equation 12-37 is much smaller than $\sin^2\frac{1}{2}\phi_0$. When the amplitude is very small, the period is T_0. Since this is less than T, the frequency is greater and the clock gains time. The difference between these two periods is

$$T - T_0 = \frac{T_0}{4}\sin^2 5°$$

$$= \frac{T_0}{4}(0.0872)^2 \approx (2 \times 10^{-3})T_0$$

The percentage change in the period is therefore

$$\frac{T - T_0}{T_0} \times 100\% = (2 \times 10^{-3}) \times 100\% = 0.2\%$$

This is a very small percentage change. However, it leads to an inaccuracy that is intolerable by today's time-keeping standards. The number of minutes in one day is

$$\frac{24\text{ h}}{1\text{ d}} \times \frac{60\text{ min}}{1\text{ h}} = 1440\text{ min/d}$$

A gain of 0.2 percent, then, leads to an accumulated error of $(2 \times 10^{-3})(1440\text{ min}) = 2.88\text{ min}$ in 1 day. The clock thus gains nearly three minutes each day. For this reason, pendulum clocks are designed to maintain a constant amplitude.

The Physical Pendulum

A rigid body that is suspended from some point other than its center of mass will oscillate when displaced from its equilibrium position. Such a system is called a **physical pendulum.** Consider a plane figure suspended from a point a distance D from its center of mass and displaced from equilibrium by the angle ϕ as shown in Figure 12-14. The torque about the point of suspension has a magnitude $MgD \sin \phi$ and tends to decrease ϕ. The angular acceleration α of the body is related to the torque by

$$\tau = I\alpha = I\frac{d^2\phi}{dt^2}$$

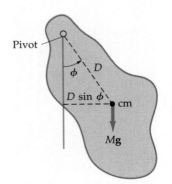

Figure 12-14   A physical pendulum. The torque about the pivot due to the force of gravity has magnitude $MgD \sin \phi$ and tends to decrease ϕ.

where I is the moment of inertia about the point of suspension. Substituting $-MgD \sin \phi$ for the net torque, we obtain

$$-MgD \sin \phi = I \frac{d^2\phi}{dt^2}$$

or

$$\frac{d^2\phi}{dt^2} = -\frac{MgD}{I} \sin \phi \qquad 12\text{-}38$$

For a simple pendulum, $I = ML^2$ and $D = L$, so Equation 12-38 is the same as Equation 12-34. Again, the motion is approximately simple harmonic motion if the angular displacements are small so that the approximation $\sin \phi \approx \phi$ holds. In this case, we have

$$\frac{d^2\phi}{dt^2} = -\frac{MgD}{I}\phi = -\omega^2\phi \qquad 12\text{-}39$$

where $\omega^2 = MgD/I$. The period is

$$T = \frac{2\pi}{\omega} = 2\pi\sqrt{\frac{I}{MgD}} \qquad 12\text{-}40 \qquad \textit{Period of a physical pendulum}$$

For large amplitudes, the period is given by Equation 12-37, where T_0 is now given by Equation 12-40.

Equation 12-40 can be used to measure the moment of inertia of a plane figure. The center of mass can be located by suspending the body from two different points, as discussed previously. Then, to find the moment of inertia about some point, we suspend the body from that point and measure the period of oscillation. The moment of inertia is obtained from

$$I = \frac{MgDT^2}{4\pi^2} \qquad 12\text{-}41$$

Example 12-9

What is the period for small angular displacements of a uniform stick of length L that is pivoted at one end?

The moment of inertia of a uniform stick of length L about an axis at one end was shown in Chapter 8 to be $I = \frac{1}{3}ML^2$. The distance from the pivot to the center of mass is $D = \frac{1}{2}L$. Substituting these values into Equation 12-40, we obtain

$$T = 2\pi\sqrt{\frac{\frac{1}{3}ML^2}{Mg(\frac{1}{2}L)}} = 2\pi\sqrt{\frac{2L}{3g}}$$

Exercise

What is the period for small angular displacements of a meterstick that is pivoted at one end? (Answer: 1.64 s)

The Torsional Pendulum

Figure 12-15 shows a **torsional pendulum**, which consists of a body suspended by a wire attached to a fixed point. When the wire is twisted through some angle ϕ, it exerts a restoring torque that is proportional to ϕ:

$$\tau = -\kappa\phi \qquad 12\text{-}42$$

Figure 12-15 A torsional pendulum. When the wire is twisted through an angle ϕ, there is a restoring torque of magnitude $-\kappa\phi$ that tends to twist the wire back to its equilibrium position.

The constant of proportionality κ is called the **torsion constant.** It can be found by applying a known torque to twist the wire and then measuring the resulting angular displacement ϕ. If I is the moment of inertia of the body about the axis along the wire, Newton's second law for rotational motion gives

$$\tau = -\kappa\phi = I\,\frac{d^2\phi}{dt^2}$$

or

$$\frac{d^2\phi}{dt^2} = -\frac{\kappa}{I}\,\phi = -\omega^2\phi \qquad\qquad 12\text{-}43$$

Equation 12-43 describes simple harmonic motion with an angular frequency $\omega = \sqrt{\kappa/I}$. The period of the motion is

Period of a torsional pendulum

$$T = 2\pi\sqrt{\frac{I}{\kappa}} \qquad\qquad 12\text{-}44$$

Note that we have not made a small-angle approximation. The motion of a torsional pendulum is simple harmonic motion as long as the restoring torque is proportional to the twist angle. Such is the case as long as the elastic limit of the wire for a shear stress is not exceeded. The balance wheel in a watch is a torsional pendulum as is the Cavendish torsional balance.

Question

7. The length of the string or wire supporting a pendulum increases slightly when its temperature is raised. How would this affect a clock operated by a simple pendulum? By a torsional pendulum?

Figure 12-16 (*a*) Graph of an arbitrary force F_x versus x. The force is zero at the equilibrium points x_1 and x_2. At x_1 the equilibrium is stable because, for small displacements away from x_1, the force is directed toward x_1. At x_2 the equilibrium is unstable because, for small displacements from x_2, the force is directed away from x_2. (*b*) Near x_1 the force can be approximated by a straight line $F_x = -k(x - x_1) = -k\epsilon$. Thus, for small displacements the motion is approximately simple harmonic motion.

12-6   General Motion near Equilibrium

Whenever a particle is displaced from a position of stable equilibrium, the motion of the particle is simple harmonic if the displacements are small enough. Examples are the various pendulums just discussed.

Figure 12-16*a* is a graph of an arbitrary net force as a function of position. At positions x_1 and x_2, the force is zero, so these are positions of equilibrium. However, the equilibrium at x_2 is unstable. If the particle is displaced slightly in the positive x direction, the force is positive, whereas if it is displaced slightly in the negative x direction, the force is negative. In either case, the force accelerates the particle away from its equilibrium position. The equilibrium at x_1 is stable. If the particle is displaced slightly from equilibrium in either direction, the force accelerates it back toward equilibrium.

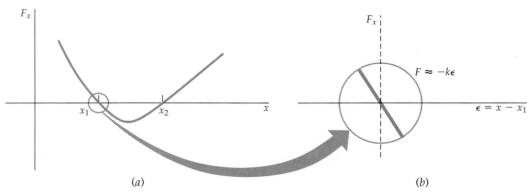

(*a*)   (*b*)

The particle therefore oscillates about the equilibrium point. Although the plot of force versus displacement is generally a curve, it will approximate a straight line if we consider it over a small enough interval, as shown in Figure 12-16b. Let $\epsilon = x - x_1$ be the displacement from equilibrium. The equation of the force for small ϵ is

$$F_x = -k\epsilon \qquad\qquad 12\text{-}45$$

where k is the magnitude of the slope of F_x versus x near x_1. Since the force is proportional to the displacement and is oppositely directed, the motion will be simple harmonic.

We can also examine the motion from the point of view of the potential-energy function U associated with the force. Figure 12-17 is a graph of U versus x for the same force as in Figure 12-16. As discussed in Chapter 6, the maximum at x_2 corresponds to unstable equilibrium, whereas the minimum at x_1 corresponds to stable equilibrium. The dashed curve in this figure is a parabolic curve that approximately fits the potential-energy curve near the stable equilibrium point. As long as the displacements from equilibrium are not too great, U is approximately parabolic. The general equation for a parabola can be written

$$U = A + B(x - x_1)^2$$

where A and B are constants. The force is related to the potential-energy curve by Equation 6-20:

$$F_x = -\frac{dU}{dx}$$

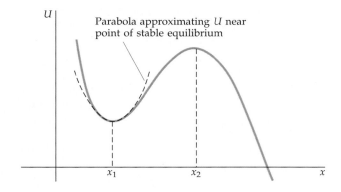

Parabola approximating U near point of stable equilibrium

Figure 12-17 Graph of the potential-energy function U corresponding to the force in Figure 12-16. The minimum at x_1 indicates stable equilibrium, whereas the maximum at x_2 indicates unstable equilibrium. Near x_1 the curve is approximately parabolic, as it is in simple harmonic motion.

Therefore,

$$F_x = -\frac{dU}{dx} = -2B(x - x_1)$$

If we set $2B = k$, this equation reduces to

$$F_x = -\frac{dU}{dx} = -k(x - x_1) = -k\epsilon$$

which is Equation 12-45.

Question

8. Give several examples of familiar motions that are either exactly or approximately simple harmonic motions.

A ball bouncing into a water tank. The motion is damped, that is, the energy of each bounce is less than that of the previous bounce because of the resistance of the water.

12-7   Damped Oscillations

In all real oscillatory motions, mechanical energy is dissipated because of some kind of frictional force. Left to itself, a spring or a pendulum eventually stops oscillating. When the mechanical energy of oscillatory motion decreases with time, the motion is said to be **damped.** If the frictional or damping forces are small, the motion is nearly periodic, except that the amplitude decreases slowly with time as shown in Figure 12-18. If the amplitude decreases slowly with time the energy does also, since the energy of an oscillator is proportional to the square of the amplitude. For the case of small damping shown in Figure 12-18, both the amplitude and the energy of the oscillations decrease by a constant percentage in a given time interval. For example, the energy of a slightly damped pendulum may decrease by 10 percent every minute. Thus, after 1 minute the energy would be 90 percent of the original energy, after 2 minutes it would be 90 percent of 90 percent or 81 percent of the original energy, and so on. This type of decrease is called exponential decrease. (We encountered exponential decrease in Section 11-3 when we studied the decrease in air pressure with altitude.)

In Figure 12-19, the oscillation of the body is damped because of the motion of the plunger immersed in the liquid. The rate of energy loss can be varied by changing the size of the plunger or the viscosity of the liquid. Although the detailed analysis of a damping force like that for this system is usually complicated, we can often represent such a force by an empirical

Figure 12-18   A graph of displacement versus time for a slightly damped oscillator. The motion is approximately simple harmonic motion with an amplitude that decreases slowly with time.

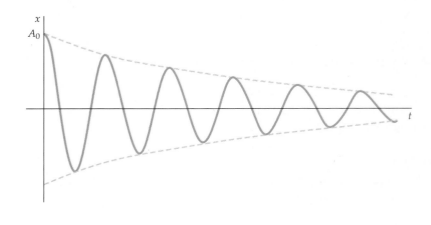

expression that is in reasonable accord with experimental results and for which the mathematics is comparatively simple. The simplest and most common representation of such a force is one that is proportional to the velocity of the object but in the opposite direction:

$$F_d = -bv$$

where b is a constant that describes the amount of damping. This is an example of a drag force, which we discussed in Section 5-2. Since a drag force is always opposite the direction of motion, the work done by this nonconservative force is always negative. Thus, it causes the mechanical energy of the system to decrease. Newton's second law applied to the motion of an object of mass m on a spring of force constant k when the damping force is $-bv$ is

$$F_x = ma_x$$

$$-kx - bv = m \frac{dv}{dt} \qquad \text{12-46}$$

We can obtain a qualitative understanding of the behavior of a damped oscillator without solving Equation 12-46 in detail. If the damping is small, we expect the body to oscillate with an angular frequency ω' that is approximately equal to the undamped frequency $\omega_0 = \sqrt{k/m}$ and we expect the amplitude to decrease slowly. In simple harmonic motion, the total mechanical energy oscillates between potential and kinetic energy. The average values of the potential and kinetic energies over one cycle are equal, and the total energy is twice the average value of either the potential or the kinetic energy. We can therefore write

$$E = 2(\tfrac{1}{2}mv^2)_{av} = m(v^2)_{av} \qquad \text{12-47}$$

For a slightly damped oscillator, only a very small fraction of the mechanical energy is lost during one cycle so that the total energy decreases slowly with time. The instantaneous rate of change of the total mechanical energy equals the power input of the damping force:

$$P = \frac{dE}{dt} = F_d \cdot v = -bv^2 \qquad \text{12-48}$$

The power input is negative, indicating that energy is leaving the system. If we replace v^2 in Equation 12-48 with its average value $(v^2)_{av} = E/m$ from Equation 12-47, we obtain

$$\frac{dE}{dt} = -\frac{b}{m} E \qquad \text{12-49}$$

Equation 12-49 describes an **exponential decrease.** The rate of decrease of the energy is proportional to the energy, so the fractional decrease, $-dE/E = (b/m)\, dt$, is the same for any given time interval.

Equation 12-49 may be solved by direct integration. We first rewrite it as

$$\frac{dE}{E} = -\frac{b}{m}\, dt \qquad \text{12-50}$$

Integrating both sides, we obtain

$$\ln E = -\frac{b}{m} t + C$$

where C is an arbitrary constant of integration. Taking the exponential of each side, we obtain

$$E = e^{-bt/m+C} = e^C e^{-bt/m} = E_0 e^{-bt/m}$$

Figure 12-19   A damped oscillator. The motion is damped by the plunger immersed in the liquid.

Shock absorbers (yellow cylinders) are used to damp the oscillations of this truck.

where $E_0 = e^C$ is another constant, which is the energy at time $t = 0$. The solution of Equation 12-49 is thus

$$E = E_0 e^{-(b/m)t} = E_0 e^{-t/\tau}$$ 12-51

where the **time constant**

$$\tau = \frac{m}{b}$$ 12-52

is the time it takes for the energy to decrease by the factor $1/e$. When the damping is small, b is small and τ is large, and the oscillator will lose only a very small fraction of its energy during one oscillation. In this case, we can obtain the energy loss per period by replacing the differentials dE and dt in Equation 12-50 by the differences ΔE and Δt and setting $\Delta t = T$, the period. We then have

$$\frac{\Delta E}{E} = -\frac{b}{m}T$$ 12-53

The damping of a slightly damped oscillator is usually described by a dimensionless quantity Q called the **quality factor** or **Q factor.** If E is the total energy and $|\Delta E|$ is the energy loss in one period, the Q factor is defined as

Q factor defined

$$Q = 2\pi \frac{E}{|\Delta E|}$$ 12-54

The Q factor is thus inversely proportional to the fractional energy loss per cycle:

$$\frac{|\Delta E|}{E} = \frac{2\pi}{Q}$$ 12-55

Using Equations 12-53 and 12-54, we can relate the Q factor to the damping constant and the time constant:

$$Q = 2\pi \frac{E}{|\Delta E|} = 2\pi \frac{m}{bT} = 2\pi \frac{\tau}{T}$$ 12-56

Example 12-10

A simple pendulum loses 1 percent of its energy during each oscillation. What is the Q factor?

Since the energy loss is 1 percent,

$$\frac{|\Delta E|}{E} = \frac{1}{100}$$

The Q factor is therefore

$$Q = 2\pi \frac{E}{|\Delta E|} = 2\pi(100) = 628$$

Since the energy of an oscillator is proportional to the square of its amplitude, we can use Equation 12-51 to find the time dependence of the amplitude of a slightly damped oscillator. If A is the amplitude at some time t and A_0 is the amplitude at $t = 0$, we have

$$\frac{E}{E_0} = \frac{A^2}{A_0^2}$$

Then, from Equation 12-51,

$$\frac{A^2}{A_0^2} = e^{-(b/m)t}$$

or

$$A = A_0 e^{-(b/2m)t} \qquad \text{12-57}$$

Thus, the amplitude decreases exponentially with time. The dashed curves in Figure 12-18 correspond to $x = A$ and $x = -A$ where A is given by Equation 12-57.

The exact solution of Equation 12-46 can be found using standard methods for solving differential equations. The solution for the case of small damping is

$$x = A_0 e^{-(b/2m)t} \cos (\omega' t + \delta) \qquad \text{12-58}$$

where A_0 is the maximum amplitude and the frequency ω' is related to the angular frequency $\omega_0 = \sqrt{k/m}$ by

$$\omega' = \omega_0 \sqrt{1 - \left(\frac{b}{2m\omega_0}\right)^2} = \omega_0 \sqrt{1 - \frac{1}{4Q^2}} \qquad \text{12-59}$$

Thus, we see that our qualitative observations were correct. For small damping, the frequency is very nearly equal to the undamped frequency, and the amplitude decreases exponentially with time. Equation 12-59 is given only to illustrate that, for all practical purposes, the frequency of a slightly damped oscillator is the same as that of the undamped oscillator. For example, for a moderately damped oscillator with $Q = 10$, ω' and ω_0 differ by only about 0.1 percent.

Equation 12-58 is the solution of Equation 12-46 only if the damping is small. If the damping is gradually increased, it eventually reaches a critical value such that no oscillation occurs. The critical value of the damping constant is

$$b_c = 2m\omega_0 \qquad \text{12-60}$$

If b is equal to or greater than this critical value, the system does not oscillate at all but merely returns to its equilibrium position. The greater the damping, the longer it takes the system to return to equilibrium. When $b = b_c$, the system is said to be **critically damped**, and it returns to equilibrium in the shortest time with no oscillation. When b is greater than b_c the system is overdamped. Figure 12-20 shows plots of the displacement versus time for a critically damped and an overdamped oscillator.

In many practical applications, critical damping or nearly critical damping is used to avoid oscillations and yet have the system return to equilibrium quickly. One example is the use of shock absorbers to damp the oscillations of an automobile on its springs. You can test the damping on your car by pushing down on the front or back of the car and releasing it. If the car moves back up to equilibrium with no oscillation, the system is critically damped or overdamped. Usually, you will note one or two oscillations, indicating that the damping is just under the critical value.

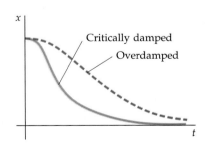

Figure 12-20 Plots of displacement versus time for a critically damped and an overdamped oscillator.

Figure 12-21   An object on a vertical spring can be driven by moving the point of support up and down.

12-8   Driven Oscillations and Resonance

We have seen that in damped oscillations, energy is continually dissipated and the amplitude of the oscillations decreases. To keep a damped system going, energy must be put into the system. When this is done, the oscillator is said to be driven or forced. For example, when you sit on a swing and keep it going by "pumping," that is, by moving your body and legs, you are driving an oscillator. If you put energy into the system at a greater rate than it is being dissipated by damping, the energy increases with time, which is indicated by an increase in the amplitude. If you put energy in at the same rate it is being dissipated, the amplitude remains constant in time.

Figure 12-21 shows a system consisting of an object on a spring that is being driven by moving the point of support up and down. Similarly, a simple pendulum can be driven by moving the support back and forth. (You should try some simple experiments with one of these systems to familiarize yourself with the behavior of driven oscillators.) If the point of support of an object on a spring or a simple pendulum is moved with simple harmonic motion of small amplitude and angular frequency ω, the system will begin oscillating. At first the motion is complicated, but eventually a steady state is reached in which the system oscillates with the same frequency as that of the driver and with a constant amplitude and, therefore, with constant energy. In the steady state, the energy put into the system by the driving force during one cycle equals the energy dissipated per cycle because of the damping.

The amplitude, and therefore the energy, of a system in the steady state depends not only on the amplitude of the driver, but also on its frequency. The **natural frequency** of an oscillator is defined as the frequency of that oscillator when no driving or damping forces are present. (The natural angular frequency of a spring, for example, is $\omega_0 = \sqrt{k/m}$.) If the driving frequency is equal (or approximately equal) to the natural frequency of the system, the system will oscillate with an amplitude much larger than the

The swaying of the Citicorp Building in New York during high winds is reduced by this tuned-mass damper, which oscillates at the same frequency as the building but 180° out of phase.

amplitude of the driving force. This phenomenon is called **resonance.** When the driving frequency equals that natural frequency of the oscillator, the energy absorbed by the oscillator is maximum. The natural frequency of the system is thus called the **resonance frequency** of the system.*

The average rate at which energy is absorbed over a cycle equals the average power delivered by the driving force. Figure 12-22 shows plots of the average power delivered to an oscillator as a function of the driving frequency for two different values of damping. These curves are called **resonance curves.** When the damping is small (large Q), the oscillator absorbs much more energy from the driving force at or near the resonance frequency than it does at other frequencies. The width of the peak of the resonance curve is correspondingly narrow, and we speak of the resonance as being sharp. When the damping is large (small Q) the oscillator still absorbs more energy at or near resonance than at other frequencies, but the difference is not as great and the resonance curve is broad. The width of each resonance curve $\Delta\omega$ is indicated in the figure. For relatively small damping, the ratio of the resonance frequency ω_0 to the width of the resonance can be shown to be equal to the Q factor (see Problem 74):

$$Q = \frac{\omega_0}{\Delta\omega} = \frac{f_0}{\Delta f} \qquad\qquad 12\text{-}61 \qquad Q \text{ factor for small damping}$$

Thus, the Q factor is a direct measure of the sharpness of resonance.

We can treat a driven oscillator mathematically by assuming that, in addition to the restoring force and a damping force, the oscillator is subject to an external force, the driving force, that varies harmonically with time:

$$F_{\text{ext}} = F_0 \cos\omega t \qquad\qquad 12\text{-}62$$

where ω is the angular frequency of the driving force, which is generally not related to the natural angular frequency of the system ω_0.

An object of mass m attached to a spring of force constant k and subject to a damping force $-bv$ and an external force $F_0 \cos\omega t$ then obeys the equation of motion given by

$$\sum F = -kx - bv + F_0 \cos\omega t = m\frac{dv}{dt}$$

or

$$m\frac{d^2x}{dt^2} + b\frac{dx}{dt} + m\omega_0^2 x = F_0 \cos\omega t \qquad\qquad 12\text{-}63$$

where we have used $k = m\omega_0^2$ and $dv/dt = d^2x/dt^2$.

We will not attempt to solve Equation 12-63. Instead, we will discuss its general solution qualitatively. The solution of Equation 12-63 consists of two parts, the **transient solution** and the **steady-state solution.** The transient part of the solution is identical to that for a damped oscillator given by Equation 12-58. The constants in this part of the solution depend on the initial conditions. Over time, this part of the solution becomes negligible because of the exponential decrease of the amplitude. We are then left with

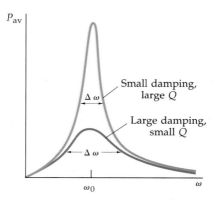

Figure 12-22 Plots of the average power delivered to an oscillator by a sinusoidal driving force versus the driving angular frequency ω for different values of damping. Resonance occurs when the (angular) frequency of the driving force equals the natural (angular) frequency of the system, ω_0. The resonance is sharp when the damping is small.

---

*When describing an oscillator mathematically, the angular frequency ω is more convenient than the frequency $f = \omega/2\pi$. Since ω and f are proportional, most statements concerning angular frequency also hold for frequency. For example, when the driving (angular) frequency ω equals the natural (angular) frequency ω_0, the driving frequency $f = \omega/2\pi$ equals the natural frequency $f_0 = \omega_0/2\pi$. In verbal descriptions we usually omit the word "angular" when such omission will not cause confusion.

the steady-state solution, which does not depend on the initial conditions. It can be written as

$$x = A \cos(\omega t - \delta)$$ 12-64

where the angular frequency ω is the same as that of the driving force and the amplitude A and phase constant δ are given by

$$A = \frac{F_0}{\sqrt{m^2(\omega_0^2 - \omega^2)^2 + b^2\omega^2}}$$ 12-65

and

$$\tan \delta = \frac{b\omega}{m(\omega_0^2 - \omega^2)}$$ 12-66

Comparing Equations 12-64 and 12-62, we can see that the displacement and the driving force oscillate with the same frequency but differ in phase by δ. The negative sign in the phase of Equation 12-64 is introduced so that the phase constant δ is positive. When the driving frequency ω is much less than the natural frequency ω_0, $\delta \approx 0$ as can be seen from Equation 12-66. At resonance, $\delta = \pi/2$, and when ω is much greater than ω_0, $\delta \approx \pi$. At resonance, $\omega = \omega_0$, and the amplitude A is maximum. (The difference in phase between the displacement and the driving force can be demonstrated by holding a meterstick at one end as a pendulum and moving the end back and forth to drive it.)

The velocity of the object in the steady state is obtained by differentiating Equation 12-64 with respect to time:

$$v = \frac{dx}{dt} = -A\omega \sin(\omega t - \delta)$$

At resonance, the velocity is in phase with the driving force:

$$v = -A\omega \sin\left(\omega t - \frac{\pi}{2}\right) = +A\omega \cos \omega t$$

Thus, at resonance, the object is always moving in the direction of the driving force, as would be expected for maximum power input.

Extended objects have more than one resonance frequency. The resonance vibration patterns of the top plate of a classical guitar are shown by these holographic interferograms. The frequency and Q value for each resonance are given.

268 Hz ($Q = 52$)

553 Hz ($Q = 66$)

672 Hz ($Q = 61$)

1010 Hz ($Q = 80$)

There are many familiar examples of resonance. When you sit on a swing and pump, the driving force is not simple harmonic. However, it is periodic, and you learn intuitively to pump with the same frequency as the natural frequency of the swing. When soldiers march across a small bridge, they often break step because it is possible that the frequency of their step might be near one of the resonance frequencies of the bridge. Many machines vibrate because they have rotating parts that are not in perfect balance. (Observe a washing machine in the spin cycle for an example.) If such a machine is attached to a structure that can vibrate, the structure becomes a driven oscillatory system that is set in motion by the machine. If the structure resonates with the applied vibration, considerable damage can occur, or it may produce unwanted sound throughout the building. Great engineering efforts are put into balancing the rotary parts of such machines, damping their vibrations, and isolating them from building supports.

A glass with low damping can be broken by an intense sound wave at a frequency equal to or very nearly equal to the natural frequency of vibration of the glass. This is often done in physics demonstrations using an audio oscillator and an amplifier. You can compare the relative damping of good crystal glasses by tapping them and listening to how long they ring. The longer the glass rings, the lower the damping. There are many stories about singers breaking a glass with a loud high note. Not one of these stories has ever been documented, however. The problem with a singer trying to break a glass is that it is nearly impossible for a singer to concentrate enough energy in the narrow frequency range near resonance. When we study the vibration of piano and violin strings in Chapter 13, we will see that such systems have a series of natural frequencies rather than just a single natural frequency and can resonate at any one of these natural frequencies.

A student laboratory apparatus for studying resonance. The inner copper rim attached to a coiled spring constitutes a torsion pendulum, which can be driven by the motor at the right. The amplitude of the oscillation is measured by the protractor which is the outer rim of the apparatus. The electromagnet at the bottom provides damping that can be varied.

Question

9. Give some examples of common systems that can be considered to be driven oscillators.

Summary

1. In simple harmonic motion, the acceleration is proportional to the displacement and is oppositely directed. If x is the displacement, the acceleration is

$$a_x = -\omega^2 x$$

where ω is the angular frequency of oscillation, which is related to the frequency f by

$$f = \frac{\omega}{2\pi}$$

2. The period of oscillation is the reciprocal of the frequency:

$$T = \frac{1}{f}$$

The period and frequency in simple harmonic motion are independent of the amplitude. For the motion of an object of mass m on a spring of force constant k, the period is given by

$$T = 2\pi\sqrt{\frac{m}{k}}$$

The period of the motion of a simple pendulum of length L is

$$T = 2\pi\sqrt{\frac{L}{g}}$$

3. The position function x for simple harmonic motion of amplitude A and angular frequency ω is given by

$$x = A \cos(\omega t + \delta)$$

where δ is the phase constant, which depends on the choice of time $t = 0$. The velocity of the particle is given by

$$v = -\omega A \sin(\omega t + \delta)$$

4. When a particle moves in a circle with constant speed, the x and y components of its position vary with simple harmonic motion.

5. The total energy in simple harmonic motion is proportional to the square of the amplitude. For a mass on a spring of force constant k, it is given by

$$E_{\text{total}} = \tfrac{1}{2}kA^2$$

The potential and kinetic energies for a mass oscillating on a spring are given by

$$U = E_{\text{total}} \cos^2(\omega t + \delta)$$

and

$$K = E_{\text{total}} \sin^2(\omega t + \delta)$$

The average value of either the potential or kinetic energy is half the total energy.

6. In the oscillations of real systems, the motion is damped because of frictional forces or other forces that dissipate energy. If the damping is greater than some critical value, the system does not oscillate but merely returns to its equilibrium position if disturbed. The motion of a slightly damped system is nearly simple harmonic with an amplitude that decreases exponentially with time. The energy of such a system also decreases exponentially with time. For a slightly damped oscillator, the damping is measured by the Q factor:

$$Q = 2\pi\frac{E}{|\Delta E|}$$

where E is the total energy and ΔE is the energy loss per period.

7. When a slightly damped system is driven by an external force that varies sinusoidally with time, the system oscillates with frequency equal to the driving frequency and an amplitude that depends on the driving frequency. If the driving frequency is equal to or nearly equal to the natural frequency of the system, the system oscillates with a large amplitude. This is called resonance. The Q factor is a measure of the sharpness of the resonance. Systems with low damping and therefore high Q factors show a sharply peaked resonance curve. The ratio of the resonance frequency to the width of the resonance curve equals the Q factor:

$$Q = \frac{\omega_0}{\Delta\omega} = \frac{f_0}{\Delta f}$$

Chaos: An Orderly Disorder

James S. Walker
Washington State University,
Pullman, Washington

The ability to predict the future of a physical system is of central importance in physics. Given the appropriate initial conditions one can calculate the trajectory of a baseball, the orbit of a satellite, or the motion of a clock pendulum. So pervasive is this type of analysis in physics that it is sometimes tempting to reiterate the claim of the 18th century French mathematician Pierre Simon de Laplace that, given the initial conditions of the universe, one could predict its future for the rest of time. Leaving aside, for the moment, the problems associated with gathering and storing the initial conditions of every particle in the universe, and leaving aside also uncertainties that are inherent in quantum mechanics, there still remains a fundamental obstacle to predicting the future—chaos.

Chaos, according to many dictionaries, is a condition of total disorder or confusion—certainly the sort of situation to cloud the future. The meaning ascribed to chaos in physics and mathematics is similar in spirit, though more precise. For example, a system undergoing chaotic motion never repeats itself, but rather, continually does something different; thus the motion may appear quite random and

and phase transitions for many years before his research turned, appropriately enough, to chaos. He received his bachelor's, master's, and doctorate degrees in physics all from the University of Washington, obtaining the Ph.D. in 1978. Before taking his current position, he did postdoctoral work at the University of Pennsylvania, the Massachusetts Institute of Technology, and the University of California at San Diego. When not rafting in Grand Canyon, or looking through his telescope on a starry night, Walker enjoys hiking, body surfing, kayaking, and juggling.

James S. Walker, an associate professor of physics at Washington State University, is a theorist who studied critical phenomena

disordered. Nonetheless, chaotic motion is far from totally disordered and instead displays a definite structure that is readily apparent.

Another aspect of chaos, as it arises in physics, is an extreme sensitivity to initial conditions. Imagine, for example, trying to stand a straight pin on end. If placed exactly vertically the pin is in equilibrium. The equilibrium is unstable, however, and even the slightest disturbance—a breath of air, the sound of a pin dropping—will upset it and cause the pin to fall in one direction or another. The vertical position of the pin is an example of a state exhibiting extreme sensitivity to initial conditions, where, if one ignores even the minute puff of wind produced by a butterfly in the room, it is no longer possible to predict with certainty in which direction the pin will fall. Chaotic motion is like this, except that in chaos every state of the system, that is, every part of its motion, is just as sensitive to disturbance as is the vertical pin. As a result, even the smallest errors in measuring the initial conditions of a chaotic system lead to enormous errors later on.

The field of chaos is one of the youngest and most rapidly developing in physics today. There are many new aspects of chaos waiting to be discovered, but already much progress has been made. The inherent unpredictability of the weather, for example, is recognized as an example of chaotic motion associated with the transfer of heat in the atmosphere. While specific long-range forecasts are impossible, the general characteristics of weather—the climate—are well represented by an underlying chaotic structure. Many mechanical systems are also known to be chaotic, ranging from such mundane examples as the dripping of water from a leaky faucet to more exotic cases, like the chaotic tumbling motion of Hyperion, a moon of Saturn. In addition, chaos occurs in fluctuations of animal populations from generation to generation, in electric circuits, and even in our own bodies; systems such as the heart and even the brain show indications of chaotic behavior.

The Bender Bouncer

Fortunately, the basic properties of chaos—order within the disorder and sensitivity to initial conditions—can be seen quite clearly even in very simple physical systems. Consider, for example, a driven, damped harmonic oscillator, as shown in Figure 1. Here a spring of force constant k is attached to a mass m and driven vertically with angular frequency ω. A strip chart is arranged so that it records the periodic driving force applied to the spring, as well as the resulting displacement x of the mass. The

Continued

equation of motion is essentially that given in Equation 12-63:

$$m\frac{d^2x}{dt^2} + b\frac{dx}{dt} + kx = F\cos(\omega t)$$

where b is the damping coefficient and F is the magnitude of the driving force. (All results presented in this essay use $m = 4$ kg, $b = 1$ N $\cdot$ s/m, $k = 1$ N/m, and $F = 1$ N.) In Section 12-8, this system was studied for the usual situation, in which the mass is allowed to oscillate freely and for which the displacement as a function of time is simply

$$x(t) = A\cos(\omega t - \delta)$$

where A is the amplitude of motion and δ is the phase shift. This result for $x(t)$ clearly leads to periodic, predictable motion.

The simple behavior of this system is directly related to the fact that the restoring force kx is linear in the displacement; in fact, chaotic motion is found only in systems that are nonlinear. For example, if the restoring force of the spring included a nonlinear term like $k'x^2$, chaotic behavior would be possible. Of course, any real spring deviates from linearity for large enough displacements, thus nonlinear behavior is virtually unavoidable in the real world. Yet in physics one often studies linear systems almost to the exclusion of others, since their solutions and behaviors are relatively straightforward. This is certainly a reasonable approach, and in many situations the results for a linear system give a good indication of what may be expected in a system that is slightly nonlinear. But this is not always the case, and as pointed out by Robert May, a pioneer in chaos research, "Not only in research, but also in the everyday world of politics and economics, we would all be better off if more people realized that simple nonlinear systems do not necessarily possess simple dynamical properties."

An interesting way to make this mass-spring system nonlinear is to introduce a barrier which blocks the motion of the mass, as indicated in Figure 1. This system I refer to as the "Bender bouncer," in honor of my friend and colleague Paul A. Bender, who first brought it to my attention. The barrier is considered to be infinitely massive and to undergo perfectly elastic collisions with the mass m; thus, it simply sends the mass back in the direction from which it came, traveling with the same speed. Clearly, the force acting on the mass is no longer a linear function of the displacement, since the barrier acts on the mass by giving it a sharp impulse.

With such simple characteristics for the mass–barrier collisions, it might seem that the barrier plays an essentially trivial role in the system; indeed, for many values of the driving frequency, the resulting motion is simple and periodic, as in the examples shown on the strip chart in Figure 1. However, for certain ranges of ω the motion is no longer periodic, but instead, never repeats. In short, the effect of the barrier is to produce chaos in this otherwise simple system.

Chaotic Bouncing

A clear indication of chaos in the system is shown in Figure 2a, where the amplitude of oscillation is plotted as a function of the driving frequency ω. The tan-colored curve shows the result for the freely swinging oscillator, as discussed in Chapter 12. Also shown are the results for the bouncing oscillator with the barrier placed at $x = 0$, which are obtained as follows: Let the system run for many cycles so that the motion settles down, as in the examples in Figure 1; then for the next 100 bounces, plot the largest value of x that occurs between successive bounces. The result is a series of peaks separated by regions of scattered points. The smooth lines tracing the peaks result from periodic motion—each of the 100 bounces is of the same height; thus, for such frequencies, a single value of x is plotted 100 times over. Between the peaks are chaotic regions where each bounce is a different height, scattering the 100 plotted points like pepper grains over a range of amplitudes.

Figure 1   A driven, damped harmonic oscillator—set up so that its motion is recorded on a strip chart. Notice that the mass is not free to oscillate, but instead bounces elastically off an infinitely massive barrier. The tracings shown on the strip chart are obtained for $\omega = 1$ rad/s.

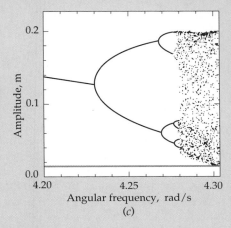

(a)

Angular frequency, rad/s

(b)

Angular frequency, rad/s

(c)

Angular frequency, rad/s

Figure 2 Amplitude versus angular frequency ω for the freely oscillating mass–spring system (tan-colored line) and the Bender bouncer (black line and scattered points). (*a*) Over a large range of angular frequency, the bouncing oscillator responds with a series of peaks (corresponding to periodic motion) separated by regions of scattered points (indicating chaotic motion). Parts (*b*) and (*c*) show details of the behavior near $\omega = 1.5$ rad/s and $\omega = 4.3$ rad/s, respectively, showing a progression of ever more closely spaced bifurcations leading to chaotic motion.

To show the chaotic behavior more clearly, the first chaotic region near $\omega = 1.5$ rad/s is shown in detail in Figure 2*b*; in Figure 2*c* the chaotic region near $\omega = 4.3$ rad/s is illustrated. As complex as this behavior is, and it is simply the response of a bouncing mass on a spring driven at different frequencies, note that the chaotic region is not simply random, but shows structure in the form of shading, as well as occasional clear "windows" where periodic motion returns. In particular, looking at Figure 2*b* one sees that simple periodic motion is observed for frequencies near 1.25 rad/s, with each bounce the same height. At roughly 1.325 rad/s the single line bifurcates, that is, it splits in two as the mass now bounces high, then low, then high, and so on. An example of such motion, for $\omega = 1.35$ rad/s, is shown in Figure 3*a*. Notice that it takes two bounces for the motion to repeat (i.e., high, then low), thus the motion is still periodic, though the period has doubled.

Another period doubling is evident at about 1.3625 rad/s, and the mass now bounces to four different heights. With each doubling of the period it takes longer to see the "repeating pattern" of the motion, which thus appears more complex. Greater resolution in this frequency range shows, in fact, that an *infinite* progression of period doublings occurs ever more closely spaced in frequency—it's like walking forward with steps that are each a fraction the size of the one before, in which case an infinite number of steps cover only a finite distance. At a frequency of roughly 1.37 rad/s, an infinite number of "frequency steps" (period doublings) has been taken, and the period is now infinite; or to put it another way, the mass now bounces in a chaotic fashion, never repeating its motion. An example of the chaotic motion is shown in Figure 3*b*, for $\omega = 1.5$ rad/s.

(a)

(b) (b)

Figure 3 Position versus time plots for the Bender bouncer. (*a*) $\omega = 1.35$ rad/s. In this case, the system's motion repeats after every two bounces, which takes two cycles of the driving force. Compare with Figure 1, where the motion repeats with every cycle of the driving force. (*b*) $\omega = 1.5$ rad/s. At this frequency the system's motion is chaotic—it never repeats.

Continued

Sensitivity to Initial Conditions

Once the system becomes chaotic, the scattering of bounce heights in Figure 2b and c is very intricate, and the dictionary definition of chaos as a condition of total disorder seems fairly appropriate. But in addition to the intricate structure, chaotic motion also displays an extreme sensitivity to initial conditions, in contrast to simple periodic motion in which all initial conditions evolve toward the same final motion. If several periodic oscillators are started off, each slightly differently, their motions quickly converge, and they spend the rest of time moving in perfect unison (as in Figure 1). Different starting conditions for a chaotic system, however, such as the bouncing oscillator at $\omega = 1.5$ rad/s, diverge to very different motions. Recall the situation with the vertical straight pin and the fact that different starting conditions arbitrarily close may still result in the pin falling in different directions. Each point in chaotic motion is just as sensitive to initial conditions, thus a set of starting points clustered together will, within a few bounces, be spread out over a range of amplitudes.

An example of this sensitivity can be seen in Figure 4, which shows five different initial conditions chosen so close together that, to begin with, they cannot be distinguished at this scale. Imagine these different starting conditions to represent a system of five different oscillators that start off moving in almost the same way. Even so, with each bounce the different motions diverge further from one another.

If each of these five starting conditions were also composed of five even more closely spaced starting conditions, it is clear that with more bounces these additional starting conditions would split off in the same way. It is like shredding a rope, splitting it apart into individual strands, then separating the strands into threads, then the threads into fibers, and so on. Chaotic motion does just this, but without limit—the separating and separating goes on to an infinitely fine level.

Since the starting conditions of an oscillator are not known exactly in any physical experiment, it follows that in a very short time the initial uncertainties will be magnified, and magnified again and again, until the mass could be located anywhere. In a practical sense, then, it is simply not meaningful to ask the question, "Where is the mass at time t?" if one is dealing with a chaotic oscillator.

Another aspect of sensitivity to initial conditions appears when the barrier is placed at a value of x that is different from zero. Consider, for example, the fantastically complex behavior in Figure 5a, which shows amplitude versus angular frequency for the barrier at $x = 0.5$. The blue line and points, taken by themselves, would seem to suggest a standard bifurcation route to chaos; but over the range of angular frequencies from about 2.22 rad/s to 2.55 rad/s, another type of motion with three amplitudes (shown in black) also exists. Beyond 2.55 rad/s, each of these three amplitudes undergoes its own bifurcations, until the system finally becomes chaotic.

Of particular interest is the range of angular frequencies where both types of motions are found. For any given frequency in this range, the final motion depends on exactly how the system is started. If the mass is always dropped from rest, then the final motion is found to depend on the height from which the mass is dropped and on the precise time at which it is dropped. In principle, the mass may be dropped from any height, and the time of the drop may occur at any point during a given cycle of the driving force; that is, one need only consider t in the interval $0 \le t \le T$, where T is the period of the driving force. What one finds is that, even though both of these motions are periodic rather than chaotic, the system is still extremely sensitive to initial condi-

Figure 4   Position versus time plot for $\omega = 1.5$ rad/s, showing the motion for a system of five bouncing oscillators with closely spaced initial conditions. For the first few bounces, the oscillators stay in step, but with time the motions become very different.

(*a*)

(*b*)

Figure 5 A different type of sensitivity to initial conditions occurs when the barrier is placed at $x = 0.5$. (*a*) The amplitude versus angular frequency plot shows two different types of motion, indicated by the blue and black points, existing in the same angular frequency range. (*b*) Initial conditions determine which motion the system displays. This plot, with $\omega = 2.4$ rad/s, shows the initial conditions that correspond to the motion indicated by the blue or the black points of part (*a*).

tions in determining which will be its final motion. This is shown clearly in Figure 5*b*, corresponding to $\omega = 2.4$ rad/s. The blue and black regions show the initial conditions that result in the blue or black motions, respectively. Again, in contrast to the apparent simplicity of this system, one finds an incredibly rich and complex structure describing its behavior.

Calculator Chaos

The Bender bouncer can be set up experimentally very easily, and it can also be simulated on a desktop microcomputer. But if you have a pocket calculator handy, you can explore chaos for yourself, right now, using a very simple mathematical relation. Consider the following nonlinear expression, $x' = wx(1 - x)$, known as the logistic equation, where $0 \leq x \leq 1$ and w is an adjustable parameter. As an example, let $w = 2.9$, then pick any allowed value of x, say $x = 0.4$. Calculate x' as defined above to find $x' = 0.696$. Next, take x' as your starting value, and calculate x'' in the same manner as above to obtain $x'' = 0.614$ as the next iteration. Continue this procedure for a few more iterations, and soon you notice the values of x settle down to approximately 0.655. Try different starting values for x, and again the same limiting value of 0.655 is found.

Do you see the analogy with the mass-spring system discussed above? The value of x is like the maximum displacement between successive bounces, as plotted in Figure 2*a*, *b*, and *c*. What you have just observed is analogous to Figure 1, where different starting conditions settle down within a few cycles to a fixed and repeating bounce height.

Now try $w = 3.3$. In a few iterations, it becomes clear that x alternates between a high value, 0.824, and a low value, 0.480; that is, the system has un-

dergone a period doubling similar to the period doublings in Figure 2*b* and 2*c*. This is analogous to the motion shown in Figure 3*a*. If one continues this procedure for different values of w and plots the stable values of x that result after several iterations, the result is as shown in Figure 6. Notice the remarkable similarity with Figure 2*c*, which emphasizes the fact that w and the angular frequency play analogous roles. As in the mass–spring system, the

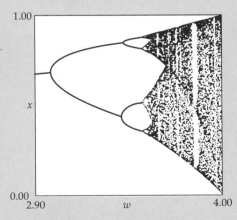

Figure 6 Results for the logistic equation $x' = wx(1 - x)$. This nonlinear equation is iterated 200 times before plotting begins, then the next 100 values of x are plotted. The bifurcations and period doublings shown here are very similar to the behavior displayed by the Bender bouncer in Figure 2*c*.

Continued

This striking image shows a small portion of a mathematical object known as the Mandelbrot set. The set occupies those regions shown in black, whereas the various colors represent regions just outside the set. Notice that the Mandelbrot set consists primarily of circular structures with the circles coming in all different sizes; from the large circle, to the smaller circles budding from it, to the even smaller circles on the buds, and so on down to circles of arbitrarily small size. This repetition of a basic shape to ever smaller sizes is one of the hallmarks of objects known as fractals. Fractals occur frequently in the study of chaos; for example, the size of the circles in the Mandelbrot set correspond exactly to the distance between successive period doublings in the logistic equation. Not surprisingly, the structure shown in Figure 5b is also a fractal.

distance between successive period doublings becomes smaller and smaller. Even more remarkably, the ratio of distances between successive period doublings approaches *exactly* the same value for both the Bender bouncer and the logistic equation, providing just one example of the many common features shared among chaotic systems. Though the logistic equation is a simple mathematical rule and may be studied without reference to a particular physical system, it nonetheless serves as a useful model for the complex situations that can actually arise in nature.

Try now a value of w slightly greater than $3.56994571\ldots$, which is where chaos sets in. Notice that the sequence of x values never repeats—you are now observing chaotic behavior and are on the road to further explorations.

Though this has been a very brief introduction to some of the unusual and extraordinary features of the orderly disorder known as chaos, it is clear that the ability to predict the future of a physical system is something that should not be taken for granted. And not only does chaos provide scientists with a new way of looking at the world, describing behavior in a wide variety of systems, but, as we have seen, there is great esthetic appeal in the fantastically complex geometrical forms associated with these systems. As the French chaotician D. Ruelle has said, "These systems of curves, these clouds of points, sometimes evoke galaxies or fireworks, other times quite weird and disturbing blossomings. There is a whole world of forms still to be explored, and harmonies still to be discovered." And, I might add, this exploration is open to us all.

When fluids are mixed the results are often chaotic, as in the experiment shown here. A small blob of yellow fluorescent tracer is placed in a container filled with glycerine, then the glycerine is stirred in a simple, repetitive fashion. As a result, the blob is repeatedly stretched and folded, stretched and folded—much like a strand of salt-water taffy—giving rise to an intricately convoluted structure. An analogous "stretching and folding" occurs with the initial conditions of the Bender bouncer, as is evident in Figure 5b.

Suggestion for Further Reading

Pierce, John R.: "Sine Waves and Resonance," in *The Science of Musical Sound*, Scientific American Books, Inc., distributed by W.H. Freeman and Company, New York, 1983.

See description of the book under Suggestions for Further Reading, Chapter 13.

Review

A. Objectives: After studying this chapter, you should:

1. Be able to describe the general characteristics of simple harmonic motion.

2. Be able to describe the relationship between simple harmonic motion and circular motion.

3. Be able to work problems involving springs and pendulums.

4. Be able to discuss the general motion of any system near equilibrium.

5. Be able to describe the motion of a damped oscillator.

6. Be able to discuss the motion of a driven oscillator and sketch a typical resonance curve for various values of the Q factor.

B. Define, explain, or otherwise identify:

Periodic motion	Damped oscillations
Simple harmonic motion	Exponential decrease
Period	Q factor
Frequency	Critical damping
Hertz	Resonance
Amplitude	Resonance frequency
Phase	Resonance curve
Phase constant	Transient solution
Angular frequency	Steady-state solution
Physical pendulum	
Torsional pendulum	
Torsion constant	

C. True or false: If the statement is true, explain why it is true. If it is false, give a counterexample.

1. In simple harmonic motion, the period is proportional to the square of the amplitude.

2. In simple harmonic motion, the frequency does not depend on the amplitude.

3. In simple harmonic motion, the total energy is proportional to the square of the amplitude.

4. The motion of a simple pendulum is simple harmonic for any initial angular displacement.

5. The motion of a simple pendulum is periodic for any initial angular displacement.

6. If the acceleration of a particle is proportional to the displacement and oppositely directed, the motion is simple harmonic.

7. The energy of a damped, undriven oscillator decreases exponentially with time.

8. Resonance occurs when the driving frequency equals the natural frequency.

9. If the Q value is high, the resonance is sharp.

Problems

Level I

12-1 Simple Harmonic Motion: A Mass on a Spring

1. A 2-kg object is attached to a horizontal spring of force constant $k = 5$ kN/m. The spring is stretched 10 cm from equilibrium and released. Find (*a*) the frequency, (*b*) the period, and (*c*) the amplitude of the motion. (*d*) What is the maximum speed? (*e*) What is the maximum acceleration? (*f*) When does the object first reach its equilibrium position? What is its acceleration at this time?

2. Answer the questions in Problem 1 for a 3-kg object attached to a spring of force constant $k = 600$ N/m when the spring is initially stretched 8 cm from equilibrium.

3. A 3-kg object attached to a horizontal spring oscillates with an amplitude $A = 10$ cm and a frequency $f = 2$ Hz. (*a*) What is the force constant of the spring? (*b*) What is the period of the motion? (*c*) What is the maximum speed of the object? (*d*) What is its maximum acceleration?

4. An object on a horizontal spring oscillates with a period of 4 s. If the object is suspended from the spring vertically, by how much is the spring stretched from its natural length when the object is in equilibrium?

5. An 80-kg person steps into a car of mass 2400 kg, causing it to sink 2.50 cm on its springs. Assuming no damping, with what frequency will the car and passenger vibrate on the springs?

6. A 5-kg object oscillates on a horizontal spring with an amplitude of 4 cm. Its maximum acceleration is 24 m/s². Find (a) the force constant k, (b) the frequency, and (c) the period of the motion.

7. An object oscillates with an amplitude of 6 cm on a horizontal spring of force constant 2 kN/m. Its maximum speed is 2.20 m/s. Find (a) the mass of the object, (b) the frequency of the motion, and (c) the period of the motion.

8. The position of a particle is given by

$$x = (5 \text{ cm}) \cos 4\pi t$$

where t is in seconds. What is (a) the frequency, (b) the period, and (c) the amplitude of the particle's motion? (d) What is the first time after t = 0 that the particle is at its equilibrium position? In what direction is it moving at that time?

9. (a) What is the maximum speed of the particle in Problem 8? (b) What is its maximum acceleration?

10. What is the phase constant δ in Equation 12-4 if the position of the oscillating particle at time t = 0 is (a) 0, (b) −A, (c) A, (d) A/2?

11. A particle has a displacement of $x = 0.4 \cos (3t + \pi/4)$ where x is in meters and t is in seconds. (a) Find the frequency f and period T of the motion. Where is the particle at (b) t = 0? (c) At t = 0.5 s?

12. (a) Find an expression for the velocity of the particle whose position is given in Problem 10. (b) What is the velocity at time t = 0? (c) What is the maximum velocity? (d) At what time after t = 0 does this maximum velocity first occur?

13. A particle of mass m begins at rest from x = +25 cm and oscillates about its equilibrium position at x = 0 with a period of 1.5 s. Write equations for (a) the position x versus the time t, (b) the velocity v versus t, and (c) the acceleration a versus t.

14. Work Problem 13 for the case in which the particle is initially at x = 25 cm and is moving with velocity $v_0 = +50$ cm/s.

15. Find (a) the maximum speed and (b) the maximum acceleration of the particle in Problem 13. (c) What is the first time that the particle is at x = 0 and moving to the right?

12-2 Simple Harmonic Motion and Circular Motion

16. A particle moves in a circle of radius 40 cm with a constant speed of 80 cm/s. Find (a) the frequency and (b) the period of the motion. (c) Write an equation for the x component of the position of the particle as a function of time t, assuming that the particle is on the x axis at time t = 0.

17. A particle moves in a circle of radius 15 cm, making 1 rev every 3 s. (a) What is the speed of the particle? (b) What is its angular velocity ω? (c) Write an equation for the x component of the position of the particle as a function of time t, assuming that the particle is on the x axis at time t = 0.

12-3 Energy in Simple Harmonic Motion

18. Find the total energy of the object oscillating on the spring in Problem 1.

19. Find the total energy of the object oscillating on the spring in Problem 2.

20. A 1.5-kg object oscillates with simple harmonic motion on a spring of force constant k = 500 N/m. Its maximum speed is 70 cm/s. (a) What is the total energy? (b) What is the amplitude of the oscillation?

21. A 3-kg object oscillating on a spring of force constant 2 kN/m has a total energy of 0.9 J. (a) What is the amplitude of the motion? (b) What is the maximum speed?

22. An object oscillates on a spring with an amplitude of 4.5 cm. Its total energy is 1.4 J. What is the force constant of the spring?

12-4 An Object on a Vertical Spring

23. A 2.5-kg object hangs from a vertical spring of force constant 600 N/m. It oscillates with an amplitude of 3 cm. When the object is at its maximum downward displacement, find (a) the total energy of the system, (b) the gravitational potential energy, and (c) the potential energy in the spring. (d) What is the maximum kinetic energy of the object? (Choose U = 0 when the object is in equilibrium.)

24. A 1.5-kg object that stretches a spring 2.8 cm from its natural length when hanging at rest oscillates with an amplitude of 2.2 cm. Find (a) the total energy of the system. At maximum downward displacement find (b) the gravitational potential energy and (c) the potential energy in the spring. (d) What is the maximum kinetic energy of the object? (Choose U = 0 when the object is in equilibrium.)

25. A 1.2-kg object hanging from a spring of force constant 300 N/m oscillates with a maximum speed of 30 cm/s. (a) What is its maximum displacement? When the object is at its maximum displacement, find (b) the total energy of the system, (c) the gravitational potential energy, and (d) the potential energy in the spring.

12-5 Pendulums

26. Find the length of a simple pendulum if the period is 5 s at a point where g = 9.81 m/s².

27. What would be the period of the pendulum in Problem 26 on the moon, where the acceleration due to gravity is one-sixth that on earth?

28. If the period of a pendulum 70 cm long is 1.68 s, what is the value of g at the location of the pendulum.

29. A pendulum set up in the stairwell of a 10-story building consists of a heavy weight suspended on a 34.0-m wire. If g = 9.81 m/s², what is the period of oscillation?

30. A small particle of mass m slides without friction in a spherical bowl of radius r. (a) Show that the motion of the particle is the same as if it were attached to a string of length r. (b) Figure 12-23 shows a particle of mass m_1 that is displaced a small distance s_1 from the bottom of the bowl, where s_1 is much smaller than r. A second particle

Figure 12-23 Problem 30.

of mass m_2 is displaced in the opposite direction a distance $s_2 = 3s_1$, where s_2 is also much smaller than r. If the particles are released at the same time, where do they meet? Explain.

31. A thin disk of mass 5 kg and radius 20 cm is suspended by a horizontal axis perpendicular to the disk through its rim. The disk is displaced slightly from equilibrium and released. Find the period of the subsequent simple harmonic motion.

32. A circular hoop of radius 50 cm is hung on a narrow horizontal rod and allowed to swing in the plane of the hoop. What is the period of its oscillation, assuming that the amplitude is small?

33. A 3-kg plane figure is suspended at a point 10 cm from its center of mass. When it is oscillating with small amplitude, the period of oscillation is 2.6 s. Find the moment of inertia I about an axis perpendicular to the plane of the figure through the pivot point.

12-6 General Motion near Equilibrium

There are no problems for this section.

12-7 Damped Oscillations

34. An oscillator has a Q factor of 200. By what percentage does its energy decrease during one period?

35. A 2-kg object oscillates with an initial amplitude of 3 cm on a spring of force constant $k = 400$ N/m. Find (*a*) the period and (*b*) the total initial energy. (*c*) If the energy decreases by 1 percent per period, find the damping constant b and the Q factor.

36. Show that the ratio of the amplitudes for two successive oscillations is constant for a damped oscillator.

37. An oscillator has a period of 3 s. Its amplitude decreases by 5 percent during each cycle. (*a*) By how much does its energy decrease during each cycle? (*b*) What is the time constant τ? (*c*) What is the Q factor?

12-8 Driven Oscillations and Resonance

38. Find the resonance frequency for each of the three systems shown in Figure 12-24.

39. A damped oscillator loses 2 percent of its energy during each cycle. (*a*) What is its Q factor? (*b*) If its resonance frequency is 300 Hz, what is the width of the resonance curve $\Delta\omega$ when the oscillator is driven?

40. A 2-kg object oscillates on a spring of force constant $k = 400$ N/m. The damping constant $b = 2.00$ kg/s. The system is driven by a sinusoidal force of maximum value

10 N and angular frequency $\omega = 10$ rad/s. (*a*) What is the amplitude of the oscillations? (*b*) If the driving frequency is varied, at what frequency will resonance occur? (*c*) Find the amplitude of oscillation at resonance. (*d*) What is the width of the resonance curve $\Delta\omega$?

Level II

41. A 3-kg object oscillates on a spring with an amplitude of 8 cm. Its maximum acceleration is 3.50 m/s². Find the total energy.

42. The period of an oscillating particle is 8 s, and its amplitude is 12 cm. At $t = 0$, it is at its equilibrium position. Find the distance traveled during the interval (*a*) $t = 0$ to $t = 2$ s, (*b*) $t = 2$ s to $t = 4$ s, (*c*) $t = 0$ to $t = 1$ s, and (*d*) $t = 1$ s to $t = 2$ s.

43. The period of an oscillating particle is 8 s. At $t = 0$, the particle is at $x = A = 10$ cm. (*a*) Sketch x as a function of t. (*b*) Find the distance traveled in the first second, the second second, the third second, and the fourth second after $t = 0$.

44. An object of mass m is supported by a vertical spring of force constant 1800 N/m. When pulled down 2.5 cm from equilibrium and released from rest, the object oscillates at 5.5 Hz. (*a*) Find m. (*b*) Find the amount the spring is stretched from its natural length when the object is in equilibrium. (*c*) Write expressions for the displacement x, the velocity v, and the acceleration a as functions of time t.

45. A spring is hanging vertically. An object of unknown mass is hung on the end of the unstretched spring and is released from rest. If it falls 3.42 cm before first coming to rest, find the period of the motion. (Assume $g = 9.81$ m/s².)

46. In heavy seas, the bow of a destroyer undergoes a simple harmonic vertical pitching motion with a period of 8.0 s and an amplitude of 2.0 m. (*a*) What is the maximum vertical velocity of the destroyer's bow? (*b*) What is its maximum acceleration? (*c*) An 80-kg sailor is standing on a scale in the bunkroom in the bow. What are the maximum and minimum readings on the scale in newtons?

47. Military specifications often call for electronic devices to be able to withstand accelerations of $10g = 98.1$ m/s². To make sure their products meet this specification, manufacturers test them using a shaking table that can vibrate a device at various specified frequencies and amplitudes. If a device is given a vibration of amplitude 1.5 cm, what should its frequency be?

Figure 12-24 Problem 38.

48. A block of wood slides on a frictionless horizontal surface. It is attached to a spring and oscillates with a period of 0.8 s. A second block rests on top of the first. The coefficient of static friction between the two blocks is 0.25. (a) If the amplitude of oscillation is 1 cm, will the block on top slip? (b) What is the greatest amplitude of oscillation for which the top block will not slip?

49. A child on a swing oscillates with a period of 3 s. The mass of the child and swing is 35 kg. The swing is pushed by a patient father such that the amplitude remains constant. At the bottom of the swing, the speed of the child and swing is 2.0 m/s. (a) What is the total energy of the child and swing? (b) If $Q = 20$, how much energy is dissipated during each oscillation of the swing? (c) What is the power input of the father? Note: Pushing a swing is usually not done sinusoidally. However, to maintain a steady amplitude, the energy lost per cycle due to damping must be replaced by an external energy source. This energy may come from the child's "pumping" or from someone else's pushing the swing.

50. A damped oscillator loses 2 percent of its energy during each cycle. (a) How many cycles elapse before half its original energy is dissipated? (b) What is the Q factor? (c) If the natural frequency is 100 Hz, what is the width of the resonance curve when the oscillator is driven?

51. An oscillator has a Q factor of 20. (a) By what fraction does the energy decrease during each cycle? (b) Use Equation 12-59 to find the percentage difference between ω' and ω_0. Hint: Use the approximation $(1 + x)^{1/2} \approx 1 + \frac{1}{2}x$ for small x.

52. For a child on a swing, the amplitude drops by a factor of $1/e$ in about eight periods if no energy is fed in. Estimate the Q factor for this system.

53. It has been stated that the vibrating earth has a resonance period of 54 min and a Q factor of about 400 and that after a large earthquake, the earth "rings" (continues to vibrate) for about 2 months. (a) Find the percentage of the energy of vibration lost to damping forces during each cycle. (b) Show that after n periods, the energy is $E_n = (0.984)^n E_0$, where E_0 is the original energy. (c) If the original energy of vibration of an earthquake is E_0, what is the energy after 2 days?

54. An object of mass m_1 sliding on a frictionless horizontal surface is attached to a spring of force constant k. It oscillates with an amplitude A. When the spring is at its greatest extension and the mass is instantaneously at rest, a second object of mass m_2 is placed on top of it. (a) What is the smallest value the coefficient of static friction μ_s can have if the second object is not to slip on the first? (b) Explain how the total energy E, the amplitude A, the angular frequency ω, and the period T of the system are changed by placing m_2 on m_1.

55. A spring of force constant $k = 250$ N/m is suspended from a rigid support. An object of mass 1 kg is attached at the bottom end and is released from rest when the spring is unstretched. (a) How far down does the object move before it starts up again? (b) How far below the starting point is the equilibrium position for the object? (c) What is the period of oscillation? (d) What is the speed of the ob-

ject when it first reaches its equilibrium position? When does it first reach its equilibrium position?

56. An object of mass 2.0 kg is attached to the top of a vertical spring that is anchored to the floor. The uncompressed length of the spring is 8.0 cm, and the equilibrium position of the object on the spring is 5.0 cm from the floor. When the object is resting at its equilibrium position, it is given a downward impulse with a hammer such that its initial speed is 0.3 m/s. (a) To what maximum height above the floor does the object eventually rise? (b) How long does it take for the object to reach its maximum height the first time? (c) Does the spring ever become uncompressed? What minimum initial velocity must be given to the object for the spring to be uncompressed at some time?

57. A spider of mass 0.36 g sits in the middle of its horizontal web, which sags 3.00 mm under its weight. Estimate the frequency of vertical vibration for this system.

58. The acceleration due to gravity g varies with location on the earth because of the earth's rotation and because the earth is not exactly spherical. This was first discovered in the seventeenth century, when it was noted that a pendulum clock carefully adjusted to keep correct time in Paris lost about 90 s/d near the equator. (a) Show that a small change in the acceleration of gravity Δg produces a small change in the period ΔT of a pendulum given by

$$\frac{\Delta T}{T} \approx -\frac{1}{2}\frac{\Delta g}{g}$$

(Use differentials to approximate ΔT and Δg.) (b) How great a change in g is needed to account for a change in the period of 90 s/d?

Level III

59. A 20-kg ball is hung on a steel wire of 1.00 mm diameter and 3.00 m length. When caused to vibrate vertically, it is found to have a frequency of 8.14 Hz. (a) Find the force constant k of the wire. (b) Show that Young's modulus for the wire is related to the force constant by $Y = kL/A$. (c) Find Young's modulus for this wire.

60. A pendulum clock that has run down to a very small amplitude gains 5 min each day. What angular amplitude should the pendulum have to keep the correct time?

61. An object of mass 2 kg resting on a frictionless horizontal surface is attached to a spring of force constant 600 N/m. A second object of mass 1 kg slides along the surface toward the first object at 6 m/s. (a) Find the amplitude of oscillation if the objects make a perfectly inelastic collision and remain together on the spring. What is the period of oscillation? (b) Find the amplitude and period of oscillation if the collision is elastic. (c) For each type of collision, write an expression for the position x as a function of time t for the object attached to the spring, assuming that the collision occurs at time $t = 0$. What is the impulse delivered to the 2-kg object in each case?

62. A small block of mass m_1 rests on a piston that is vibrating vertically with simple harmonic motion given by $y = A \sin \omega t$. (a) Show that the block will leave the piston if $\omega^2 A > g$. (*b*) If $\omega^2 A = 3g$ and $A = 15$ cm, at what time will the block leave the piston?

63. A block suspended from a spring is oscillating vertically with a frequency of 4 Hz and an amplitude of 7 cm. A very small bit of rock is placed on top of the oscillating block just as it reaches its lowest point. Assume that the rock has no effect on the oscillation. (*a*) At what distance above the block's equilibrium position does the rock lose contact with the block? (*b*) What is the speed of the rock when it leaves the block? (*c*) What is the greatest distance above the block's equilibrium position reached by the rock?

64. Show that for the situations in Figure 12-25*a* and *b*, the object oscillates with a frequency $f = (1/2\pi)\sqrt{k_{eff}/m}$, where k_{eff} is given by (*a*) $k_{eff} = k_1 + k_2$ and (*b*) $1/k_{eff} = 1/k_1 + 1/k_2$. *Hint:* Find the net force F on the object for a small displacement x and write $F = -k_{eff}x$. Note that in (*b*) the springs stretch by different amounts, the sum of which is x.

Figure 12-25   Problem 64.

(*a*)

(*b*)

65. A straight tunnel is dug through the earth as shown in Figure 12-26. Assume the walls of the tunnel are frictionless. (*a*) The gravitational force exerted by the earth on a particle of mass m at a distance r from the center of the earth when $r < R_E$ is $F_r = -(GmM_E/R_E^3)r$, where M_E is the mass of the earth and R_E is its radius. Show that the net force on a particle of mass m at a distance x from the middle of the tunnel is given by $F_x = -(GmM_E/R_E^3)x$, and that the motion of the particle is therefore simple harmonic motion. (*b*) Show that the period of the motion is given by $T = 2\pi\sqrt{R_E/g}$ and find its value in minutes. (This is the same period as that of a satellite orbiting near the surface of the earth, and it is independent of the length of the tunnel.)

Figure 12-26   Problem 65.

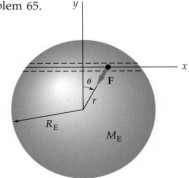

66. A simple pendulum of length L is released from rest from an angle θ_0. (*a*) Assuming the motion of the pendulum to be simple harmonic motion, find its speed as it passes through $\theta = 0$. (*b*) Using the conservation of energy, find this speed exactly. (*c*) Show that your results for parts (*a*) and (*b*) are the same when θ_0 is small. (*d*) Find the difference in your results for $\theta_0 = 0.20$ rad and $L = 1$ m.

67. A 3-kg sphere dropped in the air has a terminal speed of 25 m/s. (Assume the drag force is $-bv$.) The sphere is attached to a spring of force constant $k = 400$ N/m, and it oscillates with an initial amplitude of 20 cm. (*a*) What is Q? (*b*) When will the amplitude be 10 cm? (*c*) How much energy will have been lost when the amplitude is 10 cm?

68. A plane object has moment of inertia I about its center of mass. When pivoted at point P_1, as shown in Figure 12-27, it oscillates about the pivot with a period T. There is a second point P_2 on the opposite side of the center of mass about which the object can be pivoted so that the period of oscillation is also T. Show that $h_1 + h_2 = gT^2/4\pi^2$.

Figure 12-27   Problem 68.

69. A physical pendulum consists of a spherical bob of radius r and mass m suspended from a string (Figure 12-28). The distance from the center of the sphere to the point of support is L. When r is much less than L, such a pendulum is often treated as a simple pendulum of length L. (*a*) Show that the period for small oscillations is given by

$$T = T_0\sqrt{1 + \frac{2r^2}{5L^2}}$$

where $T_0 = 2\pi\sqrt{L/g}$ is the period of a simple pendulum of length L. (*b*) Show that when r is much smaller than L, the period is approximately $T \approx T_0(1 + r^2/5L^2)$. (*c*) If $L = 1$ m and $r = 2$ cm, find the error when the approximation $T = T_0$ is used for this pendulum. How large must the radius of the bob be for the error to be 1 percent?

Figure 12-28   Problem 69.

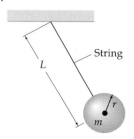

70. A damped oscillator has a frequency ω' that is 10 percent less than its undamped frequency. (*a*) By what factor is its amplitude decreased during each oscillation? (*b*) By what factor is its energy reduced during each oscillation?

71. Show by direct substitution that Equation 12-58 is a solution of Equation 12-46.

72. A block of mass m on a horizontal table is attached to a spring of force constant k as shown in Figure 12-29. The coefficient of kinetic friction between the block and the table is μ_k. The spring is stretched a distance A and released. (*a*) Apply Newton's second law to the block to obtain an equation for its acceleration d^2x/dt^2 for the first half cycle, during which the block is moving to the left. Show that the resulting equation can be written $d^2x'/dt^2 = -\omega^2 x'$, where $x = 0$ at the equilibrium position of the spring, and $x' = x - x_0$, with $x_0 = \mu_k mg/k = \mu_k g/\omega^2$. (*b*) Repeat part (*a*) for the second half cycle as the block moves to the right, and show that $d^2x''/dt^2 = -\omega^2 x''$, where $x'' = x + x_0$ and x_0 has the same value. (*c*) Sketch the first few cycles for $A = 10x_0$.

Figure 12-29   Problem 72.

73. In this problem, you will derive the expression for the average power delivered by a driving force to a driven oscillator (Figure 12-22). (*a*) Show that the instantaneous power input of the driving force is given by

$$P = Fv = -A\omega F_0 \cos \omega t \sin (\omega t - \delta)$$

(*b*) Use the trigonometric identity $\sin (\theta_1 - \theta_2) = \sin \theta_1 \cos \theta_2 - \cos \theta_1 \sin \theta_2$ to show that this can be written

$$P = A\omega F_0 \sin \delta \cos^2 \omega t - A\omega F_0 \cos \delta \cos \omega t \sin \omega t$$

(*c*) Show that the average value of the second term in your result for part (*b*) over one or more periods is zero and that therefore

$$P_{av} = \tfrac{1}{2} A\omega F_0 \sin \delta$$

(*d*) From Equation 12-66 for tan δ, construct a right triangle in which the side opposite the angle δ is $b\omega$ and the side adjacent is $m(\omega_0^2 - \omega^2)$, and use this triangle to show that

$$\sin \delta = \frac{b\omega}{\sqrt{m^2(\omega_0^2 - \omega^2)^2 + b^2\omega^2}} = \frac{b\omega A}{F_0}$$

(*e*) Use your result for part (*d*) to eliminate ωA so that the average power input can be written

$$P_{av} = \frac{1}{2}\frac{F_0^2}{b} \sin^2 \delta$$

$$= \frac{1}{2}\frac{b\omega^2 F_0^2}{m^2(\omega_0^2 - \omega^2)^2 + b^2\omega^2}$$

12-67

74. In this problem you are to use the result of Problem 73 to derive Equation 12-61, which relates the width of the resonance curve to the Q value when the resonance is sharp. At resonance, the denominator in Equation 12-67 is $b^2\omega_0^2$ and P_{av} has its maximum value. For a sharp resonance, the variation in ω in the numerator in Equation 12-67 can be neglected. Then the power input will be half its maximum value at the values of ω, for which the denominator is $2b^2\omega_0^2$. (*a*) Show that ω then satisfies

$$m^2(\omega - \omega_0)^2(\omega + \omega_0)^2 = b^2\omega_0^2$$

(*b*) Using the approximation $\omega + \omega_0 \approx 2\omega_0$, show that

$$\omega - \omega_0 \approx \pm\frac{b}{2m}$$

(*c*) Show from Equation 12-56, that Q is related to b and m by

$$Q = \frac{m\omega_0}{b}$$

(*d*) Combine the results of (*b*) and (*c*) to show that there are two values of ω for which the power input is half that at resonance, and they are given by

$$\omega_1 = \omega_0 - \frac{\omega_0}{2Q} \quad \text{and} \quad \omega_2 = \omega_0 + \frac{\omega_0}{2Q}$$

Therefore, $\omega_2 - \omega_1 = \Delta\omega = \omega_0/Q$, which is equivalent to Equation 12-61.

Chapter 13

Waves on a String

A bass viol made by Henry Jaye in London, 1619.

Wave motion can be thought of as the transport of energy and momentum from one point in space to another without the transport of matter. In mechanical waves, such as waves on a string or sound waves in air, the energy and momentum are transported by means of a disturbance in the medium. A violin string is plucked or bowed, and the disturbance in the shape of the string is propagated along the string. At the same time, the vibrating string produces a slight change in the pressure of the adjacent air, and this pressure change is propagated as a sound wave through the air. In both cases, the disturbance is propagated because of the elastic properties of the medium. On the other hand, in electromagnetic waves (such as light, radio, television, or x rays) the energy and momentum are carried by electric and magnetic fields that can propagate through a vacuum. Light waves and other electromagnetic waves, which will be studied in Chapters 29 to 33, are produced by oscillating electric charges in atoms and molecules, or perhaps a transmitting radio antenna.

Although the variety of wave phenomena observed in nature is immense, many features are common to all kinds of waves, and others are shared by a wide range of wave phenomena. In this chapter, we will study waves on strings, which are common in our everyday experience and are easily visualized. In Chapter 14, we will study sound waves, which are also common in our everyday experience but are not as easily visualized. Many of the ideas and results discussed here will be applied later when we study more abstract kinds of waves such as light and other electromagnetic waves.

(a)

(b)

Figure 13-1 (a) A wave pulse moving to the right on a stretched string. When the pulse arrives at the rigid support, it is reflected and inverted. (b) Photograph of a similar pulse on a spring.

13-1   Wave Pulses

When a string (or spring) stretched under tension is given a flip, as in Figure 13-1, its shape changes over time in a regular way. The bump produced by the flip of the string travels down the string as a <u>wave pulse</u>. The disturbance in the medium, in this case, is the change in the shape of the string from its equilibrium shape, which is that of a taut string. The wave pulse travels down the string at a definite speed that depends on the <u>tension</u> in the string and on its <u>mass density</u> (mass per unit length). As it moves, the pulse may change shape. For example, it <u>may gradually spread out</u>. This effect is called **dispersion.** We will usually neglect dispersion in this chapter and assume that the wave pulse travels with an approximately constant shape. The fate of the pulse at the other end of the string depends on how the string is fastened there. If it is tied to a rigid support, as in Figure 13-1, the pulse will be reflected and will return inverted. When the pulse arrives at the fixed support, it exerts an upward force on the support. The fixed support exerts an equal and opposite downward force on the string, causing the pulse to invert upon reflection.

In Figure 13-2, the string is tied to a smooth ring of negligible mass that is free to move vertically on a post. This arrangement approximates a free end for the string. When the pulse arrives, it exerts an upward force on the ring and the ring accelerates upward. The ring overshoots the height of the pulse, causing a reflected pulse that is not inverted.

If the string is tied to another string of different mass density, part of the pulse will be transmitted and part will be reflected. If the second string is

Figure 13-2 A free end for a string can be approximated by tying the string to a ring that moves on a frictionless post. A wave pulse traveling on the string is reflected at the free end without inversion.

Figure 13-3 (*a*) A wave pulse traveling on a light string attached to a heavier string in which the wave speed is smaller. The reflected pulse is inverted, whereas the transmitted pulse is not. (*b*) Photograph of a similar pulse on a light spring attached to a heavier spring. (*c*) A wave pulse traveling on a heavy string attached to a light string in which the wave speed is greater. In this case, the reflected pulse is not inverted. (*d*) Photograph of a similar pulse on a heavy spring attached to a lighter spring.

heavier than the first, the reflected part of the pulse will be inverted (Figure 13-3*a*). If the second string is lighter, the reflected pulse will not be inverted (Figure 13-3*c*). In either case, the transmitted pulse is not inverted.

It is important to understand that it is not the mass elements of the string that are transported in wave motion but rather the disturbance in the shape of the string caused by flipping one end. The mass elements of the string, in fact, move in a direction perpendicular to the string and thus perpendicular to the direction of motion of the pulse (Figure 13-4). A wave in which the disturbance is perpendicular to the direction of propagation is called a **transverse wave.** A wave in which the disturbance is parallel to the

Figure 13-4 Three successive drawings of a wave on a string traveling to the right. An element of the string moves up and down in simple harmonic motion.

direction of propagation is called a **longitudinal wave.** Sound consists of longitudinal waves. A longitudinal pulse in a spring analogous to a sound pulse can be produced by suddenly compressing the spring as shown in Figure 13-5a. Figure 13-5b shows a transverse wave produced in the same spring.

Figure 13-5 (a) Longitudinal wave pulse in a spring. The disturbance is in the direction of the motion of the wave. (b) Transverse wave pulse on the same spring. The disturbance is perpendicular to the direction of the motion of the wave.

(a)

(b)

Water waves are neither completely transverse nor completely longitudinal but a combination of the two. Figure 13-6 shows the motion of a water particle in a water wave. The water particle moves in a circular path with both transverse and longitudinal components.

Figure 13-6 Surface waves on water. The water particles on the surface move in circular paths having both longitudinal and transverse components.

Figure 13-7 shows a pulse on a string at time $t = 0$. The shape of the string at this time can be represented by some function $y = y(x)$. At some later time, the pulse is farther down the string, so its shape then is some other function of x. We will neglect dispersion and assume that the pulse does not vary in shape. Let us introduce a new coordinate system with an origin O' that moves with the speed of the pulse v. In this reference frame,

Figure 13-7 A wave pulse moving without change in shape in the positive x direction with a speed v relative to the origin O. In the primed coordinate system moving with the pulse, the wave function is $y' = y'(x')$ at all times. In the original, unprimed system, the wave function is $y = y(x - vt)$.

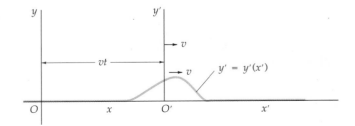

the pulse is stationary. The shape of the string is $y' = y'(x')$ for all times. The coordinates of the two reference frames are related by

$$y = y'$$

and

$$x = x' + vt$$

Thus, the shape of the string in the original frame O can be written

$$y = y(x - vt) \qquad \text{wave moving right} \qquad \text{13-1}$$

This same line of reasoning for a pulse moving to the left leads to

$$y = y(x + vt) \qquad \text{wave moving left} \qquad \text{13-2}$$

In both of these expressions, v is the speed of propagation of the wave. The function $y = y(x - vt)$ is called the **wave function**. For waves on a string, the wave function represents the transverse displacement of the string at point x and time t. Hence, the wave function is a function of the two variables x and t.

Figure 13-8 shows wave pulses moving in opposite directions on a string. The shape of the string when they meet can be found by adding the displacements produced by each pulse separately as shown in the figure. The property of wave motion that the resultant wave is the sum of two or more individual waves is called the **principle of superposition**. Mathematically, if $y_1(x - vt)$ is the wave function for the pulse moving to the right and $y_2(x + vt)$ is that for the pulse moving to the left, the total wave function when there are two pulses on the string is just the algebraic sum of the individual wave functions:

$$y(x, t) = y_1(x - vt) + y_2(x + vt) \qquad \text{13-3}$$

The principle of superposition holds only for small wave pulses for which the height of the pulse is small compared with its length. In Section 13-8, we will see that the principle of superposition follows from the fact that the wave equation is linear for small transverse displacements.

In the special case where two pulses are identical except that one is inverted relative to the other, as in Figure 13-8b, there will be one time when the pulses exactly overlap and add to zero. At this time the string is horizontal, but it is not at rest. A short time later the pulses emerge, each continuing in its original direction. Superposition is a characteristic property of wave motion. There is no analogous situation in particle motion; that is, two particles never overlap or add together in this way. Superposition is unique to wave motion.

Questions

1. Give some examples of wave pulses that occur in nature in addition to those mentioned in the text. In each case, explain what kind of disturbance from equilibrium occurs.

2. Consider a long line of cars equally spaced by one car length and moving slowly with the same speed. One car suddenly slows to avoid a dog and then speeds up until it is again one car length behind the car ahead. Discuss how the space between cars propagates back along the line. How is this like a wave pulse? Is there any transport of energy? What does the speed of propagation depend on?

3. When two waves moving in opposite directions superimpose as in Figure 13-8, does either impede the progress of the other?

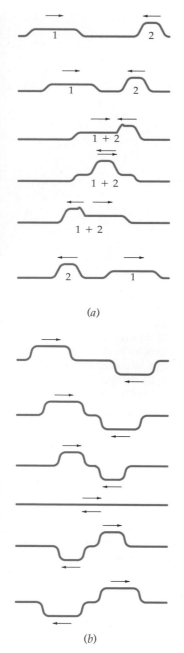

Figure 13-8 Wave pulses moving in opposite directions on a string. The shape of the string when the pulses meet is found by adding the displacements of each separate pulse. This property of wave motion is called superposition. (a) Superposition of pulses having displacements in the same direction. (b) Superposition of pulses having opposite displacements. Here the algebraic addition of the displacements amounts to the subtraction of the magnitudes.

Ocean waves breaking as they approach shallow water. The speed of a water wave depends on the depth of the water. As the depth of the water suddenly decreases, as evident here by the change in color, the back of the wave overtakes the front and the wave "breaks."

Figure 13-9 A wave pulse moving with a speed v along a string. If the pulse is small enough, the tension in the string does not change.

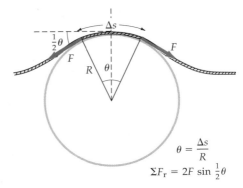

$$\theta = \frac{\Delta s}{R}$$

$$\Sigma F_r = 2F \sin \tfrac{1}{2}\theta$$

Figure 13-10 In a frame in which the wave pulse of Figure 13-9 is at rest, the string is moving with a speed v to the left. A small segment of the string of length Δs is moving in a circular arc of radius R. The centripetal acceleration of the segment is provided by the radial components of the tension.

13-2 Speed of Waves

A general property of waves is that their speed depends on the properties of the medium but is independent of the motion of the source of the waves relative to the medium. For example, the speed of a wave on a string depends only on the properties of the string. Similarly, the speed of a sound wave produced by a train whistle depends only on the properties of the air and not on the motion of the train.

If we send wave pulses down a long rope, we can easily demonstrate that the speed of propagation of the wave pulses increases as we increase the tension in the rope. Futhermore, if we have two ropes, a light rope and a heavy rope, under the same tension, wave pulses will propagate more slowly in the heavy rope. Thus, the speed of propagation v of waves on a string or rope is related to the tension F and the mass per unit length μ. (We use F for tension here rather than T so as not to confuse the tension with the period T used in the next section.) We can obtain this relation using Newton's laws.

Figure 13-9 shows a pulse traveling along a string with a speed v to the right. We assume that the pulse is small compared to the length of the string so that, to a good approximation, the tension is constant along the string and is the same as when there is no pulse. It is convenient to consider the pulse in a reference frame moving with speed v to the right. In this frame, the pulse is stationary and the string moves with a speed v to the left. Figure 13-10 shows a small segment of the string of length Δs. If the segment is small enough, we can consider it to be part of a circular arc of radius R. The segment is thus moving in a circle of radius R with a speed v, and it has a centripetal acceleration v^2/R. Let the angle subtended by the string be θ:

$$\theta = \frac{\Delta s}{R}$$

The forces acting on the segment are the tension F at each end. The horizontal components of these forces are equal and opposite and thus cancel. The vertical components of these forces point toward the center of the circular arc. The sum of these radial forces provides the centripetal acceleration. The net radial force acting on the segment is

$$\Sigma F_r = 2F \sin \tfrac{1}{2}\theta \approx 2F(\tfrac{1}{2}\theta) = F\theta$$

where we have assumed that θ is small enough to permit the approximation $\sin \frac{1}{2}\theta \approx \frac{1}{2}\theta$. If μ is the mass per unit length of the string, the mass of a segment of length $\Delta s = R\theta$ is

$$m = \mu\,\Delta s = \mu R\theta$$

Setting the net radial force equal to the mass times the centripetal acceleration gives

$$F\theta = \mu R\theta\,\frac{v^2}{R}$$

Canceling the common factor θ and solving for v, we obtain

$$v = \sqrt{\frac{F}{\mu}}$$

Since this speed is independent of R and θ, this result holds for all segments of the string. However, the derivation depends on angle θ being small, which will be true if the height of the pulse is small compared to its length or, alternatively, if the slope of the string at any point is small.

In the original frame, the string is fixed, and the pulse moves with speed

$$v = \sqrt{\frac{F}{\mu}}$$

13-4   *Speed of waves on a string*

The dependence of the speed on F/μ is not surprising when we realize that it is essentially the tension that accelerates the mass elements in a rope to provide the wave pulse.

Example 13-1

The tension in a string is provided by hanging an object of mass 3 kg at one end as shown in Figure 13-11. The length of the string is 2.5 m and its mass is 50 g. What is the speed of waves on the string?

The tension in the string is

$$F = mg = (3\ \text{kg})(9.81\ \text{N/kg}) = 29.4\ \text{N}$$

The mass per unit length is

$$\mu = \frac{m}{L} = \frac{0.05\ \text{kg}}{2.5\ \text{m}} = 0.02\ \text{kg/m}$$

The speed is therefore

$$v = \sqrt{\frac{F}{\mu}} = \sqrt{\frac{29.4\ \text{N}}{0.02\ \text{kg/m}}} = 38.3\ \text{m/s}$$

$m = 3$ kg

Figure 13-11   Example 13-1. The tension in the string is provided by the weight of the block.

Exercise

If the 3-kg mass in Example 13-1 is replaced with a 6-kg mass, what is the speed of the waves on the string?   (Answer: 54.2 m/s)

Question

4.   Two strings are stretched between the same two posts. One weighs twice as much as the other. How should their tensions be adjusted so that waves travel along each at the same speed?

13-3   Harmonic Waves

If we move one end of a string up and down in simple harmonic motion (for example, by attaching it to a vibrating tuning fork), a sinusoidal wave train propagates along the string. Such a wave is called a **harmonic wave.** The shape of the string at some instant in time is that of a sine function, as shown in Figure 13-12. (Again, whether this is a sine function or a cosine function depends merely on where the origin is chosen on the *x* axis.) Such a figure can be obtained by taking a snapshot of the string. The distance between two successive wave crests is called the **wavelength** λ. The wavelength is the distance in space after which the wave repeats itself. As the wave propagates down the string, each point on the string moves up and down, that is, perpendicular to the direction of propagation, in simple harmonic motion with the frequency *f* of the tuning fork or whatever agent is driving the end of the string. There is a simple relation among the frequency *f*, the wavelength λ, and the speed *v* of the harmonic wave. During one period $T = 1/f$, the wave moves a distance of one wavelength, so the speed is given by

$$v = \frac{\lambda}{T} = f\lambda \qquad \text{13-5}$$

Since Equation 13-5 arises simply from the definitions of wavelength and frequency, it applies to all types of harmonic waves. Because the speed of wave propagation is determined by the properties of the medium, the wavelength is determined by the frequency of the source through $\lambda = v/f$. The greater the frequency, the smaller the wavelength.

The sine function that describes the displacement shown in Figure 13-12 is

$$y(x) = A \sin kx \qquad \text{13-6}$$

where *A* is the **amplitude** and *k* is a constant called the **wave number.** The wave number is related to the wavelength. If we move from some point x_1 to another point a distance one wavelength away, $x_2 = x_1 + \lambda$, the argument of the sine function changes by 2π. We thus have

$$k(x_1 + \lambda) = kx_1 + 2\pi$$

$$k\lambda = 2\pi$$

or

$$k = \frac{2\pi}{\lambda} \qquad \text{13-7}$$

To describe a wave that is traveling to the right with a speed *v*, we replace *x* in Equation 13-6 with $x - vt$ as we did with wave pulses in Section 13-1. Thus, the wave function for a wave traveling with a speed *v* to the right is written

$$y(x, t) = A \sin k(x - vt) = A \sin (kx - kvt)$$

or

Harmonic wave function

$$y(x, t) = A \sin (kx - \omega t) \qquad \text{13-8}$$

where

$$\omega = kv \qquad \text{13-9}$$

is the angular frequency, which is related to the frequency *f* and the period *T*, as in Chapter 12, by

$$\omega = 2\pi f = \frac{2\pi}{T} \qquad \text{13-10}$$

Figure 13-12   Harmonic wave at some instant in time. *A* is the amplitude and λ is the wavelength. For waves on a string, this figure can be obtained by taking a snapshot of the string.

Substituting $\omega = 2\pi f$ into Equation 13-9 and using $k = 2\pi/\lambda$, we obtain

$$2\pi f = kv = \frac{2\pi}{\lambda} v$$

or

$$v = f\lambda$$

which is Equation 13-5. Equation 13-8 can also be written in terms of the period T and the wavelength λ. Substituting in $k = 2\pi/\lambda$ and $\omega = 2\pi/T$, we obtain

$$y(x,\, t) = A \sin \left[2\pi \left(\frac{x}{\lambda} - \frac{t}{T} \right) \right]$$ 13-11   *Harmonic wave function in terms*
λ and T

From this form of the equation we can see that if t advances by one period T or if x advances by one wavelength λ, the argument of the sine function changes by 2π and $y(x,\, t)$ has the same value as before.

Note that a string segment at a fixed point x moves up and down with simple harmonic motion with angular frequency ω.

Example 13-2

The wave function for a harmonic wave on a string is

$$y(x,\, t) = 0.03 \sin (2.2x - 3.5t)$$

where y and x are in meters and t is in seconds. Find the amplitude, the wavelength, the frequency, the period, and the speed of the wave.

Comparing this wave function with Equation 13-8, we can see that the amplitude $A = 0.03$ m, the wave number $k = 2.2$ m^{-1}, and the angular frequency $\omega = 3.5$ s^{-1}. Then the wavelength $\lambda = 2\pi/k = 2.86$ m, and the period $T = 2\pi/\omega = 1.80$ s. The speed of the wave is therefore

$$v = f\lambda = \frac{\lambda}{T} = \frac{2.86 \text{ m}}{1.80 \text{ s}} = 1.59 \text{ m/s}$$

Figure 13-13   That waves transmit energy is indicated by the upward motion of the weight when the wave pulse arrives.

13-4   Energy Transmitted by Waves

As a wave moves along a string, energy is transmitted. We can demonstrate that energy is transported by a wave pulse by hanging a weight on a string under tension and giving the string a flip at one end (Figure 13-13). When the pulse arrives at the weight, the weight is momentarily lifted. Thus, the energy introduced by the flip at the end of the string is transmitted along the string and is received by the weight.

Figure 13-14 shows a harmonic wave of amplitude A and angular frequency ω generated at the left by a vibrating tuning fork. Each element of the string is undergoing simple harmonic motion of amplitude A and angular frequency ω. In Chapter 12, we saw that the total energy of a mass oscillating on a spring is $\frac{1}{2}kA^2$, where $k = m\omega^2$ is the force constant and A is the amplitude. (Do not confuse the force constant k with the wave number k.) Applying this to a segment of the string of mass $\Delta m = \mu \, \Delta x$ oscillating perpendicular to the string with an amplitude A, we have

$$\Delta E = \tfrac{1}{2}(\Delta m)\omega^2 A^2 = \tfrac{1}{2}\mu\omega^2 A^2 \, \Delta x$$ 13-12

Figure 13-14   The vibrating tuning fork at the left generates a harmonic wave that travels to the right along the string. A segment of the string of mass $\Delta m = \mu \, \Delta x$ is oscillating perpendicular to the string with simple harmonic motion. The energy of this segment is the energy of simple harmonic motion, which is $\frac{1}{2}kA^2 = \frac{1}{2}(\Delta m)\, \omega^2 y A^2$, where A is the amplitude and $k = \Delta m \, \omega^2$ is the effective force constant.

Figure 13-15 Energy transmitted by a wave on a string. (a) At time t_1, the wave has just reached point P_1. The string to the left of this point has energy due to the simple harmonic motion of its elements, whereas the string to the right of P_1 has no energy. (b) In an additional time Δt, the wave travels a distance $v \, \Delta t$. The energy transmitted past point P_1 in time Δt is the energy in the length $v \, \Delta t$.

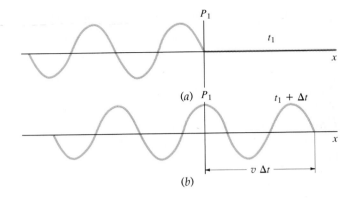

In Figure 13-15a, a harmonic wave traveling to the right has just reached point P_1 at time t_1. Each segment of string Δx to the left of P_1 has an energy ΔE given by Equation 13-12. After a time Δt, the wave travels a distance $\Delta x = v \, \Delta t$ as shown in Figure 13-15b. The energy transmitted past point P_1 during this time is thus

$$\Delta E = \tfrac{1}{2}\mu\omega^2 A^2 \, \Delta x = \tfrac{1}{2}\mu\omega^2 A^2 v \, \Delta t \qquad 13\text{-}13$$

The rate at which energy is transmitted is the power transmitted:

Power transmitted by a harmonic wave

$$P = \frac{dE}{dt} = \tfrac{1}{2}\mu\omega^2 A^2 v \qquad 13\text{-}14$$

Thus, the power transmitted is proportional to the square of the amplitude, to the square of the frequency, and to the wave speed.

Example 13-3

Waves of wavelength 35 cm and amplitude 1.2 cm move along a 15-m string that has a mass of 80 g and a tension of 12 N. (a) What is the total energy of the waves in the string? (b) Find the power transmitted past a given point on the string.

(a) The linear mass density of the string is $\mu = m/L = (0.08 \text{ kg})/(15 \text{ m}) = 5.33 \times 10^{-3}$ kg/m. The speed of the waves on the string is then

$$v = \sqrt{\frac{F}{\mu}} = \sqrt{\frac{12 \text{ N}}{5.33 \times 10^{-3} \text{ kg/m}}} = 47.4 \text{ m/s}$$

The angular frequency of the waves is

$$\omega = 2\pi f = 2\pi \frac{v}{\lambda} = 2\pi \frac{47.4 \text{ m/s}}{0.35 \text{ m}} = 851 \text{ rad/s}$$

The total energy of the waves on the string is then given by Equation 13-13:

$$\Delta E = \tfrac{1}{2}\mu\omega^2 A^2 \, \Delta x$$

$$= \tfrac{1}{2}(5.33 \times 10^{-3} \text{ kg/m})(851 \text{ rad/s})^2(0.012 \text{ m})^2(15 \text{ m}) = 4.17 \text{ J}$$

(b) The power transmitted past a point on the string is

$$P = \tfrac{1}{2}\mu\omega^2 A^2 v$$

$$= \tfrac{1}{2}(5.33 \times 10^{-3} \text{ kg/m})(851 \text{ rad/s})^2(0.012 \text{ m})^2(47.4 \text{ m/s}) = 13.2 \text{ W}$$

13-5   Superposition and Interference of Harmonic Waves

The superposition of harmonic waves is called **interference.** The interference of harmonic waves depends on the phase difference between the waves. Let $y_1(x, t)$ be the wave function for a harmonic wave traveling to the right with amplitude A, angular frequency ω, and wave number k:

$$y_1 = A \sin (kx - \omega t) \qquad\qquad 13\text{-}15$$

For this wave function, we have chosen $t = 0$ when the displacement is zero at $x = 0$. [This choice is convenient, but it is not mandatory. If, for example, we chose $t = 0$ when the displacement was maximum at $x = 0$, we would write $y_1 = A \cos (kx - \omega t) = A \sin (kx - \omega t - \pi/2)$.] If we have a second harmonic wave that is also traveling to the right with the same amplitude, frequency, and wave number and we have already chosen the zero time, the general equation for its wave function can be written

$$y_2 = A \sin (kx - \omega t + \delta) \qquad\qquad 13\text{-}16$$

where δ is the phase constant. The two waves described by Equations 13-15 and 13-16 differ in phase by δ. Figure 13-16 shows a plot of two wave functions versus position for a fixed time $t = 0$. The resultant wave is the sum

$$y_1 + y_2 = A \sin (kx - \omega t) + A \sin (kx - \omega t + \delta) \qquad 13\text{-}17$$

If the phase difference is zero, that is, if the waves are in phase, the resultant wave has an amplitude twice that of either wave:

$$y_1 + y_2 = 2A \sin (kx - \omega t) \qquad \delta = 0$$

This type of superposition is called **constructive interference** (Figure 13-17). On the other hand, if the phase difference is π rad $= 180°$, the waves are out of phase. Then

$$y_1 + y_2 = A \sin (kx - \omega t) + A \sin (kx - \omega t + \pi)$$
$$= A \sin (kx - \omega t) - A \sin (kx - \omega t)$$

or

$$y_1 + y_2 = 0 \qquad \delta = \pi \text{ rad} = 180°$$

where we have used

$$A \sin (kx - \omega t + \pi) = -A \sin (kx - \omega t)$$

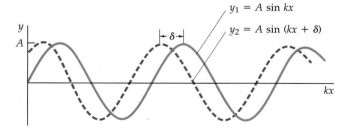

Figure 13-16   Displacement versus position for two harmonic waves having the same amplitude, frequency, and wavelength but differing in phase by δ.

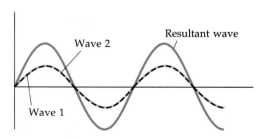

Figure 13-17   Constructive interference. When two waves are in phase, they add. The amplitude of the resultant wave is the sum of the amplitudes of the individual waves. For waves of equal amplitude, as shown here, the resultant amplitude is twice that of either wave.

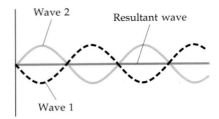

Wave 2   Resultant wave

Wave 1

Figure 13-18 Destructive interference. When two waves have a phase difference of π, the amplitude of the resultant wave is the difference of the amplitudes of the individual waves. If the original waves have equal amplitudes, they cancel completely, giving zero for the amplitude of the resultant wave.

When the waves differ in phase by 180° (π rad) or any odd integer times 180°, the waves cancel if their amplitudes are equal. This type of superposition is called **destructive interference** (Figure 13-18).

For a general phase difference δ, we can simplify Equation 13-17 by using the trigonometric identity

$$\sin \theta_1 + \sin \theta_2 = 2 \cos \tfrac{1}{2}(\theta_1 - \theta_2) \sin \tfrac{1}{2}(\theta_1 + \theta_2) \qquad 13\text{-}18$$

For this case, $\theta_1 = kx - \omega t$ and $\theta_2 = kx - \omega t + \delta$ so that

$$\tfrac{1}{2}(\theta_1 - \theta_2) = -\tfrac{1}{2}\delta$$

and

$$\tfrac{1}{2}(\theta_1 + \theta_2) = kx - \omega t + \tfrac{1}{2}\delta$$

Thus, Equation 13-17 becomes

$$y_1 + y_2 = 2A \cos \tfrac{1}{2}\delta \sin (kx - \omega t + \tfrac{1}{2}\delta) \qquad 13\text{-}19$$

where we have used $\cos (-\tfrac{1}{2}\delta) = \cos \tfrac{1}{2}\delta$. From Equation 13-19, we see that, in general, the superposition of two harmonic waves results in another harmonic wave having the same wave number and frequency. It differs in phase from both the original waves, and its amplitude is $2A \cos \tfrac{1}{2}\delta$. If the two waves are in phase, the phase difference $\delta = 0$, $\cos 0 = 1$, and the amplitude of the resultant wave is $2A$. If they are 180° out of phase, $\delta = \pi$ rad, $\cos (\pi/2) = 0$, and the amplitude of the resultant wave is zero.

Exercise

Two waves with the same frequency, wavelength, and amplitude are traveling in the same direction. If they differ in phase by $\pi/2$ and each has amplitude of 4.0 cm, what is the amplitude of the resultant wave? (Answer: 5.66 cm)

13-6   Standing Waves

When waves are confined in space, like waves on a piano string, there are reflections at both ends, and there are therefore waves traveling in both directions. These waves combine according to the principle of superposition. For a given string, there are certain frequencies for which superposition results in a stationary vibration pattern called a **standing wave.**

String Fixed at Both Ends

If we fix both ends of a long string and move a portion of the string up and down with simple harmonic motion of small amplitude, we find that at certain frequencies standing-wave patterns such as those shown in Figure 13-19 are produced. The frequencies that produce these patterns are called the **resonance frequencies** of the string system. The lowest resonance frequency is called the fundamental frequency f_1. It produces the standing-wave pattern shown in Figure 13-19a which is called the **fundamental** mode of vibration or the **first harmonic.** The second lowest frequency f_2 produces the pattern shown in Figure 13-19b. This mode of vibration has twice the fundamental frequency and is called the second harmonic. The third lowest frequency f_3 is three times the fundamental frequency, and it produces the third harmonic pattern shown in Figure 13-19c. We note from Figure 13-19 that, for each harmonic, there are certain points on the string that do not move. For example, the midpoint in Figure 13-19b does not move. Such points are called **nodes.** Midway between each pair of nodes is a point of maximum amplitude of vibration called an **antinode.** Both fixed ends of the string are, of course, nodes. If one end is attached to a tuning fork rather

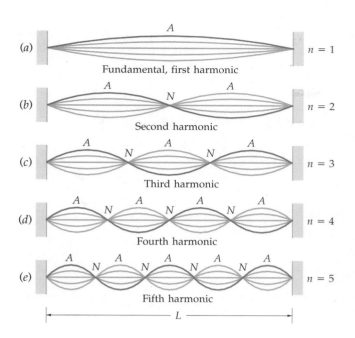

(a) *A*   *n* = 1
Fundamental, first harmonic

(b) *A*   *N*   *A*   *n* = 2
Second harmonic

(c) *A*   *N*   *A*   *N*   *A*   *n* = 3
Third harmonic

(d) *A*   *N*   *A*   *N*   *A*   *N*   *A*   *n* = 4
Fourth harmonic

(e) *A*   *N*   *A*   *N*   *A*   *N*   *A*   *N*   *A*   *n* = 5
Fifth harmonic

L

Figure 13-19   Standing waves on a string that is fixed at both ends. The points labeled *A* are antinodes and those labeled *N* are nodes. In general, the *n*th harmonic has *n* antinodes.

than being fixed, it will still be approximately a node because the amplitude of the vibration at that end is much smaller than the amplitude at the antinodes. We note that the fundamental or first harmonic has one antinode, the second harmonic has two antinodes, and so on.

We can relate the resonance frequencies to the wave speed in the string and the length of the string. We can see from Figure 13-19 that the length of the string *L* equals one-half wavelength for the first harmonic, two half wavelengths for the second harmonic, $\frac{3}{2}\lambda$ for the third harmonic, and so forth. In general, for the *n*th harmonic, we have

$$L = n \frac{\lambda_n}{2} \qquad n = 1, 2, 3, \dots$$

13-20   *Standing-wave condition, both ends fixed*

The first four harmonics of standing waves on a string fixed at both ends.

A classic Steinway grand piano. The strings vibrate when struck by the hammers, which are controlled by the keys. The longer strings (bottom of photo) vibrate at lower frequencies than the shorter strings (top of photo).

This result is known as the **standing-wave condition.** We can find the frequency of the nth harmonic from the fact that the wave speed v equals the frequency f times the wavelength. Thus,

nth harmonic =
nth nodes

$$f_n = \frac{v}{\lambda_n} = \frac{v}{(2L/n)}$$

VELOCITY
ᶻᴸᴱᴺᴳᵀᴴ / node #

or

Resonance frequencies, both ends fixed

$$f_n = n\frac{v}{2L} = nf_1 \qquad n = 1, 2, 3, \ldots \qquad 13\text{-}21$$

where

$$f_1 = \frac{v}{2L}$$

is the fundamental frequency. Since the wave speed v is $\sqrt{F/\mu}$, where F is the tension in the string and μ is its mass per unit length,

$$f_1 = \frac{1}{2L}\sqrt{\frac{F}{\mu}} \qquad 13\text{-}22$$

An easy way to remember the resonance frequencies given by Equation 13-21 is to sketch Figure 13-19 to remind yourself of the standing-wave condition $L = n\lambda_n/2$ and then use $f = v/\lambda$.

We can understand the production of standing waves in terms of resonance. Let us consider a string of length L that is attached at one end to a vibrating tuning fork, or some other vibrating body, and is fixed at the other end. The first wave sent out by the vibrating tuning fork travels down the string at a distance L to the fixed end, where it is reflected and inverted. It then travels back to the tuning fork, where it is again reflected at the tuning fork. It is again inverted and starts back down the string. If the time it takes for the wave to travel the complete distance $2L$ is exactly equal to the period of the vibrating fork, the twice-reflected wave will exactly overlap the second wave produced by the fork. The two waves will interfere constructively, which means that they will add together to produce a wave having twice the amplitude of either wave. The combined wave will travel down the string and back and add to the third wave produced by the fork, and so on. The amplitude of the combined wave will continue to grow as the string absorbs energy from the tuning fork. Various damping effects, such as the loss of energy during reflection or the imperfect flexibility of the string, put a limit on the maximum amplitude that can be reached. Still, this maximum amplitude is much larger than that of the tuning fork. Thus the tuning fork is in resonance with the string. Resonance will also occur if the time it takes for the first wave to travel the distance $2L$ is twice the period of the vibrating tuning fork or any integer n times the period. Since the time for a wave to travel the distance $2L$ is $2L/v$, where v is the wave speed, we can write the resonance condition as

$$\frac{2L}{v} = nT = \frac{n}{f}$$

or

$$f = n\,\frac{v}{2L}$$

where $T = 1/f$ is the period. This is the same result we found by fitting an integral number of half-wavelengths into the distance L.

The resonance frequencies given by Equation 13-21 are also called the **natural frequencies** of the string. When the frequency of the tuning fork is not one of the natural frequencies of the vibrating string, standing waves are not produced. After the first wave has traveled the distance $2L$ and is reflected from the fork, it differs in phase from the wave being generated at the fork (Figure 13-20). The reflected wave combines with this new wave to produce a resultant wave whose amplitude may be greater or smaller than that of the original wave, depending on the phase difference. When this resultant wave has traveled the distance $2L$ and is again reflected at the fork, it will differ in phase from the next wave generated. In some cases, the new resultant wave will have a greater amplitude than the first, whereas in other cases the new amplitude will be less. On the average, however, the amplitude will not increase but will remain on the order of the amplitude of the first wave generated, which is the amplitude of the tuning fork and is very small compared with the amplitudes attained at resonance frequencies.

Only when the frequency of the tuning fork equals one of the natural frequencies of the string given by Equation 13-21 will the waves add in phase and the amplitude build up. This resonance phenomenon is analogous to the resonance of a simple harmonic oscillator with a harmonic driving force we discussed in Section 12-8. There we saw that if the frequency of the driving force equals the natural frequency of a simple harmonic oscillator, the oscillator absorbs the maximum amount of energy from the driving force. Note, however, that a vibrating string has not just one natural frequency but a sequence of natural frequencies that are integral multiples of the fundamental frequency. This sequence is called a **harmonic series.** The first frequency f_1 is termed the first harmonic, the second frequency $f_2 = 2f_1$ the second harmonic, and so on. Each vibration pattern is called a **mode** of vibration. In the terminology often used in music, the second harmonic is called the first **overtone,** the third harmonic the second overtone, and so on.

Figure 13-20 Waves on a string produced by a tuning fork whose frequency is not in resonance with the natural frequencies of the string. The wave leaving the tuning fork for the first time (dashed line) and that leaving for the second time after being reflected twice (grey line) are not in phase and so do not interfere constructively. The resultant wave (black line) has about the same amplitude as the individual waves, which is about the amplitude of the fork.

Turbulent winds set up standing waves in the Tacoma Narrows suspension bridge, leading to its collapse on November 7, 1940, just four months after it had been opened for traffic.

Example 13-4

A string is stretched between two fixed supports 1 m apart and the tension is adjusted until the fundamental frequency of the string is 440 Hz. What is the speed of transverse waves on the string?

From the standing-wave condition (Equation 13-20), the wavelength for the first harmonic is $\lambda = 2L = 2$ m. Hence, the wave speed is

$$v = \lambda f = (2 \text{ m})(440 \text{ Hz}) = 880 \text{ m/s}$$

Exercise

The speed of transverse waves on a stretched string is 200 m/s. If the string is 5 m long, find the frequencies of the fundamental and the second and third harmonics. (Answer: $f_1 = 20$ Hz, $f_2 = 40$ Hz, $f_3 = 60$ Hz)

Example 13-5

A string of length 3 m and mass density 0.0025 kg/m is fixed at both ends. One of its resonance frequencies is 252 Hz. The next higher resonance frequency is 336 Hz. Find (a) the fundamental frequency and (b) the tension in the string.

(a) According to Equation 13-21, each resonance frequency is an integer times the fundamental frequency. The ratio of two successive resonance frequencies must then be the ratio of two successive integers. The ratio of the frequencies given is $\frac{336}{252} = 1.33$, which is the ratio of 4 to 3. Thus 336 Hz must be the fourth harmonic and 252 Hz must be the third harmonic. The fundamental frequency is thus

$$f_1 = \frac{1}{3}f_3 = \frac{252 \text{ Hz}}{3} = 84 \text{ Hz}$$

(b) We find the tension in the string by first finding the wave speed. Since the wavelength of the fundamental is $\lambda = 2L = 6$ m, the wave speed is $v = f\lambda = (84 \text{ Hz})(6 \text{ m}) = 504$ m/s. The tension is then found from

$$v = \sqrt{\frac{F}{\mu}}$$

$$F = \mu v^2 = (0.0025 \text{ kg/m})(504 \text{ m/s})^2 = 635 \text{ N}$$

String Fixed at One End

Standing waves can also be produced on a string that has one end free instead of both ends fixed (Figure 13-21). The standing-wave patterns for such a string are shown in Figure 13-22. Note that the free end is an antinode. In the fundamental mode of vibration, the length of the string L equals $\lambda/4$. In the next highest mode, $L = 3\lambda/4$. The standing-wave condition can thus be written

$$L = n\,\frac{\lambda_n}{4} \qquad n = 1, 3, 5, \ldots$$

13-23   *Standing-wave condition, one end free*

The resonance frequencies are therefore given by

$$f_n = n\,\frac{v}{4L} = nf_1 \qquad n = 1, 3, 5, \ldots$$

13-24   *Resonance frequencies, one end free*

where

$$f_1 = \frac{v}{4L}$$

13-25

is the fundamental frequency. The natural frequencies of this system occur in the ratios $1:3:5:7:\ldots$, which means that the even harmonics are missing. Again, an easy way to remember the resonance frequencies given by Equation 13-24 is to sketch Figure 13-22 to remind yourself of the standing-wave condition and use $f = v/\lambda$.

Figure 13-21 An approximation of a string fixed at one end and free at the other end can be produced by connecting the "free" end of the string to a ring that is free to move on a post. Since the amplitude of the tuning fork is very small, the end attached to the fork is approximately fixed.

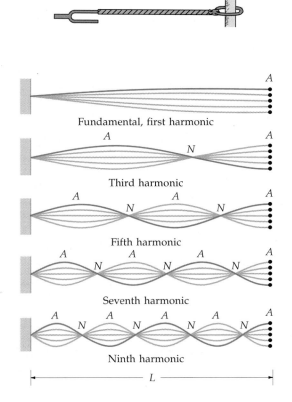

Fundamental, first harmonic

Third harmonic

Fifth harmonic

Seventh harmonic

Ninth harmonic

Figure 13-22 Standing waves on a string fixed at only one end. The free end is an antinode.

Standing-Wave Functions

We can find the equations for the wave functions illustrated in Figures 13-19 and 13-22 and also rederive the standing-wave conditions by considering the resultant wave formed by the superposition of a wave traveling to the right and a wave traveling to the left. Let the displacement of the wave traveling to the right be y_R and that of the wave traveling to the left be y_L. Assume that the amplitudes are equal. Then

DISPLACEMENT OF WAVE

$$y_R = A \sin (kx - \omega t)$$

and

$$y_L = A \sin (kx + \omega t)$$

where $k = 2\pi/\lambda$ is the wave number and $\omega = 2\pi f$ is the angular frequency. The sum of these two waves is

$$y(x, t) = y_R + y_L = A \sin (kx - \omega t) + A \sin (kx + \omega t)$$

Again, we use the identity $\sin \theta_1 + \sin \theta_2 = 2 \cos \frac{1}{2}(\theta_1 - \theta_2) \sin \frac{1}{2}(\theta_1 + \theta_2)$. If we let $\theta_1 = kx + \omega t$ and $\theta_2 = kx - \omega t$, we have

$$\tfrac{1}{2}(\theta_1 + \theta_2) = kx$$

and

$$\tfrac{1}{2}(\theta_1 - \theta_2) = \omega t$$

Then

$$y(x, t) = 2A \cos \omega t \sin kx \qquad\qquad 13\text{-}26$$

If the string is fixed at $x = 0$ and $x = L$, we have the following **boundary conditions** on the wave function:

DISPLACEMENT

$$y(x, t) = 0 \qquad \text{at } x = 0 \text{ and } x = L \qquad\qquad 13\text{-}27$$

for all times t. The boundary condition at $x = 0$ is automatically met because $\sin kx = 0$ at $x = 0$. The boundary condition at $x = L$ is met only for those particular values of the wave number that satisfy

$$\sin kL = 0 \qquad\qquad 13\text{-}28$$

The values k_n that satisfy this equation are given by

$$k_n L = n\pi \qquad n = 1, 2, 3, \ldots \qquad\qquad 13\text{-}29$$

In terms of the wavelength $\lambda = 2\pi/k$, Equation 13-29 is

$$\frac{2\pi}{\lambda_n} L = n\pi$$

or

$$n \frac{\lambda_n}{2} = L$$

which is the same as the standing-wave condition discussed earlier (Equation 13-20). The length of the string must equal an integral number of half-wavelengths. Equation 13-26 for the standing-wave function for vibration in the nth harmonic can thus be written

Standing-wave functions

$$y_n(x, t) = A_n \cos \omega_n t \sin k_n x \qquad\qquad 13\text{-}30$$

where A_n is the amplitude, k_n is given by Equation 13-29, and $\omega_n = 2\pi f_n$.

For a string with one end fixed and the other end free, the boundary conditions are that $y = 0$ at $x = 0$, which occurs automatically, and that y be a maximum or minimum at $x = L$. That is, the point $x = L$ must be an anti-node. This occurs if

$$\sin k_n L = \pm 1$$

or

$$k_n L = n\,\frac{\pi}{2} \qquad n = 1,\, 3,\, 5,\, \ldots$$

In terms of the wavelength, this condition is $2\pi L/\lambda_n = n\pi/2$ or

$$L = n\,\frac{\lambda_n}{4} \qquad n = 1,\, 3,\, 5,\, \ldots \qquad\qquad 13\text{-}31$$

which is the same as Equation 13-23.

Example 13-6

A string fixed at one end only is vibrating in its third harmonic. Its wave function is given by $y(x, t) = 0.015 \cos 189t \sin 0.262x$ where y and x are in meters and t is in seconds. (a) What is the frequency of vibration? (b) What is the length of the string?

(a) Since the angular frequency is $\omega = 189$ rad/s, the frequency is $f = \omega/2\pi = (189 \text{ rad/s})/(6.28) = 30.1$ Hz. The length of the string is related to the wavelength of the third harmonic by Equation 13-31,

$$L = n\,\frac{\lambda_n}{4} = 3\left(\frac{\lambda_3}{4}\right)$$

(b) Since the wave number is $k = 0.262$ m^{-1}, the wavelength is $\lambda_3 = 2\pi/k = 6.28/(0.262 \text{ m}^{-1}) = 24.0$ m. The length of the string is therefore

$$L = 3\left(\frac{\lambda_3}{4}\right) = 3\left(\frac{24 \text{ m}}{4}\right) = 18.0 \text{ m}$$

13-7 The Superposition of Standing Waves

In general, a vibrating system such as a string fixed at both ends does not vibrate in a single harmonic mode. Instead, the motion consists of a mixture of the allowed harmonics. For example, the general motion of a string fixed at both ends can be described by the wave function

$$y(x, t) = \sum_n A_n \cos (\omega_n t + \delta_n) \sin k_n x \qquad\qquad 13\text{-}32$$

where k_n satisfies Equation 13-29, $\omega_n = k_n v$, and A_n and δ_n are constants. The wave function in Equation 13-32 is a linear combination of the harmonic wave functions as described by Equation 13-30. Since the energy in a wave is proportional to the square of the amplitude, the quantity A_n^2 describes the fraction of the energy associated with the nth harmonic. The constants A_n and δ_n depend on the initial position and velocity of the string. A piano string, for example, is set into motion by striking it with a hammer. The location of the point of impact of the hammer on the string determines the relative values of A_n, which in turn describe the relative amounts of energy in each harmonic component. A harp string, on the other hand, is plucked. If a string is plucked at the center and released as in Figure 13-23, the initial shape of the string is symmetric about the point $x = \frac{1}{2}L$. The motion of the string after it has been released will remain symmetric about this point. Only the odd harmonics, which are also symmetric about $x = \frac{1}{2}L$, will be excited. The even harmonics, which are antisymmetric about $x = \frac{1}{2}L$, are not excited.

Figure 13-23 A string plucked at the center. When it is released, its vibration is a linear combination of standing waves.

Hologram interferograms showing standing waves in a handbell. The "bull's eyes" locate the anti-nodes.

The constant A_n is zero for every even n. (The shapes of the first four harmonics are shown in Figure 13-24.) Most of the energy of the plucked string is associated with the fundamental, but small amounts of energy are associated with the third and higher odd harmonic modes. Figure 13-25 shows an approximation to the initial shape of the string using the superpositions of only the first three odd harmonics.

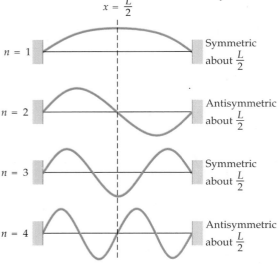

Figure 13-24 The first four harmonics for a string fixed at both ends. The odd harmonics are symmetrical about the center of the string, whereas the even harmonics are not. When a string is plucked at the center, it vibrates only in its odd harmonics.

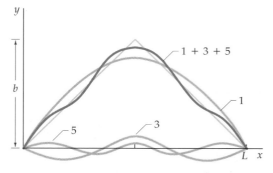

Figure 13-25 Synthesis of a string plucked at the center, as in Figure 13-23, using only the first three odd harmonics. The heavy colored line is an approximation of the original shape of the string based on the first three odd harmonics. The height of the string is exaggerated in this drawing to show the relative amplitudes of the harmonics. Most of the energy is associated with the fundamental, but there is some energy in the third, fifth, and other odd harmonics.

13-8   The Wave Equation (Optional)

A general wave function $y(x, t)$ is a solution of a differential equation called the wave equation. The wave equation follows directly from Newton's laws.

In Figure 13-26, one segment of a string has been isolated. Our derivation will apply only if the wave is small enough in amplitude for the angle between the string and the horizontal (the original direction of the string with no wave) to be small. In this case, the length of the segment is approximately Δx and its mass is $\mu \, \Delta x$. The string segment moves vertically. Its acceleration is the second derivative of $y(x, t)$ with respect to t for constant x.

A derivative of a function of two variables with respect to one of the variables with the other held constant is called a **partial derivative.** The partial derivative of y with respect to t is written $\partial y / \partial t$. The second partial derivative of y with respect to t is written $\partial^2 y / \partial t^2$. The net vertical force is

$$\sum F = F \sin \theta_2 - F \sin \theta_1$$

where θ_2 and θ_1 are the angles shown in Figure 13-26, and F is the tension in the string. Since the angles are assumed to be small, we may approximate $\sin \theta$ by $\tan \theta$. Thus, the net vertical force on the string segment can be written

$$\sum F = F(\sin \theta_2 - \sin \theta_1) \approx F(\tan \theta_2 - \tan \theta_1)$$

The tangent of the angle made by the string with the horizontal is the slope of the curve formed by the string. If we call this slope S, we have

$$S = \tan \theta = \frac{\partial y}{\partial x}$$

Then

$$\sum F = F(S_2 - S_1) = F \, \Delta S$$

where S_1 and S_2 are the slopes of either end of the string segment, and ΔS is the change in the slope. Setting this net force equal to the mass $\mu \, \Delta x$ times the acceleration $\partial^2 y / \partial t^2$ gives

$$F \, \Delta S = \mu \, \Delta x \, \frac{\partial^2 y}{\partial t^2}$$

or

$$F \, \frac{\Delta S}{\Delta x} = \mu \, \frac{\partial^2 y}{\partial t^2} \qquad \text{13-33}$$

In the limit $\Delta x \to 0$, we have

$$\lim_{\Delta x \to 0} \frac{\Delta S}{\Delta x} = \frac{\partial S}{\partial x} = \frac{\partial}{\partial x} \frac{\partial y}{\partial x} = \frac{\partial^2 y}{\partial x^2}$$

Thus, Equation 13-33 becomes

$$\frac{\partial^2 y}{\partial x^2} = \frac{\mu}{F} \frac{\partial^2 y}{\partial t^2} \qquad \qquad \text{13-34a} \qquad \textit{Wave equation}$$

Figure 13-26   Segment of a stretched string used for the derivation of the wave equation. The net vertical force on the segment is $F \sin \theta_2 - F \sin \theta_1$, where F is the tension in the string. The wave equation is derived by applying Newton's second law to the segment.

Equation 13-34 is the **wave equation** for a stretched string. It is important to remember that this equation holds only for small angles and thus for small displacements $y(x, t)$.

We now show that the wave equation is satisfied by any function of either $x - vt$ or $x + vt$. Let $\alpha = x - vt$ and consider any wave function

$$y = y(x - vt) = y(\alpha)$$

We will use y' for the derivative of y with respect to α. Then, by the chain rule for derivatives, we have

$$\frac{\partial y}{\partial x} = \frac{\partial y}{\partial \alpha} \frac{\partial \alpha}{\partial x} = y' \frac{\partial \alpha}{\partial x}$$

and

$$\frac{\partial y}{\partial t} = \frac{\partial y}{\partial \alpha} \frac{\partial \alpha}{\partial t} = y' \frac{\partial \alpha}{\partial t}$$

Since $\partial \alpha / \partial x = 1$ and $\partial \alpha / \partial t = -v$, we have

$$\frac{\partial y}{\partial x} = y'$$

and

$$\frac{\partial y}{\partial t} = -vy'$$

Taking the second derivatives, we obtain

$$\frac{\partial^2 y}{\partial x^2} = y''$$

$$\frac{\partial^2 y}{\partial t^2} = -v \frac{\partial y'}{\partial t} = -v \frac{\partial y'}{\partial \alpha} \frac{\partial \alpha}{\partial t} = +v^2 y''$$

Thus,

Wave equation

$$\frac{\partial^2 y}{\partial x^2} = \frac{1}{v^2} \frac{\partial^2 y}{\partial t^2} \qquad \text{13-34}b$$

Comparing Equations 13-34*a* and *b*, we see that the speed of propagation of the wave is

$$v = \sqrt{\frac{F}{\mu}} \qquad \text{13-35}$$

Example 13-7

Show by explicitly calculating the derivatives that $y = A \sin (kx - \omega t)$ satisfies Equation 13-34*b*.

Taking the partial derivative of y with respect to x, we obtain

$$\frac{\partial y}{\partial x} = \frac{\partial}{\partial x} [A \sin (kx - \omega t)] = A \cos (kx - \omega t) \frac{\partial(kx)}{\partial x}$$

$$= kA \cos (kx - \omega t)$$

Taking the second partial derivative with respect to x, we obtain

$$\frac{\partial^2 y}{\partial x^2} = -k^2 A \sin (kx - \omega t)$$

Similarly, the derivatives with respect to t are

$$\frac{\partial y}{\partial t} = \frac{\partial}{\partial t}[A \sin(kx - \omega t)] = A \cos(kx - \omega t)\frac{\partial(-\omega t)}{\partial t}$$

$$= -\omega A \cos(kx - \omega t)$$

and

$$\frac{\partial^2 y}{\partial t^2} = -\omega^2 A \sin(kx - \omega t)$$

Equation 13-34b then gives

$$-k^2 A \sin(kx - \omega t) = \frac{1}{v^2}[-\omega^2 A \sin(kx - \omega t)]$$

which holds if $v = \omega/k$.

An important property of the wave equation is that it is linear; that is, the function $y(x, t)$ and its derivatives occur only to the first power. An important property of linear equations is that if y_1 and y_2 are two solutions of the equation, the linear combination

$$y_3 = C_1 y_1 + C_2 y_2 \qquad\qquad 13\text{-}36$$

where C_1 and C_2 are any constants, is also a solution. This can be shown by the direct substitution of y_3 into the wave equation. This result is the mathematical statement of the principle of superposition. If any two waves satisfy a wave equation, their sum also satisfies the same wave equation. Again, the principle of superposition holds for waves on a string only if the amplitudes of the waves are small, so that the approximation used in deriving the wave equation, $\sin\theta \approx \tan\theta$, holds.

Exercise
Show that any function $y(x + vt)$ satisfies Equation 13-34b.

Exercise
Show that the function y_3 given by Equation 13-36 satisfies Equation 13-34.

Summary

1. Wave motion is the propagation of a disturbance in a medium. In transverse waves, such as waves on a string, the disturbance is perpendicular to the direction of propagation. In longitudinal waves, such as sound waves, the disturbance is along the direction of propagation. Both energy and momentum are transported by a wave.

2. The speed of a wave depends on the density and elastic properties of the medium. It is independent of the motion of the wave source. The speed of waves on a string is related to the tension F in the string and its mass per unit length μ by

$$v = \sqrt{\frac{F}{\mu}}$$

3. When two or more waves meet at the same place they superimpose; that is, the disturbances add algebraically. The principle of superposition holds for waves on a string if the transverse displacements are not too large.

4. In harmonic waves, the disturbance varies sinusoidally in time and space. In a harmonic wave on a string, a segment of the string oscillates with simple harmonic motion in a direction perpendicular to the direction of the wave. The distance between successive wave crests is the wavelength λ. The wave function $y(x, t)$ for a harmonic wave is

$$y(x, t) = A \sin (kx - \omega t)$$

where A is the amplitude, k is the wave number, which is related to the wavelength by

$$k = \frac{2\pi}{\lambda}$$

and ω is the angular frequency, which is related to the frequency f by

$$\omega = 2\pi f$$

The speed of a harmonic wave equals the frequency times the wavelength:

$$v = f\lambda = \frac{\omega}{k}$$

5. The power transmitted by a harmonic wave is proportional to the square of the amplitude of the wave and is given by

$$P = \tfrac{1}{2}\mu\omega^2 A^2 v$$

6. The superposition of harmonic waves is called interference. If the waves are in phase or differ in phase by an integer times 2π, the amplitudes of the waves add and the interference is constructive. If the waves differ in phase by π or by an odd integer times π, the amplitudes subtract and the interference is destructive.

7. When waves are confined in space, standing waves occur. For a string fixed at both ends, the standing-wave condition can be found by fitting waves into the string with a node at each end. The result is that an integral number of half-wavelengths must fit into the length of the string. The standing-wave condition in this case is

$$L = n \frac{\lambda_n}{2} \qquad n = 1, 2, 3, \ldots$$

The allowed waves form a harmonic series, with the frequencies given by

$$f_n = nf_1 \qquad n = 1, 2, 3, \ldots$$

where $f_1 = v/2L$ is the lowest frequency, which is called the fundamental. The wave functions for these standing waves are of the form

$$y_n(x, t) = A_n \cos \omega_n t \sin k_n x$$

where $k_n = 2\pi/\lambda_n$ and $\omega_n = 2\pi f_n$.

    If the string has one end fixed and one end free, there is a node at one end and an antinode at the other. The standing-wave condition in this case is

$$L = n \frac{\lambda_n}{4} \qquad n = 1, 3, 5, \ldots$$

Only the odd harmonics are present. Their frequencies are given by

$$f_n = nf_1 \qquad n = 1, 3, 5, \ldots$$

where $f_1 = v/4L$.

8. In general, a vibrating system, such as a string fixed at both ends, does not vibrate in a single harmonic mode but in a mixture of the allowed harmonics.

9. The wave function for waves on a string obeys the wave equation, which relates the spatial derivatives of the wave function to the time derivatives:

$$\frac{\partial^2 y}{\partial x^2} = \frac{1}{v^2}\frac{\partial^2 y}{\partial t^2}$$

The wave equation follows from Newton's second law applied to a segment of a vibrating string.

Suggestions for Further Reading

Pierce, John R.: "Periodicity, Pitch and Waves," in *The Science of Musical Sound*, Scientific American Books Inc., New York, 1983.

Traveling and standing waves on a stretched string form the basis of this discussion of the pitch of sound.

Schelleng, John C.: "The Bowed String," *Scientific American*, January 1974. Also in *The Physics of Music*, Scientific American Inc., New York, 1978.

Measurements of the position and velocity of points on a violin string, as the bow is drawn across it, result in new understanding of the string's peculiar standing-wave motion.

Review

A. Objectives: After studying this chapter, you should:

1. Be able to state the quantities upon which the speed of a wave on a string depend.

2. Be able to state the relationship between the speed, frequency, and wavelength of a harmonic wave.

3. Be able to sketch standing-wave patterns for a string fixed at both ends and for a string with one end free and from these sketches obtain the possible frequencies for standing waves.

4. Know that the energy in a harmonic wave is proportional to the square of the amplitude of the wave.

B. Define, explain, or otherwise identify:

Dispersion
Transverse wave
Longitudinal wave
Wave function
Principle of superposition
Harmonic wave
Wavelength
Amplitude
Wave number
Phase difference
Interference
Constructive interference
Destructive interference
Standing wave

Resonance frequencies
Fundamental
First harmonic
Node
Antinode
Standing-wave condition
Natural frequencies
Harmonic series
Mode
Overtone
Boundary condition
Partial derivative
Wave equation

C. True or false: If the statement is true, explain why it is true. If it is false, give a counterexample.

1. Wave pulses on strings are transverse waves.

2. When a wave pulse is reflected, it is always inverted.

3. The speed of a harmonic wave on a string is proportional to its wavelength.

4. The power transmitted by a wave is proportional to the square of the amplitude of the wave.

5. The frequency of the third harmonic is three times that of the first harmonic.

Problems

Level I

13-1 Wave Pulses

1. Figure 13-27 shows a wave pulse at time $t = 0$. The pulse is moving to the right without dispersion at 2 cm/s. Sketch the shape of the string at times $t = 1$, 2, and 4 s.

Figure 13-27 Problems 1, 2, and 3.

2. Figure 13-27 shows a wave pulse at time $t = 0$ moving to the right. At this particular time, which segments of the string are moving up? Which are moving down? Is there any segment of the string at the pulse that is instantaneously at rest? Answer these questions by sketching the pulse at a slightly later time and a slightly earlier time to see how the segments of the string are moving.

3. Make a sketch of the velocity of each string segment versus position for the pulse shown in Figure 13-27.

4. Two rectangular wave pulses are traveling in opposite directions along a string. At $t = 0$, the two pulses are as shown in Figure 13-28. Sketch the wave functions for $t = 1$, 2, and 3 s.

Figure 13-28 Problems 4 and 5.

5. Repeat Problem 4 for the case in which the pulse on the right in Figure 13-28 is inverted.

6. The following wave functions represent traveling waves:

(a) $y_2(x, t) = A \cos k(x + 34t)$

(b) $y_3(x, t) = Ae^{-k(x-20t)}$

(c) $y_1(x, t) = \dfrac{B}{C + (x - 10t)^2}$

where x is in meters, t is in seconds, and A, k, B, and C are constants that have the proper units for y to be in meters. Give the direction of propagation and the speed of the wave for each wave function.

7. At time $t = 0$, the shape of a wave pulse on a string is given by the function

$$y(x, 0) = \frac{0.12 \text{ m}^3}{(2.00 \text{ m})^2 + x^2}$$

where x is in meters. (a) Sketch $y(x, 0)$ versus x. Give the wave function $y(x, t)$ at a general time t if (b) the pulse is moving in the positive x direction with a speed of 10 m/s and (c) the pulse is moving in the negative x direction with a speed of 10 m/s.

13-2 Speed of Waves

8. Show that if F is in newtons and μ is in kilograms per meter, the units of $\sqrt{F/\mu}$ are meters per second.

9. A steel wire 7 m long has a mass of 100 g. It is under a tension of 900 N. What is the speed of a transverse wave pulse on this wire?

10. Transverse waves travel at 150 m/s on a wire of length 80 cm that is under a tension of 550 N. What is the mass of the wire?

11. A wave pulse propagates along a wire in the positive x direction at 20 m/s. What will the pulse velocity be (a) if we double the length of the wire but keep the tension and mass per unit length constant? (b) If we double the tension while holding the length and mass per unit length constant? (c) If we double the mass per unit length while holding the other variables constant?

12. A steel piano wire is 0.7 m long and has a mass of 5 g. It is stretched with a tension of 500 N. (a) What is the speed of transverse waves on the wire? (b) To reduce the wave speed by a factor of 2 without changing the tension, what mass of copper wire would have to be wrapped around the steel wire?

13. In a common lecture demonstration of wave pulses, a piece of rubber tubing is tied at one end to a fixed post and is passed over a pulley to a weight hanging at the other end. Suppose that the distance from the fixed support to the pulley is 10 m, the mass of this length of tubing is 0.7 kg, and the suspended weight is 110 N. If the tubing is given a transverse blow at one end, how long will it take the resulting pulse to reach the other end?

14. A long glass tube has a semicircular bend of radius 8 cm. A string of mass density 0.04 kg/m under a tension of 20 N is pulled through the tube. (a) At what speed should the string be pulled for it to pass through the tube without touching the sides? (b) What is the magnitude of the acceleration of the string segments as they pass through the semicircular bend? (c) Draw a diagram indicating the forces acting on a string segment in the bend that provide this acceleration.

15. The cable of a ski lift runs 400 m up a mountain and has a mass of 80 kg. When the cable is struck with a transverse blow at one end, the return pulse is detected 12 s later. (a) What is the speed of the wave? (b) What is the tension in the cable?

13-3 Harmonic Waves

16. One end of a string 6 m long is moved up and down with simple harmonic motion at a frequency of 60 Hz. The waves reach the other end of the string in 0.5 s. Find the wavelength of the waves on the string.

17. Equation 13-8 expresses the displacement of a harmonic wave as a function of x and t in terms of the wave parameters k and ω. Write the equivalent expressions that contain the following pairs of parameters instead of k and ω: (a) k and v, (b) λ and f, (c) λ and T, (d) λ and v, and (e) f and v.

18. Equation 13-5 applies to all types of periodic waves, including electromagnetic waves such as light waves and microwaves, which travel at 3×10^8 m/s in a vacuum. (a) The range of wavelengths of light to which the eye is sensitive is about 4×10^{-7} to 7×10^{-7} m. What are the frequencies · that correspond to these wavelengths? (b) Find the frequency of a microwave that has a wavelength of 3 cm.

19. A harmonic wave with a frequency of 60 Hz and an amplitude of 0.02 m travels along a string to the right with a speed of 10 m/s. Write a suitable wave function for this wave.

20. The wave function for a harmonic wave on a string is $y(x, t) = 0.001 \sin (62.8x + 314t)$, where y and x are in meters and t is in seconds. (a) In what direction does this wave travel, and what is its speed? (b) Find the wavelength, frequency, and period of this wave. (c) What is the maximum displacement of any string segment?

21. A harmonic wave travels along an infinite uniform rope under constant tension. The rope is marked off in meter intervals. At the 0-meter mark, the rope is observed to reach its maximum transverse displacement of 50 cm every 5 s. The distance between the maxima at any instant in time is 50 m. Give the wave function of this wave, assuming that it is harmonic, has a maximum displacement at $x = 0$ when $t = 0$, and is moving along the rope from left to right.

13-4 Energy Transmitted by Waves

22. Waves of frequency 200 Hz and amplitude 1 cm move along a 20-m string that has a mass of 0.06 kg and a tension of 50 N. (a) What is the total energy of the waves on the string? (b) Find the power transmitted past a given point on the string.

23. A harmonic wave on a string with a mass per unit length of 0.05 kg/m and a tension of 80 N has an amplitude of 5 cm. Each section of the string moves with simple harmonic motion at a frequency of 10 Hz. Find the power propagated along the string.

24. A rope 2 m long has a mass of 0.1 kg. The tension is 60 N. A power source at one end sends a harmonic wave with an amplitude of 1 cm down the rope. The wave is extracted at the other end without any reflection. What is the frequency of the power source if the power transmitted is 100 W?

13-5 Superposition and Interference of Harmonic Waves

25. Two waves traveling on a string in the same direction both have a frequency of 100 Hz, a wavelength of 2 cm, and an amplitude of 0.02 m. What is the amplitude of the resultant wave if the original waves differ in phase (a) by $\pi/6$? (b) By $\pi/3$?

26. What is the phase difference between the two waves of Problem 25 if the amplitude of the resultant wave is 0.02 m, the same as the amplitude of each original wave?

27. Two waves having the same frequency, wavelength, and amplitude are traveling in the same direction. If they differ in phase by $\pi/2$ and each has an amplitude of 0.05 m, find the amplitude of the resultant wave.

13-6 Standing Waves

28. A string fixed at both ends is 3 m long. It resonates in its second harmonic at a frequency of 60 Hz. What is the speed of transverse waves on the string?

29. A string 3 m long and fixed at both ends is vibrating in its third harmonic. The maximum displacement of any point on the string is 4 mm. The speed of transverse waves on this string is 50 m/s. (a) What are the wavelength and frequency of this wave? (b) Write the wave function for this wave.

30. A 5-g steel wire 1.4 m long is fixed at both ends and has a tension of 968 N. (a) Find the speed of transverse waves on the wire. (b) Find the wavelength and frequency of the fundamental. (c) Find the frequencies of the second and third harmonics.

31. A rope 4 m long is fixed at one end and is attached to a light string at the other end so that it is free to move there. The speed of waves on the rope is 20 m/s. Find the frequency of (a) the fundamental, (b) the first overtone, and (c) the second overtone.

32. Middle C on the equal-temperament scale used by modern instrument makers has a frequency of 261.63 Hz. If this is the fundamental frequency of a 7-g piano wire that is 80 cm long, what should the tension in the wire be?

33. A piano wire without windings has a fundamental frequency of 200 Hz. When it is wound with wire, its linear mass density is doubled. What is its new fundamental frequency assuming that the tension is unchanged?

34. The length of the B string on a certain guitar is 60 cm. It vibrates at 247 Hz. (a) What is the speed of transverse waves on the string? (b) If the linear mass density of the guitar string is 0.01 g/cm, what should its tension be when it is in tune?

35. The wave function $y(x, t)$ for a certain standing wave on a string fixed at both ends is given by $y(x, t) = 0.30 \sin 0.20x \cos 300t$, where y and x are in centimeters and t is in seconds. (a) What are the wavelength and frequency of these waves? (b) What is the speed of transverse waves on this string? (c) If the string is vibrating in its fourth harmonic, how long is it?

36. The wave function $y(x, t)$ for a certain standing wave on a string fixed at both ends is given by $y(x, t) = 0.5 \sin 0.025x \cos 500t$, where y and x are in centimeters and t is in seconds. (a) Find the speed and amplitude of the two traveling waves that result in this standing wave. (b) What is the distance between successive nodes on the string? (c) What is the shortest possible length of the string?

37. A string 2.51 m long has the wave function given in Problem 36. (a) Sketch the position of the string at the times $t = 0$, $t = \frac{1}{4}T$, $t = \frac{1}{2}T$, and $t = \frac{3}{4}T$, where $T = 1/f$ is the period of the vibration. (b) Find T in seconds. (c) At a time t when the string is horizontal, that is, $y(x) = 0$ for all x, what has become of the energy in the wave?

38. A 160-g rope 4 m long is fixed at one end and is tied to a light string at the other end. Its tension is 400 N. (a) What are the wavelengths of the fundamental and the next two harmonics? (b) What are the frequencies of these standing waves?

39. A string fixed at one end only is vibrating in its fundamental mode. The wave function is $y(x, t) = 0.02 \sin 2.36x \cos 377t$, where y and x are in meters and t is in seconds. (a) What is the wavelength of the wave? (b) What is the length of the string? (c) What is the speed of transverse waves on the string?

40. A string 5 m long that is fixed at one end only is vibrating in its fifth harmonic with a frequency of 400 Hz. The maximum displacement of any segment of the string is 3 cm. (a) What is the wavelength of this wave? (b) What is the wave number k? (c) What is the angular frequency? (d) Write the wave function for this standing wave.

41. Three successive resonance frequencies for a certain string are 75, 125, and 175 Hz. (a) Find the ratios of each pair of successive resonance frequencies. (b) How can you tell that these frequencies are for a string fixed at one end only rather than for a string fixed at both ends? (c) What is the fundamental frequency? (d) Which harmonics are these resonance frequencies? (e) If the speed of transverse waves on this string is 400 m/s, find the length of the string.

13-7 The Superposition of Standing Waves

There are no problems for this section.

13-8 The Wave Equation (Optional)

42. Show explicitly that the following functions satisfy the wave equation: (a) $y(x, t) = (x + vt)^3$; (b) $y(x, t) = Ae^{ik(x-vt)}$, where A and k are constants and $i = \sqrt{-1}$; and (c) $y(x, t) = \ln (x - vt)$.

43. Show that the function $y = A \sin kx \cos \omega t$ satisfies the wave equation.

Level II

44. The G string on a violin is 30 cm long. When played without fingering, it vibrates at a frequency of 196 Hz. The next higher notes on the scale are A (220 Hz), B (247 Hz), C (262 Hz), and D (294 Hz). How far from the end of the string must a finger be placed to play each of these notes?

45. A string with a mass density of 4×10^{-3} kg/m is under a tension of 360 N and is fixed at both ends. One of its resonance frequencies is 375 Hz. The next higher resonance frequency is 450 Hz. (a) What is the fundamental frequency of this string? (b) Which harmonics are the ones given? (c) What is the length of the string?

46. A string fastened at both ends has successive resonances with wavelengths of 0.54 m for the nth harmonic and 0.48 m for the $(n + 1)$th harmonic. (a) Which harmonics are these? (b) What is the length of the string?

47. A tuning fork attached to a stretched wire generates transverse waves. The vibration of the fork is perpendicular to the wire. Its frequency is 400 Hz, and the amplitude of its oscillation is 0.50 mm. The wire has linear mass density of 0.01 kg/m and is under a tension of 1 kN. Assume that there are no reflected waves. (a) Find the period and frequency of waves on the wire. (b) What is the speed of the waves? (c) What are the wavelength and wave number? (d) Write a suitable wave function for the waves on the wire. (e) Calculate the maximum speed and acceleration of a point on the wire. (f) At what average rate must energy be supplied to the fork to keep it oscillating at a steady amplitude?

48. In a lecture demonstration of standing waves, a string is attached to a tuning fork that vibrates at 60 Hz and sets up transverse waves of that frequency on the string. The other end of the string passes over a pulley, and the tension is varied by attaching weights to that end. The string has approximate nodes at the tuning fork and at the pulley. (a) If the string has linear mass density of 8 g/m and is 2.5 m long (from the tuning fork to the pulley), what must the tension be for the string to vibrate in its fundamental mode? (b) Find the tensions necessary for the string to vibrate in each of its first three overtones.

49. (a) For the wave function given in Problem 39, find the velocity of a string segment at some point x as a function of time. (b) Which point has the greatest speed at any time? What is the maximum speed of this point? (c) Find the acceleration of a string segment at some point x as a function of time. (d) Which point has the greatest acceleration? What is the maximum acceleration of this point?

50. Two wires of different linear mass densities are soldered together end to end and are then stretched under a tension F (the same in both wires). The wave speed in the first wire is twice that in the second wire. When a harmonic wave traveling in the first wire is reflected at the junction of the wires, the reflected wave has half the amplitude of the transmitted wave. (a) Assuming no loss in the wire, what fraction of the incident power is reflected at the junction and what fraction is transmitted? (b) If the amplitude of the incident wave is A, what are the amplitudes of the reflected and transmitted waves?

51. Power is to be transmitted along a stretched wire by means of transverse harmonic waves. The wave speed is 10 m/s, and the linear mass density of the wire is 0.01 kg/m. The power source oscillates with an amplitude of 0.50 mm. (a) What average power is transmitted along the wire if the frequency is 400 Hz? (b) The power transmitted can be increased by increasing the tension in the wire, the frequency of the source, or the amplitude of the waves. How would each of these quantities have to be changed to effect an increase in power by a factor of 100 if it is the only quantity changed? (c) Which of the quantities would probably be the easiest to change?

52. A 2-m string is fixed at one end and is vibrating in its third harmonic. The greatest displacement of any segment of the string is 3 cm. The frequency of vibration is 100 Hz. (*a*) Write the wave function for this vibration. (*b*) Write an expression for the kinetic energy of a segment of the string of length dx at a point x at some time t. At what time is this kinetic energy maximum? What is the shape of the string at this time? (*c*) Find the maximum kinetic energy of the string by integrating your expression for part (*b*) over the total length of the string. (*d*) Find the potential energy of a segment of the string, and compute the maximum potential energy of the string by integration. *Hint:* Remember that the potential energy of a mass m in simple harmonic motion of angular frequency ω is $\frac{1}{2}m\omega^2 y^2$, where y is the displacement.

53. The kinetic energy of a segment Δm of a vibrating string is given by

$$\Delta K = \tfrac{1}{2}\Delta m \left(\frac{\partial y}{\partial t}\right)^2 = \tfrac{1}{2}\mu \left(\frac{\partial y}{\partial t}\right)^2 \Delta x$$

(*a*) Find the total kinetic energy of the *n*th mode of vibration of a string of length L fixed at both ends. (*b*) Give the maximum kinetic energy of the string. (*c*) What is the wave function when the kinetic energy has its maximum value? (*d*) Show that the maximum kinetic energy in the *n*th mode is proportional to $n^2 A_n^2$.

54. (*a*) Compute the derivative of the speed of a wave on a string with respect to the tension dv/dF, and show that the differentials dv and dF obey $dv/v = \frac{1}{2}\,dF/F$. (*b*) A wave moves with a speed of 300 m/s on a wire that is under a tension of 500 N. Using dF to approximate a change in tension, find how much the tension must be changed to increase the speed to 312 m/s.

55. (*a*) Show that, if the tension in a string fixed at both ends is changed by a small amount dF, the frequency of the fundamental is changed by approximately df, where $df/f = \frac{1}{2}\,dF/F$. (See Problem 54.) Does this result apply to all harmonics? (*b*) Use this result to find the percentage change in the tension needed to increase the frequency of the fundamental of a piano wire from 260 to 262 Hz.

56. Three waves with the same frequency, wavelength, and amplitude are traveling in the same direction. The three waves are given by

$$y_1(x,\,t) = 0.05 \sin\left(kx - \omega t - \frac{\pi}{3}\right)$$

$$y_2(x,\,t) = 0.05 \sin\left(kx - \omega t\right)$$

$$y_3(x,\,t) = 0.05 \sin\left(kx - \omega t + \frac{\pi}{3}\right)$$

Find the resultant wave.

57. A very long wire can be vibrated up and down with a mechanical motor to produce waves traveling down the wire. At the far end of the wire, the traveling waves are cleverly absorbed so that no reflected waves are returned to the motor. The wave speed is observed to be 240 m/s, the maximum transverse displacement of the wire is 1 cm, and the distance between maxima is 3.0 m. (*a*) Write a wave function to represent the wave propagating down this wire. (*b*) What is the frequency of vibration of the motor? (*c*) What is the period of the transverse oscillations of the wire? (*d*) What is the maximum transverse velocity of a small insect clinging to the wire?

58. Two connected wires with linear mass densities that are related by $\mu_1 = 3\mu_2$ are under the same tension. When the wires oscillate at a frequency of 120 Hz, waves of wavelength 10 cm travel down the first wire with the linear density of μ_1. (*a*) What is the wave speed in the first wire? (*b*) What is the wave speed in the second wire? (*c*) What is the wavelength in the second wire?

59. A violin string of length 40 cm and mass 1.2 g has a frequency of 500 Hz when it is vibrating in its fundamental mode. (*a*) What is the wavelength of the standing wave on the string? (*b*) What is the tension in the string? (*c*) Where should you place your finger to increase the frequency to 650 Hz?

60. Two traveling wave pulses on a string are represented by the wave functions

$$y_1(x,\,t) = \frac{0.02 \text{ m}^3}{2 \text{ m}^2 + (x - 2t)^2}$$

and

$$y_2(x,\,t) = \frac{-0.02 \text{ m}^3}{2 \text{ m}^2 + (x + 2t)^2}$$

where x is in meters and t is in seconds. (*a*) Sketch each wave function separately as a function of x at $t = 0$, and describe the behavior of each as time increases. (*b*) Find the resultant wave function at $t = 0$. (*c*) Find the resultant wave function at $t = 1$ s. (*d*) Sketch the resultant wave function at $t = 1$ s.

61. A transverse wave of frequency 40 Hz propagates down a string. Two points 5 cm apart are out of phase by $\pi/6$. (*a*) What is the wavelength of the wave? (*b*) What is the phase difference between two displacements at a given point for times 5 ms apart? (*c*) What is the wave velocity?

62. A standing wave on a rope is represented by the following wave function:

$$y(x,\,t) = 0.02 \sin\frac{\pi x}{2} \cos 40\pi t$$

where x and y are in meters and t is in seconds. (*a*) Write wave functions for two traveling waves that when superimposed will produce the resultant standing-wave pattern. (*b*) What is the distance between the nodes of the standing wave? (*c*) What is the velocity of a segment of the rope $x = 1$ m? (*d*) What is the acceleration of a segment of the rope at $x = 1$ m?

Level III

63. The wave functions of two standing waves on a string of length L are

$$y_1(x,\,t) = A_1 \cos \omega_1 t \sin k_1 x$$

$$y_2(x,\,t) = A_2 \cos \omega_2 t \sin k_2 x$$

where $k_n = n\pi/L$ and $\omega_n = n\omega_1$. The wave function of the resultant wave is

$$y_r(x,\ t) = y_1(x,\ t) + y_2(x,\ t)$$

(a) Find the velocity of a segment dx of the string. (b) Find the kinetic energy of this segment. (c) By integration, find the total kinetic energy of the resultant wave. Notice the disappearance of the cross terms so that the total kinetic energy is proportional to $(n_1A_1)^2 + (n_2A_2)^2$.

64. A 2-m wire fixed at both ends is vibrating in its fundamental mode. The tension in the wire is 40 N and the mass of the wire is 0.1 kg. The midpoint of the wire has an amplitude of 2 cm. (a) Find the maximum kinetic energy of the wire. (b) At the instant the transverse displacement is given by (0.02 m) $\sin(\pi x/2)$, what is the kinetic energy of the wire? (c) At what position on the wire does the kinetic energy per unit length have its largest value? (d) Where does the potential energy per unit length have its maximum value?

65. If a loop of chain is spun at high speed, it will roll like a hoop without collapsing. Consider a chain of linear mass density μ that is rolling without slipping at a high speed v_0. (a) Show that the tension in the chain is $F = \mu v_0^2$. (b) If the chain rolls over a small bump, a transverse wave pulse will be generated in the chain. At what speed will it travel along the chain? (c) How far around the loop (in degrees) will a transverse wave pulse travel in the time the hoop rolls through one complete revolution?

66. A long rope with a mass per unit length of 0.1 kg/m is under a constant tension of 10 N. A motor at the point $x = 0$ drives one end of the rope with harmonic motion at 5 oscillations per second and an amplitude of 4 cm. (a) What is the wave speed? (b) What is the wavelength? (c) What is the maximum transverse linear momentum of a 1-mm segment of the rope? (d) What is the maximum net force on a 1-mm segment of the rope?

67. A heavy rope 3 m long is attached to the ceiling and is allowed to hang freely. (a) Show that the speed of transverse waves on the rope is independent of its mass and length but does depend on the distance y from the bottom according to the formula $v = \sqrt{gy}$. (b) If the bottom end of the rope is given a sudden sideways displacement, how long does it take the resulting wave pulse to go to the ceiling, reflect, and return to the bottom of the rope?

68. Two wires of different linear mass densities are joined at $x = 0$ and are under constant tension. A wave y_{in} is incident on the junction from the left at $x = 0$, where

$$y_{in} = A_{in} \cos(k_1x - \omega t)$$

This wave is partly reflected and partly transmitted at $x = 0$. The transmitted wave

$$y_t = A_t \cos(k_2x - \omega t)$$

moves to the right at $x > 0$. The reflected wave

$$y_r = A_r \cos(k_1x + \omega t)$$

moves to the left at $x < 0$. The boundary conditions at $x = 0$ are that the resultant displacement and the resultant slope $\partial y/\partial x$ are both continuous. (a) Find the amplitude of the reflected and transmitted waves. (b) Verify that mechanical energy is conserved. (c) If the wire at $x > 0$ is denser than the wire at $x < 0$, find the phase change between the incident wave and the reflected wave.

69. The linear mass density of a nonuniform wire under constant tension decreases gradually along the wire so that an incident wave is transmitted without reflection. The wire is uniform for $-\infty \le x \le 0$. In this region, a transverse wave has the form $y(x,\ t) = 0.003 \cos(25x - 50t)$ where y and x are in meters and t is in seconds. From $x = 0$ to $x = 20$ m, the linear mass decreases gradually from μ_1 to $\mu_1/4$. For $20 \le x \le \infty$, the linear mass density is constant at $\mu_1/4$. (a) What is the wave velocity for large values of x? (b) What is the amplitude of the wave for large values of x? (c) Give $y(x,\ t)$ for $20 \le x \le \infty$.

Chapter 14

Sound

Sound waves from a telephone receiver spreading out in air. The waves have been made visible by sweeping out the space in front of the receiver with a light bulb that has a brightness controlled by a microphone.

Sound waves are *longitudinal* waves of compression and rarefaction in a gas, liquid, or solid medium. They are produced when a body, such as a tuning fork or violin string, vibrates and causes a disturbance in the density of a medium. The disturbance is propagated through the medium by the interactions of its molecules. The vibration of the molecules is along the direction of propagation of the wave. As in the case of waves on a string, only the *disturbance* is propagated; the molecules themselves merely vibrate back and forth about their equilibrium positions.

In a gas, density and pressure are closely related. Therefore, sound waves in a gas, such as air, can be thought of as either density waves or pressure waves. In waves on a string, the transverse displacement of the string is described by a wave function $y(x \pm vt)$. The wave functions for sound waves analogous to the transverse displacement of a string are the longitudinal displacement of gas molecules from their equilibrium position $s(x \pm vt)$ or a related function—the change in the gas pressure $p(x \pm vt)$.

Some of the properties of sound waves discussed in this chapter, such as interference, diffraction, reflection, and refraction, are applicable to all types of waves and will be encountered again when we study optics.

14-1   Speed of Sound Waves

The speed of sound waves, like the speed of waves on strings, depends on the properties of the medium. For sound waves in a fluid such as air or water, the speed v is given by

$$v = \sqrt{\frac{B}{\rho}} \qquad\qquad 14\text{-}1$$

where ρ is the equilibrium density of the medium and B is the bulk modulus (Section 11-3). The derivation of this result from Newton's laws is given at the end of this section. For sound waves in a long, solid rod, the bulk modulus is replaced by Young's modulus Y (Section 11-2):

$$v = \sqrt{\frac{Y}{\rho}} \qquad\qquad 14\text{-}2$$

Comparing Equations 14-1 and 14-2 for the speed of sound waves with Equation 13-4 for the speed of waves on a string, we can see that, in general, the speed of waves depends on (1) an elastic property of the medium—the tension for string waves and the bulk modulus or Young's modulus for sound waves—and (2) an inertial property of the medium—the linear mass density or the volume mass density.

For sound waves in a gas such as air, the bulk modulus is proportional to the pressure, which is in turn proportional to the density ρ and to the absolute temperature T of the gas. The ratio B/ρ is thus independent of both the volume and pressure and is merely proportional to the absolute temperature T. In Chapter 16, we will show that, for sound waves in a gas, Equation 14-1 is equivalent to

Speed of sound in a gas

$$v = \sqrt{\frac{\gamma R T}{M}} \qquad\qquad 14\text{-}3$$

In this equation, T is the absolute temperature measured in kelvins (K), which is related to the Celsius temperature t_C by

$$T = t_C + 273 \qquad\qquad 14\text{-}4$$

The constant R is the universal gas constant, which has the value

$$R = 8.314 \text{ J/mol·K} \qquad\qquad 14\text{-}5$$

The constant M is the molar mass of the gas (that is, the mass of 1 mol of the gas), which for air is

$$M = 29 \times 10^{-3} \text{ kg/mol}$$

and γ is a constant that depends on the kind of gas and has the value 1.4 for air.

Example 14-1

Calculate the speed of sound in air at (*a*) 0°C and (*b*) 20°C.

(*a*) The absolute temperature corresponding to the Celsius temperature 0°C is

$$T = t_C + 273 = 0 + 273 = 273 \text{ K}$$

The speed of sound at 0°C is therefore

$$v = \sqrt{\frac{\gamma R T}{M}} = \sqrt{\frac{(1.4)(8.31 \text{ J/mol·K})(273 \text{ K})}{29.0 \times 10^{-3} \text{ kg/mol}}} = 331 \text{ m/s}$$

(b) To find the speed at 20°C = 293 K, we note that the speed of sound is proportional to the square root of the absolute temperature. Its value at 293 K, v_{293}, is therefore related to its value 273 K, v_{273}, by

$$\frac{v_{293}}{v_{273}} = \frac{\sqrt{293}}{\sqrt{273}}$$

or

$$v = \sqrt{\tfrac{293}{273}} \, (331 \text{ m/s})$$

$$= 343 \text{ m/s}$$

Exercise

For helium, $M = 4 \times 10^{-3}$ kg/mol and $\gamma = 1.67$. What is the speed of sound waves in helium at 20°C? (Answer: 1.01 km/s)

Question

1. Although the densities of most solids are more than 1000 times that of air, the speed of sound in solids is usually greater than in air. Why?

Derivation of Equation for the Speed of Sound (Optional)

Consider a fluid of density ρ and pressure P in a long tube as shown in Figure 14-1. As a source of a longitudinal wave in the fluid, we imagine a piston of area A. We suddenly compress the fluid by moving the piston to the right for a short time Δt, which increases the pressure to the left of the fluid by an amount ΔP. The piston collides with molecules in the fluid, which in turn collide with other molecules, and a disturbance in the form of a wave pulse is propagated down the tube. We make the simplifying assumption that the piston moves with a constant speed u for the time Δt. We further assume that the piston speed u is much less than the wave speed v. (It is important not to confuse these two speeds.) In time Δt the piston moves a distance $u \, \Delta t$, and the wave pulse moves a distance $v \, \Delta t$. We assume in our model that the effect of the motion of the piston is to give a speed u to all the fluid from the piston to the leading edge of the pulse, a distance $v \, \Delta t$ from the original position of the piston. This distance is based on the fact that the wave speed v is the speed at which the disturbance is propagated through the fluid. Thus, in time Δt, the greatest distance the disturbance can penetrate into the fluid ahead of the piston is $v \, \Delta t$. The assumption that all the fluid in this region moves with the same speed u amounts to assuming a rectangular shape for the wave pulse. We can calculate the speed of the pulse by setting the change in the momentum of the fluid equal to the impulse acting on the fluid due to the increased pressure for the time Δt. If A is the area of the piston, the impulse is

$$\text{Impulse} = F \, \Delta t = A \, \Delta P \, \Delta t$$

Figure 14-1 A longitudinal wave pulse is produced by suddenly moving the piston to the right with speed u. After a short time Δt, the piston has moved a distance $u \, \Delta t$ and the pulse has moved a distance $v \, \Delta t$. If a rectangular pulse is assumed, the fluid in the shaded region of length $v \, \Delta t$ is moving with speed u.

The mass of fluid set into motion is the density ρ times the volume $Av \, \Delta t$. The change in momentum is this mass times the velocity u:

$$\text{Momentum change} = \rho(Av \, \Delta t)u$$

Equating the momentum change to the impulse gives

$$A \, \Delta P \, \Delta t = \rho(Av \, \Delta t)u$$

or

$$\Delta P = \rho v u \qquad\qquad 14\text{-}6$$

The change in pressure is related to the decrease in volume of the fluid by the bulk modulus (Equation 11-12):

$$\Delta P = B \frac{-\Delta V}{V}$$

The original volume of the fluid under consideration is $V = Av\ \Delta t$, and the change in the volume swept out by the piston is $\Delta V = -Au\ \Delta t$. Thus,

$$\frac{-\Delta V}{V} = \frac{Au\ \Delta t}{Av\ \Delta t} = \frac{u}{v}$$

and

$$\Delta P = \frac{Bu}{v} \qquad\qquad 14\text{-}7$$

Using this result for ΔP in Equation 14-6, we obtain

$$\frac{Bu}{v} = \rho v u$$

or

$$v = \sqrt{\frac{B}{\rho}}$$

which is Equation 14-1.

14-2   Harmonic Sound Waves

Harmonic sound waves can be generated by a source that is vibrating with simple harmonic motion, such as a tuning fork, or a speaker driven by an audio oscillator. The vibrating source causes the air molecules next to it to oscillate with simple harmonic motion about their equilibrium position. These molecules collide with neighboring molecules, causing them to oscillate, thereby propagating the sound wave. The displacement of the molecules $s(x,t)$ for a harmonic wave can be written

$$s(x,t) = s_0 \sin (kx - \omega t) \qquad\qquad 14\text{-}8$$

where s_0 is the maximum displacement of a gas molecule from its equilibrium position, k is the wave number

$$k = \frac{2\pi}{\lambda} \qquad\qquad 14\text{-}9$$

and ω is the angular frequency

$$\omega = 2\pi f = \frac{2\pi}{T} \qquad\qquad 14\text{-}10$$

As with all harmonic waves, the speed of the wave equals the frequency times the wavelength:

$$v = f\lambda = \frac{\omega}{k} \qquad\qquad 14\text{-}11$$

The displacement from equilibrium given by Equation 14-8 is along the direction of the motion of the wave; that is, sound is a longitudinal wave. These displacements lead to variations in density of the air and in the air pressure. Figure 14-2a shows a graph of the displacement as a function of position at some fixed time. Points x_1 and x_3 are points of zero displacement at this time. Just to the left of point x_1, the displacement is negative, indicating that the gas molecules are displaced to the left, away from point x_1, at this time. Just to the right of x_1, the displacement is positive, indicating that the molecules are displaced to the right, which is again away from point x_1. Figure 14-2b shows some representative molecules near points x_1, x_2, and x_3

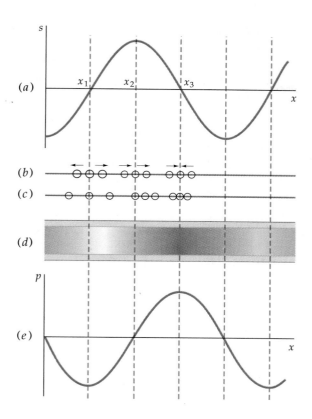

(a)

(b)

(c)

(d)

(e)

Figure 14-2 (a) Displacement from equilibrium of air molecules in a harmonic sound wave versus position at some instant. The molecules are at their equilibrium positions at points x_1 and x_3 and have maximum displacements at point x_2. (b) Some representative molecules equally spaced at their equilibrium positions before the sound wave arrives. The arrows indicate the direction of the displacement that will be caused by the sound wave. (c) Molecules near points x_1, x_2, and x_3 after the sound wave arrives. (d) Density of the air at this time. The density is maximum at x_3 and minimum at x_1, which are both points of zero displacement. (e) Pressure change versus position. The pressure change and displacement are 90° out of phase.

when there is no sound wave. These molecules are equally spaced. The arrows in this figure show the direction of displacements that will be caused by the sound wave. In Figure 14-2c, these displacements have occurred. At point x_1, the density is a minimum because the gas molecules on both sides are displaced away from that point. At point x_3 the density is a maximum because the molecules on both sides of that point are displaced toward point x_3. At point x_2 the displacement function is a maximum. The displacements to the left and to the right of this point are both positive and of equal magnitude. Thus, the density does not change near this point. Since the pressure in a gas is proportional to the density, the change in pressure is maximum where the change in density is maximum. Figure 14-2e indicates the change in pressure for this displacement wave. It could also represent the change in density for this wave. We can see from these figures that the pressure or density wave is 90° out of phase with the displacement wave. When the displacement is zero, the pressure and density changes are either maximum or minimum. When the displacement is a maximum or minimum, the pressure and density changes are zero. At the end of this section, we will show that a displacement wave given by Equation 14-8 implies a pressure wave given by*

$$p = p_0 \sin (kx - \omega t - \pi/2) \qquad\qquad 14\text{-}12$$

where p stands for the *change* in pressure from the equilibrium pressure and p_0 is the maximum value of this change. We will also see that the maximum pressure amplitude p_0 is related to the maximum displacement amplitude s_0 by

$$p_0 = \rho \omega v s_0 \qquad\qquad 14\text{-}13$$

*In the arguments of sine or cosine functions, we will always express phase angles in radians, which are dimensionless. However, in verbal descriptions it is often more convenient to describe phase differences in degrees. Thus we say that "two waves are 90° out of phase" rather than "two waves are out of phase by $\pi/2$ rad."

where v is the speed of propagation and ρ is the equilibrium density of the gas.

As a sound wave moves in time, the displacement of air molecules, the density, and the pressure at one point all vary sinusoidally with a frequency f, which equals the frequency of the vibrating source.

Example 14-2

The human ear can hear sound of frequencies that range from about 20 Hz to about 20,000 Hz (though many people have rather limited hearing above about 15,000 Hz). If the speed of sound in air is 340 m/s, what are the wavelengths that correspond to these extreme frequencies?

If the speed of sound in air is 340 m/s, the wavelength corresponding to the lowest audible frequency is

$$\lambda = \frac{v}{f} = \frac{340 \text{ m/s}}{20 \text{ Hz}} = 17 \text{ m}$$

and that corresponding to the highest audible frequency is

$$\lambda = \frac{v}{f} = \frac{340 \text{ m/s}}{20 \text{ kHz}} = 1.7 \text{ cm}$$

Exercise

If the speed of sound in air is 340 m/s, what is the frequency of a sound wave whose wavelength is 2 m?   (Answer: 170 Hz)

Derivation of Equations 14-12 and 14-13 (Optional)

For a sound wave in a gas, we can relate the change in pressure to the displacement by noting that a displacement changes the volume of a given mass of gas, and that changes in pressure and volume are related by the bulk modulus. Since we are using p for a change in pressure here, we have from the definition of the bulk modulus

$$p = -B \frac{\Delta V}{V}$$

Using $B = \rho v^2$ from Equation 14-1, we have

$$p = -\rho v^2 \frac{\Delta V}{V} \qquad 14\text{-}14$$

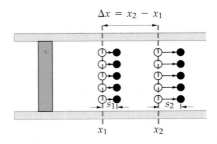

$\Delta x = x_2 - x_1$

$x_1 \qquad x_2$

Figure 14-3 Change in the volume of a given mass of gas due to a variation in the displacement with position. The original volume is $V = A \Delta x$, and the change in volume is $\Delta V = A \Delta s$, where A is the cross-sectional area of the tube.

Let us consider the mass of gas in the tube shown in Figure 14-3 that is initially between points x_1 and x_2. It occupies a volume $V = A(x_2 - x_1) = A \Delta x$, where A is the cross-sectional area of the tube and $\Delta x = x_2 - x_1$. In general, a displacement of a given mass of gas causes a change in its volume. If the displacement is the same at points x_1 and x_2, there will be no change in the volume, but if the displacement is different at these two points, the volume will change. For example, if the displacement is greater at x_2 than at x_1, as is the case in Figure 14-3, the volume will increase. If $\Delta s = s(x_2, t_0) - s(x_1, t_0)$ is the difference in the displacement at these two points, the change in the volume is

$$\Delta V = A \Delta s$$

Substituting this and $V = A \Delta x$ into Equation 14-14, we obtain

$$p = -\rho v^2 \frac{A \Delta s}{A \Delta x} = -\rho v^2 \frac{\Delta s}{\Delta x} \qquad 14\text{-}15$$

In the limit as Δx goes to zero, the ratio $\Delta s/\Delta x$ becomes the derivative of s with respect to x. This is a partial derivative because x is also a function of t, which we are holding constant. Replacing $\Delta s/\Delta x$ with the notation for the partial derivative $\partial s/\partial x$, we have

$$p = -\rho v^2 \frac{\partial s}{\partial x}$$

We compute $\partial s/\partial x$ from Equation 14-8:

$$s(x, t) = s_0 \sin (kx - \omega t)$$

$$\frac{\partial s}{\partial x} = s_0 k \cos (kx - \omega t)$$

Hence,

$$p = -\rho v^2 s_0 k \cos (kx - \omega t)$$

$$= +k\rho v^2 s_0 \sin \left(kx - \omega t - \frac{\pi}{2} \right)$$

$$= p_0 \sin \left(kx - \omega t - \frac{\pi}{2} \right) \qquad \text{14-16}$$

where

$$p_0 = k\rho v^2 s_0$$

But according to Equation 14-11, $kv = \omega$. Therefore,

$$p_0 = \rho \omega v s_0 \qquad \text{14-17}$$

and Equations 14-16 and 14-17 are identical to Equations 14-12 and 14-13.

14-3 Waves in Three Dimensions: Intensity

So far, we have discussed one-dimensional waves, that is, waves that propagate in a straight line. Figure 14-4 shows two-dimensional circular waves on the surface of water in a ripple tank. These waves are generated by a point source moving up and down with simple harmonic motion. The wavelength, in this case, is the distance between successive wave crests, which are concentric circles. These circles are called **wavefronts.** For a point source of sound, the waves are emitted in three dimensions. They move out in all directions, and the wavefronts are concentric spherical surfaces.

The motion of any set of wavefronts can be indicated by **rays,** which are directed lines perpendicular to the wavefronts (Figure 14-5). For circular or spherical waves, the rays are radial lines.

If a point source emits waves uniformly in all directions, the energy at a distance r from the source will be distributed uniformly on a spherical shell of radius r and area $4\pi r^2$. If P is the power emitted by the source, the power per unit area at a distance r from the source will be $P/4\pi r^2$. The average power per unit area that is incident perpendicular to the direction of propagation is called the **intensity:**

$$I = \frac{P_{av}}{A} \qquad \text{14-18}$$

The units of intensity are watts per square meter. At a distance r from a point source, the intensity is

$$I = \frac{P_{av}}{4\pi r^2} \qquad \text{14-19}$$

Figure 14-4 Circular wavefronts diverging from a point source in a ripple tank.

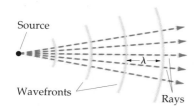

Figure 14-5 The motion of wavefronts can be represented by rays drawn perpendicular to the wavefronts. For a point source, the rays are radial lines diverging from the source.

Intensity due to a point source

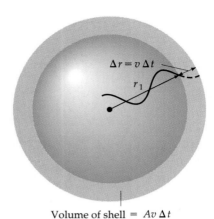

Volume of shell = $Av\,\Delta t$

Figure 14-6   A spherical wave that has just reached the distance r_1 at this time. All the energy is contained in the sphere of this radius. In some time Δt, the wave moves out a distance $v\,\Delta t$ and occupies the additional volume $Av\,\Delta t$ of the spherical shell. The energy transmitted past r_1 is $\eta Av\,\Delta t$, where η is the energy per unit volume and A is the average surface area of the spherical shell.

The intensity of a three-dimensional wave varies inversely with the square of the distance from a point source.

There is a simple relation between the intensity of a wave and the energy per unit volume in the medium carrying the wave. Consider the spherical wave that has just reached the radius r_1 in Figure 14-6. The volume inside the radius r_1 contains energy because the particles in that region are oscillating with simple harmonic motion. The region outside r_1 contains no energy because the wave has not yet reached that region. After a short time Δt, the wave moves out past r_1 a short distance $\Delta r = v\,\Delta t$. The total energy in the medium is increased by the energy in the spherical shell of surface area A, thickness $v\,\Delta t$, and volume $\Delta V = Av\,\Delta t$. The additional energy in the spherical shell is

$$\Delta E = \eta\,\Delta V = \eta Av\,\Delta t$$

where η is the average energy per unit volume in the shell that now contains energy. The rate of increase of energy is the power passing into the shell. The source of this energy is at the center of the sphere from which the wave is radiating. Thus, the average incident power is

$$P_{av} = \frac{\Delta E}{\Delta t} = \eta Av$$

and the intensity of the wave is

$$I = \frac{P_{av}}{A} = \eta v \qquad\qquad 14\text{-}20$$

Thus, the intensity equals the product of the wave speed v and the average energy per unit volume η. This result applies to all waves.

The energy of a sound wave in air is the energy of oscillation of air molecules that vibrate in simple harmonic motion along the direction of wave propagation. As found in Chapter 12, the total energy of a mass m oscillating in simple harmonic motion with angular frequency ω and amplitude A is $E = \frac{1}{2}m\omega^2 A^2$. Writing s_0 for the amplitude and $\Delta m = \rho\,\Delta V$ for the mass, where ρ is the density, we have for the energy of a sound wave in volume element ΔV

$$\Delta E = \tfrac{1}{2}\rho\omega^2 s_0^2\,\Delta V$$

The energy density η is thus

$$\eta = \frac{\Delta E}{\Delta V} = \tfrac{1}{2}\rho\omega^2 s_0^2 \qquad\qquad 14\text{-}21$$

and the intensity is

$$I = \eta v = \tfrac{1}{2}\rho\omega^2 s_0^2 v = \frac{1}{2}\frac{p_0^2}{\rho v} \qquad\qquad 14\text{-}22$$

where we have used $s_0 = p_0/\rho\omega v$ from Equation 14-13 to relate the displacement amplitude s_0 to the pressure amplitude p_0. This result—that the intensity of a sound wave is proportional to the square of the amplitude—is a general property of harmonic waves.

The human ear can accommodate a rather large range of sound-wave intensities, from about 10^{-12} W/m^2 (which is usually taken to be the threshold of hearing) to about 1 W/m^2 (which produces a sensation of pain in most people). The pressure variations that correspond to these extreme intensities are about 3×10^{-5} Pa for the hearing threshold and 30 Pa for the pain threshold. (Recall that a pascal is a newton per square meter.) These small pressure variations are superimposed on the normal atmospheric pressure of about 101 kPa.

Loudness and Intensity Level

Because of the enormous range of intensities to which the ear is sensitive and because the psychological sensation of loudness does not vary directly with intensity but more nearly logarithmically, a logarithmic scale is used to describe the intensity level of a sound wave. The **intensity level** β measured in **decibels** (dB) is defined by

$$\beta = 10 \log \frac{I}{I_0}$$

14-23   *Intensity level in dB*

where I is the intensity of the sound and I_0 is a reference level, which we will take to be the threshold of hearing:

$$I_0 = 10^{-12} \text{ W/m}^2$$

14-24

On this scale, the threshold of hearing is

$$\beta = 10 \log \frac{I_0}{I_0} = 0 \text{ dB}$$

and the pain threshold is

$$\beta = 10 \log \frac{1}{10^{-12}} = 10 \log 10^{12} = 120 \text{ dB}$$

Thus, the range of sound intensities from 10^{-12} W/m^2 to 1 W/m^2 corresponds to a range of intensity levels from 0 dB to 120 dB. Table 14-1 lists the intensity levels of some common sounds.

Exercise

What is the ratio of the intensity of a 90-dB sound to that of a 60-dB sound?   (Answer: 1000)

Table 14-1   Intensity and Intensity Level of Some Common Sounds
($I_0 = 10^{-12}$ W/m^2)

Source	I/I_0	dB	Description
	10^0	0	Hearing threshold
Normal breathing	10^1	10	Barely audible
Rustling leaves	10^2	20	
Soft whisper (at 5 m)	10^3	30	Very quiet
Library	10^4	40	
Quiet office	10^5	50	Quiet
Normal conversation (at 1 m)	10^6	60	
Busy traffic	10^7	70	
Noisy office with machines; average factory	10^8	80	
Heavy truck (at 15 m); Niagara Falls	10^9	90	Constant exposure endangers hearing
Old subway train	10^{10}	100	
Construction noise (at 3 m)	10^{11}	110	
Rock concert with amplifiers (at 2 m); jet takeoff (at 60 m)	10^{12}	120	Pain threshold
Pneumatic riveter; machine gun	10^{13}	130	
Jet takeoff (nearby)	10^{15}	150	
Large rocket engine (nearby)	10^{18}	180	

Example 14-3

A barking dog delivers about 1 mW of power. (*a*) If this power is uniformly distributed in all directions, what is the sound intensity level at a distance of 5 m? (*b*) What would the intensity level be of two dogs barking at the same time if each delivered 1 mW of power?

(*a*) The intensity at a distance of 5 m is the power divided by the area (Equation 14-19):

$$I = \frac{P}{4\pi r^2} = \frac{10^{-3}\text{ W}}{4\pi(5\text{ m})^2} = 3.18 \times 10^{-6}\text{ W/m}^2$$

The intensity level at this distance is

$$\beta = 10 \log \frac{I}{I_0} = 10 \log \frac{3.18 \times 10^{-6}}{10^{-12}}$$

$$= 10 \log (3.18 \times 10^6)$$

$$= 10 (\log 3.18 + \log 10^6)$$

$$= 10(0.50 + 6) = 65.0\text{ dB}$$

(*b*) If there are two dogs barking at the same time, the intensity will be 2 times as great, or

$$I = 2 \times (3.18 \times 10^{-6}\text{ W/m}^2) = 6.36 \times 10^{-6}\text{ W/m}^2$$

Then I/I_0 will be

$$\frac{I}{I_0} = 6.36 \times 10^6$$

and the intensity level will be

$$\beta = 10 \log (6.36 \times 10^6) = 68.0\text{ dB}$$

We can see from this example that if the intensity is doubled, the intensity level increases by 3 dB.

Example 14-4

A sound absorber attenuates the sound level by 30 dB. By what factor is the intensity decreased?

From Table 14-1, we can see that for every 10-dB drop in the intensity level, the intensity decreases by a factor of 10. Thus, a drop of 30 dB corresponds to a decrease in intensity by a factor of $10^3 = 1000$.

The sensation of loudness depends on the frequency as well as the intensity of a sound. Figure 14-7 is a plot of intensity level versus frequency for sounds of equal loudness to the human ear. (In this figure, the frequency is plotted on a logarithmic scale to display the wide range of frequencies from 20 Hz to 10 kHz.) The lowest curve corresponds to the threshold of hearing for someone with a very good ear. We can see from this curve that the hearing threshold is 0 dB at 1 kHz, but it is about 50 dB at 60 Hz. About 1 percent of the population have a hearing threshold this low. The second lowest curve is a more typical hearing threshold curve that applies to about 50 percent of the population. The upper curve represents the pain threshold. Note that it does not vary as much with frequency as do the lower-level curves. We note from this figure that the human ear is most sensitive at about 4 kHz for all intensity levels.

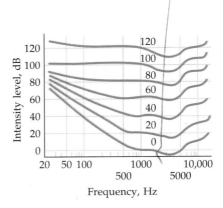

Figure 14-7 Intensity level versus frequency for sounds of equal loudness. The lowest curve is below the threshold for hearing of all but about 1 percent of the population. The second lowest curve is approximately the hearing threshold for about 50 percent of the population.

14-4   Interference: Beats

In Section 13-5, where we discussed the interference of harmonic waves, we saw that the superposition of two harmonic waves of equal amplitude results in a third harmonic wave whose amplitude depends on the phase difference between the original waves. If the two waves are in phase, the interference is constructive and the amplitude of the resultant wave is twice that of either of the original waves. If the two waves differ in phase by 180°, the interference is destructive and the waves cancel. For a phase difference δ, the pressure of the resultant wave is (Equation 13-19)

$$p_1 + p_2 = 2p_0 \cos (\tfrac{1}{2}\delta) \sin (kx - \omega t + \tfrac{1}{2}\delta) \qquad 14\text{-}25$$

A common cause of a phase difference between two sound waves is different path lengths between the sources of waves and the point of interference. Suppose we have two sources that are emitting harmonic waves of the same frequency and wavelength and are oscillating in phase; that is, when a positive crest leaves one source, a positive crest leaves the other source at the same time. If the path difference to some point is one wavelength, as is the case in Figure 14-8, the interference is constructive. The two waves fall on top of one another with their maxima occurring at the same time and their minima also occurring at the same time. If the waves have equal amplitudes, the amplitude of the resulting wave will be twice that of either original wave. We can see from the figure that a path difference of one wavelength or any integral number of wavelengths is equivalent to no path difference at all. If the path difference is one-half wavelength or an odd number of half wavelengths, as in Figure 14-9, the maximum of one wave falls at the minimum of the other and the interference is destructive. For the general case, we can write the wave functions

$$p_1 = p_0 \sin (kx_1 - \omega t)$$

and

$$p_2 = p_0 \sin (kx_2 - \omega t)$$

The phase difference for these two wave functions is

$$\delta = (kx_2 - \omega t) - (kx_1 - \omega t) = k(x_2 - x_1) = k \, \Delta x$$

Using $k = 2\pi/\lambda$, we have

$$\delta = 2\pi \frac{\Delta x}{\lambda} = (360°) \frac{\Delta x}{\lambda} \qquad 14\text{-}26 \qquad \textit{Phase difference due to path difference}$$

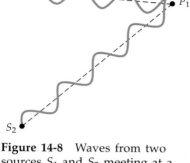

Figure 14-8 Waves from two sources S_1 and S_2 meeting at a point P_1 that is farther from one source than the other by one wavelength λ. The waves are in phase at P_1 and interfere constructively. They will also interfere constructively at any other point for which the path difference is zero or an integral number of wavelengths.

Figure 14-9 Waves from two sources S_1 and S_2 meeting at a point P_2 that is farther from one source than the other by $\tfrac{1}{2}\lambda$. The sources are in phase, but the waves at P_2 are out of phase by 180° and therefore interfere destructively. If the waves are of equal amplitude at P_2, they will cancel completely at this point. They will also interfere destructively at any other point for which the path difference is $\tfrac{1}{2}\lambda$ or any odd number of half wavelengths.

Example 14-5

Two sound sources oscillate in phase. At a point 5.00 m from one source and 5.17 m from the other, the amplitude of the sound from each source separately is p_0. Find the amplitude of the resultant wave if the frequency of the sound waves is (a) 1000 Hz, (b) 2000 Hz, and (c) 500 Hz. (Use 340 m/s for the speed of sound.)

In all the cases, the path difference is $\Delta x = 5.17$ m $- 5.00$ m $= 0.17$ m $= 17$ cm.

(a) If the frequency is 1000 Hz, the wavelength is

$$\lambda = \frac{v}{f} = \frac{340 \text{ m/s}}{1000 \text{ Hz}} = 0.34 \text{ m} = 34 \text{ cm} = 2 \Delta x$$

Since the path difference is one-half the wavelength, the phase difference is π and the resultant amplitude will be 0.

(b) When the frequency is 2000 Hz, the wavelength is

$$\lambda = \frac{340 \text{ m/s}}{2000 \text{ Hz}} = 17 \text{ cm} = \Delta x$$

In this case, the path difference is one wavelength and the resultant amplitude is $2p_0$.

(c) For $f = 500$ Hz, the wavelength is

$$\lambda = \frac{340 \text{ m/s}}{500 \text{ Hz}} = 68 \text{ cm} = 4 \Delta x$$

The phase difference is then

$$\delta = 2\pi \frac{\Delta x}{\lambda} = 2\pi \frac{\Delta x}{4 \Delta x} = \frac{\pi}{2}$$

The factor $2p_0 \cos \frac{1}{2}\delta$ in Equation 14-25 is the amplitude of the resulting wave. For $\delta = \pi/2$, the amplitude is

$$2p_0 \cos \tfrac{1}{2}\left(\frac{\pi}{2}\right) = 2p_0 \cos \left(\frac{\pi}{4}\right) = 1.41 \, p_0$$

Figure 14-10 (a) Interference pattern of two point sources oscillating in phase in a ripple tank. (b) Geometric construction of the interference pattern in (a). The waves interfere constructively at the points of intersection. These points occur wherever the path lengths from the two sources differ by an integral number of wavelengths.

Figure 14-10a shows the wave pattern produced by two point sources separated by a small distance that are oscillating in phase, each producing circular waves of wavelength λ. We can construct a similar pattern with a compass by drawing circular arcs representing the wave crests from each source at some particular time (Figure 14-10b). At the points where the crests from each source overlap, the waves interfere constructively. At these points, the paths for the waves from the two sources are either equal in length or differ by an integral number of wavelengths. The dashed lines

(a)

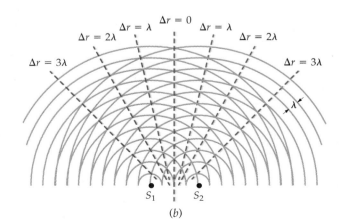

(b)

indicate the points that are equidistant from the sources or whose path differences are one wavelength, two wavelengths, or three wavelengths. If we assume that the two sources generate waves of equal amplitude, the resulting amplitude at points where the waves interfere constructively will be twice the amplitude of either wave separately. Since the energy of a wave is proportional to the square of the amplitude, the energy of the resultant wave where the waves interfere constructively will be 4 times that of either wave separately. Between each pair of interference maxima is an interference minimum for which the path difference is an odd number of half wavelengths. The lines along which the waves completely cancel are called nodes or nodal lines. At any point between the maxima and minima, the amplitude of the resultant wave is given by $A = 2p_0 \cos \frac{1}{2}\delta$, where p_0 is the amplitude of each wave separately, and δ is related to the path difference by Equation 14-26.

Figure 14-11 shows the intensity of the resultant wave from two sources as a function of path difference. At points where the interference is constructive, the intensity is 4 times that due to either source alone because the amplitude of the resultant wave is twice that of either wave alone. At points of destructive interference, the intensity is zero. The average intensity, shown by the dashed line in the figure, is twice the intensity due to either source alone, a result required by the conservation of energy. We can see that, as a result of the interference of the waves from the two sources, the energy is redistributed in space.

The interference of two sound sources can be demonstrated by driving two separated speakers with the same amplifier (so that they are always in phase) fed by an audio-signal generator. Moving about the room, one can detect by ear the positions of constructive and destructive interference. (In this demonstration, the sound intensity will not be quite zero at the points of destructive interference of the direct sound waves because of sound reflections from the walls or objects in the room.)

Two sources need not be in phase to produce an interference pattern. Figure 14-12 shows a plot of the intensity versus path difference produced by two sources that are 180° out of phase. (Two sound sources that are in phase can be made to be out of phase by 180° merely by switching the leads to one of the speakers.) The pattern is the same as that in Figure 14-11 except that the maxima and minima are interchanged. At points equidistant from the sources or those for which the distance differs by an integral number of wavelengths the interference is destructive because the waves are 180° out of phase and cancel (Figure 14-13). At points where the path difference is an odd number of half wavelengths, the waves are now in phase because the 180° phase difference of the sources is offset by the 180° phase difference due to the path difference.

No matter what the phase difference between the sources might be, similar interference patterns will be produced as long as the phase difference is constant over time. Two sources that are in phase or have a constant phase difference are said to be **coherent sources.** Coherent sources of water waves in a ripple tank are easy to produce by driving both sources with the same motor. Coherent sound sources are obtained by driving two speakers with the same signal source and amplifier. Wave sources whose phase difference is not constant over time but varies randomly are said to be incoherent sources. There are many examples of incoherent sources, such as two speakers driven by different amplifiers or two violins bowed by different players. In general, any two separate light sources, such as two candles, are incoherent. For incoherent sources, the interference at a particular point varies rapidly back and forth from constructive to destructive and no inererence pattern is observed. The resultant intensity of waves from two or more incoherent sources is simply the sum of the intensities due to the individual sources.

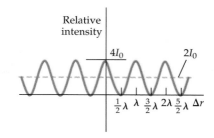

Figure 14-11   Plot of relative intensity versus path difference for two sources that are in phase. When the path difference is an integral number of wavelengths, the intensity is $4I_0$, where I_0 is the intensity due to each source individually. When the path difference is an odd number of half wavelengths, the intensity is zero. The dashed line indicates the average intensity, which is twice that due to either source alone.

Figure 14-12   Plot of relative intensity versus path difference for two sources that are 180° out of phase. The pattern is the same as that in Figure 14-11 except that it is shifted by one-half wavelength.

Figure 14-13   At a point equidistant from two sources that are out of phase by 180°, the waves cancel.

Beats

The interference of two waves of different but nearly equal frequencies produces the interesting phenomenon known as **beats.** A familiar example is the beats produced by the sound waves from two tuning forks or two guitar strings of nearly equal but not identical frequencies. What is heard is a tone whose intensity varies alternately between loud and soft. The frequency of this variation in intensity is called the **beat frequency.**

Let us consider two sound waves with angular frequencies of ω_1 and ω_2 having the same pressure amplitude p_0. We consider how the resultant wave behaves over time at a single point in space. We therefore neglect the spatial dependence of the wave, which merely contributes a phase constant. The pressure at the ear due to either wave acting alone will be a simple harmonic function of the type

$$p_1 = p_0 \sin \omega_1 t$$

or

$$p_2 = p_0 \sin \omega_2 t$$

where we have chosen sine functions for convenience and have assumed that the waves are in phase at time $t = 0$. Using

$$\sin \theta_1 + \sin \theta_2 = 2 \cos \tfrac{1}{2}(\theta_1 - \theta_2) \sin \tfrac{1}{2}(\theta_1 + \theta_2)$$

for the sum of two sine functions, we obtain for the resultant wave

$$p = p_0 \sin \omega_1 t + p_0 \sin \omega_2 t$$
$$= 2p_0 \cos \tfrac{1}{2}(\omega_1 - \omega_2)t \sin \tfrac{1}{2}(\omega_1 + \omega_2)t$$

If we write $\omega_{av} = \tfrac{1}{2}(\omega_1 + \omega_2)$ for the average angular frequency and $\Delta\omega = \omega_1 - \omega_2$ for the difference in angular frequencies, the resultant wave function is

$$p = 2p_0 \cos (\tfrac{1}{2} \Delta\omega \, t) \sin \omega_{av}t$$
$$= 2p_0 \cos (2\pi\tfrac{1}{2} \Delta f \, t) \sin 2\pi f_{av}t \qquad \text{14-27}$$

where $\Delta f = \Delta\omega/2\pi$ and $f_{av} = \omega_{av}/2\pi$.

Figure 14-14 shows a plot of the pressure variation as a function of time. The waves are originally in phase and add constructively at time $t = 0$. Because their frequencies differ, the waves gradually become out of phase, and at time t_1 they are 180° out of phase and interfere destructively. (Complete cancellation occurs only when the pressure amplitudes of the two waves are equal.) An equal time interval later (time t_2 in the figure), the two waves will again be in phase and will interfere constructively. The greater the difference in frequencies of the two waves, the sooner they become out of phase and then back in phase again.

Figure 14-14 Beats. (a) Two waves of different but nearly equal frequencies that are in phase at $t_0 = 0$ are 180° out of phase at some later time t_1. At a still later time t_2, they are back in phase. (b) The resultant of the two waves shown in (a). The frequency of the resultant wave is about the same as those of the original waves, but the amplitude is modulated as indicated by the dashed envelope. The amplitude is maximum at times t_0 and t_2 and zero at times t_1 and t_3.

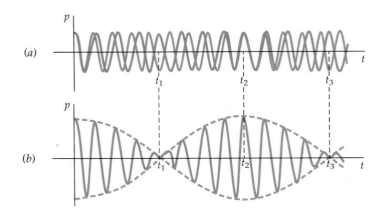

The ear hears the average frequency $f_{av} = \frac{1}{2}(f_1 + f_2)$ with an amplitude of $2p_0 \cos(2\pi \frac{1}{2}\Delta f\, t)$. The amplitude thus oscillates with the frequency $\frac{1}{2}\Delta f$. Since the intensity is proportional to the square of the amplitude, the sound is loud whenever the amplitude is either maximum or minimum. Since the maximum amplitude occurs with frequency $\frac{1}{2}\Delta f$, the frequency of maximum and minimum amplitudes is just twice this or Δf. Thus, the **beat frequency** equals the difference in the individual frequencies of the two waves:

$$f_{beat} = \Delta f \qquad\qquad 14\text{-}28$$

The closer the two waves are in frequency, the longer it takes for them to move out of phase and back in phase, so the closer the waves are in frequency, the smaller the beat frequency. For example, if we simultaneously strike two tuning forks having the frequencies 241 Hz and 243 Hz, we will hear a pulsating tone at the average frequency of 242 Hz that has a maximum intensity 2 times per second; that is, the beat frequency will be 2 Hz.

The phenomenon of beats is often used to compare an unknown frequency with a known frequency, as when a tuning fork is used to tune a piano string. The ear can detect up to about 15 to 20 beats per second. Above this frequency, the fluctuations in loudness are too rapid to be heard. Beats are also used to detect small frequency changes such as those that are produced when a radar beam is reflected from a moving car. The shift in frequency of the reflected beam is due to the doppler effect (Section 14-9), which depends on the speed of the car. The speed can be determined by measuring the beats produced when the reflected radar beam is combined with the original radar beam.

An interesting phenomenon somewhat related to beats is the moiré pattern that is produced when two sets of parallel lines of slightly different spacing overlap (Figure 14-15). Moiré patterns can often be observed when one looks through two parallel fences along a road or through two window screens.

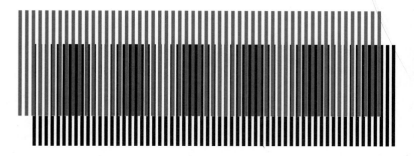

Figure 14-15 Moiré pattern produced by the overlap of two sets of parallel lines where the spacing of one set differs slightly from that of the other.

Example 14-6

When a 440-Hz tuning fork is struck simultaneously with the playing of the note A on a guitar, 3 beats per second are heard. After the guitar string is tightened slightly to increase its frequency, the beat frequency increases to 6 beats per second. What is the frequency of the guitar string after it is tightened?

Since 3 beats per second were heard initially, the original frequency of the guitar string was either 443 Hz or 437 Hz. Had it been 437 Hz, increasing the string's frequency by slightly tightening it would decrease the beat frequency because the frequency of the string then would be closer to that of the tuning fork. Since the beat frequency increases to 6 beats per second, the original frequency has to have been 443 Hz and the new frequency is 446 Hz. To tune the string to 440 Hz, the string should be loosened slightly to lower its frequency.

Questions

2. When musical notes are sounded together to make chords, beats are not heard. Why not?

3. About how accurately do you think you can tune a piano string to a tuning fork?

14-5   Standing Sound Waves

Much of what we learned about standing waves on strings in Chapter 13 can be applied to standing sound waves. Figure 14-16 shows a tube of air closed at the right end and fitted with a movable piston at the left end. Because the air cannot vibrate past the closed end, this point must be a displacement node. If the piston at the left end vibrates with an amplitude that is small relative to the amplitude of the waves in the air, then that end is also approximately a displacement node. (This approximation is similar to that for the driven end of a vibrating string.)

Figure 14-16   Air confined to a tube closed at both ends. There is a displacement node at each end if the amplitude of oscillation of the piston is small. The condition for standing sound waves is the same as that for standing waves on a string fixed at both ends.

The standing-wave condition for this system is the same as that for a string fixed at both ends, and all the same equations apply. The length of the tube L must contain an integral number of half wavelengths. That is, the allowed wavelengths are those that can be fitted into the length of the tube such that there are displacement nodes at each end. Thus,

$$L = n \frac{\lambda_n}{2} \qquad n = 1, 2, 3, \ldots \qquad \text{14-29}$$

The allowed frequencies are then

$$f_n = \frac{v}{\lambda_n} = \frac{v}{(2L/n)}$$

or

Standing-wave frequencies,
both ends closed

$$f_n = n \frac{v}{2L} = n f_1 \qquad n = 1, 2, 3, \ldots \qquad \text{14-30}$$

where

$$f_1 = \frac{v}{2L}$$

is the fundamental frequency.

As we have seen, a sound wave can be thought of as a pressure wave; that is, the longitudinal vibrations of the air molecules back and forth cause variations in the air pressure. These pressure variations are sinusoidal if the displacement is sinusoidal, as in a harmonic wave. However, the pressure and displacement variations in a sound wave are 90° out of phase. Thus, in a standing sound wave, the displacement nodes are pressure antinodes and vice versa.

There are no musical instruments based on tubes that are closed at both ends. However, there are electrical devices—antennas and cavity resonators, for example—that are analogous to this case.

Example 14-7

If the speed of sound is 340 m/s, what are the allowed frequencies and wavelengths for standing sound waves in a closed tube 1 m long?

The fundamental frequency is the lowest allowed frequency, which corresponds to the longest wavelength

$$\lambda_1 = 2L = 2 \text{ m}$$

Thus, the fundamental frequency is

$$f_1 = \frac{v}{\lambda_1} = \frac{340 \text{ m/s}}{2 \text{ m}} = 170 \text{ Hz}$$

The frequencies of the other harmonics are

$$f_2 = 2f_1 = 2(170 \text{ Hz}) = 340 \text{ Hz}$$

$$f_3 = 3f_1 = 3(170 \text{ Hz}) = 510 \text{ Hz}$$

$$f_n = nf_1 = 170n \text{ Hz}$$

The other wavelengths are

$$\lambda_2 = \frac{2L}{2} = 1 \text{ m}$$

$$\lambda_n = \frac{2L}{n} = \frac{2 \text{ m}}{n}$$

If the end of the tube at the right in Figure 14-16 is not closed but is open to the atmosphere, that open end is a displacement antinode. (It is also a pressure node since the pressure is fixed at atmospheric pressure.) The standing-wave condition for this system is the same as that for a string with one end fixed and one end free. The length of the tube must equal an odd integer times $\lambda/4$. That is, the wavelength of the fundamental mode is 4 times the length of the tube, and only the odd harmonics are present. Thus,

$$L = n\frac{\lambda_n}{4} \qquad n = 1, 3, 5, \ldots \qquad \text{14-31}$$

The allowed frequencies are therefore

$$f_n = n\frac{v}{4L} = nf_1 \qquad n = 1, 3, 5, \ldots \qquad \text{14-32} \qquad \textit{Standing-wave frequencies, one end open}$$

where

$$f_1 = \frac{v}{4L} \qquad \text{14-33}$$

is the fundamental frequency.

Exercise

Find the three lowest frequencies for a tube of length 1 m that is closed at one end and open at the other.
(Answer: $f_1 = 85$ Hz, $f_3 = 3f_1 = 255$ Hz, $f_5 = 5f_1 = 425$ Hz)

The result that the open end of the tube is a displacement antinode (and pressure node) is based on the assumption that the sound wave in the tube is a one-dimensional wave, which is approximately true if the diameter of the tube is very small compared with the wavelength of the sound wave. In practice, the displacement antinode and pressure node lie slightly beyond the open end of the tube. The effective length of the tube is therefore somewhat longer than the true length. If L is the true length of the tube and ΔL is

Figure 14-17 Apparatus for determining the speed of sound in air. Sound waves of the frequency of the tuning fork are excited in the tube on the left, whose length L can be adjusted by adjusting the level of the water. Resonance occurs when the effective length of the tube equals $\frac{1}{4}\lambda$, $\frac{3}{4}\lambda$, $\frac{5}{4}\lambda$, and so on, where λ is the wavelength of the sound.

the end correction for the actual position of the displacement antinode or pressure node, the effective length of the tube is

$$L_{\text{eff}} = L + \Delta L$$

The end correction ΔL is of the order of the radius of the tube. The distance between two successive nodes or antinodes is still $\frac{1}{2}\lambda$, but the distance from the open end of the tube to the first displacement node is slightly less than $\frac{1}{4}\lambda$ because of the end correction.

Figure 14-17 shows an apparatus for measuring the speed of sound in air. A narrow vertical tube on the left is partially filled with water. Since the level of the water can be adjusted, the length of the air column can be adjusted. A tuning fork of frequency f_0 vibrating at the open end of the tube sends sound waves of wavelength $\lambda_0 = v/f_0$ down the tube. They are reflected at the adjustable bottom, the water level. When the water level is such that the standing-wave condition is met, resonance occurs and the increased energy of the sound wave can be detected by ear. If ΔL is the end correction for the tube, the tube will resonate when its length is L_1, which is given by

$$L_1 + \Delta L = \tfrac{1}{4}\lambda_0$$

It will also resonate at lengths L_3, L_5, and so on, which are given by

$$L_3 + \Delta L = \tfrac{3}{4}\lambda_0$$
$$L_5 + \Delta L = \tfrac{5}{4}\lambda_0$$

and so on. The wavelength can be measured by measuring the distance between two consecutive resonance lengths. For example,

$$L_3 - L_1 = \tfrac{1}{2}\lambda_0$$

Then the speed can be found from

$$v = f_0\lambda_0 = 2f_0(L_3 - L_1)$$

Example 14-8

When a tuning fork of frequency 500 Hz is held above the tube in Figure 14-17, resonances are found when the water level is at distances 16.0, 50.5, 85.0, and 119.5 cm from the top of the tube. (a) What is the speed of sound in air? (b) How far from the end of the tube is the pressure node?

(a) The distance between the water levels for two successive resonances is 34.5 cm. The wavelength is thus twice this distance or 69.0 cm = 0.690 m. From these data, the speed of sound in air is

$$v = f\lambda = (500 \text{ Hz})(0.690 \text{ m}) = 345 \text{ m/s}$$

(b) Since the distance to the first position of resonance is 16.0 cm and $\lambda/4 = (69.0 \text{ cm})/4 = 17.25$ cm, the distance from the end of the tube to the pressure node or displacement antinode is

$$\Delta L = \tfrac{1}{4}\lambda - L = 17.25 \text{ cm} - 16.0 \text{ cm} = 1.25 \text{ cm}$$

Figure 14-18 Flue-type organ pipe. A stream of air is blown against the edge causing a swirling motion of the air near A that excites standing waves in the pipe. There is a pressure node near point A, which is open to the atmosphere. The resonance frequencies of the pipe depend on the length of the pipe and on whether the top of the pipe is open or closed.

An organ pipe is a familiar example of the use of standing waves in air columns. In the flue-type organ pipe (Figure 14-18), a stream of air is directed against the sharp edge of an opening (point A in the figure). The complicated swirling motion of the air near the edge sets up vibrations in the air column. The resonance frequencies of the pipe depend on the length of the pipe and on whether the top is closed or open.

In a closed organ pipe, there is a displacement node at the closed end and an antinode near the opening (point *A* in Figure 14-18). The resonance frequencies for such a pipe are therefore those for a tube open at one end and closed at the other. The wavelength of the fundamental is approximately 4 times the length of the pipe, and only the odd harmonics are present. The musical instruments in the clarinet family are effectively open–closed cylinders with odd harmonics for the lowest notes. Figure 14-19 shows the standing-wave patterns in a pipe open at one end and closed at the other.

In an open organ pipe, there is a displacement antinode (and pressure node) near both ends of the pipe. The resonance frequencies for a pipe open at both ends are the same as for a pipe closed at both ends except that there is an end correction at each end. The wavelength of the fundamental is 2 times the effective length of the pipe, and all harmonics are present. Flutes and recorders behave as open–open pipes with all the harmonics present. The standing-wave patterns for a pipe open at both ends are shown in Figure 14-20.

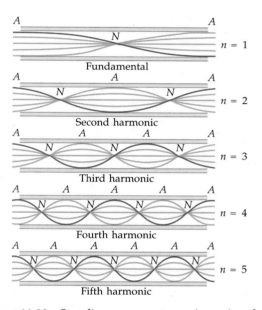

Figure 14-19 Standing-wave patterns in pipe that is open at one end and closed at the other. These plots are for the displacement, which has a node at the closed end and an antinode at the open end. As is the case for a string fixed at one end, only the odd harmonics are present.

Figure 14-20 Standing-wave patterns in a pipe that is open at both ends. The displacement has an antinode at each end. As is the case for a string fixed at both ends, all harmonics are present.

As with vibrating strings, the general motion of the air in an organ pipe contains a mixture of harmonics. In addition, most musical instruments are much more complicated than simple cylindrical tubes. The conical tube, which is the basis for the oboe, bassoon, English horn, and saxophone, has a complete harmonic series with its fundamental wavelength equal to twice the length of the cone. Brass instruments are combinations of cones and cylinders. The fact that they have nearly harmonic series is a triumph of educated trial and error rather than mathematical calculation. The analysis of these instruments is extremely complex.

Questions

4. How do the resonance frequencies of an organ pipe change when the air temperature increases?

14-6   Harmonic Analysis and Synthesis

When an oboe and a violin play the same note, say, concert A, they sound quite different. Both notes have the same pitch, a physiological sensation of the highness or lowness of the note that is strongly correlated with frequency—the higher the frequency, the higher the pitch. However, the notes differ in what is called **tone quality**. The principal reason for the difference in tone quality is that, although both the violin and oboe are producing vibrations at the same fundamental frequency, 440 Hz for concert A, each instrument is also producing harmonics whose relative intensities depend on the instrument and how it is played. If each instrument produced only the fundamental frequency, the sound would be the same for each.

Figure 14-21   Waveforms of (*a*) a tuning fork, (*b*) a clarinet, and (*c*) a cornet, each at a fundamental frequency of 440 Hz and all at approximately the same intensity.

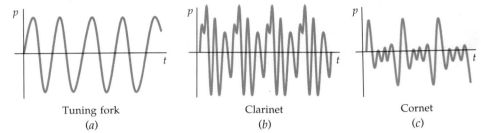

Tuning fork
(*a*)

Clarinet
(*b*)

Cornet
(*c*)

Figure 14-21 shows plots of the pressure variations versus time for a tuning fork, a clarinet, and a cornet, each playing the same note. These patterns are called **waveforms**. The waveform for the tuning fork is very nearly a pure sine wave, but those for the clarinet and the cornet are clearly not simple sine waves. Waveforms can be analyzed in terms of the harmonics that comprise them. Such analysis is called **harmonic analysis.** (It is also sometimes called **Fourier analysis** after the French mathematician Fourier who developed the mathematics for analyzing periodic functions.) Figure 14-22 shows a plot of the relative intensities of the harmonics of the waveforms in Figure 14-21. The waveform of the tuning fork contains only the fundamental frequency. That of the clarinet contains the fundamental, large amounts of the third, fifth, and seventh harmonics, and lesser amounts of the second, fourth, and sixth harmonics. For the cornet, there is more energy in the second and third harmonics than in the fundamental.

Although the harmonic content of steady tones is important in identifying the instrument, other factors are also important. These include

1. The attack (the way the tone begins)

2. The presence or absence of vibrato and tremolo (variations in pitch and loudness)

3. The rate of buildup of the harmonics

4. The decay (the rate at which the tone decreases from its peak level)

5. The release (the rate at which the sound level drops off at the end of the note)

Figure 14-22   Relative intensities of the harmonics in the waveforms shown in Figure 14-21 for (*a*) the tuning fork, (*b*) the clarinet, and (*c*) the cornet.

Tuning fork
(*a*)

Clarinet
(*b*)

Cornet
(*c*)

Figure 14-23 (*a*) A square wave and the first three odd harmonics used to synthesize it. (*b*) Synthesis of the square wave using the first three odd harmonics.

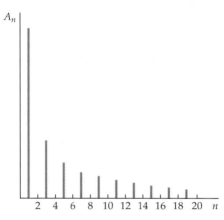

The inverse of harmonic analysis is **harmonic synthesis,** which is the construction of a periodic wave from its harmonic components. Figure 14-23 shows a square wave that can be produced by an electronic square-wave generator and the approximate synthesis of the square wave using only three harmonics. The more harmonics used in a synthesis, the closer the approximation is to the actual waveform. The relative amplitudes of the harmonics needed to synthesize the square wave are shown in Figure 14-24.

An electronic music synthesizer is a device that produces a series of harmonics whose relative amplitudes can be adjusted so that the combination produces the desired waveform. In addition, each note can be shaped by adjusting such parameters as the attack, decay, vibrato, tremolo, and release. Sophisticated modern synthesizers can produce tones that sound like any instrument in an orchestra.

Figure 14-24   Relative amplitudes A_n of the harmonics needed to synthesize the square wave shown in Figure 14-23. The more harmonics that are used, the closer the approximation is to the square wave.

(a)

(b)

(a) Synthesizer designed by Heinrich von Helmholtz and constructed by Rudolph Koenig. Each electrically driven tuning fork has a resonator that can be mechanically opened or closed to vary the intensity of that harmonic. (b) In the more modern SY 77 Yamaha synthesizer, the harmonics are produced and mixed electronically.

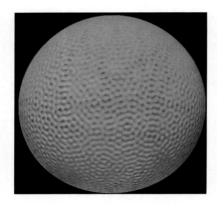

Standing sound waves on the surface of the sun. The surface of the sun is observed to oscillate with a period of about 5 min. Out of the ten million modes of oscillation known to exist, a combination of approximately 100 modes is illustrated here. The displacements of the surface are exaggerated by a factor of 1000. The period of oscillation for each mode contains information about the structure and dynamics of the solar interior.

14-7   Wave Packets and Dispersion

The complicated waveforms discussed in the previous section are periodic in time. Functions that are not periodic, such as pulses, can also be represented by harmonic (sinusoidal) wave functions, but a continuous distribution of frequencies is needed.

The characteristic feature of a wave pulse that distinguishes it from a periodic wave of a single frequency is that the pulse has a beginning and an end, whereas a harmonic wave repeats over and over. To send a signal with a wave, a pulse consisting of a group of waves of different frequencies is needed rather than a harmonic wave of a single frequency. Such a group of waves is called a **wave packet.** There is an important relation between the distribution of frequencies of the harmonic functions that make up a wave packet describing a pulse, and the time duration of the pulse. If the duration Δt is very short, the range of frequencies $\Delta \omega$ is very large. The general relation between Δt and $\Delta \omega$ is

$$\Delta \omega \, \Delta t \sim 1 \qquad \qquad 14\text{-}34$$

The exact value of this product depends on just how the quantities $\Delta \omega$ and Δt are defined. For any reasonable definitions, $\Delta \omega$ is of the order of $1/\Delta t$. We can understand this relation qualitatively from our discussion of beats (Section 14-4). The envelope in Figure 14-14b resembles a wave pulse, in that it has a beginning and an end; however, it repeats over and over in time. If the two waves that produce the envelope are very close in frequency, that is, if $\Delta \omega$ is very small, the time interval $\Delta t = t_3 - t_1$ (between the first time t_1 when the waves are 180° out of phase and the next time t_3) will be large. On the other hand, if the frequency difference between the two waves is large, they will quickly be out of phase again, and Δt will be small.

A wave pulse produced by a source of short duration Δt has a narrow width in space $\Delta x = v \, \Delta t$, where v is the wave speed. Each harmonic wave of frequency ω has a wave number $k = \omega/v$. A range of frequencies $\Delta \omega$ implies a range of wave numbers $\Delta k = \Delta \omega/v$. Substituting $v \, \Delta k$ for $\Delta \omega$ in Equation 14-34 gives

$$v \, \Delta k \, \Delta t \sim 1$$

or

$$\Delta k \, \Delta x \sim 1 \qquad \qquad 14\text{-}35$$

where again the exact value for the product depends on the precise definitions of Δk and Δx.

The relations expressed by Equations 14-34 and 14-35 are important characteristics of wave packets that apply to all types of waves. Since information cannot be transported by a single harmonic wave, which has no beginning or end in time, the transmission of short pulses depends on the ability to transmit a wide range of frequencies.

If a wave packet is to maintain its shape as it travels, all the component harmonic waves that make up the packet must travel the same speed. This occurs if the speed of harmonic waves does not depend on wavelength or frequency. A medium for which wave speed is independent of wavelength or frequency is called a **nondispersive medium.** Air is a nondispersive medium for sound waves, but solids and liquids generally are not.

Sometimes the wave speed in a dispersive medium depends only slightly on the frequency and wavelength. Under such conditions a wave packet will change shape as it travels, but it will travel a considerable distance as a recognizable packet. However, the speed of the packet, for example, the speed of the center of the packet, is not the same as the (average) speed of the individual component harmonic waves. The average speed of the harmonic waves is called the **phase velocity,** and the speed of the packet is called the **group velocity.**

14-8   Reflection, Refraction, and Diffraction

In a homogeneous medium, such as air at constant density, a wave travels in straight lines in the direction of the rays. At a great distance from a point source, a small part of the wavefront can be approximated by a plane, and the rays are approximately parallel lines; such a wave is called a **plane wave** (Figure 14-25). The two-dimensional analog of a plane wave is a line wave, which is a small part of a circular wavefront at a great distance from the source. Such waves can also be produced in a ripple tank by a line source as in Figure 14-26. A plane wave propagates in a straight line in the direction of the rays—much like a beam of particles.

Figure 14-25   Plane waves. At great distances from a point source, the wavefronts are approximately parallel planes, and the rays are parallel lines perpendicular to the wavefronts.

Figure 14-26   A two-dimensional analog of a plane wave can be generated in a ripple tank by a flat board that oscillates up and down in the water to produce wavefronts that are straight lines.

When a wave is incident on a boundary surface that separates two regions of differing wave speed, part of the wave is reflected and part is transmitted. This occurs, for example, when a sound wave in air strikes a solid or liquid surface. The reflected ray makes an angle with the normal to the surface equal to that of the incident ray, whereas the transmitted ray is bent toward or away from the normal—depending on whether the wave speed in the second medium is less or greater than that in the incident medium (Figure 14-27). The bending of the transmitted ray is called **refraction.** We will study refraction when we study light in Chapter 30.

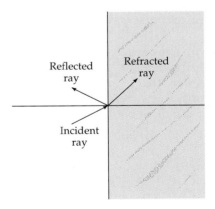

Figure 14-27   A wave striking a boundary surface between two media in which the wave speed differs. Part of the wave is reflected and part is transmitted. The change in direction of the transmitted ray is called refraction.

The amount of sound energy reflected from a surface depends on the surface. Flat walls, floors, and ceilings make good reflectors; whereas less rigid and porous materials, such as cloth in draperies and furniture coverings, absorb much of the incident sound. The reflection of sound waves plays an important role in the design of a lecture hall, a library, or a music auditorium. In a lecture hall with many flat reflecting surfaces, speech is difficult to understand because of the many echoes that arrive at different times at the listener's ear. Absorbent material is often placed on the walls

(*Left*) This anechoic chamber at the Naval Research Laboratory is used in the testing of electronic equipment. (*Right*) Davies Symphony Hall in San Francisco. The plastic reflectors above the orchestra reflect the sound waves out toward the audience.

Figure 14-28 Plane waves in a ripple tank meeting a barrier with a small opening. The waves to the right of the barrier are circular waves that are concentric about the opening just as if there were a point source at the opening.

Figure 14-29 Comparison of the transmission through a narrow opening in a barrier of (*a*) a beam of particles and (*b*) a wave. In (*a*), the transmitted particles are confined to a narrow angle. In (*b*), the opening acts as a point source of circular waves that are radiated to the right in a much wider angle than are the particles in (*a*).

and ceiling to reduce such reflections. In a concert hall, a reflecting shell is placed behind the orchestra, and reflecting panels are hung from the ceiling to reflect and direct the sound back toward the listeners.

When a portion of a wave is cut off by an obstruction, the propagation of the wave is complicated. The portion of the wavefront that is not obstructed does not simply propagate in the direction of the straight rays as might be expected. Figure 14-28 shows plane waves in a ripple tank meeting a barrier with a small opening. The waves to the right of the barrier are not confined to the narrow angle of the rays from the source that can pass through the opening; instead, they are circular waves, just as if there were a point source at the opening.

The propagation of a wave is thus quite different from the propagation of a stream of particles. In Figure 14-29*a*, the arrows indicate streams of particles that hit a barrier with a small opening, or aperture. Those particles that get through the opening will be confined to a small angle. In Figure 14-29*b*, the arrows indicate the rays that describe the propagation of circular waves toward the barrier. In passing beyond the barrier, the rays bend around the edges of the small opening. This bending of the rays, which always occurs to some extent when part of a wavefront is limited, is called **diffraction.**

In Figure 14-28, the opening in the barrier is smaller than the wavelength and could be considered to be a point. Figure 14-30 shows plane waves hitting a barrier with an opening that is much larger than the wavelength. Near the edges of the opening, the wavefronts are distorted, and the waves appear to bend slightly. For the most part, however, the wavefronts are not affected.

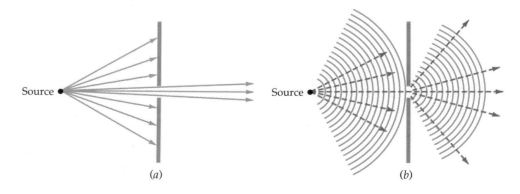

(*a*)   (*b*)

If an aperture or obstacle is large compared with the wavelength, the bending of the wavefront is not noticeable and the wave propagates in straight lines or rays, much as a beam of particles does.

Ray approximation

The result is known as the **ray approximation.** Because the wavelengths of audible sound range from a few centimeters to several meters (Example 14-2) and are often large compared with apertures and obstacles (doors or windows, for example), the bending of sound waves around corners is a common phenomenon. On the other hand, the wavelengths of visible light range from about 4×10^{-7} to 7×10^{-7} m. Because these wavelengths are so small compared with the size of ordinary objects and apertures, the diffraction of light is not easily noticed, so light appears to travel in straight lines.

The effects of diffraction place a limitation on the ability to locate small objects or identify their details by reflecting waves off of them. No appreciable reflection of waves occurs unless an object is at least as large as the wavelength. Thus, no detail can be observed on a scale smaller than the wavelength used. For example, when an object is located using waves of any kind, the position of the object is uncertain by an amount that is of the order of the wavelength used. Since the smallest wavelength of audible sound is about 2 cm (corresponding to a frequency of 17 kHz), an object cannot be located to better than ±2 cm using audible sound. Sound waves with frequencies above 20,000 Hz are called **ultrasonic waves.** Ships use ultrasonic waves to detect submarines and other submerged objects with a device called sonar (from *sound navigation and ranging*). Ultrasonic waves are used in medicine for diagnostic purposes. For example, a "sonogram," a picture constructed from ultrasonic waves, is taken of a pregnant woman to measure the size of the fetus or to detect any abnormalities. Because of their very small wavelengths, narrow beams of ultrasonic waves can be sent out and reflected from small objects. Bats emit ultrasonic waves and detect

Figure 14-30 Plane waves in a ripple tank meeting a barrier with an opening that is much larger than the wavelength. The barrier has a noticeable effect only near the edges of the opening.

Water waves on the surface of the ocean diffract as they encounter an aperture formed by the breakwater.

(a)

(b)

(a) Sonogram of a pregnant woman showing the fetus in the womb. (b) Remains of the American schooner Hamilton sunk in 300 feet of water in Lake Ontario in the War of 1812. The black outline of the ship results from processing of low-intensity signals. The red, yellow, and green are false colors resulting from the processing of moderate to high intensity signals.

their reflection to locate prey such as moths. They can hear frequencies up to about 120,000 Hz, which corresponds to a wavelength of (340 m/s)/(120 kHz) = 2.8 mm. The shortest wavelength of visible light is about 4×10^{-7} m, which is much smaller than most objects but is about 4000 times larger than the diameter of a typical atom. Thus, atoms can never be seen with visible light.

Question

5. You may have noticed that when you hear music coming from another room around a corner, there is a loss in the high frequencies compared to the rest of the sound. Why does this occur?

14-9   The Doppler Effect

When a wave source and a receiver are moving relative to each other, the frequency observed by the receiver is not the same as that of the source. When they are moving toward each other, the observed frequency is greater than the source frequency; when they are moving away from each other, the observed frequency is less than the source frequency. This is called the **doppler effect**. A familiar example is the change in the pitch of a car horn as the car approaches or recedes.

The change in frequency of a sound wave is slightly different depending on whether the source or the receiver is moving relative to the medium. When the source moves, the wavelength changes, and the new frequency f' is found by first finding the new wavelength λ and then computing $f' = v/\lambda$. On the other hand, when the source is stationary and the receiver moves, the frequency is different simply because the receiver moves past more or fewer waves in a given time.

We will first consider the case of a moving source. Figure 14-31a shows waves in a ripple tank produced by a source moving to the right with a speed less than the wave speed. We can see that the waves in front of the source are compressed so that the wavefronts are closer together than they would be for a stationary source, whereas behind the source, they are farther apart. We can calculate the wavelength in front of the source λ_f and that behind the source λ_b as follows. Let the frequency of the source be f_0. In some time Δt, the source produces a number of waves N given by $N = f_0\,\Delta t$. The first wavefront moves a distance $v\,\Delta t$ while the source moves a distance $u_s\,\Delta t$, where u_s is the speed of the source relative to the medium (Figure 14-31b). Since these N waves are contained in the distance $(v - u_s)\,\Delta t$, the wavelength in front of the source is obtained by dividing this distance by N:

$$\lambda_f = \frac{(v - u_s)\,\Delta t}{N} = \frac{(v - u_s)\,\Delta t}{f_0\,\Delta t}$$

or

$$\lambda_f = \frac{v - u_s}{f_0} = \frac{v}{f_0}\left(1 - \frac{u_s}{v}\right) \qquad\qquad 14\text{-}36$$

Behind the source, the N waves are contained in a distance $(v + u_s)\,\Delta t$, so the wavelength behind the source is

$$\lambda_b = \frac{(v + u_s)\,\Delta t}{f_0\,\Delta t}$$

or

$$\lambda_b = \frac{v + u_s}{f_0} = \frac{v}{f_0}\left(1 + \frac{u_s}{v}\right) \qquad 14\text{-}37$$

The speed v of the waves depends only on the properties of the medium and not on the motion of the source. For a source approaching the receiver, the frequency f' at which the waves pass a point at rest relative to the medium is thus

$$f' = \frac{v}{\lambda_f} = \frac{f_0}{1 - u_s/v} \qquad \text{source approaching} \qquad 14\text{-}38a$$

For a source receding from the receiver, the frequency is

$$f' = \frac{v}{\lambda_b} = \frac{f_0}{1 + u_s/v} \qquad \text{source receding} \qquad 14\text{-}38b$$

When the source is at rest and the receiver is moving relative to the medium, there is no change in the wavelength, but the frequency with which waves pass the receiver is increased when the receiver moves toward the source and decreased when the receiver moves away from the source. The number of waves passing a stationary receiver in time Δt is the number in the distance $v\,\Delta t$, which is $v\,\Delta t/\lambda$. When the receiver moves toward the source with speed u_r, it passes an additional number of waves $u_r\,\Delta t/\lambda$ (Figure 14-32). The total number of waves passing the receiver in time Δt is then

$$N = \frac{v\,\Delta t + u_r\,\Delta t}{\lambda} = \frac{v + u_r}{\lambda}\,\Delta t$$

The frequency observed is this number of waves divided by the time interval:

$$f' = \frac{N}{\Delta t} = \frac{v + u_r}{\lambda}$$

or

$$f' = f_0\left(1 + \frac{u_r}{v}\right) \qquad \text{receiver approaching} \qquad 14\text{-}39a$$

If the receiver moves away from the source with speed u_r, similar reasoning leads to the observed frequency

$$f' = f_0\left(1 - \frac{u_r}{v}\right) \qquad \text{receiver receding} \qquad 14\text{-}39b$$

When both the source and the receiver are moving relative to the medium, the results expressed in Equations 14-38 and 14-39 can be combined:

$$f' = \frac{(1 \pm u_r/v)}{(1 \pm u_s/v)}\,f_0 \qquad 14\text{-}40$$

(a)

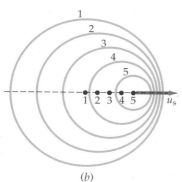

(b)

Figure 14-31   (a) Waves in a ripple tank produced by a point source moving to the right with a speed that is less than the wave speed. The wavefronts are closer together in front of the source and farther apart behind the source than they would be for a stationary source. (b) Successive wavefronts emitted by a point source moving with speed u_s to the right. Each numbered wavefront was emitted when the source was at the correspondingly numbered position.

Figure 14-32   The number of waves passing a stationary receiver in time Δt is the number in the distance $v\,\Delta t$, where v is the wave speed. If the receiver moves toward the source with speed u_r, he passes the additional number of waves in the distance $u_r\,\Delta t$. The frequency of the waves received is therefore increased.

The correct choices for the plus or minus signs are most easily obtained by remembering that the frequency increases when the source and receiver are moving toward the other, whereas it decreases when they are moving away from each other. Thus, for example, if the source is moving toward the receiver and the receiver is moving toward the source, the plus sign is used in the numerator and the minus sign is used in the denominator.

In general, the observed frequency f' can always be written

$$f' = \frac{v'}{\lambda} \qquad \qquad 14\text{-}41$$

where v' is the speed of the waves relative to the observer, and λ is the wavelength in the medium, that is, the distance between successive wave crests as measured by an observer at rest relative to the medium.

If a medium is moving (for example, if there is a wind blowing), the wave speed v is replaced by $v' = v \pm u_w$, where u_w is the speed of the medium.

Example 14-9

The frequency of a car horn is 400 Hz. What frequency is observed if the car moves through still air toward a stationary receiver with a speed $u_s = 34$ m/s (about 76 mi/h)? Take the speed of sound in air to be 340 m/s.

According to Equation 14-36, the wavelength in front of the car is

$$\lambda_f = \frac{v - u_s}{f_0} = \frac{340 - 34 \text{ m/s}}{400 \text{ s}^{-1}} = \frac{306}{400} \text{ m} = 0.765 \text{ m}$$

The frequency observed is thus

$$f' = \frac{v}{\lambda_f} = \frac{340 \text{ m/s}}{0.765 \text{ m}} = 444 \text{ Hz}$$

Note that the observed frequency is 11 percent higher than the original frequency of the car horn.

Example 14-10

The horn of a stationary car has a frequency of 400 Hz. What frequency is observed by a receiver moving through still air toward the car at 34 m/s?

For a moving receiver, the wavelength does not change; the receiver merely passes more waves in a given time. The observed frequency is (Equation 14-39a):

$$f' = f_0\left(1 + \frac{u_r}{v}\right) = (400 \text{ Hz})\left(1 + \frac{34}{340}\right) = (400 \text{ Hz})(1.10) = 440 \text{ Hz}$$

In this case, the increase in the observed frequency is 10 percent.

Exercise

As a train moving at 90 km/h is approaching a stationary listener, it blows its horn, which has a frequency of 630 Hz. (a) What is the wavelength of the sound waves in front of the train? (b) What frequency is heard by the listener? (Use 340 m/s for the speed of sound.) [Answers: (a) $\lambda_f = 0.5$ m, (b) $f' = 680$ Hz]

A familiar example of the use of the doppler effect is the use of radar by the police to measure the speed of a car. Electromagnetic waves emitted by the radar transmitter strike the moving car. The car acts as both a moving receiver and a moving source as the waves reflect off it back to the radar receiver. Another application involves the famous red shift in the light from distant galaxies. Because the galaxies are moving away from us, the light they emit is shifted toward the longer, red wavelengths. The speed of the galaxies relative to us can be determined by measuring this shift.

In Examples 14-9 and 14-10, we saw that the magnitude of the doppler shift in frequency depends on whether it is the source or the receiver that is moving relative to the medium. In Example 14-9, the source moved with a speed of 34 m/s relative to still air, and the frequency shifted from 400 to 444 Hz, a difference of 11 percent. In Example 14-10, the receiver moved with a speed of 34 m/s relative to still air, and the frequency shifted from 400 to 440 Hz, a difference of 10 percent. Though these shifts are approximately the same, they are not exactly equal.

The frequency f' observed for a source moving with a speed u toward a receiver is, from Equation 14-38a,

$$f' = \frac{f_0}{1 - u/v} = f_0 \left(1 - \frac{u}{v}\right)^{-1} \qquad 14\text{-}42a$$

where we have dropped the subscripts on the u for simplicity. We can compare this with Equation 14-39a for an approaching receiver by using the binomial expansion:

$$(1 + x)^n = 1 + nx + \frac{n(n - 1)}{2} x^2 + \cdots$$

For values of x much less than 1, we can approximate using only the first few terms. For $x = -u/v$ and $n = -1$, we have

$$\left(1 - \frac{u}{v}\right)^{-1} = 1 + (-1)\left(-\frac{u}{v}\right) + \frac{(-1)(-2)}{2}\left(-\frac{u}{v}\right)^2 + \cdots$$

$$\approx 1 + \frac{u}{v} + \frac{u^2}{v^2}$$

Thus, Equation 14-42a is approximately

$$f' \approx f_0 \left(1 + \frac{u}{v} + \frac{u^2}{v^2}\right) \qquad 14\text{-}42b$$

On the other hand, if the receiver moves toward a stationary source with a speed u, the observed frequency from Equation 14-39a is

$$f' = f_0 \left(1 + \frac{u}{v}\right) \qquad 14\text{-}43$$

Comparing Equations 14-42b and 14-43, we can see that the *difference* in the frequency shift between the source moving with a speed u and the receiver moving with a speed u is of the order of $(u/v)^2$. Often this difference can be neglected because u is usually much smaller than v. Then the doppler shift in frequency can be written

$$\frac{\Delta f}{f_0} \approx \pm \frac{u}{v} \qquad 14\text{-}44$$

where $\Delta f = f' - f_0$. The plus sign is used when the source and receiver are approaching each other with a relative speed u and the minus sign is used when they are receding.

The small difference in the doppler shift for a moving source or moving receiver is real, and it is of theoretical importance because it shows that these two situations are really different. Not only is the motion of the source and receiver relative to each other important, but so is the "absolute" motion of both relative to the medium. Thus, if we can measure the doppler shift in frequency to the order of $(u/v)^2$, we can tell whether it is the source or receiver that is moving relative to the medium. If you move relative to still air, you feel air rushing past you. In your reference frame, there is a wind. For sound waves in air, therefore, we can tell whether the source or receiver is moving by noting if there is a wind in the reference frame of the source or the receiver.

A problem arises when we apply the doppler effect to light and other electromagnetic waves that propagate through a vacuum. Our equations for the doppler effect for sound seem to imply that we could detect absolute motion relative to a vacuum if we could measure the doppler shift accurately enough. This contradicts the principle of relativity enunciated by Albert Einstein that states that it is impossible to detect absolute motion. It turns out that for light and other electromagnetic waves, our equations for the doppler effect are only approximately correct. When the proper relativistic corrections are made, the doppler shift for electromagnetic waves is the same for a moving source as for a moving receiver; that is, it depends only on the relative motion of the source and receiver.

In our derivations of the doppler-shift expressions, we have assumed that the speed u of the source or receiver is less than the wave speed v. If the receiver moves toward the source faster than the wave speed, there is no problem. Equation 14-39a for the observed frequency still holds. If the receiver moves away from the source faster than the wave speed, the waves never reach the receiver. (According to the special theory of relativity, a receiver cannot move faster than the speed of light c in a vacuum.) If a sound source moves with speed greater than the wave speed, there will be no waves in front of the source. Behind the source, the waves pile up on top of one another to form a shock wave that is heard as a sonic boom when it arrives at the receiver. The shock wave is confined to a cone that narrows as u increases. We can calculate the angle of this cone from Figure 14-33a, which shows a source originally at point P_1 moving to the right with speed u. After some time t, the wave emitted from point P_1 will have traveled a distance vt. The source will have traveled a distance ut and will be at point P_2. The line from this new position of the source to the wavefront emitted when the source was at P_1 makes an angle θ with the path of the source, given by

$$\sin \theta = \frac{vt}{ut} = \frac{v}{u} \qquad 14\text{-}45$$

The ratio of the source speed u to the wave speed v is called the **Mach number**:

$$\text{Mach number} = \frac{u}{v} \qquad 14\text{-}46$$

Equation 14-45 also applies to the electromagnetic radiation given off when a charged particle moves in a medium with speed u that is greater than the speed of light v in that medium.* This radiation, called Čerenkov radiation, is confined to a cone with an angle given by Equation 14-45.

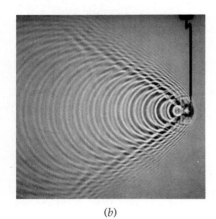

(a)

(b)

Figure 14-33 (a) A source moving from P_1 to P_2 with a speed u that is greater than the wave speed v. The envelope of the wavefronts forms a cone with the source at the apex. The angle θ of this cone is given by $\sin \theta = v/u$. (b) Waves in a ripple tank produced by a source moving with a speed greater than the wave speed.

---

*It is impossible for a particle to move faster than c, the speed of light in a vacuum. In a medium such as glass, however, electrons and other particles can move faster than the speed of light in that medium.

(a)

(b)

(c)

(d)

(a) Shock waves from a supersonic airplane. (b) Bow waves from a boat. (c) Shock waves produced by a bullet traversing a helium balloon. (d) Glow of a small reactor produced by Čerenkov radiation.

Question

6. If the source and receiver are at rest relative to each other but the wave medium is moving relative to them, will there be any doppler shift in frequency?

Summary

1. Sound waves are longitudinal waves of compression and rarefaction. In a fluid, they travel with speed

$$v = \sqrt{\frac{B}{\rho}}$$

where B is the bulk modulus and ρ is the equilibrium density of the fluid. The speed of sound in a gas is related to the absolute temperature by

$$v = \sqrt{\frac{\gamma RT}{M}}$$

where the absolute temperature T is related to the Celsius temperature t_C by

$$T = t_C + 273$$

and $R = 8.314$ J/mol·K is the universal gas constant, M is the molar mass (mass per mole), and γ is a constant that depends on the kind of gas and has the value 1.4 for air. In a solid, the speed of sound is related to Young's modulus Y and the density ρ by

$$v = \sqrt{\frac{Y}{\rho}}$$

2. Sound waves can be considered to be either displacement waves or pressure waves. In a harmonic sound wave, the pressure amplitude p_0 is related to the displacement amplitude s_0 by

$$p_0 = \rho \omega v s_0$$

where ω is the angular frequency, ρ is the density of the medium, and v is the speed of the wave. The human ear is sensitive to sound waves of frequencies in the range of about 20 Hz to 20 kHz.

3. The intensity of a wave is the power divided by the area. The intensity of spherical waves from a point source varies inversely as the square of the distance from the source:

$$I = \frac{P_{av}}{4\pi r^2}$$

The intensity of a harmonic wave is proportional to the square of the amplitude of the wave.

4. Sound intensity levels are measured on a logarithmic scale. The sound intensity level β in decibels (dB) is related to the intensity I by

$$\beta = 10 \log \frac{I}{I_0}$$

where $I_0 = 10^{-12}$ W/m², which is approximately the threshold of hearing. On this scale, the threshold of hearing is 0 dB and the pain threshold is 120 dB.

5. Two sources that are in phase or have a constant phase difference are coherent. Interference is observed only for waves from coherent sources. Waves from incoherent sources have a phase difference that varies randomly in time, so the interference at a particular point varies back and forth from constructive to destructive and no interference pattern is observed. A common cause of a phase difference between two waves is a difference in the path length traveled by the waves. A path difference Δx produces a phase difference δ given by

$$\delta = 2\pi \frac{\Delta x}{\lambda}$$

6. Beats are the result of the interference of two waves of slightly different frequencies. The beat frequency equals the difference in the frequencies of the two waves:

$$f_{beat} = \Delta f$$

7. When sound waves are confined in space, as in an organ pipe, standing waves occur. For a pipe that is either open or closed at both ends, the standing-wave condition can be found by fitting waves into the pipe with a node at each end (closed at both ends) or an antinode at each end (open at both ends). An integral number of half wavelengths must fit into the length of the pipe. The standing-wave condition is thus

$$n \frac{\lambda_n}{2} = L \qquad n = 1, 2, 3, \ldots$$

The allowed frequencies are

$$f_n = n f_1 \qquad n = 1, 2, 3, \ldots$$

where $f_1 = v/2L$ is the fundamental frequency. If the pipe is closed at one end and open at the other end, there is a node at one end and an antinode at the other. The standing-wave condition in this case is

$$n\,\frac{\lambda_n}{4} = L \qquad n = 1, 3, 5, \ldots$$

Only the odd harmonics are present. The allowed frequencies are

$$f_n = nf_1 \qquad n = 1, 3, 5, \ldots$$

where $f_1 = v/4L$.

8. Sounds of different tone quality contain different mixtures of harmonics. The analysis of a particular tone in terms of its harmonic content is called harmonic analysis. Harmonic synthesis is the construction of a tone by adding the proper mixture of harmonics.

9. A wave pulse can be represented by a continuous distribution of harmonic waves. If the duration of the pulse in time is small, a wide range of frequencies is needed. The range of frequencies $\Delta\omega$ is related to the width in time Δt by

$$\Delta\omega\,\Delta t \sim 1$$

Similarly, the range of wave numbers Δk is related to the width in space Δx by

$$\Delta k\,\Delta x \sim 1$$

10. In a nondispersive medium, the wave speed does not depend on the frequency or wavelength, and a pulse travels without change in shape. In a dispersive medium, the wave speed does depend on the wavelength and frequency, and a pulse changes shape as it moves. In a dispersive medium, the velocity of the pulse, called the group velocity, is not the same as the phase velocity, which is the average velocity of the harmonic components of the pulse.

11. Waves can be reflected, refracted, and diffracted. Refraction is the change in direction of a wave that occurs when the wave speed changes because the medium changes. Diffraction is the bending of a wave around an obstacle or an edge of an aperture. It occurs whenever the wavefront is limited. When the obstacles or apertures are large compared with wavelength, diffraction is negligible and the wave propagates in straight lines like a beam of particles. This is known as the ray approximation. Because of diffraction, waves can be used to locate an object only to within a wavelength or so.

12. When a sound source and receiver are in relative motion, the observed frequency is increased if they move toward each other and decreased if they move away from each other. This is known as the doppler effect. The observed frequency f' is related to the source frequency f_0 by

$$f' = \frac{1 \pm u_{\mathrm{r}}/v}{1 \pm u_{\mathrm{s}}/v}\,f_0$$

When the relative speed of the source or receiver u is much less than the wave speed v, the doppler shift is approximately the same whether it is the source or the receiver that is moving and is given by

$$\frac{\Delta f}{f_0} \approx \pm\frac{u}{v}$$

Seismic Waves

Jack L. Flinner
Mankato State University

A bat swoops on a moth in the dark of night. A minesweeper hunts for submerged mines. A doctor carefully examines a fetus being "scanned" in its mother's womb. In each of these cases sound waves are emitted, reflected, analyzed, and interpreted. In a way, they resemble the method used to estimate the distance to a large reflecting wall, such as a cliff across a valley, by timing the echo of a loud yell "Hello-o-o-o." To find the distance to the cliff, you simply multiply the speed of sound in air by one-half the time that elapses between the shout and the arrival of its echo. Reflection of a sound wave by an obstacle creates a pattern rich in information. Along with its distance, the shape or the speed of the object relative to the observer can be determined from a careful analysis of this pattern. Correct interpretation of the information is to the clear advantage of both bat and minesweeper bent on survival.

The center of the earth is approximately 6370 km below where you are sitting. The deepest hole ever drilled managed to get only 15 km of the way there. Our information about the structure and composition of the material along the remaining 6355 km has been derived primarily from the analysis of the behavior of seismic waves. Seismic waves are waves of very low frequency that travel through the earth. Sudden displacements of crustal segments that are bounded by fault zones can produce seismic waves; these are often of catastrophic dimensions affecting both people and structures. Seismic waves can also be generated in the earth's upper crust by explosives and other artificial devices. These enable geologists and geophysicists to gain knowledge of the earth's interior and assist in the search for new fossil fuel resources.

The hundreds of thousands of earthquakes that occur each year (most, thankfully, too minor for us to notice—see Table 1) provide ample opportunity for the production and study of seismic waves. The device used to detect these waves is called a seismograph. As you might guess, the calculations needed to extract information from these data are not as simple as those done when you holler across a valley. Like all waves, the speed of seismic waves depends on the properties of the medium through

Jack L. Flinner is Professor of Physics at Mankato State University in Mankato, Minnesota. He earned his Bachelor's degree from Wittenberg University, Springfield, Ohio in 1953; his M.S. from the University of Illinois, Urbana–Champaign, in 1955; and his Ph.D. from Ohio State University in 1965. His research was in low-energy nuclear physics. He taught at Wittenberg University for twelve years and has been at Mankato State University since 1969. He served as chairperson of the department from 1975 to 1981 and then returned to teaching and to research with the Van de Graaff accelerator using the PIXE method to investigate the L-shell spectrum of high-Z elements. He has served on such projects as the establishments of an East Asia studies program at Wittenberg, as the Minnesota State University representative on the committee to propose a state school of optometry, and then as a representative on the Minnesota State Legislature ad hoc committee on science and technology.

Table 1   Earthquake Magnitudes and Frequencies for the Entire Earth and their Damaging Effects

Richter magnitude	Number per year	Characteristic effects of shocks in populated areas
<3.4	800,000	Recorded only by seismographs
3.5–4.2	30,000	Felt by some people
4.3–4.8	4800	Felt by many people
4.9–5.4	1400	Felt by everyone
5.5–6.1	500	Slight building damage
6.2–6.9	100	Much building damage
7.0–7.3	15	Serious damage, bridges twisted, walls fractured
7.4–7.9	4	Great damage, buildings collapse
>8.0	One every 5–10 yr	Total damage, waves seen on ground surface, objects thrown in the air

Source: From Skinner and Porter (1987) *Physical Geology*, p. 456, John Wiley & Sons, Inc.

which they propagate—in particular, on the rigidity (for example, the bulk modulus) and on the density of the medium. Also, at a boundary surface between two media in which the wave speed differs, a seismic wave is partially reflected and partially refracted. Further complications arise due to the fact that the media through which seismic waves travel can be solid, liquid, or gaseous. The physical properties of the medium have profound effects on the transmission pattern of seismic waves. The rigidity and density of the materials that make up the earth vary with depth, and their values are of particular importance for an accurate analysis. Furthermore, the intensity of the reflected (or deflected) waves can be extremely small, so the sensitivity of the detectors used must be correspondingly large.

There are four types of seismic waves that fall into two categories: body waves (P for primary, and S for secondary) and surface waves (Love and Rayleigh). All four types of waves are produced by an earthquake. Body waves travel through the earth outward from the focus of an earthquake (Figure 1). P waves are longitudinal waves—like sound waves (refer to Figure 13-5 in the main text). They travel at speeds up to 14 km/s and can pass through solids, liquids, and gases. Since they travel faster than S waves they are the first to arrive at an earthquake detector (hence "primary"). S waves are transverse shear waves that travel at speeds up to 3.5 km/s (refer to Figure 13-5 in the main text). They can travel only through solids because liquids and gases cannot support shear stresses. Love and Rayleigh waves remain at the earth's surface. Love waves are torsional waves. They produce momentary distortions of the earth's surface like those you could produce on a spherical balloon globe if you grasped it at the poles and gave a twist with each hand in opposite directions. Rayleigh waves are similar to the rise and swell of ocean waves. They produce momentary distortions like those you could produce on a spherical balloon globe by pushing the poles in or pulling them out.

P and S waves propagate through the earth at substantially different speeds, so their separate arrival times at seismograph stations can be used to locate earthquake epicenters with great precision. Travel-time graphs for P and S waves have been constructed (Figure 2), first using earthquake epicenters whose locations were relatively well established from physical evidence, and later by the use of data from nuclear explosions whose detonation times and locations were precisely known. When P and S seismic waves are recorded at a seismograph station, the time interval between the early arrival of the P wave and the later arrival of the S wave is

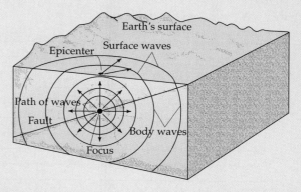

Figure 1 The body waves move radially outward from the focus of an earthquake submerged within the earth. The epicenter, which lies at the earth's crust directly above the focus, is the center of the surface waves that radiate from that point.

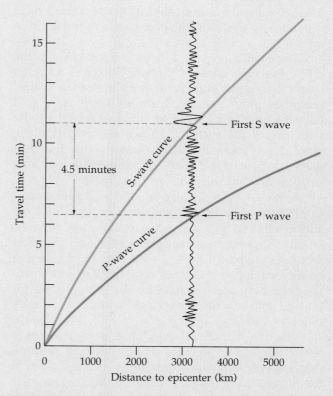

Figure 2 A travel-time graph can be used to determine the distance of a seismograph station from an earthquake's epicenter.

Continued

Figure 3 A typical seismograph record.

easily read from the seismograph (Figure 3). This time interval can be entered on the travel-time graph, allowing the distance to the recording site to be read directly from the graph. The precise location of the epicenter is determined from the intersection point provided by distance data from three or more stations. Clustered seismograph arrays and high-speed computers, used to discriminate between nuclear explosions and natural earthquakes, have proved very useful in the accurate determination of epicenters.

The earth's average density is 5.5 g/cm³. At the crust, the rock density is about 3 g/cm³, so some parts of the earth must have much greater densities. Seismographic readings from all over the world have enabled geologists to build a model of the structure of the earth. It shows not only that the most dense rocks are at the center with the least dense at the surface, but also that the interior of the earth is divided into regions (Figure 4).

Seismic waves can also be produced deliberately to map the subsurface rock strata of geological basins in order to find likely sites for oil or gas accumulations. These induced seismic waves (and their correct interpretation) have a vital role to play in our energy-intensive industrial society; they often provide the only cost-effective means of exploring for oil and gas reserves concealed within the earth's crust.

The use of artificially produced seismic waves to search for fuel deposits is called reflection seismology, or seismic prospecting. It has been practiced in the United States since 1923. Reflection seismology simply stated is the generation of minor, artificial earthquakes to produce seismic waves so that their reflection and refraction patterns can be studied.

The method used to produce the waves depends on whether the survey is taking place on land or at sea. Options include using dynamite blasts in shallow boreholes; deploying large, low-intensity hydraulic vibrators (particularly useful in populated or ecologically sensitive areas); dropping heavy weights;

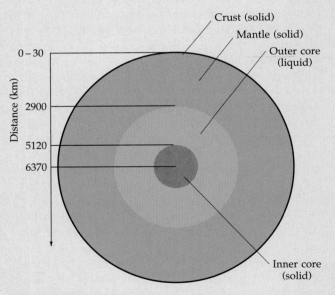

Figure 4 Geologists have constructed this model of the interior of the earth on the basis of the refraction and transmission patterns of seismic body waves (S and P). The inability of S waves to penetrate the large interior of the earth indicates that the outer core is liquid. The transmission patterns of those P waves that are not too strongly refracted by the outer core indicate that the inner core is solid.

striking the ground with an air hammer; or using an array of air guns. Similar effects are produced by each method. The waves that are produced travel through the earth's crust and are reflected back toward the surface at the boundaries of various strata. The arrival times of seismic waves at the earth's surface is recorded by detectors (called geophones). The arrival times, amplitudes, and frequencies of the seismic waves are then stored on computers (Figure 5).

The very fast, large-memory computers, which have become available in recent years, not only make it possible to store a seismic reflection profile (Figure 6) that depicts a two-dimensional slice through the earth's crust, but also make it possible to move the location of the seismic energy sources—thus producing a series of profiles that can be used to generate a three-dimensional representation. The two-dimensional profiles indicate the discontinuities encountered in the density and rigidity in the different layers of the earth's crust and can be interpreted to reveal the nature of geologic configurations, some of which are known to be particularly

Figure 5 Seismic reflection studies at sea involve a survey ship towing an energy source nearby, which is used to generate seismic waves. It also tows cables up to 5 km long that carry hydrophones that detect the reflected seismic waves returning from the crust.

Figure 6 A seismic reflection profile together with an interpretation of the various layers revealed.

Continued

likely to contain oil or gas (Figure 7). The three-dimensional representation shows even more clearly the various rock layers and makes it possible to study carefully a volume of crust of particular interest. Once a pocket of gas has been found, for instance, its size can be determined by rotating its image and "viewing" it from different angles. The computers and software used today by geologists have become very sophisticated and make such rotations possible. It is also possible, by digital processing, to reduce the effect of the scattering of the wavefronts that emerge from the features being analyzed. This process, called migration, is roughly equivalent to focusing the lens of a camera or microscope, thus enabling one to examine finer geological features. This "focusing" becomes more important as the larger oil fields are depleted, and interest is shifted to smaller reserves.

Computers are also used to store data about local geology, such as rock porosity for water and oil, that are important in determining the ease with which oil can be extracted and the likelihood of oil or gas in a new find being economically retrievable. The geologist is thereby able to process and utilize vast amounts of data rapidly and effectively and to search for those features that are characteristic of a promising new fuel source before any investment is made in an expensive exploratory drilling. (The present state of the art in reflection seismology is, indeed, a beautiful example of equipment and technology being available soon enough to help solve an increasingly difficult problem.)

Reflection seismology is one of many applications of the reflection and refraction of waves as they travel through matter. As the power and speed of computers grow, all aspects of seismology are expected to play increasingly important and sophisticated roles in studies of the earth's crust—its features, its often unpredictable behavior, and its hidden resources.

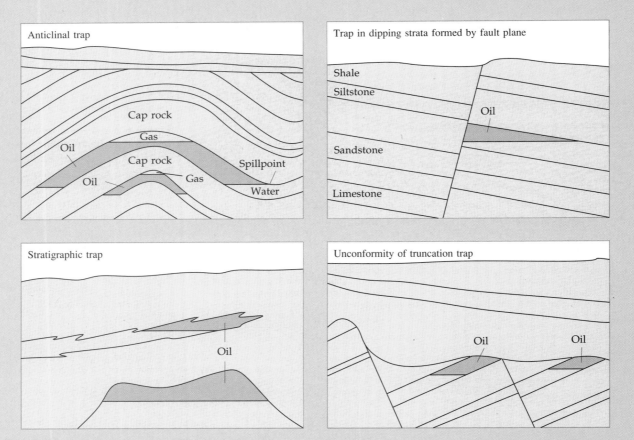

Figure 7   Some of the rock structures that can trap hydrocarbons, and schematic versions of their appearance in profiles.

Suggestions for Further Reading

Boore, David M.: "The Motion of the Ground in Earthquakes," *Scientific American*, December 1977, p. 68.

Sudden slippage of a continental plate relative to its neighbor along a fault line excites mechanical waves in the earth. The study of this phenomenon may lead to improved building designs for earthquake zones.

Devey, Gilbert B., and Peter N. T. Wells: "Ultrasound in Medical Diagnosis," *Scientific American*, May 1978, p. 98.

The physics and technology of imaging using ultrahigh-frequency sound waves and the interpretation of the medical images produced are discussed.

Fletcher, Neville H., and Suzanne Thwaites: "The Physics of Organ Pipes," *Scientific American*, January 1983, p. 94.

Modern measurements show that the competing theories of nineteenth-century physicists Helmholtz and Rayleigh to explain the sound production of organ pipes are both correct but apply to different pressure ranges.

The Physics of Music, readings from *Scientific American* with an introduction by Carleen M. Hutchins, Scientific American, Inc., distributed by W. H. Freeman and Co., New York, 1978.

This collection of articles includes analyses of the acoustics of the singing voice, the piano, woodwinds, brasses, violins, and the bowed string as well as architectural acoustics.

Jeans, Sir James: *Science and Music*, The University Press, Cambridge, 1953.

The basic features of sound production by vibrating strings and columns of air as well as harmony are well treated for the general reader in this book originally published in 1937.

Matthews, Max V. and John R. Pierce: "The Computer as a Musical Instrument," *Scientific American*, February 1981, p. 126.

Analysis and synthesis of waveforms has helped to determine what constitutes a "pleasing" musical sound. This article describes some of the history of computers in music generation.

Pierce, John R.: *The Science of Musical Sound*, Scientific American Books, Inc., distributed by W. H. Freeman and Co., New York, 1983.

This excellent book discusses the physical and mathematical aspects of sound waves that underlie our experience of music as well as the quantifiable aspects of sound perception.

Quate, Calvin F.: "The Acoustic Microscope," *Scientific American*, October 1979, p. 62.

The development of an acoustic microscope employing sound with a frequency 1000 times higher than that used in medical imaging technology is described. The images are comparable in quality to light micrographs.

Review

A. Objectives: After studying this chapter, you should:

1. Know how the speed of sound in air depends on the absolute temperature, and know the approximate value in meters per second of the speed of sound in air at ordinary temperatures.

2. Be able to calculate the intensity level in decibels from the intensity in watts per square meter and vice versa.

3. Be able to sketch the interference pattern produced by two coherent sources.

4. Know that the beat frequency equals the difference between the two frequencies producing the beats.

5. Be able to explain what is meant by the terms coherent sources and diffraction.

6. Be able to sketch standing-wave patterns for sound waves in a tube open at both ends, closed at both ends, and closed at one end and from these sketches determine the possible frequencies for standing waves.

7. Be able to derive expressions for the doppler frequency shift for a moving source or a moving receiver and use these expressions in working problems.

B. Define, explain, or otherwise identify:

Wavefront	Wave packet
Ray	Nondispersive medium
Intensity	Dispersive medium
Intensity level	Phase velocity
Decibel	Group velocity
Coherent sources	Plane wave
Beats	Reflection
Beat frequency	Refraction
Tone quality	Diffraction
Waveform	Ray approximation
Harmonic analysis	Ultrasonic waves
Fourier analysis	Doppler effect
Harmonic synthesis	Mach number

C. True or false: If the statement is true, explain why it is true. If it is false, give a counterexample.

1. Sound waves in air are transverse waves of compression and rarefaction.

2. The speed of sound at 20°C is twice that at 5°C.

3. A 60-dB sound has twice the intensity of a 30-dB sound.

4. The waves from two coherent sources that are in phase interfere constructively everywhere in space.

5. Two wave sources that are out of phase by 180° are incoherent.

6. Interference patterns are observed only for coherent sources.

7. Diffraction occurs only for transverse waves.

8. The beat frequency between two sound waves of nearly equal frequencies equals the difference in the frequencies of the individual sound waves.

9. The frequency of the fifth harmonic is 5 times that of the fundamental.

10. In a pipe that is open at one end and closed at the other, the even harmonics are not excited.

11. The doppler shift in sound waves depends only on the relative motion of the source and receiver.

Problems

Unless a problem specifies otherwise, take the speed of sound to be 340 m/s in air and 1500 m/s in water.

Level I

14-1 Speed of Sound Waves

1. Show that $\sqrt{B/\rho}$ has the units of meters per second if B has the correct SI units and ρ is in kilograms per cubic meter.

2. The bulk modulus for water is 2.0×10^9 N/m². Use it to find the speed of sound in water.

3. Young's modulus for aluminum is 7.0×10^{10} N/m². The density of aluminum is 2.7×10^3 kg/m³. Find the speed of sound in an aluminum rod.

4. Calculate the speed of sound waves in hydrogen gas at $T = 300$ K. (Take $M = 2$ g/mol and $\gamma = 1.4$.)

5. The speed of sound in mercury is 1410 m/s. What is the bulk modulus for mercury ($\rho = 13.6 \times 10^3$ kg/m³)?

14-2 Harmonic Sound Waves

6. A sound wave in air produces a pressure variation given by

$$p(x, t) = 0.75 \cos \frac{\pi}{2} (x - 340t)$$

where p is in pascals, x is in meters, and t is in seconds. (a) What is the pressure amplitude, (b) the wavelength, (c) the frequency and (d) the speed of the sound wave?

7. (a) Middle C on the musical scale has a frequency of 262 Hz. What is the wavelength of this note in air? (b) The frequency of the C an octave above middle C is twice that of middle C. What is the wavelength of this note in air?

8. What is the displacement amplitude for a sound wave having a frequency of 100 Hz and a pressure amplitude of 10^{-4} atm?

9. The displacement amplitude of a sound wave of frequency 300 Hz is 10^{-7} m. What is the pressure amplitude of this wave?

10. (a) Find the displacement amplitude of a sound wave of frequency 500 Hz at the pain-threshold pressure amplitude of 29 Pa. (b) Find the displacement amplitude of a sound wave with the same pressure amplitude but a frequency of 1 kHz.

11. A typical loud sound wave with a frequency of 1 kHz has a pressure amplitude of about 10^{-4} atm. (a) At $t = 0$, the pressure is a maximum at some point x_1. What is the displacement at that point at $t = 0$? (b) What is the maximum value of the displacement at any time and place? (Take the density of air to be 1.29 kg/m³.)

12. (a) Find the displacement amplitude of a sound wave of frequency 500 Hz at the threshold-of-hearing pressure amplitude of 2.9×10^{-5} Pa. (b) Find the displacement amplitude of a wave of the same pressure amplitude but a frequency of 1 kHz.

14-3 Waves in Three Dimensions: Intensity

13. A piston at one end of a long tube filled with air at room temperature and normal pressure oscillates with a frequency of 500 Hz and an amplitude of 0.1 mm. The area of the piston is 100 cm². (a) What is the pressure amplitude of the sound waves generated in the tube? (b) What is the intensity of the waves? (c) What average power is required to keep the piston oscillating (neglecting friction)?

14. What is the intensity level in decibels of a sound wave of intensity (a) 10^{-10} W/m² and (b) 10^{-2} W/m²?

15. Find the intensity of a sound wave if (a) $\beta = 10$ dB and (b) $\beta = 3$ dB. (c) Find the pressure amplitudes of sound waves in air for each of these intensities.

16. Show that if the intensity is doubled, the intensity level increases by 3.0 dB.

17. What fraction of the acoustic power of a noise would have to be eliminated to lower its sound intensity level from 90 to 70 dB?

14-4 Interference: Beats

18. Two sound sources oscillate in phase with the same amplitude A. They are separated in space by $\frac{1}{3}\lambda$. What is the amplitude of the resultant wave from the two sources at a point that is on the line that passes through the sources but is not between the sources?

19. Two sound sources oscillate in phase with a frequency of 100 Hz. At a point 5.00 m from one source and 5.85 m from the other, the amplitude of the sound from each source separately is A. (a) What is the phase difference in the sound waves from the two sources at that point? (b) What is the amplitude of the resultant wave at that point?

20. With a compass, draw circular arcs representing wave crests originating from each of two point sources a distance $d = 6$ cm apart for $\lambda = 1$ cm. Connect the intersections corresponding to points of constant path difference, and label the path difference for each line. (See Figure 14-10.)

21. Two loudspeakers are separated by a distance of 6 m. A listener sits directly in front of one speaker at a distance of 8 m from it such that the two speakers and the listener form a right triangle. (*a*) Find the two lowest frequencies for which the path difference from the speakers to the listener is an odd number of half wavelengths. (*b*) Why can these frequencies be heard even if the speakers are driven in phase by the same amplifier?

22. Two violinists standing a few feet apart are playing the same notes. Are there places in the room where certain notes are not heard because of destructive interference? Explain.

23. Two speakers separated by some distance emit sound waves of the same frequency. At some point P, the intensity due to each speaker separately is I_0. The path distance from P to one of the speakers is $\frac{1}{2}\lambda$ greater than that from P to the other speaker. What is the intensity at P if (*a*) the speakers are coherent and in phase, (*b*) the speakers are incoherent, and (*c*) the speakers are coherent but have a phase difference of π rad?

24. Answer the questions of Problem 23 for a point P' for which the distance to the far speaker is 1λ greater than the distance to the near speaker. Assume that the intensity at point P' due to each speaker separately is again I_0.

25. Two speakers separated by some distance emit sound waves of the same frequency, but the waves from speaker 1 lead those from speaker 2 in phase by 90°. Let r_1 be the distance from some point to speaker 1 and r_2 be the distance from that point to speaker 2. Find the smallest value of $r_2 - r_1$ such that the sound at the point will be (*a*) maximum and (*b*) minimum. (Express your answers in terms of the wavelength.)

26. Two tuning forks have frequencies of 256 and 260 Hz. If the forks are vibrating at the same time, what is the beat frequency?

27. When a violin string is played (without fingering) simultaneously with a tuning fork of frequency 440 Hz, beats are heard at the rate of ±3 per second. When the tension in the string is increased slightly, the beat frequency decreases. What was the initial frequency of the violin string?

28. Two tuning forks are struck simultaneously, and 4 beats per second are heard. The frequency of one fork is 500 Hz. (*a*) What are the possible values for the frequency of the other fork? (*b*) A piece of wax is placed on the 500-Hz fork to lower its frequency slightly. Explain how the measurement of the new beat frequency can be used to determine which of your answers to part (*a*) is the correct frequency of the second fork.

14-5 Standing Sound Waves

29. Calculate the fundamental frequency for a 10-m organ pipe that is (*a*) open at both ends and (*b*) closed at one end.

30. The normal range of hearing is about 20 to 20,000 Hz. What is the greatest length of an organ pipe that would have its fundamental note in this range if (*a*) it is closed at one end and (*b*) it is open at both ends?

31. The shortest pipes used in organs are about 7.5 cm long. (*a*) What is the fundamental frequency of a pipe this long that is open at both ends? (*b*) For such a pipe, what is the highest harmonic that is within the audible range (see Problem 30)?

32. The space above the water in a tube like that shown in Figure 14-17 is 120 cm long. Near the open end, there is a loudspeaker that is driven by an audio oscillator whose frequency can be varied from 10 to 5000 Hz. (*a*) What is the lowest frequency of the oscillator that will produce resonance within the tube? (*b*) What is the highest frequency that will produce resonance? (*c*) How many different frequencies of the oscillator will produce resonance? (Neglect the end correction.)

33. The ear canal, which is about 2.5 cm long, roughly approximates a pipe that is open at one end and closed at the other. (*a*) What are the resonance frequencies of the ear canal? (*b*) Describe the possible effect of the resonance modes of the ear canal on the threshold of hearing.

14-6 Harmonic Analysis and Synthesis

There are no problems for this section.

14-7 Wave Packets and Dispersion

34. Information for use by computers is transmitted along a cable in the form of short electric pulses at the rate of 10^5 pulses per second. (*a*) What is the maximum duration of each pulse if no two pulses overlap? (*b*) What is the range of frequencies to which the receiving equipment must respond?

35. A tuning fork of frequency f_0 begins vibrating at time $t = 0$ and is stopped after a time interval Δt. The waveform of the sound at some later time is shown as a function of x in Figure 14-34. Let N be the (approximate) number of cycles in this waveform. (*a*) How are N, f_0, and Δt related? (*b*) If Δx is the length in space of this wave packet, what is the wavelength in terms of Δx and N? (*c*) What is the wave number k in terms of N and Δx? (*d*) The number N is uncertain by about ±1 cycle. Use Figure 14-34 to explain why. (*e*) Show that the uncertainty in the wave number due to the uncertainty in N is $2\pi/\Delta x$.

Figure 14-34   Problem 35

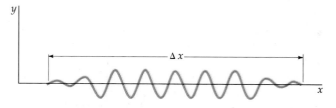

14-8 Reflection, Refraction, and Diffraction

36. If a wavelength is much larger than the diameter of a loudspeaker, the speaker radiates in all directions, much like a point source. On the other hand, if the wavelength is much smaller than the diameter, the sound travels in approximately a straight line in front of the speaker. Find the frequency of a sound wave that has a wavelength (a) 10 times and (b) one-tenth the diameter of a 30-cm speaker. (c) Do the same for a 6-cm speaker.

14-9 The Doppler Effect

37. This problem is a doppler-effect analogy. A conveyor belt moves to the right with a speed $v = 300$ m/min. A very fast piemaker puts pies on the belt at a rate of 20 per minute, and they are received at the other end by a pie eater. (a) If the piemaker is stationary, find the spacing λ between the pies and the frequency f with which they are received by the stationary pie eater. (b) The piemaker now walks with a speed of 30 m/min toward the receiver while continuing to put pies on the belt at 20 per minute. Find the spacing of the pies and the frequency with which they are received by the stationary pie eater. (c) Repeat your calculations for a stationary piemaker and pie eater who moves toward the piemaker at 30 m/min.

38. For the situation described in Problem 37, derive general expressions for the spacing of the pies λ and the frequency f with which they are received by the pie eater in terms of the speed of the belt v, the speed of the sender u_s, the speed of the receiver u_r, and the frequency f_0 with which the piemaker places pies on the belt.

In Problems 39 through 44, a source emits sounds of frequency 200 Hz that travel through still air at 340 m/s.

39. The source moves with a speed of 80 m/s relative to still air toward a stationary listener. (a) Find the wavelength of the sound between the source and the listener. (b) Find the frequency heard by the listener.

40. Consider the situation in Problem 39 from the reference frame in which the source is at rest. In this frame, the listener moves toward the source with a speed of 80 m/s, and there is a wind blowing at 80 m/s from the listener to the source. (a) What is the speed of the sound from the source to the listener in this frame? (b) Find the wavelength of the sound between the source and the listener. (c) Find the frequency heard by the listener.

41. The source moves away from the stationary listener at 80 m/s. (a) Find the wavelength of the sound waves between the source and the listener. (b) Find the frequency heard by the listener.

42. The listener moves at 80 m/s relative to still air toward the stationary source. (a) What is the wavelength of the sound between the source and the listener? (b) What is the frequency heard by the listener?

43. Consider the situation in Problem 42 in a reference frame in which the listener is at rest. (a) What is the wind velocity in this frame? (b) What is the speed of the sound from the source to the listener in this frame, that is, relative to the listener? (c) Find the wavelength of the sound between the source and the listener in this frame. (d) Find the frequency heard by the listener.

44. The listener moves at 80 m/s relative to the still air away from the stationary source. Find the frequency heard by the listener.

45. A whistle of frequency 500 Hz moves in a circle of radius 1 m at 3 rev/s. What are the maximum and minimum frequencies heard by a stationary listener in the plane of the circle and 5 m away from its center?

46. Ocean waves move toward the beach with a speed of 8.9 m/s and a crest-to-crest separation of 15.0 m. You are in a small boat anchored off shore. (a) What is the frequency of the ocean waves? (b) You now lift anchor and head out to sea at a speed of 15 m/s. What frequency of the waves do you observe?

47. A boat traveling at 10 m/s on a still lake makes a bow wave at an angle of 20° with its direction of motion. What is the speed of the bow wave?

48. A jet is traveling at Mach 2.5 at an altitude of 5000 m. (a) What is the angle that the shock wave makes with the track of the jet? (Assume the speed of sound at this altitude is still 340 m/s.) (b) Where is the jet when a person on the ground hears the shock wave?

Level II

49. Two parallel tubes with the same diameter are filled with gas at the same pressure and temperature. One tube contains H_2, and the other contains O_2. (a) If sound waves traveling in the tubes have the same displacement amplitude and frequency, how do their intensities compare? (b) If the waves have the same frequency and pressure amplitude, how do their intensities compare? (c) If the waves have the same frequency and intensity, how do their pressure amplitudes and displacement amplitudes compare?

50. A common rule of thumb for finding the distance to a lightning flash is to begin counting when the flash is observed and stop when the thunder clap is heard. The number of seconds counted is then divided by 3 to get the distance in kilometers. Why is this justified? How accurate is this procedure? What is the velocity of sound in kilometers per second? Is a correction for the time it takes for the light to reach you important? (The speed of light is about 3×10^8 m/s.)

51. A method for measuring the speed of sound using an ordinary watch with a second hand is to stand some distance from a large flat wall and clap your hands rhythmically in such a way that the echo from the wall is heard halfway between every two claps. Show that the speed of sound is given by $v = 4LN$, where L is the distance to the wall and N is the number of claps per second. What is a reasonable value for L for this experiment to be feasible? (If you have access to a flat wall outdoors somewhere, try this method and compare your result with the standard value for the speed of sound.)

52. A man drops a stone from a high bridge and hears it strike the water below exactly 4 s later. (a) Estimate the distance to the water based on the assumption that the travel time for the sound to reach the man is negligible. (b) Improve your estimate by using your result from part (a) for the distance to the water to estimate the time it takes for sound to travel this distance and then calculate the distance the rock falls in 4 s minus this time. (c) Calculate the exact distance and compare your result with your previous estimates.

53. Kundt's method was an early procedure for determining the speed of sound in different gases. A cylindrical glass tube is placed in a horizontal position, and a quantity of light powder is spread along the bottom of the tube. One end of the tube is closed by a piston that is attached to an oscillator of a known frequency f such as a tuning fork. The other end is closed by a piston whose position can be varied. While the first piston is made to oscillate at the frequency f, the movable piston's position is adjusted until resonance occurs. When this happens, the powder collects in equally spaced piles along the bottom of the tube. (a) Explain why the powder collects in this way. (b) Derive a formula that gives the speed of sound in the gas in terms of f and the distance between the piles of powder. (c) Give suitable values for the frequency f and the distance between the piles of powder. (d) Give suitable values of the frequency f and the length L of the tube for which the speed of sound could be measured in either air or helium.

54. Three successive resonance frequencies in an organ pipe are 1310, 1834, and 2358 Hz. (a) Is the pipe closed at one end or open at both ends? (b) What is the fundamental frequency? (c) What is the length of the pipe?

55. A wire of mass 1 g and length 50 cm is stretched with a tension of 440 N. It is placed near the open end of the tube in Figure 14-17 and is stroked with a violin bow so that it oscillates at its fundamental frequency. The water level in the tube is lowered until a resonance is first obtained at 18 cm below the top of the tube. Use the data given to determine the speed of sound in air. Why is this method not very accurate?

56. A 460-Hz tuning fork causes resonance in the tube in Figure 14-17 when the top of the tube is 18.3 cm and 55.8 cm above the water surface. (a) Find the speed of sound in air. (b) What is the end correction to adjust for the fact that the antinode does not occur exactly at the end of the open tube?

57. (a) Compute the derivative of the velocity of sound with respect to the absolute temperature, and show that the differentials dv and dT obey $dv/v = \frac{1}{2} dT/T$. (b) Use this result to compute the percentage change in the velocity of sound when the temperature changes from 0 to 27°C. (c) If the speed of sound is 331 m/s at 0°C, what is it (approximately) at 27°C? How does this approximation compare with the result of an exact calculation?

58. (a) Show that if the temperature changes by a small amount ΔT, the fundamental frequency of an organ pipe changes by approximately Δf, where $\Delta f/f = \frac{1}{2} \Delta T/T$. (b) Suppose an organ pipe that is closed at one end has a fundamental frequency of 200 Hz when the temperature is 20°C. What will its fundamental frequency be when the temperature is 30°C? (Ignore any change in the length of the pipe due to thermal expansion.)

59. Two loudspeakers emit sound waves in phase at 170 Hz. An observer sits at 8 m from one speaker and 11 m from the other. The intensity level from either speaker acting alone is 60 dB. (a) Find the observed inten-

sity level when both speakers are on together. (b) Find the observed intensity level when both speakers are on together but the leads to one have been reversed so that the speakers are 180° out of phase. (c) Find the observed intensity level when both speakers are on together and in phase but the frequency is 85 Hz.

60. Instructions for connecting stereo speakers to an amplifier correctly so that they are in phase are as follows: "After both speakers are connected, play a monophonic record or program with the bass control turned up and the treble control turned down. While listening to the speakers, turn the balance control so that first one speaker is heard separately, then the two together, and then the other separately. If the bass is stronger when both speakers play together, they are connected properly. If the bass is weaker when both play together than when each plays separately, reverse the connections on one speaker." Explain why this method works. In particular, explain why a stereo source is not used and why only the bass is compared.

61. Two identical speakers emit sound waves of frequency 680 Hz uniformly in all directions. The total audio output of each speaker is 1 mW. A point P is 2.00 m from one speaker and 3.00 m from the other. (a) Find the intensities I_1 and I_2 from each speaker separately at point P. (b) If the speakers are driven coherently and are in phase, what is the intensity at point P? (c) If they are driven coherently but are 180° out of phase, what is the intensity at point P? (d) If the speakers are incoherent, what is the intensity at point P?

62. A radar device emits microwaves with a frequency of 2.00 GHz. When the waves are reflected from a car moving directly away from the emitter, the beat frequency is 293 beats per second. Find the speed of the car.

63. Hovering over the pit of hell, the devil observes that as a student falls past (with the terminal velocity), the frequency of his scream decreases from 842 to 820 Hz. (a) Find the speed of descent of the student. (b) The student's scream generates beats when it is mixed with its echo from the bottom of the pit. Find the number of beats per second heard by the student. (c) Find the number of beats per second heard by the devil after the student has passed by.

64. The ratio of the frequencies of a note to those of the semitone above it on the diatonic scale is about 15:16. Find the speed of a car for which the tone of its horn will drop a semitone as it passes you.

65. A loudspeaker diaphragm 30 cm in diameter is vibrating at 1 kHz with an amplitude of 0.020 mm. Assuming that the air molecules in the vicinity have this same amplitude of vibration, find (a) the pressure amplitude immediately in front of the diaphragm, (b) the sound intensity, and (c) the acoustic power being radiated.

66. A stationary destroyer is equipped with sonar that sends out pulses of sound at 40 MHz. Reflected pulses are received from a submarine directly below with a time delay of 80 ms at a frequency of 39.958 MHz. If the speed of sound in seawater is 1.54 km/s, find (a) the depth of the submarine and (b) its vertical speed.

67. Two students with vibrating 440-Hz tuning forks walk away from each other with equal speeds. How fast must they walk to hear a beat frequency of 2 Hz?

68. A physics student walks down a long hall carrying a vibrating 512-Hz tuning fork. The end of the hall is closed so that sound reflects from it. The student hears 4 beats per second. How fast is the student walking?

69. A spherical source radiates sound uniformly in all directions. At a distance of 10 m, the sound intensity level is 80 dB, with 0 dB corresponding to 10^{-12} W/m^2. (a) At what distance from the source is the intensity level 60 dB? (b) What power is radiated by this source?

70. A plane, harmonic, acoustical wave that oscillates in air with an amplitude of 10^{-6} m has an intensity of 100 dB. What is the frequency of the sound wave?

71. A loudspeaker at a rock concert generates 10^{-2} W/m^2 at 20 m at a frequency of 1 kHz. Assume that the speaker spreads its energy uniformly in the forward hemisphere and that no energy is reflected from the ground or elsewhere so that the intensity is given by $I = P/2\pi r^2$. (a) What is the intensity level at 20 m? (b) What is the total acoustic power output of the speaker? (c) At what distance will the intensity level be at the pain threshold of 120 dB? (d) What is the intensity level at 30 m?

72. When a pin of mass 0.1 g is dropped from a height of 1 m, 0.05 percent of its energy is converted into a sound pulse with a duration of 0.1 s. (a) Estimate the range at which the dropped pin can be heard if the minimum audible intensity is 10 pW/m^2. (b) Your result in (a) is much too large in practice because of background noise. If you assume that the intensity level must be at least 40 dB for the sound to be heard, estimate the range at which the dropped pin can be heard. (In both parts, assume that the intensity is $P/4\pi r^2$.)

73. Normal human speech has a sound intensity level of about 65 dB at 1 m. Estimate the power of human speech.

74. Three noise sources produce intensity levels of 70, 73, and 80 dB when acting separately. When the sources act together, their intensities add. (There is no interference between the sound waves from the different sources because the relative phase differences vary randomly.) (a) Find the sound intensity level in decibels when the three sources act at the same time. (b) Discuss the effectiveness of eliminating the two least intense sources in reducing the intensity level of the noise.

75. An article on noise pollution claims that sound intensity levels in large cities have been increasing by about 1 dB annually. (a) To what percentage increase in intensity does this correspond? Does this increase seem reasonable? (b) In about how many years will the intensity of sound double if it increases at 1 dB annually?

76. A column of precision marchers keeps in step by listening to the band positioned at the head of the column. The beat of the music is for 100 paces per minute. A television camera shows that only the marchers at the front and the rear of the column are actually in step. The marchers in the middle section are striding forward with the left foot when those at the front and rear are striding forward

with the right foot. The marchers are so well trained, however, that they are all certain they are in proper step with the music. Explain the source of the problem, and calculate the length of the column.

Level III

77. Everyone at a party is talking equally loudly. If only one person were talking, the sound level would be 72 dB. Find the sound level when all 38 people are talking.

78. The noise level in an empty examination hall is 40 dB. When 100 students are writing an exam, the sounds of heavy breathing and pens traveling rapidly over paper cause the noise level to rise to 60 dB (not counting the occasional groans). Assuming that each student contributes an equal amount of noise power, find the noise level to the nearest decibel when 50 students have left.

79. Two point sources that are in phase are separated by a distance d. An interference pattern is detected along a line parallel to that through the sources and a large distance D from the sources as shown in Figure 14-35. (a) Show that the path difference from the two sources to some point on the line at an angle θ is given approximately by

$$\Delta x = d \sin \theta$$

(b) Show that the distance y_m from the central maximum point to the mth interference maximum is given approximately by

$$y_m = m\,\frac{D\lambda}{d}$$

Figure 14-35   Problems 79–81.

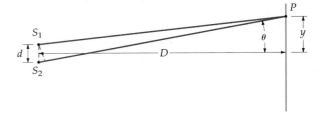

80. Two loudspeakers are driven in phase by an audio amplifier at a frequency of 600 Hz. The speakers are on the y axis, one at $y = +1.00$ m and the other at $y = -1.00$ m. A listener begins at $y = 0$ and walks along a line parallel to the y axis at a very large distance D away. (See Problem 79.) (a) At what angle θ will she first hear a minimum in the sound intensity? (b) At what angle will she first hear a maximum (after $\theta = 0$)? (c) How many maxima can she possibly hear if she keeps walking in the same direction?

81. Two sound sources, driven in phase by the same amplifier, are 2 m apart on the y axis. At a point a very large distance from the y axis, constructive interference is first heard at an angle $\theta_1 = 0.140$ rad with the x axis and is next heard at $\theta_2 = 0.283$ rad. (See Figure 14-35.) (a) What is the wavelength of the sound waves from the sources? (b) What is the frequency of the sources? (c) At what other angles is constructive interference heard? (d) What is the smallest angle for which the sound waves completely cancel?

82. Two sources have a phase difference δ_0 that is proportional to time: $\delta_0 = Ct$, where C is a constant. The amplitude of the wave from each source at some point P is A_0. (a) Write the wave functions for each of the two waves at point P, assuming this point to be a distance x_1 from one source and $x_1 + \Delta x$ from the other. (b) Find the resultant wave function, and show that its amplitude is $2A_0 \cos \frac{1}{2}(\delta + \delta_0)$, where δ is the phase difference at P due to the path difference. (c) Sketch the intensity at point P versus time for a zero path difference. (Let I_0 be the intensity due to each wave separately.) What is the time average of the intensity? (d) Make the same sketch for the intensity at a point for which the path difference is $\frac{1}{2}\lambda$.

83. In this problem, you will derive a convenient formula for the speed of sound in air at temperature t in Celsius degrees. Begin by writing the temperature as $T = T_0 + \Delta T$, where $T_0 = 273$ K corresponds to 0°C and $\Delta T = t$, the Celsius temperature. The speed of sound is a function of T, $v(T)$. To a first-order approximation, you can write

$$v(T) \approx v(T_0) + \left(\frac{dv}{dT}\right)_{T_0} \Delta T$$

where $(dv/dT)_{T_0}$ is the derivative evaluated at $T = T_0$. Compute this derivative, and show that the result leads to

$$v \approx (331 \text{ m/s})\left(1 + \frac{t}{2T_0}\right) = (331 + 0.606t) \text{ m/s}$$

84. While studying physics in her dorm room, a student is listening to the live radio broadcast of a baseball game. She is 1.6 km due south of the baseball field. Over her radio, the student hears a noise generated by the electromagnetic pulse of a lightning bolt. Two seconds later, she hears over the radio the thunder picked up by the microphone at the baseball field. Four seconds after she hears the noise of the electromagnetic pulse over the radio,

thunder rattles her windows. Where, relative to the ballpark, did the lightning bolt occur?

85. Two identical sound sources have a frequency of 500 Hz. The coordinates of the sources are (0, 1) m and (0, −1) m. A detector 80 m from the origin is free to revolve in the xy plane with a radius of 80 m. The first maxima in intensity is detected when the detector is at (80, 0) m. (a) Find the coordinates of the detector for the first five maxima in order of decreasing positive values of x. (b) Find the coordinates for the first four minima in order of decreasing positive values of x subject to the condition $\sqrt{x^2 + y^2} = 80$ m.

86. A radio telescope consists of two antennas separated by a distance of 200 m. Both antennas are tuned to a particular frequency, such as 20 MHz. The signals from each antenna are fed into a common amplifier, but one signal first passes through a phase adjuster that delays its phase by a chosen amount so that the telescope can "look" in different directions. When the phase delay is zero, plane radio waves that are incident vertically on the antennas produce signals that add constructively at the amplifier. What should the phase delay be so that signals coming from an angle $\theta = 10°$ with the vertical (in the plane formed by the vertical and the line joining the antennas) will add constructively at the amplifier?

87. A coiled spring, such as a Slinky, is stretched to a length L. It has a force constant k and a mass m. (a) Show that the velocity of longitudinal compression waves along the spring is given by $v = L\sqrt{k/m}$. (b) Show that this is also the velocity of transverse waves along the spring if the natural length of the spring is much less than L.

88. A physics student drops a vibrating 440-Hz tuning fork down an elevator shaft of a tall building. When the student hears a frequency of 400 Hz, how far has the tuning fork fallen?

Part 3 Thermodynamics

Two red intersecting laser beams are used here to study the combustion of coal–water slurries in a conventional power conversion device. The test material is injected into the combustion reactor (blue flame), giving off a yellow-orange emission as it ignites and burns. The laser light is used to measure the particle size of combustible material.

Chapter 15

Temperature

Sunspots appear on the surface of the sun when streams of gases slowly erupt from deep within the star. The solar "flower" is 10,000 miles in diameter. The temperature variation, indicated by color changes made by computer color enhancement, is not fully understood. The central portion of the sunspot is cooler than the outer regions as indicated by the dark area. The temperature at the sun's core is of the order of 10^6 K, whereas at the surface the temperature is only about 6000 K.

Thermodynamics is the study of temperature, heat, and the exchange of energy. It has practical applications in all branches of science and engineering as well as in many aspects of daily life, from dealing with the weather to cooking. In this chapter, we will study temperature and some of the thermal properties of matter.

Temperature is familiar to us all as the measure of the hotness or coldness of objects. More precisely, it is a measure of the average internal molecular kinetic energy of an object. The definition and determination of temperature are subtle. For example, it is quite difficult to define temperature so that different thermometers will agree with each other in the measurement of the temperature of a substance. However, the properties of gases at low densities do allow us to define a temperature scale and to construct gas thermometers that do agree. In Chapter 17, we will see that the second law of thermodynamics can be used to define a temperature scale that is independent of the properties of any substance and can be chosen to coincide with the temperature scale defined using the properties of gases.

15-1   The Celsius and Fahrenheit Temperature Scales

We are born with built-in thermometers. Our sense of touch can usually tell us if an object is hot or cold. Early in childhood we learn that to make a cold object warmer we place it in contact with a hot object. To warm water, we put the pot on a flame. To make a warm object cool, we place it in contact with a cold object. To chill a drink, we place the bottle in an ice bucket.

When an object is heated or cooled, some of its physical properties change. For example, most solids and liquids expand when they are heated. A gas, if it is allowed to, will also expand when it is heated, or, if its volume is kept constant, its pressure will rise. If an electrical conductor is heated, its electrical resistance changes. A physical property that changes with temperature is called a **thermometric property.** A change in a thermometric property indicates a change in the temperature of the object.

Suppose we place a warm copper bar in close contact with a cold iron bar. The copper bar contracts slightly, indicating that it is cooling, and the iron bar expands slightly, indicating that it is warming. The two bars are said to be in **thermal contact.** Eventually this process stops; that is, neither bar changes anymore in length. When this happens the two bars are said to be in **thermal equilibrium** with each other.

Now, suppose that instead of placing the bars in contact with each other, we first place the warm copper bar in a cool lake. The bar cools down and the water in the lake warms slightly, but the lake is large enough that the change in its temperature is negligible. Again, this process eventually stops and the bar and water are in thermal equilibrium. Now suppose we place the cold iron bar in the lake far away from the copper bar so that the two bars are not in thermal contact. The iron bar warms until it and the lake water are also in thermal equilibrium. If we remove the iron and copper bars and place them in contact with each other, we find that their lengths do not change. They are in thermal equilibrium with each other. This is such a common experience that you are probably not surprised. However, there is no logical way to deduce this result. It is an important statement of the thermal behavior of all objects.

> If two objects are in thermal equilibrium with a third, then they are in thermal equilibrium with each other.

This is often called the **zeroth law of thermodynamics.** It is illustrated in Figure 15-1. The zeroth law of thermodynamics allows us to define a temperature scale. Two objects in thermal equilibrium with each other are said to have the same temperature.

Any thermometric property can be used to establish a temperature scale and construct a thermometer. Figure 15-2 shows a common mercury thermometer, which consists of a glass bulb and tube containing a fixed amount of mercury. When the mercury is heated by placing the thermometer in contact with a warmer body, the mercury expands more than the glass, and the length of the mercury column increases. Temperatures are measured by comparing the end of the mercury column with markings on the glass tube. These markings are determined as follows. The thermometer is first placed in ice and water in equilibrium at a pressure of 1 atm.* If the thermometer is

---

*Water and ice in equilibrium provide a constant-temperature bath. When ice is placed in warm water, the water cools as some of the ice melts. Eventually, thermal equilibrium is reached and no more ice melts. If the system is heated slightly, some more of the ice melts, but the temperature does not change as long as some ice remains.

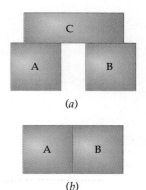

Figure 15-1 The zeroth law of thermodynamics. (*a*) Systems *A* and *B* are in thermal contact with system *C* but not with each other. When *A* and *B* are each in thermal equilibrium with *C*, they are in thermal equilibrium with each other, which can be checked by placing them in contact with each other as in (*b*).

Figure 15-2 A simple mercury thermometer. When in thermal contact with a warmer body, the mercury expands and the column rises.

(a)

(b)

(c)

(a) The temperature in the interior of a furnace is measured by a radiation thermometer. (b) Various temperature probes for use with a laboratory thermocouple digital thermometer. (c) Thermometers built around 1660 in Florence, Italy.

initially warmer than the ice water, the length of the mercury column will decrease, but it will eventually cease changing. The thermometer is now in thermal equilibrium with the ice water. The position of the mercury column is marked on the glass tube. This is the **ice-point temperature** (also called the **normal freezing point** of water). Next, the thermometer is placed in boiling water at a pressure of 1 atm, and the length of the mercury column increases until the thermometer is in thermal equilibrium with the boiling water. The new position of the column is marked. This is the **steam-point temperature** (also called the **normal boiling point** of water).

The **Celsius temperature scale** (formerly called the centigrade scale) is constructed by defining the ice-point temperature to be zero degrees Celsius (0°C) and the steam-point temperature to be 100°C. The space on the glass tube between the ice-point mark and the steam-point mark is then divided into 100 equal intervals or degrees, and the degree markings are extended below the ice-point mark and above the steam-point mark. The temperature of another system can now be measured by placing the mercury thermometer in thermal contact with it, waiting for thermal equilibrium to be established, and noting the position of the mercury column. If L_t is the length of the mercury column, the Celsius temperature t_C is given by

$$t_C = \frac{L_t - L_0}{L_{100} - L_0} \times 100°$$ 15-1

where L_0 is the length of the mercury column when the thermometer is in an ice bath and L_{100} is its length when the thermometer is in a steam bath. The normal temperature of the human body measured on the Celsius scale is about 37°C.

The **Fahrenheit temperature scale** is constructed by defining the ice-point temperature to be 32°F and the steam-point temperature to be 212°F.* Because the Fahrenheit scale is in common use in the United States and the Celsius scale is used in scientific work and throughout the rest of the world, we often need to convert temperatures between these two scales. We note that there are 100 Celsius degrees and 180 Fahrenheit degrees between the ice point and the steam point. A temperature change of one Fahrenheit degree is therefore smaller than a change of one Celsius degree. A temperature change of one Celsius degree (written 1 C° to distinguish it from a temperature of 1°C) equals a change of $\frac{9}{5}$ Fahrenheit degrees. To convert a temperature given on one scale to that of the other scale, we must also take

*Fahrenheit wanted all measurable temperatures to be positive. Originally, he chose 0°F for the coldest temperature he could obtain with a mixture of ice and saltwater, and 96°F (a convenient number with many factors for subdivision) for the temperature of the human body. He then modified his scale slightly to make the ice-point and steam-point temperatures whole numbers. This resulted in the temperature of the human body being between 98° and 99°F.

into account the fact that the zero temperatures of the two scales are not the same. The general relation between a Fahrenheit temperature t_F and a Celsius temperature t_C is

$$t_C = \tfrac{5}{9}(t_F - 32°)$$

15-2   *Fahrenheit–Celsius conversion*

Example 15-1

Find the temperature on the Celsius scale equivalent to 41°F.

From Equation 15-2, we obtain

$$t_C = \tfrac{5}{9}(41° - 32°) = \tfrac{5}{9}(9°) = 5°C$$

We can understand this calculation as follows. A temperature of 41°F is 9 Fahrenheit degrees above the ice point (41 − 32 = 9). But 9 Fahrenheit degrees above ice point is equivalent to 5 Celsius degrees above the ice point. Since the ice point on the Celsius scale is 0°C, a temperature of 41°F corresponds to 5°C.

Example 15-2

Find the temperature on the Fahrenheit scale equivalent to −10°C.

The temperature −10°C is 10 C° below the ice point. This is equivalent to 18 F° below the ice point on the Fahrenheit scale. Since the Fahrenheit ice point is 32°F, −10°C is equivalent to 14°F. This result can also be obtained directly from Equation 15-2 by substituting $t_C = -10°C$.

Exercise

(*a*) Find the Celsius temperature equivalent to 68°F. (*b*) Find the Fahrenheit temperature equivalent to −40°C.   [Answers: (*a*) 20°C, (*b*) −40°F]

Other thermometric properties can be used to set up thermometers and construct temperature scales. Figure 15-3 shows a bimetallic strip consisting of two different metals bonded together. When the strip is heated or cooled, it bends to accommodate the difference in the thermal expansion of the two metals. One application of a bimetallic strip is in a thermostat to regulate a furnace. The bending of the strip is used to make or break an electrical contact to turn the furnace on or off (Figure 15-4). Figure 15-5 shows a thermometer consisting of a bimetallic coil with a pointer attached to indicate the temperature. When the thermometer is heated, the coil bends and the pointer moves to the right. The thermometer is calibrated in the same way as a mercury thermometer, that is, by determining the ice point and the steam point and dividing the interval between them into 100 Celsius degrees (or 180 Fahrenheit degrees).

Figure 15-3 A bimetallic strip. When heated or cooled the two metals expand or contract by different amounts, causing the strip to bend.

To relay

Figure 15-4 A bimetallic strip can be used in a thermostat to make or break an electrical contact.

Figure 15-5 (*Left*) A thermometer using a bimetallic coil. (*Right*) An office thermostat. The coil on the right controls the heater. When the coil expands or contracts, the tube mounted on it tilts, and mercury in the tube slides to open or close an electrical switch. A slide lever (at the top) mounted on the coil is used to set the desired temperature. The coil on the left (partially obscured by two mercury switches) is used in a similar way to control the air conditioner.

Questions

1. How can you determine if two bodies are in thermal equilibrium with each other if it is impossible to put them into thermal contact with each other?

2. Which is greater, an increase in temperature of 1 C° or one of 1 F°?

3. One body is at −2°C and another is at +20°F. Which body is colder?

4. Distinguish between 1°C and 1 C°.

15-2   Gas Thermometers and the Absolute Temperature Scale

Mercury thermometers and bimetallic thermometers are commonly used. If they are calibrated in ice water and steam as described in Section 15-1, they must agree at 0°C and 100°C. But there is no reason to expect them to agree at any other temperature such as 50°C. Indeed, if they did it would mean that the thermal expansion of mercury, glass, and the metals used in the bimetallic thermometer varied with temperature in precisely the same way. In fact, at temperatures between 0 and 100°C different thermometers give slightly different readings, though the agreement is good enough for most purposes. But at temperatures well above the steam point or below the ice point, the agreement becomes very poor. Furthermore, mercury freezes at very low temperatures and vaporizes at very high temperatures. So the question of the measurement of temperatures other than 0°C or 100°C is more than just one of accuracy. How are such temperatures defined? Which thermometer is right?

There is one group of thermometers for which measured temperatures are in very close agreement even far from the calibration points. These are the gas thermometers. In one version, the **constant-volume gas thermometer,** the volume of the gas is kept constant, and the pressure of the gas is used as the thermometric property. Figure 15-6 shows a constant-volume gas thermometer. The volume of the gas in tube B_1 is kept constant by raising or lowering tube B_3 so that the mercury in tube B_2 remains at the zero mark. The pressure of the gas is found by reading the height h of the mercury column in tube B_3. As we learned in Chapter 11, a column of mercury 760 mm high is equivalent to a pressure of 1 atm. The ice-point pressure and the steam-point pressure are determined in the usual way, and the interval between them is divided into 100 equal degrees (for the Celsius scale).

If the pressure in a constant-volume gas thermometer is P_0 in an ice bath, P_{100} in a steam bath, and P_t in a bath whose temperature is to be determined, that temperature in degrees Celsius is defined to be

$$t_C = \frac{P_t - P_0}{P_{100} - P_0} \times 100° \qquad 15\text{-}3$$

Figure 15-7 shows the results of measurements of the boiling point of sulfur using constant-volume gas thermometers filled with various gases. In this figure, the measured temperature is plotted as a function of the steam-point pressure, P_{100}, which is varied by varying the amount of gas in the thermometer. As the amount of gas is reduced, the density of the gas decreases and steam-point pressure decreases. From this figure, we see that the agreement among the thermometers is very close no matter what kind of gas is used, as long as the density of the gas is low. Indeed, in the limit as the gas density goes to zero, all gas thermometers give exactly the same value for the temperature, independent of the gas used. Because this temperature measurement does not depend on the properties of some particular

Figure 15-6 A constant-volume gas thermometer. The volume is kept constant by raising or lowering tube B_3 such that the mercury in tube B_2 remains at the zero mark. The temperature is chosen to be proportional to the pressure of the gas in tube B_1, which is indicated by the height h of the mercury column in tube B_3.

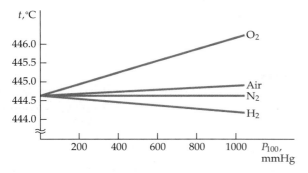

Figure 15-7 Temperature of the boiling point of sulfur measured with constant-volume gas thermometers filled with various gases. The pressure at the steam point of water P_{100} is varied by varying the amount of gas in the thermometers. As the amount of gas is reduced, the temperatures measured by all the thermometers approach the value 444.60°C.

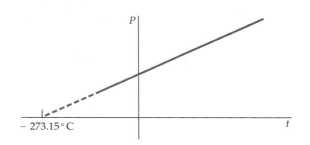

Figure 15-8 Plot of pressure versus temperature as measured by a constant-volume gas thermometer. When extrapolated to zero pressure, the plot intersects the temperature axis at the value −273.15°C.

substance, like mercury, low-density gas thermometers are used to define temperature. This choice is reinforced by theoretical considerations, which we will discuss later.

Let us now consider a series of temperature measurements with a constant-volume gas thermometer that has a very small but fixed amount of gas. According to Equation 15-3, the pressure in the thermometer P_t varies linearly with the measured temperature t_C. Figure 15-8 shows a plot of measured temperature versus pressure in a constant-volume gas thermometer. When we extrapolate this straight line to zero pressure, the temperature approaches −273.15°C. This limit is the same no matter what kind of gas is used, but there is often considerable experimental uncertainty in the measurement of this number because of the difficulty in duplicating the ice-point and steam-point states with high precision in different laboratories. Because of these difficulties, a temperature scale based on a single fixed point was adopted in 1954 by the International Committee on Weights and Measures. A reference state that is much more precisely reproducible than either the ice point or the steam point is the triple point of water—the single temperature and pressure at which water, water vapor, and ice coexist in equilibrium. If we put water, ice, and water vapor in a sealed container with no air, the system will eventually reach a state of equilibrium where no ice will melt or evaporate, no water will freeze or evaporate, and no vapor will condense or freeze. This equilibrium state, called the **triple point of water,** occurs at a specific pressure and temperature. The triple-point pressure is 4.58 mmHg, and the temperature is 0.01°C. The **ideal-gas temperature scale** is defined so that the temperature of the triple point is 273.16 kelvins (K) (the **kelvin,** formerly called the degree kelvin, is a degree unit that is the same size as the Celsius degree). The temperature T of any other state is defined to be proportional to the pressure in a constant-volume gas thermometer:

$$T = \frac{273.16 \text{ K}}{P_3} P \qquad \text{15-4}$$

where P is the pressure of the gas in the thermometer when the thermometer is in thermal equilibrium with the system whose temperature is being measured and P_3 is the pressure of the gas in the thermometer when the thermometer is immersed in a water–ice–vapor bath at its triple point. The value of P_3 depends on the amount of gas in the thermometer. The ideal-gas

H$_2$O at its triple point. The container is a hollow cylindrical shell that is sealed and evacuated. It contains water, ice, and water vapor in equilibrium. The cylindrical well in the center is filled with water and contains an aluminum bushing for inserting a thermometer. The container, stored at the National Institute of Standards and Technology, is used in an ice-water bath that is just slightly below the triple-point temperature so that the water in the cell freezes very slowly.

temperature scale, defined by Equation 15-4, has the advantage that the measured temperature of any state is the same no matter what kind of gas is used. This temperature scale depends on the properties of gases but not on the properties of any particular gas.

The lowest temperature that can be measured with a gas thermometer using helium for the gas is about 1 K. Below this temperature helium liquefies. (All other gases liquefy at higher temperatures.) In Chapter 17, we will see that the second law of thermodynamics can be used to define the **absolute temperature scale** independent of the properties of any substance, and with no limitations on the range of temperatures that can be measured. As a result, it has been possible to measure temperatures as low as a millionth of a kelvin. The absolute scale so defined is identical to that defined by Equation 15-4 for the range of temperatures for which gas thermometers can be used. The symbol T is used when referring to the absolute temperature.

Because the Celsius degree and the kelvin are the same size, temperature *differences* are the same on both the Celsius scale and the absolute temperature scale (also called the **Kelvin scale**). That is, a temperature *change* of 1 K is identical to a temperature *change* of 1 C°. As a result, the only difference between the two scales lies in the choice of the zero temperature. To convert from degrees Celsius to kelvins, we merely add 273.15:

Celsius–absolute conversion

$$T = t_C + 273.15 \qquad\qquad 15\text{-}5$$

(For most purposes, we can round off the temperature of absolute zero to $-273°C$, so we can simply add 273 to the Celsius temperature to get the absolute temperature.) To convert from kelvins to degrees Fahrenheit, it is usually easiest to convert first to degrees Celsius by subtracting 273 and then use Equation 15-2. Although the Celsius and Fahrenheit scales are convenient for everyday use, the absolute scale is much more convenient for scientific purposes, partly because many formulas are more simply expressed when the absolute scale is used, and partly because the absolute temperature can be given a more fundamental interpretation.

Example 15-3

What is the Kelvin temperature corresponding to 70°F?

From Equation 15-2, the Celsius temperature equal to 70°F is

$$t_C = \tfrac{5}{9}(70° - 32°) = \tfrac{5}{9}(38°) = 21°C$$

The Kelvin temperature is then obtained from Equation 15-5:

$$T = t_C + 273 = 21 + 273 = 294 \text{ K}$$

Exercise
Find the Fahrenheit temperature corresponding to 300 K.
(Answer: 80.6°F)

Questions

5. Why might the Celsius and Fahrenheit scales be more convenient than the absolute scale for ordinary, nonscientific purposes?

6. The temperature of the interior of the sun is said to be about 10^7 degrees. Do you think that this is degrees Celsius or kelvins, or does it matter?

15-3 Thermal Expansion

When the temperature of an object increases, the object usually expands. Consider a long rod of length L at a temperature T. When the temperature changes by ΔT, the change in length ΔL is proportional to ΔT and to the original length L:

$$\Delta L = \alpha L \, \Delta T \qquad\qquad 15\text{-}6$$

where α is called the **coefficient of linear expansion.** It is the ratio of the fractional change in length to the change in temperature:

$$\alpha = \frac{\Delta L/L}{\Delta T} \qquad\qquad 15\text{-}7$$

Its units are reciprocal Celsius degrees ($1/C°$) or reciprocal kelvins ($1/K$). The coefficient of linear expansion for a solid or liquid usually does not vary much with pressure, but it may vary with temperature. Equation 15-7 gives the average value of α over the temperature interval ΔT. The coefficient of linear expansion at a particular temperature T is found by taking the limit as ΔT approaches zero:

$$\alpha = \lim_{\Delta T \to 0} \frac{\Delta L/L}{\Delta T} = \frac{1}{L}\frac{dL}{dT} \qquad\qquad 15\text{-}8$$

In most cases, sufficient accuracy is obtained by using the average value of α over a wide temperature range.

The **coefficient of volume expansion** β is defined similarly as the ratio of the fractional change in volume to the change in temperature (at constant pressure):

$$\beta = \lim_{\Delta T \to 0} \frac{\Delta V/V}{\Delta T} = \frac{1}{V}\frac{dV}{dT} \qquad\qquad 15\text{-}9$$

Like α, β does not usually vary with pressure for solids and liquids but may vary with temperature. Average values for α and β for various substances are given in Table 15-1.

We can show that for a given material, the coefficient of volume expansion is 3 times the coefficient of linear expansion. Consider a box of dimensions L_1, L_2, and L_3. Its volume at a temperature T is

$$V = L_1 L_2 L_3$$

The rate of change of the volume with respect to temperature is

$$\frac{dV}{dT} = L_1 L_2 \frac{dL_3}{dT} + L_1 L_3 \frac{dL_2}{dT} + L_2 L_3 \frac{dL_1}{dT}$$

Dividing each side of the equation by the volume, we obtain

$$\beta = \frac{1}{V}\frac{dV}{dT} = \frac{1}{L_3}\frac{dL_3}{dT} + \frac{1}{L_2}\frac{dL_2}{dT} + \frac{1}{L_1}\frac{dL_1}{dT}$$

Since each term on the right side of the above equation equals α, we have

$$\beta = 3\alpha \qquad\qquad 15\text{-}10$$

The increase in size of any part of a body for a given temperature change is proportional to the original size of that part of the body. Thus, if we increase the temperature of a steel ruler, for example, the effect will be similar to that of a (very slight) photographic enlargement. Lines that were previously equally spaced will still be equally spaced, but the spaces will be larger. Likewise, the width of the ruler will be slightly larger. If the ruler has a hole in it, the hole will get larger, just as it would in a photographic enlargement.

Table 15-1 Approximate Values of the Coefficients of Thermal Expansion for Various Substances

Material	α, K^{-1}
Aluminum	24×10^{-6}
Brass	19×10^{-6}
Carbon	
Diamond	1.2×10^{-6}
Graphite	7.9×10^{-6}
Copper	17×10^{-6}
Glass	
Ordinary	9×10^{-6}
Pyrex	3.2×10^{-6}
Ice	51×10^{-6}
Invar	1×10^{-6}
Steel	11×10^{-6}

Material	β, K^{-1}
Acetone	1.5×10^{-3}
Air	3.67×10^{-3}
Alcohol	1.1×10^{-3}
Mercury	0.18×10^{-3}
Water (20°C)	0.207×10^{-3}

Expansion joint allows the bridge
to expand or contract without
stress as the temperature changes.

Though most materials expand when heated, water between 0 and 4°C
is an important exception. Figure 15-9 shows the volume occupied by 1 g of
water as a function of temperature. The volume is minimum, and therefore
the density is maximum, at 4°C. Thus, when water is heated at temperatures
below 4°C, it contracts rather than expands. This property has important
consequences for the ecology of lakes. At temperatures above 4°C, the water
in a lake becomes denser as it cools and sinks to the bottom. But at temperatures below 4°C, the water becomes less dense as it cools, so it stays at the
surface. Ice therefore forms first on the surface of a lake and, being less
dense than water, it remains there and acts as a thermal insulator for the
water below. If water contracted when it freezes, as most substances do, ice
would sink and new water would be exposed at the surface to freeze. Lakes
would fill with ice from the bottom up and would be much more likely to
become completely frozen in the winter, killing the fish and other aquatic
life.

Figure 15-9   Volume of 1 g of
water at atmospheric pressure
versus temperature. The minimum volume, which corresponds
to the maximum density, occurs
at 4°C.

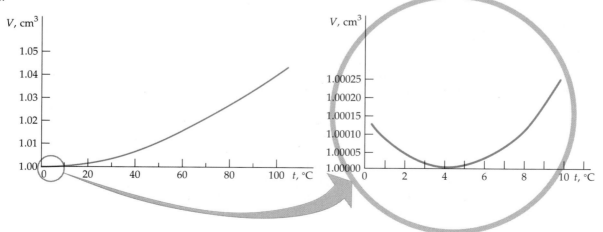

Example 15-4

A steel bridge is 1000 m long. By how much does it expand when the temperature rises from 0 to 30°C?

From Table 15-1, the coefficient of linear expansion for steel is 11×10^{-6} K^{-1}. The change in length for a 30-C° = 30-K rise in temperature is thus

$$\Delta L = \alpha L\, \Delta T = (11 \times 10^{-6}\ \text{K}^{-1})(1000\ \text{m})(30\ \text{K})$$

$$= 0.33\ \text{m} = 33\ \text{cm}$$

To relieve the stresses that would occur if such expansions were not allowed for, expansion joints must be included in the bridge. We can calculate the stress that would result if the bridge could not expand by using Equation 11-7 for Young's modulus:

$$Y = \frac{\text{stress}}{\text{strain}} = \frac{F/A}{\Delta L/L}$$

$$\frac{F}{A} = Y\frac{\Delta L}{L}$$

$$= (2 \times 10^{11}\ \text{N/m}^2)\frac{0.33\ \text{m}}{1000\ \text{m}} = 6.6 \times 10^7\ \text{N/m}^2$$

where we have used the value of Young's modulus for steel from Table 11-2. This stress is about one-third of the breaking stress for steel under compression. A compression stress of this magnitude would cause a steel bridge to buckle and become permanently deformed.

Example 15-5

A 1-L glass flask is filled to the brim with alcohol at 10°C. If the temperature warms to 30°C, how much alcohol spills out of the flask?

Using $\Delta T = 20$ C° = 20 K and the expansion coefficients from Table 15-1, we obtain for the change in the volume of the glass flask

$$\Delta V_g = \beta V\, \Delta T = 3\alpha V\, \Delta T$$

$$= 3(9 \times 10^{-6}\ \text{K}^{-1})(1\ \text{L})(20\ \text{K})$$

$$= 5.4 \times 10^{-4}\ \text{L} = 0.54\ \text{mL}$$

and for the change in the volume of the alcohol

$$\Delta V_a = \beta V\, \Delta T = (1.1 \times 10^{-3}\ \text{K}^{-1})(1\ \text{L})(20\ \text{K})$$

$$= 2.2 \times 10^{-2}\ \text{L} = 22.0\ \text{mL}$$

The amount of alcohol that spills out is thus 22.0 mL − 0.54 mL ≈ 21.5 mL.

Example 15-6

A copper bar is heated to 300°C and is then clamped rigidly between two fixed points so that it can neither expand nor contract. If the breaking stress of copper is 230 MN/m², at what temperature will the bar break as it cools?

In this example, the change in length ΔL that would occur if the bar contracted as it cooled is offset by an equal stretching due to tensile stress in the bar. From Equation 11-7, for Young's modulus, we have for the stretching ΔL caused by a tensile stress F/A

$$\Delta L = L \frac{F/A}{Y}$$

Setting this equal to the change in length that would occur if the bar could contract, we have

$$\Delta L = L\alpha \, \Delta T = L \frac{F/A}{Y}$$

Solving for ΔT and using $Y = 110 \text{ GN/m}^2$ from Table 11-2, $\alpha = 17 \times 10^{-6} \text{ K}^{-1}$ from Table 15-1, and $F/A = 230 \text{ MN/m}^2$ for the breaking stress of copper, we obtain

$$\Delta T = \frac{F/A}{\alpha Y}$$

$$= \frac{230 \times 10^6 \text{ N/m}^2}{(17 \times 10^{-6} \text{ K}^{-1})(110 \times 10^9 \text{ N/m}^2)}$$

$$= 123 \text{ K} = 123 \text{ C}°$$

Since the original temperature is 300°C, the temperature at which the bar will break is

$$t = 300°C - 123 \text{ C}° = 177°C$$

Question

7. If mercury and glass had the same coefficient of thermal expansion, could a mercury thermometer work?

15-4   The Ideal-Gas Law

If we compress a gas while keeping its temperature constant, we find that the pressure increases as the volume decreases. Similarly, if we cause a gas to expand at constant temperature, its pressure decreases as its volume increases. To a good approximation, the pressure of a gas varies inversely with the volume. This implies that, at a constant temperature, the product of the pressure and the volume of a gas is constant. This result was discovered experimentally by Robert Boyle (1627–1691), a younger contemporary of Galileo and an older one of Newton, and is known as **Boyle's law:**

$$PV = \text{constant} \qquad \text{constant temperature}$$

This law holds approximately for all gases at low densities. But according to Equation 15-4, the absolute temperature of a gas at low densities is proportional to the pressure at constant volume. Similarly, the absolute temperature is proportional to the volume of a gas if the pressure is kept constant, a result discovered experimentally by Jacques Charles (1746–1823) and Gay-Lussac (1778–1850). Thus, at low densities, the product PV is very nearly proportional to the temperature T:

$$PV = CT \qquad\qquad 15\text{-}11$$

where C is a constant of proportionality appropriate to a particular body of gas.

Now, suppose we have two identical containers, each holding the same amount of the same kind of gas at the same temperature. Each has a volume V described by Equation 15-11. If we combine the two containers, we end up with twice the volume of gas at the same pressure P and temperature T. It follows from Equation 15-11 that C must have increased by a factor of 2. In other words, C is proportional to the amount of gas. We can therefore write

$$C = kN$$

where N is the number of molecules of the gas and k is a constant. Equation 15-11 then becomes

$$PV = NkT \qquad\qquad 15\text{-}12$$

The constant k is called **Boltzmann's constant.** It is found experimentally to have the same value for any kind or amount of gas. Its value in SI units is

$$k = 1.381 \times 10^{-23} \text{ J/K} \qquad\qquad 15\text{-}13$$

It is often convenient to write the amount of a gas in terms of the number of moles. A **mole** (mol) of any substance is the amount of that substance that contains Avogadro's number of atoms or molecules. **Avogadro's number** N_A is defined as the number of carbon atoms in 12 grams of ^{12}C. The value of Avogadro's number is

$$N_A = 6.022 \times 10^{23} \text{ molecules/mol} \qquad\qquad 15\text{-}14$$

If we have n moles of a substance, the number of molecules is

$$N = nN_A \qquad\qquad 15\text{-}15$$

Equation 15-12 is then

$$PV = nN_A kT = nRT \qquad\qquad 15\text{-}16$$

where

$$R = kN_A \qquad\qquad 15\text{-}17$$

is called the **universal gas constant.** Its value, which is the same for all gases, is

$$R = 8.314 \text{ J/mol·K} = 0.08206 \text{ L·atm/mol·K} \qquad\qquad 15\text{-}18$$

Figure 15-10 shows plots of PV/nT versus the pressure P for several gases. (In these plots, the pressure is varied by varying the amount of gas.) We can see that for real gases, PV/nT is very nearly constant over a rather large range of pressures. (Even for oxygen, which has the largest variation in this graph, the variation is only about 1 percent between 0 and 5 atm.) We define an **ideal gas** as one for which PV/nT is constant for all pressures. For an ideal gas, the pressure, volume, and temperature are related by

$$PV = nRT \qquad\qquad 15\text{-}19 \qquad \textit{Ideal-gas law}$$

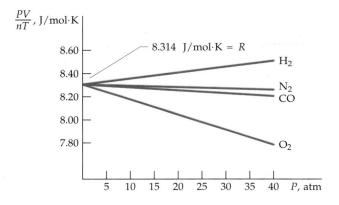

Figure 15-10 Plot of PV/nT versus P for real gases. As the density of the gas is reduced, thereby reducing the pressure, the ratio PV/nT approaches the same value, 8.314 J/mol·K, for all gases. This value is the universal gas constant R. The ideal-gas equation, $PV = nRT$, is a good approximation for all real gases at low pressures, that is, at pressures up to a few atmospheres.

The mass of 1 mol is called the **molar mass** M. (The terms *molecular weight* or *molecular mass* are also sometimes used.) The molar mass of ^{12}C is, by definition, 12 g/mol or 12×10^{-3} kg/mol. Molar masses of the elements are given in the periodic table in Appendix E. The molar mass of a molecule, such as CO_2, is the sum of the molar masses of the elements in the molecule. Since the molar mass of oxygen is 16 g/mol (actually 15.999 g/mol), the molar mass of O_2 is 32 g/mol and that of CO_2 is $12 + 32 = 44$ g/mol.

The mass of n moles of gas is given by

$$m = nM$$

The density ρ of an ideal gas is

$$\rho = \frac{m}{V} = \frac{nM}{V}$$

Using $n/V = P/RT$ from Equation 15-19, we can write

$$\rho = \frac{M}{RT}P \qquad\qquad 15\text{-}20$$

At a given temperature, the density of an ideal gas is proportional to the pressure.

Equation 15-19, which relates P, V, and T for a given amount of gas, is called an **equation of state.** The state of a gas of constant mass is determined by any two of the three variables P, V, and T. For example, if P and V are known, then the temperature T is determined by some function $T(P, V)$. The particular function, the equation of state, for an ideal gas is given by Equation 15-19. The concept of an ideal gas is an extrapolation of the behavior of real gases at low densities and pressures to ideal behavior. As can be seen from Figure 15-10, at reasonably low pressures real gases differ little from an ideal gas. Thus, Equation 15-19 is quite useful for describing the properties of real gases. At higher gas densities and pressures, corrections must be made to Equation 15-19 if it is to apply to real gases. One set of these corrections is discussed in Section 15-6.

Figure 15-11 shows plots of P versus V for various temperatures T. These curves are called **isotherms.** The isotherms for an ideal gas are hyperbolas.

For a fixed amount of gas, we can see from Equation 15-19 that the quantity PV/T is constant. If we use the subscripts 1 for the initial values and 2 for the final values, we have

$$\frac{P_2 V_2}{T_2} = \frac{P_1 V_1}{T_1} \qquad\qquad 15\text{-}21$$

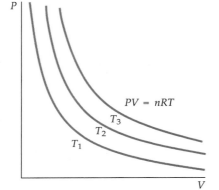

Figure 15-11 Isotherms on the PV diagram for a gas. For an ideal gas, these curves are hyperbolas given by $PV = nRT$.

Example 15-7

What volume is occupied by 1 mol of gas at a temperature of 0°C and a pressure of 1 atm?

The absolute temperature corresponding to 0°C is 273 K. From the ideal-gas law (Equation 15-19), we have

$$V = \frac{nRT}{P} = \frac{(1 \text{ mol})(0.0821 \text{ L·atm/mol·K})(273 \text{ K})}{1 \text{ atm}} = 22.4 \text{ L}$$

The temperature of 0°C = 273 K and the pressure of 1 atm are often referred to as **standard conditions.** From Example 15-7, we can see that under standard conditions, 1 mol of any gas occupies a volume of 22.4 L.

Exercise

Find (*a*) the number of moles *n* and (*b*) the number of molecules *N* in 1 cm³ of a gas under standard conditions.
[Answers: (*a*) $n = 4.46 \times 10^{-5}$ mol, (*b*) $N = 2.68 \times 10^{19}$ molecules]

Example 15-8

The molar mass of hydrogen is 1.008 g/mol. What is the mass of one hydrogen atom?

Since there are N_A hydrogen atoms in 1 mol, the mass *m* of one atom is

$$m = \frac{1.008 \text{ g/mol}}{6.022 \times 10^{23} \text{ atoms/mol}} = 1.67 \times 10^{-24} \text{ g/atom}$$

We can see from this example that Avogadro's number is approximately the reciprocal of the mass of the hydrogen atom measured in grams.

Example 15-9

A gas has a volume of 2 L, a temperature of 30°C, and a pressure of 1 atm. It is heated to 60°C and compressed to a volume of 1.5 L. Find its new pressure.

From Equation 15-21, we have

$$P_2 = \frac{T_2 V_1}{T_1 V_2} P_1$$

Since we are dealing with ratios, we can express the pressure and volume in any units, but we must remember that the ideal-gas equation holds only if the temperatures are absolute temperatures. Using $T_1 = 273 + 30 = 303$ K and $T_2 = 273 + 60 = 333$ K, we have

$$P_2 = \frac{(333 \text{ K})(2 \text{ L})}{(303 \text{ K})(1.5 \text{ L})}(1 \text{ atm}) = 1.47 \text{ atm}$$

Example 15-10

One hundred grams of CO_2 occupies a volume of 55 L at a pressure of 1 atm. (*a*) Find the temperature. (*b*) If the volume is increased to 80 L and the temperature is kept constant, what is the new pressure?

(*a*) We can find the temperature from the ideal-gas equation (Equation 15-19) if we first find the number of moles. Since the molar mass of CO_2 is 44 g/mol, the number of moles is

$$n = \frac{m}{M} = \frac{100 \text{ g}}{44 \text{ g/mol}} = 2.27 \text{ mol}$$

The absolute temperature is then

$$T = \frac{PV}{nR} = \frac{(1 \text{ atm})(55 \text{ L})}{(2.27 \text{ mol})(0.0821 \text{ L} \cdot \text{atm/mol} \cdot \text{K})} = 295 \text{ K}$$

(*b*) Using Equation 15-21 with $T_2 = T_1$, we have

$$P_2 V_2 = P_1 V_1$$

$$P_2 = \frac{V_1}{V_2} P_1 = \frac{55 \text{ L}}{80 \text{ L}}(1 \text{ atm}) = 0.688 \text{ atm}$$

15-5   The Molecular Interpretation of Temperature: The Kinetic Theory of Gases

In previous sections of this chapter, the behavior of gases has been described in terms of P, V, and T, quantities that are called **macroscopic variables.** Such variables describe the macroscopic state of a substance. To describe the *microscopic* state of a gas would require giving the coordinates and velocities of all its molecules—a truly impossible task. In this section, we will show how the macroscopic description can be related to simple averages of microscopic quantities. In particular we will show that the absolute temperature of a gas is a measure of the average kinetic energy of the gas molecules. We do this by using a simple model to calculate the pressure exerted by the gas on the walls of its container.

In this microscopic point of view, called the **kinetic theory of gases,** the pressure of a gas is the result of the collisions between the gas molecules and the walls of their container. We can calculate this pressure by calculating the rate of change of momentum of the gas molecules due to collisions with a wall of the container. By Newton's second law, the rate of change of momentum equals the force exerted by the wall on the gas molecules:

$$\mathbf{F} = \frac{d\mathbf{p}}{dt}$$

By Newton's third law, this force equals the force exerted by the molecules on the wall. The force per unit area equals the pressure.

We begin by making the following assumptions:

1. The gas consists of a large number of molecules that make elastic collisions with each other and with the walls of the container.

2. The molecules are separated, on the average, by distances that are large compared with their diameters, and they exert no forces on each other except when they collide.

3. In the absence of external forces (the molecules are moving fast enough that we can neglect gravity), there is no preferred position for a molecule in the container, and there is no preferred direction for the velocity vector.

The second assumption, that the gas molecules are, on the average, far apart, is equivalent to assuming a very low gas density, which as we saw in the last section is the same as assuming that the gas is an ideal gas. Since momentum is conserved, the collisions the molecules make with each other have no effect on the total momentum in any direction, so we can neglect these collisions.

Let us assume that we have a rectangular container of volume V containing N molecules, each of mass m and moving with a speed v. We wish to calculate the force exerted by these molecules on the right wall, which is perpendicular to the x axis. The x component of the momentum of a molecule before it hits the wall is $+mv_x$, and after it makes an elastic collision with the wall it is $-mv_x$. The magnitude of the change in momentum during the collision of one molecule with the wall is thus $2mv_x$. The total change in momentum of all the molecules during some time interval Δt is $2mv_x$ times the number of molecules that hit the wall during the interval. Figure 15-12 depicts gas molecules in a rectangular container. The number of molecules that hit the right wall of area A is the number that are within the distance $v_x \Delta t$ and are moving to the right. This number is the number of molecules

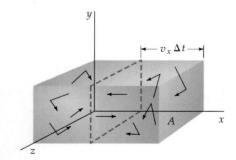

Figure 15-12 Gas molecules in a rectangular container. In a time interval Δt, the molecules at a distance $v_x \Delta t$ from the right wall will hit the right wall if they are moving to the right. The number of molecules within this distance is proportional to v_x and to the number of molecules per unit volume.

per unit volume N/V times the volume $v_x \, \Delta t \, A$ times $\frac{1}{2}$, because on the average, half the molecules will be moving to the right and half to the left. The total change in the momentum Δp of the gas molecules in the time interval Δt is this number times $2mv_x$, the change in momentum per molecule:

$$\Delta p = \frac{1}{2}\frac{N}{V}(v_x \, \Delta t \, A)2mv_x = \frac{N}{V}mv_x^2 \, A \, \Delta t$$

The force exerted by the wall on the molecules and by the molecules on the wall is this change in momentum divided by the time interval Δt. The pressure is this force divided by the area:

$$P = \frac{F}{A} = \frac{1}{A}\frac{\Delta p}{\Delta t}$$

Dividing the change in momentum by the time Δt and by the area A, we obtain

$$P = \frac{N}{V}mv_x^2 \qquad\qquad 15\text{-}22$$

and

$$PV = Nmv_x^2 \qquad\qquad 15\text{-}23$$

To allow for the fact that all the molecules in a container do not have the same speed, we merely replace v_x^2 with the average value $(v_x^2)_{\mathrm{av}}$. Then, if we write Equation 15-23 in terms of the kinetic energy $\frac{1}{2}mv_x^2$ associated with motion along the x axis, we have

$$PV = 2N(\tfrac{1}{2}mv_x^2)_{\mathrm{av}} \qquad\qquad 15\text{-}24$$

Comparing this with Equation 15-12, which was obtained experimentally for any gas at very low densities, we can see that

$$PV = NkT = 2N(\tfrac{1}{2}mv_x^2)_{\mathrm{av}}$$

or

$$(\tfrac{1}{2}mv_x^2)_{\mathrm{av}} = \tfrac{1}{2}kT \qquad\qquad 15\text{-}25$$

Thus, the average kinetic energy associated with motion along the x axis is $\frac{1}{2}kT$. But there is nothing special about the x direction. On the average

$$(v_x^2)_{\mathrm{av}} = (v_y^2)_{\mathrm{av}} = (v_z^2)_{\mathrm{av}}$$

and

$$(v^2)_{\mathrm{av}} = (v_x^2)_{\mathrm{av}} + (v_y^2)_{\mathrm{av}} + (v_z^2)_{\mathrm{av}} = 3(v_x^2)_{\mathrm{av}}$$

Writing $(v_x^2)_{\mathrm{av}} = \frac{1}{3}(v^2)_{\mathrm{av}}$ and K_{av} for the average kinetic energy of the molecules, Equation 15-25 becomes

$$K_{\mathrm{av}} = (\tfrac{1}{2}mv^2)_{\mathrm{av}} = \tfrac{3}{2}kT \qquad\qquad 15\text{-}26$$

The average kinetic energy of the molecules is $\frac{3}{2}kT$. The absolute temperature is thus a measure of the average translational kinetic energy of the molecules. We include the word "translational" because the molecules may also have rotational or vibrational kinetic energy. Only the translational kinetic energy is relevant to the calculation of the pressure exerted by a gas on the walls of its container. The total translational kinetic energy of n moles of a gas containing N molecules is

$$K = N(\tfrac{1}{2}mv^2)_{\mathrm{av}} = \tfrac{3}{2}NkT = \tfrac{3}{2}nRT \qquad\qquad 15\text{-}27$$

Thus, the translational kinetic energy is $\frac{3}{2}kT$ per molecule and $\frac{3}{2}RT$ per mole.

We can use these results to estimate the order of magnitude of the speeds of the molecules in a gas. The average value of v^2 is, by Equation 15-26,

$$(v^2)_{av} = \frac{3kT}{m} = \frac{3N_A kT}{N_A m} = \frac{3RT}{M}$$

where $M = N_A m$ is the molar mass. The square root of $(v^2)_{av}$ is the **root mean square** (rms) speed:

$$v_{rms} = \sqrt{(v^2)_{av}} = \sqrt{\frac{3kT}{m}} = \sqrt{\frac{3RT}{M}} \qquad\qquad 15\text{-}28$$

Note that Equation 15-28 is similar to Equation 14-3 for the speed of sound in a gas:

$$v_{sound} = \sqrt{\frac{\gamma RT}{M}} \qquad\qquad 15\text{-}29$$

where $\gamma = 1.4$ for air. This should not be surprising since a sound wave in air is a pressure disturbance that is propagated by collisions between air molecules.

Example 15-11

Oxygen gas (O_2) has a molar mass of about 32 g/mol and hydrogen gas (H_2) has a molar mass of about 2 g/mol. Calculate (*a*) the rms speed of an oxygen molecule and (*b*) the rms speed of a hydrogen molecule when the temperature is 300 K.

(*a*) To make the units come out right, we must express the molecular mass of O_2 in Equation 15-28 in kilograms per mole. We then have

$$v_{rms} = \sqrt{\frac{3RT}{M}} = \sqrt{\frac{3(8.31 \text{ J/mol·K})(300 \text{ K})}{32 \times 10^{-3} \text{ kg/mol}}} = 483 \text{ m/s}$$

(*b*) Since the molar mass of hydrogen is one-sixteenth that of oxygen and v_{rms} is proportional to $1/\sqrt{M}$, the rms speed of hydrogen is 4 times that of oxygen or about 1.93 km/s.

We can see from Example 15-11 that the rms speed of oxygen molecules is somewhat greater than but of the same order of magnitude as the speed of sound in air, which at 300 K is about 347 m/s.

The Distribution of Molecular Speeds

As we have noted, all the molecules in a gas do not have the same speed. The distribution of the molecular speeds of a gas can be measured directly using the apparatus illustrated in Figure 15-13. A substance is vaporized in an oven, and the vapor molecules are allowed to escape through a hole in the oven wall into a vacuum chamber. The molecules are collimated into a narrow beam by a series of slits (not shown in the figure). The beam is aimed at a detector that counts the number of molecules that are incident on it in a given period of time. Most of the beam is stopped by a rotating cylinder. Small helical slits in the cylinder (only one of which is depicted in the figure) allow the passage of molecules that have a narrow range of speeds that is determined by the angular velocity of rotation of the cylinder. The number of molecules in each range of speeds is measured by varying the angular velocity of the cylinder and counting the number of molecules that reach the detector for each angular velocity.

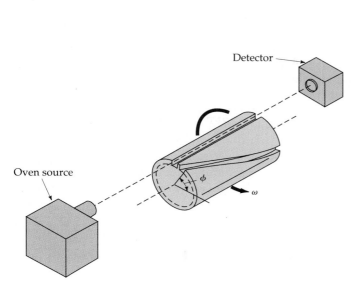

Figure 15-13 Schematic diagram of the apparatus for determining the speed distribution of the molecules of a gas. The molecules emerge from the oven and are collimated into a narrow beam by a series of slits (not shown). If the cylinder is not rotating, the beam passes through the straight slit and hits the detector. When the cylinder is rotating, only those molecules in a narrow speed range determined by the angular speed of rotation ω pass through the helical slit and reach the detector.

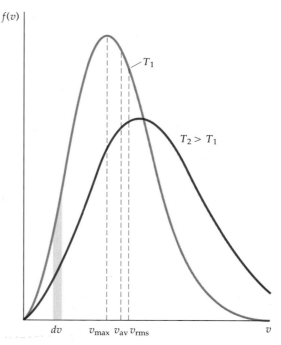

Figure 15-14 Distributions of molecular speeds in a gas at two temperatures, T_1 and $T_2 > T_1$. The shaded area $f(v) \, dv$ equals the fraction of the number of molecules having a particular speed in a narrow range of speeds dv. The mean speed v_{av} and the rms speed v_{rms} are both slightly greater than the most probable speed v_{max}.

Figure 15-14 shows the distributions of molecular speeds in a gas for two different temperatures. In this figure the quantity $f(v)$, called the **Maxwell–Boltzmann distribution function,** is defined as follows: If N is the total number of molecules, the number that have speeds in the range between v and $v + dv$ is dN, given by

$$dN = N f(v) \, dv \qquad\qquad 15\text{-}30$$

The fraction $dN/N = f(v) \, dv$ in a particular range dv is illustrated by the shaded region in the figure. The speed v_{max} at which the distribution is maximum is the most probable speed. The mean speed v_{av} and the rms speed v_{rms} are somewhat greater than the most probable speed.

The Maxwell–Boltzmann distribution function can be derived using statistical mechanics. The result is

$$f(v) = \frac{4}{\sqrt{\pi}} \left(\frac{m}{2kT} \right)^{3/2} v^2 e^{-mv^2/2kT} \qquad\qquad 15\text{-}31$$

The most probable speed v_{max} is that speed for which $f(v)$ is maximum. It is left as a problem to show that

$$v_{max} = \sqrt{\frac{2kT}{m}} \qquad\qquad 15\text{-}32$$

Comparing Equation 15-32 with Equation 15-28, we see that the most probable speed is slightly less than the rms speed.

Example 15-12

Use the Maxwell-Boltzmann distribution function (Equation 15-30) to calculate the average value of v^2 for the molecules in a gas.

To find the average value of a quantity, we sum all the values of that quantity and then divide by the number of terms in the sum. For example, to find the average score on a test given to N students, we add all the scores and divide by N. An alternative to summing each score obtained is to multiply each possible score by the number of students that received that score and sum. The number of molecules that have a certain speed v in the range dv is $Nf(v)\,dv$. We obtain the average value of v^2 by multiplying this number by v^2, summing over all the possible speeds, and dividing by the total number of molecules N. The average square speed is thus given by

$$(v^2)_{av} = \frac{1}{N} \int_0^\infty v^2\, Nf(v)\, dv = \int_0^\infty v^2\, f(v)\, dv$$

Using Equation 15-31 for $f(v)$, we obtain

$$(v^2)_{av} = \int_0^\infty v^2\, \frac{4}{\sqrt{\pi}} \left(\frac{m}{2kT}\right)^{3/2} v^2 e^{-mv^2/2kT}\, dv$$

$$= \frac{4}{\sqrt{\pi}} \left(\frac{m}{2kT}\right)^{3/2} \int_0^\infty v^4 e^{-mv^2/2kT}\, dv$$

This integral can be found in tables. Its value is

$$\int_0^\infty v^4 e^{-mv^2/2kT}\, dv = \frac{3}{8}\sqrt{\pi} \left(\frac{2kT}{m}\right)^{5/2}$$

Combining these results, we obtain

$$(v^2)_{av} = \frac{4}{\sqrt{\pi}} \left(\frac{m}{2kT}\right)^{3/2} \frac{3}{8}\sqrt{\pi} \left(\frac{2kT}{m}\right)^{5/2} = \frac{3kT}{m}$$

which agrees with $v_{rms} = \sqrt{3kT/m}$ from Equation 15-28.

In Example 15-11, we found that the rms speed of hydrogen molecules is about 1.93 km/s. This is about one-sixth of the escape speed at the surface of the earth, which we found to be 11.2 km/s in Section 10-5. So why is there no free hydrogen in the earth's atmosphere? As we can see from Figure 15-14, a considerable fraction of the molecules of a gas in equilibrium at some temperature has speeds greater than the rms speed. When the rms speed of the molecules of a particular gas is only about one-sixth of the escape speed for a planet, enough of the molecules have speeds greater than the escape speed so that the gas cannot exist in the atmosphere of that planet. Thus, no hydrogen. The rms speed of oxygen molecules, on the other hand, is about one-fourth that of hydrogen molecules, which makes it much less than the escape speed at the surface of the earth. Therefore, very few oxygen molecules have speeds greater than the escape speed, so oxygen is found in the earth's atmosphere.

The Maxwell–Boltzmann distribution of speed as given by Equation 15-31 can also be written as an energy distribution. We write the number of molecules with energy E in the range between E and $E + dE$ as

$$dN = NF(E)\, dE$$

where $F(E)$ is the energy distribution function. This will be the same number as given by Equation 15-30 if the energy E is related to the speed v by $E = \frac{1}{2}mv^2$. Then

$$dE = mv\, dv$$

(a)

(b)

and

$$Nf(v)\ dv = NF(E)\ dE$$

We can write

$$f(v)\ dv = Cv^2 e^{-mv^2/2kT}\ dv = Cve^{-E/kT}v\ dv = C\left(\frac{2E}{m}\right)^{1/2}e^{-E/kT}\frac{dE}{m}$$

where $C = (4/\sqrt{\pi})(m/2kT)^{3/2}$ from Equation 15-31. The energy distribution function $F(E)$ is thus given by

$$F(E) = \frac{4}{\sqrt{\pi}}\left(\frac{m}{2kT}\right)^{3/2}\left(\frac{2}{m}\right)^{1/2}\frac{1}{m}E^{1/2}e^{-E/kT}$$

Simplifying yields the **Maxwell–Boltzmann energy distribution function:**

$$F(E) = \frac{2}{\sqrt{\pi}}\left(\frac{1}{kT}\right)^{3/2}E^{1/2}e^{-E/kT} \qquad\qquad 15\text{-}33$$

The energy distribution function is thus proportional to $E^{1/2}e^{-E/kT}$.

(a) The atmosphere of Venus is almost entirely CO_2. However, measurements by the Pioneer Venus Orbiter show an atomic hydrogen cloud surrounding Venus. The crescent-shaped image shows atomic oxygen, and the bars show hydrogen extending far above the atmosphere. Since the escape speed on Venus is 10.3 km/s, slightly smaller than that on earth, and the temperature of its atmosphere is considerably warmer, all of the hydrogen in the atmosphere at the time of the formation of Venus should have escaped by now. (b) Jupiter as seen from about twelve million miles. Because the escape speed at the surface of Jupiter is about 60 km/s, Jupiter easily retains hydrogen in its atmosphere.

Questions

8. By what factor must the absolute temperature of a gas be increased to double the rms speed of its molecules?

9. How does the average translational kinetic energy of a molecule of a gas change if the pressure is doubled while the volume is kept constant? If the volume is doubled while the pressure is kept constant?

10. Why wouldn't you expect all molecules in a gas to have the same speed?

11. Two different gases are at the same temperature. What can you say about the rms speeds of the gas molecules? What can you say about the average kinetic energies of the molecules?

12. If a gas is heated at constant volume, explain in terms of molecular motion why its pressure on the walls of its container increases.

13. If the volume of a gas is reduced at constant temperature, explain in terms of molecular motion why its pressure on the walls of its container increases.

15-6   The van der Waals Equation and Liquid–Vapor Isotherms

Although most gases behave like an ideal gas at ordinary pressures, this ideal behavior breaks down at sufficiently high pressures or low temperatures when the density of the gas is high and the molecules are not, on the average, far apart. An equation of state called the **van der Waals equation** describes the behavior of many real gases over a wide range of pressures more accurately than does the ideal-gas equation of state ($PV = nRT$). The van der Waals equation for n moles of gas is

The van der Waals equation of state

$$\left(P + \frac{an^2}{V^2}\right)(V - bn) = nRT \qquad\qquad 15\text{-}34$$

The constant b in this equation arises because the gas molecules are not point particles but have a finite size; therefore, the volume available for the molecules to move about in is reduced. The magnitude of b is the volume of one mole of gas molecules. The term an^2/V^2 arises from the attraction of the gas molecules for each other. As a molecule approaches the wall of the container, it is pulled back by the molecules surrounding it with a force that is proportional to the density of those molecules n/V. Since the number of molecules that hit the wall in a given time is also proportional to the density of the molecules, the decrease in pressure due to the attraction of the molecules for each other is proportional to the square of the density and therefore to n^2/V^2. The constant a depends on the gas and is small for inert gases that have small chemical interactions. The terms bn and an^2/V^2 are both negligible when the volume V is large, that is, at low densities. Thus, at low densities the van der Waals equation approaches the ideal-gas law, whereas at high densities it provides a much better description of the behavior of real gases.

Figure 15-15 shows P-versus-V isothermal curves for a real substance at various temperatures. For temperatures above some critical temperature T_c, these curves are described quite accurately by the van der Waals equation and can be used to determine the constants a and b. For example, the values of these constants that give the best fit to the experimental curves for nitrogen are $a = 0.14$ Pa·m^6/mol^2 and $b = 39.1$ cm^3/mol. This volume of 39.1 cm^3 per mole is about 0.2 percent of the volume of 22,400 cm^3 occupied by 1 mol of nitrogen under standard conditions. Since the molar mass of nitrogen is 28 g/mol, if 1 mol of nitrogen molecules were packed into a volume of 39.1 cm^3, the density would be

$$\rho = \frac{M}{V} = \frac{28 \text{ g}}{39.1 \text{ cm}^3} = 0.72 \text{ g/cm}^3$$

which compares favorably with the density of liquid nitrogen, 0.80 g/cm^3.

Since the constant b is the actual volume of 1 mol of gas molecules, its value can be used to estimate the size of a molecule. For instance, if 1 mol = N_A molecules of nitrogen has a volume of 39.1 cm^3, the volume of one nitrogen molecule is

$$V = \frac{b}{N_A} = \frac{39.1 \text{ cm}^3/\text{mol}}{6.02 \times 10^{23} \text{ molecules/mol}}$$

$$= 6.50 \times 10^{-23} \text{ cm}^3/\text{molecule}$$

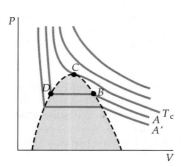

Figure 15-15 Isotherms on the PV diagram for a real substance. For temperatures above the critical temperature T_c, the substance remains a gas at all pressures and is described by the van der Waals equation. The pressure for the horizontal portions of the curves in the shaded region is the vapor pressure, which is the pressure at which the vapor and liquid are in equilibrium. To the left of the shaded region for temperatures below the critical temperature, the substance is a liquid and is nearly incompressible.

If we assume that each molecule is a sphere of diameter d occupying a cubic volume of side d, we obtain

$$d^3 = 6.50 \times 10^{-23} \text{ cm}^3$$

or

$$d = 4.0 \times 10^{-8} \text{ cm}$$

which is a reasonable estimate for the diameter of a molecule.

At temperatures below T_c, the van der Waals equation describes those portions of the isotherms outside the shaded region in Figure 15-15 but not those portions inside the shaded region. Suppose we have a gas at a temperature below T_c that initially has a low pressure and a large volume. We begin to compress the gas while holding the temperature constant (isotherm A in the figure). At first the pressure rises, but when we reach point B on the dashed curve, the pressure ceases to rise and the gas begins to liquefy at constant pressure. Along the horizontal line BD in the figure, the gas and liquid are in equilibrium. As we continue to compress the gas, more and more gas liquefies until at point D on the dashed curve we have only liquid. Then, if we try to compress the substance further, the pressure rises sharply because a liquid is nearly incompressible.

The constant pressure at which a liquid is in equilibrium with its own vapor at a given temperature is called the **vapor pressure.** When this equilibrium exists and the liquid is heated slightly (or the pressure is reduced slightly), the liquid boils. We can see from the figure that the vapor pressure for a gas depends on its temperature. If we had started compressing the gas at a lower temperature, as with isotherm A' in Figure 15-15, the vapor pressure would be lower, as is indicated by the horizontal constant-pressure line for A' at a lower value of pressure. The temperature for which the vapor pressure for a substance equals 1 atm is the **normal boiling point** of that substance. For example, the temperature at which the vapor pressure of water is 1 atm is 373 K = 100°C, so this temperature is the normal boiling point of water. At high altitudes, such as on the top of a mountain, the pressure is less than 1 atm, and water then boils at a temperature lower than 373 K. Table 15-2 gives the vapor pressures of water at various temperatures.

At temperatures greater than the critical temperature T_c, a gas will not liquefy at any pressure. The critical temperature for water vapor is 647 K = 374°C. The point at which the critical isotherm intersects the dashed curve in Figure 15-15 (point C) is called the critical point.

Table 15-2   Vapor Pressure of Water versus Temperature

t, °C	P, mmHg	P, kPa
0	4.581	0.611
10	9.209	1.23
15	12.653	1.69
20	17.535	2.34
30	31.827	4.24
40	55.335	7.38
50	92.55	12.3
60	149	19.9
70	233.8	31.2
80	355	47.4
90	526	70.1
100	760	101.3
110	1074	143.3
120	1489	198.5
130	2026	270.1

Critical temperature demonstration. (a) Below the critical temperature, the liquid is in equilibrium with its saturated vapor, and a clear meniscus separating the two is visible. (b) At the critical point, there is no distinction between the liquid and gas phases. (c) As the gas cools, clusters of molecules of increasing size form and redissolve. Eventually, the liquid–gas interface with a meniscus returns as in the first picture.

(a)   (b)   (c)

15-7   Phase Diagrams

Consider what happens if we put a liquid such as water in an evacuated container that is sealed so that its volume remains constant. At first, some of the water will evaporate, and water-vapor molecules will fill the previously empty space in the container. Some of the water-vapor molecules will hit the liquid surface and condense again into liquid water. Initially, the rate of evaporation will be greater than the rate of condensation, and the density of the water-vapor molecules will increase. But as the number of water-vapor molecules increases, the condensation rate will increase until it equals the rate of evaporation and equilibrium is established. The pressure of the water vapor at equilibrium is the vapor pressure for that temperature. If we now heat the container to a greater temperature, more liquid will evaporate and a new equilibrium will be established at a higher vapor pressure.

Figure 15-16 is a plot of pressure versus temperature at a constant volume for water. Such a plot is called a **phase diagram**. The portion of the diagram between points O and C shows the vapor pressure versus the temperature. As we continue to heat the container, the density of the liquid decreases and the density of the vapor increases. At point C on the diagram, these densities are equal. Point C is the critical point, which was also shown in Figure 15-15 on the isotherm labeled T_c. At this point and above it, there is no distinction between the liquid and the gas.* Critical temperatures for various substances are listed in Table 15-3.

If we now cool our container, some of the vapor condenses into a liquid as we move back down the curve OC until the substance reaches point O in Figure 15-16. At this point, the liquid begins to solidify. Point O is the triple point, that one point at which the vapor, liquid, and solid phases of a substance can coexist in equilibrium. Every substance has a unique triple point, with a specific temperature and pressure. As we saw in Section 15-2, the triple-point temperature for water is 273.16 K = 0.01°C and the triple-point pressure is 4.58 mmHg.

At temperatures and pressures below the triple point, the liquid cannot exist. The curve OA in the phase diagram of Figure 15-16 is the locus of pressures and temperatures for which the solid and vapor coexist in equilibrium. The direct change from a solid to a vapor is called **sublimation.** You can observe sublimation if you put ice cubes in the freezer compartment of a refrigerator (especially a self-defrosting refrigerator). The ice cubes eventually disappear due to sublimation. Because atmospheric pressure is well above the triple-point pressure of water, equilibrium is never established

Table 15-3 Critical Temperatures T_c for Various Substances	
Substance	T_c, K
Argon	150.8
Carbon dioxide	304.2
Chlorine	417.2
Helium	5.3
Hydrogn	33.3
Neon	44.4
Nitric oxide	180.2
Oxygen	154.8
Sulfur dioxide	430.9
Water	647.4

*Often the word "vapor" is used if the temperature is below the critical temperature, and the word "gas" is used if the temperature is above the critical temperature, though there is no need for such a distinction.

Figure 15-16   Phase diagram for water. The pressure and temperature scales are not linear but are compressed to show the interesting points. Curve OC is the curve of vapor pressure versus temperature. Curve OB is the melting curve, and curve OA is the sublimation curve.

between the ice and water vapor. The triple-point temperature and pressure of carbon dioxide (CO_2) are 216.55 K and 3880 mmHg, which means that liquid CO_2 can exist only at pressures above 3880 mmHg = 5.1 atm. Thus, at ordinary atmospheric pressures, liquid carbon dioxide cannot exist at any temperature. Solid carbon dioxide sublimates directly into gaseous CO_2 without going through the liquid phase; hence, the name "dry ice."

The curve OB in Figure 15-16 is the melting curve separating the liquid and solid phases. For a substance such as water for which the melting temperature decreases as the pressure increases, curve OB slopes upward to the left from the triple point, as in this figure. For most other substances, the melting temperature increases as the pressure increases. For such a substance, curve OB slopes upward to the right from the triple point.

Boiling is a rapid form of vaporization in which bubbles of vapor form inside the liquid. To form a bubble, the vapor must have enough pressure to push the liquid back against the pressure applied to the top of the liquid (normally atmospheric pressure) plus the pressure due to the liquid above the bubble. If the applied pressure is increased, as in a pressure cooker, the vapor pressure required for boiling is increased. Therefore, the temperature of the liquid must be raised further before boiling can occur.

For a molecule to escape from the surface of a liquid, energy is required to break the molecular bonds responsible for surface tension, which we discussed in Chapter 11. Vaporization is therefore a cooling process for the liquid left behind. If water is boiled in the usual way through the input of heat, the cooling effect keeps the temperature of the liquid constant at the boiling point. This is the reason that the boiling point of a substance can be used to calibrate thermometers. However, water can also be caused to boil without adding heat by evacuating the air above it and thereby lowering the applied pressure. The energy needed for vaporization is then taken from the liquid left behind. As a result, the liquid will cool down, even to the point that ice forms on the top of the boiling water.

Humidity

Air is 78 percent nitrogen and 21 percent oxygen plus small amounts of other gases, such as argon, carbon dioxide, and water vapor. The pressure exerted by air is the sum of the partial pressures exerted by each of the various gases that comprise it. In a given volume of air, the partial pressure of any particular gas—nitrogen, oxygen, water vapor, or whatever—is the same as if that gas occupied that volume alone. That is, the presence of other gases does not alter the partial pressure of any given gas. This result is known as **Dalton's law.** Dalton's law can be understood in terms of the kinetic-theory model discussed in Section 15-5. The collisions of one type of gas molecule with the walls of a container are not affected by the presence of other types of gas molecules.

If more water vapor is added to a given volume of air at a given temperature, the partial pressure of the water vapor is increased. When this partial pressure equals the vapor pressure for that temperature, the air is said to be saturated. The water vapor then begins to condense into liquid water if the temperature is above the melting point or into ice crystals (snow or frost) if the temperature is below the melting point. The ratio of the partial pressure of water vapor to the vapor pressure for a particular temperature is called the **relative humidity,** which is usually expressed as a percentage:

$$\text{Relative humidity} = \frac{\text{partial pressure}}{\text{vapor pressure}} \times 100\% \qquad \text{15-35} \qquad \textit{Relative humidity defined}$$

Dew drops form on these spiderwebs when the dew point is above 0°C.

Frost forms on a window when the dew point is below 0°C.

The relative humidity can be increased either by increasing the amount of water vapor in the air at a given temperature or by lowering the temperature and thereby lowering the vapor pressure. The temperature at which air becomes saturated with water vapor, that is, the temperature at which relative humidity equals 100 percent, is called the **dew point.** When the surface of the earth cools below the dew point at night due to radiation (which will be discussed in the next chapter), dew forms if the dew point is above 0°C and frost forms if it is below 0°C.

Example 15-13

On a humid, 20°C day, the dew point is measured by cooling a metal container until moisture forms on its surface. This happens when the temperature of the container is 15°C. What is the relative humidity?

At the dew point of 15°C, the partial pressure of the water vapor in the air equals the vapor pressure of water, which according to Table 15-2 is 1.69 kPa. This, then, is the original partial pressure of the water vapor at the original temperature of 20°C. Since the vapor pressure at 20°C is 2.34 kPa (Table 15-2, again), the relative humidity by Equation 15-35 is

$$\text{Relative humidity} = \frac{1.69 \text{ kPa}}{2.34 \text{ kPa}} \times 100\% = 72.2\%$$

Questions

14. At high altitudes, as in the mountains, cooking things in boiling water takes longer than at sea level. Why?

15. What is the advantage of a pressure cooker?

16. When a pan of water is heated, just before the water really boils small bubbles form on the bottom but disappear as they rise. Later, when the water is really boiling, the bubbles grow in size as they rise. Explain.

Summary

1. A temperature scale can be constructed by choosing some thermometric property and defining the temperature to vary linearly with that property using two fixed points, such as the ice point and steam point of water. On the Celsius scale, the ice point is defined to be 0°C and the steam point 100°C. On the Fahrenheit scale, the ice point is 32°F and the steam point 212°F. Temperatures on the Fahrenheit and Celsius scales are related by

$$t_C = \tfrac{5}{9}(t_F - 32°)$$

2. Different thermometers do not always agree with each other on the measurement of a temperature except at the fixed points. Gas thermometers have the property that they all agree with each other in the measurement of any temperature as long as the density of the gas in the thermometers is very low. The ideal-gas temperature T is defined by

$$T = \frac{273.16 \text{ K}}{P_3} P$$

   where P is the pressure of the gas in the thermometer when the thermometer is in thermal equilibrium with the system whose temperature is being measured, and P_3 is the pressure when the thermometer is immersed in a water–ice–vapor bath at its triple point. The absolute or kelvin temperature scale is the same as the ideal-gas scale in the range of temperatures for which gas thermometers can be used. The absolute temperature is related to the Celsius temperature by

$$T = t_C + 273.15 \text{ K}$$

3. The coefficient of linear expansion is the ratio of the fractional change in length to the change in temperature:

$$\alpha = \frac{\Delta L/L}{\Delta T}$$

   The coefficient of volume expansion, which is the ratio of the fractional change in volume to the change in temperature, is three times the coefficient of linear expansion:

$$\beta = \frac{\Delta V/V}{\Delta T} = 3\alpha$$

4. At low densities, all gases obey the ideal-gas law:

$$PV = nRT$$

   where

$$R = 8.314 \text{ J/mol·K}$$

   is the universal gas constant, which is related to Avogadro's number N_A and Boltzmann's constant k by

$$R = kN_A$$

Avogadro's number is

$$N_A = 6.022 \times 10^{23} \text{ molecules/mol}$$

and Boltzmann's constant is

$$k = 1.381 \times 10^{-23} \text{ J/K}$$

A form of the ideal-gas law that is useful for solving problems involving a fixed amount of gas is

$$\frac{P_2 V_2}{T_2} = \frac{P_1 V_1}{T_1}$$

5. The absolute temperature T is a measure of the average molecular energy. For an ideal gas, the average translational kinetic energy of the molecules is

$$K_{av} = (\tfrac{1}{2}mv^2)_{av} = \tfrac{3}{2}kT$$

The total translational kinetic energy of n moles of a gas containing N molecules is given by

$$K = N(\tfrac{1}{2}mv^2)_{av} = \tfrac{3}{2}NkT = \tfrac{3}{2}nRT$$

The rms speed of a molecule of a gas is related to the absolute temperature by

$$v_{rms} = \sqrt{(v^2)_{av}} = \sqrt{\frac{3kT}{m}} = \sqrt{\frac{3RT}{M}}$$

where m is the mass of the molecule and M is the molar mass.

6. The van der Waals equation of state describes the behavior of real gases over a wide range of temperatures and pressures:

$$\left(P + \frac{an^2}{V^2}\right)(V - bn) = nRT$$

It takes into account the space occupied by the gas molecules themselves and the attraction of the molecules for one another.

7. Vapor pressure is the pressure at which the liquid and gas phases of a substance are in equilibrium at a given temperature. The liquid boils at that temperature for which the external pressure equals the vapor pressure.

8. The triple point is the particular temperature and pressure at which the gas, liquid, and solid phases of a substance can coexist. At temperatures and pressures below the triple point, the liquid phase of a substance cannot exist.

9. Relative humidity is the ratio of the partial pressure of water vapor in air to its vapor pressure at a particular temperature.

Suggestions for Further Reading

Hall, Marie Boas: "Robert Boyle," *Scientific American*, August 1967, p. 96.

This English investigator of the seventeenth century performed careful experiments that led to his discovery of the inverse relationship between pressure and volume in a gas, and he obtained numerous results in chemistry, guided by his views on the "mechanical" nature of matter.

Romer, Robert H: "Temperature Scales: Celsius, Fahrenheit, Kelvin, Reamur, and Romer," *The Physics Teacher*, vol. 20, no. 7, 1982, p. 450.

A compact and well-researched history of the development of temperature scales.

Review

A. Objectives: After studying this chapter, you should:

1. Be able to state the zeroth law of thermodynamics.

2. Be able to discuss how a mercury thermometer is calibrated using two fixed points for either the Celsius or Fahrenheit temperature scales.

3. Be able to convert temperatures from Celsius to Fahrenheit and from Fahrenheit to Celsius.

4. Be able to define the ideal-gas temperature scale.

5. Be able to calculate the linear expansion and volume expansion of a substance given the change in its temperature.

6. Be able to work problems using the ideal-gas equation, $PV = nRT$.

7. Be able to discuss the molecular interpretation of temperature.

8. Be able to discuss the van der Waals equation.

9. Be able to sketch the liquid–vapor isotherms for a real substance and indicate the critical point on a PV diagram.

10. Be able to sketch the phase diagram for a real substance.

11. Be able to interpret the parts of a phase diagram.

B. Define, explain, or otherwise identify:

Thermodynamics	Celsius temperature scale
Temperature	Fahrenheit temperature scale
Thermal contact	scale
Thermometric property	Constant-volume gas
Thermal equilibrium	thermometer
Zeroth law of	Triple point of water
thermodynamics	Ideal-gas temperature
Ice-point temperature	scale
Normal freezing point	Kelvin
Steam-point temperature	Absolute temperature
Normal boiling point	scale

Coefficient of linear expansion	RMS speed
Coefficient of volume expansion	Maxwell–Boltzmann distribution function
Boyle's law	Maxwell–Boltzmann energy distribution function
Boltzmann's constant	
Mole	Van der Waals equation
Avogadro's number	
Universal gas constant	Vapor pressure
Ideal gas	Normal boiling point
Molar mass	Phase diagram
Equation of state	Sublimation
Isotherm	Dalton's law
Standard conditions	Relative humidity
Macroscopic variables	Dew point
Kinetic theory of gases	

C. True or false: If the statement is true, explain why it is true. If it is false, give a counterexample.

1. Two objects in thermal equilibrium with each other must be in thermal equilibrium with a third object.

2. All materials expand when they are heated.

3. The Fahrenheit and Celsius temperature scales differ only in the choice of the zero temperature.

4. The kelvin is the same size as the Celsius degree.

5. All thermometers give the same result when measuring the temperature of a particular system.

6. If the pressure of a gas increases, the temperature must increase.

7. The absolute temperature of a gas is a measure of the average translational kinetic energy of the gas molecules.

8. The temperature at which water boils depends on the pressure.

9. The vapor pressure of a gas depends on the temperature.

Problems

Level I

15-1 The Celsius and Fahrenheit Temperature Scales

1. A certain ski wax is rated for use between -12 and $-7°C$. What is this temperature range on the Fahrenheit scale?

2. The melting point of gold (Au) is $1945.4°F$. Express this temperature in degrees Celsius.

3. The highest and lowest temperatures ever recorded in the United States are $134°F$ (in California in 1913) and $-80°F$ (in Alaska in 1971). Express these temperatures on the Celsius scale.

4. What is the Celsius temperature corresponding to the normal temperature of the human body of $98.6°F$?

5. The length of the column of a mercury thermometer is 4.0 cm when the thermometer is immersed in ice water and 24.0 cm when the thermometer is immersed in boiling water. (a) What should the length be at room temperature, $22.0°C$? (b) The mercury column is 25.4 cm long when the thermometer is immersed in a chemical solution. What is the temperature of the solution?

15-2 Gas Thermometers and the Absolute Temperature Scale

6. The temperature of the interior of the sun is about 10^7 K. What is this temperature on (a) the Celsius scale and (b) the Fahrenheit scale?

7. The boiling point of nitrogen N_2 is 77.35 K. Express this temperature in degrees Fahrenheit.

8. The pressure of a constant-volume gas thermometer is 0.400 atm at the ice point and 0.546 atm at the steam point. (a) When the pressure is 0.100 atm, what is the temperature? (b) What is the pressure at 444.6°C, the boiling point of sulfur?

9. A constant-volume gas thermometer reads 50 torr at the triple point of water. (a) What will the pressure be when the thermometer measures a temperature of 300 K? (b) What ideal-gas temperature corresponds to a pressure of 678 torr?

10. A constant-volume gas thermometer has a pressure of 30 torr when it reads a temperature of 373 K. (a) What is its triple-point pressure P_3? (b) What temperature corresponds to a pressure of 0.175 torr?

15-3 Thermal Expansion

11. A steel ruler has a length of 30 cm at 20°C. What is its length at 100°C?

12. A bridge 100 m long is built of steel. If it is built as a single, continuous structure, how much will its length change from the coldest winter days (−30°C) to the hottest summer days (40°C)?

13. A steel tape is placed around the earth at the equator when the temperature is 0°C. What will the clearance between the tape and the ground (assumed to be uniform) be if the temperature of the tape rises to 30°C? Neglect the expansion of the earth.

15-4 The Ideal-Gas Law

14. (a) If 1 mol of a gas in a container occupies a volume of 10 L at a pressure of 1 atm, what is the temperature of the gas in kelvins? (b) The container is fitted with a piston so that the volume can change. When the gas is heated at constant pressure, it expands to a volume of 20 L. What is the temperature of the gas in kelvins? (c) The volume is fixed at 20 L, and the gas is heated at constant volume until its temperature is 350 K. What is the pressure of the gas?

15. A gas is kept at constant pressure. If its temperature is changed from 50 to 100°C, by what factor does the volume change?

16. A pressure as low as 1×10^{-8} torr can be achieved using an oil diffusion pump. How many molecules are there in 1 cm^3 of a gas at this pressure if its temperature is 300 K?

17. A 10-L vessel contains gas at 0°C and a pressure of 4 atm. How many moles of gas are there in the vessel? How many molecules?

18. How many moles of lead ($M = 207.2$ g/mol) are there in a lead brick 5 cm by 10 cm by 20 cm?

19. A room is 6 m by 5 m by 3 m. (a) If the air pressure in the room is 1 atm and the temperature is 300 K, find the number of moles of air in the room. (b) If the temperature rises by 5 K and the pressure remains constant, how many moles of air leave the room?

15-5 The Molecular Interpretation of Temperature: The Kinetic Theory of Gases

20. (a) Find v_{rms} for an argon atom if 1 mol of the gas is confined to a 1-L container at a pressure of 10 atm. (For argon, $M = 40 \times 10^{-3}$ kg/mol.) (b) Compare this with v_{rms} for a helium atom under the same conditions. (For helium, $M = 4 \times 10^{-3}$ kg/mol.)

21. Find the total translational kinetic energy of 1 L of oxygen gas held at a temperature of 0°C and a pressure of 1 atm.

22. At what temperature will the rms speed of an H$_2$ molecule equal 332 m/s?

23. Find the rms speed and the average kinetic energy of a hydrogen atom at a temperature of 10^7 K. (At this temperature, which is of the order of the temperature in the interior of stars, the hydrogen is ionized and consists of a single proton.)

15-6 The van der Waals Equation and Liquid–Vapor Isotherms

24. (a) Calculate the volume of 1 mol of steam at 100°C and a pressure of 1 atm, assuming that it is an ideal gas. (b) Find the temperature at which the steam will occupy the volume found in part (a) if it obeys the van der Waals equation with $a = 0.55$ Pa·m^6/mol^2 and $b = 30$ cm^3.

25. Use the values in Table 15-2 to draw a graph of the vapor pressure of water versus temperature. From your graph, find (a) the temperature at which water boils on a mountain where the atmospheric pressure is 70 kPa, (b) the temperature at which water will boil in a container in which the pressure has been reduced to 0.5 atm, and (c) the pressure at which water will boil at 115°C.

15-7 Phase Diagrams

26. Which gases in Table 15-3 can not be liquefied by applying pressure at 20°C?

27. On a certain day, the temperature is 30°C and the relative humidity is 80 percent. What is the partial pressure of the water vapor in the air?

28. If the temperature is 20°C and the dew point is 10°C, what is the relative humidity?

29. If the partial pressure of water vapor is 3.00 kPa when the temperature is 30°C, what is the relative humidity?

Level II

30. A cubic metal box of sides 20 cm contains air at a pressure of 1 atm and a temperature of 300 K. The box is sealed so that the volume is constant, and it is heated to a temperature of 400 K. Find the net force on each wall of the box.

31. The escape velocity on Mars is 5.0 km/s, and the surface temperature is typically 0°C. Calculate the rms speeds for (a) H$_2$, (b) O$_2$, and (c) CO$_2$ at this temperature. (d) If the rms speed of a gas is greater than about one-sixth of the escape velocity of a planet, virtually all of the molecules of that gas will have escaped the atmosphere of the planet by the present time. Based on this criterion, are H$_2$, O$_2$, and CO$_2$ likely to be found in Mars's atmosphere?

32. Repeat Problem 31 for Jupiter, whose escape velocity is 60 km/s and whose temperature is typically −150°C.

33. At what temperature do the Fahrenheit and Celsius temperature scales give the same reading?

34. A solid-state temperature transducer is essentially a linear amplifier whose gain is linearly temperature dependent. If the gain is 25 at 20°C and 60 at 70°C, what is the temperature when the gain is 45?

35. An aluminum rod is measured with a steel tape measure at 25°C. Its length is found to be 75 cm. What will the tape indicate the length of the rod to be when both the tape and the rod are at (a) 0°C and (b) 50°C?

36. (a) From the definition of β, the coefficient of volume expansion (at constant pressure), show that $\beta = 1/T$ for an ideal gas. (b) The experimentally determined value of β for N_2 gas at 0°C is 0.003673 K^{-1}. Compare this value with the theoretical value $\beta = 1/T$, assuming that N_2 is an ideal gas.

37. One way to construct a device with two points whose separation remains the same in spite of temperature changes is to bolt together one end of two rods having different coefficients of linear expansion as in the arrangement shown in Figure 15-17. (a) Show that the distance L will not change with temperature if the lengths L_A and L_B are chosen such that $L_A/L_B = \alpha_B/\alpha_A$. (b) If material B is steel, material A is brass, and $L_A = 250$ cm at 0°C, what is the value of L?

Figure 15-17 Problem 37.

38. For the phase diagram given in Figure 15-18, state what changes (if any) occur for each line segment—AB, BC, CD, and DE—in (a) volume and (b) phase. (c) For what type of substance would OH be replaced by OG? (d) What is the significance of point F?

Figure 15-18 Problems 38 and 40.

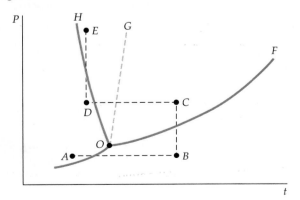

39. The van der Waals constants for helium are $a = 0.03412$ L^2·atm/mol and $b = 0.0237$ L/mol. Use these data to find the volume in cubic centimeters occupied by one helium atom and to estimate the radius of the atom.

40. The phase diagram in Figure 15-18 can be interpreted to yield information on how the boiling and melting points of water change with altitude. (a) Explain how this information can be obtained. (b) How might this information affect cooking procedures in the mountains?

41. An automobile tire is filled to a gauge pressure of 200 kPa when its temperature is 20°C. (Gauge pressure is the difference between the actual pressure and atmospheric pressure.) After the car has been driven at high speeds, the tire temperature has increased to 50°C. (a) Assuming that the volume of the tire has not changed, find the gauge pressure of the air in the tire (assuming air is an ideal gas). (b) Calculate the gauge pressure if the tire expands such that the volume increases by 10 percent.

42. A 1-km steel railroad rail is fastened securely at both ends when the temperature is 20°C. As the temperature increases, the rail begins to buckle. When the temperature is 25°C, how high is the buckle, assuming it is triangularly shaped?

43. A car has a 60-L steel gas tank filled to the top with gasoline when the temperature is 10°C. The coefficient of volume expansion of gasoline is $\beta = 0.900 \times 10^{-3}$ K^{-1}. Taking the expansion of the steel tank into account, how much gasoline spills out of the tank when the car is parked in the sun and its temperature rises to 25°C?

44. (a) Define a coefficient of area expansion. (b) Calculate it for a square and a circle, and show that it is 2 times the coefficient of linear expansion.

45. A container fitted with a piston holds 1 mol of gas. Its initial pressure and temperature are 2 atm and 300 K. The gas is allowed to expand at constant temperature until the pressure is 1 atm. The gas is then compressed and heated at the same time until it is back to its original volume, at which point the pressure is 2.5 atm. What is its final temperature?

46. A scuba diver at 40 m below the surface of a lake, where the temperature is 5°C, releases an air bubble with a volume of 15 cm^3. The bubble rises to the surface, where the temperature is 25°C. What is the volume of the bubble right before it breaks the surface? *Hint:* Remember that the pressure also changes.

47. Oxygen (O_2) is confined to a cubic container, 15 cm on a side, at a temperature of 300 K. Compare the average kinetic energy of a molecule of the gas to the change in its gravitational potential energy if it falls from the top of the container to the bottom.

48. Ultra-high-vacuum (UHV) laboratory systems can produce pressures of 10^{-12} torr. At a temperature of 20°C, how many molecules are there in 1 cm^3 at this pressure?

49. Ten objects have the following speeds:

Speed, m/s	2	5	6	8
Number of objects	3	3	3	1

Calculate (a) the average speed and (b) the rms speed of the objects.

50. Show that $f(v)$ given by Equation 15-31 is maximum when $v = \sqrt{2kT/m}$. *Hint:* Set $df/dv = 0$ and solve for v.

51. Since $f(v)\,dv$ gives the fraction of molecules that have speeds in the range dv, the integral of $f(v)\,dv$ over all the possible ranges of speeds must equal 1. Given the integral

$$\int_0^\infty v^2 e^{-av^2}\,dv = \frac{\sqrt{\pi}}{4}\,a^{-3/2}$$

show that

$$\int_0^\infty f(v)\,dv = 1$$

where $f(v)$ is given by Equation 15-31.

52. Given the integral

$$\int_0^\infty v^3\,e^{-av^2}\,dv = \frac{a^{-2}}{2}$$

calculate the average speed v_{av} of molecules in a gas using the Maxwell–Boltzmann distribution function (Equation 15-31).

Level III

53. At the critical point on the critical isotherm, dP/dV and d^2P/dV^2 are both zero. Show that for a van der Waals gas, the critical volume $V_c = 3b$ per mole.

54. (a) For a van der Waals gas, show that the critical temperature is $8a/27Rb$ and the critical pressure is $a/27b^2$. (b) Rewrite the van der Waals equation of state in terms of the reduced variable $V_r = V/V_c$, $P_r = P/P_c$, and $T_r = T/T_c$.

55. A thermistor is a solid-state device whose resistance varies greatly with temperature. Its temperature dependence is given approximately by $R = R_0 e^{B/T}$, where R is in ohms, T is in kelvins, and R_0 and B are constants that can be determined by measuring R at calibration points such as the ice point and the steam point. (a) If $R = 7360$ ohms at the ice point and 153 ohms at the steam point, find R_0 and B. (b) What is the resistance of the thermistor at $t = 98.6°F$? (c) What is the rate of change of the resistance with temperature (dR/dT) at the ice point and the steam point? At which temperature is the thermistor most sensitive?

56. A grandfather's clock is calibrated at a temperature of 20°C. (a) On a hot day, when the temperature is 30°C, does the clock "run" fast or slow? (b) How much does it gain or lose in a 24-h period? Assume that the pendulum is a thin brass rod with a heavy mass attached to the end.

57. At a temperature of 20°C, a steel bar of radius 2.2 cm and length 60 cm is jammed horizontally perpendicular between two vertical concrete walls. With a blow torch, the temperature of the bar is raised to 60°C. Find the force exerted by the bar on each wall.

58. A steel tube has an outside diameter of 3.000 cm at room temperature (20°C). A brass tube has an inside diameter of 2.997 cm at the same temperature. To what temperature must the ends of the tubes be heated if the steel tube is to be inserted into the brass tube?

59. A constant-volume gas thermometer with a triple-point pressure $P_3 = 500$ torr is used to measure the boiling point of some substance. When the thermometer is placed in thermal contact with the boiling substance, its pressure is 734 torr. Some of the gas in the thermometer is then allowed to escape so that its triple-point pressure is 200 torr. When it is again placed in thermal contact with the boiling substance, its pressure is 293.4 torr. Again, some of the gas is removed from the thermometer so that its triple-point pressure is 100 torr. When it is placed in thermal contact with the boiling substance once again, its pressure is 146.65 torr. Find the ideal-gas temperature of the boiling substance.

60. At the surface of the sun, the temperature is about 6000 K, and all the substances present are gaseous. From data given by the light spectrum of the sun, it is known that most elements are present. (a) What is the average kinetic energy of translation of an atom at the surface of the sun? (b) What is the range of rms speeds at the surface of the sun if the atoms present range from hydrogen ($M = 1$ g/mol) to uranium ($M = 238$ g/mol)?

61. In Section 10-5, we found that the escape speed at the surface of a planet of radius R is $v_e = \sqrt{2gR}$, where g is the acceleration of gravity. (a) At what temperature is v_{rms} for O_2 equal to the escape speed for the earth? (b) At what temperature is v_{rms} for H_2 equal to the escape speed for the earth? (c) Temperatures in the upper atmosphere reach 1000 K. How does this account for the low abundance of hydrogen in the earth's atmosphere? (d) Compute the temperatures for which the rms speeds of O_2 and H_2 are equal to the escape velocity at the surface of the moon, where g is about one-sixth of its value on earth and $R = 1738$ km. How does this account for the absence of an atmosphere on the moon?

Chapter 16

Heat and the First Law of Thermodynamics

Steel ingots in a twin-tube tunnel furnace. The three 53-cm diameter carbon steel ingots seen here have been heated for about 7 hours to approximately 1340°C. Each 3200-kg ingot sits on a furnace car that transports it through the 81-m furnace, which is divided into twelve separate heating zones so that the temperature of the ingot is increased gradually to prevent cracking. The ingots, glowing a yellow-whitish color, exit the furnace to be milled into large, heavy-walled pipes.

Heat is energy that is transferred from one object to another because of a difference in temperature. In the seventeenth century, Galileo, Newton, and other scientists generally supported the theory of the ancient Greek atomists, who considered heat to be a manifestation of molecular motion. In the next century, methods were developed for making quantitative measurements of the amount of heat that leaves or enters a body, and it was found that when two bodies are in thermal contact, the amount of heat that leaves one body equals the amount that enters the other. This discovery led to the development of an apparently successful theory of heat as a conserved material substance—an invisible fluid called "caloric"—that was neither created nor destroyed but merely flowed out of one body and into another.

The caloric theory of heat served quite well in the description of heat transfer but eventually was discarded when it was observed that caloric apparently could be created endlessly by friction with no corresponding disappearance of caloric somewhere else. In other words, the principle of conservation of caloric, which had been the experimental foundation of this theory of heat, proved to be false.

The first clear observations showing that caloric cannot be conserved were made at the end of the eighteenth century by Benjamin Thompson, an American who emigrated to Europe, became the director of the Bavarian arsenal, and was given the title Count Rumford. Thompson supervised the boring of cannon for Bavaria. Because of the heat generated by the boring tool, water was used for cooling. The water had to be replaced continually because it boiled away during the boring. According to the caloric theory, as the metal from the bore was cut into small chips, its ability to retain caloric was decreased. It therefore released caloric to the water, heating it and causing it to boil. Thompson noticed, however, that even when the drill was too dull to cut the metal, the water still boiled away as long as the drill was turning. Apparently, heat was being produced merely by friction and could be produced endlessly. Based on this observation, Thompson suggested that heat is not a substance that is conserved but rather is some form of motion that was communicated from the bore to the water. He showed, in fact, that the heat produced was approximately proportional to the work done by the boring tool.

The caloric theory continued to be the leading theory of heat for some 40 years after Thompson's work, but it gradually weakened as more and more examples of the nonconservation of heat were observed. The modern, mechanical theory of heat did not emerge until the 1840s. In this view, heat is energy that is transferred from one object to another because of a difference in temperature. (The internal energy of a system is often referred to as thermal energy. When a warm system is in contact with a cooler system, internal energy is transferred from the warm system to the cooler system in the form of heat.) The most varied and precise early experiments demonstrating this were performed, starting in the late 1830s, by James Joule (1818–1889), after whom the SI unit of energy is named. Joule showed that the appearance or disappearance of a given quantity of thermal energy is always accompanied by the disappearance or appearance of an equivalent quantity of mechanical energy. The experiments of Joule and others showed that neither thermal energy nor mechanical energy is conserved independently but that the mechanical energy lost always equals the thermal energy produced. What is conserved is the total mechanical energy plus thermal energy.

16-1 Heat Capacity and Specific Heat

When heat energy is added to a substance, the temperature of the substance usually rises. (An exception occurs during a change of phase, such as when water freezes or evaporates, which we will discuss in the next section.) The amount of heat energy Q needed to raise the temperature of a substance is proportional to the temperature change and to the mass of the substance:

$$Q = C\,\Delta T = mc\,\Delta T \qquad \text{16-1}$$

where C is the **heat capacity** of the substance, which is defined as the heat energy needed to raise the temperature of a substance by one degree. The **specific heat** c is the heat capacity per unit mass:

$$c = \frac{C}{m} \qquad \text{16-2}$$

The historical unit of heat energy, the **calorie,** was originally defined to be the amount of heat energy needed to raise the temperature of one gram of water one Celsius degree (or one kelvin since the Celsius degree and the kelvin are the same size). The kilocalorie is then the amount of heat energy needed to raise the temperature of one kilogram of water by one Celsius degree. (The "calorie" used in measuring the energy equivalent of foods is actually the kilocalorie.) Since we now recognize that heat is just another form of energy, we do not need any special units for heat that differ from other energy units. The calorie is now defined in terms of the SI unit of energy, the joule:

$$1 \text{ cal} = 4.184 \text{ J} \qquad 16\text{-}3$$

The U.S. customary unit of heat is the **Btu** (for British thermal unit), which was originally defined to be the amount of energy needed to raise the temperature of one pound of water by one Fahrenheit degree. The Btu is related to the calorie and to the joule by

$$1 \text{ Btu} = 252 \text{ cal} = 1.054 \text{ kJ} \qquad 16\text{-}4$$

From the original definition of the calorie, the specific heat of water is

$$c_{\text{water}} = 1 \text{ cal/g·C°} = 1 \text{ kcal/kg·C°}$$

$$= 1 \text{ kcal/kg·K} = 4.184 \text{ kJ/kg·K} \qquad 16\text{-}5a$$

Similarly from the definition of the Btu, the specific heat of water in U. S. customary units is

$$c_{\text{water}} = 1 \text{ Btu/lb·F°} \qquad 16\text{-}5b$$

Careful measurement shows that the specific heat of water varies slightly with temperature, but this variation is only about 1 percent over the entire temperature range from 0 to 100°C. We will usually neglect this small variation and take the specific heat of water to be 1 kcal/kg·K = 4.18 kJ/kg·K.

The heat capacity per mole is called the **molar heat capacity** C_m. The molar heat capacity equals the specific heat (heat capacity per unit mass) times the molar mass M (the mass per mole):

$$C_m = Mc \qquad 16\text{-}6$$

The heat capacity of n moles of a substance is then

$$C = nC_m \qquad 16\text{-}7$$

Table 16-1 lists the specific heats and molar heat capacities of some solids and liquids. (Note that the molar heat capacities of all the metals are about the same. We will discuss the significance of this in Section 16-7.)

Table 16-1 **Specific Heat and Molar Heat Capacities for Various Solids and Liquids at 20°C**			
Substance	c, kJ/kg·K	c, kcal/kg·K or Btu/lb·F°	C_m, J/mol·K
Aluminum	0.900	0.215	24.3
Bismuth	0.123	0.0294	25.7
Copper	0.386	0.0923	24.5
Gold	0.126	0.0301	25.6
Ice (−10°C)	2.05	0.49	36.9
Lead	0.128	0.0305	26.4
Silver	0.233	0.0558	24.9
Tungsten	0.134	0.0321	24.8
Zinc	0.387	0.0925	25.2
Alcohol (ethyl)	2.4	0.58	111
Mercury	0.140	0.033	28.3
Water	4.18	1.00	75.2

Example 16-1

How much heat is needed to raise the temperature of 3 kg of copper by 20 C°?

From Table 16-1, the specific heat of copper is 0.386 kJ/kg·K. Equation 16-1 then gives

$$Q = mc \, \Delta T = (3 \text{ kg})(0.386 \text{ kJ/kg·K})(20 \text{ K}) = 23.2 \text{ kJ}$$

Note we use 20 C° = 20 K. Alternatively, we could express the specific heat as 0.386 kJ/kg·C° and write the temperature change as 20 C°.

Exercise

A 2-kg aluminum block is originally at 10°C. If 36 kJ of heat energy are added to the block, what is its final temperature?   (Answer: 30°C)

We can see from Table 16-1 that the specific heat of water is considerably larger than that of other substances. For example, it is more than 10 times the specific heat of copper. Because of its very large heat capacity, water is an excellent material for storing thermal energy, as in a solar heating system. It is also an excellent coolant. A large body of water, such as a lake or an ocean, tends to moderate the variations of temperature near it because it can absorb or release large quantities of thermal energy while undergoing only a very small change in temperature.

Since the specific heat of water is practically constant over a wide range of temperatures, the specific heat of an object can be conveniently measured by heating the object to some easily measured temperature, placing it in a water bath of known mass and temperature, and measuring the final equilibrium temperature. If the whole system is isolated from its surroundings, the heat leaving the object equals the heat entering the water and its container. This procedure is called **calorimetry,** and the insulated water container is called a **calorimeter.** Let m be the mass of the object, c its specific heat, and T_{io} its initial temperature. If T_f is the final temperature of the object in its water bath, the heat leaving the object is

$$Q_{out} = mc(T_{io} - T_f)$$

Similarly, if T_{iw} is the initial temperature of the water and container, and T_f is their final temperature (the final temperature of the object and water are the same, since they eventually come to equilibrium), the heat absorbed by the water and container is

$$Q_{in} = m_w c_w (T_f - T_{iw}) + m_c c_c (T_f - T_{iw})$$

where m_w and $c_w = 4.18$ kJ/kg·K are the mass and specific heat of the water, and m_c and c_c are the mass and specific heat of the container. Note that in these equations we have chosen to write the temperature differences so that both the heat in and the heat out are positive quantities. (We know that if we place a heated object in cooler water, the final temperature T_f will be greater than the initial temperature of the water bath and less than the initial temperature of the object.) Since these amounts of heat are equal, the specific heat c of the object can be calculated by setting the heat out of the object equal to the heat into the water and its container.

$$Q_{out} = Q_{in}$$

$$mc(T_{io} - T_f) = m_w c_w (T_f - T_{iw}) + m_c c_c (T_f - T_{iw}) \qquad 16\text{-}8$$

Since only temperature differences occur in Equation 16-8 and since the Kelvin and Celsius degree are the same size, the temperatures can all be measured on either the Celsius or Kelvin scale without affecting the result.

Example 16-2

Lead shot of mass 600 g is heated to 100°C and placed in an aluminum can of mass 200 g which contains 500 g of water initially at 17.3°C. The specific heat of the aluminum can is 0.900 kJ/kg·K. The final equilibrium temperature of the mixture is 20.0°C. What is the specific heat of lead?

Since the temperature change of the water is 20.0°C − 17.3°C = 2.7 C° = 2.7 K, the heat absorbed by the water is

$$Q_w = m_w c_w \, \Delta T_w = (0.5 \text{ kg})(4.18 \text{ kJ/kg·K})(2.7 \text{ K}) = 5.64 \text{ kJ}$$

Similarly, the heat absorbed by the aluminum container is

$$Q_c = m_c c_c \, \Delta T_c = (0.2 \text{ kg})(0.900 \text{ kJ/kg·K})(2.7 \text{ K}) = 0.486 \text{ kJ}$$

The temperature change of the lead is 100°C − 20.0°C = 80.0 C° = 80.0 K, and the heat given off by the lead is

$$Q_{Pb} = m_{Pb} c_{Pb} \, \Delta T_{Pb} = (0.6 \text{ kg})(c_{Pb})(80.0 \text{ K}) = (48.0 \text{ kg·K}) c_{Pb}$$

Setting the heat given off by the lead equal to that absorbed by the water and its container, we obtain

$$Q_{Pb} = Q_w + Q_c$$

$$(48.0 \text{ kg·K}) c_{Pb} = 5.64 \text{ kJ} + 0.486 \text{ kJ} = 6.13 \text{ kJ}$$

Solving for c_{Pb} gives

$$c_{Pb} = \frac{6.13 \text{ kJ}}{48.0 \text{ kg·K}} = 0.128 \text{ kJ/kg·K}$$

Note that the specific heat of lead is considerably less than that of water.

The amount of heat needed to raise the temperature of a substance a given amount depends on whether the substance is allowed to expand while it is being heated. If it is kept from expanding, all the added heat will go into increasing the temperature. If it is allowed to expand, however, it will do work on its surroundings—on the air, if on nothing else. Therefore, more heat energy will be needed to do the work as well as raise the temperature. Thus, for all substances that expand when they are heated, the heat capacity or specific heat at constant volume c_v is less than the heat capacity or specific heat at constant pressure c_p. It is very difficult to prevent a solid or a liquid from expanding when it is heated. However, the expansion, and therefore the work done, is usually very small, so the difference between c_v and c_p can be ignored in most cases. Since solids and liquids are usually heated at constant (atmospheric) pressure, the values listed in Table 16-1 are those for constant pressure.

For gases, the situation is quite different. Gases expand significantly when they are heated at constant pressure, so they do a significant amount of work. Thus, there is a large difference between the specific heat or the heat capacity of a gas at constant pressure and that at constant volume. (It is quite easy to keep the volume of a gas constant while it is being heated, as when a gas is heated in a closed container in a laboratory.) We will calculate the difference in the heat capacities of an ideal gas in Section 16-7.

Questions

1. A potato wrapped in aluminum foil is baked in an oven; it is taken out and its foil removed. The foil cools much faster than the potato. Why?

2. Body A has twice the mass and twice the specific heat of body B. If they are supplied with equal amounts of heat, how do the changes in their temperatures compare?

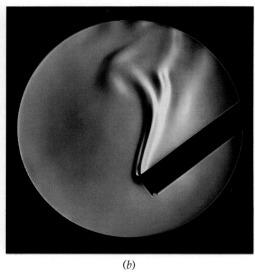

(a)   (b)

(a) Crystal spires of KCl condensed from gas at 550°C.
(b) Gasoline vapor made visible by Schlieren photography, which is sensitive to very slight changes in density. (c) Color changes produced by changes in temperature are characteristic of liquid crystals such as those shown here. A liquid crystal is a phase between a solid and a liquid—having fluid properties of a liquid and optical properties of a solid. They are observed in organic compounds and occur widely in living matter. (d) Ice floating on a thawing lake is an example of a liquid and a solid in thermal equilibrium. (e) Carbon dioxide subliming from solid to gas.

16-2   Change of Phase and Latent Heat

When heat is supplied to a substance at constant pressure, the result is usually an increase in the temperature of the substance. Sometimes, though, a substance can absorb large amounts of heat without undergoing any change in its temperature. This happens during a **phase change,** that is, when the physical condition of the substance is changing from one form to another. Types of phase changes are **fusion,** the change of a liquid to a solid (as in the freezing of water into ice); **vaporization,** the change of a liquid to a vapor or gas (as in the evaporation of water); and **sublimation,** the change of a solid directly into a gas (as in the vaporization of moth balls or solid carbon dioxide, frequently called dry ice). There are also other types of phase changes, as when a solid changes from one crystalline form to another.

Phase changes can be understood in terms of molecular theory. An increase in the temperature of a substance reflects an increase in the kinetic energy of motion of its molecules. When a substance changes from liquid to gaseous form, its molecules, which are close together in the liquid form, are moved far apart from each other. This requires that work be done against the attractive forces that hold the molecules close together; that is, energy must be supplied to the molecules to separate them. This energy goes into an increase in the potential energy of the molecules rather than an increase in their kinetic energy. Therefore, the temperature of the substance, which is a measure of the average kinetic energy of its molecules, does not change.

For a pure substance, a change in phase at a given pressure occurs only at a particular temperature. For example, pure water at a pressure of 1 atm changes from solid to liquid at 0°C and from liquid to gas at 100°C. The first temperature is the normal melting point of water and the second is the normal boiling point.

A specific quantity of heat energy is required to change the phase of a given amount of substance. The heat required is proportional to the mass of the substance. The heat required to melt a substance of mass m with no change in its temperature is

$$Q = mL_f \qquad \text{16-9}$$

(c)

(d)

(e)

where L_f is called the **latent heat of fusion** of the substance. For the melting of ice to water at a pressure of 1 atm, the latent heat of fusion is 333.5 kJ/kg = 79.7 kcal/kg. When the phase change is from liquid to gas, the heat required is

$$Q = mL_v \qquad\qquad 16\text{-}10$$

where L_v is the **latent heat of vaporization**. For water at a pressure of 1 atm, the latent heat of vaporization is 2.26 MJ/kg = 540 kcal/kg. Table 16-2 gives the normal melting and boiling points and the latent heats of fusion and vaporization at 1 atm for various substances.

Table 16-2   **Normal Melting Point (MP), Latent Heat of Fusion L_f, Normal Boiling Point (BP), and Latent Heat of Vaporization L_v for Various Substances at 1 atm**

Substance	MP, K	L_f, kJ/kg	BP, K	L_v, kJ/kg
Alcohol, ethyl	159	109	351	879
Bromine	266	67.4	332	369
Carbon dioxide	—	—	194.6†	573†
Copper	1356	205	2839	4726
Gold	1336	62.8	3081	1701
Helium	—	—	4.2	21
Lead	600	24.7	2023	858
Mercury	234	11.3	630	296
Nitrogen	63	25.7	77.35	199
Oxygen	54.4	13.8	90.2	213
Silver	1234	105	2436	2323
Sulfur	388	38.5	717.75	287
Water	273.15	333.5	373.15	2257
Zinc	692	102	1184	1768

†These values are for sublimation. Carbon dioxide does not have a liquid state at 1 atm.

Example 16-3

If 1 kg of ice at $-20°C$ is heated at a pressure of 1 atm until all the ice has been changed into steam, how much heat is required?

Provided that the heat capacity of ice is constant and is equal to 2.05 kJ/kg·K (Table 16-1), the heat energy needed to raise the temperature of the ice from -20 to $0°C$ is

$$Q_1 = mc \ \Delta T = (1 \text{ kg})(2.05 \text{ kJ/kg·K})(20 \text{ K}) = 41 \text{ kJ}$$

The latent heat of fusion for ice is 334 kJ/kg (Table 16-2), so the heat needed to melt 1 kg of ice is

$$Q_2 = mL_f = (1 \text{ kg})(334 \text{ kJ/kg}) = 334 \text{ kJ}$$

The heat needed to raise the temperature of the resulting 1 kg of water from 0 to $100°C$ is

$$Q_3 = mc \ \Delta T = (1 \text{ kg})(4.18 \text{ kJ/kg·K})(100 \text{ K}) = 418 \text{ kJ}$$

where we have neglected any variation in the heat capacity of water over this range of temperatures. Finally, the heat needed to vaporize 1 kg of water at $100°C$ is

$$Q_4 = mL_v = (1 \text{ kg})(2.26 \times 10^3 \text{ kJ/kg}) = 2.26 \text{ MJ}$$

The total amount of heat required is

$$Q = Q_1 + Q_2 + Q_3 + Q_4$$
$$= 0.041 \text{ MJ} + 0.334 \text{ MJ} + 0.418 \text{ MJ} + 2.26 \text{ MJ}$$
$$= 3.05 \text{ MJ}$$

Note that most of the heat supplied in this process was needed to change the phases of the water, not to raise its temperature.

Figure 16-1 Plot of temperature versus time when heat is added to 1 kg of ice originally at $-20°C$ at a constant rate of 1 kJ/s. The temperature first rises to $0°C$, at which point the ice melts at constant temperature. When the ice has melted, the temperature of the water rises to $100°C$. Then the water vaporizes at constant temperature. If heat were added after all the water has turned to steam, the temperature of the steam would rise.

Figure 16-1 shows a graph of the temperature versus time for Example 16-3, provided that the heat is added at a constant rate of 1 kJ/s. The temperature of the ice increases at a steady rate until it reaches $0°C$, the normal melting point. The temperature then remains constant at $0°C$ as the ice melts. Note that it takes longer to melt the ice than it does to raise its temperature from -20 to $0°C$. As calculated in the example, 334 kJ is needed to melt the ice whereas only 41 kJ is needed to raise its temperature by 20 C°. So at 1 kJ/s, it takes only 41 s to raise the temperature of the ice 20 C°, but it takes 334 s to melt the ice. When all the ice is melted to water at $0°C$, the temperature of the water begins to rise. Since the specific heat of water is about twice that of ice, the temperature of the water increases at a slower rate than does the temperature of the ice. As calculated, it takes 418 kJ to raise the temperature by 100 C°, so at 1 kJ/s, it takes 418 s for the water to reach the boiling point. When the temperature reaches $100°C$, the normal boiling point, the water vaporizes (boils), and the temperature remains constant until all of the water has turned to vapor (steam). Since 2.26 MJ are needed to vaporize 1 kg of water, it takes considerably longer to vaporize the water than it does to melt the ice or to raise the temperature of the water by 100 C°. When all of the water has vaporized, the temperature will again rise as heat is added.

Exercise

An 830-g piece of lead is heated to its melting point of 600 K. How much additional heat energy must be added to melt the lead? (Answer: 20.5 kJ)

16-3 The Transfer of Thermal Energy

Thermal energy is transferred from one place to another by three processes: conduction, convection, and radiation. In **conduction,** thermal energy is transferred by interactions among atoms or molecules, though there is no transport of the atoms or molecules themselves. For example, if one end of a solid bar is heated, the atoms in the heated end vibrate with greater energy than those at the cooler end. Because of the interaction of the more energetic atoms with their neighbors, this energy is transported along the bar. If the solid is a metal, the transport of thermal energy is helped by free electrons, which move throughout the metal, receiving and giving off thermal energy as they collide with the atoms of the metal. In a gas, heat is conducted by the direct collisions of the gas molecules. Molecules in the warmer part of the gas have a higher average energy than the molecules in the cooler part of the gas. When the high-energy molecules collide with the low-energy molecules, some of the energy of the high-energy molecules is transferred to the low-energy molecules.

In **convection,** heat is transported by direct mass transport. For example, when air near the floor is warmed, it expands and rises because of its lower density. The thermal energy in this warm air is thus transported from the floor to the ceiling along with the mass of warm air.

In **radiation,** energy is emitted and absorbed by bodies in the form of electromagnetic radiation. This radiation moves through space at the speed of light. Thermal radiation, light waves, radio waves, television waves, and x rays are all forms of electromagnetic radiation that differ from one another only in their wavelengths and frequencies. All bodies emit and absorb electromagnetic radiation. When a body is in thermal equilibrium with its surroundings, it emits and absorbs energy at the same rate. However, if it is warmed to a higher temperature than its surroundings, it radiates away more energy than it absorbs, thereby cooling down as its surroundings warm up.

Conduction and Convection

Figure 16-2 shows a solid bar of cross-sectional area A. If we keep one end of the bar at a high temperature (for example, in a steam bath) and the other end at a low temperature (for example, in an ice bath), thermal energy is continually conducted down the bar from the hot end to the cold end. In the steady state, the temperature varies uniformly (if the bar is uniform) from the hot end to the cold end. The rate of change of the temperature along the bar $\Delta T/\Delta x$ is called a **temperature gradient.** Let us consider a small portion of the bar, a section of thickness Δx, and let ΔT be the temperature difference across the section (Figure 16-2b). If ΔQ is the amount of thermal energy conducted through the section in some time Δt, the rate of conduction of thermal energy $\Delta Q/\Delta t$ is called the **thermal current** I. Experimentally, it is found that the thermal current is proportional to the temperature gradient and to the cross-sectional area A:

$$I = \frac{\Delta Q}{\Delta t} = kA\,\frac{\Delta T}{\Delta x}$$

16-11

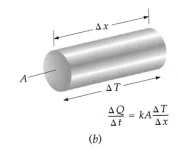

$$\frac{\Delta Q}{\Delta t} = kA\frac{\Delta T}{\Delta x}$$

(b)

Figure 16-2 (a) A conducting bar with its ends at two different temperatures. (b) A section of the bar of length Δx. The rate at which thermal energy is conducted across the section is proportional to the cross-sectional area and the temperature difference and is inversely proportional to the thickness of the section.

Thermal conduction

The proportionality constant k, called the **coefficient of thermal conductivity** or just the **thermal conductivity,** depends on the composition of the bar. In SI units, thermal current is expressed in watts (joules per second), and the thermal conductivity has units of watts per meter-kelvin (though in some tables, the energy may be given in calories or kilocalories and the

Table 16-3   **Thermal Conductivities k for Various Materials**

Material	k, W/m·K	k, Btu·in/h·ft²·F°
Air (27°C)	0.026	0.18
Ice	0.592	4.11
Water (27°C)	0.609	4.22
Aluminum	237	1644
Copper	401	2780
Gold	318	2200
Iron	80.4	558
Lead	353	2450
Silver	429	2980
Steel	46	319
Oak	0.15	1.02
White Pine	0.11	0.78
Concrete	0.19–1.3	6–9
Glass	0.7–0.9	5–6

thickness in centimeters). In practical calculations in the United States—for finding the heat conducted through the walls of a room, for example—the thermal current is usually expressed in Btu per hour, the area in square feet, the thickness in inches, and the temperature in degrees Fahrenheit. The thermal conductivity is then given in Btu·in/h·ft²·F°. Table 16-3 lists thermal conductivities for various materials in both SI and U.S. customary units.

If we solve Equation 16-11 for the temperature difference, we obtain

$$\Delta T = \frac{\Delta x}{kA} I \qquad \text{16-12}$$

or

$$\Delta T = IR \qquad \text{16-13}$$

where $\Delta x/kA$ is called the **thermal resistance R**:

Thermal resistance

$$R = \frac{\Delta x}{kA} \qquad \text{16-14}$$

Equation 16-13 is of the same form as Equation 11-29 for the viscous flow of a fluid through a pipe, except that I now stands for the flow of thermal energy rather than the flow of a fluid, R is now the thermal resistance, and the pressure difference ΔP is replaced by the temperature difference ΔT.

Exercise

Calculate the thermal resistance of an aluminum slab of cross-sectional area 15 cm² and thickness 2 cm. (Answer: 0.0563 K/W)

In many practical problems, we are interested in the flow of heat through two or more conductors (or insulators) in series. For example, we may wish to know the effect of adding insulating material of a certain thickness and thermal conductivity to the space between two layers of plasterboard. Figure 16-3 shows two thermally conducting slabs of the same cross-sectional area but of different materials and different thicknesses. Let T_1 be the temperature on the warm side, T_2 be that at the interface between the slabs, and T_3 be that on the cool side. Under the conditions of steady-state heat flow, the thermal current must be the same through both slabs. This follows from energy conservation; in the steady state, the energy going in

T_1 T_2 T_3

Figure 16-3 Two thermally conducting slabs of different materials in series. The equivalent thermal resistance of the slabs in series is the sum of their individual thermal resistances. The thermal current is the same through both slabs.

must equal that coming out. If R_1 and R_2 are the thermal resistances of the two slabs, we have from Equation 16-13 for each slab

$$T_1 - T_2 = IR_1$$

and

$$T_2 - T_3 = IR_2$$

Adding these equations gives

$$\Delta T = T_1 - T_3 = (R_1 + R_2)I = R_{eq}I \qquad \text{16-15}$$

where R_{eq} is the equivalent resistance. Thus, for thermal resistances in series, the equivalent resistance is the sum of the individual resistances:

$$R_{eq} = R_1 + R_2 + \cdots \qquad \text{16-16}$$

Resistances in series

This result can be applied to any number of resistances in series. It is the same result we will obtain later for electrical resistances in series.

In the building industry, the thermal resistance in U.S. customary units for a square foot of material is called the **R factor,** which we will designate by R_f. The R factor is merely the thickness of the material divided by its thermal conductivity:

$$R_f = \frac{\Delta x}{k} = RA \qquad \text{16-17}$$

The reciprocal of the thermal conductivity in U.S. customary units is the R factor per inch of material. Table 16-4 lists R factors for several materials. To find the rate of heat conduction per unit area in Btu/h·ft^2 through several slabs of materials, we divide temperature difference in Fahrenheit degrees by the sum of the R factors. From Equation 16-15, we have

$$\Delta T = R_{eq}I = R_{eq}A\left(\frac{I}{A}\right) = R_{f,eq}\left(\frac{I}{A}\right)$$

or

$$\frac{I}{A} = \frac{\Delta T}{R_{f,eq}} \qquad \text{16-18}$$

Table 16-4 R Factors $\Delta x/k$ for Various Building Materials

Material	Thickness, in	R_f, h·ft^2·F°/Btu
Building board		
Gypsum or plasterboard	0.375	0.32
Plywood (Douglas fir)	0.5	0.62
Plywood or wood panels	0.75	0.93
Particle board, medium		
density	1.0	1.06
Finish flooring materials		
Carpet and fibrous pad	1.0	2.08
Tile		0.5
Wood, hardwood finish	0.75	0.68
Roof insulation	1.0	2.8
Roofing		
Asphalt roll roofing		0.15
Asphalt shingles		0.44
Windows		
Single-pane		0.9
Double-pane		1.8

Figure 16-4 Two thermally conducting cubes in series for Example 16-4.

Example 16-4

Figure 16-4 shows two metal cubes, with 2-cm sides, between two walls, one held at 100°C and the other at 0°C. The cubes are lead and silver. Find (a) the total thermal current through the cubes and (b) the temperature at the interface.

(a) We first find the thermal resistance of each cube using Equation 16-14 and values for the thermal conductivity k from Table 16-3. For the lead cube, we have

$$R_{Pb} = \frac{\Delta x}{kA} = \frac{0.02 \text{ m}}{(353 \text{ W/m·K})(0.02 \text{ m})^2} = 0.142 \text{ K/W}$$

Similarly, the resistance of the silver cube is

$$R_{Ag} = \frac{0.02 \text{ m}}{(429 \text{ W/m·K})(0.02 \text{ m})^2} = 0.117 \text{ K/W}$$

The equivalent thermal resistance for the two cubes in series is thus

$$R_{eq} = R_{Pb} + R_{Ag} = 0.142 \text{ K/W} + 0.117 \text{ K/W} = 0.259 \text{ K/W}$$

The thermal current is then given by

$$I = \frac{\Delta T}{R_{eq}} = \frac{100 \text{ K}}{0.259 \text{ K/W}} = 386 \text{ W}$$

(b) We find the temperature at the interface by noting that the temperature drop between the end at 100°C and the interface must equal the current I times the thermal resistance of the lead cube. Calling the temperature at the interface t_{if}, we have

$$100°C - t_{if} = IR_{Pb} = (386 \text{ W})(0.142 \text{ K/W}) = 54.8 \text{ K} = 54.8 \text{ C}°$$

The temperature at the interface is thus

$$t_{if} = 100°C - 54.8 \text{ C}° = 45.2°C$$

Example 16-5

A 60- by 20-ft roof is made of 1-in pine board covered with asphalt shingles. (a) If the overlap in the shingles is neglected, at what rate is heat conducted through the roof when the temperature inside is 70°F and that outside is 40°F? (b) By what factor is the heat loss reduced if 2 in of roof insulation are added?

(a) From Table 16-3, the thermal conductivity of pine board in U.S. customary units is 0.78 Btu·in/h·ft²·F°. The R factor for a board 1 in thick is therefore

$$R_f = \frac{\Delta x}{k} = \frac{1 \text{ in}}{0.78 \text{ Btu·in/h·ft}^2\text{·F}°} = 1.28 \frac{\text{h·ft}^2\text{·F}°}{\text{Btu}}$$

From Table 16-4, the R factor for asphalt shingles is 0.44 h·ft²·F°/Btu. The R factor for the combination is therefore $1.28 + 0.44 = 1.72$ h·ft²·F°/Btu. For a temperature difference of 30 F°, the heat conducted per square foot is

$$\frac{I}{A} = \frac{\Delta T}{R_{f,eq}} = \frac{30 \text{ F}°}{1.72 \text{ h·ft}^2\text{·F}°/\text{Btu}} = 17.4 \frac{\text{Btu}}{\text{h·ft}^2}$$

Thus, the rate of conduction through the area of 60 by 20 ft = 1200 ft² is

$$I = (17.4 \text{ Btu/h·ft}^2)(1200 \text{ ft}^2) = 21,000 \text{ Btu/h}$$

(*b*) From Table 16-4, the *R* factor for roof insulation is 2.8 h·ft²·F°/Btu per inch, so for 2 in of insulation, the *R* factor is 5.6 h·ft²·F°/Btu. Adding the insulation thus increases the *R* factor from 1.72 h·ft²·F°/Btu to 1.72 + 5.6 = 7.32 h·ft²·F°/Btu. Since 7.32/1.72 = 4.26, the rate at which heat is conducted through the roof is decreased by a factor of 4.26. Thus, for the temperature difference given, the rate of heat conduction through the roof with the insulation will be reduced to (21,000 Btu/h)/4.26 = 4900 Btu/h.

To calculate the amount of heat leaving a room by conduction in a given time, we need to know how much leaves through the walls, the windows, the floor, and the ceiling. Such a problem involves what are called "parallel paths" for heat flow. The walls, windows, floor, and ceiling all represent independent paths for heat loss to the outside. The temperature difference is the same for each path, but the thermal current is different. The total thermal current is the sum of the thermal currents through each of the independent or parallel paths:

$$I_{total} = I_1 + I_2 + \cdots = \frac{\Delta T}{R_1} + \frac{\Delta T}{R_2} + \cdots$$

or

$$I_{total} = \frac{\Delta T}{R_{eq}} \qquad \text{16-19}$$

where the equivalent thermal resistance is given by

$$\frac{1}{R_{eq}} = \frac{1}{R_1} + \frac{1}{R_2} + \cdots \qquad \text{16-20} \qquad \textit{Resistances in parallel}$$

We will encounter this equation again when we study electric conduction through parallel resistances.

The thermal conductivity of air is very small compared with that of solid materials, so air is a very good insulator. However, when there is a large air gap—say, between a storm window and the inside window—the insulating

The effect of insulation can be seen by the snow on the roof of this van. The roof over the driver's compartment is insulated but the rest of the roof is not. Three large braces across the back roof evidently act to insulate the cool metal roof from the warm air in the van.

efficiency of air is greatly reduced because of convection. Whenever there is a temperature difference between different parts of the air space, convection currents act quickly to equalize the temperature, so the effective conductivity is greatly increased. For storm windows, air gaps of about 1 to 2 cm are optimal. Wider air gaps actually reduce the thermal resistance of a double-pane window because of convection.

The insulating properties of air can be used very effectively if the air can be trapped in small pockets that are separated from each other so that convection cannot take place. Goose down, for example, is a good thermal insulator because it traps air so that it cannot circulate and transport heat by convection. Another example is Styrofoam, a cellular material in which tiny pockets of air are separated by poorly conducting cell walls that prevent convection. The thermal conductivity of such a material is essentially the same as that of air.

If you touch the inside glass surface of a window when it is cold outside, you will observe that the surface is considerably colder than the inside air. The thermal resistance of a window is due mainly to thin insulating air films that adhere to either side of the glass surface. The thickness of the glass has little effect on the overall thermal resistance. An air film typically adds an R factor of about 0.45. Thus, the R factor of a window with N separated glass layers is approximately $0.9N$ because of the two sides of each layer. Under windy conditions, the outside air film may be greatly decreased, leading to a smaller R factor for the window.

It is possible to write an equation for the thermal energy transported by convection and to define a coefficient of convection, but analyses of practical problems involving convection are very difficult, so they will not be discussed here. The heat transferred from a body to its surroundings by convection is approximately proportional to the area of the body and to the difference in temperature between the body and the surrounding fluid.

Convection from a heated cylinder. This interferometer photograph shows isotherms as the cylinder cools by convection.

Radiation

The third mechanism for the transfer of thermal energy is radiation in the form of electromagnetic waves. The rate at which an object radiates thermal energy is proportional to the area of the object and to the fourth power of its absolute temperature. This result, found empirically by Josef Stefan in 1879 and derived theoretically by Ludwig Boltzmann about five years later, is called the **Stefan–Boltzmann law:**

Stefan–Boltzmann law

$$P = e\sigma AT^4$$

16-21

where P is the power radiated in watts, A is the area, e is the **emissivity** of the object, and σ is a universal constant called **Stefan's constant,** which has the value

$$\sigma = 5.6703 \times 10^{-8} \text{ W/m}^2\cdot\text{K}^4$$

16-22

The emissivity e is a fraction that varies from 0 to 1 and depends on the composition of the surface of the object.

When radiation falls on an opaque object, part of the radiation is reflected and part is absorbed. Light-colored objects reflect most visible radiation, whereas dark objects absorb most of it. The rate at which an object absorbs radiation is given by

$$P_a = e\sigma AT_0^4$$

16-23

where T_0 is the temperature of the surroundings.

(a)

(b)

(a) Thermograph of a boy and his dog. The bright regions are those of high temperature. Note the dog's cold nose. (b) Thermograph of Niagara falls taken from an altitude of 762 m. The brighter regions indicate vegetation. The land mass to the right (West) of the spectacular Canadian (Horseshoe) Falls is Ontario, Canada.

If an object emits more radiation than it absorbs, it cools down while the surroundings absorb radiation from the object and warm up. If the object absorbs more than it emits, the object warms and the surroundings cool. When an object is in thermal equilibrium with its surroundings, $T = T_0$, and the object emits and absorbs radiation at the same rate. The net power radiated by an object at temperature T in an environment at temperature T_0 is

$$P_{\text{net}} = e\sigma A(T^4 - T_0^4) \qquad \text{16-24}$$

An object that absorbs all the radiation incident upon it has an emissivity equal to 1 and is called a **blackbody.** A blackbody is also an ideal radiator. The concept of an ideal blackbody is important because the characteristics of the radiation emitted by such an object can be calculated theoretically. Materials such as black velvet come close to being ideal blackbodies. The best practical approximation of an ideal blackbody is a small hole leading into a cavity, such as a keyhole in a closet door (Figure 16-5). Radiation incident on the hole has little chance of being reflected back out the hole before it is absorbed by the walls of the cavity. The radiation emitted through the hole is thus characteristic of the temperature of the object.

The radiation emitted by an object at temperatures below about 600°C is not visible. Most of it is concentrated at wavelengths much longer than those of visible light. (When we study light, we will see that visible light is electromagnetic radiation with wavelengths between about 400 and 700 nm.) As an object is heated, the rate of energy emission increases (Equation 16-21), and the energy radiated extends to shorter and shorter wavelengths. Between about 600 and 700°C, enough of the radiated energy is in the visible spectrum for the body to glow a dull red. At higher temperatures,

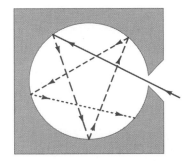

Figure 16-5   A hole in a cavity approximating an ideal blackbody. Radiation entering the cavity has little chance of leaving the cavity before it is completely absorbed. The radiation emitted through the hole is therefore characteristic of the temperature of the object.

it may become bright red or even "white hot." Figure 16-6 shows the power radiated by a blackbody as a function of wavelength for several different temperatures. The wavelength at which the power is a maximum varies inversely with the temperature, a result known as **Wien's displacement law:**

Wien's displacement law

$$\lambda_{max} = \frac{2.898 \text{ mm·K}}{T}$$

16-25

This law is used to determine the temperatures of stars from analyses of their radiation. It can also be used to map out the variation in temperature over different regions of the surfaces of an object. Such a map is called a **thermograph.** Thermographs can be used to detect cancer because cancerous material is at a slightly higher temperature than surrounding tissue.

Figure 16-6   Radiated power versus wavelength for radiation emitted by a blackbody. The wavelength of the maximum power varies inversely with the absolute temperature of the blackbody.

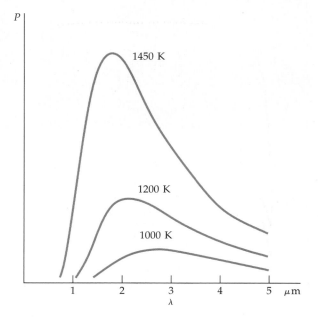

The spectral-distribution curves shown in Figure 16-6 played an important role in the history of physics. It was the discrepancy between theoretical calculations of these curves using classical thermodynamics and the experimental measurements of them that led to Max Planck's first ideas about the quantization of energy in 1897. We will discuss this problem further in Chapter 35.

Example 16-6

The surface temperature of the sun is about 6000 K. If the sun is assumed to be a blackbody radiator, at what wavelength λ_{max} would its spectrum peak?

From Wien's displacement law (Equation 16-25), we have

$$\lambda_{max} = \frac{2.898 \text{ mm·K}}{6000 \text{ K}} = 483 \times 10^{-9} \text{ m} = 483 \text{ nm}$$

This wavelength is in the visible spectrum. The blackbody radiation spectrum describes the sun's radiation fairly well, so the sun is indeed a good example of a blackbody.

Example 16-7

Calculate $\lambda_{\max}$ for a blackbody at room temperature, $T = 300$ K.

From Equation 16-25, we have

$$\lambda_{\max} = \frac{2.898 \text{ mm·K}}{300 \text{ K}} = 9.66 \times 10^{-6} \text{ m} = 9660 \text{ nm}$$

This is an infrared wavelength, far longer than the wavelengths visible to the eye. This obscures the fact that surfaces that are not black to our eyes may act as blackbodies for infrared radiation and absorption. For example, it has been found experimentally that the skin of human beings of all races is black to infrared radiation; hence, the emissivity of skin is 1.00 for its own radiation process.

Example 16-8

Calculate the net loss in radiated energy for a naked person in a room at 20°C, assuming the person to be a blackbody with a surface area of 1.4 m^2 and a surface temperature of 33°C = 306 K. (The surface temperature of the human body is slightly less than the internal temperature, about 37°C, because of the thermal resistance of the skin.)

Using Equation 16-24 with $T = 306$ K and $T_0 = 293$ K, we have

$$P_{\text{net}} = (1)(5.67 \times 10^{-8} \text{ W/m}^2\text{·K}^4)(1.4 \text{ m}^2)[(306)^4 - (293)^4] = 111 \text{ W}$$

This is a large energy loss. It is approximately equal to the basal metabolic rate of about 120 W for a food intake of 2500 kcal per day. (See Problem 1.) We protect ourselves from such a great energy loss by wearing clothing, which, because of its low thermal conductivity, has a much lower outside temperature and therefore a much lower rate of thermal radiation.

When the absolute temperature of a body differs only slightly from that of its surroundings, the net power radiated is approximately proportional to the temperature difference. We can see this from Equation 16-24 by noting that $T^4 - T_0^4 = (T^2 + T_0^2)(T^2 - T_0^2) = (T^2 + T_0^2)(T + T_0)(T - T_0)$, and also noting that in the sums we can replace T_0 by T without changing the result much. Replacing T_0 by T in these sums, we have

$$T^4 - T_0^4 \approx (T^2 + T^2)(T + T)(T - T_0) = 4T^3 \, \Delta T$$

(This result can also be obtained by computing the differential $d(T^4) = 4T^3 \, dT$.) Thus for all mechanisms of heat transfer

The rate of cooling of a body is approximately proportional to the temperature difference between the body and its surroundings.

Newton's law of cooling

This result is known as **Newton's law of cooling.**

In many real situations, all three mechanisms for the transfer of thermal energy occur simultaneously, though one may be more effective than the others. For example, an ordinary space heater with a metal heating element heats partly by direct radiation and partly by convection. Because the emissivity of metals is fairly small, the main mechanism is convection. The heater warms the nearby air, which then rises and is replaced by cooler air. Often a fan is included in the heater to speed the convection process. (This is called "forced convection.") On the other hand, a space heater with a quartz filament heats mainly by radiation because the filament has a large emissivity.

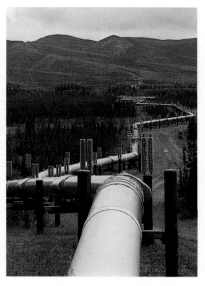

Temperature regulation along the Alaska pipeline. The 1.2-m diameter pipeline carrying hot oil (up to 145°F) is rigidly attached to vertical support members that are anchored in the ground, which is typically at 30°F. A typical support member is 15 m long of which only 4 m is above ground. Because the ground unfreezes enough to lose its structural rigidity at temperatures greater than 31°F, thermal energy must be transported away from the pipe and ground. Inside the support members are heat pipes containing liquid ammonia that absorbs heat, turns to gas, and rises. The ammonia then liquefies, releasing the heat to the air via the aluminum radiator fins. It then runs back down the walls of the heat pipe.

Similarly, coffee in a ceramic coffee cup cools by radiation as well as conduction (and evaporation) because the cup has a relatively large emissivity, whereas coffee in a metal cup cools mainly by conduction (and evaporation).

Questions

3. In a cool room, a metal or marble table top feels much colder to the touch than a wood surface does even though they are at the same temperature. Why?

4. Which heat-transfer mechanisms are most important in the warming effect of a fire in a fireplace?

5. Which heat-transfer mechanism is important in the transfer of energy from the sun to the earth?

6. Do you think that it is a coincidence that the radiant energy from the sun is concentrated in the range of visible wavelengths? If not, what is the reason?

16-4   The First Law of Thermodynamics

The first law of thermodynamics is a statement of the conservation of energy. It reflects the results of many experiments relating the work done on a system, the heat added to or subtracted from the system, and the internal energy of the system. From the original definition of the calorie, we know that it takes one calorie of heat to raise the temperature of 1 g of water by 1 C°. But we can also raise the temperature of water or any other system by doing work on it without adding any heat.

Figure 16-7a is a schematic diagram of the apparatus Joule used in his most famous experiment in which he determined the amount of work equivalent to a given amount of heat, that is, the amount of work needed to raise the temperature of one gram of water by one Celsius degree. Once the experimental equivalence of heat and energy has been established, Joule's experiment can be described as determining the size of the calorie in the usual energy units. The water in Figure 16-7 is enclosed by insulating walls so that the temperature of the system cannot be affected by heat entering or leaving it. As the weights fall at constant speed, they turn a paddle wheel,

James Prescott Joule (1818–1889).

Figure 16-7 (a) Schematic diagram of the apparatus for Joule's experiment to determine the amount of work required to produce a given temperature increase in a given amount of water. The work done on the water is determined by calculating the loss in the potential energy of the weights. (b) Photograph of the apparatus for Joule's experiment showing the paddle wheel and its container.

(a)

(b)

which does work against the water. If the energy loss through friction in the bearings, and so forth, is negligible, the work done by the paddle wheel against the water equals the loss of mechanical energy of the falling weights. The latter is easily measured by determining the distance through which the weights fall. The result of Joule's experiment and of those of many others after his is that it takes about 4.18 units of mechanical work (joules) to raise the temperature of 1 g of water by 1 C°. The result that 4.18 J of mechanical energy is equivalent to 1 cal of heat energy was known as the mechanical equivalence of heat. Historically, it was customary to express heat energy in calories and then use the mechanical equivalence of heat to convert to the standard units of mechanical energy. Today, all forms of energy are usually expressed in joules.

There are other ways of doing work on this system. For example, we could merely drop the insulated container of water from some height h, letting the system make an inelastic collision with the ground, or we could do mechanical work to generate electricity and then use the electricity to heat the water (Figure 16-8). In all such experiments, the same amount of work is required to produce a given temperature change in a given system. By the conservation of energy, this work done must go into an increase in internal energy U of the system.* The change in the internal energy of the system is usually indicated by the change in the temperature of the system or by a phase change.

Figure 16-8   Another method of doing work on a thermally insulated container of water. Electrical work is done on the system by the generator, which is driven by the falling weight.

Example 16-9

If a thermally insulated container of water is dropped from a height h and collides inelastically with the ground, what must h be for the temperature of the water to increase by 1 C°?

For the temperature to increase by 1 C°, the internal energy of the water must increase by 4.18 J for each gram of water or 4.18 kJ for each kilogram. If the mass of the water is m, the increase in its internal energy must be $m(4.18 \text{ kJ/kg})$. Setting this equal to the loss in potential energy mgh, we have

$$mgh = m(4.18 \text{ kJ/kg})$$

$$h = \frac{4.18 \text{ kJ/kg}}{9.81 \text{ N/kg}} = 0.426 \text{ km} = 426 \text{ m}$$

Note that this result is independent of the mass of the water because the loss in potential energy and the energy needed to raise the temperature by one degree are both proportional to the mass. Note also that the distance is rather large. This result illustrates one of the difficulties with Joule's experiment. A large amount of work must be done to produce a measurable change in temperature of the water.

Now, suppose we perform Joule's experiment but replace the insulating walls of the container with conducting walls. We find that the work needed to produce a given change in the temperature of the system is not always the same but rather depends on how much heat is added to or subtracted from the system by conduction through the walls. However, if we measure the heat added to or subtracted from the system and the work done on the system, we find that the sum of the work done on the system and the net heat added to the system is always the same for a given temperature change.

*The symbol U, which we used in previous chapters to denote potential energy, is now used for the internal energy of a system, which may include both kinetic and potential energy of the molecules in the system.

$$\Delta U = Q - W$$

Figure 16-9 Sign conventions for quantities in the first law of thermodynamics. Heat Q is positive when it goes into the system, whereas the work W is positive when it is done by the system.

First law of thermodynamics

That is, the sum of the heat added and the work done on the system equals the change in the internal energy of the system. This is a statement of the first law of thermodynamics. It is customary to write W for the work done *by* the system on its surroundings. Then $-W$ is the work done *on* the system. For example, if a gas expands against a piston, doing work on the surroundings, W is positive. Conversely, if work is done to compress the gas, W is negative. The heat Q is taken to be positive if it is put *into* the system and negative if it is taken *out of* the system (Figure 16-9). Using these conventions and writing U for the internal energy of a system, the **first law of thermodynamics** is

$$Q + (-W) = \Delta U$$

or, as it is usually written:

$$Q = \Delta U + W \qquad\qquad 16\text{-}26$$

The net heat added to a system equals the change in the internal energy of the system plus the work done by the system.

The first law of thermodynamics is a statement of the conservation of energy. The heat energy put into a system is accounted for by the work done by the system or by the increase in internal energy of the system or by some combination of the two.

Example 16-10

A system consists of 3 kg of water at 80°C. Twenty-five kilojoules of work is done on the system by stirring it with a paddle wheel, while 15 kcal of heat is removed. (*a*) What is the change in the internal energy of the system? (*b*) What is the final temperature of the system?

(*a*) We first change the units of the heat removed from calories to joules so that both the work and heat are expressed in the same units. We then calculate the change in the internal energy of the system from the first law of thermodynamics as expressed in Equation 16-26. Since 1 kcal = 4.18 kJ, 15 kcal = 15 × 4.18 kJ = 62.7 kJ. Thus 25 kJ of work is done on the system and 62.7 kJ of heat is removed. The net heat added to the system is $Q = -62.7$ kJ, and the work done by the system is $W = -25$ kJ. Equation 16-26 gives

$$Q = \Delta U + W$$

$$-62.7 \text{ kJ} = \Delta U + (-25 \text{ kJ})$$

$$\Delta U = -62.7 \text{ kJ} + 25 \text{ kJ} = -37.7 \text{ kJ}$$

The net change in the internal energy of the system is negative because more energy is removed from the system as heat than is added to the system by doing work on it.

(*b*) Since 4.18 kJ will change the temperature of 1 kg of water 1 C°, the change in temperature of the 3 kg of water will be

$$\Delta T = \frac{-37.7 \text{ kJ}}{(4.18 \text{ kJ/kg·C°})(3 \text{ kg})} = -3.01 \text{ C°}$$

The final temperature is therefore 80°C − 3.01 C° = 76.99°C ≈ 77°C.

It is important to understand that the internal energy U of a system is a property of the system that depends on the state of the system. That is, U is a function of the state of the system, just as P, V, and T are. As we have mentioned, the state of a gas is determined by any two properties such as P and V. Suppose that a gas is in some initial state (P_1, V_1). If we compress the gas or let it expand, add or remove heat from it, or do work on it or let it do work, the gas will move through a sequence of states; that is, it will have different values of the state functions P, V, T, and U. If the gas is then returned to its original state (P_1, V_1), the temperature T and the internal energy U must equal their original values.

On the other hand, the net heat input Q and the work W done by the gas are not functions of the state of the system. It is correct to say that a system has a large amount of internal energy, but it is not correct to say that a system has a large amount of heat or a large amount of work. In Example 16-10, for instance, it is not correct to say that the system contains 62.7 kJ less heat because we removed that much heat from the system. Nor can we say that the system contains 25 kJ more work because we did that much work on the system. What we can say is that after doing that much work on the system and removing that much heat, the internal energy of the system is 37.7 kJ less than it was originally. To increase the internal energy of a system by 2 J, we may add 2 J of heat, or we may add no heat but instead do 2 J of work on the system, or we may do 3 J of work on the system and remove 1 J of heat. In other words, heat is not something that is contained in a system. Rather, it is a measure of the energy that flows from one system to another because of a difference in temperature.

For very small amounts of heat added, work done, and changes in internal energy, it is customary to write Equation 16-26 as

$$dQ = dU + dW \qquad \text{16-27}$$

In this equation, dU is the differential of the internal-energy function. However, neither dQ nor dW is a differential of any function. Instead, dQ merely represents a small amount of heat added to the system, and dW represents a small amount of work done by the system.

Questions

7. The experiment of Joule discussed in this section involved the conversion of mechanical energy into internal energy. Give some examples in which the internal energy of a system is converted into mechanical energy.

8. Can a system absorb heat with no change in its internal energy?

16-5   The Internal Energy of an Ideal Gas

In Chapter 15, we saw that, according to a simple molecular model of a gas, the temperature T of the gas is associated with the kinetic energy K of translation of the molecules of the gas by the relation

$$K = \tfrac{3}{2}nRT$$

where n is the number of moles of gas and R is the universal gas constant. If this translational energy is taken to be the total internal energy of the gas, the internal energy will depend only on the temperature of the gas and not on its volume or pressure. Writing U for K, we have

$$U = \tfrac{3}{2}nRT \qquad \text{16-28}$$

If the internal energy of the gas includes other kinds of energy, its value will be different from that given by Equation 16-28, and it may or may not depend on the pressure and volume of the gas. Suppose, for example, that nearby gas molecules exert attractive forces on each other. Work is then required to increase the separation of the molecules. Thus, if the average distance between the molecules is increased, the potential energy associated with the molecular attraction will increase. The internal energy of the gas will then depend on the volume of the gas as well as on its temperature.

Joule, using an apparatus like that shown in Figure 16-10*a*, performed an interesting experiment to determine whether the internal energy of a gas depends on its volume. Initially, the compartment on the left contains a gas and the compartment on the right has been evacuated. The compartments are connected by a stopcock that is initially closed. The whole system is thermally insulated from its surroundings by rigid walls so that no heat can go into or out of the system and no work can be done. When the stopcock is opened, the gas rushes into the evacuated chamber. This process is called a **free expansion.** Eventually, the gas reaches thermal equilibrium with itself. Since no work has been done and no heat has been transferred, the final internal energy of the gas must equal its initial internal energy. If the gas molecules exert attractive forces on one another, the potential energy associated with these forces will increase when the volume increases. Since energy is conserved, the kinetic energy of translation must therefore decrease, which will result in a decrease in the temperature of the gas.

When Joule did this experiment, he found the final temperature to be equal to the initial temperature. Subsequent experiments verified this result for low gas densities. That is, the result held when the amount of gas originally in the left compartment was small. This implies that for a gas at low density—that is, for an ideal gas—the temperature depends only on the internal energy, or as we usually think of it, the internal energy depends only on the temperature. However, when the experiment is done with a large amount of gas initially in the left compartment so that the density is high, the temperature after the expansion is slightly lower than that before the expansion. This indicates that there is a small attraction between the gas molecules of a real gas at high density.

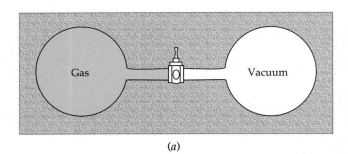

(a)

Figure 16-10 (a) Free expansion of a gas. When the stopcock is opened, the gas expands rapidly into the evacuated chamber. Since no work is done and the whole system is thermally insulated, the initial and final internal energies of the gas are equal. (b) The planetary nebula NGC 7293 (also known as the Helix nebula). This large-scale expansion of a hot gas occurs when an unstable star blows off its outer layers as its core shrinks to form a white dwarf star. In contrast to the free expansion in (a), in which the gas is insulated from its surroundings, the shell of hot gas in (b) is heated by ultraviolet radiation from the star.

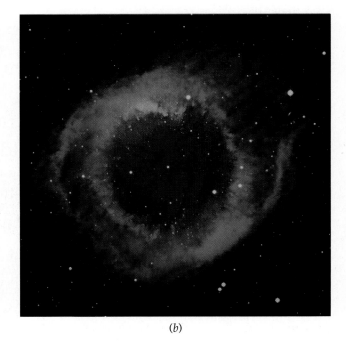

(b)

16-6   Work and the *PV* Diagram for a Gas

In the rest of this chapter, we will look at some applications of the first law of thermodynamics. In this section, we will discuss the work done by a gas under various conditions. In our discussion, we will assume that we have an ideal gas because the equation of state for an ideal gas is simple, and it provides a fairly accurate description of real gases in many situations. Understanding the work done by a gas is important for an understanding of engines, which are used to convert thermal energy into usable work. In many engines, work is done by a gas expanding against a movable piston. For example, in a steam engine, water is heated in a boiler to produce steam. The steam then does work as it expands and drives the piston. In an automobile engine, a mixture of gasoline vapor and air is ignited, causing it to explode. The resulting high temperatures and pressures cause the gas to expand rapidly, driving a piston and doing work. Heat engines will be considered in more detail in Chapter 17 when we discuss the second law of thermodynamics.

Figure 16-11 shows a gas confined in a container with a tightly fitting piston that we will assume to be frictionless. When the piston moves, the volume of the gas changes. The temperature or pressure or both must also change since these three variables are related by the equation of state $PV = nRT$. If we suddenly push the piston in to compress the gas, the pressure will initially be greater near the piston than far from it. Until the gas settles down into thermal equilibrium with itself, we cannot determine such macroscopic variables as T, P, or U for the entire gas system. However, if we move the piston slowly in small steps and wait after each step for equilibrium to be reestablished, we can compress or expand the gas in such a way that the gas is never far from an equilibrium state. In this kind of process, called a **quasi-static process,** the gas moves through a series of equilibrium states. In practice, it is possible to approximate quasi-static processes fairly well.

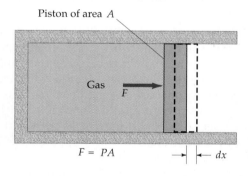

Piston of area A

Gas F

$F = PA$ ← | ← dx

Figure 16-11   Gas confined in a thermally insulated cylinder with a movable piston. When the piston moves a distance dx, the volume of the gas changes by $dV = A\,dx$. The work done by the gas is $PA\,dx = P\,dV$, where P is the pressure.

Let us begin with a gas at a fairly high pressure, and let it expand quasi-statically. The force exerted by the gas on the piston is PA, where A is the area of the piston and P is the gas pressure, which usually changes as the gas expands. (We could keep the pressure constant by heating the gas as it expands.) We are assuming that the piston does not accelerate, so there must be an external force pushing against the piston that is also equal to PA. The piston does work on the agent providing this external force. If the piston moves a small distance dx, the work done by the gas on the piston is

$$dW = F\,dx = PA\,dx = P\,dV \qquad\qquad 16\text{-}29$$

where $dV = A\,dx$ is the increase in the volume of the gas. Equation 16-29 also holds for a compression. In this case, dV is negative, indicating that work is done on the gas. To calculate the work done by the gas during an expansion from a volume of V_1 to a volume of V_2, we need to know how the pressure varies during the expansion.

Since the state of a gas is determined by two variables such as P and V, we can represent the states of a gas on a diagram of P versus V. Each point on the PV diagram indicates a particular state of the gas. Since P and V are related to the absolute temperature T by the equation of state for an ideal gas, $PV = nRT$, there is also a temperature T associated with each point. There is also a value of U associated with each point on the PV diagram. Figure 16-12 shows a PV diagram with a horizontal line representing a series of states that all have the same value of P. This line represents an expansion at constant pressure. Such a process is called an **isobaric expansion**. For a volume change of ΔV, the work done is $P \, \Delta V$ which is equal to the shaded area under the curve in the figure. In general, the work done by the gas is equal to the area under the P-versus-V curve:

Work done by a gas

$$W = \int P \, dV = \text{area under the } P\text{-versus-}V \text{ curve} \qquad 16\text{-}30$$

Since pressures are often given in atmospheres and volumes are often given in liters, it is convenient to have a conversion factor between liter-atmospheres and joules:

$$1 \text{ L·atm} = (10^{-3} \text{ m}^3)(101.3 \times 10^3 \text{ N/m}^2) = 101.3 \text{ J} \qquad 16\text{-}31$$

Exercise

If 3 L of an ideal gas at a pressure of 2 atm is heated so that it expands at constant pressure until its volume is 5 L, what is the work done by the gas? (Answer: 405.2 J)

Figure 16-13 shows three different possible paths on a PV diagram for a gas that is originally in state (P_1, V_1) and is finally in state (P_2, V_2). We assume the gas is ideal and have chosen the original and final states to have the same temperature so that $P_1V_1 = P_2V_2 = nRT$. In Figure 16-13a, the gas is heated at constant pressure until its volume is V_2, after which it is cooled at constant volume until its pressure is P_2. The work done along path A is $P_1(V_2 - V_1)$ for the horizontal part of the path and zero for the constant-volume part. This work is indicated by the shaded area under the curve. In Figure 16-13b, the gas is first cooled at constant volume until its pressure is P_2, after which it is heated at constant pressure until its volume is V_2. The work done along path B is $P_2(V_2 - V_1)$, which is much less than that done along the path A as can be seen by comparing the shaded regions in Figures 16-13a and b. In Figure 16-13c, both the pressure and the volume vary along the path. Path C is one of constant temperature, $PV = $ constant, so the process is called an **isothermal expansion**. The work done is again indicated by

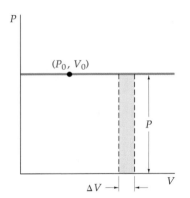

Figure 16-12 Each point on a PV diagram, such as (P_0, V_0), represents a particular state of the gas. The horizontal line represents states with a constant pressure P_0. The work done by a gas as it expands an amount ΔV is represented by the shaded area, $P_0 \, \Delta V$.

Figure 16-13 Three paths on PV diagrams connecting an initial state (P_1, V_1) and a final state (P_2, V_2). The work done along each path is indicated by the shaded area.

(a)

(b)

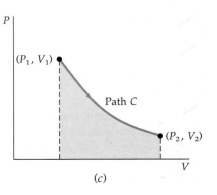

(c)

the shaded area. We can calculate this work using integral calculus. Using $P = nRT/V$, we have

$$dW = P \, dV = \frac{nRT}{V} \, dV$$

Hence, the work done by the gas as it expands isothermally from V_1 to V_2 is

$$W_{\text{isothermal}} = \int_{V_1}^{V_2} P \, dV = \int_{V_1}^{V_2} \frac{nRT}{V} \, dV$$

Since T is constant, we can remove it from the integral. We then have

$$W_{\text{isothermal}} = nRT \int_{V_1}^{V_2} \frac{dV}{V} = nRT \ln \frac{V_2}{V_1} \qquad 16\text{-}32$$

where the symbol "ln" stands for the natural logarithm, the logarithm to the base e.

In Figure 16-13, we can see that the amount of work done by the gas is different for each process illustrated. Since the change in the internal energy must be the same for each of the paths that begin at one state and end at another state, the net amount of heat added must also be different for each of the processes illustrated. In this case, the final internal energy equals the initial internal energy because we have an ideal gas and the initial and final temperatures are the same. Therefore, the net amount of heat added for each process equals the net work done by the gas. This discussion illustrates the fact that both the work done and the heat added depend on just how a system moves from one state to another, but the change in the internal energy of the system does not.

In Figure 16-14, an ideal gas undergoes a cyclic process from point A to point B to point C to point D and back to point A. (Such cyclic processes have important applications for heat engines, as we will see in the next chapter.) The gas begins at a volume of 1 L and a pressure of 2 atm and expands at constant pressure until the volume is doubled to 2 L, after which it is cooled at constant volume until its pressure is 1 atm. It is then compressed at constant pressure until its volume is again 1 L, after which it is heated at constant volume until it is back in its original state. We wish to find the total work done by the gas and the total heat added during this cycle.

As the gas expands from point A to point B at constant pressure, the work done by the gas is

$$W_{AB} = P(V_B - V_A) = (2 \text{ atm})(2 \text{ L} - 1 \text{ L}) = 2 \text{ L·atm}$$

This work is equal to the shaded area under curve AB in Figure 16-14. Using Equation 16-31 to convert from liter-atmospheres to joules, we find this work equals 202.6 J. As the gas cools from point B to point C, the volume is constant, so no work is done. As the gas is compressed at constant pressure from point C to point D, it does negative work; that is, work must be done on the gas. The work done *by* the gas is

$$W_{CD} = P(V_D - V_C) = (1 \text{ atm})(1 \text{ L} - 2 \text{ L}) = -1 \text{ L·atm} = -101.3 \text{ J}$$

The magnitude of this work is equal to the area under curve CD in Figure 16-14, which is the shaded area with the cross hatching. As the gas is heated back to its original state A, the volume is again constant, so no work is done. The net work done by the gas during the entire cycle is therefore 202.6 J − 101.3 J = 101.3 J. This net work is represented in the figure by the enclosed, shaded area without cross hatching. Since the gas is back in its original state, all the state properties, P, V, T, and U, are at their original values. There has thus been no net change in the internal energy of the gas. However, 101.3 J of work has been done by the gas, so by the first law of thermodynamics, 101.3 J of heat must have been added to the gas during the cycle.

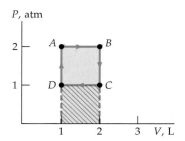

Figure 16-14   A *PV* diagram for a cyclic process in which a gas expands at constant pressure from state A to state B, is then cooled at constant volume from state B to state C, is then compressed at constant pressure from state C to state D, and is finally heated at constant volume back to state A. The work done by the gas as it expands from A to B minus that done on the gas as it is compressed from C to D is the net work done by the gas during the cycle. This net work is represented by the shaded area (without cross hatching) enclosed by the cycle.

16-7   Heat Capacities and the Equipartition Theorem

The determination of the heat capacity of a substance provides information about its internal energy, which in turn provides information about its molecular structure. For all substances that expand when heated, the heat capacity at constant pressure C_p is greater than the heat capacity at constant volume C_v because of the work done by the substance as it expands at constant pressure. As was mentioned in Section 16-1, this work is usually negligible for solids and liquids, so for them C_p and C_v are approximately equal. However, for gases we must distinguish between C_p and C_v because a gas heated at constant pressure readily expands and does a significant amount of work.

When heat is added to a gas at constant volume, no work is done by or on the gas, so the heat added equals the increase in the internal energy of the gas. Writing Q_v for the heat added at constant volume, we have

$$Q_v = C_v \, \Delta T$$

Since $W = 0$, we have from the first law of thermodynamics

$$Q_v = \Delta U + W = \Delta U$$

Thus,

$$\Delta U = C_v \, \Delta T$$

Taking the limit as ΔT approaches zero, we obtain

$$dU = C_v \, dT \qquad\qquad 16\text{-}33$$

and

$$C_v = \frac{dU}{dT} \qquad\qquad 16\text{-}34$$

Thus, the heat capacity at constant volume is the rate of change of the internal energy with temperature.

If we add heat at constant pressure, the gas will expand and do work on its surroundings. Therefore, only part of the added heat will go into an increase in the internal energy of the gas. Since the absolute temperature of the gas T is a measure of its internal energy, more heat must be added at constant pressure than at constant volume to achieve the same increase in temperature. In other words, the heat capacity at constant pressure is greater than that at constant volume.

We will now calculate the difference $C_p - C_v$ for an ideal gas. If we use Q_p for the heat added at constant pressure, we have from the definition of C_p

$$Q_p = C_p \, \Delta T$$

From the first law of thermodynamics,

$$Q_p = \Delta U + W = \Delta U + P \, \Delta V$$

Therefore,

$$C_p \, \Delta T = \Delta U + P \, \Delta V$$

For infinitesimal changes, this becomes

$$C_p \, dT = dU + P \, dV$$

Using Equation 16-33 for dU, we obtain

$$C_p \, dT = C_v \, dT + P \, dV \qquad \text{16-35}$$

The pressure, volume, and temperature for an ideal gas are related by

$$PV = nRT$$

Taking the differentials of both sides with $dP = 0$ for constant pressure, we obtain

$$P \, dV = nR \, dT$$

Substituting this into Equation 16-35, we obtain

$$C_p \, dT = C_v \, dT + nR \, dT$$

Therefore,

$$C_p = C_v + nR \qquad \text{16-36}$$

For an ideal gas, the heat capacity at constant pressure is greater than that at constant volume by the amount nR.

Table 16-5 lists measured molar heat capacities C_{mp} and C_{mv} for several gases. We note from this table that the ideal gas prediction, $C_{mp} - C_{mv} = R$, holds quite well for all gases. We also note that C_{mv} is approximately $1.5R$ for all monatomic gases, $2.5R$ for all diatomic gases, and greater than $2.5R$ for gases consisting of more complex molecules. We can understand these results by considering the molecular model of a gas discussed in Section 15-5. In that section, we saw that the total translational kinetic energy of n moles of a gas is $K = \frac{3}{2}nRT$ (Equation 15-27). Thus, if the internal energy of a gas consists of translational kinetic energy only, we have

$$U = \tfrac{3}{2}nRT \qquad \text{16-37}$$

Table 16-5 Molar Heat Capacities in J/mol·K of Various Gases at 25°C

Gas	C_{mp}	C_{mv}	$\dfrac{C_{mv}}{R}$	$C_{mp} - C_{mv}$	$\dfrac{C_{mp} - C_{mv}}{R}$
Monatomic					
He	20.79	12.52	1.51	8.27	0.99
Ne	20.79	12.68	1.52	8.11	0.98
Ar	20.79	12.45	1.50	8.34	1.00
Kr	20.79	12.45	1.50	8.34	1.00
Xe	20.79	12.52	1.51	8.27	0.99
Diatomic					
N_2	29.12	20.80	2.50	8.32	1.00
H_2	28.82	20.44	2.46	8.38	1.01
O_2	29.37	20.98	2.52	8.39	1.01
CO	29.04	20.74	2.49	8.30	1.00
Polyatomic					
CO_2	36.62	28.17	3.39	8.45	1.02
N_2O	36.90	28.39	3.41	8.51	1.02
H_2S	36.12	27.36	3.29	8.76	1.05

The heat capacities are then

$$C_v = \frac{dU}{dT} = \tfrac{3}{2}nR$$ 16-38

and

$$C_p = C_v + nR = \tfrac{5}{2}nR$$ 16-39

(handwritten margin note: dith $\frac{5}{2}$R ... For MONOTOMIC)

(handwritten margin note: C_v)

We can see that the experimental results in Table 16-5 agree well with these predictions for monatomic gases, but for other gases, the heat capacities are greater than those predicted by Equations 16-38 and 16-39. The internal energy for a gas consisting of diatomic or more complicated molecules is evidently greater than $\tfrac{3}{2}nRT$. The reason is that such molecules can have other types of energy, such as rotational or vibrational energy, in addition to translational kinetic energy.

In Section 15-5, we saw that the average kinetic energy associated with translational motion in the x direction is $\tfrac{1}{2}kT$ per molecule (Equation 15-25), where k is Boltzmann's constant, or, equivalently, $\tfrac{1}{2}RT$ per mole. Since x was an arbitrary direction, equivalent relations must also hold for the average kinetic energy of translation associated with the motion of a molecule in the y and z directions. If the energy of a molecule associated with its motion in the x direction is momentarily increased, say, by a collision between the molecule and a moving piston during a compression (Figure 16-15), collisions between that molecule and other molecules will quickly redistribute the additional energy so that, when the gas is again in equilibrium, the energy will be equally partitioned among the translational kinetic energies associated with motion in the x, y, and z directions. This sharing of the energy equally between the three terms in the translational kinetic energy is a special case of the **equipartition theorem,** a result that follows from statistical mechanics. Each coordinate, velocity component, angular velocity component, and so forth that appears in the expression for the energy of a molecule* is called a **degree of freedom.** Typical degrees of freedom are associated with the kinetic energy of translation, rotation, and vibration, and with the potential energy of vibration. The equipartition theorem states that

Equipartition theorem

> When a substance is in equilibrium, there is an average energy of $\tfrac{1}{2}kT$ per molecule or $\tfrac{1}{2}RT$ per mole associated with each degree of freedom.

According to Table 16-5 nitrogen, oxygen, hydrogen, and carbon monoxide all have molar heat capacities at constant volume of about $\tfrac{5}{2}R$. The internal energy of n moles of any of these gases must therefore be

$$U = \tfrac{5}{2}nRT$$ 16-40

*Strictly speaking, the coordinate must appear as a squared term in the expression for energy.

Figure 16-15 Molecules bouncing off a moving piston during a compression of the gas. The magnitude of the x component of momentum of each molecule is increased because of the collision. This results in an increase in the average kinetic energy associated with the translational motion of each molecule in the x direction, $\tfrac{1}{2}mv_x^2$. Collisions with other molecules quickly distribute this additional energy equally among other forms such as $\tfrac{1}{2}mv_y^2$ and $\tfrac{1}{2}mv_z^2$.

Thus, each of these gases has five degrees of freedom. About 1880, Clausius speculated that these gases must be diatomic gases that can rotate about two axes, giving them two additional degrees of freedom. Figure 16-16 shows a rigid-dumbbell model of a diatomic molecule. The two degrees of freedom besides the three for translation are now known to be associated with their rotation about each of the two axes, x' and y', perpendicular to the line joining the atoms. The kinetic energy of a diatomic molecule is therefore

$$K = \tfrac{1}{2}mv_x^2 + \tfrac{1}{2}mv_y^2 + \tfrac{1}{2}mv_z^2 + \tfrac{1}{2}I_{x'}\omega_{x'}^2 + \tfrac{1}{2}I_{y'}\omega_{y'}^2$$

From the equipartition theorem, an average energy of $\tfrac{1}{2}RT$ per mole is associated with each term, resulting in the total internal energy given by Equation 16-40. The heat capacity at constant volume for a diatomic gas is thus

$$C_v = \tfrac{5}{2}nR \qquad\qquad 16\text{-}41$$

In Section 16-1, we noted that all of the metals listed in Table 16-1 have approximately equal molar heat capacities. For instance, the molar heat capacity of copper is 24.5 J/mol·K, that for gold is 25.6 J/mol·K, and that for lead is 26.4 J/mol·K. Indeed, it is found experimentally that most solids have molar heat capacities approximately equal to $3R$:

$$C_m = 3R = 24.9 \text{ J/mol·K} \qquad\qquad 16\text{-}42$$

This result is known as the **Dulong–Petit law.** We can understand it by applying the equipartition theorem to the simple model for a solid shown in Figure 16-17. According to this model, a solid consists of a regular array of atoms in which each of the atoms has a fixed equilibrium position and is connected by springs to its neighbors. Each atom can vibrate in the x, y, and z directions. The total energy of an atom in a solid is thus

$$E = \tfrac{1}{2}mv_x^2 + \tfrac{1}{2}mv_y^2 + \tfrac{1}{2}mv_z^2 + \tfrac{1}{2}Kx^2 + \tfrac{1}{2}Ky^2 + \tfrac{1}{2}Kz^2$$

where K is the effective force constant of the hypothetical springs. Each atom thus has six degrees of freedom. The equipartition theorem states that a substance in equilibrium has an average energy of $\tfrac{1}{2}RT$ per mole for each degree of freedom. Thus, the internal energy of n moles of a solid is

$$U = 6 \times \tfrac{1}{2}nRT = 3nRT \qquad\qquad 16\text{-}43$$

which means that the molar heat capacity is $3R$.

Although the equipartition theorem has had spectacular successes in explaining the measured heat capacities of gases and solids, it has had equally spectacular failures. If a diatomic gas molecule rotates about the line joining the atoms, there should be an additional degree of freedom. Similarly, if a diatomic molecule is not rigid, the two atoms should vibrate along the line joining them. We would then have two more degrees of freedom corresponding to kinetic and potential energies of vibration. According to the measured values of the molar heat capacities in Table 16-5, however, diatomic gases apparently do not rotate about the line joining them nor do they vibrate. The equipartition theorem gives no explanation for this, nor for the fact that monatomic atoms apparently do not rotate about any of the three possible perpendicular axes in space. Furthermore, the equipartition theorem predicts constant values for the heat capacities of gases, but careful measurement shows that these quantities are somewhat dependent on temperature. Finally, the equipartition theorem predicts a constant value of $3R$ for the heat capacity of solids. This result holds for many, though not all, solids at high temperatures, but it does not hold at very low temperatures.

The equipartition theorem fails because classical mechanics itself breaks down when applied to atomic and molecular systems and must be replaced by quantum mechanics. We will discuss this in more detail in Chapter 35. It is interesting to note that the successes of the equipartition theorem in ex-

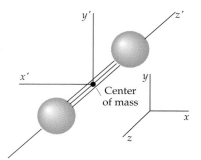

Figure 16-16 Rigid-dumbbell model of a diatomic gas molecule. If there is no rotation about the z' axis, such a molecule has five degrees of freedom, three associated with translation in the x, y, and z directions and two associated with rotation about the x' and y' axes.

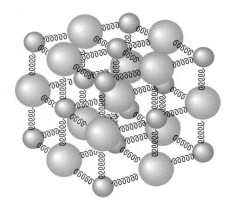

Figure 16-17 Model of a solid in which the atoms are connected to each other by springs. The internal energy of the solid consists of the kinetic and potential energies of vibration.

plaining the measured heat capacities of gases and solids led to the first real understanding of molecular structure in the nineteenth century, whereas its failures played an important role in the development of quantum mechanics in the twentieth century.

Example 16-11

The molar mass of copper is 63.5 g/mol. Use the Dulong–Petit law to calculate the specific heat of copper.

According to the Dulong–Petit law, the molar heat capacity of a solid is

$$C_m = 3R = 3(8.31 \text{ J/mol·K}) = 24.9 \text{ J/mol·K}$$

The specific heat of copper should therefore be

$$c = \frac{C_m}{M} = \frac{24.9 \text{ J/mol·K}}{63.5 \text{ g/mol}} = 0.392 \text{ J/g·K} = 0.392 \text{ kJ/kg·K}$$

This agrees fairly closely with the measured value of 0.386 kJ/kg·K given in Table 16-1.

Exercise

The specific heat of a certain metal is measured to be 1.02 kJ/kg·K. (*a*) Calculate the molar mass of this metal, assuming that the metal obeys the Dulong–Petit law. (*b*) What is the metal? [Answers: (*a*) M = 24.4 g/mol; (*b*) the metal must be magnesium, which has a molar mass of 24.31 g/mol.]

Example 16-12

One mole of oxygen gas is heated from a temperature of 20°C and a pressure of 1 atm to a temperature of 100°C. Assume oxygen is an ideal gas. (*a*) How much heat must be supplied if the volume is kept constant during the heating? (*b*) How much heat must be supplied if the pressure is kept constant? (*c*) How much work does the gas do in part (*b*)?

(*a*) Using $n = 1$ mol and $R = 8.31$ J/K, the heat capacity of oxygen at constant volume is

$$C_v = \tfrac{5}{2}nR = \tfrac{5}{2}(1)(8.31 \text{ J/K}) = 20.8 \text{ J/K}$$

The heat that must be added to raise the temperature by 80 C° = 80 K is

$$Q_v = C_v \, \Delta T = (20.8 \text{ J/K})(80 \text{ K}) = 1.66 \text{ kJ}$$

No work is done when the volume is kept constant, so the internal energy of the gas must increase by 1.66 kJ. Since the internal energy depends only on the temperature, this is the increase in the internal energy when the temperature of 1 mol of oxygen changes from 20°C to 100°C by any process.

(*b*) If we keep the pressure constant, the heat that must be added to raise the temperature by 80 K is

$$Q_p = C_p \, \Delta T = \tfrac{7}{2}(8.31 \text{ J/K})(80 \text{ K}) = 2.33 \text{ kJ}$$

where we have used $C_p = C_v + R = \tfrac{7}{2}R$ for 1 mol.

(*c*) Since the increase in internal energy is again 1.66 kJ, the work done by the gas is

$$W = Q - \Delta U = 2.33 \text{ kJ} - 1.66 \text{ kJ} = 0.67 \text{ kJ}$$

We can check this result by calculating directly the work done by the gas when the heat is added at constant pressure. Since the volume of 1 mol of any gas is 22.4 L at the standard conditions of 1 atm and 0°C = 273 K, and since the volume is proportional to the absolute temperature if the pressure is constant, the initial volume V_1 at 20°C = 293 K is

$$V_1 = (22.4 \text{ L})\frac{293 \text{ K}}{273 \text{ K}} = 24.0 \text{ L}$$

When the temperature is raised by 80 K to 373 K at constant pressure, the volume increases to V_2, which is given by

$$V_2 = (22.4 \text{ L})\frac{373 \text{ K}}{273 \text{ K}} = 30.6 \text{ L}$$

The work done is thus

$$W = P \, \Delta V = (1 \text{ atm})(30.6 \text{ L} - 24.0 \text{ L}) = 6.6 \text{ L·atm}$$

Converting this to joules, we obtain

$$W = (6.6 \text{ L·atm})\frac{101.3 \text{ J}}{1 \text{ L·atm}} = 0.67 \text{ kJ}$$

which agrees with our previous result.

16-8   The Quasi-Static Adiabatic Expansion of a Gas

A process in which no heat flows into or out of a system is called an **adiabatic process.** In this section, we will consider the quasi-static adiabatic expansion of a gas in which the gas in a thermally insulated container expands slowly against a piston, doing work on it. Since no heat enters or leaves the gas, the work done by the gas equals the decrease in the internal energy of the gas, and the temperature of the gas decreases. The curve representing this process on a PV diagram is shown in Figure 16-18. This curve is steeper than that for an isothermal expansion, as shown. That is, when the volume increases by some amount, the pressure decreases by more in an adiabatic expansion than it does in an isothermal expansion. This is to be expected since the temperature decreases during an adiabatic expansion.

We can find the equation for the adiabatic curve for an ideal gas using the equation of state and the first law of thermodynamics. We have

$$dQ = dU + dW = C_v \, dT + P \, dV = 0 \qquad \text{16-44}$$

where we have used $dU = C_v \, dT$ from Equation 16-34. Then using $P = nRT/V$, we obtain,

$$C_v \, dT + nRT \, \frac{dV}{V} = 0$$

Rearranging slightly, we obtain

$$\frac{dT}{T} + \frac{nR}{C_v} \frac{dV}{V} = 0 \qquad \text{16-45}$$

Equation 16-45 can be further simplified by noting that $C_p - C_v = nR$, so

$$\frac{nR}{C_v} = \frac{C_p - C_v}{C_v} = \frac{C_p}{C_v} - 1 = \gamma - 1$$

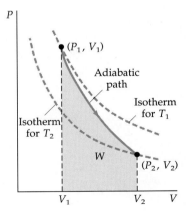

Figure 16-18   Quasi-static adiabatic expansion of an ideal gas. The dashed lines are the isotherms for the initial and final temperatures. The curve connecting the initial and final states of the adiabatic expansion is steeper than the isotherms because the temperature drops during the expansion.

where γ is the ratio of the heat capacities:

$$\gamma = \frac{C_p}{C_v} \qquad \text{16-46}$$

We then have

$$\frac{dT}{T} + (\gamma - 1)\frac{dV}{V} = 0$$

Integrating, we obtain

$$\ln T + (\gamma - 1)\ln V = \text{constant}$$

Using the properties of logarithms (Appendix A), we obtain

$$\ln (TV^{\gamma-1}) = \text{constant}$$

Then

$$TV^{\gamma-1} = \text{constant} \qquad \text{16-47}$$

(The constants in the two preceding equations are not the same.) We can eliminate T from Equation 16-47 using $PV = nRT$. We then have

$$\frac{PV}{nR}V^{\gamma-1} = \text{constant}$$

Quasi-static adiabatic expansion   or

$$PV^\gamma = \text{constant} \qquad \text{16-48}$$

[handwritten annotation: FOR THAT ONE CURVE]

Equation 16-48 relates P and V for a quasi-static adiabatic expansion and for a quasi-static adiabatic compression in which the piston does work on the gas.

We can use Equation 16-48 to calculate the adiabatic bulk modulus for an ideal gas, which is related to the speed of sound waves in air. Differentiating Equation 16-48, we obtain

$$P\gamma V^{\gamma-1}\,dV + V^\gamma\,dP = 0$$

$$dP = -\frac{\gamma P\,dV}{V}$$

Recalling that the bulk modulus is the ratio of the change in pressure to the fractional change in volume, we have for the adiabatic bulk modulus

$$B_{\text{adiabatic}} = -\frac{dP}{dV/V} = \gamma P \qquad \text{16-49}$$

The speed of sound is given by Equation 14-1

$$v = \sqrt{\frac{B_{\text{adiabatic}}}{\rho}}$$

where the mass density ρ is related to the number of moles n and the molecular mass M by $\rho = m/V = nM/V$. Using the ideal gas law, $PV = nRT$, we can eliminate V from the density

$$\rho = \frac{nM}{V} = \frac{nM}{(nRT/P)} = \frac{MP}{RT}$$

Using this result and γP for $B_{\text{adiabatic}}$, we obtain

$$v = \sqrt{\frac{B_{\text{adiabatic}}}{\rho}} = \sqrt{\frac{\gamma P}{(MP/RT)}} = \sqrt{\frac{\gamma RT}{M}}$$

which is Equation 14-3.

The work done by the gas in the quasi-static adiabatic expansion in the PV diagram of Figure 16-18 is equal to the area under the curve. It is simply related to the change in the temperature of the gas. Setting $dQ = 0$ in Equation 16-44, we have

$$P \, dV = -C_v \, dT$$

Then

$$W_{\text{adiabatic}} = \int P \, dV = -\int C_v \, dT$$

or

$$W_{\text{adiabatic}} = -C_v \, \Delta T \qquad \text{16-50}$$

Adiabatic work

where we have assumed that C_v is constant. (For an ideal gas, U is proportional to the absolute temperature, and therefore $C_v = dU/dT$ is a constant.) We note from Equation 16-50 that the work done by the gas depends only on the change in the absolute temperature of the gas. In a quasi-static adiabatic expansion, the gas does work, and its internal energy and temperature decrease. In a quasi-static adiabatic *compression*, work is done *on* the gas, and the internal energy and temperature increase.

We can use the ideal gas law to write Equation 16-50 in terms of the initial and final values of the pressure and volume. If T_1 is the initial temperature and T_2 is the final temperature, we have for the work done

$$W_{\text{adiabatic}} = -C_v \, \Delta T = -C_v(T_2 - T_1) = C_v(T_1 - T_2)$$

Using $PV = nRT$, we obtain

$$W_{\text{adiabatic}} = C_v\left(\frac{P_1 V_1}{nR} - \frac{P_2 V_2}{nR}\right) = \frac{C_v}{C_p - C_v}(P_1 V_1 - P_2 V_2)$$

where we have used $nR = C_p - C_v$. Dividing the numerator and denominator by C_v and writing γ for C_p/C_v, we obtain

$$W_{\text{adiabatic}} = \frac{P_1 V_1 - P_2 V_2}{\gamma - 1} \qquad \text{16-51}$$

Clouds are formed when rising moist air cools below the dew point due to adiabatic expansion of the air.

Example 16-13

A quantity of air ($\gamma = 1.4$) expands adiabatically and quasi-statically from an initial pressure of 2 atm and volume of 2 L at temperature 20°C to twice its original volume. Find (*a*) the final pressure, (*b*) the final temperature, and (*c*) the work done by the gas.

(*a*) According to Equation 16-48, the quantity PV^γ remains unchanged during a quasi-static adiabatic expansion. Thus, if P_1 and V_1 are the initial pressure and volume and P_2 and V_2 are the final pressure and volume, we have

$$P_1 V_1^\gamma = P_2 V_2^\gamma \qquad \text{or} \qquad P_2 = P_1\left(\frac{V_1}{V_2}\right)^\gamma$$

For the given values we have

$$P_2 = (2 \text{ atm})\left(\frac{2 \text{ L}}{4 \text{ L}}\right)^{1.4} = 0.758 \text{ atm}$$

(*b*) The change in temperature is found most easily from Equation 16-47, which implies that

$$T_1 V_1^{\gamma - 1} = T_2 V_2^{\gamma - 1}$$

or

$$T_2 = T_1\left(\frac{V_1}{V_2}\right)^{\gamma - 1} = (293 \text{ K})\left(\frac{2 \text{ L}}{4 \text{ L}}\right)^{0.4} = 222 \text{ K} = -51°\text{C}$$

(c) The work done by the gas is most easily found using Equation 16-51:

$$W_{\text{adiabatic}} = \frac{P_1 V_1 - P_2 V_2}{\gamma - 1}$$

$$= \frac{(2 \text{ atm})(2 \text{ L}) - (0.758 \text{ atm})(4 \text{ L})}{1.4 - 1}$$

$$= 2.42 \text{ L·atm}$$

Converting to joules, we obtain

$$W_{\text{adiabatic}} = 2.42 \text{ L·atm} \times \frac{101.3 \text{ J}}{1 \text{ L·atm}}$$

$$= 245 \text{ J}$$

Summary

1. Heat is energy that is transferred from one object to another because of a temperature difference. The heat capacity of a substance is the heat energy needed to raise the temperature of the substance by one degree. The specific heat is the heat capacity per unit mass. The calorie, originally defined as the heat necessary to raise the temperature of one gram of water by one Celsius degree, is now defined to be 4.184 joules. The specific heat of water is 4.184 kJ/kg·K.

2. The heat needed to melt a substance is the product of the mass of the substance and its latent heat of fusion L_f:

$$Q = mL_f$$

The heat needed to vaporize a liquid is the product of the mass of the liquid and its latent heat of vaporization L_v:

$$Q = mL_v$$

Both melting and vaporization occur at a constant temperature. For water, $L_f = 333.5$ kJ/kg and $L_v = 2257$ kJ/kg. The heat required to melt 1 g of ice or to vaporize 1 g of water is large compared with the heat required to raise the temperature of 1 g of water by one degree.

3. The three mechanisms by which thermal energy is transferred are conduction, convection, and radiation.

4. The rate of conduction of thermal energy is given by

$$I = \frac{\Delta Q}{\Delta t} = kA \frac{\Delta T}{\Delta x}$$

where I is the thermal current and k is the coefficient of thermal conductivity. This equation can be written

$$\Delta T = IR$$

where R is the thermal resistance:

$$R = \frac{\Delta x}{kA}$$

The thermal resistance for a unit area of a slab of material is called the R factor R_f:

$$R_f = RA = \frac{\Delta x}{k}$$

The equivalent thermal resistance of a series of thermal resistances equals the sum of the individual resistances:

$$R_{eq} = R_1 + R_2 + \cdots \qquad \text{resistances in series}$$

The equivalent thermal resistance for thermal resistances in parallel is given by

$$\frac{1}{R_{eq}} = \frac{1}{R_1} + \frac{1}{R_2} + \cdots \qquad \text{resistances in parallel}$$

5. The rate at which an object radiates thermal energy is given by

$$P = e\sigma A T^4$$

where $\sigma = 5.6703 \times 10^{-8}$ $\text{W/m}^2 \cdot \text{K}^4$ is Stefan's constant, and e is the emissivity, which varies between 0 and 1 depending on the composition of the surface of the object. Materials that are good heat absorbers are good heat radiators. A blackbody has an emissivity of 1. It is a perfect radiator, and it absorbs all the radiation incident upon it. The net thermal power radiated by an object at a temperature T in an environment at a temperature T_0 is given by

$$P_{net} = e\sigma A(T^4 - T_0^4)$$

The spectrum of electromagnetic energy radiated by a blackbody has a maximum at a wavelength λ_{max}, which varies inversely with the absolute temperature of the body:

$$\lambda_{max} = \frac{2.898 \text{ mm} \cdot \text{K}}{T}$$

This is known as Wien's displacement law.

6. For all mechanisms of heat transfer, if the temperature difference between the body and its surroundings is small, the rate of cooling of a body is approximately proportional to the temperature difference. This result is known as Newton's law of cooling.

7. The first law of thermodynamics is a statement of the conservation of energy. It says that the net heat added to a system equals the change in the internal energy of the system plus the work done by the system:

$$Q = \Delta U + W$$

The internal energy of a system is a property of the state of the system, as are the pressure, volume, and temperature, but heat and work are not.

8. The internal energy of an ideal gas depends only on the absolute temperature T.

9. A quasi-static process is one that occurs slowly so that the system moves through a series of equilibrium states. A process is isobaric if the pressure remains constant, isothermal if the temperature remains constant, and adiabatic if no heat is transferred. For a quasi-static adiabatic expansion of an ideal gas, the pressure and volume are related by

$$PV^\gamma = \text{constant}$$

where γ is the ratio of the heat capacity at constant pressure to that at constant volume:

$$\gamma = \frac{C_p}{C_v}$$

10. When a system expands quasi-statically, the work done by the system is given by

$$W = \int P \, dV$$

The work done by a gas can be represented graphically as the area under the P-versus-V curve. It can be calculated if P is known as a function of V for the expansion. For an isothermal expansion of an ideal gas, the work done by the gas is

$$W_{\text{isothermal}} = nRT \ln \frac{V_2}{V_1}$$

For an adiabatic expansion of an ideal gas, the work done by the gas is

$$W_{\text{adiabatic}} = -C_v \, \Delta T = \frac{P_1 V_1 - P_2 V_2}{\gamma - 1}$$

11. The heat capacity at constant volume is related to the change in internal energy by

$$C_v = \frac{dU}{dT}$$

For an ideal gas, the heat capacity at constant pressure is greater than that at constant volume by the amount nR:

$$C_p = C_v + nR$$

The heat capacity at constant pressure is greater because a gas that is heated at constant pressure expands and does work, so it takes a greater amount of heat to achieve the same change in temperature. The heat capacity at constant volume for a monatomic gas is

$$C_v = \tfrac{3}{2}nR \qquad \text{monatomic gas}$$

For a diatomic gas, it is

$$C_v = \tfrac{5}{2}nR \qquad \text{diatomic gas}$$

12. The equipartition theorem states that when a system is in equilibrium, there is an average energy of $\tfrac{1}{2}kT$ per molecule or $\tfrac{1}{2}RT$ per mole associated with each degree of freedom. Monatomic gases have three degrees of freedom, which are associated with the kinetic energy of translation in three dimensions. Diatomic gases have two additional degrees of freedom, which are associated with their rotation about axes perpendicular to the line joining the atoms.

13. The molar heat capacity of most solids is $3R$, a result known as the Dulong–Petit law. This result can be understood by applying the equipartition theorem to a model of a solid in which each atom in the solid can vibrate in three dimensions and therefore has a total of six degrees of freedom, three associated with the kinetic energy of vibration and three with the potential energy of vibration.

The Earth's Energy Balance and Global Warming

Jerrold H. Krenz
University of Colorado

Humans have inhabited this planet for millions of years, but it is only during the past hundred years or so that—as a consequence of both population growth and industrialization—we have begun to affect our climate. The global temperature has risen by 0.5 K since 1900, the ocean levels have risen, inland glaciers are receding, polar ice caps are shrinking, and the depth to Alaskan permafrost has decreased. The combustion of fossil fuels and deforestation combine to produce an increase in atmospheric levels of carbon dioxide (CO_2); combined with other gaseous products of industrialization (for example, nitrous oxide, chlorofluorocarbons, and ozone) this promotes global warming, due to a process commonly called the "greenhouse effect."

There is still debate as to whether current climatological models prove conclusively that increased global warming has already occurred. There is, however, a consensus in the scientific community that a continuation of present human activities will result in a significant additional warming of the earth (1.5 to 5.5 K) over the next 50–100 years. Global warming of this magnitude would have profound and long-lasting effects on agriculture, wildlife, and human society.

The Earth's Temperature

If the earth merely absorbed solar radiation, its temperature would continually increase. However, the earth also radiates energy into space. Through this process of absorption and radiation, an equilibrium condition is maintained that is generally referred to as the earth's energy balance. It is the *rate* at which energy arrives and leaves the earth that is balanced so, to be precise, it is actually a power balance.

To appreciate the process by which the earth is being warmed, it is necessary to understand the mechanism by which the temperature of the earth is maintained. At the earth's average distance from the sun, radiant energy from the sun arrives at a rate of 1353 W/m² (the solar constant, *S*). Since the surface temperature of the sun is approximately 6000 K, its radiation spectrum consists of very short wavelengths centered around approximately 0.5 μm. This is observed experimentally; and it also follows from Wien's displacement law if the sun is approximated as a blackbody (Section 16-3).

The solar radiation absorbed by the earth depends on the cross-sectional area of the earth as viewed from the sun, πR_E^2 (where R_E is the radius of the earth, 6.4×10^6 m). Not all the solar radiation intercepted by the earth is absorbed; a portion ($r \approx 0.3$) is reflected (this is the "earthlight" that astronauts see from space). We may therefore write the following equation:

$$\text{Power absorbed by the earth} = (1 - r)\pi R_E^2 SW \quad (1)$$

The earth reradiates some of this absorbed power back into space. Although the average surface temperature of the earth is about 13°C (286 K), the average radiative temperature of the earth's atmosphere is only about −22°C (251 K). A relative emissivity constant *e* is used to account for this difference in the blackbody radiation expression ($e \approx 0.6$). The power radiated by each square meter of the earth's surface is described by the following equation:

$$\text{Radiated power density} = e\sigma T_E^4 \quad \text{W/m}^2 \quad (2)$$

where σ is the Stefan radiation constant (Equation 16-22), and T_E is the surface temperature of the earth. The earth's surface is largely covered with water with a relatively stable temperature; this means that T_E varies little from day to night. Because of this, and due to the fact that solar energy is *absorbed* only by the portion of the earth's surface

Dr. Krenz did his undergraduate work at the State University of New York at Buffalo and his graduate work in electrical engineering (MS and PhD) at Stanford University. Before joining the faculty of the University of Colorado–Boulder, he worked several years at the General Electric Microwave Research Laboratory in Palo Alto, California.

At the University of Colorado his teaching interests have included the interaction of technology and society. During the 1970s, in response to the "energy crisis," he offered a senior-level engineering course on energy which dealt with both the potentialities and limitations of various existing and proposed energy systems. He has published several books, including two that address energy issues and include discussions of the "greenhouse effect."

In addition to trying to understand how societies use energy, Dr. Krenz has been deeply concerned with the interaction of technological developments and the nuclear arms race.

Continued

that is facing the sun but is *radiated* from the entire surface of the globe, the radiative behavior of the earth may be approximated by a spherical radiator with a uniform temperature:

Power radiated by the earth = $4\pi R_E^2 e\sigma T_E^4$ W (3)

For an equilibrium condition, the power absorbed and radiated by the earth must be equal:

$$(1 - r)\pi R_E^2 S = 4\pi R_E^2 e\sigma T_E^4 \quad \text{W} \qquad (4)$$

This equation, for the approximations introduced, is the energy or power balance of the earth. Dividing by the surface area of the earth, $4\pi R_E^2$ we have

$$\frac{(1 - r)S}{4} = e\sigma T_E^4 \quad \text{W/m}^2 \qquad (5)$$

which is the average rate at which energy is absorbed and subsequently radiated by each square meter of the earth's surface (≈ 237 W/m², about the strength of four 60-W light bulbs for every square meter of the earth's surface).

Other energy inputs (see Table 1) have only a slight effect on the earth's energy/power balance. Ignoring these, we may rewrite Equation 5 to solve for temperature of the earth, T_E:

$$T_E = \left[\frac{(1 - r)S}{4e\sigma}\right]^{1/4}$$

A decrease in either the reflectivity of the earth, r, or its relative emissivity, e, will result in an increase in its average temperature. It is *primarily* the properties of the gases that make up the earth's atmosphere that affect these quantities.

Some Atmospheric Gases Absorb Thermal Radiation

The atmosphere is fairly transparent to *incoming* solar radiation from the sun. Although the principal gases of the atmosphere, oxygen (O_2) and nitrogen (N_2), are transparent to the long-wavelength thermal radiation from the earth, this is *not* the case for all gases in the atmosphere. Solar radiation can therefore penetrate the earth's atmosphere, but a portion is trapped when it is reradiated as longer wavelength thermal radiation, resulting in a warming of the earth (Figure 1).

You are probably familiar with the terms "greenhouse effect" to describe global warming and "greenhouse gases" to characterize the gases that contribute to global warming by absorbing thermal radiation. It is interesting to note that these names are somewhat inappropriate as greenhouses operate by another mechanism altogether (by the reduction of convective cooling; see Section 16-3).

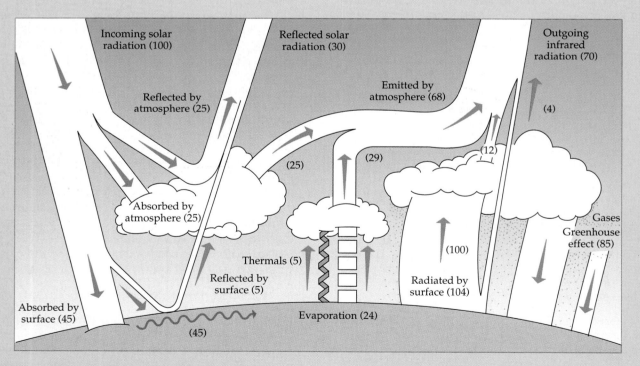

Figure 1 The earth's energy/power balance, including the effect of the atmospheric system. The quantities indicated are relative to the earth-averaged solar constant $S/4$. From Stephen H. Schneider, "The Green-house Effect: Science and Policy," *Science*, vol. 243, pp. 771–781. (Reprinted with the permission of the American Association for the Advancement of Science.)

The most prevalent gases that absorb thermal radiation are water vapor and carbon dioxide. Without the *benefit* of the thermal effects of water vapor and CO_2, the earth's temperature would actually be insufficient for most existing life forms. The relationship between the "greenhouse gases" is complex. As their total levels build up and the global temperature increases, the rate of water evaporation from the oceans will be enhanced. While this would appear to create a positive feedback loop by boosting the atmospheric concentrations of water vapor, it is also likely to influence the earth's cloud cover. Clouds have an as-yet poorly understood role to play in the overall equation: They increase the earth's reflectivity, thereby reducing global warming; at the same time, they reduce the rate at which thermal energy can be radiated into space thereby enhancing global warming. The effect of clouds may be further complicated by changing seasonal conditions.

Table 1   **Average Surface Contribution to the Earth's Energy Balance from Several Minor Sources of Thermal Energy**

Source	Contribution (W/m^2)
Radioisotope decay	0.06
Fuel consumption	0.018
Tidal friction	0.005

Carbon Dioxide

Contemporary levels of atmospheric CO_2 are being influenced by the combustion of fossil fuels and by the net result of global biomass accumulation or destruction. Since 1957, accurate measurements of atmospheric CO_2 have been made at Mauna Loa Observatory, Hawaii (Figure 2). Overshadowing the seasonal fluctuations, which are due primarily to seasonal variations in the total levels of plants' pho-

tosynthetic activity, there can be seen a steady and continuing increase in the concentration of atmospheric CO_2. This now stands at 350 parts per million by volume (ppmv). Based on analyses of air trapped long ago in glacial ice, the concentration of atmospheric CO_2 in 1750 was about 280 ppmv. There has therefore been a 25 percent increase since the beginning of modern industrialization.

Carbon dioxide emissions are a direct consequence of the energy conversion process, the oxidation of the carbon of the fuel. Consider the case of coal, which is primarily carbon:

$$C + O_2 \rightarrow CO_2$$

Combustion of hydrocarbon fuels, petroleum, and natural gas produces a similar effect, also creating water vapor (another "greenhouse gas") by the oxidation of the hydrogen of the fuel.

Every year approximately 5×10^{12} kg of carbon in the form of CO_2 is released to the atmosphere as a result of global consumption of fossil fuels. In 1988, on average each of us in the United States consumed about 3.6 tons of coal, 73,000 cubic feet of natural gas, and 1070 gallons of petroleum. With less than 5 percent of the world's population, the United States accounts for approximately 33 percent of the world's fossil fuel consumption. The majority of the world's inhabitants have low levels of economic activity and concurrent low energy-usage rates but have aspirations for higher standards of living. An increase in global fossil fuel usage (and hence CO_2 emissions) is therefore considered inevitable. Although increased energy efficiencies and the use of alternative energy sources may moderate the growth in fossil fuel usage, atmospheric concentrations of CO_2 of 440 to 550 ppmv are nevertheless anticipated by the year 2100. As the level of CO_2 dissolved in the oceans increases and as deforestation and acid rain take their toll, the capacity of the earth to absorb increased atmospheric CO_2 levels diminishes.

Figure 2   Concentration of atmospheric carbon dioxide at Mauna Loa Observatory, Hawaii. (Courtesy John Chin.)

Continued

Other Greenhouse Gases

Although present in trace quantities only, other gases (methane, ozone, nitrous oxide, and chloro-fluorocarbons) also absorb long-wavelength thermal radiation. The concentration of methane (CH_4) in the atmosphere is less than 1 percent that of CO_2, however, each molecule of CH_4 is about twenty times as effective as a molecule of CO_2 at absorbing radiant heat. Moreover, its concentration has been increasing at a rate of 1 percent per year and has doubled over the past 250 years. A positive feedback loop could develop because increases in global temperature enhance net rates of organic decay (for example, in swamps), thereby accelerating the overall increase in atmospheric concentrations of CH_4.

Ozone (O_3) is formed as a result of photochemical processes primarily involving sunlight, methane, carbon monoxide, and nitrogen oxides (typical pollutants). The concentration of O_3 in the troposphere has increased by 10 percent (notwithstanding its *decreased* concentration in the stratosphere over the poles). A similar increase (5–10 percent) has occurred for nitrous oxide; this has arisen primarily due to the use of nitrogen-based fertilizers, deforestation, and the burning of biomass—all consequences of cultivation of new land areas.

Chlorofluorocarbons (CFCs) are not naturally present in the atmosphere. These compounds are manufactured for use as solvents, refrigerants, and spray-can propellants. They act as CO_2 does in trapping radiant energy, in addition to destroying ozone in the stratosphere. While the manufacture of CFCs is being restricted as a result of an international agreement, their combined atmospheric concentration is nevertheless expected to reach levels of 1 ppbv, owing to their long lifetimes.

Global Warming

The average global temperature has already increased by about 0.5 K since 1900 (Figure 3). The six warmest recorded years to date this century have been since 1980. More data are necessary before it will be possible to state without equivocation that a climatic change is occurring and that increases in concentrations of "greenhouse gases" are indeed

Figure 3 Global surface air temperature relative to 1950–1979 average. The data from the University of East Anglia and the UK Meteorological Office indicate a global warming of about 0.5 K this century. From Richard A. Kerr, "1988 Ties for Warmest Year," *Science*, vol. 243, p. 891. (Reprinted with the permission of the American Association for the Advancement of Science.)

the cause of the change. However, there is no doubt that their continued increase would certainly lead at some point to this effect. Deferring our response to this until analyses are conclusive will mean that the damage is more pronounced and more difficult to control or undo.

The net effect that global warming will have on a particular region is uncertain. Global precipitation patterns are expected to change dramatically, with consequent disruption of agriculture, trade, and the global economy. Thermal expansion of the ocean due to overall global warming and partial melting of land-based ice are expected to increase sea levels. Low-lying coastal and inland areas, where a large portion of the world's population lives, would be threatened with severe social, economic, and environmental repercussions.

The difference between this and previous climatic shifts is that these changes are a direct result of our own activities, not the effects of natural occurrences over which we have no control. By changing our activities we can moderate and perhaps even prevent the anticipated climatic changes. Given the economic costs, we must decide what should be done, taking into account both the present levels of uncertainty and the scale of the possible consequences of our choices. Will we have the wisdom to deal with this challenge effectively?

Sunset over the Amazon Basin as seen from the space shuttle Challenger, showing the thinness of the atmosphere.

Suggestions for Further Reading

Allen, Philip B.: "Conduction of Heat," *The Physics Teacher*, vol. 21, no. 9, 1983, p. 582.

This is an advanced article on our present understanding of heat conduction by solids and gases.

"Heat," *Scientific American* (Special Issue), September 1954.

This issue discusses many aspects of heat. It includes articles on the use of fire by early human beings, the temperature range in which life is possible, the temperatures of flames, materials that are resistant to high temperatures, chemistry at high temperatures, and the ultrahigh temperatures found in the interiors of stars.

Velarde, Manuel G., and Christiane Normand: "Convection," *Scientific American*, July 1980, p. 92.

Interesting illustrations and a discussion of Lord Rayleigh's theory of convection make this article useful, though the modern theories are quite mathematical.

Walker, Jearl: "The Amateur Scientist: Icicles Ensheathe a Number of Puzzles: Just How Does the Water Freeze?" *Scientific American*, May 1988, p. 114.

A discussion of research into a complex and beautiful change of phase.

Weaver, Jefferson Hane: "Energy," in *The World of Physics: A Small Library of the Literature of Physics from Antiquity to the Present*, Simon and Schuster, New York, 1987.

Includes "The Source of the Heat Which Is Excited by Friction" by Benjamin Thompson, Count Rumford, and writings on the mechanical equivalent of heat by J. R. Mayer and J. P. Joule.

Wilson, Mitchell: "Count Rumford," *Scientific American*, October 1960, p. 158.

This is an article about Benjamin Thompson, the highly unusual man whose experiments dealt heavy blows to the caloric theory of heat.

Review

A. Objectives: After studying this chapter, you should:

1. Be able to work calorimetry problems.

2. Be able to calculate the rate of conduction of thermal energy for various thermal resistances in series or in parallel.

3. Be able to state the Stefan–Boltzmann law of radiation and use it to calculate the power radiated by an object at a given temperature.

4. Be able to state Wien's displacement law.

5. Be able to state the first law of thermodynamics and use it in solving problems.

6. Be able to calculate the work done by a gas during various quasi-static processes and sketch the processes on a *PV* diagram.

7. Be able to state the equipartition theorem and use it to explain the molar heat capacities of gases and solids.

8. Be able to state the Dulong–Petit law and use it to estimate the heat capacity of a given solid or to calculate the molar mass of a solid from its specific heat.

B. Define, explain, or otherwise identify:

Heat
Heat capacity
Specific heat
Calorie
Btu
Molar heat capacity
Calorimetry
Calorimeter
Phase change
Fusion
Vaporization
Sublimation
Latent heat of fusion
Latent heat of vaporization
Conduction
Convection
Radiation
Temperature gradient
Thermal current
Coefficient of thermal conductivity
Thermal resistance
R factor
Stefan–Boltzmann law
Emissivity
Stefan's constant
Blackbody
Wien's displacement law
Thermograph
Newton's law of cooling
First law of thermodynamics
Free expansion
Quasi-static process
Isobaric expansion
Isothermal expansion
Equipartition theorem
Degree of freedom
Dulong–Petit law
Adiabatic process

C. True or false: If the statement is true, explain why it is true. If it is false, give a counterexample.

1. During a phase change, the temperature of a substance remains constant.

2. The heat capacity of a body is the amount of heat it can store at a given temperature.

3. The rate of conduction of thermal energy is proportional to the temperature gradient.

4. The rate at which an object radiates energy is proportional to the square of its absolute temperature.

5. When a system goes from state 1 to state 2, the amount of heat added to the system is the same for all processes.

6. When a system goes from state 1 to state 2, the work done on the system is the same for all processes.

7. When a system goes from state 1 to state 2, the change in the internal energy of the system is the same for all processes.

8. The internal energy of a given amount of an ideal gas depends only on its absolute temperature.

9. A quasi-static process is one in which there is no motion.

10. For any material that expands when heated, C_P is greater than C_V.

Problems

Level I

16-1 Heat Capacity and Specific Heat

1. A man typically consumes food having a total energy value of 2500 kcal each day. (*a*) How many joules of energy is this? (*b*) Assuming that this energy is dissipated at a steady rate during 24 h, calculate the person's power output in watts.

2. A 200-g piece of lead is heated to 90°C and is then dropped into a calorimeter containing 500 g of water that is initially at 20°C. Neglecting the heat capacity of the container, find the final temperature of the lead and water.

3. The specific heat of a certain metal can be determined by measuring the temperature change that occurs when a heated piece of the metal is placed in an insulated container made of the same material and containing water. The piece of metal has a mass of 100 g and is initially at 100°C. The container has a mass of 200 g and contains 500 g of water at an initial temperature of 20.0°C. The final temperature is 21.4°C. What is the specific heat of the metal?

4. A solar home contains 10^5 kg of concrete (specific heat = 1.00 kJ/kg·K). How much heat is given off by the concrete when it cools from 25 to 20°C?

16-2 Change of Phase and Latent Heat

5. A 200-g piece of ice at 0°C is placed in 500 g of water at 20°C. The system is in a container of negligible heat capacity and insulated from its surroundings. (*a*) What is the final equilibrium temperature of the system? (*b*) How much of the ice melts?

6. How much heat must be removed when 100 g of steam at 150°C is cooled and frozen into 100 g of ice at 0°C? (Take the specific heat of steam to be 2.01 kJ/kg·K.)

7. If 500 g of molten lead at 327°C is poured into a cavity in a large block of ice at 0°C, how much of the ice melts?

16-3 The Transfer of Thermal Energy

8. A copper bar 2 m long has a circular cross section of radius 1 cm. One end is kept at 100°C and the other end is kept at 0°C. The surface of the bar is insulated so that there is negligible heat loss through it. Find (*a*) the thermal resistance of the bar, (*b*) the thermal current I, (*c*) the temperature gradient $\Delta T/\Delta x$, and (*d*) the temperature of the bar 25 cm from the hot end.

9. A 20- by 30-ft slab of insulation has an R factor of 11. How much heat (in Btu per hour) is conducted through the slab if the temperature on one side is 68°F and that on the other side is 30°F?

10. Two metal cubes, one copper (Cu) and one aluminum (Al), with 3-cm edges are arranged as shown in Figure 16-19. Find (*a*) the thermal resistance of each cube, (*b*) the thermal resistance of the two-cube system, (*c*) the thermal current I, and (*d*) the temperature at the interface of the two cubes.

Figure 16-19 Two metal cubes in series for Problem 10.

11. The same cubes as in Problem 10 are arranged as shown in Figure 16-20. Find (*a*) the thermal current carried by each cube from one side to the other, (*b*) the total thermal current, and (*c*) the equivalent thermal resistance of the two-cube system.

Figure 16-20 Two metal cubes in parallel for Problem 11.

12. The heating wires of a 1-kW electric heater are "red hot" at a temperature of 900°C. Assuming that 100 percent of the heat output is due to radiation and that the wires act as blackbody radiators, what is the effective area of the radiating surface? (Assume a room temperature of 20°C.)

13. Calculate $\lambda_{\max}$ for a human blackbody radiator, assuming the surface temperature of the skin to be 33°C.

16-4 The First Law of Thermodynamics

14. A lead bullet initially at 30°C just melts upon striking a target. Assuming that all of the initial kinetic energy of the bullet goes into the internal energy of the bullet to raise its temperature and melt it, calculate the speed of the bullet upon impact.

15. If 400 kcal is added to a gas that expands and does 800 kJ of work, what is the change in the internal energy of the gas?

16. A lead bullet moving at 200 m/s is stopped in a block of wood. Assuming that all of the energy change goes into heating the bullet, find the final temperature of the bullet if its initial temperature is 20°C.

17. (a) At Niagara Falls, the water drops 50 m. If the change in potential energy goes into the internal energy of the water, compute the increase in its temperature. (b) Do the same for Yosemite Falls, where the water drops 740 m. (These temperature rises are not observed because the water cools by evaporation as it falls.)

16-5 The Internal Energy of an Ideal Gas

18. A certain gas consists of ions that repel each other. The gas undergoes a free expansion with no heat exchange and no work done. How does the temperature of the gas change? Why?

16-6 Work and the *PV* Diagram for a Gas

In Problems 19 through 22, the initial state of 1 mol of an ideal gas is $P_1 = 3$ atm, $V_1 = 1$ L, and $U_1 = 456$ J. Its final state is $P_2 = 2$ atm, $V_2 = 3$ L, and $U_2 = 912$ J. All the processes are quasi-static.

19. The gas is allowed to expand at constant pressure to a volume of 3 L. It is then cooled at constant volume until its pressure is 2 atm. (a) Show this process on a *PV* diagram, and calculate the work done by the gas. (b) Find the heat added during this process.

20. The gas is first cooled at constant volume until its pressure is 2 atm. It is then allowed to expand at constant pressure until its volume is 3 L. (a) Show this process on a *PV* diagram, and calculate the work done by the gas. (b) Find the heat added during this process.

21. The gas is allowed to expand isothermally until its volume is 3 L and its pressure is 1 atm. It is then heated at constant volume until its pressure is 2 atm. (a) Show this process on a *PV* diagram, and calculate the work done by the gas. (b) Find the heat added during this process.

22. The gas is heated and is allowed to expand such that it follows a straight-line path on a *PV* diagram from its initial state to its final state. (a) Show this process on a *PV* diagram, and calculate the work done by the gas. (b) Find the heat added during this process.

23. One mole of an ideal gas initially at a pressure of 1 atm and a temperature of 0°C is compressed isothermally and quasi-statically until its pressure is 2 atm. Find (a) the work needed to compress the gas and (b) the heat removed from the gas during the compression.

24. An ideal gas initially at 20°C and 200 kPa has a volume of 4 L. It undergoes a quasi-static, isothermal expansion until its pressure is reduced to 100 kPa. Find (a) the work done by the gas and (b) the heat added to the gas during the expansion.

16-7 Heat Capacities and the Equipartition Theorem

25. The specific heat of air at 0°C is listed in a handbook as having the value of 1.00 J/g·K measured at constant pressure. (a) Assuming air is an ideal gas with a molar mass $M = 29.0$ g/mol, what is its specific heat at 0°C at constant volume? (b) How much internal energy is there in 1 L of air at 0°C?

26. The heat capacity at constant volume of a certain amount of a monatomic gas is 49.8 J/K. (a) Find the number of moles of the gas. (b) What is the internal energy of this gas at $T = 300$ K? (c) What is its heat capacity at constant pressure?

27. One mole of an ideal diatomic gas is heated at constant volume from 300 to 600 K. (a) Find the increase in internal energy, the work done, and the heat added. (b) Find the same quantities if this gas is heated from 300 to 600 K at constant pressure. Use the first law of thermodynamics and your results for part (a) to calculate the work done. (c) Calculate the work done in part (b) directly from $dW = P\,dV$.

28. The Dulong–Petit law was originally used to determine the molecular mass of a substance from its measured heat capacity. The specific heat of a certain solid is measured to be 0.447 kJ/kg·K. (a) Find the molecular mass of the substance. (b) What element is this?

16-8 The Quasi-Static Adiabatic Expansion of a Gas

29. One mole of an ideal gas ($\gamma = \frac{5}{3}$) expands adiabatically and quasi-statically from a pressure of 10 atm and a temperature of 0°C to a pressure of 2 atm. Find (a) the initial and final volumes, (b) the final temperature, and (c) the work done by the gas.

30. An ideal gas at a temperature of 20°C is compressed quasi-statically and adiabatically to half its original volume. Find its final temperature if (a) $C_v = \frac{3}{2}nR$ and (b) $C_v = \frac{5}{2}nR$.

Level II

31. A 200-g aluminum calorimeter can contains 500 g of water at 20°C. Aluminum shot of mass 300 g is heated to 100°C and is then placed in the calorimeter. (a) Using the value of the specific heat of aluminum given in Table 16-1, find the final temperature of the system, assuming that no heat is lost to the surroundings. (b) The error due to heat transfer to or from the surroundings can be minimized if the initial temperature of the water and calorimeter is chosen to be $\frac{1}{2}\Delta t_w$ below room temperature, where Δt_w is the temperature change of the calorimeter and water during the measurement. Then the final temperature is $\frac{1}{2}\Delta t_w$ above room temperature. What should the initial temperature of the water and container be if the room temperature is 20°C?

32. A thermally insulated system consists of 1 mol of a diatomic ideal gas at 100 K and 2 mol of a solid at 200 K that are separated by a rigid insulating wall. Find the equilibrium temperature of the system after the insulating wall is removed, assuming that the solid obeys the Dulong–Petit law.

33. At very low temperatures, the specific heat of a metal is given by $c = aT + bT^3$. For copper, $a = 0.0108$ J/kg·K^2 and $b = 7.62 \times 10^{-4}$ J/kg·K^4. (a) What is the specific heat of copper at 4 K? (b) How much heat is required to heat copper from 1 to 3 K?

34. A 50-g piece of aluminum at 20°C is cooled to −196°C by placing it in a large container of liquid nitrogen at that temperature. How much nitrogen is vaporized? (Assume that the specific heat of aluminum is constant and is equal to 0.90 kJ/kg·K.)

35. A piece of ice is dropped from a height H. (a) Find the minimum value of H such that the ice melts when it makes an inelastic collision with the ground. Assume that all the mechanical energy lost goes into melting the ice. (b) Is it reasonable to neglect the variation in the acceleration of gravity in doing this problem? (c) Comment on the reasonableness of neglecting air resistance. What effect would air resistance have on your answer?

36. An insulated cylinder fitted with a movable piston to maintain constant pressure initially contains 100 g of ice at −10°C. Heat is supplied to the contents at a constant rate by a 100-W heater. Make a graph showing the temperature of the cylinder contents as a function of time starting at $t = 0$, when the temperature is −10°C, and ending when the temperature is 110°C. (Use $c = 2.0$ kJ/kg·K for the average specific heat of ice from −10 to 0°C and of steam from 100 to 110°C.)

37. A 200-g aluminum calorimeter can contains 500 g of water at 20°C. A 100-g piece of ice cooled to −20°C is placed in the calorimeter. (a) Find the final temperature of the system, assuming no heat losses. (Assume that the specific heat of ice is 2.0 kJ/kg·K.) (b) A second 200-g piece of ice at −20°C is added. How much ice remains in the system after it reaches equilibrium? (c) Would your answer to part (b) be different if both pieces of ice were added at the same time?

38. A well-insulated bucket contains 150 g of ice at 0°C. (a) If 20 g of steam at 100°C is injected into the bucket; what is the final equilibrium temperature of the system? (b) Is any ice left?

39. A six-pack of 12-oz aluminum cans of diet cola originally at 80°F is placed in a well-insulated Styrofoam container. How many 30-g ice cubes must you add to the container so that the final temperature of the cola is 40°F? (Neglect heat losses through the container and the heat removed from the aluminum. The cola is mostly water.)

40. The surface temperature of the filament of an incandescent lamp is 1300°C. If the electric power input is doubled, what will the temperature become? *Hint:* Show that you can neglect the temperature of the surroundings.

41. A copper-bottomed saucepan containing 0.8 L of boiling water boils dry in 10 min. Assuming that all the heat flows through the flat copper bottom, which has a diameter of 15 cm and a thickness of 3.0 mm, calculate the temperature of the outside of the copper bottom while some water is still in the pan.

42. For a boiler at a power station, heat must be transferred to boiling water at the rate of 3 GW. The boiling water passes through copper pipes having a wall thickness of 4.0 mm and a surface area of 0.12 m^2 per meter length of pipe. Find the total length of pipe (actually there are many pipes in parallel) that must pass through the furnace if the steam temperature is 225°C and the external temperature of the pipes is 600°C.

43. A hemispherical igloo made of compacted snow has an inside radius of 2 m. It is desired to maintain an inside temperature of 20°C when the outside temperature is −20°C. The heat generated by the igloo inhabitants is 38 MJ/day. How thick must the igloo walls be if the thermal conductivity of the compacted snow is 0.209 W/m·K. As an approximation, assume that the inner surface area of the igloo is equal to the outer surface area.

44. A blackened, solid copper sphere of radius 4.0 cm hangs in a vacuum in an enclosure whose walls have a temperature of 20°C. If the sphere is initially at 0°C, find the rate at which its temperature changes, assuming that heat is transferred by radiation only.

45. Liquid helium is stored at its boiling point (4.2 K) in a spherical can that is separated by a vacuum space from a surrounding shield that is maintained at the temperature of liquid nitrogen (77 K). If the can is 30 cm in diameter and is blackened on the outside so it acts as a blackbody, how much helium boils away per hour?

46. The solar constant is the power received from the sun per unit area perpendicular to the sun's rays at the mean distance of the earth from the sun. Its value at the upper atmosphere of the earth is about 1.35 kW/m^2. Calculate the effective temperature of the sun if it radiates like a blackbody. (The radius of the sun is 6.96×10^8 m.)

47. A 20-kg block of ice at 0°C slides down a 5-m plane inclined at 30°. The coefficient of sliding friction between the ice and the plane is 0.05. Calculate the amount of ice that melts due to friction. (Assume that all the mechanical energy lost goes into melting the ice.)

48. It is possible to warm your hands on a cold day by rubbing them together. (a) Assume that the coefficient of friction between your hands is 0.5, that the normal force between your hands is 35 N, and that you rub them together at an average speed of 35 cm/s. What is the rate at which heat is generated? (b) Assume further that the mass of each of your hands is approximately 350 g, that the specific heat of your hands is about 4 kJ/kg·K, and that all the heat generated goes into raising the temperature of your hands. How long must you rub your hands together to produce a 5-C° increase in their temperature?

49. When an ideal gas undergoes a temperature change at constant volume, its energy changes by $\Delta U = C_v \, \Delta T$. (a) Explain why the result $\Delta U = C_v \, \Delta T$ holds for an ideal gas for any temperature change independent of the process. (b) Show explicitly that this result holds for the expansion of an ideal gas at constant pressure by first calculating the work done and showing that it can be written as $W = nR \, \Delta T$, and then by using $\Delta U = Q - W$ where $Q = C_p \, \Delta T$.

50. The heat capacity of a certain amount of a particular gas at constant pressure is greater than that at constant volume by 29.1 J/K. (*a*) How many moles of the gas are there? (*b*) If the gas is monatomic, what are C_v and C_p? (*c*) If the gas consists of diatomic molecules that rotate but do not vibrate, what are C_v and C_p?

51. One mole of a monatomic ideal gas is initially at 273 K and 1 atm. (*a*) What is its initial internal energy? Find its final internal energy and the work done by the gas when 500 J of heat are added (*b*) at constant pressure and (*c*) at constant volume.

52. One mole of an ideal monatomic gas is heated at constant volume from 300 to 600 K. (*a*) Find the heat added, the work done by the gas, and the change in its internal energy. (*b*) Find these same quantities if the gas is heated from 300 to 600 K at constant pressure.

53. One-half mole of helium is expanded adiabatically and quasi-statically from an initial pressure of 5 atm and temperature of 500 K to a final pressure of 1 atm. Find (*a*) the final temperature, (*b*) the final volume, (*c*) the work done by the gas, and (*d*) the change in the internal energy of the gas.

54. One mole of N_2 ($C_v = \frac{5}{2}R$) gas is originally at room temperature (20°C) and a pressure of 5 atm. It is allowed to expand adiabatically and quasi-statically until its pressure equals the room pressure of 1 atm. It is then heated at constant pressure until its temperature is again 20°C. During this heating, the gas expands. After it reaches room temperature, it is heated at constant volume until its pressure is 5 atm. It is then compressed at constant pressure until it is back to its original state. (*a*) Construct an accurate *PV* diagram showing each process in the cycle. (*b*) From your graph, determine the work done by the gas during the complete cycle. (*c*) How much heat is added or subtracted from the gas during the complete cycle? (*d*) Check your graphical determination of the work done by the gas in part (*b*) by calculating the work done during each part of the cycle.

55. Two moles of an ideal monatomic gas have an initial pressure $P_1 = 2$ atm and an initial volume $V_1 = 2$ L. The gas is taken through the following quasi-static cycle: It is expanded isothermally until it has a volume $V_2 = 4$ L. It is then heated at constant volume until it has a pressure $P_3 = 2$ atm. It is then cooled at constant pressure until it is back to its initial state. (*a*) Show this cycle on a *PV* diagram. (*b*) Calculate the heat added and the work done by the gas during each part of the cycle. (*c*) Find the temperatures T_1, T_2, and T_3.

56. Heat in the amount of 500 J is supplied to 2 mol of an ideal diatomic gas. (*a*) Find the change in temperature if the pressure is kept constant. (*b*) Find the work done by the gas. (*c*) Find the ratio of the final volume of the gas to the initial volume if the initial temperature is 20°C.

57. A 100-g piece of copper is heated in a furnace to a temperature *t*. The copper is then inserted into a 150-g copper calorimeter containing 200 g of water. The initial temperature of the water and calorimeter is 16°C, and the final temperature after equilibrium is established is 38°C. When the calorimeter and its contents are weighed, 1.2 g of water are found to have evaporated. What was the temperature *t*?

Level III

58. A small pond has a layer of ice 1 cm thick floating on its surface. (*a*) If the air temperature is −10°C, find the rate in centimeters per hour at which ice is added to the bottom of the layer. The density of ice is 0.917 g/cm³. (*b*) How long does it take for a 20-cm layer to be built up?

59. A hand pump is used to inflate a bicycle tire to a gauge pressure of 70 lb/in². How much work must be done if each stroke of the pump is an adiabatic process? Atmospheric pressure is 1 atm, the air temperature is initially 20°C, and the volume of the air in the tire remains constant at 1 L.

60. A steam pipe of length *L* is insulated with a layer of material of thermal conductivity *k*. Find the rate of heat transfer if the temperature outside the insulation is t_1, that on the inside is t_2, the outside radius of the insulation is r_1, and the inside radius is r_2.

61. In an isothermal expansion, an ideal gas at an initial pressure P_0 expands until its volume is twice its initial volume. (*a*) Find its pressure after the expansion. (*b*) The gas is the compressed adiabatically and quasi-statically back to its original volume, at which point its pressure is $1.32P_0$. Is the gas monatomic, diatomic, or polyatomic? (*c*) How does the translational kinetic energy of the gas change in these processes?

62. Brine at −16°C circulating through copper pipes with walls 1.5 mm thick is used to keep a cold room at 0°C. The diameter of each pipe is very large compared to the thickness of its walls. By what fraction is the transfer of heat reduced when the pipes are coated with a 5-mm layer of ice?

63. A body initially at a temperature T_i cools in a room where the temperature is T_0 by convection and radiation. The body obeys Newton's law of cooling, which can be written

$$\frac{dQ}{dt} = hA(T - T_0)$$

where *A* is the area of the body and *h* is a constant called the surface coefficient of heat transfer. Show that the temperature *T* at any time *t* is given by

$$T = T_0 + (T_i - T_0)e^{-hAt/mc}$$

where *m* is the mass of the body and *c* is its specific heat.

64. Two 200-g copper containers, each holding 0.7 L of water, are connected by a 10-cm copper rod of cross-sectional area 1.5 cm². Initially, one container is at 60°C; the second is maintained at 0°C. (*a*) Show that the temperature t_c of the first container changes over time *t* according to

$$t_c = t_{c0}e^{-t/RC}$$

where t_{c0} is the initial temperature of the first container, *R* is the thermal resistance of the rod, and *C* is the total heat capacity of the container plus the water. (*b*) Evaluate *R*, *C*, and the "time constant" *RC*. (*c*) Show that the total amount of heat *Q* conducted in time *t* is

$$Q = Ct_{c0}(1 - e^{-t/RC})$$

(*d*) Find the time it takes for the temperature of the first container to be reduced to 30°C.

65. An ideal gas at initial volume V_1 and pressure P_1 expands quasi-statically and adiabatically to volume V_2 and pressure P_2. Calculate the work done by the gas directly by integrating $P\,dV$, and show that your result is the same as given by Equation 16-51.

66. Prove that the slope of the adiabatic curve passing through a point on the PV diagram is γ times the slope of the isothermal curve passing through the same point.

67. An ideal gas of n mol is initially at pressure P_1, volume V_1, and temperature T_h. It expands isothermally until its pressure and volume are P_2 and V_2. It then expands adiabatically until its temperature is T_c and its pressure and volume are P_3 and V_3. It is then compressed isothermally until it is at a pressure P_4 and a volume V_4, which is related to its initial volume V_1 by

$$T_c V_4^{\gamma-1} = T_h V_1^{\gamma-1}$$

The gas is then compressed adiabatically until it is back in its original state. (a) Assuming that each process is quasi-static, plot this cycle on a PV diagram. (This cycle is known as the Carnot cycle for an ideal gas.) (b) Show that the heat absorbed Q_h during the isothermal expansion at T_h is

$$Q_h = nRT_h \ln \frac{V_2}{V_1}$$

(c) Show that the heat rejected Q_c by the gas during the isothermal compression at T_c is

$$Q_c = nRT_c \ln \frac{V_3}{V_4}$$

(d) Using the result that, for an adiabatic expansion, $TV^{\gamma-1}$ is constant, show that $V_2/V_1 = V_3/V_4$. (e) The efficiency of a Carnot cycle is defined to be the net work done divided by the heat absorbed Q_h. Using the first law of thermodynamics, show that the efficiency is $1 - Q_c/Q_h$. (f) Using your results from the previous parts of this problem, show that $Q_c/Q_h = T_c/T_h$.

68. According to the Einstein model of a crystalline solid, the molar heat capacity at constant volume is given by

$$C_{mv} = 3R \left(\frac{T_E}{T}\right)^2 \frac{e^{T_E/T}}{(e^{T_E/T} - 1)^2}$$

where T_E is a characteristic temperature called the Einstein temperature and T is the temperature of the solid in kelvins. (a) Show that the Dulong–Petit law, $C_{mv} \approx 3R$, holds for the Einstein model when $T \gg T_E$. (b) For diamond, T_E is approximately 1060 K. If 1 mol of diamond is heated from 300 to 600 K, numerically integrate

$$\Delta U = \int C_{mv}\,dT$$

to find the increase in the internal energy of the diamond.

Chapter 17

The Availability of Energy

Solar energy is directed toward the solar oven at the center by this circular array of reflectors at Barstow, California.

We are continually told to conserve energy. Yet, according to the first law of thermodynamics, energy is always conserved. In a closed system, the total amount of energy cannot change. What, then, does it mean to conserve energy if the total amount of energy in the universe does not change no matter what we do? In this chapter, we will see that the first law of thermodynamics does not tell the whole story. Energy is always conserved, but some forms of energy are more useful than others. The possibility or impossibility of putting energy to use is the subject of the second law of thermodynamics. Simply stated, it is easy to convert mechanical work or the internal energy of a system completely into heat with no other changes, but it is impossible to remove heat or internal energy from a system and convert it completely into mechanical work with no other changes. This experimental fact is one statement of the second law of thermodynamics. There is thus a lack of symmetry in the roles played by heat and work that is not evident from the first law. This lack of symmetry is related to the fact that some processes are irreversible.

A common example of the conversion of mechanical energy into internal energy is motion with kinetic friction. Consider, for instance, a block with some initial kinetic energy that is sliding along a rough table. In this process, mechanical energy, the initial kinetic energy of the block, is converted into the internal energy of the block and the table. The reverse process never occurs, however. The internal energy of the block and table is never spontaneously converted into kinetic energy to send the block sliding along the table while the table and block cool. Yet, such an amazing occurrence would not violate the conservation of energy or any other physical laws we have encountered so far. It does, however, violate the second law of thermodynamics, as we will discuss shortly.

There are many irreversible processes, seemingly quite different from one another, but all related to the second law. Suppose, for example, that we pour a layer of black sand into a jar and cover it with a layer of white sand. If we now shake the jar, the sand becomes mixed. However, we cannot "unshake" the jar. That is, we cannot reverse the shaking until the black sand is again on the bottom and the white on top. Another example of an irreversible process is heat conduction. If we place a hot body in contact with a cold body, heat will flow from the hot body to the cold body until they are at the same temperature. However, the reverse does not occur. Two bodies in contact at the same temperature remain at the same temperature. Heat does not flow from one to the other, making one colder and the other warmer. The second law of thermodynamics summarizes the fact that processes of this type do not occur. There are many different ways of stating the second law. We will study several and see that they are all equivalent.

17-1   Heat Engines and the Second Law of Thermodynamics

The study of the efficiency of heat engines gave rise to the first clear statements of the second law of thermodynamics. The first practical heat engine was the steam engine, invented in the eighteenth century for pumping water out of coal mines. Today, the primary use of steam engines is in generating electrical energy. In a typical steam engine (Figure 17-1), water is heated under high pressure (typically several hundred atmospheres) until it vaporizes into steam at a high temperature (normally around 500°C). The steam expands and does work against a piston. The steam exits at a much lower temperature and is cooled further until it condenses. The water is then pumped back into the boiler to be heated again.

Figure 17-1 Schematic drawing of a steam engine. The high-pressure steam from the boiler does work against the piston. The steam is then exhausted at lower pressure and condenses back into water, which is recycled back to the boiler.

Figure 17-2 illustrates schematically how a simple heat engine performs work in a cyclic process. A cylinder containing gas at pressure P_1 is fitted with a piston. In step (a) the piston is held fixed as the gas is heated to a new pressure P_2. The pressure increases at constant volume as shown in the PV diagram in part (e). A weight w is then added to balance the increase in the force on the piston due to the increase in pressure. In step (b), more heat is added, and the gas is allowed to expand at constant pressure, lifting the

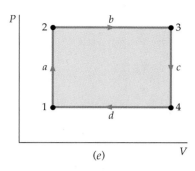

weight to some height h. The volume increases at constant pressure, as shown in part (e). In step (c), the piston is again held fixed as heat is removed, reducing the pressure back to P_1. The pressure decreases at constant volume as shown in part (e). The weight is then removed so that the piston is again in equilibrium. In step (d), the gas is compressed at constant pressure, and heat is exhausted until the gas is back to its original volume. The net result is that an amount of heat $Q_{in} = Q_1 + Q_2$ has been put into the system, some heat $Q_{out} = Q_3 + Q_4$ has been exhausted, and work $W = mgh = Q_{in} - Q_{out}$ has been performed.

Figure 17-3 is a schematic diagram of a practical engine, the internal-combustion engine used in most automobiles. With the exhaust valve closed, a mixture of gasoline vapor and air enters the combustion chamber as the piston moves down during the intake stroke. The mixture is then compressed (compression stroke), after which it is ignited by a spark from the spark plug. The hot gases then expand against the piston, doing work during the power stroke. The gases are then exhausted through the exhaust

Figure 17-2 A simple heat engine that does work in a cycle. (a) A gas enclosed in a cylinder is heated while the piston is held fixed. The pressure increases from P_1 to P_2, as shown in the PV diagram of part (e). A weight w is added so that the piston is in equilibrium. (b) More heat is added, and the gas is allowed to expand at constant pressure, lifting the weight. (c) The piston is held fixed while the gas is cooled back to its original pressure. (d) The weight is removed, and the gas is compressed at constant pressure back to its original state. The net result of the cycle is that work has been done in lifting the weight to some height h. (e) The PV diagram for the simple heat engine in this figure. Work is done by the gas during step b, and on the gas in step d. The net work done by the gas is the shaded area of the diagram.

Figure 17-3 Internal-combustion engine. In (a), a mixture of gasoline vapor and air enters the combustion chamber as the piston moves down. The piston then moves up in (b), compressing the gas for ignition in (c). The hot gases expand, moving the piston down in (d), the power stroke. In (e), the piston moves up again to exhaust the burned gases. The cycle then repeats.

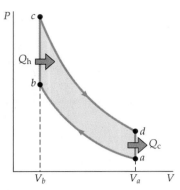

Figure 17-4 Otto cycle, representing the internal-combustion engine. The gasoline–air mixture enters at a and is adiabatically compressed to b. It is then heated (by ignition from the spark plug) at constant volume to c. The power stroke is represented by the adiabatic expansion from c to d. The cooling at constant volume from d to a represents the exhausting of the burned gases and the intake of a fresh gasoline–air mixture.

valve, and the cycle repeats. Figure 17-4 shows an *Otto cycle*, which is an idealized model of the processes that occur in the internal-combustion engine. The gasoline–air mixture enters the cylinder at point a. The compression ab is adiabatic; that is, there is no heat exchange. The combustion of the gases is represented by the heat input Q_h that raises the temperature at constant volume from b to c. The power stroke is an adiabatic expansion from c to d. Heat Q_c leaves the system during the exhaust from d to a. Since an equivalent amount of gasoline vapor and air then enters at a, we can view the process as though the same gasoline–air mixture were used again. The net work done during the cycle is indicated by the shaded area in the figure.

The important features of any **heat engine** are that a substance or system called the **working substance** (water for a steam engine, air and gasoline vapor for an internal-combustion engine) absorbs a quantity of heat Q_h at a high temperature T_h, does work W, and rejects heat $|Q_c|^*$ at a lower temperature T_c. The working substance then returns to its original state. The heat engine is therefore a cyclic device.

A heat engine's purpose is to convert as much heat input Q_h into work as possible. Figure 17-5 shows a heat engine in schematic form. The heat input Q_h comes from a heat reservoir at temperature T_h; the exhaust $|Q_c|$ goes into a heat reservoir at a lower temperature T_c. A **heat reservoir** is an idealized system with a heat capacity so large it can absorb or give off heat with no appreciable change in temperature. In practice, the surrounding atmosphere or lakes often act as heat reservoirs. Since the initial and final states of the engine and working substance are the same, the final internal energy of the working substance must equal its initial internal energy. Then, by the first law of thermodynamics, the work done equals the net heat absorbed:

$$W = Q_h - |Q_c| \qquad \text{17-1}$$

The **efficiency** ϵ is defined as the ratio of the work done to the heat absorbed from the hot reservoir:

Efficiency defined

$$\epsilon = \frac{W}{Q_h} = \frac{Q_h - |Q_c|}{Q_h} = 1 - \frac{|Q_c|}{Q_h} \qquad \text{17-2}$$

Since the heat Q_h is usually produced by burning coal, oil, or some other kind of fuel that must be paid for, heat engines are designed to have the greatest possible efficiency. The efficiencies of steam engines are typically about 40 percent, and those of internal-combustion engines may be as high as 50 percent. We can see from Equation 17-2 that we want to reject as small a fraction of the heat absorbed as possible. For perfect efficiency ($\epsilon = 1 = 100$ percent), $|Q_c| = 0$. In other words, all the heat absorbed from the hot reservoir would be converted into work and no heat would be rejected to the cold reservoir.

The development and propagation of the flame in a piston engine is studied in this series of computer-enhanced schlieren photographs. The ignition begins at the left, and the flame kernel develops to the right.

*According to our sign convention for the first law of thermodynamics, the heat rejected by a system is negative. Since we are interested here only in the magnitudes of the heat absorbed or rejected, we will use absolute value signs where needed.

Although the efficiency of heat engines has been greatly increased since the early steam engines, it is impossible to make a perfect heat engine, that is, an engine with 100 percent efficiency.

> It is impossible for a heat engine working in a cycle to produce no other effect than that of extracting heat from a reservoir and performing an equivalent amount of work.

Second law of thermodynamics: heat-engine statement

This experimental result is the Kelvin–Planck or **heat-engine statement of the second law of thermodynamics.** It is important to include the word "cycle" in this statement because it is possible to convert heat completely into work in a noncyclic process. An ideal gas undergoing an isothermal expansion does just this. But after the expansion, the gas is not in its original state. To bring the gas back to its original state, work must be done on the gas, and some heat will be exhausted. Essentially, the second law tells us that if we want to extract energy from a heat reservoir to do work, we must have available a colder reservoir to which part of the energy can be exhausted. If this were not true, we could design a heat engine for a ship that would extract energy from the ocean (a convenient heat reservoir with an enormous supply of energy) and use it to power the ship. Unfortunately, this enormous reservoir of energy is not available for such use unless a colder reservoir is available for the exhaust.

Figure 17-5   Schematic representation of a heat engine. The engine removes heat energy Q_h from a hot reservoir at a temperature T_h, does work W, and rejects heat $|Q_c|$ to a cold reservoir at a temperature T_c.

Example 17-1

A heat engine absorbs 200 J of heat from a hot reservoir, does work, and exhausts 160 J to a cold reservoir. What is its efficiency?

By the first law, the work done is

$$W = Q_h - |Q_c| = 200 \text{ J} - 160 \text{ J} = 40 \text{ J}$$

The efficiency is thus

$$\epsilon = \frac{W}{Q_h} = \frac{40 \text{ J}}{200 \text{ J}} = 0.20 = 20\%$$

Exercise

A heat engine has an efficiency of 35 percent. (*a*) How much work does it perform in a cycle if it extracts 150 J of heat from a hot reservoir per cycle? (*b*) How much heat is exhausted per cycle?
[Answers: (*a*) 52.5 J, (*b*) 97.5 J]

Questions

1. Where does the energy come from in an internal-combustion engine? In a steam engine?

2. How does friction in an engine affect its efficiency?

17-2 Refrigerators and the Second Law of Thermodynamics

A **refrigerator** is essentially a heat engine run backwards. Work is put into the refrigerator to extract heat from a cold reservoir and transfer it to a hot reservoir. Figure 17-6 is a schematic representation of a refrigerator. It is desirable to remove as much heat Q_c as possible while doing as little work W as possible. It is found from experience that some work must always be done. This result is the Clausius or **refrigerator statement of the second law of thermodynamics:**

Second law of thermodynamics: refrigerator statement

> It is impossible for a refrigerator working in a cycle to produce no other effect than the transfer of heat from a cold object to a hot object.

If this statement were not true, it would be possible in principle to cool our homes in the summer with a refrigerator that pumps heat to the outside without using any electricity or other energy.

A measure of the performance of a refrigerator is the ratio Q_c/W, which is called the **coefficient of performance** COP:

Coefficient of performance defined

$$\text{COP} = \frac{Q_c}{W} \qquad\qquad 17\text{-}3$$

The greater the coefficient of performance, the better the refrigerator. Typical refrigerators have coefficients of performance of about 5 or 6. In terms of this ratio, the refrigerator statement of the second law says that the coefficient of performance of a refrigerator cannot be infinite (because W cannot be zero).

Example 17-2

A refrigerator has a coefficient of performance of 5.5. How much work is needed to make ice cubes from 1 L of water at 10°C?

The mass of 1 L of water is 1 kg. The heat energy that must be removed to reduce the temperature of the water by 10 C° = 10 K is

$$Q_1 = mc\,\Delta T = (1\ \text{kg})(4.18\ \text{kJ/kg·K})(10\ \text{K}) = 41.8\ \text{kJ}$$

The heat of fusion of water is 333.5 kJ/kg. The heat that must be removed to freeze 1 L of water into ice cubes is therefore $Q_2 = 333.5$ kJ. The total amount of heat that must be removed is thus

$$Q_c = Q_1 + Q_2 = 41.8\ \text{kJ} + 333.5\ \text{kJ} = 375.3\ \text{kJ} \approx 375\ \text{kJ}$$

From Equation 17-3, the work needed is

$$W = \frac{Q_c}{\text{COP}} = \frac{375\ \text{kJ}}{5.5} = 68.2\ \text{kJ}$$

Figure 17-6 Schematic representation of a refrigerator. The refrigerator removes heat energy Q_c from a cold reservoir and rejects heat $|Q_h|$ to a hot reservoir using work W.

Exercise

A refrigerator has a coefficient of performance of 4.0. How much heat is exhausted to the hot reservoir when 200 kJ of heat are removed from the cold reservoir?   (Answer: 250 kJ)

Question

3.   To cool his kitchen, a man leaves his refrigerator door open. Explain why this does not have the desired effect and, in fact, heats the kitchen.

17-3   Equivalence of the Heat-Engine and Refrigerator Statements

Although the heat-engine and refrigerator statements of the second law of thermodynamics seem quite different, they are actually equivalent. That is, if one of the statements is true, the other must also be true. This equivalence can be proven by showing that if either statement is assumed to be false, the other must also be false. We will use a numerical example to show that if the heat-engine statement is false, the refrigerator statement must also be false.

Figure 17-7a shows an ordinary refrigerator that uses 50 J of work to remove 100 J of energy from a cold reservoir and reject 150 J of energy to a hot reservoir. Its coefficient of performance is 2.0. If the heat-engine statement of the second law were not true, we could have a perfect heat engine that could remove energy from a single reservoir and convert it completely into work with 100 percent efficiency. We could use this perfect heat engine to remove 50 J of energy from the hot reservoir and do 50 J of work (Figure 17-7b). Then, if we used the perfect heat engine in conjunction with the ordinary refrigerator, we could construct a perfect refrigerator that would transfer 100 J of energy from the cold reservoir to the hot reservoir without requiring any work, as illustrated in Figure 17-7c. This would violate the refrigerator statement of the second law. Thus, if the heat-engine statement is false, the refrigerator statement must also be false. It can be similarly shown that if a perfect refrigerator existed, it could be used in conjunction with an ordinary heat engine to construct a perfect heat engine; see Problem 6. Thus, if the refrigerator statement is false, the heat-engine statement must also be false. It then follows that if either statement is true, the other must also be true. Therefore, the two statements are equivalent.

Figure 17-7   (a) An ordinary refrigerator that removes 100 J from a cold reservoir and rejects 150 J to the hot reservoir, requiring the input of 50 J of work. (b) A perfect heat engine that violates the heat-engine statement of the second law by removing 50 J from the hot reservoir and converting it completely into work. (c) The ordinary refrigerator in (a) and the perfect heat engine in (b) working together make a perfect refrigerator that violates the refrigerator statement of the second law by transferring 100 J of energy from the cold reservoir to the hot reservoir with no other effect.

17-4   The Carnot Engine

We have seen that, according to the second law of thermodynamics, it is impossible for a heat engine working between two heat reservoirs to be 100 percent efficient. What, then, is the maximum possible efficiency for such an engine? This question was answered in 1824 by a young French engineer, Sadi Carnot, before the first law of thermodynamics was established. Carnot found that all reversible engines working between two heat reservoirs have the same efficiency and that no engine could have a greater efficiency than that of a *reversible engine*. This result is known as the **Carnot theorem:**

No engine working between two given heat reservoirs can be more efficient than a reversible engine working between those reservoirs.

Carnot theorem

A reversible engine working between two heat reservoirs is called a **Carnot engine.**

Let us look at what makes a process reversible or irreversible. We have already seen that the conversion of mechanical energy into heat by friction is irreversible. The reverse process of converting heat into mechanical energy with no other effects violates the heat-engine statement of the second law. Similarly, the conduction of heat from a hot object to a cold object is irreversible. The reverse process, the conduction of heat from a cold body to a hot body with no other effect, violates the refrigerator statement of the second law. A third type of irreversibility occurs when a system passes through nonequilibrium states, such as when there is turbulence in a gas or an explosion. The free expansion of a gas discussed in Section 16-5 is an example of this type of irreversibility. We never observe a free compression in which a gas in a large container moves by itself into the left or right half of the container. For a process to be reversible, we must be able to move the system back through the same equilibrium states in the reverse order.

From these considerations and our statements of the second law of thermodynamics, we can list some conditions that are necessary for a process to be reversible:

1. No mechanical energy can be lost due to friction, viscous forces, or other dissipative forces that produce heat.

Conditions for reversibility   2. There can be no heat conduction due to a finite temperature difference.

3. The process must be quasi-static so that the system is always in an equilibrium state (or infinitesimally near an equilibrium state).

Any process that violates any of the above conditions is irreversible. Most processes in nature are irreversible. To have a reversible process, great care must be taken to eliminate frictional and other dissipative forces and to make the process quasi-static. Since this can never be done completely, a reversible process seems impossible in practice. Nevertheless, one can come very close to a reversible process, and the concept is very important in theory.

We now illustrate the Carnot theorem with a numerical example in which we show that if we assume that an engine exists that is more efficient than a reversible engine, the second law of thermodynamics is violated.

The engine in Figure 17-8a is defined to be reversible. It removes 100 J from the hot reservoir, does 40 J of work, and exhausts 60 J to the cold reservoir. Its efficiency is therefore 40 percent. In Figure 17-8b, the engine is reversed and run backwards as a refrigerator. Now, 40 J of work are done to remove 60 J from the cold reservoir and exhaust 100 J to the hot reservoir. (If

Figure 17-8 Illustration of the Carnot theorem. (a) A reversible heat engine with 40 percent efficiency. (b) The same engine runs backwards as a refrigerator. (c) An assumed heat engine working between the same two reservoirs with an efficiency of 45 percent, which is greater than that of the reversible engine in (a). (d) The net effect of running the engine in (c) in conjunction with the refrigerator in (b) is the same as that of a perfect heat engine that removes 5 J from the cold reservoir and converts it completely into work with no other effect. This is impossible because it violates the heat-engine statement of the second law. Thus, the reversible engine in (a) is the most efficient engine that can operate between these two reservoirs.

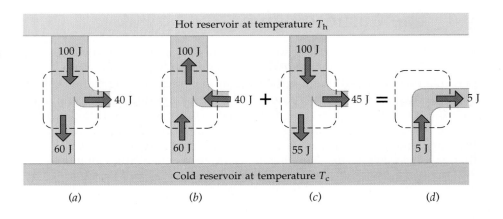

the engine were not reversible, more than 40 J of work would have to be done to remove 60 J from the cold reservoir.) Now, let us assume that there is another (nonreversible) engine working between the same two reservoirs that is more than 40 percent efficient. Figure 17-8c shows an engine with an efficiency of 45 percent. If we run this engine in conjunction with the refrigerator in Figure 17-8b, the net effect would be the same as that of a perfect engine that removes 5 J from the cold reservoir and converts it completely into work (Figure 17-8d). We have already seen that this is not possible because it violates the heat-engine statement of the second law of thermodynamics. Therefore, the reversible engine is the most efficient engine that can work between these two reservoirs.

If no engine can have a greater efficiency than a reversible engine, it follows that all reversible engines working between the same two reservoirs must have the same efficiency, which is called the **Carnot efficiency.** Because this efficiency is the same for all reversible engines working between these reservoirs, it must be independent of the working substances of the engines. Therefore, the Carnot efficiency depends only on the temperatures of the reservoirs.

We can calculate the Carnot efficiency in terms of the temperatures of the reservoirs by considering a reversible engine that has an ideal gas as the working substance. A reversible cycle is called a **Carnot cycle.** Figure 17-9 shows a Carnot cycle for an ideal gas. The cycle starts at state 1 with a quasi-static isothermal absorption of heat from the hot reservoir at a temperature T_h. Since the heat is absorbed isothermally, the process could be reversed without violating the second law of thermodynamics. Work is done by the gas as it expands to state 2. From state 2 to state 3, the gas expands adiabatically, that is, with no heat exchange. If this expansion is done quasistatically, it will be reversible. More work is done by the gas, and the temperature of the gas is reduced to T_c. The third part of the cycle is an isothermal compression at temperature T_c from state 3 to state 4. During this part of the cycle, work is done on the gas, and heat $|Q_c|$ is rejected to the cold reservoir at a temperature T_c. The last part of the cycle is an adiabatic compression from state 4 back to the original state 1. Work is again done on the gas during this compression. The net work done during the cycle is represented by the shaded area in the figure.

To calculate the efficiency of this cycle, we calculate the heat absorbed from the hot reservoir and that rejected to the cold reservoir. We first consider the isothermal expansion from state 1 to state 2. Since the internal energy of an ideal gas remains constant during an isothermal expansion, the work done by the gas equals the heat absorbed from the hot reservoir. The heat absorbed is thus

$$Q_h = W = \int_1^2 P\,dV = \int_1^2 \frac{nRT_h}{V}\,dV$$

or

$$Q_h = nRT_h \ln \frac{V_2}{V_1}$$

Similarly, the heat rejected to the cold reservoir equals the work done on the gas during the isothermal compression at temperature T_c from state 3 to state 4. This work is of the same magnitude as that done by the gas if it expands from state 4 to state 3. The heat rejected is thus

$$|Q_c| = nRT_c \ln \frac{V_3}{V_4}$$

Exhaust manifolds in the 8-cylinder Dusenberg carry heat from the engine to reduce its temperature. A typical automobile engine has an efficiency of less than 15 percent.

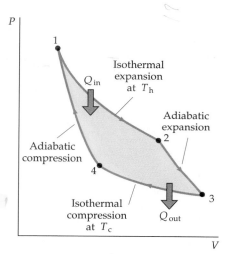

17-9 A Carnot cycle for an ideal gas. Heat is absorbed during the isothermal expansion from state 1 to state 2 at a temperature T_h. The gas expands adiabatically from state 2 to state 3 until its temperature is reduced to T_c. It then rejects heat as it is compressed isothermally to state 4, after which it is compressed adiabatically until its temperature is again T_h. All the processes are reversible.

The ratio of these heats is

$$\frac{|Q_c|}{Q_h} = \frac{T_c \ln (V_3/V_4)}{T_h \ln (V_2/V_1)} \qquad \text{17-4}$$

We can relate the volumes V_1, V_2, V_3, and V_4 using Equation 16-47 for an quasi-static adiabatic expansion:

$$TV^{\gamma-1} = \text{constant}$$

Applying this equation to the expansion from state 2 to state 3, we obtain

$$T_h V_2{}^{\gamma-1} = T_c V_3{}^{\gamma-1}$$

Similarly, for the adiabatic compression from state 4 to state 1, we have

$$T_h V_1{}^{\gamma-1} = T_c V_4{}^{\gamma-1}$$

Dividing these two equations, we obtain

$$\left(\frac{V_2}{V_1}\right)^{\gamma-1} = \left(\frac{V_3}{V_4}\right)^{\gamma-1}$$

and so $V_2/V_1 = V_3/V_4$. Then $\ln (V_2/V_1) = \ln (V_3/V_4)$, so we can cancel the logarithmic terms in Equation 17-4. We thus obtain

$$\frac{|Q_c|}{Q_h} = \frac{T_c}{T_h} \qquad \text{17-5}$$

The Carnot efficiency ϵ_C is thus

Carnot efficiency

$$\epsilon_C = 1 - \frac{|Q_c|}{Q_h} = 1 - \frac{T_c}{T_h} \qquad \text{17-6}$$

Equation 17-6 holds for any reversible engine working between reservoirs at the temperatures T_h and T_c. It gives the greatest possible efficiency for an engine working between these temperatures. No engine can have an efficiency greater than the Carnot efficiency. As we have seen, an engine with a greater efficiency than this would violate the second law of thermodynamics.

In Section 15-2, the ideal-gas temperature scale was defined in terms of the properties of gases at low densities. Since the Carnot efficiency depends only on the temperatures of the two heat reservoirs, it can be used to define the ratio of the temperatures of the reservoirs independent of any substance. We define the ratio of the absolute temperatures of the hot and cold reservoirs to be

$$\frac{T_c}{T_h} = 1 - \epsilon_C = \frac{|Q_c|}{Q_h} \qquad \text{17-7}$$

To use this definition to measure the ratio of two temperatures, we have to set up a reversible engine to operate between two reservoirs at the two temperatures and carefully measure the heats absorbed from or rejected to those reservoirs during one cycle. Temperatures as low as a millionth of a kelvin have been measured in this way. The absolute temperature is completely determined by defining the ratio of two temperatures as in Equation 17-7 and choosing one fixed point, as was done for the ideal-gas temperature scale. If the absolute temperature of the triple point of water is defined to be 273.16 K, then the absolute temperature scale is identical to the ideal-gas temperature scale for the range of temperatures for which a gas thermometer can be used.

Example 17-3

A steam engine works between a hot reservoir at 100°C = 373 K and a cold reservoir at 0°C = 273 K. What is the maximum possible efficiency of this engine?

From Equation 17-6, the Carnot efficiency of the engine is

$$\epsilon_C = 1 - \frac{T_c}{T_h}$$

$$= 1 - \frac{273\ K}{373\ K}$$

$$= 0.268 = 26.8\%$$

Even though this efficiency seems to be quite low, it is the greatest efficiency possible for any engine working between these temperatures. Real engines will have lower efficiencies because of friction, heat conduction, and other irreversible processes.

We can see that it is very useful to know the Carnot efficiency. Suppose, for example, that we have a real engine that removes 100 J from a hot reservoir at 373 K, does 25 J of work, and exhausts 75 J to a cold reservoir at 273 K. Though its efficiency is only 25 percent, this is a very good engine. If we want to improve it by reducing friction and eliminating various other irreversibilities, the best efficiency we could obtain is 26.8 percent, as we calculated in Example 17-3. That is, the most work we could possibly get using an ideal engine between these reservoirs would be 26.8 J of work for every 100 J removed from the hot reservoir.

The ratio of the actual efficiency ϵ of an engine to the Carnot efficiency ϵ_C is called the **second-law efficiency** ϵ_{SL}:

$$\epsilon_{SL} = \frac{\text{actual efficiency}}{\text{Carnot efficiency}} = \frac{\epsilon}{\epsilon_C} \qquad 17\text{-}8$$

The second-law efficiency of a real engine that has an efficiency of 25 percent when working between 373 K and 273 K is

$$\epsilon_{SL} = \frac{25\%}{26.8\%} = 0.93 = 93\%$$

Example 17-4

An engine removes 200 J from a hot reservoir at 373 K, does 48 J of work, and exhausts 152 J to a cold reservoir at 273 K. How much work is "lost" per cycle due to the irreversibilities of this engine?

The efficiency of this engine is ϵ = (48 J)/(200 J) = 24 percent. As we just saw in Example 17-3, the Carnot efficiency for an engine working between these two reservoirs is 26.8 percent, so this is a very good engine. Its second-law efficiency is 24/26.8 = 90 percent. A Carnot engine working between these reservoirs that removes 200 J from the hot reservoir could do 0.268 × 200 = 53.6 J of work per cycle. Since the real engine does 48 J of work per cycle, it could do only 5.6 J more work if it were an ideal engine. Thus, only 5.6 J of work is lost per cycle by this engine. This energy is, of course, not lost completely. The total energy is conserved. However, the real engine exhausts 5.6 J of energy into the cold reservoir that could have been converted into useful work if the engine were ideal, that is, reversible.

(a)   (b)

(a) Coal-fueled electric generating plant at Four Corners, New Mexico. (b) Power plant at Wairakei, New Zealand, that converts geothermal energy into electricity. (c) Solar energy is focused and collected individually to produce electricity by these heliostats that are being tested at Sandia National Laboratory. (d) Control rods are inserted into this nuclear reactor at Tihange, Belgium, (e) An experimental wind-powered electric generator at Sandia National Laboratory. The propeller is designed for optimum transfer of wind energy to mechanical energy.

Example 17-5

If 200 J of heat are conducted from a heat reservoir at 373 K to one at 273 K, how much work capability is "lost" in this process?

We saw in the previous example that a Carnot engine working between these two reservoirs could do 53.6 J of work if it extracted 200 J from the 373-K reservoir and exhausted to a 273-K reservoir. Thus, if 200 J is conducted directly from the hot reservoir to the cold reservoir without any work being done, 53.6 J of this energy has been "lost" in the sense that it could have been converted into useful work.

Exercise

A Carnot engine works between heat reservoirs at 500 K and 300 K. (a) What is its efficiency? (b) If it removes 200 kJ of heat from the hot reservoir, how much work does it do?   [Answers: (a) 40%, (b) 80 kJ]

Exercise

A real engine works between heat reservoirs at 500 K and 300 K. It removes 500 kJ of heat from the hot reservoir and does 150 kJ of work during each cycle. (a) What is its efficiency? (b) What is its second-law efficiency?   [Answers: (a) 30%, (b) 75%]

Questions

4. Why do power-plant designers try to increase the temperature of the steam fed to engines as much as possible?

5. Some think that electrical heat for the home is more efficient because all the electrical energy goes into heat in the home, whereas with a gas or oil furnace, some of the heat goes out the chimney. However, electrical energy is usually generated in a steam power plant, which is only about 40 percent efficient. In light of this fact, discuss the relative merits of electrical heat versus oil or gas heat for homes.

(c)

(d)

17-5   The Heat Pump

About 20 percent of our energy use is for the heating of our homes and office buildings. Much of this heating is done by the on-site combustion of fossil fuels (oil, coal, or natural gas). Although such combustion is relatively efficient, it is a major source of pollution in our cities. The alternative of using electric heat seems at first glance to be more efficient and less polluting, but this is actually far from the truth. Although some of our electricity is generated in hydroelectric and nuclear power plants, most comes from power plants that burn fossil fuels. This leads to just as much air pollution as the combustion of fossil fuels in homes and office buildings. In addition, the heat exhausted by power plants results in the thermal pollution of our lakes and rivers and the atmosphere. Furthermore, the Carnot efficiency of a power plant is about 50 percent, and the actual efficiency is only about 40 percent. An efficient alternative to the direct or indirect combustion of fossil fuels as a method of heating homes and office buildings in moderate climates is the heat pump.

(e)

A **heat pump** is essentially a refrigerator that is used to pump thermal energy from a cold reservoir (for example, the cold air outside a house) to a hot reservoir (for example, the hot air inside the house). If work W is done to remove heat Q_c from the cold reservoir and reject heat $|Q_h| = W + Q_c$ to the hot reservoir, the coefficient of performance (Equation 17-3) is

$$\text{COP} = \frac{Q_c}{W} \qquad\qquad 17\text{-}9$$

Using $W = |Q_h| - Q_c$, this can be written

$$\text{COP} = \frac{Q_c}{|Q_h| - Q_c} = \frac{Q_c/|Q_h|}{1 - Q_c/|Q_h|} \qquad\qquad 17\text{-}10$$

The maximum coefficient of performance is obtained using a reversible or Carnot heat pump. Then Q_c and Q_h are related by Equation 17-5:

$$\frac{Q_c}{|Q_h|} = \frac{T_c}{T_h}$$

Substituting $Q_c/|Q_h| = T_c/T_h$ into Equation 17-10, we obtain for the maximum coefficient of performance

$$\text{COP}_{max} = \frac{T_c/T_h}{1 - T_c/T_h} = \frac{T_c}{T_h - T_c}$$

or

$$\text{COP}_{max} = \frac{T_c}{\Delta T} \qquad\qquad 17\text{-}11$$

where ΔT is the difference in temperature between the hot and cold reservoirs. Real heat pumps and refrigerators have COPs less than the Carnot COP because of friction, heat conduction, and other irreversible processes.

We are usually interested in the work that must be done to exhaust a given amount of heat $|Q_h|$ into the hot reservoir. (For a house, the hot reservoir may be the hot-air supply for the heating fan.) Using $|Q_h| = Q_c + W$, we can write Equation 17-9 as

$$\text{COP} = \frac{Q_c}{W} = \frac{|Q_h| - W}{W}$$

$$W(\text{COP}) = |Q_h| - W$$

or

$$W = \frac{|Q_h|}{1 + \text{COP}} \qquad\qquad 17\text{-}12$$

Example 17-6

An ideal heat pump is used to pump heat from the outside air at $-5°C$ to the hot-air supply for the heating fan in a house, which is at $40°C$. The pump is run by electric energy. How much work is required to pump 1 kJ of heat into the house?

From Equation 17-11, the Carnot coefficient of performance for a heat pump operating at $T_c = -5°C = 268$ K and $\Delta T = 40°C - (-5°C) = 45$ C° = 45 K is

$$\text{COP}_{max} = \frac{T_c}{\Delta T} = \frac{268 \text{ K}}{45 \text{ K}} = 5.96$$

We can calculate the work needed from Equation 17-12 with $|Q_h| = 1$ kJ:

$$W = \frac{|Q_h|}{1 + \text{COP}} = \frac{1 \text{ kJ}}{1 + 5.96} = 0.144 \text{ kJ}$$

Thus, only 0.144 kJ of work is needed to pump 1 kJ of heat into the hot-air supply in the house.

We see from this example and Equation 17-12 that the heat pump essentially multiplies the energy needed to run the pump by 1 + COP. That is, if we use 1 kJ to run a heat pump, with a COP = 6.83, we can exhaust 7.83 kJ of heat into the house.

17-6   Entropy and Disorder

We have seen that the second law of thermodynamics is related to the fact that some processes are irreversible; that is, they go in one direction only. However, there are many irreversible processes that are not easily described by the heat-engine or refrigerator statements of the second law. Examples are the free expansion of a gas or a glass falling off a table and shattering when it hits the floor.

All irreversible processes have one thing in common—the system plus its surroundings moves toward a less ordered state. This can be illustrated with a simple example. Consider a box containing a gas of mass M at a temperature T that is moving along a frictionless table with a velocity V_{cm} (Figure 17-10a). The total kinetic energy of the gas has two components. The first is the kinetic energy associated with the movement of the center of mass of the gas, $\frac{1}{2}MV_{cm}^2$. This is ordered mechanical energy that could be converted directly into work. For example, if a weight were attached with a string to the moving box, this energy could be used to lift the weight. The other component of the gas's kinetic energy is the energy of the motion of its molecules relative to its center of mass. This is the gas's internal thermal energy, which is related to its temperature T; it is random, nonordered energy that cannot be converted directly into work.

Now, suppose the box of gas collides inelastically with a fixed wall (Figure 17-10b), which is an irreversible process. Disregarding the box itself, we note that all of the ordered mechanical energy of the gas is converted into random internal energy, causing the temperature of the gas to rise. The gas still has the same total energy, but now all of it is associated with the random motion of its molecules about its center of mass, which is now at rest. Thus, the gas has become less ordered, or more disordered, and it has lost the ability to do work.

 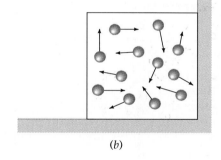

(a)   (b)

Figure 17-10 (a) A box of gas moving to the right with a speed V_{cm}. The kinetic energy of the molecules associated with the motion of the gas's center of mass is ordered energy. It can be converted directly into work. The kinetic energy of the molecules associated with their motion relative to the center of mass is their thermal energy, which is related to the temperature of the gas. This is disordered energy and cannot be converted directly into work. (b) After the box has collided inelastically with the wall, the gas molecules have the same total energy as before, but all of it is now disordered. This is indicated by the higher temperature of the gas.

There is a thermodynamic function called **entropy** S that is a measure of the disorder of a system. Like the pressure P, volume V, temperature T, and internal energy U, entropy is a function of the state of a system. As with internal energy, it is the change in entropy that is important. The change in entropy ΔS of a system when it goes from one state to another is defined as

$$\Delta S = \int \frac{dQ_{rev}}{T}$$

17-13   *Entropy change defined*

where dQ_{rev} is the heat that must be added to the system in a reversible process to bring it from its initial state to its final state. If heat is subtracted from the system, dQ_{rev} is negative and the entropy change of the system is negative.

We will illustrate the existence of the entropy function for an ideal gas. Let us consider an arbitrary reversible quasi-static process in which a system consisting of an ideal gas absorbs an amount of heat dQ. According to the first law, dQ is related to the change in the internal energy dU of the gas and the work done $dW = P\,dV$ by

$$dQ = dU + dW = dU + P\,dV$$

For an ideal gas, we can write dU in terms of the heat capacity, $dU = C_v \, dT$, and we can substitute nRT/V for P from the equation of state. Then

$$dQ = C_v \, dT + nRT \, \frac{dV}{V} \qquad\qquad 17\text{-}14$$

Because W and Q are not state functions, both the work done and the heat absorbed by a system when it goes from one state to another depend on the process by which the change of state occurs. We must therefore know the path of the system on a PV diagram in order to calculate the heat absorbed by the system. This can also be seen mathematically from the fact that we cannot integrate Equation 17-14. There is no problem with the first term on the right side of the equation because $C_v = dU/dT$ is a function of temperature only. The integration of this term gives the change in the internal energy ΔU. However, we cannot integrate the second term unless we know how T depends on V. This is just another way of saying that dQ and dW are not differentials of state functions because Q and W are not state functions. But if we divide each term in Equation 17-14 by T, we obtain

$$\frac{dQ}{T} = C_v \, \frac{dT}{T} + nR \, \frac{dV}{V} \qquad\qquad 17\text{-}15$$

Since C_v depends only on T, the first term on the right can still be integrated and now the second term can be integrated also. (Mathematically, the factor $1/T$ is called an integrating factor for Equation 17-14.) For simplicity, we will assume that C_v is constant. Integrating Equation 17-15, we obtain

$$\Delta S = \int \frac{dQ}{T} = C_v \ln \frac{T_2}{T_1} + nR \ln \frac{V_2}{V_1} \qquad\qquad 17\text{-}16$$

The dQ_{rev} term in Equation 17-13 does not mean that a reversible heat transfer must take place for the entropy of a system to change. Indeed, there are many situations in which the entropy of a system changes when there is no transfer of heat whatsoever. (We will look at an example shortly.) Equation 17-13 simply gives us a method of calculating the entropy difference between two states of a system. Like internal energy, entropy is a state function. Therefore, the change in the entropy of a system when it moves from one state to another depends only on the system's initial and final states, not on the process by which the change occurs. However, to *calculate* the change in entropy for an irreversible process, we must first find a reversible process that connects the initial and final states, then find the heat added or subtracted during this reversible process, and use Equation 17-13. Calculating the entropy changes for some simple examples should clarify this somewhat subtle point.

We first consider a substance that is heated at constant pressure from temperature T_1 to temperature T_2. The amount of heat dQ absorbed by a substance at constant pressure, is related to its temperature change dT by

$$dQ = C_p \, dT$$

As we have seen, heat conduction between two systems that have a finite temperature difference is irreversible. We can approximate reversible heat conduction if we have a large number of heat reservoirs with temperatures ranging from T_1 to T_2 in very small steps. We then place our substance whose initial temperature is T_1 in contact with the first reservoir at a temperature just slightly greater than T_1 and let the substance absorb a small amount of heat. Since the heat transfer is approximately isothermal, the process will be approximately reversible. We then place the substance in contact with the next reservoir at a slightly higher temperature, and so on, until the final temperature T_2 is reached. When the heat is absorbed reversibly, the entropy change of the substance is

$$dS = \frac{dQ}{T} = C_p \frac{dT}{T}$$

Integrating from T_1 to T_2, we obtain the total entropy change of the substance

$$\Delta S = C_p \int_{T_1}^{T_2} \frac{dT}{T} = C_p \ln \frac{T_2}{T_1} \qquad 17\text{-}17$$

Since entropy is a state function, its change is independent of the process. If the final pressure equals the initial pressure, the entropy change of a substance is given by Equation 17-17 when it is heated by any process— whether it is reversible or irreversible. Equation 17-17 also gives the entropy change of a substance that is cooled. In this case, T_2 is less than T_1, and $\ln (T_2/T_1)$ will be negative, giving a negative entropy change.

Exercise

Find the change in entropy of 1 kg of water that is heated from 0°C to 100°C.   (Answer: $\Delta S = 1.31$ kJ/K)

Next we consider a reversible isothermal expansion of an ideal gas at a temperature T from an initial volume V_1 to a final volume V_2. As we have seen before, work W is done by the gas and heat $Q = W$ is absorbed from a reservoir at temperature T. The entropy change of the gas is given by Equation 17-16 with $T_2 = T_1$:

$$\Delta S = S_2 - S_1 = nR \ln \frac{V_2}{V_1} \qquad 17\text{-}18$$

The entropy change of the gas is positive because V_2 is greater than V_1. In this process, an amount of heat $|Q|$ leaves the reservoir and enters the gas. This amount of heat equals the work done by the gas

$$|Q| = W = \int_1^2 P \, dV = nRT \int_{V_1}^{V_2} \frac{dV}{V} = nRT \ln \frac{V_2}{V_1} \qquad 17\text{-}19$$

The entropy change of the gas is $+|Q|/T$. Since the same amount of heat leaves the reservoir at temperature T, the entropy change of the reservoir is $-|Q|/T$. The net entropy change of the gas plus the reservoir is zero. This example illustrates a general result:

> In a reversible process, the entropy change of the universe is zero. By "universe," we mean the system plus its surroundings.

We now consider the free expansion of an ideal gas—the irreversible process we first discussed in Section 16-5. Let the gas be at a temperature T and let it expand freely from an initial volume V_1 to a final volume V_2. As we have seen, no work is done and no heat is transferred in a free expansion. Because the process is not reversible, we cannot use $\int dQ/T$ to find the change in entropy of the gas. However, the initial and final states of the gas in the free expansion are the same as those of the gas in the isothermal expansion in the previous example. Therefore, since entropy is a state function and since the change in the entropy of a system for any process depends only on the initial and final states of the system, the entropy change for the free expansion is the same as that for the isothermal expansion. (As was noted earlier, to find the entropy change for an irreversible process, we must first find a reversible process that connects the same initial and final states.) The entropy change for the free expansion is thus given by Equation 17-18. Since the gas is a system that is isolated from its surroundings, this is also the entropy change for the universe:

$$\Delta S_{\mathrm{u}} = S_2 - S_1 \qquad\qquad 17\text{-}20$$

$$= nR \ln \frac{V_2}{V_1}$$

Note that because V_2 is greater than V_1, the change in entropy of the universe for this irreversible process is positive; that is, the entropy of the universe increases. This is also a general result:

> In an irreversible process, the entropy of the universe increases.

If the final volume in the free expansion were less than the initial volume the entropy of the universe would decrease, but this does not happen. A gas does not freely contract by itself into a smaller volume. This leads us to another statement of the second law of thermodynamics:

> For any process, the entropy of the universe never decreases.

This statement is equivalent to the heat-engine and refrigerator statements.

Next, we consider an example of a block of mass m falling from a height h and making an inelastic collision with the ground. Clearly this is an irreversible process. Let the block, ground, and atmosphere all be at a temperature T, which is not significantly changed by the process. If we consider the block, ground, and atmosphere as our isolated system, there is no heat conducted into or out of the system. The state of the system has been changed because its internal energy has been increased by an amount mgh. This change is the same as if we added heat $Q = mgh$ to the system at constant temperature T. To calculate the change in entropy of the system, we thus consider a reversible process in which heat $Q_{\mathrm{rev}} = mgh$ is added at a constant temperature T. According to Equation 17-13, the change in entropy is then

$$\Delta S = \frac{Q}{T} = \frac{mgh}{T}$$

This is also the entropy change of the universe ΔS_{u}.

In this process, energy is "wasted." When the block of mass m was at a height h, its potential energy mgh could have been used to do useful work, such as lifting a weight. After the inelastic collision this energy is no longer available because it has become the disordered internal energy of the block and its surroundings. (Some of it would be available if there were a reservoir at a temperature lower than that of the atmosphere T so that the energy mgh could be extracted and used to power a heat engine running between temperature T and that of the colder reservoir.) The energy that has become unavailable is equal to $T\,\Delta S_{\mathrm{u}}$. This is a general result:

> In an irreversible process, energy equal to the entropy change of the universe times the temperature of the coldest available reservoir becomes unavailable for doing work.

For simplicity, we will call the energy that becomes unavailable for doing work the "work lost":

$$W_{lost} = T \, \Delta S_u \qquad\qquad 17\text{-}21$$

In the free expansion discussed earlier the ability to do work was also lost. In that case, the entropy change of the universe was $nR \ln (V_2/V_1)$, so the work lost was $nRT \ln (V_2/V_1)$. This is the amount of work that could have been done if the gas had expanded quasi-statically and isothermally from V_1 to V_2, as given by Equation 17-19.

Lastly, we consider heat conduction from one heat reservoir to another. Let heat $|Q|$ be conducted from a hot reservoir at a temperature T_h to a cold reservoir at a temperature T_c. The state of a heat reservoir is determined only by its temperature and its internal energy. The change in entropy of a heat reservoir due to a heat exchange is the same whether the heat exchange is reversible or not. If heat $|Q|$ is put into a reservoir at temperature T, the entropy of the reservoir increases by $|Q|/T$. If the heat is removed, the entropy of the reservoir decreases by $-|Q|/T$. In the case of heat conduction, the hot reservoir loses heat, so its entropy change is

$$\Delta S_h = -\frac{|Q|}{T_h}$$

The cold reservoir absorbs heat, so its entropy change is

$$\Delta S_c = +\frac{|Q|}{T_c}$$

The net entropy change of the universe is

$$\Delta S_u = \frac{|Q|}{T_c} - \frac{|Q|}{T_h}$$

According to the general result that the work lost equals the entropy change times the temperature of the coldest reservoir available, the work lost in this process is

$$W_{lost} = T_c \, \Delta S_u = |Q|\left(1 - \frac{T_c}{T_h}\right)$$

We can see that this is just the work that could have been done by a Carnot engine running between these reservoirs, removing heat $|Q|$ from the hot reservoir and doing work $W = \epsilon_C |Q|$, where $\epsilon_C = 1 - T_c/T_h$.

Example 17-7

During each cycle, a Carnot engine removes 100 J of energy from a reservoir at 400 K, does work, and exhausts heat to a reservoir at 300 K. Compute the entropy change of each reservoir for each cycle, and show that the entropy change of the universe is zero for this reversible process.

The efficiency of a Carnot engine working between these two reservoirs is

$$\epsilon_C = 1 - \frac{T_c}{T_h} = 1 - \frac{300 \text{ K}}{400 \text{ K}} = 0.25$$

The engine therefore exhausts 75 J to the cold reservoir during each cycle. Since the hot reservoir loses heat, its entropy change is negative:

$$\Delta S_{400} = -\frac{|Q_h|}{T_h} = -\frac{100\ J}{400\ K} = -0.250\ J/K$$

The cold reservoir gains heat, so its entropy increases:

$$\Delta S_{300} = \frac{|Q_c|}{T_c} = \frac{75\ J}{300\ K} = +0.250\ J/K$$

Since the engine works in a cycle, its entropy change is zero. The entropy change of the universe is therefore the sum of the entropy changes of the hot and cold reservoirs, which is zero.

Example 17-8

During each cycle, a real heat engine with a second-law efficiency of 60 percent removes 100 J from a reservoir at 400 K, does work, and exhausts heat to a reservoir at 300 K. Compute the entropy change of each reservoir for each cycle.

Since the second-law efficiency is equal to the actual efficiency divided by the Carnot efficiency, the efficiency ϵ of this engine is

$$\epsilon = \epsilon_{SL}\epsilon_C = (0.60)(0.25) = 0.15$$

where we have used the Carnot efficiency we calculated in Example 17-7. The engine therefore does $\epsilon|Q_h| = (0.15)(100\ J) = 15\ J$ of work and exhausts 85 J to the cold reservoir during each cycle. The entropy change of the hot reservoir is the same as in Example 17-7:

$$\Delta S_{400} = -\frac{100\ J}{400\ K} = -0.250\ J/K$$

The entropy change of the cold reservoir is

$$\Delta S_{300} = \frac{85\ J}{300\ K} = +0.283\ J/K$$

In this case, the entropy increase of the cold reservoir is greater than the decrease for the hot reservoir. The entropy change of the universe for one cycle is therefore

$$\Delta S_u = \Delta S_{400} + \Delta S_{300}$$
$$= -0.250\ J/K + 0.283\ J/K = 0.033\ J/K$$

17-7   Entropy and Probability

In the previous section, we saw that entropy is a measure of the disorder of a system and that, even though the entropy of a given system can decrease during an irreversible process, the net change in entropy of the universe (the system plus its surroundings) is always positive. In other words, the universe always moves toward a state of less order. In this section, we will see that the entropy of a state is related to probability. Essentially, a state of high order has a low probability, whereas a state of low order has a high probability. Thus, in an irreversible process, the universe moves from a state of low probability to one of high probability.

Let us consider a free expansion in which a gas expands from an initial volume V_1 to a final volume $V_2 = 2V_1$. The entropy change of the universe for this process is given by Equation 17-20:

$$\Delta S_\mathrm{u} = nR \ln \frac{V_2}{V_1} = nR \ln 2 \qquad\qquad 17\text{-}22$$

As we have mentioned, this process wastes energy in that the gas could have been used to absorb heat Q from a reservoir and change it completely into work W. After the free expansion, this capability is no longer there because the gas is at a greater volume and a lower pressure.

Why is this process irreversible? Why can't the gas compress by itself back into its original volume? Since there is no energy change involved, a compression would not violate the first law of thermodynamics. The reason is merely that such a compression is extremely improbable. To see this, let us begin by assuming that the gas consists of only 10 molecules, and that, initially, these molecules occupy the entire volume of their container. If we assume that each molecule moves about the total volume of the container randomly, there will be an equal chance of a given molecule being in the left or right half of the container. That is, the chance that any one particular molecule will be in the left half of the container at any given time is $\frac{1}{2}$. The chance that any two particular molecules will both be in the left half is $\frac{1}{2} \times \frac{1}{2} = \frac{1}{4}$. (This is the same as the chance that a coin flipped twice will come up heads both times.) The chance that three particular molecules will be in the left half is $\frac{1}{2} \times \frac{1}{2} \times \frac{1}{2} = (\frac{1}{2})^3 = \frac{1}{8}$. The chance that all 10 molecules will be in the left half is $(\frac{1}{2})^{10} = \frac{1}{1024}$. That is, there is 1 chance in 1024 that all 10 molecules will be in the left half of the container at any given time.

© 1987 Sidney Harris

Though the probability of all 10 molecules being on one side of the container is very small, we would not be too surprised to see it occur. For example, if we look at the gas once each second, we could expect to see it happen once in every 1024 seconds, which is about once every 17 minutes. If we started with the 10 molecules randomly distributed and then found them all in the left half of the original volume, the entropy of the universe would have decreased by $nR \ln 2$. However, this decrease is extremely small, since the number of moles n corresponding to 10 molecules is only about 10^{-23}. Still, it would violate the entropy statement of the second law of thermodynamics, which says that for any process, the entropy of the universe never decreases. Therefore, we must consider the second law to be a statement of probability.

It may be disturbing to learn that the second law of thermodynamics seems to say that irreversible processes, such as the spontaneous compression of a gas or the spontaneous conduction of heat from a cold body to a hot body, are only improbable, not impossible. But, as we have just seen, if we consider a system consisting of a very small number of molecules, there is a reasonable chance that an irreversible process will occur. However, thermodynamics itself is applicable only to macroscopic systems, that is, to systems that have a very large number of molecules. Consider, for example, trying to measure the pressure of a gas consisting of only 10 molecules. The pressure would vary wildly depending on whether no or one or two molecules were colliding with the wall of the container at the time of measurement.

As we increase the number of molecules in a system, the chance of an irreversible process occurring decreases dramatically. For example, if we have 50 molecules in a container, the chance that they will all wind up in the left half of the volume is $(\frac{1}{2})^{50} \approx 10^{-15}$. Thus, if we look at the gas once each second, we could expect to see all 50 molecules in the left half of the volume about once in every 10^{15} seconds or once in every 36 million years! For 1 mole $= 6 \times 10^{23}$ molecules, the chance that all will wind up in half of the volume is extremely small, so small that it is essentially zero. So, for macroscopic systems, the probability of a process resulting in a decrease in the entropy of the universe is so extremely small that the distinction between improbable and impossible becomes blurred.

We can relate the probability of a gas spontaneously compressing itself into a smaller volume to the change in its entropy. If the original volume is V_1, the probability p of finding N molecules in a smaller volume V_2 is

$$p = \left(\frac{V_2}{V_1}\right)^N$$

Taking the natural logarithm of both sides of this equation, we obtain

$$\ln p = N \ln \left(\frac{V_2}{V_1}\right) \qquad\qquad \text{17-23}$$
$$= nN_A \ln \left(\frac{V_2}{V_1}\right)$$

where n is the number of moles and N_A is Avogadro's number. The entropy change of the gas is

$$\Delta S = nR \ln \left(\frac{V_2}{V_1}\right) \qquad\qquad \text{17-24}$$

(The entropy change is negative if V_2 is less than V_1). Comparing Equations 17-23 and 17-24, we see that

$$\Delta S = \frac{R}{N_A} \ln p = k \ln p \qquad\qquad \text{17-25}$$

where k is Boltzmann's constant.

Summary

1. A heat engine removes heat Q_h from a hot reservoir, does work W, and exhausts heat $|Q_c|$ to a cold reservoir. The efficiency of a heat engine is the ratio of the work done to the total heat absorbed:

$$\epsilon = \frac{W}{Q_h} = 1 - \frac{|Q_c|}{Q_h}$$

2. A refrigerator uses work W to remove heat Q_c from a cold reservoir and exhausts heat $|Q_h| = Q_c + W$ to a warmer reservoir. The coefficient of performance of a refrigerator is the ratio of the heat removed to the work that must be added:

$$\text{COP} = \frac{Q_c}{W}$$

3. According to the heat-engine statement of the second law of thermodynamics, it is impossible for a heat engine working in a cycle to remove heat from a reservoir and convert it completely into work with no other effects. According to the refrigerator statement of the second law, it is impossible to transfer heat from a cold reservoir to a hot reservoir with no other effects. These two statements of the second law of thermodynamics are equivalent.

4. The Carnot theorem states that no engine working between two heat reservoirs can be more efficient than a reversible engine. A reversible engine is called a Carnot engine. The efficiency of a Carnot engine is

$$\epsilon_C = 1 - \frac{T_c}{T_h}$$

For a Carnot engine, the ratio of the heat exhausted to the cold reservoir to the heat removed from the hot reservoir is

$$\frac{|Q_c|}{Q_h} = \frac{T_c}{T_h}$$

The second-law efficiency of a heat engine is the ratio of the actual efficiency to the Carnot efficiency:

$$\epsilon_{SL} = \frac{\epsilon}{\epsilon_C}$$

5. A heat pump is a device that removes energy from a cold reservoir, such as the air outside a house, and exhausts heat into a hot reservoir, such as the air inside a house. The coefficient of performance of a Carnot refrigerator or heat pump is

$$COP_{max} = \frac{T_c}{\Delta T}$$

where $\Delta T = T_h - T_c$.

The work W needed to exhaust heat $|Q_h|$ into a hot reservoir is related to the coefficient of performance by

$$W = \frac{|Q_h|}{1 + COP}$$

6. During an irreversible process, the universe moves toward a state of less order. A measure of the disorder of a system and of the reversibility of a process is the entropy function. The change in entropy of a system is given by

$$\Delta S = \int \frac{dQ_{rev}}{T}$$

where dQ_{rev} is the heat added in a reversible process connecting the initial and final states of the system.

7. The entropy of a system can increase or decrease, but the entropy of the universe or of any other isolated system never decreases. For a reversible process, the entropy of the universe remains constant; for an irreversible process, it increases. The statement that for any process the entropy of the universe never decreases is a statement of the second law of thermodynamics that is equivalent to the heat-engine and refrigerator statements.

8. During an irreversible process, the entropy of the universe S_u increases and a certain amount of energy

$$W_{lost} = T \Delta S_u$$

becomes unavailable for doing work.

9. Entropy is related to probability. A highly ordered system is one of low probability and low entropy. An isolated system moves towards a state of low order, high probability, and high entropy.

Exponential Growth and Decay

Albert A. Bartlett
University of Colorado at Boulder

If a town with a population of 10,000 has a steady annual population growth of 5 percent, how large will the population be in 10, 30, and 100 y?* If I invest $100 in an account that pays 15 percent annual interest (compounded continuously), how large will the account be in 10, 30, and 100 y? These questions are real-life applications of what may be the single most important differential equation:

$$\frac{dN}{dt} = kN \qquad (1)$$

This equation describes the situation in which the rate of change of a quantity dN/dt is proportional to the size N of the quantity. In our first example, N is the population of the town and $k = 0.05$ y^{-1}; in the second example N is the number of dollars in the account and $k = 0.15$ y^{-1}. In both cases t is the time in years. The meaning of k can be seen by rearranging Equation 1 to read

$$k = \frac{1}{N}\frac{dN}{dt} \qquad (2)$$

Thus k is the fractional change (dN/N) in N per unit time dt. It follows that $100k$ is P, the percent growth per unit time. These two examples are cases of steady growth in which the quantity always in-

*These figures are for continuous compounding. Annual compounding will give growth at slightly slower rates.

Albert A. Bartlett is a professor of physics at the University of Colorado in Boulder (80309-0390) where he has been a member of the faculty since 1950. His undergraduate degree is from Colgate University and he has a Ph.D. from Harvard (nuclear physics). In 1978 he was President of the American Association of Physics Teachers. He has lectured widely on the "Forgotten Fundamentals of the Energy Crisis."

creases by a fixed fraction (for example, 5 percent) in a fixed length of time (each year).

The solution of Equation 1 is the exponential function

$$N = N_0 e^{kt} \qquad (3)$$

where e is the base of natural logarithms (2.718 . . .) and N_0 is the size of N at the time $t = 0$. Equation 3 describes steady growth at a fixed percent per unit time. This is sometimes called exponential growth and it occurs whenever the rate of increase in a quantity is proportional to that quantity.

In our two examples in the first paragraph, the town's population will be

$$N = 10{,}000 e^{0.05t}$$

and the size of the bank account will be

$$N = 100 e^{0.15t}$$

where t is the time in years.

Using Equation 3, one can show that for a given value of k, a constant time T_c is required for N to increase in size from its present value N_0 to CN_0, where C is a constant:

$$T_c = \frac{1}{k}\ln C \qquad (4)$$

The time for N to double in size ($C = 2$) is

$$T_2 = \frac{1}{k}\ln 2 = \frac{0.693}{k} \qquad (5)$$

which can be expressed as

$$T_2 = \frac{69.3}{P} \approx \frac{70}{P} \qquad (6)$$

where $P = 100k$ is the percent growth per unit time. Table 1 shows doubling times for different steady rates of growth calculated from Equation 6.

From Table 1 we see that a quantity growing steadily at 5 percent per year will double in size in 13.9 y (steady growth at a rate of 5 percent per month will cause doubling to take place in 13.9 months). A quantity growing steadily at 15 percent per year will double in 4.62 y. We should develop the habit of doing the mental arithmetic of Equation 6 every time we see a percent growth rate in the news or elsewhere. People who do not grasp the meaning of 15 percent annual inflation will understand when you tell them that this means that prices will double every 4.6 years!

Equation 3 can be rewritten in a very useful form using T_2:

$$N = N_0 e^{kt} = N_0 2^{t/T_2} \qquad (7)$$

Table 1   **Doubling Times for Various Rates of Steady Growth**

k, y^{-1}	P, %/y	T_2, y
0.01	1	69.3
0.02	2	34.7
0.03	3	23.1
0.05	5	13.9
0.07	7	9.90
0.10	10	6.93
0.15	15	4.62
0.20	20	3.47
0.30	30	2.31

In this form we can estimate the answers to the questions that opened this essay. Since the doubling time for an annual growth rate of 5 percent is roughly 14 y, it can be seen that 10 y is less than one doubling time, 30 y is approximately two doubling times, and 100 y is approximately 7 doubling times. Thus in 10 y the population will increase from 10,000 to roughly 16,000; in 30 y it will have doubled twice (to 40,000); and in 100 y it will have doubled 7 times (a factor of $2^7 = 128$), and so the population will be 1,280,000. For a 15 percent annual interest rate compounded continuously ($T_2 = 4.62$ y) the figures are shown in Table 2.

We can estimate the value of large powers of 2 by noting that

$$2^{10} = 1024 \approx 10^3$$

For example,

$$2^{22} = 2^{10} \times 2^{10} \times 2^2 \approx 10^3 \times 10^3 \times 4 \approx 4 \times 10^6 \tag{8}$$

Some very important properties of steady growth can be illustrated by the following example. Legend has it that the game of chess was invented by a mathematician who worked for a king. As a reward the mathematician asked the king for the quantity of wheat determined by placing one grain of wheat on the first square of the chessboard, on the next square doubling the one grain to make two, on the next square doubling the two grains to make four, and continuing until the doubling has been done for all the squares (see Table 3). We can use Equation 8 to get an approximate expression for this total number of grains:

$$2^{64} \approx 10^3 \times 10^3 \times 10^3 \times 10^3 \times 10^3 \times 10^3 \times 16$$
$$\approx 1.6 \times 10^{19} \text{ grains}$$

The actual value of 2^{64} is 1.84×10^{19}. This is approximately 500 times the 1980 worldwide harvest of wheat, which may be more wheat than people have harvested in the entire history of the earth. How did we get such a large number? It was simple. We started with one grain and let the number grow until it had doubled a mere 63 times. A study of Table 3 reveals another important point. The growth in any doubling time is greater than the sum of all of the preceding growth. Thus when we put 32 grains on the sixth square, the 32 is larger than the total of 31 already on the board.

In his famous speech on energy (April 18, 1977) President Carter said, "In each of those decades [the 1950s and 1960s] more oil was consumed [worldwide] than in all of mankind's previous history." This statement sounds incredible, but a look at Table 3 shows that this is a simple consequence of steady growth with a doubling time of 10 years, that is, an annual growth rate of 7 percent. This was the growth rate of world oil consumption for nearly 100 years before 1970. Oil prices increased dramatically in the 1970s and this led people to seek improved efficiency in the use of oil. As a result, the growth in consumption stopped. In the 1980s it started to rise again but at a rate less than 7 percent.

Table 2   **The Value of $100 Deposited at 15% Annual Interest (Compounded Continuously) after a Time t**

t, y	t/T_2	$2^{t/T_2}$	Size of account
10	$10/4.62 = 2.16$	4.48	$448
30	$30/4.62 = 6.49$	90	$9000
100	$100/4.62 = 21.6$	3.27×10^6	$327,000,000

Table 3   **Grains of Wheat on a Chessboard**

No. of square	No. on square	Total No. on board
1	1	1
2	2	3
3	4	7
4	8	15
5	16	31
6	32	63
. . . .		
64	2^{63}	$2^{64} - 1$

Continued

This feature of exponential growth is extremely important. Suppose, for example, that the annual growth rate of 7 percent continued and that just as the world's supply of oil is to be exhausted, a new source of oil is discovered equal to total amount of oil consumed worldwide in the entire history of mankind. If oil consumption continued to grow at the annual rate of 7 percent, this fantastic new supply will last approximately one doubling time, or 10 years. This example makes it clear that, contrary to what prominent people often suggest, exploration and production of fossil fuels cannot meet the needs of continued growth for very long. It is interesting to calculate how long a quantity M_0 of a resource will last if it is now being consumed at a rate R_0 and if its rate of consumption is growing steadily in accord with Equation 3. It is easy to show[1,2] that the expiration time T_e is

$$T_e = \frac{1}{k} \ln \left(\frac{kM_0}{R_0} + 1 \right) \qquad (9)$$

We can use Equation 9 to estimate the lifetime T_e of United States coal reserves for different steady rates of growth of coal production. The U.S. Geological Survey[3] estimates the reserve base of United States coal to be $M_0 = 379 \times 10^9$ tonnes (1 tonne = 10^3 kg). The rate of production in 1978 was $R_0 = 0.6 \times 10^9$ tonnes/y. Using these numbers in Equation 10 we find the results shown in Table 4.

If we have zero growth of coal production, United States coal would last $T_e = M_0/R_0 = 632$ y. This is the basis for news stories that say "at present rates of consumption" United States coal will last about 600 y.[4] These stories then go on to stress the national urgency of achieving rapid growth of United States coal production, but the stories don't tell how growth reduces T_e (see Table 4).

In late 1989, the U.S Geological Survey released its estimates of the amount of petroleum (discovered *and undiscovered*) that remains in the ground in the United States. If the rate of extraction of oil from the ground does not change from the 1989 level (no growth), this petroleum will last 32 years.[5] This 1989 rate of extraction was supplying only about half of our domestic consumption. The other half is imported.

If the rate of change of N is negative, the right sides of Equations 1 and 2 are negative and Equation 3 becomes

$$N = N_0 e^{-kt} \qquad (10)$$

Table 4 Lifetime of United States Coal Reserves for Different Steady Rates of Growth of Coal Production

Annual growth rate, %	Expiration time of the reserve base
0	632
1	199
2	131
3	100
5	70
10	42
15	30

Exponential decay occurs in Newton's law of cooling (Chapter 16), where N is the temperature difference between an object and its surroundings; in the study of RC circuits (Chapter 23), where N is the charge on the capacitor; in LR circuits (Chapter 26), where N is the current in the circuit; and in radioactive decay (Chapter 40), where N is the number of radioactive nuclei that have not yet decayed.

A common example of this is the decay of radioactive substances. In this case N_0 is the number of atoms of a radioactive substance at time $t = 0$, and N is the number that have not decayed after a time t. The "half-life" is the time $T_{1/2}$ required for N to decay to $N_0/2$:

$$T_{1/2} = \frac{\ln 2}{k} = \frac{0.693}{k} \qquad (11)$$

After one half-life, half the substance remains, after two half-lives, one quarter remains, and after n half-lives the fraction remaining is $(\frac{1}{2})^n$. Table 5 lists half-lives of a few radioactive species.

The exponential arithmetic that arises so naturally in physics and in many other real-life situations is incredibly important—in part because it is so poorly understood by the people who make our policies (both local and national), who follow the dictum that "growth is good," and who never ask how long growth can continue. One can see that the energy shortage was caused by growth in consumption; the energy crisis was caused by our failure to understand the arithmetic and consequences of this growth. If we try to continue to have growth in our national and global patterns of energy consumption, we can ask, "Is the next energy crisis far away?" In the words of Aldous Huxley, "Facts do not cease to exist because they are ignored."

Table 5   **Half-Lives of Some Radioactive Substances**

Substance	Half-life
Carbon-14	5735 years
Potassium-40	1.26×10^9 years
Cobalt-60	5.26 years
Strontium-90	27.7 years
Iodine-131	8.05 days
Cesium-137	30.0 years
Lead-210	22 years
Polonium-211	0.52 second
Radium-215	10^{-6} second
Radon-222	3.823 days
Radium-226	1620 years
Uranium-238	4.51×10^9 years
Plutonium-239	2.44×10^4 years

Carbon-14 is used to date archaeological remains. Potassium-40 and uranium-238 have half-lives that are comparable to the age of the earth, so they can be used in establishing the age of geological samples. Cobalt-60 is used as a source of gamma rays in medical therapy. Strontium-90 and cesium-137 are produced in nuclear reactors and in nuclear explosions. Because of their half-lives they are a significant biological hazard. Radon-222 is the radioactive gas that is found in homes and other buildings. Radium-226 was discovered by Marie Curie and is used as a source of gamma rays for medical therapy. Plutonium-239 is a fissionable material that is used in nuclear weapons.

Let us summarize the points we have made about the arithmetic of steady growth (the exponential function):

1. Steady growth happens whenever the rate of increase of a quantity is proportional to the quantity (Equation 1).

2. It is characterized by a fixed doubling time (Equations 5 and 6).

3. In every doubling time, the amount of growth is greater than the total of all the preceding growth.

4. Growth for a modest number of doubling times gives astronomically large numbers.

5. If consumption of a resource is growing steadily, enormous new supplies of the resource will allow only very short extensions of the period of steady growth in the rate of consumption.

The important points of exponential decay are

1. It happens whenever the rate at which decay events take place is proportional to the quantity of material that has not yet decayed.

2. It is characterized by a constant half-life.

3. In a modest number of half-lives, the amount of material remaining undecayed will become very small and will approach zero.

It has been said that the greatest shortcoming of the human race is our inability to understand the exponential function.[6]

These topics have been developed in more detail by **1.** A. A. Bartlett, *The American Journal of Physics*, vol. 46, Sept. 1978, p. 876. **2.** A. A. Bartlett, *Journal of Geological Education*, vol. 28, Jan. 1980, p. 4. **3.** P. Averitt, "Coal Resources of the United States," U.S. Geological Survey, Bulletin 1412, Jan. 1, 1974. **4.** *Newsweek Magazine*, July 16, 1979, p. 23. **5.** *Science*, vol. 245, Sept. 22, 1989, p. 1330. **6.** A. A. Bartlett, *The Physics Teacher*, a series of articles on "The Exponential Function" was published between October 1976 and January 1979. A videotaped lecture, "The Forgotten Fundamentals of the Energy Crisis" by A. A. Bartlett is available from the Media Center, University of Colorado, Boulder 80309-0379

Suggestions for Further Reading

Angrist, Stanley W.: "Perpetual Motion Machines," *Scientific American*, January 1968, p. 114.

An amusing account of some of these machines, all of which would violate either the first or second law of thermodynamics, is presented in this article.

Atkins, P. W.: *The Second Law*, Scientific American Books, Inc., New York, 1984.

Thermodynamics as a link between the steam engine and the processes of life—a clear account with many diagrams and little mathematics.

Bennett, Charles H: "Demons, Engines and the Second Law," *Scientific American*, November 1987, p. 108.

This article gives an answer to the puzzle of Maxwell's Demon. The answer comes from the theory of computing.

Dyson, Freeman J.: "What Is Heat?" *Scientific American*, September 1954, p. 58.

This article includes a discussion of the relation of energy and entropy to heat and a short history of our evolving concepts of heat.

Ehrenberg, W.: "Maxwell's Demon," *Scientific American*, November 1967, p. 103.

Maxwell's thought experiment involving a violation of the second law of thermodynamics by a small, intelligent being has troubled physicists for over 100 years.

Penney, Terry R., and Desikan Bharathan: "Power from the Sea," *Scientific American*, January 1987, p. 86.

The developing technology of electrical power generation using heat engines that operate between warm surfaces and cooler waters below.

Shinbrot, Marvin: "Things Fall Apart," *The Sciences*, May/June 1987, p. 32.

A history of the controversies over Boltzmann's assertion that the entropy of a gas always increases—from 1872 to present-day research.

Tribus, Myron, and Edward C. McIrvine: "Energy and Information," *Scientific American*, September 1971, p. 179.

The concept of entropy can also be applied to the transmission of information, forming the basis of modern information theory.

von Baeyer, Hans Christian: "Physika: On the Cold Front," *The Sciences*, July/August 1989, p. 2.

The third law of thermodynamics states that the absolute zero of temperature is unattainable. This article describes the significance of a current experiment that tests the law.

Weaver, Jefferson Hane: "Entropy," in *The World of Physics: A Small Library of the Literature of Physics from Antiquity to the Present*, Simon and Schuster, New York, 1987.

Includes writings by Carnot, Clausius, Kelvin, Gibbs, Boltzmann, Planck, and Bridgman.

Wilson, S. S.: "Sadi Carnot," *Scientific American*, August 1981, p. 134.

A brief biography of this French investigator, who was concerned mainly with the practical improvement of steam engines and other heat engines, is presented in this article.

Review

A. Objectives: After studying this chapter, you should:

1. Be able to give the definitions of the efficiency of a heat engine and the coefficient of performance of a refrigerator.

2. Be able to give both the heat-engine and refrigerator statements of the second law of thermodynamics and illustrate their equivalence with a numerical example.

3. Be able to state the Carnot theorem and illustrate it with a numerical example.

4. Be able to give the expression for the Carnot efficiency of a heat engine.

5. Be able to discuss the concept of entropy, including its interpretation in terms of the unavailability of energy.

B. Define, explain, or otherwise identify:

Heat engine
Working substance
Heat reservoir
Efficiency
Heat-engine statement of the second law
Refrigerator
Refrigerator statement of the second law
Coefficient of performance
Carnot theorem
Carnot engine
Carnot efficiency
Carnot cycle
Second-law efficiency
Heat pump
Entropy

C. True or false: If the statement is true, explain why it is true. If it is false, give a counterexample.

1. Work can never be converted completely into heat.

2. Heat can never be converted completely into work.

3. All heat engines have the same efficiency.

4. It is impossible to transfer a given quantity of heat from a cold reservoir to a hot reservoir.

5. The coefficient of performance of a refrigerator cannot be greater than 1.

6. All Carnot engines are reversible.

7. The entropy of a system can never decrease.

8. The entropy of the universe can never decrease.

Problems

Level I

17-1 Heat Engines and the Second Law of Thermodynamics

1. An engine with 20 percent efficiency does 100 J of work in each cycle. (*a*) How much heat is absorbed in each cycle? (*b*) How much heat is rejected in each cycle?

2. An engine absorbs 400 J of heat and does 120 J of work in each cycle. (*a*) What is its efficiency? (*b*) How much heat is rejected in each cycle?

3. An engine absorbs 100 J and rejects 60 J in each cycle. (*a*) What is its efficiency? (*b*) If each cycle takes 0.5 s, find the power output of this engine in watts.

4. An engine with an output of 200 W has an efficiency of 30 percent. It works at 10 cycles/s. (*a*) How much work is done in each cycle? (*b*) How much heat is absorbed and how much is rejected in each cycle?

17-2 Refrigerators and the Second Law of Thermodynamics

5. A refrigerator absorbs 5 kJ of energy from a cold reservoir and rejects 8 kJ to a hot reservoir. (*a*) Find the coefficient of performance of the refrigerator. (*b*) The refrigerator is reversible and is run backward as a heat engine between the same two reservoirs. What is its efficiency?

17-3 Equivalence of the Heat-Engine and Refrigerator Statements

6. A certain engine running at 30 percent efficiency draws 200 J of heat from a hot reservoir. Assume that the refrigerator statement of the second law of thermodynamics is false, and show how this engine combined with a perfect refrigerator can violate the heat-engine statement of the second law.

7. A certain refrigerator takes in 500 J of heat from a cold reservoir and rejects 800 J to a hot reservoir. Assume that the heat-engine statement of the second law of thermodynamics is false, and show how a perfect engine working with this refrigerator can violate the refrigerator statement of the second law.

17-4 The Carnot Engine

8. A reversible engine working between two reservoirs at temperatures T_h and T_c has an efficiency of 30 percent. Working as a heat engine, it rejects 140 J of heat to the cold reservoir. A second engine working between the same two reservoirs also rejects 140 J to the cold reservoir. Show that if the second engine has an efficiency greater than 30 percent, the two engines working together would violate the heat-engine statement of the second law.

9. A reversible engine working between two reservoirs at temperatures T_h and T_c has an efficiency of 20 percent. Working as a heat engine, it does 100 J of work in each cycle. A second engine working between the same two reservoirs also does 100 J of work in each cycle. Show that if the efficiency of the second engine is greater than 20 percent, the two engines working together would violate the refrigerator statement of the second law.

10. A Carnot engine works between two heat reservoirs as a refrigerator. It does 50 J of work in removing 100 J from the cold reservoir and rejecting 150 J to the hot reservoir during each cycle. Its coefficient of performance COP $= Q_c/W = (100 \text{ J})/(50 \text{ J}) = 2$. (*a*) What is the efficiency of the Carnot engine when it works as a heat engine between the same two reservoirs? (*b*) Show that no other engine working as a refrigerator between the same two reservoirs can have a COP greater than 2.

11. A Carnot engine works between two heat reservoirs at temperatures $T_h = 300$ K and $T_c = 200$ K. (*a*) What is its efficiency? (*b*) If it absorbs 100 J from the hot reservoir during each cycle, how much work does it do? (*c*) How much heat does it reject during each cycle? (*d*) What is the COP of this engine when it works as a refrigerator between the same two reservoirs?

12. Which has a greater effect on increasing the efficiency of a Carnot engine, a 5-K increase in the temperature of the hot reservoir or a 5-K decrease in the temperature of the cold reservoir?

13. A refrigerator works between an inside temperature of 0°C and a room temperature of 20°C. (*a*) What is the largest possible coefficient of performance it can have? (*b*) If the inside of the refrigerator is to be cooled to −10°C, what is the largest possible coefficient of performance it can have, assuming the same room temperature of 20°C?

14. In each cycle, an engine removes 150 J from a reservoir at 100°C and rejects 125 J to a reservoir at 20°C. (*a*) What is the efficiency of this engine? (*b*) What is its second-law efficiency?

15. An engine has a second-law efficiency of 85 percent. In each cycle, it removes 200 kJ of heat from a hot reservoir at 500 K and exhausts heat to a cold reservoir at

200 K. (*a*) What is the efficiency of this engine? (*b*) How much work is done in each cycle? (*c*) How much heat is exhausted in each cycle?

16. An engine removes 250 J from a reservoir at 300 K and exhausts 200 J to a reservoir at 200 K. (*a*) What is its efficiency? (*b*) What is the second-law efficiency of the engine? (*c*) How much more work could be done if the engine were reversible?

17-5 The Heat Pump

17. A heat pump delivers 20 kW to heat a house. The outside temperature is $-10°C$ and the inside temperature of the hot-air supply for the heating fan is 40°C. (*a*) What is the coefficient of performance of a Carnot heat pump operating between these temperatures? (*b*) What must be the minimum power of the engine needed to run the heat pump? (*c*) If the second-law efficiency of the heat pump is 60 percent, what must the minimum power of the engine be?

18. Rework Problem 17 for an outside temperature of $-20°C$.

19. A refrigerator is rated at 370 W. (*a*) What is the maximum amount of heat it can remove in 1 min if the inside temperature of the refrigerator is 0°C and it exhausts into a room at 20°C? (*b*) If the second-law efficiency of the refrigerator is 70 percent, how much heat can it remove in 1 min?

20. Rework Problem 19 for a room temperature of 35°C.

17-6 Entropy and Disorder

21. Two moles of an ideal gas at $T = 400$ K expand quasi-statically and isothermally from an initial volume of 40 L to a final volume of 80 L. (*a*) Find the entropy change of the gas. (*b*) What is the entropy change of the universe for this process?

22. The gas in Problem 21 is taken from the same initial state ($T = 400$ K, $V_1 = 40$ L) to the same final state ($T = 400$ K, $V_2 = 80$ L) by a process that is not quasi-static. (*a*) What is the entropy change of the gas? (*b*) What can be said about the entropy change of the universe?

23. A system absorbs 200 J of heat reversibly from a reservoir at 300 K and rejects 100 J reversibly to a reservoir at 200 K as it moves from state *A* to state *B*. During this process, the system does 50 J of work. (*a*) What is the change in the internal energy of the system? (*b*) What is the change in entropy of the system? (*c*) What is the change in entropy of the universe? (*d*) If the system goes from state *A* to state *B* by a nonreversible process, how would your answers for parts (*a*), (*b*), and (*c*) differ?

24. A system absorbs 300 J from a reservoir at 300 K and 200 J from a reservoir at 400 K. It then returns to its original state, doing 100 J of work and rejecting 400 J of heat to a reservoir at a temperature *T*. (*a*) What is the entropy change of the system for the complete cycle? (*b*) If the cycle is reversible, what is the temperature *T*?

25. Two moles of an ideal gas originally at $T = 400$ K and $V = 40$ L undergo a free expansion to twice their initial volume. What is (*a*) the entropy change of the gas and (*b*) the entropy change of the universe?

26. A 5-kg block is dropped from rest at a height of 6 m above the ground. It hits the ground and comes to rest. The block, the ground, and the atmosphere are all initially at 300 K. What is the entropy change of the universe for this process?

27. If 500 J of heat is conducted from a reservoir at 400 K to one at 300 K, what is the change in entropy of the universe?

28. A 200-kg block of ice at 0°C is placed in a large lake. The temperature of the lake is just slightly higher than 0°C, and the ice melts. (*a*) What is the entropy change of the ice? (*b*) What is the entropy change of the lake? (*c*) What is the entropy change of the universe (the ice plus the lake)?

29. What is the change in entropy of 1.0 kg of water when it changes to steam at 100°C and a pressure of 1 atm?

30. How much of the 500 J of heat conducted from the 400-K reservoir to the 300-K reservoir in Problem 27 could have been converted into work using a cold reservoir at 300 K?

17-7 Entropy and Probability

There are no problems for this section.

Level II

31. (*a*) Which process is more wasteful: (1) a block moving with 500 J of kinetic energy being slowed to rest by friction when the temperature of the atmosphere is 300 K or (2) 1 kJ of heat being conducted from a reservoir at 400 K to one at 300 K? *Hint:* How much of the 1 kJ of heat could be converted into work in an ideal situation? (*b*) What is the change in entropy of the universe for each process?

32. An engine operates with 1 mol of an ideal gas for which $C_v = \frac{3}{2}R$ and $C_p = \frac{5}{2}R$ as its working substance. The cycle begins at $P_1 = 1$ atm and $V_1 = 24.6$ L. The gas is heated at constant volume to $P_2 = 2$ atm. It then expands at constant pressure until $V_2 = 49.2$ L. During these two steps, heat is absorbed by the gas. The gas is then cooled at constant volume until its pressure is again 1 atm. It is then compressed at constant pressure to its original state. During the last two steps, heat is rejected by the gas. All the steps are quasi-static and reversible. (*a*) Show this cycle on a *PV* diagram. Find the work done, the heat added, and the change in the internal energy of the gas for each step of the cycle. (*b*) Find the efficiency of the cycle.

33. An engine using 1 mol of a diatomic ideal gas performs a cycle consisting of three steps: (1) an adiabatic expansion from an initial pressure of 2.64 atm and an initial volume of 10 L to a pressure of 1 atm and a volume of 20 L, (2) a compression at constant pressure to its original volume of 10 L, and (3) heating at constant volume to its original pressure of 2.64 atm. Find the efficiency of this cycle.

34. An engine using 1 mol of an ideal gas initially at $V_1 = 24.6$ L and $T = 400$ K performs a cycle consisting of four steps: (1) an isothermal expansion at $T = 400$ K to twice its initial volume, (2) cooling at constant volume to $T = 300$ K, (3) an isothermal compression to its original volume, and (4) heating at constant volume to its original temperature of 400 K. Assume that $C_v = 21$ J/K. Sketch the cycle on a PV diagram, and find its efficiency.

35. A steam engine takes in superheated steam at 270°C and discharges condensed steam from its cylinder at 50°C. Its efficiency is 30 percent. (*a*) How does this efficiency compare with the maximum possible efficiency for these temperatures? (*b*) If the useful power output of the engine is 200 kW, how much heat does the engine discharge to its surroundings in 1 h?

36. In the cycle shown in Figure 17-11, 1 mol of an ideal gas ($\gamma = 1.4$) is initially at a pressure of 1 atm and a temperature of 0°C. The gas is heated at constant volume to $t_2 = 150$°C and is then expanded adiabatically until its pressure is again 1 atm. It is then compressed at constant pressure back to its original state. Find (*a*) the temperature t_3 after the adiabatic expansion, (*b*) the heat entering or leaving the system during each process, (*c*) the efficiency of this cycle, and (*d*) the efficiency of a Carnot cycle operating between the temperature extremes of this cycle.

Figure 17-11 Problem 36.

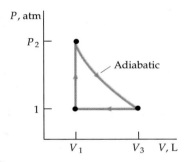

37. Helium gas ($\gamma = 1.67$) is initially at a pressure of 16 atm, a volume of 1 L, and a temperature of 600 K. It is expanded isothermally until its volume is 4 L and is then compressed at constant pressure until its volume and temperature are such that an adiabatic compression will return the gas to its original state. (*a*) Sketch this cycle on a PV diagram, and find (*b*) the volume and temperature after the isobaric compression, (*c*) the work done during each cycle, and (*d*) the efficiency of the cycle.

38. A heat engine that does the work of blowing up a balloon at a pressure of 1 atm extracts 4 kJ from a hot reservoir at 120°C. The volume of the balloon increases by 4 L, and heat is exhausted to a cold reservoir at a temperature t_c. If the second-law efficiency of the heat engine is 50 percent, find the temperature t_c.

39. A 100-g piece of ice at 0°C is placed in an insulated container with 100 g of water at 100°C. (*a*) When thermal equilibrium is established, what is the final temperature of the water? Ignore the heat capacity of the container. (*b*) Find the entropy change of the universe for this process.

40. A 1-kg block of copper at 100°C is placed in a calorimeter of negligible heat capacity containing 4 L of water at 0°C. Find the entropy change of (*a*) the copper block, (*b*) the water, and (*c*) the universe.

41. If a 2-kg piece of lead at 100°C is dropped into a lake at 10°C, find the entropy change of the universe.

42. Find the net change in entropy of the universe when 10 g of steam at 100°C and a pressure of 1 atm are introduced into a calorimeter of negligible heat capacity containing 150 g of water and 150 g of ice at 0°C.

43. A 1500-kg car traveling at 100 km/h crashes into a concrete wall. If the temperature of the air is 20°C, calculate the entropy change of the universe.

44. A Carnot engine works between two heat reservoirs at temperatures $T_h = 300$ K and $T_c = 200$ K. (*a*) What is its efficiency? (*b*) If it absorbs 100 J from the hot reservoir during each cycle, how much work does it do? (*c*) How much heat does it reject in each cycle? (*d*) What is the coefficient of performance of this engine when it works as a refrigerator between these two reservoirs?

45. A heat engine works in a cycle between reservoirs at 400 and 200 K. The engine absorbs 1000 J of heat from the hot reservoir and does 200 J of work in each cycle. (*a*) What is the efficiency of this engine? (*b*) Find the entropy change of the engine, each reservoir, and the universe for each cycle. (*c*) What would be the efficiency of a Carnot engine working between the same two reservoirs? (*d*) How much work could be done in each cycle by a Carnot engine if it absorbs 1000 J from the hot reservoir? (*e*) Show that the difference in the work done by the Carnot engine and that done by the original engine is $T_c \Delta S_u$, where ΔS_u is the change in entropy of the universe.

46. One mole of an ideal gas first undergoes a free expansion from $V_1 = 12.3$ L and $T_1 = 300$ K to $V_2 = 24.6$ L and $T_2 = 300$ K. It is then compressed isothermally and quasi-statically back to its original state. (*a*) What is the entropy change of the universe for the complete cycle? (*b*) How much work is wasted in this cycle? (*c*) Show that the work wasted is $T \Delta S_u$.

47. The cooling compartment of a refrigerator and its contents are at 5°C and have an average heat capacity of 84 kJ/K. The refrigerator exhausts heat to the room, which is at 25°C. What minimum power must the motor used to operate the refrigerator have if it is to reduce the temperature of the cooling compartment and its contents by 1 C° in 1 min?

48. (*a*) Show that the coefficient of performance of a Carnot refrigerator working between two reservoirs at temperatures T_h and T_c is related to the efficiency of a Carnot engine by $\text{COP} = T_c/(\epsilon_c T_h)$.

Level III

49. The behavior of a gasoline engine can be approximated by an *Otto cycle*. See Figure 17-4. (*a*) Compute the heat input and the heat output, and show that the efficiency can be written as

$$\epsilon = 1 - \frac{T_d - T_a}{T_c - T_b}$$

where T_a is the temperature of state a, and so forth.

(b) Using the relation for an adiabatic expansion or compression, $TV^{\gamma-1} = $ a constant, show that

$$\epsilon = 1 - \left(\frac{V_2}{V_1}\right)^{\gamma-1}$$

where $V_1 = V_a = V_d$ and $V_2 = V_b = V_c$. (c) The ratio V_1/V_2 is called the *compression ratio*. Find the efficiency of this cycle for a compression ratio of 8, using $\gamma = 1.4$. (Compression ratios much greater than this can not be achieved in real engines because of preignition problems.) (d) Explain why the efficiency of a real gasoline engine might be much less than that calculated in part (c).

50. The *diesel cycle* shown in Figure 17-12 approximates the behavior of a diesel engine. Process *ab* is an adiabatic compression, process *bc* is an expansion at constant pressure, process *cd* is an adiabatic expansion, and process *da* is cooling at constant volume. Find the efficiency of this cycle in terms of the volumes V_a, V_b, V_c, and V_d.

Figure 17-12   Diesel cycle for Problem 50.

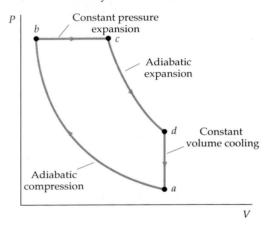

51. In the *Stirling cycle* shown in Figure 17-13, process *ab* is an isothermal compression, process *bc* is heating at constant volume, process *cd* is an isothermal expansion, and process *da* is cooling at constant volume. Find the efficiency of the Stirling cycle in terms of the temperatures T_h and T_c, and the volumes V_a and V_b.

Figure 17-13   Stirling cycle for Problem 51.

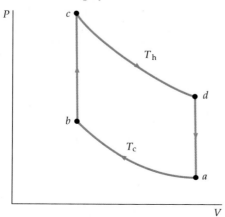

52. If two adiabatic curves intersect on a *PV* diagram, a cycle could be completed by an isothermal path between the two adiabatic curves as shown in Figure 17-14. Show that such a cycle could violate the second law of thermodynamics.

Figure 17-14   Problem 52.

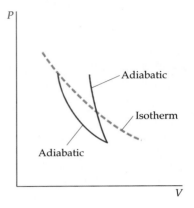

53. Using the equation for the entropy change of an ideal gas when the volume and temperature change and $TV^{\gamma-1} = $ a constant, show explicitly that the entropy change is zero for a quasi-static adiabatic expansion from state (V_1,T_1) to state (V_2,T_2).

54. (a) Show that if the refrigerator statement of the second law of thermodynamics were not true, the entropy of the universe could decrease. (b) Show that if the heat-engine statement of the second law were not true, the entropy of the universe could decrease. (c) An alternative statement of the second law is that the entropy of the universe can not decrease. Have you just proved that this statement is equivalent to the refrigerator and heat-engine statements?

55. One mole of an ideal monatomic gas at an initial volume $V_1 = 25$ L follows the cycle shown in Figure 17-15. All the processes are quasi-static. Find (a) the temperature of each state of the cycle, (b) the heat flow for each part of the cycle, and (c) the efficiency of the cycle.

Figure 17-15   Problem 55.

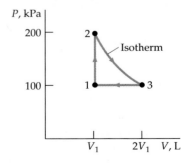

56. Suppose that two heat engines are connected in "series" such that the heat exhaust of the first engine is used as the heat input of the second engine as shown in Figure 17-16. The efficiencies of the engines are ϵ_1 and ϵ_2, respec-

tively. Show that the net efficiency of the combination is given by

$$\epsilon_{net} = \epsilon_1 + (1 - \epsilon_1)\epsilon_2$$

57. Suppose that each engine in Figure 17-16 is an ideal reversible heat engine. Engine 1 operates between temperatures T_h and T_m and engine 2 operates between T_m and T_c, where $T_h > T_m > T_c$. Show that

$$\epsilon_{net} = 1 - \frac{T_c}{T_h}$$

This means that two reversible heat engines in "series" are equivalent to one reversible heat engine operating between the hottest and coldest reservoirs.

Figure 17-16   Problems 56 and 57.

58. An ideal gas ($\gamma = 1.4$) follows the cycle shown in Figure 17-17. The temperature of state 1 is 200 K. Find (*a*) the temperatures of the other three states of the cycle and (*b*) the efficiency of the cycle.

Figure 17-17   Problem 58.

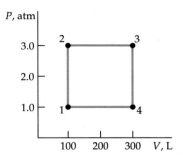

59. The Clausius equation of state is $P(V - bn) = nRT$ where b is a constant. Show that the efficiency of a Carnot cycle is the same for a gas that obeys this equation of state as it is for one that obeys the ideal-gas equation of state, $PV = nRT$.

Appendix A

Review of Mathematics

In this appendix, we will review some of the basic results of algebra, geometry, trigonometry, and calculus. In many cases, we will merely state results without proof. Table A-1 lists some mathematical symbols.

Equations

The following operations can be performed on mathematical equations to facilitate their solution:

1. The same quantity can be added to or subtracted from each side of the equation.

2. Each side of the equation can be multiplied or divided by the same quantity.

3. Each side of the equation can be raised to the same power.

It is important to understand that the preceding rules apply to each *side* of the equation and not to each *term* in the equation.

Table A-1	Mathematical Symbols
$=$	is equal to
$\neq$	is not equal to
$\approx$	is approximately equal to
$\sim$	is of the order of
$\propto$	is proportional to
$>$	is greater than
$\geq$	is greater than or equal to
$\gg$	is much greater than
$<$	is less than
$\leq$	is less than or equal to
$\ll$	is much less than
Δx	change in x
$|x|$	absolute value of x
$n!$	$n(n-1)(n-2)\cdots 1$
Σ	sum
lim	limit
$\Delta t \to 0$	Δt approaches zero
$\dfrac{dx}{dt}$	derivative of x with respect to t
$\dfrac{\partial x}{\partial t}$	partial derivative of x with respect to t
$\int$	integral

Example A-1

Solve the following equation for x:

$$(x - 3)^2 + 7 = 23$$

We first subtract 7 from each side of the equation to obtain $(x - 3)^2 = 16$. Taking the square root of each side, which is the same as raising each side to the $\frac{1}{2}$ power, we obtain $\pm(x - 3) = \pm4$. We have included the plus-or-minus signs because either $(+4)^2 = 16$ or $(-4)^2 = 16$. We do not need to write $\pm$ on both sides of the equation as all the possibilities are included in $x - 3 = \pm4$. We can now solve for x by adding 3 to each side. There are two solutions: $x = 4 + 3 = 7$ and $x = -4 + 3 = -1$. These values can be checked by substituting them into the original equation.

Example A-2

Solve the following equation for x:

$$\frac{1}{x} + \frac{1}{4} = \frac{1}{3}$$

This type of equation occurs both in geometric optics and in analyses of electric circuits. Although it is easy to solve, errors are often made. We solve it by first subtracting $\frac{1}{4}$ from each side to obtain

$$\frac{1}{x} = \frac{1}{3} - \frac{1}{4} = \frac{4}{12} - \frac{3}{12} = \frac{1}{12}$$

We then multiply each side by $12x$ to obtain $x = 12$. Note that this is equivalent to taking the reciprocal of each side of the equation. A typical mistake in handling this type of equation is to take the reciprocal of each term first to obtain $x + 4 = 3$. This operation is not allowed; it changes the relative values of each side of the equation and leads to incorrect results.

Direct and Inverse Proportion

The relationships of direct proportion and inverse proportion are so important in physics that they deserve special consideration. Often much algebraic manipulation can be avoided through a simple knowledge of these relationships. Suppose, for example, that you work for 5 days at a certain pay rate and earn $400. How much would you earn at the same pay rate if you worked 8 days? In this problem, the money earned is *directly proportional* to the time worked. We can write an equation relating the money earned M to the time worked t using a constant of proportionality R:

$$M = Rt$$

The constant of proportionality in this case is the pay rate. We can express R in dollars per day. Since $400 was earned in 5 d, the value of R is $400/(5 \text{ d}) = \$80/\text{d}$. In 8 d, the amount earned is therefore

$$M = (\$80/\text{d})(8 \text{ d}) = \$640$$

However, we do not have to find the pay rate explicitly to work the problem. Since the amount earned in 8 d is $\frac{8}{5}$ times that earned in 5 d, this amount is

$$M = \tfrac{8}{5}(\$400) = \$640$$

We can use a similar example to illustrate inverse proportion. If you get a 25 percent raise, how long would you need to work to earn $400? Here we consider R to be a variable and we wish to solve for t:

$$t = \frac{M}{R}$$

In this equation, the time t is *inversely proportional* to the pay rate R. Thus, if the new rate is $\frac{5}{4}$ times the old rate, the new time will be $\frac{4}{5}$ times the old time or 4 d.

There are some situations in which one quantity varies as the square or some other power of another quantity where the ideas of proportionality are also very useful. Suppose, for example, that a 10-in diameter pizza costs $8.50. How much would you expect a 12-in diameter pizza to cost? We expect the cost of a pizza to be approximately proportional to the amount of its contents, which is proportional to the area of the pizza. Since the area is in turn proportional to the square of the diameter, the cost should be proportional to the square of the diameter. If we increase the diameter by a factor of 12/10, the area increases by a factor of $(12/10)^2 = 1.44$, so we should expect the cost to be $(1.44)(\$8.50) = \$12.24$.

Example A-3

The intensity of light from a point source varies inversely with the square of the distance from the source. If the intensity is 3.20 W/m^2 at 5 m from a source, what is it at 6 m from the source?

The equation expressing the fact that the intensity varies inversely with the square of the distance can be written

$$I = \frac{C}{r^2}$$

where C is some constant. Then, if $I_1 = 3.20$ W/m^2 at $r_1 = 5$ m and I_2 is the unknown intensity at $r_2 = 6$ m, we have

$$\frac{I_2}{I_1} = \frac{C/r_2^2}{C/r_1^2} = \frac{r_1^2}{r_2^2} = (\frac{5}{6})^2 = 0.694$$

The intensity at 6 m from the source is thus

$$I_2 = 0.694(3.20 \ W/m^2) = 2.22 \ W/m^2$$

Linear Equations

An equation in which the variables occur only to the first power is said to be linear. A linear equation relating y and x can always be put into the standard form

$$y = mx + b \qquad\qquad \text{A-1}$$

where m and b are constants that may be either positive or negative. Figure A-1 shows a graph of the values of x and y that satisfy Equation A-1. The constant b, called the **intercept,** is the value of y at $x = 0$. The constant m is the **slope** of the line, which equals the ratio of the change in y to the corresponding change in x. In the figure, we have indicated two points on the line, x_1, y_1 and x_2, y_2, and the changes $\Delta x = x_2 - x_1$ and $\Delta y = y_2 - y_1$. The slope m is then

$$m = \frac{y_2 - y_1}{x_2 - x_1} = \frac{\Delta y}{\Delta x}$$

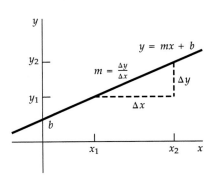

Figure A-1 Graph of the linear equation $y = mx + b$, where b is the intercept and $m = \Delta y/\Delta x$ is the slope.

If x and y are both unknown, there is no unique solution for their values. Any pair of values x_1, y_1 on the line in Figure A-1 will satisfy the equation. If we have two equations, each with the same two unknowns x and y, the equations can be solved simultaneously for the unknowns.

Example A-4

Find the values of x and y that satisfy

$$3x - 2y = 8 \qquad\qquad \text{A-2}$$

and

$$y - x = 2 \qquad\qquad \text{A-3}$$

Figure A-2 shows a graph of each of these equations. At the point where the lines intersect, the values of x and y satisfy both equations. We can solve two simultaneous equations by first solving either equation for one variable in terms of the other variable and then substituting the result into the other equation. From Equation A-3, we have

$$y = x + 2$$

Substituting this value for y in Equation A-2, we obtain

$$3x - 2(x + 2) = 8$$
$$3x - 2x - 4 = 8$$
$$x = 12$$

Then

$$y = x + 2 = 14$$

An alternative method that is sometimes easier is to multiply one equation by a constant such that one of the unknown terms is eliminated

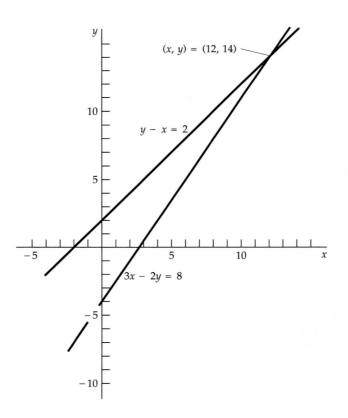

Figure A-2 Graph of Equations A-2 and A-3. At the point where the lines intersect, the values of x and y satisfy both equations.

when the equations are added or subtracted. If we multiply Equation A-3 by 2, we can add the resulting equation to Equation A-2 and eliminate y:

$$3x - 2y = 8$$

$$2y - 2x = 4$$

Adding, we obtain $3x - 2x = 12$ or $x = 12$, as before.

Factoring

Equations can often be simplified by factoring. Three important examples are

1. Common factor: $2ax + 3ay = a(2x + 3y)$
2. Perfect square: $x^2 \pm 2xy + y^2 = (x \pm y)^2$
3. Difference of squares: $x^2 - y^2 = (x + y)(x - y)$

The Quadratic Formula

An equation that contains a variable to the second power is called a *quadratic equation*. The standard form for a quadratic equation is

$$ax^2 + bx + c = 0 \qquad \text{A-4}$$

where a, b, and c are constants. The general solution of this equation is

$$x = -\frac{b}{2a} \pm \frac{1}{2a}\sqrt{b^2 - 4ac} \qquad \text{A-5}$$

When b^2 is greater than $4ac$, there are two solutions corresponding to the + and − signs. Figure A-3 shows a graph of y versus x where $y = ax^2 + bx + c$. The curve, called a **parabola,** crosses the x axis twice. The values of x for which $y = 0$ are the solutions to Equation A-4. When $b^2 < 4ac$, the graph of y versus x does not intersect the x axis, as is shown in Figure A-4, and there are no real solutions to Equation A-4. When $b^2 = 4ac$, the graph of y versus x is tangent to the x axis at the point $x = -b/2a$.

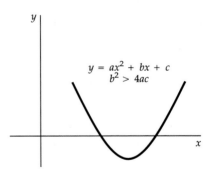

Figure A-3   Graph of y versus x when $y = ax^2 + bx + c$ for the case $b^2 > 4ac$. The two values of x for which $y = 0$ satisfy the quadratic equation (Equation A-4).

Exponents

The notation x^n stands for the quantity obtained by multiplying x times itself n times. For example, $x^2 = x \cdot x$ and $x^3 = x \cdot x \cdot x$. The quantity n is called the **power,** or the **exponent,** of x. When two powers of x are multiplied, the exponents are added:

$$(x^m)(x^n) = x^{m+n} \qquad \text{A-6}$$

This can be readily seen from an example:

$$x^2x^3 = (x \cdot x)(x \cdot x \cdot x) = x^5$$

Any number raised to the 0 power is defined to be 1:

$$x^0 = 1 \qquad \text{A-7}$$

Then

$$x^n x^{-n} = x^0 = 1$$

$$x^{-n} = \frac{1}{x^n} \qquad \text{A-8}$$

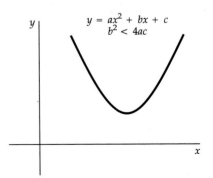

Figure A-4   Graph of y versus x when $y = ax^2 + bx + c$ for the case $b^2 < 4ac$. In this case, there are no (real) values of x for which $y = 0$.

When two powers are divided, the exponents are subtracted:

$$\frac{x^n}{x^m} = x^n x^{-m} = x^{n-m} \qquad \text{A-9}$$

Using these rules, we have

$$x^{1/2} \cdot x^{1/2} = x$$

so

$$x^{1/2} = \sqrt{x}$$

When a power is raised to another power, the exponents are multiplied:

$$(x^n)^m = x^{nm} \qquad \text{A-10}$$

Logarithms

When y is related to x by $y = a^x$, the number x is said to be the logarithm of y to the base a and is written

$$x = \log_a y$$

If $y_1 = a^n$ and $y_2 = a^m$, then

$$y_1 y_2 = a^n a^m = a^{n+m}$$

and

$$\log_a y_1 y_2 = n + m = \log_a y_2 + \log_a y_1 \qquad \text{A-11}$$

It then follows that

$$\log_a y^n = n \log_a y \qquad \text{A-12}$$

Since $a^1 = a$ and $a^0 = 1$,

$$\log_a a = 1 \qquad \text{A-13}$$

and

$$\log_a 1 = 0 \qquad \text{A-14}$$

There are two bases in common use: base 10, called **common logarithms,** and base e ($e = 2.728 \ldots$), called **natural logarithms.** When no base is specified, the base is understood to be 10. Thus, $\log 100 = \log_{10} 100 = 2$ since $100 = 10^2$.

The symbol ln is used for natural logarithms. Thus,

$$y = \ln x \qquad \text{A-15}$$

implies

$$x = e^y \qquad \text{A-16}$$

Logarithms can be changed from one base to another. Suppose that

$$z = \log x \qquad \text{A-17}$$

Then

$$10^z = x \qquad \text{A-18}$$

Taking the natural logarithm of both sides of Equation A-18, we obtain

$$z \ln 10 = \ln x$$

or

$$\ln x = (\ln 10)\log x \qquad \text{A-19}$$

The Exponential Function

When the rate of change of a quantity is proportional to the quantity itself, the quantity increases or decreases exponentially. An example of *exponential decrease* is nuclear decay. If N is the number of radioactive nuclei at some time, then the change dN in some very small time interval dt will be proportional to N and to dt:

$$dN = -\lambda N \, dt$$

where the constant of proportionality λ is the decay rate. The function N satisfying this equation is

$$N = N_0 \, e^{-\lambda t} \qquad \text{A-20}$$

where N_0 is the number at time $t = 0$. Figure A-5 shows N versus t. A characteristic of exponential decay is that N decreases by a constant factor in a given time interval. The time interval for N to decrease to half its original value is its half-life $t_{1/2}$, which is related to the decay rate by

$$t_{1/2} = \frac{\ln 2}{\lambda} = \frac{0.693}{\lambda} \qquad \text{A-21}$$

An example of *exponential increase* is population growth. If the number of organisms is N, the change in N after a small time interval dt is given by

$$dN = +\lambda N \, dt$$

where λ is a constant that characterizes the rate of increase. The function N satisfying this equation is

$$N = N_0 \, e^{\lambda t} \qquad \text{A-22}$$

A graph of this function is shown in Figure A-6. An exponential increase is characterized by a doubling time T_2, which is related to λ by

$$T_2 = \frac{\ln 2}{\lambda} = \frac{0.693}{\lambda} \qquad \text{A-23}$$

If the rate of increase λ is expressed as a percentage, $r = \lambda / 100\%$, the doubling time is

$$T_2 = \frac{69.3}{r} \qquad \text{A-24}$$

For example, if the population increases by 2 percent per year, the population will double every $69.3/2 \approx 35$ years. Table A-2 lists some useful relations for exponential and logarithmic functions.

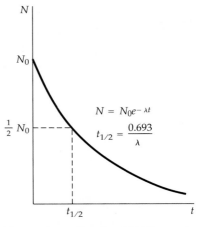

Figure A-5 Graph of N versus t when N decreases exponentially. The time $t_{1/2}$ is the time it takes for N to decrease by one-half.

Table A-2 **Exponential and Logarithmic Functions**

$e = 2.71828 \qquad e^0 = 1$
If $y = e^x$, then $x = \ln y$.
$e^{\ln x} = x$
$e^x e^y = e^{(x+y)}$
$(e^x)^y = e^{xy} = (e^y)^x$
$\ln e = 1 \qquad \ln 1 = 0$
$\ln xy = \ln x + \ln y$
$\ln \dfrac{x}{y} = \ln x - \ln y$
$\ln e^x = x \qquad \ln a^x = x \ln a$
$\ln x = (\ln 10) \log x$
$\qquad = 2.3026 \log x$
$\log x = \log e \ln x = 0.43429 \ln x$
$e^x = 1 + x + \dfrac{x^2}{2!} + \dfrac{x^3}{3!} + \cdots$
$\ln(1 + x) = x - \dfrac{x^2}{2} + \dfrac{x^3}{3}$
$\qquad\qquad - \dfrac{x^4}{4} + \cdots$

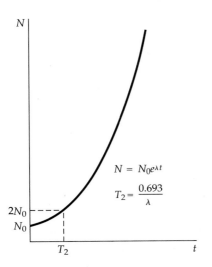

Figure A-6 Graph of N versus t when N increases exponentially. The time T_2 is the time it takes for N to double.

Area of parallelogram
$A = bh$

Figure A-7 Area of a parallelogram.

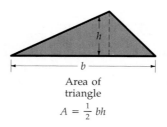

Area of triangle
$A = \frac{1}{2} bh$

Figure A-8 Area of a triangle.

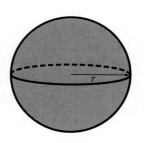

Spherical surface area
$A = 4\pi r^2$
Spherical volume
$V = \frac{4}{3}\pi r^3$

Figure A-9 Surface area and volume of a sphere.

Cylindrical surface area
$A = 2\pi rL$
Cylindrical volume
$V = \pi r^2 L$

Figure A-10 Surface area (not including the end faces) and volume of a cylinder.

Geometry

The ratio of the circumference of a circle to its diameter is a natural number π, which has the approximate value

$$\pi = 3.141592$$

The circumference of a circle C is thus related to its diameter d and its radius r by

$$C = \pi d = 2\pi r \qquad \text{circumference of circle} \qquad \text{A-25}$$

The area of a circle is

$$A = \pi r^2 \qquad \text{area of circle} \qquad \text{A-26}$$

The area of a parallelogram is the base b times the height h (Figure A-7) and that of a triangle is one-half the base times the height (Figure A-8). A sphere of radius r (Figure A-9) has a surface area given by

$$A = 4\pi r^2 \qquad \text{spherical surface area} \qquad \text{A-27}$$

and a volume given by

$$V = \frac{4}{3}\pi r^3 \qquad \text{spherical volume} \qquad \text{A-28}$$

A cylinder of radius r and length L (Figure A-10) has surface area (not including the end faces) of

$$A = 2\pi rL \qquad \text{cylindrical surface} \qquad \text{A-29}$$

and volume of

$$V = \pi r^2 L \qquad \text{cylindrical volume} \qquad \text{A-30}$$

Trigonometry

The angle between two intersecting straight lines is measured as follows. A circle is drawn with its center at the intersection of the lines, and the circular arc is divided into 360 parts called **degrees.** The number of degrees in the arc between the lines is the measure of angle between the lines. For very small angles, the degree is divided into minutes (') and seconds ('') with $1' = 1°/60$ and $1'' = 1'/60 = 1°/3600$. For scientific work, a more useful measure of an angle is the radian (rad), which is defined as the length of the circular arc between the lines divided by the radius of the circle (Figure A-11). If s is the arc length and r is the radius of the circle, the angle θ measured in radians is

$$\theta = \frac{s}{r} \qquad \text{A-31}$$

Since the angle measured in radians is the ratio of two lengths, it is dimensionless. The relation between radians and degrees is

$$360° = 2\pi \text{ rad}$$

or

$$1 \text{ rad} = \frac{360°}{2\pi} = 57.3° \qquad \text{A-32}$$

Figure A-11 The angle θ in radians is defined to be the ratio s/r, where s is the arc length intercepted on a circle of radius r.

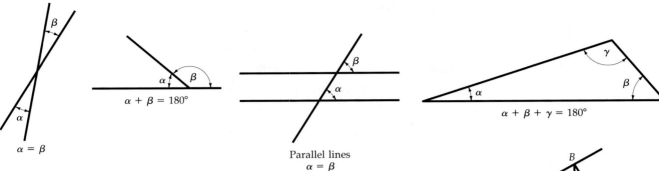

Figure A-12 shows some useful relations for angles.

Figure A-13 shows a right triangle formed by drawing the line BC perpendicular to AC. The lengths of the sides are labeled a, b, and c. The trigonometric functions $\sin\theta$, $\cos\theta$, and $\tan\theta$ for an acute angle θ are defined as

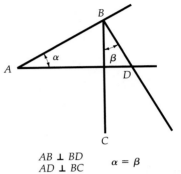

$$\sin\theta = \frac{a}{c} = \frac{\text{opposite side}}{\text{hypotenuse}} \qquad \text{A-33}$$

$$\cos\theta = \frac{b}{c} = \frac{\text{adjacent side}}{\text{hypotenuse}} \qquad \text{A-34}$$

$$\tan\theta = \frac{a}{b} = \frac{\text{opposite side}}{\text{adjacent side}} = \frac{\sin\theta}{\cos\theta} \qquad \text{A-35}$$

Three other trigonometric functions are the reciprocals of these functions and are defined as

$$\sec\theta = \frac{c}{b} = \frac{1}{\cos\theta} \qquad \text{A-36}$$

$$\csc\theta = \frac{c}{a} = \frac{1}{\sin\theta} \qquad \text{A-37}$$

$$\cot\theta = \frac{b}{a} = \frac{1}{\tan\theta} = \frac{\cos\theta}{\sin\theta} \qquad \text{A-38}$$

The angle θ whose sine is x is called the arcsine of x, and is written $\sin^{-1}x$. That is, if

$$\sin\theta = x$$

then

$$\theta = \arcsin x = \sin^{-1} x \qquad \text{A-39}$$

The arcsine is the inverse of the sine. The inverse of the cosine and tangent are defined similarly. The angle whose cosine is y is the arccosine of y. That is, if

$$\cos\theta = y$$

then

$$\theta = \arccos y = \cos^{-1} y \qquad \text{A-40}$$

The angle whose tangent is z is the arctangent of z. That is, if

$$\tan\theta = z$$

$$\theta = \arctan z = \tan^{-1} z \qquad \text{A-41}$$

The pythagorean theorem

$$a^2 + b^2 = c^2 \qquad \text{A-42}$$

Figure A-12 Some useful relations for angles.

Figure A-13 A right triangle with sides of length a and b and a hypotenuse of length c.

gives some useful identities. If we divide each term in this equation by c^2, we obtain

$$\frac{a^2}{c^2} + \frac{b^2}{c^2} = 1$$

or, from the definitions of $\sin\theta$ and $\cos\theta$,

$$\sin^2\theta + \cos^2\theta = 1 \qquad\qquad\qquad \text{A-43}$$

Similarly, we can divide each term in Equation A-42 by a^2 or b^2 and obtain

$$1 + \cot^2\theta = \csc^2\theta \qquad\qquad\qquad \text{A-44}$$

and

$$1 + \tan^2\theta = \sec^2\theta \qquad\qquad\qquad \text{A-45}$$

These and other useful trigonometric formulas are listed in Table A-3.

Table A-3   Trigonometric Formulas

$\sin^2\theta + \cos^2\theta = 1 \qquad \sec^2\theta - \tan^2\theta = 1 \qquad \csc^2\theta - \cot^2\theta = 1$

$\sin 2\theta = 2\sin\theta\cos\theta$

$\cos 2\theta = \cos^2\theta - \sin^2\theta = 2\cos^2\theta - 1 = 1 - 2\sin^2\theta$

$$\tan 2\theta = \frac{2\tan\theta}{1 - \tan^2\theta}$$

$$\sin\frac{1}{2}\theta = \sqrt{\frac{1 - \cos\theta}{2}} \qquad \cos\frac{1}{2}\theta = \sqrt{\frac{1 + \cos\theta}{2}} \qquad \tan\frac{1}{2}\theta = \sqrt{\frac{1 - \cos\theta}{1 + \cos\theta}}$$

$\sin(A \pm B) = \sin A\cos B \pm \cos A\sin B$

$\cos(A \pm B) = \cos A\cos B \mp \sin A\sin B$

$$\tan(A \pm B) = \frac{\tan A \pm \tan B}{1 \mp \tan A\tan B}$$

$\sin A \pm \sin B = 2\sin[\frac{1}{2}(A \pm B)]\cos[\frac{1}{2}(A \mp B)]$

$\cos A + \cos B = 2\cos[\frac{1}{2}(A + B)]\cos[\frac{1}{2}(A - B)]$

$\cos A - \cos B = 2\sin[\frac{1}{2}(A + B)]\sin[\frac{1}{2}(B - A)]$

$$\tan A \pm \tan B = \frac{\sin(A \pm B)}{\cos A\cos B}$$

Example A-5

Use the isosceles right triangle shown in Figure A-14 to find the sine, cosine, and tangent of 45°.

It is clear from the figure that the two acute angles of this triangle are equal. Since the sum of the three angles in a triangle must equal 180° and the right angle is 90°, each acute angle must be 45°. If we multiply each side of any triangle by a common factor, we obtain a similar triangle with the same angles as the first. Since the trigonometric functions involve the ratios of only two sides of a triangle, we can choose any convenient length for one side. Let the equal sides of this triangle have a length of 1 unit. The length of the hypotenuse can then be found from the pythagorean theorem:

$$c = \sqrt{a^2 + b^2} = \sqrt{1^2 + 1^2} = \sqrt{2} \text{ units}$$

The sine, cosine, and tangent for the angle 45° are then given by Equations A-33, A-34, and A-35, respectively:

$$\sin 45° = \frac{1}{\sqrt{2}} = 0.707 \qquad \cos 45° = \frac{1}{\sqrt{2}} = 0.707 \qquad \tan 45° = \frac{1}{1} = 1$$

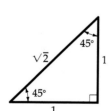

Figure A-14 An isosceles right triangle for Example A-5.

Example A-6

The sine of 30° is exactly $\frac{1}{2}$. Find the ratios of the sides of a 30–60° right triangle.

This common triangle is shown in Figure A-15. We choose a length of 1 unit for the side opposite the 30° angle. The hypotenuse is then obtained from the fact that sin 30° = 0.5:

$$c = \frac{a}{\sin 30°} = \frac{1}{0.5} = 2$$

The length of the side opposite the 60° angle is found from the pythagorean theorem:

$$b = \sqrt{c^2 - a^2} = \sqrt{2^2 - 1^2} = \sqrt{3}$$

From these results, we can obtain the following trigonometric functions for the angles 30° and 60°:

$$\cos 30° = \frac{b}{c} = \frac{\sqrt{3}}{2} = 0.866$$

$$\tan 30° = \frac{a}{b} = \frac{1}{\sqrt{3}} = 0.577$$

$$\sin 60° = \frac{b}{c} = \cos 30° = \frac{\sqrt{3}}{2} = 0.866$$

$$\cos 60° = \frac{a}{c} = \sin 30° = 0.500$$

$$\tan 60° = \frac{b}{a} = \frac{\sqrt{3}}{1} = 1.732$$

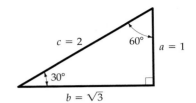

Figure A-15 A 30–60° right triangle for Example A-6.

For small angles, the length a is nearly equal to the arc length s, as can be seen in Figure A-16. The angle $\theta = s/c$ is therefore nearly equal to sin $\theta = a/c$:

$$\sin \theta \approx \theta \qquad \text{for small values of } \theta \qquad \text{A-46}$$

Similarly, the lengths c and b are nearly equal, so tan $\theta = a/b$ is nearly equal to both θ and sin θ for small values of θ:

$$\tan \theta \approx \sin \theta \approx \theta \qquad \text{(for small values of } \theta) \qquad \text{A-47}$$

Figure A-16 For small angles, sin $\theta = a/c$, tan $\theta = a/b$, and the angle $\theta = s/c$ are all approximately equal.

Equations A-46 and A-47 hold only if θ is measured in radians. Since cos $\theta = b/c$ and these lengths are nearly equal for small values of θ, we have

$$\cos \theta \approx 1 \qquad \text{for small values of } \theta \qquad \text{A-48}$$

Example A-7

By how much do sin θ, tan θ, and θ differ when $\theta = 15°$?

This angle in radians is

$$\theta = 15° \frac{2\pi \text{ rad}}{360°} = 0.262 \text{ rad}$$

Using a calculator or a table of trigonometric functions, we obtain

$$\sin 15° = 0.259$$

and

$$\tan 15° = 0.268$$

Thus, sin θ and θ (in radians) differ by 0.003 or about 1 percent, and tan θ and θ differ by 0.006 or about 2 percent. For small angles, the approximation $\theta \approx \sin \theta \approx \tan \theta$ is even more accurate.

Example A-7 shows that if accuracy of a few percent is needed, small angle approximations can be used only for angles of about 15° or less. Figure A-17 shows graphs of θ, sin θ, and tan θ versus θ, for small values of θ.

Figure A-17 Graphs of tan θ, θ, and sin θ versus θ for small values of θ.

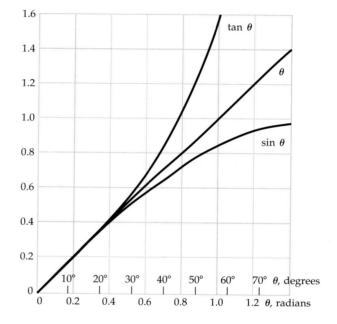

Figure A-18 shows an obtuse angle with its vertex at the origin and one side along the x axis. The trigonometric functions for a general angle such as this are defined by

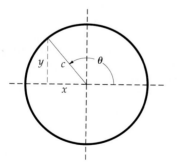

Figure A-18 Diagram for defining the trigonometric functions for an obtuse angle.

$$\sin \theta = \frac{y}{c} \qquad \text{A-49}$$

$$\cos \theta = \frac{x}{c} \qquad \text{A-50}$$

$$\tan \theta = \frac{y}{x} \qquad \text{A-51}$$

Figure A-19 shows plots of these functions versus θ. All trigonometric functions have a period of 2π. That is, when an angle changes by 2π rad, the function returns to its original value. Thus, sin $(\theta + 2\pi) = $ sin θ and so forth. Some other useful relations are

$$\sin (\pi - \theta) = \sin \theta \qquad \text{A-52}$$

$$\cos (\pi - \theta) = -\cos \theta \qquad \text{A-53}$$

$$\sin (\pi/2 - \theta) = \cos \theta \qquad \text{A-54}$$

$$\cos (\pi/2 - \theta) = \sin \theta \qquad \text{A-55}$$

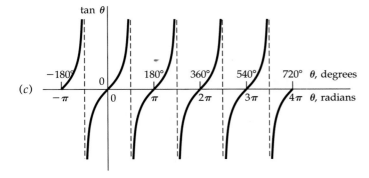

Figure A-19 The trigonometric functions sin θ, cos θ, and tan θ versus θ.

The trigonometric functions can be expressed as power series in θ. The series for sin θ and cos θ are

$$\sin \theta = \theta - \frac{\theta^3}{3!} + \frac{\theta^5}{5!} - \frac{\theta^7}{7!} + \cdots \qquad \text{A-56}$$

$$\cos \theta = 1 - \frac{\theta^2}{2!} + \frac{\theta^4}{4!} - \frac{\theta^6}{6!} + \cdots \qquad \text{A-57}$$

When θ is small, good approximations are obtained using only the first few terms in the series.

The Binomial Expansion

The binomial theorem is very useful for making approximations. One form of this theorem is

$$(1 + x)^n = 1 + nx + \frac{n(n - 1)}{2!} x^2 + \frac{n(n - 1)(n - 2)}{3!} x^3$$

$$+ \frac{n(n - 1)(n - 2)(n - 3)}{4!} x^4 + \cdots \qquad \text{A-58}$$

If n is a positive integer, there are $n + 1$ terms in this series. If n is a real number other than a positive integer, there are an infinite number of terms. The series is valid for any value of n if x^2 is less than 1. It is also valid for $x^2 = 1$ if n is positive. The series is particularly useful if $|x|$ is much less than 1. Then each term is much smaller than the previous term and we can drop all but the first two or three terms in the equation. If $|x|$ is much less than 1, we have

$$(1 + x)^n \approx 1 + nx \qquad |x| \ll 1 \qquad \text{A-59}$$

Example A-8

Use Equation A-59 to find an approximate value for the square root of 101.

We first restate the problem in the form $(1 + x)^n$ with x much less than 1:

$$(101)^{1/2} = (100 + 1)^{1/2} = (100)^{1/2}(1 + 0.01)^{1/2} = 10(1 + 0.01)^{1/2}$$

We can now use Equation A-59 with $n = \frac{1}{2}$ and $x = 0.01$:

$$(1 + 0.01)^{1/2} \approx 1 + \tfrac{1}{2}(0.01) = 1.005$$

Then

$$(101)^{1/2} \approx 10(1.005) = 10.05$$

We can get an idea of the accuracy of this approximation by looking at the first term in Equation A-58 that has been neglected. This term is

$$\frac{n(n - 1)}{2}x^2$$

Since x is 0.01, x^2 is 0.0001 and

$$\frac{n(n - 1)}{2}x^2 = \frac{\tfrac{1}{2}(-\tfrac{1}{2})}{2}(0.0001) = -\frac{0.0001}{8} \approx -0.00001 = -0.001\%$$

We therefore expect our answer to be correct to within about 0.001 percent. The value of $(101)^{1/2}$ to eight significant figures is 10.049876, which differs from our approximate value of 10.05 by 0.000124 or about 0.001% of 10.05.

Complex Numbers

A general complex number z can be written

$$z = a + bi \qquad\qquad \text{A-60}$$

where a and b are real numbers and $i = \sqrt{-1}$. The quantity a is called the real part and the quantity ib is called the imaginary part of z. We can represent a complex number in a plane as shown in Figure A-20, where the x axis is the real axis and the y axis is the imaginary axis. We can use the relations $a = r \cos\theta$ and $b = r \sin\theta$ from Figure A-20 to write the complex number z in polar coordinates:

$$z = r \cos\theta + (r \sin\theta)\,i \qquad\qquad \text{A-61}$$

where $r = \sqrt{a^2 + b^2}$ is called the magnitude of z.

When complex numbers are added or subtracted, the real and imaginary parts are added or subtracted separately:

$$z_1 + z_2 = (a_1 + b_1 i) + (a_2 + b_2 i) = (a_1 + a_2) + (b_1 + b_2)i \qquad \text{A-62}$$

However, when two complex numbers are multiplied, each part of one number is multiplied by each part of the other number:

$$z_1 z_2 = (a_1 + b_1 i)(a_2 + b_2 i) = a_1 a_2 + b_1 b_2 i^2 + (a_1 b_2 + a_2 b_1)i \qquad \text{A-63}$$
$$= a_1 a_2 - b_1 b_2 + (a_1 b_2 + a_2 b_1)i$$

where we have used $i^2 = -1$.

The complex conjugate of a complex number z^* is that number obtained by replacing i with $-i$:

$$z^* = (a + bi)^* = a - bi \qquad\qquad \text{A-64}$$

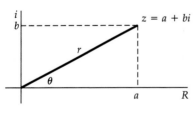

$$z = a + bi$$
$$= r \cos\theta + (r \sin\theta)i$$
$$= r(\cos\theta + i \sin\theta)$$

Figure A-20 Representation of a complex number in a plane. The real part of the complex number is plotted along the horizontal axis, and the imaginary part is plotted along the vertical axis.

The product of a complex number and its complex conjugate equals the square of the magnitude of the number:

$$zz^* = (a + bi)(a - bi) = a^2 + b^2 \qquad \text{A-65}$$

A particularly useful function of a complex number is the exponential $e^{i\theta}$. Using the expansion for e^x given in Table A-2, we have

$$e^{i\theta} = 1 + i\theta + \frac{(i\theta)^2}{2!} + \frac{(i\theta)^3}{3!} + \frac{(i\theta)^4}{4!} + \cdots$$

Using $i^2 = -1$, $i^3 = -i$, $i^4 = +1$, and so forth and separating the real parts from the imaginary parts, this expansion can be written

$$e^{i\theta} = 1 - \frac{\theta^2}{2!} + \frac{\theta^4}{4!} + \cdots + i(\theta - \frac{\theta^3}{3!} + \cdots)$$

Comparing this result with Equations A-56 and A-57, we can see that

$$e^{i\theta} = \cos\theta + i\sin\theta \qquad \text{A-66}$$

Using this result, we can express a general complex number as an exponential:

$$z = a + bi = r\cos\theta + (r\sin\theta)i = re^{i\theta} \qquad \text{A-67}$$

where $r = \sqrt{a^2 + b^2}$.

Differential Calculus

When we say that x is a function of t, we mean that for each value of t there is a corresponding value of x. An example is $x = At^2$, where A is a constant. To indicate that x is a function of t, we sometimes write $x(t)$ for x. Figure A-21 is a graph of x versus t for a typical function $x(t)$. At a particular value $t = t_1$, x has the value of x_1 as indicated. At another value t_2, x has the value x_2. The change in t, $t_2 - t_1$, is written $\Delta t = t_2 - t_1$ and the corresponding change in x is written $\Delta x = x_2 - x_1$. The ratio $\Delta x/\Delta t$ is the slope of the straight line connecting (x_1, t_1) and (x_2, t_2). If we make Δt smaller and smaller, the line connecting (x_1, t_1) and (x_2, t_2) approaches the line that is tangent to the curve at the point (x_1, t_1). The slope of this tangent line is called the derivative of x with respect to t and is written dx/dt:

$$\frac{dx}{dt} = \lim_{\Delta t \to 0} \frac{\Delta x}{\Delta t} \qquad \text{A-68}$$

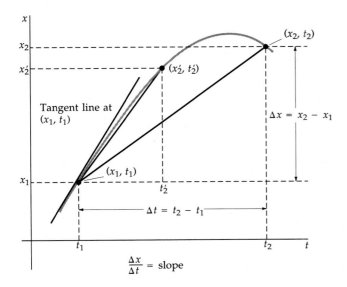

Figure A-21 Graph of a typical function $x(t)$. The points (x_1, t_1) and (x_2, t_2) are connected by a straight line. The slope of this line is $\Delta x/\Delta t$. As the time interval beginning at t_1 is decreased, the slope for that interval approaches the slope of the line tangent to the curve at time t_1, which is the derivative of x with respect to t.

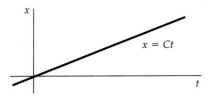

Figure A-22 Graph of the linear function $x = Ct$. This function has a constant slope C.

The derivative of a function of t is another function of t. If x is a constant, the graph of x versus t is a horizontal line with zero slope. The derivative of a constant is thus zero. In Figure A-22, x is proportional to t:

$$x = Ct$$

This function has a constant slope equal to C. Thus the derivative of Ct is C. Table A-4 lists some properties of derivatives and the derivatives of some particular functions that occur often in physics. It is followed by comments aimed at making these properties and rules clearer. More detailed discussion can be found in most calculus books.

Table A-4   Properties of Derivatives and Derivatives of Particular Functions

Linearity

1. The derivative of a constant times a function equals the constant times the derivative of the function:

$$\frac{d}{dt}[Cf(t)] = C\frac{df(t)}{dt}$$

2. The derivative of a sum of functions equals the sum of the derivatives of the functions:

$$\frac{d}{dt}[f(t) + g(t)] = \frac{df(t)}{dt} + \frac{dg(t)}{dt}$$

Chain rule

3. If f is a function of x and x is in turn a function of t, the derivative of f with respect to t equals the product of the derivative of f with respect to x and the derivative of x with respect to t:

$$\frac{d}{dt}f(x) = \frac{df}{dx}\frac{dx}{dt}$$

Derivative of a product

4. The derivative of a product of functions $f(t)g(t)$ equals the first function times the derivative of the second plus the second function times the derivative of the first:

$$\frac{d}{dt}[f(t)g(t)] = f(t)\frac{dg(t)}{dt} + \frac{df(t)}{dt}g(t)$$

Reciprocal derivative

5. The derivative of t with respect to x is the reciprocal of the derivative of x with respect to t, assuming that neither derivative is zero:

$$\frac{dx}{dt} = \left(\frac{dt}{dx}\right)^{-1} \quad \text{if} \quad \frac{dt}{dx} \neq 0$$

Derivatives of particular functions

6. $\dfrac{dC}{dt} = 0$ where C is a constant

7. $\dfrac{d(t^n)}{dt} = nt^{n-1}$

8. $\dfrac{d}{dt}\sin \omega t = \omega \cos \omega t$

9. $\dfrac{d}{dt}\cos \omega t = -\omega \sin \omega t$

10. $\dfrac{d}{dt}\tan \omega t = \omega \sec^2 \omega t$

11. $\dfrac{d}{dt}e^{bt} = be^{bt}$

12. $\dfrac{d}{dt}\ln bt = \dfrac{1}{t}$

Example A-9

Find the derivative of $x = at^2 + bt + c$, where a, b, and c are constants.

From rule 2, we can differentiate each term separately and add the results. Using rules 1 and 7, we have

$$\frac{d(at^2)}{dt} = 2at^1 = 2at$$

Similarly, $d(bt)/dt = b$ and $dc/dt = 0$. Adding these results, we obtain

$$\frac{dx}{dt} = 2at + b$$

Comments on Rules 1 through 5

Rules 1 and 2 follow from the fact that the limiting process is linear. We can understand rule 3, the chain rule, by multiplying $\Delta f/\Delta t$ by $\Delta x/\Delta x$ and noting that, since x is a function of t, both Δx and Δf approach zero as Δt approaches zero. Since the limit of a product of two functions equals the product of their limits, we have

$$\lim_{\Delta t \to 0} \frac{\Delta f}{\Delta t} = \lim_{\Delta t \to 0} \frac{\Delta f}{\Delta x}\frac{\Delta x}{\Delta t} = \left(\lim_{\Delta x \to 0}\frac{\Delta f}{\Delta x}\right)\left(\lim_{\Delta t \to 0}\frac{\Delta x}{\Delta t}\right) = \frac{df}{dx}\frac{dx}{dt}$$

Rule 4 is not immediately apparent. The derivative of a product of functions is the limit of the ratio

$$\frac{f(t + \Delta t)g(t + \Delta t) - f(t)g(t)}{\Delta t}$$

If we add and subtract the quantity $f(t + \Delta t)g(t)$ in the numerator, we can write this ratio as

$$\frac{f(t + \Delta t)g(t + \Delta t) - f(t + \Delta t)g(t) + f(t + \Delta t)g(t) - f(t)g(t)}{\Delta t}$$

$$= f(t + \Delta t)\left[\frac{g(t + \Delta t) - g(t)}{\Delta t}\right] + g(t)\left[\frac{f(t + \Delta t) - f(t)}{\Delta t}\right]$$

As Δt approaches zero, the terms in brackets become $dg(t)/dt$ and $df(t)/dt$, respectively, and the limit of the expression is

$$f(t)\frac{dg(t)}{dt} + g(t)\frac{df(t)}{dt}$$

Rule 5 follows directly from the definition:

$$\frac{dx}{dt} = \lim_{\Delta t \to 0}\frac{\Delta x}{\Delta t} = \lim_{\Delta t \to 0}\left(\frac{\Delta t}{\Delta x}\right)^{-1} = \lim_{\Delta x \to 0}\left(\frac{\Delta t}{\Delta x}\right)^{-1} = \left(\lim_{\Delta x \to 0}\frac{\Delta t}{\Delta x}\right)^{-1}$$

Comments on Rule 7

We can obtain this important result using the binomial expansion. We have

$$f(t) = t^n$$

$$f(t + \Delta t) = (t + \Delta t)^n = t^n\left(1 + \frac{\Delta t}{t}\right)^n$$

$$= t^n\left[1 + n\frac{\Delta t}{t} + \frac{n(n - 1)}{2!}\left(\frac{\Delta t}{t}\right)^2 + \frac{n(n - 1)(n - 2)}{3!}\left(\frac{\Delta t}{t}\right)^3 + \cdots\right]$$

Then

$$f(t - \Delta t) - f(t) = t^n \left[n \frac{\Delta t}{t} + \frac{n(n - 1)}{2!} \left(\frac{\Delta t}{t} \right)^2 + \cdots \right]$$

and

$$\frac{f(t - \Delta t) - f(t)}{\Delta t} = nt^{n-1} + \frac{n(n - 1)}{2!} t^{n-2} \Delta t + \cdots$$

The next term omitted from the last sum is proportional to $(\Delta t)^2$, the following to $(\Delta t)^3$, and so on. Each term except the first approaches zero as Δt approaches zero. Thus

$$\frac{df}{dt} = \lim_{\Delta x \to 0} \frac{f(t + \Delta t) + f(t)}{\Delta t} = nt^{n-1}$$

Comments on Rules 8 to 10

We first write $\sin \omega t = \sin \theta$ with $\theta = \omega t$ and use the chain rule,

$$\frac{d \sin \theta}{dt} = \frac{d \sin \theta}{d\theta} \frac{d\theta}{dt} = \omega \frac{d \sin \theta}{d\theta}$$

We then use the trigonometric formula for the sine of the sum of two angles θ and $\Delta\theta$:

$$\sin (\theta + \Delta\theta) = \sin \Delta\theta \cos \theta + \cos \Delta\theta \sin \theta$$

Since $\Delta\theta$ is to approach zero, we can use the small-angle approximations

$$\sin \Delta\theta \approx \Delta\theta \qquad \text{and} \qquad \cos \Delta\theta \approx 1$$

Then

$$\sin (\theta + \Delta\theta) \approx \Delta\theta \cos \theta + \sin \theta$$

and

$$\frac{\sin (\theta + \Delta\theta) - \sin \theta}{\Delta\theta} \approx \cos \theta$$

Similar reasoning can be applied to the cosine function to obtain rule 9.

Rule 10 is obtained by writing $\tan \theta = \sin \theta / \cos \theta$ and applying Rule 4 along with Rules 8 and 9.

$$\frac{d}{dt} (\tan \theta) = \frac{d}{dt} (\sin \theta)(\cos \theta)^{-1} = \sin \theta \frac{d}{dt} (\cos \theta)^{-1} + \frac{d (\sin \theta)}{dt} (\cos \theta)^{-1}$$

$$= \sin \theta (-1)(\cos \theta)^{-2}(-\sin \theta) + (\cos \theta)(\cos \theta)^{-1}$$

$$= \frac{\sin^2 \theta}{\cos^2 \theta} + 1 = \tan^2 \theta + 1 = \sec^2 \theta$$

Comments on Rule 11

Again we use the chain rule

$$\frac{de^\theta}{dt} = b \frac{de^\theta}{d\theta} \qquad \text{with} \qquad \theta = bt$$

and the series expansion for the exponential function:

$$e^{\theta + \Delta\theta} = e^\theta e^{\Delta\theta} = e^\theta \left[1 + \Delta\theta + \frac{(\Delta\theta)^2}{2!} + \frac{(\Delta\theta)^3}{3!} + \cdots \right]$$

Then

$$\frac{e^{\theta + \Delta\theta} - e^\theta}{\Delta\theta} = e^\theta + e^\theta \frac{\Delta\theta}{2!} + e^\theta \frac{(\Delta\theta)^2}{3!} + \cdots$$

As $\Delta\theta$ approaches zero, the right side of the equation above approaches e^θ.

Comments on Rule 12

Let

$$y = \ln bt$$

Then

$$e^y = bt \quad \text{and} \quad \frac{dt}{dy} = \frac{1}{b} e^y = t$$

Then using rule 5, we obtain

$$\frac{dy}{dt} = \left(\frac{dt}{dy}\right)^{-1} = \frac{1}{t}$$

Integral Calculus

Integration is related to the problem of finding the area under a curve. It is also the inverse of differentiation. Figure A-23 shows a function $f(t)$. The area of the shaded element is approximately $f_i \, \Delta t_i$, where f_i is evaluated anywhere in the interval Δt_i. This approximation improves if Δt_i is very small. The total area from t_1 to t_2 is found by summing all the area elements from t_1 to t_2 and taking the limit as each Δt_i approaches zero. This limit is called the integral of f over t and is written

$$\int_{t_1}^{t_2} f \, dt = Area = \lim_{\Delta t_i \to 0} \sum_i f_i \, \Delta t_i$$

If we integrate some function $f(t)$ from t_1 to some general value of t, we obtain another function of t. Let us call this function y:

$$y = \int_{t_1}^{t} f \, dt$$

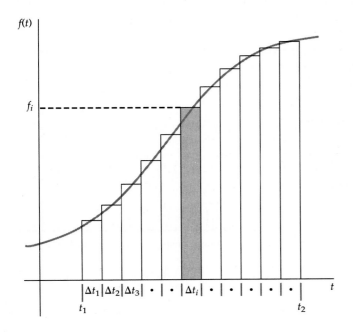

Figure A-23   A general function $f(t)$. The area of the shaded element is approximately $f_i \, \Delta t_i$, where f_i is evaluated anywhere in the interval.

The function y is the area under the f-versus-t curve from t_1 to a general value t. For a small interval Δt, the change in the area Δy is approximately $f \Delta t$.

$$\Delta y \approx f \Delta t$$

$$f \approx \frac{\Delta y}{\Delta t}$$

If we take the limit as Δt approaches 0, we can see that f is the derivative of y:

$$f = \frac{dy}{dt}$$

The relation between y and f is often written

$$y = \int f\, dt$$

where $\int f\, dt$ is called an **indefinite integral**. To evaluate an indefinite integral, we find the function y whose derivative is f. The definite integral of f from t_1 to t_2 is $y(t_1) - y(t_2)$, where $df/dt = y$:

$$\int_{t_1}^{t_2} f\, dt = y(t_2) - y(t_1)$$

Example A-10

Find the indefinite integral of $f(t) = t$.

The function whose derivative is t is $\frac{1}{2}t^2$ plus any constant. Thus,

$$\int t\, dt = \tfrac{1}{2}t^2 + C$$

where C is any constant.

Table A-5 lists some important integration formulas. More extensive lists of differentiation and integration formulas can be found in handbooks such as Herbert Dwight's *Tables of Integrals and Other Mathematical Data*, fourth edition, Macmillan Publishing Company, Inc., New York, 1961.

Table A-5 **Integration Formulas†**	
1. $\int A\, dt = At$	5. $\int e^{bt}\, dt = \dfrac{1}{b} e^{bt}$
2. $\int At\, dt = \frac{1}{2}At^2$	6. $\int \cos \omega t\, dt = \dfrac{1}{\omega} \sin \omega t$
3. $\int At^n\, dt = A\,\dfrac{t^{n+1}}{n+1} \quad n \neq -1$	7. $\int \sin \omega t\, dt = -\dfrac{1}{\omega} \cos \omega t$
4. $\int At^{-1}\, dt = A \ln t$	

†In these formulas, A, b, and ω are constants. An arbitrary constant C can be added to the right side of each equation.

Appendix B

SI Units

Basic Units	
Length	The *meter* (m) is the distance traveled by light in a vacuum in 1/299,792,458 s
Time	The *second* (s) is the duration of 9,192,631,770 periods of the radiation corresponding to the transition between the two hyperfine levels of the ground state of the ^{133}Cs atom
Mass	The *kilogram* (kg) is the mass of the international standard body preserved at Sèvres, France
Current	The *ampere* (A) is that current in two very long parallel wires 1 m apart that gives rise to a magnetic force per unit length of 2×10^{-7} N/m
Temperature	The *kelvin* (K) is 1/273.16 of the thermodynamic temperature of the triple point of water
Luminous intensity	The *candela* (cd) is the luminous intensity, in the perpendicular direction, of a surface of area 1/600,000 m^2 of a blackbody at the temperature of freezing platinum at a pressure of 1 atm

Derived Units		
Force	newton (N)	$1 \text{ N} = 1 \text{ kg·m/s}^2$
Work, energy	joule (J)	$1 \text{ J} = 1 \text{ N·m}$
Power	watt (W)	$1 \text{ W} = 1 \text{ J/s}$
Frequency	hertz (Hz)	$1 \text{ Hz} = \text{s}^{-1}$
Charge	coulomb (C)	$1 \text{ C} = 1 \text{ A·s}$
Potential	volt (V)	$1 \text{ V} = 1 \text{ J/C}$
Resistance	ohm (Ω)	$1 \text{ }\Omega = 1 \text{ V/A}$
Capacitance	farad (F)	$1 \text{ F} = 1 \text{ C/V}$
Magnetic field	tesla (T)	$1 \text{ T} = 1 \text{ N/A·m}$
Magnetic flux	weber (Wb)	$1 \text{ Wb} = 1 \text{ T·m}^2$
Inductance	henry (H)	$1 \text{ H} = 1 \text{ J/A}^2$

Appendix C

Numerical Data

Terrestrial Data	
Acceleration of gravity g	
Standard value	9.80665 m/s^2
	32.1740 ft/s^2
At sea level, at equator†	9.7804 m/s^2
At sea level, at poles†	9.8322 m/s^2
Mass of earth M_E	5.98×10^{24} kg
Radius of earth R_E, mean	6.37×10^6 m
	3960 mi
Escape speed $\sqrt{2R_E g}$	1.12×10^4 m/s
	6.95 mi/s
Solar constant‡	1.35 kW/m^2
Standard temperature and pressure (STP):	
Temperature	273.15 K
Pressure	101.325 kPa
	1.00 atm
Molar mass of air	28.97 g/mol
Density of air (STP), ρ_{air}	1.293 kg/m^3
Speed of sound (STP)	331 m/s
Heat of fusion of H_2O (0°C, 1 atm)	333.5 kJ/kg
Heat of vaporization of H_2O (100°C, 1 atm)	2.257 MJ/kg

†Measured relative to the earth's surface.
‡Average power incident normally on 1 m^2 outside the earth's atmosphere at the mean
distance from the earth to the sun.

Astronomical Data	
Earth	
Distance to moon†	3.844×10^8 m
	2.389×10^5 mi
Distance to sun, mean†	1.496×10^{11} m
	9.30×10^7 mi
	1.00 AU
Orbital speed, mean	2.98×10^4 m/s
Moon	
Mass	7.35×10^{22} kg
Radius	1.738×10^6 m
Period	27.32 d
Acceleration of gravity at surface	1.62 m/s^2
Sun	
Mass	1.99×10^{30} kg
Radius	6.96×10^8 m

†Center to center.

Physical Constants

Gravitational constant	G	$6.672\ 6 \times 10^{-11}$ N·m^2/kg^2
Speed of light	c	$2.997\ 924\ 58 \times 10^8$ m/s
Electron charge	e	$1.602\ 177 \times 10^{-19}$ C
Avogadro's number	N_A	$6.022\ 137 \times 10^{23}$ particles/mol
Gas constant	R	$8.314\ 51$ J/mol·K
		$1.987\ 22$ cal/mol·K
		$8.205\ 78 \times 10^{-2}$ L·atm/mol·K
Boltzmann's constant	$k = R/N_A$	$1.380\ 658 \times 10^{-23}$ J/K
		$8.617\ 385 \times 10^{-5}$ eV/K
Unified mass unit	$u = (1/N_A)$ g	$1.660\ 540 \times 10^{-24}$ g
Coulomb constant	$k = 1/4\pi\epsilon_0$	$8.987\ 551\ 788 \times 10^9$ N·m^2/C^2
Permittivity of free space	ϵ_0	$8.854\ 187\ 817 \times 10^{-12}$ C^2/N·m^2
Permeability of free space	μ_0	$4\pi \times 10^{-7}$ N/A^2
		$1.256\ 637 \times 10^{-6}$ N/A^2
Planck's constant	h	$6.626\ 076 \times 10^{-34}$ J·s
		$4.135\ 669 \times 10^{-15}$ eV·s
	$\hbar = h/2\pi$	$1.054\ 573 \times 10^{-34}$ J·s
		$6.582\ 122 \times 10^{-16}$ eV·s
Mass of electron	m_e	$9.109\ 390 \times 10^{-31}$ kg
		$510.999\ 1$ keV/c^2
Mass of proton	m_p	$1.672\ 623 \times 10^{-27}$ kg
		$938.272\ 3$ MeV/c^2
Mass of neutron	m_n	$1.674\ 929 \times 10^{-27}$ kg
		$939.565\ 6$ MeV/c^2
Bohr magneton	$m_B = e\hbar/2m_e$	$9.274\ 015\ 4 \times 10^{-24}$ J/T
		$5.788\ 382\ 63 \times 10^{-5}$ eV/T
Nuclear magneton	$m_n = e\hbar/2m_p$	$5.050\ 786\ 6 \times 10^{-27}$ J/T
		$3.152\ 451\ 66 \times 10^{-8}$ eV/T
Magnetic flux quantum	$\phi_0 = h/2e$	$2.067\ 834\ 6 \times 10^{-15}$ T·m^2
Quantized Hall resistance	$R_K = h/e^2$	$2.581\ 280\ 7 \times 10^4$ Ω
Rydberg constant	R_H	$1.097\ 373\ 153\ 4 \times 10^7$ m^{-1}
Josephson frequency–voltage quotient	$2e/h$	$4.835\ 979 \times 10^{14}$ Hz/V
Compton wavelength	$\lambda_C = h/m_e c$	$2.426\ 310\ 58 \times 10^{-12}$ m

For additional data, see the front and back endpapers and the following tables in the text

Appendix D

Conversion Factors

Conversion factors are written as equations for simplicity; relations marked with an asterisk are exact.

Length

1 km = 0.6215 mi

1 mi = 1.609 km

1 m = 1.0936 yd = 3.281 ft = 39.37 in

*1 in = 2.54 cm

*1 ft = 12 in = 30.48 cm

*1 yd = 3 ft = 91.44 cm

1 lightyear = 1 $c \cdot$y = 9.461 × 10^{15} m

*1 Å = 0.1 nm

Area

*1 m^2 = 10^4 cm^2

1 km^2 = 0.3861 mi^2 = 247.1 acres

*1 in^2 = 6.4516 cm^2

1 ft^2 = 9.29 × 10^{-2} m^2

1 m^2 = 10.76 ft^2

*1 acre = 43,560 ft^2

1 mi^2 = 640 acres = 2.590 km^2

Volume

*1 m^3 = 10^6 cm^3

*1 L = 1000 cm^3 = 10^{-3} m^3

1 gal = 3.786 L

1 gal = 4 qt = 8 pt = 128 oz = 231 in^3

1 in^3 = 16.39 cm^3

1 ft^3 = 1728 in^3 = 28.32 L = 2.832 × 10^4 cm^3

Time

*1 h = 60 min = 3.6 ks

*1 d = 24 h = 1440 min = 86.4 ks

1 y = 365.24 d = 31.56 Ms

Speed

1 km/h = 0.2778 m/s = 0.6215 mi/h

1 mi/h = 0.4470 m/s = 1.609 km/h

1 mi/h = 1.467 ft/s

Angle and Angular Speed

*π rad = 180°

1 rad = 57.30°

1° = 1.745 × 10^{-2} rad

1 rev/min = 0.1047 rad/s

1 rad/s = 9.549 rev/min

Mass

*1 kg = 1000 g

*1 tonne = 1000 kg = 1 Mg

1 u = 1.6606 × 10^{-27} kg

1 kg = 6.022 × 10^{23} u

1 slug = 14.59 kg

1 kg = 6.852 × 10^{-2} slug

1 u = 931.50 MeV/c^2

Density

*1 g/cm^3 = 1000 kg/m^3 = 1 kg/L

(1 g/cm^3)g = 62.4 lb/ft^3

Force

1 N = 0.2248 lb = 10^5 dyn

1 lb = 4.4482 N

(1 kg)g = 2.2046 lb

Pressure

*1 Pa = 1 N/m^2

*1 atm = 101.325 kPa = 1.01325 bars

1 atm = 14.7 lb/in^2 = 760 mmHg

      = 29.9 inHg = 33.8 ftH$_2$O

1 lb/in^2 = 6.895 kPa

1 torr = 1 mmHg = 133.32 Pa

1 bar = 100 kPa

Energy

*1 kW·h = 3.6 MJ

*1 cal = 4.1840 J

1 ft·lb = 1.356 J = 1.286 × 10^{-3} Btu

*1 L·atm = 101.325 J

1 L·atm = 24.217 cal

1 Btu = 778 ft·lb = 252 cal = 1054.35 J

1 eV = 1.602 × 10^{-19} J

1 u·c^2 = 931.50 MeV

*1 erg = 10^{-7} J

Power

1 horsepower = 550 ft·lb/s = 745.7 W

1 Btu/min = 17.58 W

1 W = 1.341 × 10^{-3} horsepower

      = 0.7376 ft·lb/s

Magnetic Field

*1 G = 10^{-4} T

*1 T = 10^4 G

Thermal Conductivity

1 W/m·K = 6.938 Btu·in/h·ft^2·F°

1 Btu·in/h·ft^2·F° = 0.1441 W/m·K

Appendix E Periodic Table

1	2	3	4	5	6	7	8	9	10	11	12	13	14	15	16	17	18
1 **H** 1.00797																	2 **He** 4.003
3 **Li** 6.941	4 **Be** 9.012											5 **B** 10.81	6 **C** 12.011	7 **N** 14.007	8 **O** 15.9994	9 **F** 19.00	10 **Ne** 20.179
11 **Na** 22.990	12 **Mg** 24.31											13 **Al** 26.98	14 **Si** 28.09	15 **P** 30.974	16 **S** 32.064	17 **Cl** 35.453	18 **Ar** 39.948
19 **K** 39.102	20 **Ca** 40.08	21 **Sc** 44.96	22 **Ti** 47.88	23 **V** 50.94	24 **Cr** 52.00	25 **Mn** 54.94	26 **Fe** 55.85	27 **Co** 58.93	28 **Ni** 58.69	29 **Cu** 63.55	30 **Zn** 65.38	31 **Ga** 69.72	32 **Ge** 72.59	33 **As** 74.92	34 **Se** 78.96	35 **Br** 79.90	36 **Kr** 83.80
37 **Rb** 85.47	38 **Sr** 87.62	39 **Y** 88.906	40 **Zr** 91.22	41 **Nb** 92.91	42 **Mo** 95.94	43 **Tc** (98)	44 **Ru** 101.1	45 **Rh** 102.905	46 **Pd** 106.4	47 **Ag** 107.870	48 **Cd** 112.41	49 **In** 114.82	50 **Sn** 118.69	51 **Sb** 121.75	52 **Te** 127.60	53 **I** 126.90	54 **Xe** 131.29
55 **Cs** 132.905	56 **Ba** 137.33	57–71 **Rare Earths**	72 **Hf** 178.49	73 **Ta** 180.95	74 **W** 183.85	75 **Re** 186.2	76 **Os** 190.2	77 **Ir** 192.2	78 **Pt** 195.09	79 **Au** 196.97	80 **Hg** 200.59	81 **Tl** 204.37	82 **Pb** 207.19	83 **Bi** 208.98	84 **Po** (210)	85 **At** (210)	86 **Rn** (222)
87 **Fr** (223)	88 **Ra** (226)	89–103 **Actinides**	104 **Rf** (261)	105 **Ha** (260)	106 (263)	107 (262)	108 (265)	109 (266)									

Rare Earths (Lanthanides)

57 **La** 138.91	58 **Ce** 140.12	59 **Pr** 140.91	60 **Nd** 144.24	61 **Pm** (147)	62 **Sm** 150.36	63 **Eu** 152.0	64 **Gd** 157.25	65 **Tb** 158.92	66 **Dy** 162.50	67 **Ho** 164.93	68 **Er** 167.26	69 **Tm** 168.93	70 **Yb** 173.04	71 **Lu** 174.97

Actinides

89 **Ac** 227.03	90 **Th** 232.04	91 **Pa** 231.04	92 **U** 238.03	93 **Np** 237.05	94 **Pu** (244)	95 **Am** (243)	96 **Cm** (247)	97 **Bk** (247)	98 **Cf** (251)	99 **Es** (252)	100 **Fm** (257)	101 **Md** (258)	102 **No** (259)	103 **Lr** (260)

The 1–18 group designation has been recommended by the International Union of Pure and Applied Chemistry (IUPAC).

Atomic Numbers and Atomic Masses

Actinium	Ac	89	227.03	Mercury	Hg	80	200.59
Aluminum	Al	13	26.98	Molybdenum	Mo	42	95.94
Americium	Am	95	(243)	Neodymium	Nd	60	144.24
Antimony	Sb	51	121.75	Neon	Ne	10	20.179
Argon	Ar	18	39.948	Neptunium	Np	93	237.05
Arsenic	As	33	74.92	Nickel	Ni	28	58.69
Astatine	At	85	(210)	Niobium	Nb	41	92.91
Barium	Ba	56	137.3	Nitrogen	N	7	14.007
Berkelium	Bk	97	(247)	Nobelium	No	102	(259)
Beryllium	Be	4	9.012	Osmium	Os	76	190.2
Bismuth	Bi	83	208.98	Oxygen	O	8	15.9994
Boron	B	5	10.81	Palladium	Pd	46	106.4
Bromine	Br	35	79.90	Phosphorus	P	15	30.974
Cadmium	Cd	48	112.41	Platinum	Pt	78	195.09
Calcium	Ca	20	40.08	Plutonium	Pu	94	(244)
Californium	Cf	98	(251)	Polonium	Po	84	(210)
Carbon	C	6	12.011	Potassium	K	19	39.102
Cerium	Ce	58	140.12	Praseodymium	Pr	59	140.91
Cesium	Cs	55	132.905	Promethium	Pm	61	(147)
Chlorine	Cl	17	35.453	Protactinium	Pa	91	231.04
Chromium	Cr	24	52.00	Radium	Ra	88	(226)
Cobalt	Co	27	58.93	Radon	Rn	86	(222)
Copper	Cu	29	63.55	Rhenium	Re	75	186.2
Curium	Cm	96	(247)	Rhodium	Rh	45	102.905
Dysprosium	Dy	66	162.50	Rubidium	Rb	37	85.47
Einsteinium	Es	99	(252)	Ruthenium	Ru	44	101.1
Erbium	Er	68	167.26	Rutherfordium	Rf	104	(261)
Europium	Eu	63	152.0	Samarium	Sm	62	150.36
Fermium	Fm	100	(257)	Scandium	Sc	21	44.96
Fluorine	F	9	19.00	Selenium	Se	34	78.96
Francium	Fr	87	(223)	Silicon	Si	14	28.09
Gadolinium	Gd	64	157.25	Silver	Ag	47	107.870
Gallium	Ga	31	69.72	Sodium	Na	11	22.990
Germanium	Ge	32	72.59	Strontium	Sr	38	87.62
Gold	Au	79	196.97	Sulfur	S	16	32.064
Hafnium	Hf	72	178.49	Tantalum	Ta	73	180.95
Hahnium	Ha	105	(260)	Technetium	Tc	43	(98)
Helium	He	2	4.003	Tellurium	Te	52	127.60
Holmium	Ho	67	164.93	Terbium	Tb	65	158.92
Hydrogen	H	1	1.00797	Thallium	Tl	81	204.37
Indium	In	49	114.82	Thorium	Th	90	232.04
Iodine	I	53	126.90	Thulium	Tm	69	168.93
Iridium	Ir	77	192.2	Tin	Sn	50	118.69
Iron	Fe	26	55.85	Titanium	Ti	22	47.88
Krypton	Kr	36	83.80	Tungsten	W	74	183.85
Lanthanum	La	57	138.91	Uranium	U	92	238.03
Lawrencium	Lr	103	(260)	Vanadium	V	23	50.94
Lead	Pb	82	207.19	Xenon	Xe	54	131.29
Lithium	Li	3	6.941	Ytterbium	Yb	70	173.04
Lutetium	Lu	71	174.97	Yttrium	Y	39	88.906
Magnesium	Mg	12	24.31	Zinc	Zn	30	65.38
Manganese	Mn	25	54.94	Zirconium	Zr	40	91.22
Mendelevium	Md	101	(258)				

Illustration Credits

Part Openers
Part 1 p. xix, 18 © James Sugar/Black Star; **p. 19** © D. Cavagnaro/Peter Arnold, Inc.
Part 2 p. xxii, 366–367 © Michael Freeman.
Part 3 p. xxiii, 484–485 Sandia National Laboratories.
Part 4 p. xxiv, 596–597 © 1988 Richard Megna/Fundamental Photos.
Part 5 p. xxvii, 972–973 Courtesy AT&T Archives.
Part 6 p. xxviii, 1098–1099 Courtesy AT&T Archives.

Chapter 1
Opener p. 1 The Granger Collection; **p. 3 (a)** The Granger Collection; **(b)** Frequency Electronics, Inc.; **p. 4 (a)** McDonald Observatory; **(b)** Bruce Coleman; **p. 5** Owen Franken/Stock, Boston; **p. 9 (a)** IBM Almaden Research Center; **(b)** © Lennart Nilsson, from *The Body Victorious*, Delacourt Press, 1985; **(c)** Kent and Donnan Dannen/Photo Researchers; **(d)** N.A.S.A. (68-HC-74); **(e)** Smithsonian Institution; **p. 11** B. A. Watkinson; **p. 12** Fermi Institute, University of Chicago; **p. 14** AIP Niels Bohr Library.

Chapter 2
Opener p. 20 © George D. Lepp/Comstock; **p. 24** © Gunter Ziesler/Peter Arnold, Inc.; **p. 29** Estate of Harold E. Edgerton/Courtesy Palm Press, Inc.; **p. 30** © Sidney Harris; **p. 34 (a)** Courtesy Stanford Linear Accelerator Center, U.S. Department of Energy; **(b)** John Seeman, SLAC; **p. 35** Courtesy General Motors Corp.

Chapter 3
Opener p. 47 © Lewis Portnoy/Spectra-Action, Inc.; **p. 60** © 1968 Fundamental Photographs; **p. 63** Scala/Art Resource; **p. 64** *PSSC Physics*, 2nd ed., 1965. D. C. Heath & Co. and Education Development Center, Inc., Newton, Massachusetts; **p. 65** N.A.S.A. (82-HC-78); **Figure 3-27** AIP Niels Bohr Library; **p. 68** N.A.S.A. (81-HC-408); **p. 75** Courtesy New York Public Library.

Chapter 4
Opener p. 77 The Granger Collection; **p. 79** (*left*) © George Hall/Woodfin Camp and Assoc.; (*right*) © Arthur Tilley/FPG International; **p. 83** © Berenice Abbott/Commerce Graphics, Ltd., Inc.; **p. 84** N.A.S.A. (76-HC-6); **p. 87** Culver Pictures; **p. 88 (a)** © Cotton Coulson/Woodfin Camp and Assoc.; **(b)** © 1972 Gary Ladd; **p. 89 (c)** Los Alamos Scientific Laboratory;

(d) © Science Photo Library/Photo Researchers; **p. 90** © Fundamental Photographs; **Figure 4-7 (b)** © Michael Abbey/Photo Researchers; **p. 95** Museum of Modern Art, New York; **p. 105** © David de Lossy/The Image Bank.

Chapter 5
Opener p. 107 Courtesy Center for Engineering Design, University of Utah, Salt Lake City; **p. 108 (a)** F. P. Bowden and D. Tabor, *Friction and Lubrication of Solids*, Oxford University Press, 1950. Reprinted by permission of the publisher; **(b)** Uzi Landman, W. David Luedtke, Georgia Institute of Technology; **Figure 5-5** © Jean-Claude Lejeune/Stock, Boston; **Figure 5-10 (b)** © Malraux Photography/The Image Bank; **p. 115** © Robert Alexander/Photo Researchers; **p. 116** (*top*) © Fundamental Photographs; **Figure 5-12 (b)** © Tom Walker/Stock, Boston; **p. 117** © **(a)** N.A.S.A. (81-HC-945); **(b)** © Jonathan Blair/Woodfin Camp and Assoc.; **p. 123** Sandia National Laboratories; **p. 124** N.A.S.A. (85-HC-345); **p. 131** © P. Thomann/The Image Bank.

Chapter 6
Opener p. 135 © 1990 Estate of Harold E. Edgerton/ Courtesy Palm Press, Inc.; **p. 139** Naples National Museum/Scala/Art Resource; **p. 142** © L. Grant/FPG International; **p. 148** © Steve Leonard/Black Star; **Figure 6-12 (b)** © T. Alvez/FPG International; **p. 156** © 1989 M. C. Escher Heirs/Cordon Art, Baarn, Holland; **p. 157** © 1973 Berenice Abbott/Photo Researchers; **p. 167** (*left*) USGS; (*right*) New York State Commerce Department; **p. 168 (a)** © Stan Sholik/FPG International; **(b)** © Leicester University/Science Photo Library/Photo Researchers; **p. 169** © Visual Horizons/FPG International.

Chapter 7
Opener p. 182 © David Parker/Science Photo Library/Photo Researchers; **Figure 7-1** © Estate of Harold E. Edgerton/Courtesy Palm Press, Inc.; **p. 198** © Estate of Harold E. Edgerton/Courtesy Palm Press, Inc.; **Figure 7-23** © Estate of Harold E. Edgerton/Courtesy Palm Press, Inc.; **Figure 7-24 (b)** © Estate of Harold E. Edgerton/Courtesy Palm Press, Inc.; **p. 205** Courtesy of Mercedes-Benz of North America, Montvale, New Jersey; **p. 209 (a)** © 1973 Berenice Abbott/ Photo Researchers; **(b)** Brookhaven National Laboratory; **(c) (1–4)** Alar and Juri Toomre, from *Cambridge*

Atlas of Astronomy, Cambridge University Press; **(d)** Kitt Peak National Observatory, from *Cambridge Atlas of Astronomy*, Cambridge University Press, 1985; **p. 211** © Dr. David Jones/Science Photo Library/Photo Researchers; **p. 215 (a)** N.A.S.A. (393-479)/Shostal Superstock; **(b)** © Bill Curtsinger/Photo Researchers; **p. 220** Courtesy Central Scientific Company.

Chapter 8

Opener p. 227 © David F. Malin/Anglo-Australian Telescope Board; **p. 229** Courtesy Center for Engineering Design, University of Utah, Salt Lake City; **p. 232** Culver Pictures; **p. 236** PAR/NYC, Inc. Archives; **p. 237 (a)** and **(b)** © David Malin, Anglo-Australian Telescope Board; **Figure 8-21** Focus on Sports; **Figure 8-22** © Estate of Harold E. Edgerton/Courtesy Palm Press, Inc.; **p. 247** © Dick Luria/FPG International; **p. 248 (a)** and **(b)** N.A.S.A. (84-HC-582); **p. 254** National Oceanographic and Atmospheric Administration; **p. 257** © Dick Luria/Photo Researchers; **p. 261** (*left*) Richard Minnix; (*right*) Courtesy Central Scientific Company.

Chapter 9

Opener p. 275 © Tom McHugh/Photo Researchers; **p. 276** Manitowoc Engineering Co., Manitowoc, Wisconsin; **p. 277** © Mike and Carol Werner/Comstock; **p. 278** © Malcolm Hanes/Bruce Coleman; **p. 281** © Marc Romanelli/The Image Bank; **p. 286** © Rick Rickman/Black Star; **p. 287** (*left*) © Yulsman/The Image Bank; (*right*) Demetrios Zangos.

Chapter 10

Opener p. 295 Collection of Historical Scientific Instruments, Harvard University; **p. 296** © Erich Lessing/Magnum; **p. 298 (a)** N.A.S.A.; **(b)** © Dennis di Cicco/Peter Arnold, Inc.; **p. 302** © Erich Lessing, Culture and Fine Arts Archive, Vienna; **Figure 10-8 (a)** Courtesy Deutsches Museum, Munich; **p. 306** Courtesy Central Scientific Company; **p. 308 (a)** *Physics Today*, May 1988; **(b)** N.A.S.A.; **p. 313** N.A.S.A. (69-HC-905); **p. 317 (b)** N.A.S.A. (80-HC-627); **Figure 1 p. 322** By permission of the Syndics of Cambridge University Library, Additional MS 3975, p. 15; **Figure 2 p. 323** From Voltaire, *The Elements of Sir Isaac Newton's Philosophy*, 1738, New York: Frank Cass, 1967, p. 97; **Figure 3 p. 323** From Newton, *Opticks*, 1730; **p. 327** Art Resource.

Chapter 11

Opener p. 331 Dr. Tai Il Mah, from *Popular Science*, June 1990; **p. 332** © Daemmrich/Stock, Boston; **p. 335 (a)** Scala/Art Resource; **(b)** © Jonathan Watts/Science Photo Library/Photo Researchers; **p. 341 (a)** © David Burnett/Woodfin Camp and Assoc.; **(b)** PAR/NYC, Inc. Archives; **p. 342** © Chuck O'Rear/Woodfin Camp and Assoc.; **p. 344 (a)** and **(b)** © Estate of Harold E. Edgerton/Courtesy Palm Press, Inc.; **p. 346 (a)** © Jerry Howard/Stock, Boston; **(b)** © Alan Oddie/Photo Edit; **Figure 11-16 (a)** © Richard Megna/Fundamental Photographs; **(b)** © Estate of Harold E. Edgerton/Courtesy

Palm Press, Inc.; **p. 348** Picker International; **p. 350 (a)** and **(b)** Office National d'Etudes et de Rechèrches Aérospatiales; **p. 353** Dr. Owen M. Griffin, Naval Research Laboratory; **Figure 3 p. 359** © Jack Zohrt/FPG International; **Figure 4 p. 359** Dave Custer, 1986, photo courtesy of Milliken Research Associates, Inc.

Chapter 12

Opener p. 368 © Fundamental Photographs; **p. 371** N.A.S.A. (73-HC-787); **p. 376** Institute for Marine Dynamics; **p. 383** © Berenice Abbott/Commerce Graphics Ltd., Inc.; **p. 388** © Estate of Harold E. Edgerton/Courtesy Palm Press, Inc.; **p. 390** Courtesy of Monroe Auto Equipment, Monroe, Michigan; **p. 392** Courtesy Citicorp Building, New York City; **p. 394** Royal Swedish Academy of Music, courtesy Prof. Thomas D. Rossing, Northern Illinois University, DeKalb; **p. 395** Courtesy Central Scientific Company **p. 402** (*top*) from H.-O. Peitgen, P. H. Richter, *The Beauty of Fractals*, Springer-Verlag, New York, 1986; (*bottom*) Prof. Julio M. Ottino, University of Massachusetts, Amherst.

Chapter 13

Opener p. 409 (c) © Hashi; **Figure 13-1 (b)** *PSSC Physics*, 2nd ed., 1965. D. C. Heath & Co. and Education Development Center, Inc., Newton, Massachusetts; **Figure 13-3 (b)** and **(d)** *PSSC Physics* 2nd ed., 1965. D. C. Heath & Co. and Education Development Center Inc., Newton, Massachusetts; **Figure 13-5 (a)** and **(b)** © Berenice Abbott/Photo Researchers; **p. 414 (a)** and **(b)** © Four By Five, Inc.; **p. 421 (a,b,c,d)** *PSSC Physics*, 2nd ed., 1965. D. C. Heath & Co. and Education Development Center, Inc., Newton, Massachusetts; **p. 422** © David Yost/Courtesy Steinway & Sons; **p. 424 (a,b,c)** University of Washington; **p. 428** Prof. Thomas D. Rossing, Northern Illinois University, DeKalb.

Chapter 14

Opener p. 439 From Winston E. Kock, *Lasers and Holography*, Dover Publications, New York, 1981; **Figure 14-4** *PSSC Physics*, 2nd ed., 1965. D. C. Heath & Co. and Education Development Center, Inc., Newton, Massachusetts; **Figure 14-10 (a)** © 1973 Berenice Abbott/Photo Researchers; **p. 459 (a)** from Adolphe Ganot, *Elementary Treatise on Physics*, William Wood & Publishers Ltd., 1910; **(b)** © 1989 Mike Chesser, Yamaha BBDO Worldwide Network, Los Angeles; **p. 460** Davis Hathaway/N.A.S.A./Courtesy of National Optical Astronomy Observatories; **p. 462 (a)** Courtesy Naval Research Laboratory; **(b)** Courtesy San Francisco Symphony; **Figure 14-28** and **Figure 14-30** *PSSC Physics*, 2nd ed., 1965. D. C. Heath & Co. and Education Development Center, Inc., Newton, Massachusetts; **p. 463** Dr. John S. Shelton; **p. 464 (a)** © 1986 Howard Sochurek/Medichrome; **(b)** N.A.S.A. (81-HC-670); **Figure 14-31 (a)** *PSSC Physics*, 2nd ed., 1965. D. C. Heath & Co. and Education Development Center, Inc., Newton, Massachusetts; **Figure 14-33 (b)** *PSSC Physics*,

Answers

These answers are calculated using $g = 9.81$ m/s^2 unless otherwise specified in the exercise or problem. The results are usually rounded to three significant figures. Differences in the last figure can easily result from differences in rounding the input data and are not important.

Chapter 1

True or False

1. True

2. False; for example, $x = vt$ where the speed v and the time t have different dimensions

3. True

Problems

1. (*a*) 1 MW   (*b*) 2 mg   (*c*) 3 μm   (*d*) 30 ks

3. (*a*) 1 picoboo   (*b*) 1 gigalow
(*c*) 1 microscope   (*d*) 1 attoboy   (*e*) 1 megaphone
(*f*) 1 nanogoat   (*g*) 1 terabull

5. (*a*) C_1, ft; C_2, ft/s   (*b*) C_1, ft/s^2 (*c*) C_1, ft/s^2
(*d*) C_1, ft; C_2, s^{-1} (*e*) C_1, ft/s; C_2, s^{-1}

7. (*a*) 62.1 mi/h   (*b*) 23.6 in   (*c*) 91.4 m

9. (1.61 km/h)/(1 mi/h)

11. (*a*) 36.00 km/h·s   (*b*) 10.00 m/s^2 (*c*) 88 ft/s
(*d*) 27 m/s

13. 4046.9 m^2

15. (*a*) C_1, L; C_2, L/T   (*b*) C_1, L/T^2 (*c*) C_1, L/T^2
(*d*) C_1, L; C_2, T^{-1} (*e*) C_1, L/T; C_2, T^{-1}

17. $F = mv^2/r$

19. (*a*) 30,000   (*b*) 0.0062   (*c*) 0.000004   (*d*) 217,000

21. (*a*) 1.14×10^5 (*b*) 2.25×10^{-8} (*c*) 8.27×10^3
(*d*) 6.27×10^2

23. 1 in = 3.63×10^6 cell membranes

25. (*a*) 1690   (*b*) 5   (*c*) 5.6   (*d*) 10

27. 3.51 Mm

29. (*a*) 6000 km   (*b*) 8760 tankers/y
(*c*) 43.8 billion dollars

31. (*a*) 1.41×10^{17} kg/m^3 (*b*) 216 m

33. (*a*) 4.92×10^9 m   (*b*) 12.8 r_{e-m}

35. (*a*) $T = Cm^n$, with $n = 0.5$ and $C = 1.81$ s/kg$^{1/2}$.
(*b*) Those points with the greatest deviation are at $T = 1.05$ and 1.75 s.

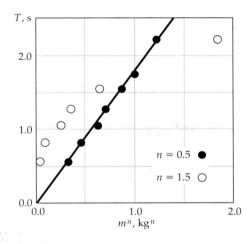

37. (*a*) $T = C\sqrt{L/g}$ (*b*) example results: $L = 1$ m, $T = 2$ s; $L = 0.5$ m, $T = 1.4$ s   (*c*) $C = 2\pi$

39. (*a*) R increases with increasing H and with increasing v (*b*) $R = v\sqrt{2H/g}$

41. (*a*) 8×10^8 bits = 1×10^8 bytes   (*b*) ≈ 500 books

Chapter 2

True or False   **1.** False; this equation holds only for constant acceleration   **2.** False; a ball at the top of its flight is momentarily at rest but has acceleration g **3.** False; for example, motion with constant velocity   **4.** True
5. True; by definition of average velocity

Problems

1. (*a*) 24 km/h   (*b*) -12 km/h   (*c*) 0 km/h
(*d*) 16 km/h

3. (*a*) -2 m/s   (*b*) 2.25 m/s   (*c*) -0.3 m/s

5. (*a*) 260 km   (*b*) 65 km/h

7. (*a*) 1.82 h   (*b*) 4.85 h   (*c*) 945 km/h   (*d*) 621 km/h

9. (*a*) 8.33 min   (*b*) 1.28 s   (*c*) 1 lightyear = 9.47×10^{12} km = 5.88×10^{12} mi

11. $66\frac{2}{3}$ km/h

13. (*a*) 1 m/s   (*b*) 2 m/s

15. (*a*) 1 m/s   (*b*) 0.7 m/s (*c*) 8 s

17. -2 m/s^2

19. (*a*) 80 m/s   (*b*) 400 m   (*c*) 40 m/s

21. 15.6 m/s^2

23. $t = 4.47$ s, $v = 44.7$ m/s, $v_{av} = 22.4$ m/s

25. 4.59 km

27. (*a*) 25.1 s   (*b*) 3.08 km

29. (*a*) The area under the curve in the indicated time interval is -36 m.

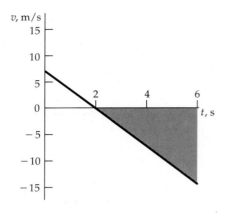

(b) $x(t) = 7t - 2t^2 + c$, $\Delta x = -36$ m   (c) -9 m/s
31. (a) 0 m/s   (b) $\frac{1}{3}$ m/s   (c) -2 m/s   (d) 1 m/s
33. (a)

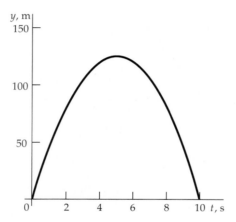

(b) $t = 0$ to 1 s, $v_{av} = 45$ m/s;
$t = 1$ to 2 s, $v_{av} = 35$ m/s;
$t = 2$ to 3 s, $v_{av} = 25$ m/s; $t = 3$ to 4 s, $v_{av} = 15$ m/s;
$t = 4$ to 5 s, $v_{av} = 5$ m/s; $t = 5$ to 6 s, $v_{av} = -5$ m/s;
$t = 6$ to 7 s, $v_{av} = -15$ m/s;
$t = 7$ to 8 s, $v_{av} = -25$ m/s;
$t = 8$ to 9 s, $v_{av} = -35$ m/s;
$t = 9$ to 10 s, $v_{av} = -45$ m/s.

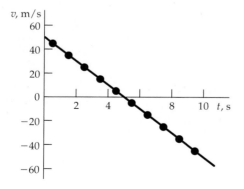

(c) $v(t) = -10t + 50$

35. (a) Velocity is greatest for $t \approx 2$ to 3 s

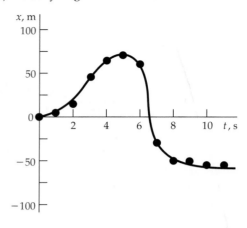

(b) Speed is least near $t = 5$ s, and for $t \approx$ 8 s and $t \approx 11$ s   (c) Same as part (b).   (d) Speed is constant for $t > 10$ s   (e) Acceleration is positive for $t \approx 0$ s to 3 s and for $t \approx 7$ s to 10 s.   (f) Acceleration is negative for $t \approx 3$ s to 7 s.
37. (a) -7.67 m/s   (b) 6.26 m/s   (c) 697 m/s^2 upward
39. (a) $x(t) = (C/6)t^3 + Dt + E$, where E is a constant in meters.   (b) At $t = 5$ s, we find $v = 37.5$ m/s and $x = 62.5$ m.
41. $\Delta x = 4\frac{1}{3}$ m. The average velocity is not equal to the mean of the initial and final values when the acceleration is not constant, as is the case here.
43.

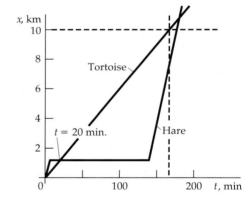

Tortoise passes Hare at $t = 20$ min. When Tortoise finishes, Hare is 2.4 km behind. Hare could nap up to 125 min without losing the race.
45. (a) t_0, t_1 (b) t_3, t_4, t_6, t_7 (c) t_2, t_5 (d) t_4
(e) t_2, t_6 (f) t_0, t_1, t_3, t_5, t_7
47. For Figure 2–22a,   (a) $a > 0$ for $t \approx 3$ to 6 s; $a < 0$ for $t \approx 1$ to 3 s; $a = 0$ near $t = 3$ and for $t = 6$ s to 7 s.
(b) $a = $ constant where the curve is straight.   (c) $v = 0$ at $t \approx 8.75$ s. For Figure 2–22b,   (a) $a > 0$ for $t \approx 5$ to $t \approx 7$ s; $a < 0$ for $t \approx 0$ to 3 s and $t > 7$ s; $a = 0$ for $t \approx 3$ to 5 s   (b) $a = $ constant on the parabolic and straight-line segments of the curve.   (c) $v = 0$ near $t = 2$, 6, and 8 s.

49. (*a*) 34.7 s   (*b*) 1.21 km
(*c*)

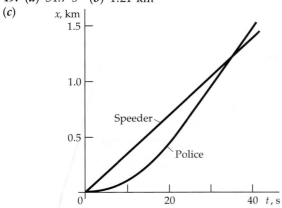

51. $v_0 = 40$ cm/s; $a = -6.88$ cm/s^2
53. (*a*) $6\frac{2}{3}$ s   (*b*) $6\frac{2}{3}$ s
55. 18.4 m
57. (*a*) 26.1 m   (*b*) 23.4 m/s
59.

61. (*a*) Yes
(*b*)

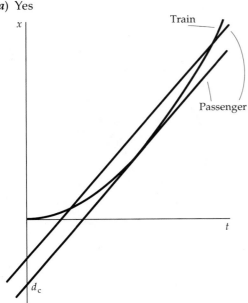

(*c*) The train's speed is 8 m/s, its average speed is 4 m/s and $d_c = 32$ m.
63. The balls collide two-thirds of the way to the top of the building.
65. (*a*)

(*b*)

67. (*a*) 0.875 cm/s

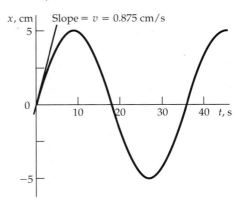

(*b*) The average velocities are 0.723, 0.835, 0.857, 0.871, 0.874, and 0.875 cm/s for the time intervals ending at 6, 3, 2, 1, 0.5, and 0.25 s, respectively.   (*c*) $dx/dt = A\omega \cos \omega t = 0.875$ cm/s at $t = 0$ s.

69. (*a*)

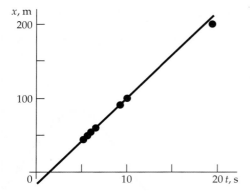

(*b*) $x = \frac{1}{2}v_0T + aT(t - T)$, or $x = v_0(t - \frac{1}{2}T)$ (*c*) The slope is ≈ 11.6 m/s $= v_0$, and the intercept with the time axis is ~ 1.3 s. Thus, $T = 2.6$ s and $a = 4.46$ m/s^2 (*d*) The sprint is run at top speed. For longer races, top speed cannot be maintained and a runner must run at a slower, but sustainable, pace

71. (*a*) Initially the particle accelerates with the acceleration of gravity, but as its velocity increases, its acceleration diminishes. When $v = g/b$, the acceleration becomes zero.   (*b*) The figure shows a rough sketch of v versus t

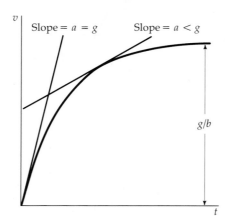

73. $x(t) = x_0 e^{A(t-t_0)}$, where $A = 1$ s^{-1}
75. (*a*)

(*b*) 295 m/s

Chapter 3

True or False   1. False; for example, equal and opposite vectors **2.** True **3.** False; for example, projectile motion **4.** False; for example, circular motion **5.** True **6.** True **7.** True

Problems
1.

18.5 m, 22.5° north of east.
3. (*a*) $t = 15$ s, $\Delta r = 14.1$ m, and $\theta = 45°$; $t = 30$ s, $\Delta r = 20$ m, and $\theta = 90°$; $t = 45$ s, $\Delta r = 14.1$ m, and $\theta = 135°$; $t = 60$ s, $\Delta r = 0$ m   (*b*) $\Delta r = 14.1$ m, $\theta = 45°$; $\Delta r = 14.1$ m, $\theta = 135°$; $\Delta r = 14.1$ m, $\theta = 225°$; $\Delta r = 14.1$ m, $\theta = 315°$ (*c*) same magnitude, but rotated by 90° (*d*) same magnitude but in the opposite direction
5. (*a*) 1 km   (*b*) 57.3° north of east   (*c*) 1/5
7. (*a*) $A_x = 8.66$ m, $A_y = 5$ m   (*b*) $A_x = 3.54$ m, $A_y = 3.54$ m   (*c*) $A_x = 3.5$ km, $A_y = 6.06$ km (*d*) $A_x = 0$ km, $A_y = 5$ km   (*e*) $A_x = -13$ km/s, $A_y = 7.5$ km/s   (*f*) $A_x = -5$ m/s, $A_y = -8.66$ m/s   (*g*) $A_x = 0$ m/s^2, $A_y = -8$ m/s^2
9. (*a*) $A_x = 1.41$ m $= A_y$, $B_x = 1.73$ m, $B_y = -1$ m (*b*) $C_x = 3.15$ m, $C_y = 0.414$ m, $C = 3.17$ m, $\theta = 7.5°$ (*c*) $C_x = -0.318$ m, $C_y = 2.41$ m, $C = 2.44$ m, $\theta = 97.5°$
11. (*a*) $A = 8.06$, $\theta = 240°$; $B = 3.61$, $\theta = -33.7°$; $C = 9.06$, $\theta = 264°$ (*b*) $A = 4.12$, $\theta = -76°$; $B = 6.32$, $\theta = 71.6°$; $C = 3.61$, $\theta = 33.7°$
13. (*a*) $\mathbf{v} = 5$ m/s $\mathbf{i} + 8.66$ m/s $\mathbf{j}$ (*b*) $\mathbf{A} = (-3.54$ m$)(\mathbf{i} + \mathbf{j})$ (*c*) $\Delta\mathbf{r} = 14$ m $\mathbf{i} - 6$ m $\mathbf{j}$
15. $A_y/A_x = \pm B_y/B_x$
17.

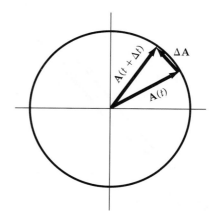

$\Delta\mathbf{A}$ becomes perpendicular to $\mathbf{A}(t)$.
19. $v_{av} = 14.7$ km/h, $\theta = -16.3°$

21. (*a*)

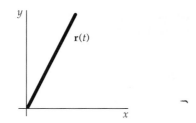

(*b*) $\mathbf{v} = d\mathbf{r}/dt = 5$ m/s $\mathbf{i} + 10$ m/s $\mathbf{j}$; $v = 11.2$ m/s
23. (*a*) *AB*, north; *BC*, northeast; *CD*, east; *DE*, southeast; *EF*, south (*b*) *AB*, north; *BC*, southeast; *CD*, acceleration = 0; *DE*, southwest; *EF*, north
(*c*) The acceleration is greater on segment *DE*
25. (*a*) $\mathbf{v}_{av} = 33\frac{1}{3}$ m/s $\mathbf{i} + 26\frac{2}{3}$ m/s $\mathbf{j}$
(*b*) $\mathbf{a}_{av} = -3$ m/s^2 $\mathbf{i} - 1.77$ m/s^2 $\mathbf{j}$
27. (*a*) west of north by $\theta = 13.1°$ (*b*) 300 km/h
29. 0.533 s
31. (*a*) 2294 m (*b*) 43.2 s (*c*) 9174 m
33. $\mathbf{v} = 15$ m/s $\mathbf{i}$, where $\mathbf{i}$ is taken to be horizontal; $a = 9.81$ m/s^2 downward
35. 45 m/s^2
37. (*a*) $a_r \rightarrow 4a_r$ (*b*) $a_r \rightarrow \frac{1}{2}a_r$ (*c*) A perfectly sharp turn implies $r = 0$, which would require an infinitely large acceleration for any finite speed.
39. 29.9 rev/min
41. (*a*) $\mathbf{a} = 2$ m/s^2 $\mathbf{i} - 3.5$ m/s^2 $\mathbf{j}$ (*b*) $\mathbf{v} = (1 + 2t)$ m/s $\mathbf{i} + (1 - 3.5t)$ m/s $\mathbf{j}$ (*c*) $\mathbf{r} = (4 + t + t^2)$ m $\mathbf{i} + (3 + t - 1.75t^2)$ m $\mathbf{j}$
43. (*a*) 31.6 m/s^2
(*b*) 26.1 m/s^2 (*c*) 23.9 m/s^2
45. $\mathbf{v} = 30\mathbf{i} + (40 - 10t)\mathbf{j}$, $\mathbf{a} = d\mathbf{v}/dt = -10\mathbf{j}$
47. (*a*) 8.14 m/s (*b*) 23.2 m/s
49. The projectile covers a horizontal distance $x = 408$ m
51. (*a*) -120 m $\mathbf{i} + 4$ m $\mathbf{j}$ (*b*) -20 m/s $\mathbf{i} - 12$ m/s $\mathbf{j}$
(*c*) -2 m/s^2 $\mathbf{j}$
53. 3.20 m/s in the direction of the thrower
55. 14.2 m/s = 31.7 mi/h
57. 4.29 m
59. 0.785 m
61. (*a*) 2 s (*b*) π m/s (*c*) $-\mathbf{j}$ (*d*) $a_r = \pi^2$ m/s^2, $a_t = \pi/2$ m/s^2 (*e*) $a = 9.99$ m/s^2, $\theta = 189°$
63. 34.6 m/s
65. (*a*) 20.0 m/s (*b*) 2.87 m
67. (*a*) Note that $r = 4$ m for all time; thus, the path is a circle of radius 4 m centered on the origin.
(*b*) $\mathbf{v} = (8\pi$ m/s$)(\cos 2\pi t \, \mathbf{i} - \sin 2\pi t \, \mathbf{j})$ (*c*) $\mathbf{a} = (-16\pi^2$ m/s$^2)(\sin 2\pi t \, \mathbf{i} + \cos 2\pi t \, \mathbf{j}) = (-4\pi^2$ s$^{-2})\mathbf{r}$; hence $\mathbf{a}$ is radial, and in particular, is radially inward as indicated by the minus sign.
69. (*a*) $H = W \tan \theta$ (*b*)
$$v_0 = \frac{W}{\cos \theta} \sqrt{\frac{g}{2(W \tan \theta - H)}}$$
71. (*a*) 54.3 m/s (*b*) 269 m (*c*) 6.56 s
73. (*a*) $\mathbf{v} = (10 - 10 \sin 2t)\mathbf{i} - (10 \cos 2t)\mathbf{j}$
(*b*) The particle is in contact with the surface at $5\pi/2$, $25\pi/2$, $\ldots$ = 7.85, 39.2, $\ldots$.

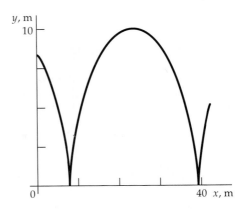

(*c*) $\mathbf{a} = -20 \cos 2t \, \mathbf{i} + 20 \sin 2t \, \mathbf{j}$ (*d*) $\mathbf{v} = 0$ when $t = (4n + 1)\pi/4$, where $n = 0, 1, 2, \ldots$. The point in contact with the ground is instantaneously at rest.

Chapter 4

True or False **1.** True **2.** False; the forces acting sum to zero **3.** False; the *acceleration* is in the direction of the unbalanced force **4.** True **5.** False **6.** False

Problems
1. (*a*) 8 m/s^2 (*b*) $m_2 = \frac{1}{2}m_1$ (*c*) 8/3 m/s^2
3. (*a*) $a = 7.07$ m/s^2 in a direction bisecting the angle between the two forces (*b*) $a = 14.0$ m/s^2 in a direction 14.6° away from $2\mathbf{F}_0$ toward $\mathbf{F}_0$
5. 12 m/s^2
7. (*a*) 6 m/s (*b*) 9 m
9. 10 kg
11.

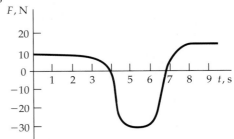

13. *AB*, force vertically upward; *BC*, force radial toward center of circular arc; *CD*, no force; *DE*, force is toward center of circular arc; *EF*, force acts vertically upward
15. (*a*) 79.4 kg (*b*) 7.94×10^4 g
17. (*a*) $W = 784.8$ N = 176 lb (*b*) $W = 716$ N = 161 lb (*c*) $m = 80$ kg = 5.48 slugs
19.

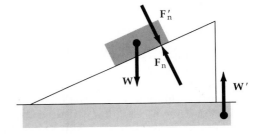

21. (a) 1.28 s   (b) 8.7 y
23. $5\frac{1}{3}$ m/s^2
25. (a) 98.1 N   (b) 98.1 N   (c) 49.05 N
(d) 49.05 N
27. (a) 8.77 m/s^2 (b) 26.3 N
29. (a) 132 lb   (b) 105 lb   (c) Same as part (b)
31. (a) $T = w/(2 \sin \theta)$, $T_{min} = \frac{1}{2}w$, $T_{max} \rightarrow \infty$
(b) 19.6 N
33. (a) 19.6 N   (b) 19.6 N   (c) 39.6 N   (d) For $t = 0$
to 2 s, $T = 19.6$ N; for $t = 2$ to 4 s, $T = 9.62$ N
35. (a) 11,810 N   (b) 9810 N   (c) $T = 7810$ N
37. (a) 3821 N   (b) 5732 N
39. 500 N
41. (a) 149 N   (b) 3.41 m/s^2
43. $\mathbf{F} = 329$ N = 73.9 lb directed toward the man.
45. (a) 17 m/s^2 (b) If a is made greater, the block will
accelerate up the wedge.
47. (a) 398 N   (b) 368 N
49. (a) The direction of deflection is opposite to the
direction of acceleration.   (c) $\theta = 9.31°$ in the forward
direction.
51. $\theta = 51.6°$

Chapter 5
True or False   1. False   **2.** False; it opposes the relative
motion of the surfaces   **3.** True   **4.** True   **5.** False;
round-off errors increase and may be important

Problems
1. 0.417
3. (a) Since $\mu_s > \mu_k$ in general, the car's acceleration is
greater when the wheels do not spin.   (b) 0.212
5. (a) 0.289   (b) 600 N
7. 4.57°
9. (a) M/T, kg/s   (b) M/L, kg/m
11. 2.79×10^{-4} kg/m
13. 205 km/h
15. (a) $a = 2.45$ m/s^2, $T = 36.8$ N
(b) $T = m_1 m_2 g(1 + \sin \theta)/(m_1 + m_2)$
17. Consider first the mass m_2. If we take radially
inward as positive, then $T_1 = 4\pi^2[m_1 L_1 + m_2(L_1 +$
$L_2)]/T^2$ and $T_2 = m_2 v_2^2/(L_1 + L_2)$.
19. (a) In the inertial frame, the object moves with
constant speed. In the boxcar frame, the object has an
acceleration $a = -5$ m/s^2.   (b) 4 s
21. In the inertial frame, the object slows with time,
then sticks to the floor and accelerates with the boxcar.
In the boxcar frame, the object slows, then comes to
rest and remains at rest.
23. 6.61 cm
25. 24.5 m/s
27. 83.9 m
29. (a) 10 m/s^2 (b) $T = (100 - 8x)$ N
31. (a) 400 N   (b) 625 N
33. 23.6 rev/min
35. (a) 16.4 m/s^2 (b) 19.6 N   (c) The frictional force
will not increase if the acceleration is increased.
37. (a) The force initially decreases with θ since the
vertical component of $\mathbf{F}$ will decrease the normal force
F_n exerted by the table. As θ increases further,
however, the horizontal component of $\mathbf{F}$ decreases so

much that $\mathbf{F}$ must become larger again in order to
move the box   (b) $F = \mu_s mg/(\cos \theta + \mu_s \sin \theta)$; F is a
minimum near 31°.

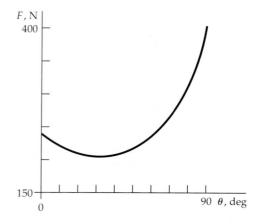

39. (a) 51.7 km/h   (b) 27.6 km/h
41. (a) 80 N   (b) $F_{net} = 600$ N, $F = 680$ N
(c) 6.8 m/s^2
43. (a) $a = \frac{1}{2}$ m/s^2, $F_c = 2$ N
(b) $F_c = m_2 F/(m_1 + m_2)$
45. (a) The maximum m is 3.39 kg; the minimum m is
0.614 kg.   (b) 9.81 N
47. (a) 17.7 N   (b) $a = 1.47$ m/s^2, $f = 2.94$ N
(c) $a_2 = 1.96$ m/s^2, $a_4 = 7.85$ m/s^2, $f = 3.92$ N
49. (a) $F(\theta) = \mu_s mg/(\cos \theta + \mu_s \sin \theta)$ (b) $dF/d\theta =$
$\mu_s mg(\sin \theta - \mu_s \cos \theta)/(\cos \theta + \mu_s \sin \theta)^2$, $\theta =$
$\tan^{-1}(0.6) = 31°$
51. (a) 7.25 m/s   (b) 0.536
53. $v(t) = v_0/[1 + (\mu_k/r)v_0 t]$

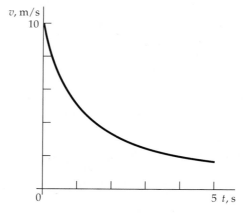

55. $v_{max} = 56.0$ km/h, $v_{min} = 20.1$ km/h
57. $a = -1.12$ m/s^2, $T = 44.7$ N
59. (a) $F_n = 4\pi^2 mrB^2/3600$ (b) Relative to the
compartment, the mass falls to the floor. From the
point of view of the rotating frame of reference, this is
caused by a pseudoforce of magnitude mv^2/r acting
radially outward.   (c) In an inertial frame, no forces
act on the mass; therefore, it travels in a straight line
with the same velocity it had when dropped. This
straight-line path intercepts the circular path of the
compartment floor; thus, the mass is seen to drop to
the floor.

Chapter 6

True or False 1. False 2. True 3. False 4. True
5. False; it is a unit of energy 6. False 7. False
8. True

Problems

1. (a) 7200 J (b) 1800 J (c) 28,800 J
3. 1 J = 0.738 ft·lb
5. (a) 9 J (b) 12 J (c) 4.58 m/s
7. $\frac{81}{4}$ C
9. (a) F = 327, 245, and 196 N (b) 981 J, independent of L (c) The work is the same in each case, but the force required is less as L becomes greater
11. 18 m^2
13. (a) 142° (b) 101° (c) 90°
15. (a) $\mathbf{A}\cdot\mathbf{i} = A_x(\mathbf{i}\cdot\mathbf{i}) + A_y(\mathbf{j}\cdot\mathbf{i}) + A_z(\mathbf{k}\cdot\mathbf{i}) = A_x$
(b) $\mathbf{A}/\sqrt{A_x^2 + A_y^2 + A_z^2}$ (c) 2
17. (a) nonconservative (b) conservative
(c) conservative (d) nonconservative
19. (a) 392 J (b) x = 2.45 m, v = 4.91 m/s
(c) K = 24.1 J, U = 368 J (d) K = 392 J, v = 19.8 m/s
21. (a) $U(x) - U(x_0) = -4(x - x_0)$ (b) $U(x) = -4x + 24$
(c) $U(x) = -4x + 36$
23. (a) A, negative; B, zero; C, positive; D, zero; E, negative; F, zero (b) Point C would appear to have the force with greatest magnitude, followed closely by point A. (c) Points B, D, and F are equilibrium points. B is unstable, D is stable, and F is neutral.
25.

(a) C/x^2 (b) The force points away from the origin when $C > 0$. (c) U decreases as x increases for $C > 0$.
(d) With $C < 0$, the force is toward the origin and U increases with increasing x.
27. (a) The work done by the woman is 1.22 J. The work done by the spring is -1.22 J. (b) 1.10 m/s
29. (a) 0.858 m. (b) After coming to rest, the object is accelerated by the spring. The kinetic energy gained by the object is just enough to allow it to rise again to the original height of 5 m. Assuming no nonconservative forces act on the object, it will continue to oscillate back and forth indefinitely.
31. (a) The object stops 2.5 m below the point Q, slides back to the left, rising 2.5 m above the point P, and continues this motion. (b) The object passes the point Q with a speed of 2.10 m/s. (c) 9.90 m/s
33. 5.05 m
35. (a) 104 J (b) -70.2 J (c) 33.7 J
(d) 2.90 m/s
37. (a) 7.67 m/s (b) -58.9 J (c) 1/3
39. (a) 3.14×10^{16} J (b) 7.47 megatons TNT
41. (a) No work is done on the skater (b) 560 J
(c) The kinetic energy of the skater results from the conversion of chemical energy in the muscles
43. (a) 1.67 m/s (b) 15 J

45. 1.37×10^6 kW
47. (a) 8.64×10^6 J (b) 2065 kcal
49. (a) 60° (b) $v^2 = \frac{2}{3}gL(2 - \cos\theta_0)$
51. (a) -11 J, -10 J, -7 J, -3 J, 0 J, 1 J, 0 J, -2 J, and -3 J
(b)

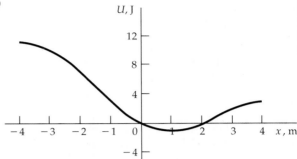

53. (a) 11/4 J (b) 6.13 J (c) 2.02 m/s
(d) 14.25/4 J (e) 2.15 m/s
55. (a) 57.6 J (b) U = 14.4 J, K = 43.2 J
(c) 175 J (d) 41.9 m/s
57. $U(x) = \frac{1}{3}ax^3$

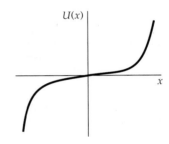

59 (a) 0.989 m (b) 0.783 m (c) 1.54 m
61. (a) $M(y) = 40 - \frac{1}{2}y$ (b) 1.18×10^4 J
63. (a) The forces and the work done by them are: gravity, $-mgL \sin 60°$; friction, $-\mu_k(mg \cos 60°)L$; the normal force, no work. (b) 0.451 m (c) The work done by gravity, friction, and the normal force is $mgL \sin 60°$, $-\mu_k mgL \cos 60°$, and zero, respectively.
(d) 2.52 m/s
65. (a) $W_{nc} = -\frac{3}{4}(\frac{1}{2}mv_0^2)$ (b) $\mu_k = -W_{nc}/(2\pi rmg)$
(c) $\frac{1}{3}$
67. y_0
69. 16.0 hp
71. (a) 0 (b) 78 J
73. (a) $U(y) = 2Mg[(d^2 + y^2)^{1/2} - d] - mgy$
(b)

$$y_0 = d\frac{\left(\dfrac{m}{2M}\right)}{\sqrt{1 - \dfrac{m^2}{4M^2}}}$$

75. 1.45 hp
77. (a) $mg(h - 2R)$ (b) $2g(h - 2R)/R$ (c) 2.5R
79. $T_{rest} = mg - K/L$, $T_{bottom} = mg + 2K/L = T_{rest} + 3K/L$
81. (a)
$$v(\theta) = \sqrt{v_0^2 + 2gR(1 - \cos\theta)}$$
(b) $\cos\theta = \frac{2}{3} + v_0^2/(3gR)$

83. (*a*) $x = 0$ m, $x = 2$ m, and $x > 3$ m   (*b*) A plot of U versus x is presented below

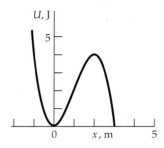

(*c*) The equilibria at $x = 0$ m, $x = 2$ m, and $x > 3$ m are stable, unstable, and neutral, respectively   (*d*) 2 m/s
85. For the path from $x = 4$ m, $y = 1$ m to $x = 4$ m, $y = 4$ m, $W = 36A$. For a path consisting of three parts—horizontal from $x = 4$ m to $x = 5$ m with $y = 1$ m; vertical from $y = 1$ m to $y = 4$ m with $x = 5$ m; horizontal from $x = 5$ m to $x = 4$ m with $y = 4$ m—
$W = 45A$
87. (*a*)

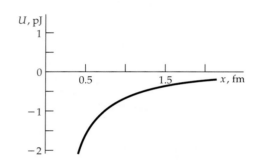

(*b*)

$$F_x = -\frac{dU}{dx} = -U_0 e^{-x/a}\left[\frac{a}{x^2} + \frac{1}{x}\right]$$

(*c*) $F_x(x = 2a)/F_x(x = a) = 0.138$
(*d*) $F_x(x = 5a)/F_x(x = a) = 0.00220$
89. (*a*)

$$\frac{dx}{\sqrt{E - U}} = \sqrt{\frac{2}{m}}\, dt$$

(*b*)

$$\frac{dx}{\sqrt{(A^2 - x^2)}} = \sqrt{\frac{k}{m}}\, dt$$

(*c*)

$$x = A \cos\left(\sqrt{\frac{k}{m}}t\right)$$

Chapter 7

True or False 1. True **2.** False; it is true only in the center-of-mass reference frame **3.** True **4.** True
6. True

Problems

1. 0.233 m
3. $X_{cm} = 6\frac{2}{3}$ m, $Y_{cm} = 6\frac{2}{3}$ m
5. $X_{cm} = 52.5$ cm, $Y_{cm} = 22.5$ cm
7. $X_{cm} = 2$ cm, $Y_{cm} = 1.4$ cm
9. 4 m/s to the east
11. (*a*) 11.8 m/s **i**   (*b*) 32.9 m

13. 1.5 m/s, in the direction opposite the girl's motion
15. 0.192 m/s
17. (*a*) The ball moves in pure translational motion
(*b*) The ball rotates about the center of mass
19. (*a*) 43.5 J   (*b*) 1.5 m/s   (*c*) $u_1 = 3.5$ m/s, $u_2 = -3.5$ m/s, taking motion to the right to be positive.
(*d*) 36.75 J   (*e*) $K_{tot} - K_{rel} = 6.75$ J. For the center of mass, $K_{cm} = \frac{1}{2}MV^2_{cm} = 6.75$ J.
21. 0.652 m/s
23. 3.13 m/s
25. (*a*) $v_1 = v_2 = 3.75$ m/s   (*b*) $v_{1f} = 2.5$ m/s, $v_{2f} = 4.5$ m/s
27. (*a*) 0.588 m/s   (*b*) The collision is inelastic
29. (*a*) One-fifth of the mechanical energy is lost with each bounce.   (*b*) 0.894
31. (*a*) The cue ball moves at an angle of 60° on the opposite side of the original direction from the 8 ball.
(*b*) $v_c = 2.5$ m/s, $v_8 = 4.33$ m/s
33. (*a*) 10.75 kg·m/s   (*b*) 1.34×10^3 N
35. (*a*) 6 kg·m/s   (*b*) 4.62×10^3 N
37. 3290 N
39. 1.2×10^6 N
41. (*a*) 4.83 km/s   (*b*) 6.91 km/s   (*c*) 13.8 km/s
43. 450 m/s
45. (*a*) 212 m from the launch site   (*b*) $K_i = 3600$ J, $K_f = \frac{1}{2}(2$ kg$)(34.6$ m/s$)^2 + \frac{1}{2}(4$ kg$)(69.2$ m/s$)^2 = 10,800$ J. The explosion released at least $K_f - K_i = 7200$ J of energy
47. $v_{1f} = -1.2$ m/s; $v_{2f} = 0.8$ m/s
49. (*a*) 1700 km below the earth's surface   (*b*) The predominant external force is that exerted by the sun, though the other planets and every other body in the universe also act on the system.   (*c*) To a very good approximation, the center of mass accelerates directly toward the sun.   (*d*) 9340 km radially every 14 days

51. $r_{cm} = \dfrac{r}{14}$, where positive is taken as the direction away from the hole
53. (*a*) 1.22 m/s   (*b*) 1.17 m/s
55. (*a*) 58.8 N   (*b*) Yes. The weight of the ball is $w = 1.47$ N, only about 2.5 percent of the force exerted during the throw
57.

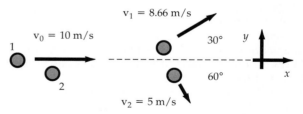

$v_1 = 8.66$ m/s; $v_2 = 5$ m/s
59. (*a*) 4.8 kg·m/s   (*b*) 1600 N   (*c*) 2.4 kg·m/s
(*d*) 19.2 N
61. 36.4 m/s
63. $m_1^2 v^2/(8gm_2^2)$
65. (*a*) 3.6×10^5 N   (*b*) 120 s   (*c*) 1720 m/s
67. (*a*) 0.0165 N   (*b*) The weight of a single water droplet is $mg = 0.0981$ N
69. 2252 m/s

71. $v_{1f} = v_{2f} = 0.693v_{ci}$, $v_{cf} = -0.2v_{ci}$. The cue ball returns along its original direction of motion; the other two balls move at 30° above and below the original motion of the cue ball.

73.

$E_2 = (\sin^2 \theta) \, E_0$

75. $(X_{cm}, Y_{cm}) = (6.67 \text{ m}, 3.33 \text{ m})$

77. (a) 9.66 km/s (b) 8.68 km/s (c) The rocket rises ≈ 400 km. Since this is only ≈ 7 percent of the earth's radius, the variation of g can be safely neglected.

Chapter 8

True or False 1. False **2.** True **3.** True **4.** True **5.** False **6.** False; if the net torque is zero, the *rate of change* of the angular momentum is zero **7.** False; it is true for a rigid body.

Problems

1. (a) 0.2 rad/s (b) 0.955 rev
3. (a) -0.0291 rad/s^2 (b) 1.75 rad/s (c) $33\frac{1}{3}$ rev
5. 0.625 s
7. (a) 0.628 rad/s^2 (b) 6.28 rad/s (c) 11.3 m
9. (a) -7.33 rad/s^2 (b) -0.0359 N·m
11. (a) $\tau = FR = 1.5$ N·m, $\alpha = 100$ rad/s^2
(b) $\omega = 400$ rad/s
13. (a) 56 kg·m^2 (b) 112 J
15. (a) 0.136 J (b) 356 rev/min
17. 28 kg·m^2
19. (a) $I_x = 28$ kg·m^2 (b) $I_y = 28$ kg·m^2
(c) $I_z = 56$ kg·m^2
21. $\frac{1}{4}MR^2$
23. (a) 60 kg·m^2/s (b) 75 kg·m^2 (c) 0.8 rad/s
25. (a) L is doubled (b) L is doubled
27. (a) zero (b) $v_1/v_2 = r_2/r_1$
29. (a) K_{rot} is 28.6 percent of the total kinetic energy
(b) K_{rot} is 33.3 percent of the total kinetic energy
(c) K_{rot} is 50 percent of the total kinetic energy
31. (a) $\frac{5}{7}g \sin \theta$ (b) $(2/7) \, mg \sin \theta$ (c) $\tan \theta = (7/2)\mu_s$
33. $\mathbf{F} = -F\mathbf{i}$; $\mathbf{r} = R\mathbf{j}$; $\boldsymbol{\tau} = FR\mathbf{k}$
35. (a) $36\mathbf{k}$ (b) $-36\mathbf{j}$ (c) $12\mathbf{k}$
37. (a) -25.8 kg·m^2/s $\mathbf{k}$ (b) 12.9 N·m $\mathbf{k}$
39. (a) 20.8 kg·m^2/s (b) 0.520 rad/s (c) 12.1 s
(d) 0.172 kg·m^2/s, upward (downward) when the angular momentum due to the spin of the wheel is away (toward) the pivot
41. (a) 2.5 N (b) 25 cm from the axis in a direction away from the center of the disk

43. (a) 2.4 N·m (b) 66.7 rad/s^2 (c) 200 rad/s
(d) 720 J (e) 7.2 kg·m^2/s (f) 300 rad (g) The work done by the torque is $W = 720$ J
45. (a) 283 kg·m^2/s (b) 9.43 N·m (c) 9.43 N·m
(d) 31.4 N
47. (a) 1.33×10^{-5} kg·m^2/s
(b) 1.33×10^{-5} kg·m^2/s (c) 1.33×10^{-5} kg·m^2/s
(d) $L_{cm} = 8.83 \times 10^{-5}$ kg·m^2/s or -6.18×10^{-5} kg·m^2/s, depending on the direction of $\mathbf{v}_{cm}$.
49. $K_{rot} = 2.60 \times 10^{29}$ J; $K_{orb} = 2.67 \times 10^{33}$ J
51. (a) 13.6 kg·m^2 (b) 7.14 N·m
53. 1.95 m
55. (a) 5.42 rad/s (b) $(5/2)mg$
57. $a = 0.2$ m/s^2; $v = at$; $x = x_0 + \frac{1}{2}at^2$; $\alpha = 3.33$ rad/s^2; $\omega = \alpha t$
59. (a)

$$a = \frac{g}{\left(1 + \dfrac{2}{5}\dfrac{M}{m}\right)}$$

(b) $T = \frac{2}{5}Ma$
61. (a) $2.7R$ (b) $2.5R$
63. (b) $\mathbf{T} = \frac{1}{3}M\mathbf{g}$
65. (a) 5 rev/s (b) 592 J (c) The additional energy is supplied by the man's muscles
67. (a) 5.36 rad/s (b) 9.99 J
69. (a) 552 N (b) 1.77 rad/s^2 (c) 2.17 rad/s
71. 1.13×10^{-4} s
73. (a) 0.0948 m/s^2 (b) $T_1 = 4.952$ N; $T_2 = 4.955$ N; $T_2 - T_1 = 2.37 \times 10^{-3}$ N (c) If you neglect the motion of the pulley, that is, if you set $a = -0.0971$ m/s^2, $T_1 = T_2 = 4.95$ N
75. (b)

$$W = -\frac{L_0^2}{m} \int_{r_0}^{r_f} \frac{1}{r^3} \, dr = \frac{L_0^2}{2m}(r_f^{-2} - r_0^{-2})$$

(c) $v_f = L_0/mr_f$, $\Delta K = L_0^2(r_f^{-2} - r_0^{-2})/2m = W$
(d) 0.282 m
77. The day would lengthen by 0.554 s
79. (a) $I = 0.04$ kg·m^2 (b) $I = 0.0415$ kg·m^2, about 3.5 percent greater than the approximate result
81. (a) $s_1 = (12/49)v_0^2/\mu_k g$; $t_1 = (2/7)v_0/\mu_k g$; $v_1 = (5/7)v_0$ (b) $s_1 = 3.99$ m; $t_1 = 0.583$ s; $v_1 = 5.71$ m/s
83. (a) $v_{cm} = (L/2)\omega = (3/2) \, P_0 x/ML$ (b) $P = P_0(3x/2L - 1)$
85. (a) 0.0443 m/s^2 (b) 2.21 rad/s^2 (c) 3.63 J
(d) 399 J
87. (a) 200 m/s (b) 8000 rad/s (c) 257 m/s
(d) 11.7 s
89.

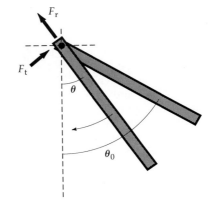

$F_r = \frac{1}{2}Mg(5\cos\theta - 3\cos\theta_0)$; $F_t = \frac{1}{4}Mg\sin\theta$
91. (a) $(11/7)v_0$ (b) $4v_0/(7\mu_k g)$ (c) $(36/49)v_0^2/\mu_k g$
93. 54.0°

Chapter 9
True or False   **1.** False; the resultant torque must also be zero   **2.** True   **3.** True   **4.** False   **5.** False

Problems
1. 1.4 m away from the center
3. $F_B = 417$ N; $F_J = 270$ N
5. $F_1 = 117$ N; $F_2 = 333$ N
7. (b) Part (a) gives the maximum angle the leg may make with the vertical before the foot slips.   (c) On ice μ_s is very small; thus the maximum θ is also small. This implies that short steps must be taken
9. The U-shaped figure has its center of gravity on the vertical line about which the shape is symmetric. The height of the center of gravity above the base of the "U" is $Y_{cg} = 4L/5$. For the L-shaped wire, choosing the origin at the bend in the "L" gives $X_{cg} = L/6$, $Y_{cg} = 2L/3$. For the triangular shape, the center of gravity is on the vertical line of symmetry, and $Y_{cg} = \sqrt{15}L/5$
11. 99.0 cm from the man's feet
13. Let F_1 and F_2 denote the forces exerted by the left and right supports, respectively. $F_1 = 245$ N; $F_2 = 736$ N
15. (a) 5400 N   (b) 9
17. $\frac{1}{2}F(\sqrt{3}b - a)$
19. (a) $\tau = Fa$ (b) The effective point of application of the normal force is a distance $a/6$ in from the corner of the cube.   (c) $\frac{1}{2}Mg$
21. $x_{cg} = 0.944a$, $y_{cg} = a$
23. Let F_1 and F_2 be the forces exerted on the cylinder by the left and right planes, respectively. $F_1 = 0.866W$; $F_2 = 0.5W$
25. (a) 42.4 N   (b) 57.0 N, along a direction 7.13° above the strut
27. 26.6°
29. (a) 450 N   (b) 241 N   (c) The tension decreases as the brace is moved downward
31.
$$F = \frac{Mg\sqrt{h(2R - h)}}{(R - h)}$$
33. 9 m
35. (a)
$$N = Mg - F\sqrt{\frac{(2R - h)}{h}}$$
(b) $-F$
(c)
$$F_{cy} = F\sqrt{\frac{(2R - h)}{h}}$$
37. (a) 49.1 N   (b) 73.6 N·m   (c) Let F_1 and F_2 refer to the forces exerted on the end of the rod and 0.1 m in from the end, respectively. With upward taken as positive, $F_2 = 736$ N and $F_1 = -687$ N
39. F_1 and F_2 correspond to the forces exerted by the left and right supports, respectively. $F_1 = -1938$ N, a tension; $F_2 = 2919$ N, a compression
41. $a = g/3 = 3.27$ m/s²
43. $x_{cg} = 0$, $y_{cg} = 2R/\pi$
45. 566 N

47. 61.6°
49. (a) $0.361mg$ (b) $0.313mg$ (c) $0.820mg$

Chapter 10
True or False   **1.** False   **2.** True   **3.** True

Problems
1. 2.92 AU
3. 10^4 AU
5. 248 y
7. 2.63×10^6 m/s
9. (a) 1.16×10^6 s   (b) 8.79×10^{25} kg
11. 6.02×10^{24} kg
13. 5.97×10^{24} kg
15. 10 times as much
17. (a) 4.80×10^{-9} N   (b) 8.64×10^{-10} N·m
19. (a) 5.77 kg   (b) gravitational
21. 5330 mi/h
23. (a) -6.25×10^9 J   (b) -3.13×10^9 J (c) 7.92×10^3 m/s
25. (a) 0   (b) 0   (c) 3.20×10^{-9} m/s²
27.
$$W = \frac{GMm_0}{R}$$
29. (a) $GM_1 m/(3a)^2 + GM_2 m/(2.2a)^2$
(b) $GM_1 m/(1.1a)^2$ (c) 0
31. $\sqrt{3}v_e = 1.94 \times 10^4$ m/s
33. 51.2 km/s
35. $h = 0.147\, R_E = 935$ km
37. (a) 3.31 y   (b) 1.49×10^{29} kg   (c) Planet 2. Planet 2 also has the greater total energy.   (d) $v_A = v_P/1.8$
39. (b) $\mathbf{g} = \mathbf{F}/m_0 = [-2Gmx/(x^2 + a^2)^{3/2}]\mathbf{i}$
41. 9.66×10^{-8} N
43. (a) 7.38 m   (b) 0.0319 mm
45. (a) 2.74×10^{-3} N   (b) 0.480 N, a value 175 times that due to the moon   (c) -0.06 percent
47. 1.88 N
49.
$$\frac{GMm}{a(a + L)}$$
51. (a) $F_r = GMm_0/r$ for $r > R$ and $F_r = 0$ for $r < R$.   (b) $U(r) = -GMm_0/r$ (c) Since $dU = -F_r\, dr = 0$ everywhere inside the shell, it follows that U is constant in value there. From the results of part (c), we conclude that $U(r) = -GMm_0/R$ for $r \le R$
(d)

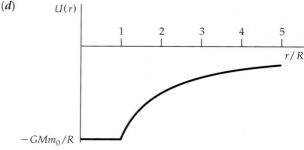

53. (b) $U(x) = -(GMm_0/L)\ln[(x + L/2)/(x - L/2)]$ for $x > \frac{1}{2}L$.   (c) $F_x = -GMm_0/(x^2 - L^2/4)$
55.
$$g = F/m_0 = GM\left[\frac{1}{x^2} - \frac{1}{8(x - R/2)^2}\right]$$

57. $\omega = (GM/R^3)^{1/2} = (\tfrac{4}{3}G\pi\rho_0)^{1/2}$

59. (a) $F_s/F_m = (M_s/M_m)(r_m/r_s)^2$. Substituting the appropriate numerical values yields 169 for this ratio, showing that the sun exerts by far the greater force. (c) Numerically, this is equal to 0.422, showing that the moon has the greater effect in causing the tides in earth's oceans

61. (a) $\mathbf{F} = -(GMm/d^2)[1 - \tfrac{1}{4}d^3/(d^2 + R^2/4)^{3/2}]\mathbf{i}$
(b) $\mathbf{F} = -0.821(GMm/R^2)\mathbf{i}$

Chapter 11

True or False 1. False 2. True 3. False; it can float because of surface tension 4. True

Problems

1. 0.676 kg
3. 103 kg
5. 54 kg
7. 0.976 mm
9. 5.01°
11. (a) 10.3 m (b) 0.757 m
13. $F = 6.46 \times 10^4$ N The table doesn't collapse because air pressure also exerts an upward force on the bottom surface of the card table
15. 230 N
17. (a) $F = 8.97 \times 10^4$ N (b) $F = 5.02 \times 10^4$ N
(c) If a window is rolled down the pressure on the two sides of the door will equalize, and it may then be opened
19. 1060 kg/m^3
21. 15.8 cm
23. 4.36 N
25. (a) 11.1×10^3 kg/m^3 (b) lead
27. 0.0233 N/m
29. (a) 4.58 L/min (b) 763 cm^2
31. (a) 12 m/s (b) 133 kPa (c) The flow rates are identical in the two sections
33. $F = 1.31 \times 10^5$ N = 29,400 lb
35. 1.43 mm
37. 1.49 ft
39. 183 m^3
41. $2[h(H - h)]^{1/2}$
43. 0.441 Pa
45. 3.89 kg
47. The readings on the upper and lower scales are 12.4 N and 36.7 N, respectively
49. 0.25×10^3 kg/m^3
51. (a) 21.2 kg (b) 636 N·s
(c)

$\Delta p = 900$ N·s, $F = 900$ N

53. (a) $dF = (\rho g y)(w\, dy)$
(b)
$$F = \tfrac{1}{2}\rho g w d^2 = 9.20 \times 10^7 \text{ N}$$
(c) The reason atmospheric pressure may be neglected is that it acts equally on both sides of the dam and thus gives rise to no net force
55. (a) 34.7 cm (b) 7.76×10^{-2} J
57. (a) 14.7 g (b) 1.03×10^3 kg/m^3
59. (b) 1.12 mm
61. 5.02 h
63. (c)
$$t = \frac{1}{\sqrt{2g}}\,\frac{2A_1}{A_2}(\sqrt{H} - \sqrt{h})$$
or
$$h = \left(\frac{\sqrt{H} - \sqrt{g}\,A_2 t}{\sqrt{2}\,A_1}\right)^2$$
(d) $t = (2H/g)^{1/2}(A_1/A_2) = 1.77$ h

Chapter 12

True or False 1. False; it is independent of the amplitude 2. True 3. True 4. False; it is simple harmonic only for small angular displacements 5. True 6. True 7. True 8. True 9. True

Problems

1. (a) 7.96 Hz (b) 0.126 s (c) 0.1 m
(d) 5 m/s (e) 250 m/s^2 (f) $t = 0.0314$ s; $a = 0$
3. (a) 474 N/m (b) 0.5 s (c) 1.26 m/s
(d) 15.8 m/s^2
5. 0.801 Hz
7. (a) 1.49 kg (b) 5.84 Hz (c) 0.171 s
9. (a) 62.8 cm/s (b) 790 cm/s^2
11. (a) $f = 0.477$ Hz, $T = 2.09$ s (b) 0.283 m
(c) -0.262 m
13. (a) $x(t) = (0.25$ m$)\cos(4\pi/3)t$
(b) $v = -(\pi/3$ m/s$)\sin(4\pi/3)t$
(c) $a = -(4\pi^2/9$ m/s$^2)\cos(4\pi/3)t$
15. (a) 1.05 m/s (b) 4.39 m/s^2 (c) 1.13 s
17. (a) 0.314 m/s (b) $2\pi/3$ rad/s (c) $x = (0.15$ m$)\cos 2\pi t/3$
19. 1.92 J
21. (a) 3 cm (b) 77.5 cm/s
23. (a) 0.270 J (b) -0.736 J (c) 1.01 J
(d) 0.270 J
25. (a) 1.90 cm (b) 0.054 J (c) 0.223 J
(d) 0.277 J
27. 12.3 s
29. 11.7 s
31. 1.10 s
33. 0.504 kg·m^2
35. (a) 0.444 s (b) 0.180 J (c) $Q = 628$; $b = 0.045$ kg/s
37. (a) The energy decreases by 10 percent each cycle
(b) 30 s (c) 62.8
39. (a) 314 (b) 6 rad/s
41. 0.420 J

43. (*a*)

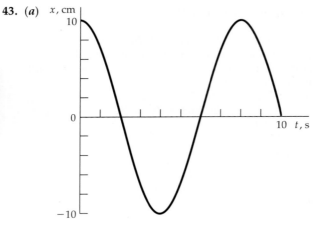

(*b*) 2.93 cm, 7.07 cm, 7.07 cm, 2.93 cm
45. 2.62 s
47. 12.9 Hz
49. (*a*) 70 J   (*b*) 22 J   (*c*) 7.33 W
51. (*a*) 0.314   (*b*) 0.000313 percent
53. (*a*) 1.57 percent   (*b*) $E_n = (0.984)^n E_0$
(*c*) $(0.430)E_0$
55. (*a*) 7.85 cm   (*b*) 3.92 cm   (*c*) 0.397 s
(*d*) 62.0 cm/s; $t = 0.0993$ s
57. 9.10 Hz
59. (*a*) 5.23×10^4 N/m   (*c*) 2×10^{11} N/m
61. (*a*) $A = 0.141$ m; $T = 0.444$ s   (*b*) $A = 0.231$ m; $T =$
0.363 s   (*c*) For the inelastic collision, we have $x(t) =$
$(0.141$ m$)$ cos $[(14.1$ rad/s$)t - \pi/2]$; $I_2 = 4$ kg·m/s. For
the elastic collision, $x(t) = (0.231$ m$)$ cos $[(17.3$ rad/s$)t -$
$\pi/2]$; $I_2 = 8$ kg·m/s.
63. (*a*) 1.55 cm   (*b*) 1.72 m/s   (*c*) 16.6 cm
65. (*a*) $F_r = -(GM_E m/R_E^3)r = -kr$, so $F_r \sin \theta =$
$-kr \sin \theta = -kx$. The motion is simple harmonic, with
a force constant $k = GM_E m/R_E^3$.
(*b*) $T = 2\pi(m/k)^{1/2} = 2\pi(R_E/g)^{1/2} = 84.4$ min
67. (*a*) 29.4   (*b*) 3.53 s   (*c*) 6 J
69. (*c*) The error is 0.008 percent. For the error to be 1
percent the requirement is $r = 22.4$ cm
71. answer given in problem
73. answer given in problem

Chapter 13

True or False   1. True   **2.** False   **3.** False; even though
$\nu = f\lambda$, the speed depends only on the tension and mass
density. When λ changes, f also changes so that ν remains
the same.   **4.** True   **5.** True

Problems
1.

3.

5.

7. (*a*)

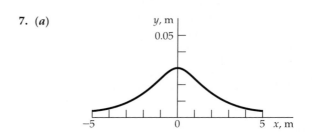

(*b*) $y(x, t) = (0.12$ m$^3)/[(2.00)^2 + (x - 10t)^2]$ m^2
(*c*) $y(x, t) = (0.12$ m$^3)/[(2.00)^2 + (x + 10t)^2]$ m^2
9. 251 m/s
11. (*a*) 20 m/s   (*b*) 28.3 m/s   (*c*) 14.1 m/s
13. 0.252 s
15. (*a*) 66.7 m/s   (*b*) 889 N
17. (*a*) $y = y_0 \sin k(x - vt)$ (*b*) $y = y_0 \sin 2\pi(x/\lambda - ft)$
(*c*) $y = y_0 \sin 2\pi(x/\lambda - t/T)$ (*d*) $y = y_0 \sin (2\pi/\lambda)(x -$
$vt)$ (*e*) $y = y_0 \sin 2\pi f(x/v - t)$
19. $y = (0.02$ m$)$ sin $12\pi(x - 10t)$
21. $y = (0.5$ m$)$ cos $(2\pi/50)(x - 10t)$
23. 9.87 W
25. (*a*) 3.86 cm   (*b*) 3.46 cm
27. $A_{resul} = 2A \cos \frac{1}{2}\delta = 2(0.05$ m$)$ cos $\pi/4 = 7.07$ cm
29. (*a*) $\lambda = 2$ m, $f = 25$ Hz   (*b*) $y =$
$(4 \times 10^{-3}$ m$)$ cos $50\pi t \sin \pi x$
31. (*a*) 1.25 Hz   (*b*) 3.75 Hz   (*c*) 6.25 Hz
33. 141 Hz
35. (*a*) $\lambda = 31.4$ cm, $f = 47.7$ Hz   (*b*) 1500 cm/s
(*c*) 62.8 cm

37. (a)

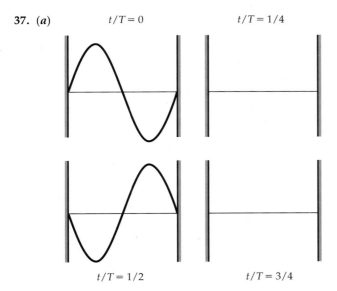

t/T = 0   t/T = 1/4

t/T = 1/2   t/T = 3/4

(b) 0.0126 s   (c) The energy of the wave is kinetic
39. (a) 2.66 m   (b) 0.666 m   (c) 160 m/s
41. (a) 5/3, 7/5   (b) Since these are successive
frequencies, it is clear that the even harmonics are
missing, as is the case for a string fixed on one end
only.   (c) 25 Hz   (d) The frequencies 75, 125, and 175
Hz are the third, fifth, and seventh harmonics,
respectively.   (e) 4 m
43. For $y(x, t) = A \sin kx \cos \omega t$, it is clear that
$\partial^2 y/\partial t^2 = -\omega^2 y$ and $\partial^2 y/\partial x^2 = -k^2 y$. Therefore,
$\partial^2 y/\partial x^2 = (k^2/\omega^2)\partial^2 y/\partial t^2 = (1/\omega^2)\partial^2 y/\partial t^2$, and thus $y(x,t)$
satisfies the wave equation
45. (a) 75 Hz   (b) 375 Hz is the fifth harmonic, and
450 Hz is the sixth harmonic.   (c) 2 m
47. (a) $f = 400$ Hz, $T = 2.5 \times 10^{-3}$ s   (b) 316 m/s
(c) $\lambda = 0.791$ m, $k = 7.95$ m^{-1} (d) $(0.5 \times 10^{-3}$ m)
$\sin 2\pi(x/\lambda - ft)$ (e) $v_{max} = 1.26$ m/s, $a_{max} =$
3.16×10^3 m/s^2 (f) 2.50 W
49. (a) $v = -(7.54$ m/s$) \sin 2.36x \sin 377t$
(b) The maximum speed occurs at the point
where $\sin 2.36x = 1$ or $x = 0.666$ m; $v_{max} = 7.54$ m/s
(c) $a = -(2.84 \times 10^3$ m/s$^2) \sin 2.36x \cos 377t$ (d) The
greatest acceleration also occurs at $x = 0.666$ m; $a_{max} =$
2.84×10^3 m/s^2
51. (a) 1.11 W   (b) The amplitude must be increased
by a factor of 10 or the frequency must be increased by
a factor of 10, which implies that the tension must be
increased by 10,000 times its original value.   (c) The
frequency of the power source is probably easiest to
change
53. (a)

$$K = \frac{1}{2}\mu n^2 \omega_1^2 A^2 \sin^2 n\omega_1 t \int_0^L \sin^2 \frac{n\pi x}{L} \, dx$$

(b) $K = \frac{1}{4}\mu n^2 \omega_1^2 A_n^2 L$ (c) $y(x, t) = 0$ for all x (d) $K =$
$(\frac{1}{2}\mu \omega_1^2 L)(n^2 A_n^2)$
55. (a) $df_n/f_n = \frac{1}{2}dF/F$ (b) A 1.54 percent increase in
the tension is required
57. (a) $y(x, t) = (0.01$ m$) \sin (2\pi x/3 - 160\pi t)$
(b) 80 Hz   (c) 0.0125 s   (d) 5.03 m/s
59. (a) 0.8 m   (b) 480 N   (c) The finger should be
placed 9.23 cm from the end

61. (a) 60 cm   (b) $2\pi/5$ (c) 24 m/s
63. (a)
$v = -A_1\omega_1 \sin \omega_1 t \sin k_1 x - A_2\omega_2 \sin \omega_2 t \sin k_2 x$
(b)

$$dK = \frac{1}{2}\mu v^2 \, dx =$$
$$\frac{1}{2}\mu[A_1^2 \omega_1^2 \sin^2 \omega_1 t \sin^2 k_1 x +$$
$$A_2^2 \omega_2^2 \sin^2 \omega_2 t \sin^2 k_2 x +$$

$2A_1 A_2 \omega_1 \omega_2 \sin \omega_1 t \sin \omega_2 t \sin k_1 x \sin k_2 x] \, dx$
(c) $\frac{1}{4}\mu L\omega_1[(n_1 A_1)^2 + (n_2 A_2)^2]$
65. (a)

$2F \sin \dfrac{\theta}{2} \approx F\theta = mv_0^2/R = \mu R\theta v_0^2/R$, so $F = \mu v_0^2$.   (b) v_0

(c) A forward-moving pulse always remains in contact
with the ground (where it was generated), whereas a
backward-moving pulse appears to make two
revolutions for one rotation of the loop
67. (b) 2.21 s
69. (a) 4 m/s   (b) 0.00424 m   (c) $y(x, t) =$
$(0.00424$ m$) \cos (12.5x - 50t)$

Chapter 14

True or False 1. False; sound waves are longitudinal
2. False 3. False; it has 1000 times the intensity
4. False 5. False 6. True 7. False 8. True 9. True
10. True 11. False

Problems
1. $[(B/\rho)^{1/2}] = [(kg/m \cdot s^2)/(kg/m^3)]^{1/2} = (m/s)$
3. 5.09×10^3 m/s
5. 2.70×10^{10} N/m^2
7. (a) 1.30 m   (b) 0.649 m
9. 8.27×10^{-2} Pa
11. (a) zero   (b) 3.67×10^{-6} m
13. (a) 138 Pa   (b) 21.6 W/m^2 (c) 0.216 W
15. (a) 10^{-11} W/m^2 (b) 2×10^{-12} W/m^2
(c) For $\beta = 3$ dB, $p_0 = 4.18 \times 10^{-5}$ Pa; for $\beta = 10$ dB,
$p_0 = 9.37 \times 10^{-5}$ Pa
17. 99/100 of the acoustic power must be eliminated to
reduce β by 20 dB
19. (a) 90°   (b) $\sqrt{2}A$
21. (a) 85 Hz; 255 Hz   (b) Some sound will reflect off
walls and other surfaces, thus giving a variety of path
differences
23. (a) zero   (b) $2I_0$ (c) $4I_0$
25. (a) $\lambda/4$ (b) The smallest magnitude of $r_2 - r_1$
occurs for $r_2 - r_1 = -\lambda/4$, and the smallest positive
value is $r_2 - r_1 = 3\lambda/4$
27. 437 Hz
29. (a) 17 Hz   (b) 8.5 Hz
31. (a) 2267 Hz   (b) The eighth harmonic
33. (a) 3400 Hz, 10,200 Hz and 17,000 Hz
(b) We might expect the ear to be more sensitive near
the above resonant frequencies, since these special

frequencies produce standing waves within the ear canal
35. (a) $N \approx \Delta t/T = f_0\, \Delta t$ (b) $\lambda \approx \Delta x/N$
(c) $k = 2\pi/\lambda \approx 2\pi N/\Delta x$ (d) N is uncertain because the waveform dies out gradually, rather than stopping abruptly at some time.   (e) $\Delta k = (2\pi/\Delta x)\,\Delta N$
37. (a) $\lambda = 15$ m and $f = 20$ pies/min   (b) $\lambda = 13.5$ m; $f = 22.2$ pies/min   (c) $\lambda = 15$ m; $f = 22$ pies/min
39. (a) 1.3 m   (b) 262 Hz
41. (a) 2.1 m   (b) 162 Hz
43. (a) 80 m/s   (b) 420 m/s   (c) 1.7 m
(d) 247 Hz
45. 529 Hz; 474 Hz
47. 3.42 m/s
49. (a) $I_1/I_2 = (M_1/M_2)^{1/2} = 4$ (b) $I_1/I_2 = s_{01}/s_{02} = \frac{1}{4}$
(c) $s_{01}/s_{02} = [\rho_2 v_2/\rho_1 v_1]^{1/2} = \frac{1}{2}$; $p_{01}/p_{02} = \rho_1 v_1 s_{01}/\rho_2 v_2 s_{02} = 2$
51. A reasonable time for Δt might be $\frac{1}{4}$ to $\frac{1}{2}$ s, giving a distance to the wall of 20 to 40 m
53. (a) At a node, the powder is not vibrated; hence, it collects there.   (b) $v = \lambda f = 2Sf$ (c) $S = 5$ cm, $f = v/\lambda = 3400$ Hz   (d) Suppose the tube is 1 m long. The lowest frequency in air is $f = 85$ Hz, the highest frequency is 8500 Hz. For the same tube, the appropriate frequencies in helium would be 35 to 3500 Hz
55. $v = 338$ m/s. This method lacks accuracy since the antinode generally does not occur exactly at the end of an open pipe
57. (a) $dv/dT = \frac{1}{2}(\gamma R/MT)^{1/2} = \frac{1}{2}(v/T)$
(b) 4.95 percent   (c) $v = 347.4$ m/s. An exact calculation gives $v = (331\text{ m/s})[(273 + 27)/273]^{1/2} = 347$ m/s, showing the approximate result to be quite accurate
59. (a) 0 dB   (b) 66 dB   (c) 63 dB
61. (a) $I_1 = 1.99 \times 10^{-5}$ W/m²; $I_2 = 0.884 \times 10^{-5}$ W/m²   (b) $[(I_1)^{1/2} + (I_2)^{1/2}]^2 = 5.53 \times 10^{-5}$ W/m²   (c) $[(I_1)^{1/2} - (I_2)^{1/2}] = 0.221 \times 10^{-5}$ W/m²   (d) $I_1 + I_2 = 2.87 \times 10^{-5}$ W/m²
63. (a) 4.5 m/s   (b) 22.1 beats/s   (c) 22 beats/s
65. (a) 55.1 N/m²   (b) 3.46 W/m²   (c) 0.245 W
67. 0.773 m/s
69. (a) $r = 100$ m   (b) 0.126 W
71. (a) 100 dB   (b) 25.1 W   (c) 2 m   (d) 96.5 dB
73. 3.97×10^{-5} W
75. (a) 26 percent increase per year. This seems to be an unreasonably large increase.   (b) 3.01 years
77. 87.8 dB
79. answer given in problem
81. (a) 0.279 m   (b) 1.22×10^3 Hz   (c) $\theta_3 = 24.7°$; $\theta_4 = 33.9°$; $\theta_5 = 44.2°$; $\theta_6 = 56.9°$; $\theta_7 = 77.6°$ (d) 4°
83. answer given in problem
85. (a) (80, 0) m; (75.2, ±27.2) m; (58.7, ±54.4) m
(b) (78.8, ±13.6) m; (68.8, ±40.8) m
87. answer given in problem

Chapter 15

True or False   **1.** False   **2.** False   **3.** False; the degree size is also different   **4.** True   **5.** False   **6.** False; it could remain the same or even decrease depending on the change in the volume   **7.** True   **8.** True   **9.** True

Problems
1. 10.4°F to 19.4°F
3. 56.7°C and −62.2°C
5. (a) 8.4 cm   (b) 107°C
7. −320.44°F
9. (a) 54.9 torr   (b) 3704 K
11. 30.0264 cm
13. 2.10 km
15. $V_2 = 1.16V_1$
17. 1.79 mol; 1.07×10^{24} molecules
19. (a) 3.66×10^3 moles   (b) 59.9 moles
21. 152 J
23. $v_{\text{rms}} = 4.99 \times 10^5$ m/s; $K_{\text{av}} = \frac{3}{2}kT = 2.07 \times 10^{-16}$ J
25.

(a) 88.2°C   (b) 80.4°C   (c) 173 kPa
27. 3.39 kPa
29. 70.8
31. (a) 1846 m/s   (b) 461 m/s   (c) 393 m/s
(d) One-sixth of the escape speed on Mars is 833 m/s; hence, O_2 and CO_2 should still be found on Mars, but not H_2
33. If $t_F = -40°$F, then $t_C = -40°$C.
35. (a) 74.98 cm   (b) 75.02 cm
37. (a) $L'_B - L'_A = L_B - L_A$ if $\alpha_B L_B = \alpha_A L_A$ or $L_A/L_B = \alpha_B/\alpha_A$ (b) 182 cm
39. $V = 3.94 \times 10^{-23}$ cm³; $r = 2.11$ Å
41. (a) 231 kPa   (b) 201 kPa
43. 0.78 L
45. 375 K
47. $K_{\text{av}} = 6.21 \times 10^{-21}$ J. The change in gravitational potential energy is $mgh = 7.82 \times 10^{-26}$ J
49. (a) 4.7 m/s   (b) 5.09 m/s
51.
$$\int_0^\infty f(v)\, dv = \frac{4}{\sqrt{\pi}}\left(\frac{m}{2kT}\right)^{3/2}\frac{\sqrt{\pi}}{4}\left(\frac{m}{2kT}\right)^{-3/2} = 1$$
53. $dP/dV = -nRT/(V - nb)^2 + 2an^2/V^3$ and $d^2P/dV^2 = 2nRT/(V - nb)^3 - 6an^2/V^4$. Setting these two derivatives to zero yields $V_c = 3nb$, which is $3b$ per mole
55. (a) $B = 3.95 \times 10^3$ K, $R_0 = 3.89 \times 10^{-3}$ ohms
(b) 1320 ohms   (c) For the ice point and the steam point, dR/dT is −393 ohms/K and −4.37 ohms/K, respectively. The thermistor is more sensitive at the ice point since the resistance changes more there for a given change in temperature
57. 1.34×10^5 N
59. 400.5 K

61. (*a*) 1.61×10^5 K   (*b*) 1.01×10^4 K   (*c*) For H_2, $f(v) \approx 10^{-9}$; for O_2, $f(v) \approx 10^{-106}$ (*d*) On the moon, the escape speed is $v_m = 2.38 \times 10^3$ m/s. O_2 and H_2 have an rms speed equal to v_m when the temperature is 7.28×10^3 K and 455 K, respectively

Chapter 16

True or False 1. True **2.** False **3.** True **4.** False; it is proportional to the fourth power of T **5.** False **6.** False **7.** True **8.** True **9.** False **10.** True

Problems

1. (*a*) 1.046×10^7 J   (*b*) 121 W
3. 0.0924 kcal/kg·K
5. (*a*) 0°C   (*b*) 125 g
7. 99.8 g
9. 2073 Btu/h
11. (*a*) $I_{Al} = 569$ W, $I_{Cu} = 962$ W   (*b*) $I_{total} = I_{Al} + I_{Cu} = 1531$ W   (*c*) 0.0522 K/W
13. 9470 nm
15. 874 kJ
17. (*a*) 0.117 K   (*b*) 1.74 K
19. (*a*)

$W = 608$ J   (*b*) 254 cal
21. (*a*)

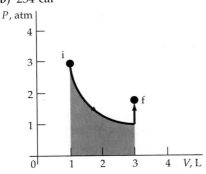

$W = 3.30$ L·atm   (*b*) 189 cal
23. (*a*) 1.57 kJ   (*b*) 1.57 kJ
25. (*a*) 0.713 J/g·K   (*b*) 252 J/L
27. (*a*) $\Delta U = 6.24$ kJ; $W = 0$; $Q = 6.24$ kJ
(*b*) $\Delta U = 6.24$ kJ; $Q = 8.73$ kJ; $W = 2.49$ kJ
(*c*) $W = 2.49$ kJ
29. (*a*) $V_i = 2.24$ L; $V_f = 5.89$ L   (*b*) $T_f = 143$ K
(*c*) 1.62 kJ
31. (*a*) 28.5°C   (*b*) 15.5°C
33. (*a*) 0.0920 J/kg·K   (*b*) 0.0584 J/kg
35. (*a*) 34.0 km   (*b*) yes   (*c*) The potential energy released is the same with or without air resistance, so there should be little net effect

37. (*a*) 2.99°C   (*b*) 199.85   (*c*) no
39. 18
41. 101.3°C
43. 47.8 cm
45. 96.6 g
47. 0.127 g
49. (*a*) For an ideal gas, U depends only on T; hence, for a given ΔT there is a unique ΔU.   (*b*) $\Delta U = Q - W = (C_p - nR) \Delta T = C_v \Delta T$
51. (*a*) 3.40 kJ   (*b*) $U_f = 3.70$ kJ, $W = 200$ J
(*c*) $U_f = 3.90$ kJ, $W = 0$
53. (*a*) 263 K   (*b*) 10.8 L   (*c*) 1.48 kJ
(*d*) -1.48 kJ
55. (*a*)

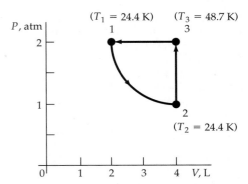

(*b*) For 1 to 2, $W = 2.77$ L·atm; $Q = 2.77$ L·atm. For 2 to 3, $W = 0$; $Q = 6.0$ L·atm. For 3 to 1, $W = -4$ L·atm; $Q = -10.0$ L·atm.   (*c*) $T_1 = 24.4$ K; $T_2 = T_1$; $T_3 = 48.7$ K
57. 626°C = 1159°F
59. 577 J
61. (*a*) $\frac{1}{2}P_0$ (*b*) diatomic   (*c*) No change occurs during the isothermal expansion; during the adiabatic compression, the kinetic energy increases
63. answer given in problem
65. answer given in problem
67. (*a*)

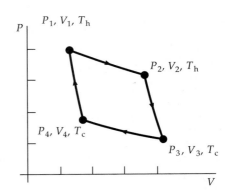

Chapter 17

True or False 1. False; it can be in a noncyclic process
2. False **3.** False **4.** False **5.** False **6.** True
7. False **8.** True

Problems

1. (*a*) 500 J   (*b*) 400 J
3. (*a*) 0.4   (*b*) 80 W

5. (*a*) 1.67   (*b*) 0.375
7. answer given in problem
9. answer given in problem
11. (*a*) $\frac{1}{3}$ (*b*) 33.3 J   (*c*) 66.7 J   (*d*) 2
13. (*a*) 13.7   (*b*) 8.77
15. (*a*) 0.51   (*b*) 102 kJ   (*c*) 98 kJ
17. (*a*) 5.26   (*b*) 3.19 kW   (*c*) 5.32 kW
19. (*a*) 303 kJ   (*b*) 212 kJ
21. (*a*) 11.5 J/K   (*b*) zero
23. (*a*) 50 J   (*b*) 0.167 J/K   (*c*) zero   (*d*) The answers to parts (*a*) and (*b*) stay the same; the answer to part (*c*) is $\Delta S_u > 0$
25. (*a*) 11.5 J/K   (*b*) 11.5 J/K
27. 0.417 J/K
29. 6.05 kJ/K
31. (*a*) Process 1 is more wasteful.   (*b*) $\Delta S_1 =$ 1.67 J/K, $\Delta S_2 = 0.833$ J/K
33.

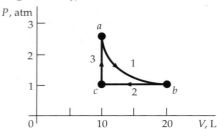

$\epsilon = W/Q_3 = 0.146$
35. (*a*) The maximum possible efficiency is 40.5 percent.   (*b*) 1.68×10^9 J

37. (*a*)

(*b*) $V_3 = 2.30$ L; $T_3 = 345$ K   (*c*) 5.15 L·atm   (*d*) 0.232
39. (*a*) 10.1°C   (*b*) 22 J/K
41. 10.7 J/K
43. 1.97×10^3 J/K
45. (*a*) 0.2   (*b*) $\Delta S_{engine} = 0$; $\Delta S_h = -2.5$ J/K; $\Delta S_c = 4.0$ J/K; $\Delta S_u = 1.5$ J/K   (*c*) 0.5   (*d*) 500 J
47. 93.9 W
49. (*a*) The heat input is $Q_h = C_v (T_c - T_b)$, and the magnitude of the heat output is $|Q_c| = C_v(T_d - T_a)/(T_c - T_b)$.   (*c*) 0.565   (*d*) In a real engine, the processes are not truly adiabatic, nor are they quasi-static
51. $\epsilon = (T_h - T_c)/\{T_h + c_v(T_h - T_c)/[R \ln (V_a/V_b)]\}$
53. answer given in problem
55. (*a*) $T_1 = 301$ K; $T_2 = T_3 = 601$ K   (*b*) $Q_{12} = 3.75$ kJ; $Q_{23} = 3.47$ kJ; $Q_{31} = -6.25$ kJ   (*c*) 0.134
57. answer given in problem
59. answer given in problem

Index

Some Conversion Factors

1 m = 39.37 in = 3.281 ft = 1.094 yd

1 m = 10^{15} fm = 10^{10} Å = 10^9 nm

1 km = 0.6215 mi

1 mi = 5280 ft = 1.609 km

1 lightyear = 1 $c \cdot y$ = 9.461×10^{15} m

1 in = 2.540 cm

1 L = 10^3 cm^3 = 10^{-3} m^3 = 1.057 qt

1 h = 3.6 ks

1 y = 365.24 d = 3.156×10^7 s

1 km/h = 0.278 m/s = 0.6215 mi/h

1 ft/s = 0.3048 m/s = 0.6818 mi/h

1 rev = 2π rad = 360°

1 rad = 57.30°

1 rev/min = 0.1047 rad/s

1 slug = 14.59 kg

1 tonne = 10^3 kg = 1 Mg

1 atm = 101.3 kPa = 1.013 bar = 76.00 cmHg = 14.70 lb/in^2

1 N = 10^5 dyn = 0.2248 lb

1 lb = 4.448 N

1 Pa·s = 10 poise

1 J = 10^7 erg = 0.7373 ft·lb = 9.869×10^{-3} L·atm

1 kW·h = 3.6 MJ

1 cal = 4.184 J = 4.129×10^{-2} L·atm

1 L·atm = 101.3 J = 24.22 cal

1 eV = 1.602×10^{-19} J

1 Btu = 778 ft·lb = 252 cal = 1054 J

1 horsepower = 550 ft·lb/s = 746 W

1 W/m·K = 6.938 Btu·in/h·ft^2·°F

1 T = 10^4 G

1 kg weighs about 2.205 lb